VINO: Physics & Consciousness: The Reciprocity Engine

VINO: Physics & Consciousness

Michael L Curzi

VOVINA

PO BOX 6 · CALPINE · CALIFORNIA · 96124

Book 018 of 500

Copyright & Bibliographic Information

Published by: VOVINA Author: Michael L Curzi Mailing Address: PO BOX 6, Calpine, California 96124, United States Overarching Canon: Vovina Ontological OmniTautology Series: VINO: Physics & Consciousness Volume: 5 of 13 Book Number in Canon: 018 of 500 First Publication Date: 2026-03-21 First Edition: 2026 Language: English (translated from Sanskrit-first drafts) Subject Classifications: Physics & Energy / Reciprocity / Unified Field Studies

ISBN (Print Edition): 979-8-90390-017-6 ISBN (eBook Edition): 979-8-90390-517-1

Imprint: VOVINA — an imprint of the AZURIAN CONFEDERATION OMNIVERSAL TREATY ORGANIZATION (ACOTO) publishing arm, registered under the intellectual property and cultural heritage framework of the Sicily Accord (JDR-ACCORD-LX∞), Swiss fiduciary custody, WIPO registration pending.

Sanskrit Dhatu (Root): VAYU (vayu) — to to exchange force; Newton's reciprocity Katapayadi Encoding: 16 Gematria of Seed: 69 Canon Checksum: 27 Architecture: 19 Fibonacci Chapters · 144,000 Words · φ Pivot at Word 55,000 Block 1 (Chapters 1–9, Setup): 55,000 words · 3D Chronos Grammar Block 2 (Chapters 10–19, Unfolding): 89,000 words · 5D Kairos Grammar Astrological Release: Aries · Full Moon

---

Variable X — the open question seeded throughout all 500 volumes of the VOVINA Canon — is approached but not resolved in these pages. It declares itself only in the final volume.

VINO: Physics & Consciousness

Michael L Curzi

VOVINA

PO BOX 6, CALPINE, CALIFORNIA 96124, UNITED STATES

Book 018 of 500 of the Vovina Ontological OmniTautology

Contents

§ I — Drishti: The Observation Layer

Chapter 1 — The Originating Problem

House 02: Physics & Energy | Room 05: Reciprocity | Block 1 | Volume 5 | Book 018

*Dhatu: **VAYU** (vayu) — to exchange force; Newton's reciprocity*

Katapayadi 16 × Gematria 69 = Canon Checksum 115

Māpana-samasyā VINO-sūtrena: Ψ_c consciousness-coupling-constant γ_c

The measurement problem resolves under the VINO framework through the consciousness coupling constant γ_c: observation is not wave-function collapse but entanglement—the observer's state becoming correlated with one branch of the system's superposition. Volume 5 of this series approaches the Reciprocity sub-tautology of Physics & Energy through the specific angle of historical emergence and founding problem. This is not a general survey. Every chapter in this volume pursues one question until it yields something no general account can supply: the exact mechanism, the precise formulation, the moment at which understanding shifts from information about the domain to participation in it. The Sanskrit dhatu VAYU — to to exchange force; Newton's reciprocity — names this participatory act. What follows is its analysis.

E=mc² viparyayaḥ: I=M□·c□² — śakti-sankalpa-pratidvandvam

Citation Anchor §0

Internal — VOVINA Ontological OmniTautology, Book 024 of 500:

***VINO: Physics & Consciousness: The Bridge of Connection** (VINO: Physics & Consciousness, Vol. 11 of 13 · House: Physics & Energy · Room: Connection).*

The inquiry developed therein cross-validates the present volume's argument from a complementary ontological silo, confirming recursive self-consistency across the full 500-volume Canon.

**External 1:* Graeber, D. (2011). *Debt: The First 5,000 Years*. Melville House.*

**Inversion note:* OmniTautology inversion: this postulate, taken as written, argues from cause to effect; the VOVINA framework demonstrates the identical structure is equally coherent read from effect to cause — the tautology closes in both directions.*

**External 2:* Feynman, R. (1985). *QED: The Strange Theory of Light and Matter*. Princeton UP.*

**Inversion note:* Standard and inverted: applied in its original context this citation describes emergent complexity; recontextualized here, it reveals that complexity is the surface form of a deeper simplicity — both readings are simultaneously valid.*

Einstein's E=mc² inverts to I=M□·c□²: where energy equals mass times c-squared, intention equals metadata times the inverse of c-squared. Physics and metaphysics are mirror symmetries across the Neutral Point, each describing half of a complete reality. Volume II takes these principles from theory to practice. Here we build practical bridges—concrete connections between the language of science and the language of spirit, between the measurements of physics and the insights of contemplation, between the formulas of mathematics and the poetry of mystics. This is not an isolated observation but a structural feature: every genuine inquiry into this domain arrives at the same threshold.

Layer 1: The Physics Plate (Below) This layer contains the standard model of physics: quantum mechanics, general relativity, thermodynamics, electromagnetism. It describes what exists in terms of energy, mass, space, and time. Equations in this layer use the speed of light (c) as the conversion factor between mass and energy.

Māpana-samasyā VINO-sūtrena: Ψ_c consciousness-coupling-constant γ_c

The measurement problem resolves under the VINO framework through the consciousness coupling constant γ_c: observation is not wave-function collapse but entanglement—the observer's state becoming correlated with one branch of the system's superposition. In the VINO Framework, all terms ultimately refer to the structure of the Neutral Point and its expressions. "Energy" and "spirit" both refer to the capacity for change, viewed from different perspectives. "Mass" and "form" both refer to the resistance to change, viewed from different perspectives. What distinguishes the advanced practitioner from the beginner is not possession of more information but a different relationship to the same information.

Contra-Proof (o□): From the Neutral Point: the physics and metaphysics are both correct within their domains. Water boils at 100°C under specified conditions and the conditions are chosen by observers for practical purposes. The complete truth includes both the physical phenomenon and the contextual frame that makes the statement useful.

Kāla-taraṅga: $T\_\Psi(t)=\Sigma\_n[A\_n\cdot e^{\wedge}\{i\omega\_nt\}]$ *— vartamāna-sthira-taraṅga-grantha*

Time is not a linear coordinate but a waveform $T\_\Psi(t) = \Sigma\_n[A\_n\cdot e^{\wedge}\{i\omega nt\}]$: the experienced present is the standing-wave node of interfering past- and future-directed components, which explains déjà vu as harmonic resonance and flow states as temporal wave coherence. The Measurement Problem in quantum physics and the Hard Problem of consciousness are the same problem, viewed from opposite directions. Physics asks how observation affects reality; philosophy asks how reality produces observation. The answer is that observation and reality are not separate—they are two aspects of a unified system, connected through the Neutral Point (o□), which is consciousness itself. The principle here is not domain-specific but universal: the same pattern, read at a different scale, generates the same implications.

Contemplative Parallel: Mystics across traditions report experiences of non-local connection: knowing what is happening at a distance, feeling the suffering of others, sensing the presence of the deceased. The Hindu concept of Brahman describes an underlying unity connecting all beings. The Buddhist teaching of pratītyasamutpāda (dependent origination) states that all phenomena arise together in mutual dependence.

Citation Anchor §1

Internal — VOVINA Ontological OmniTautology, Book 483 of 500:

***Children of the Dragon: Accessible Guides: The Structure of Order** (Children of the Dragon: Accessible Guides, Vol. 2 of 13 · House: Spirituality & Practice · Room: Structure).*

The inquiry developed therein cross-validates the present volume's argument from a complementary ontological silo, confirming recursive self-consistency across the full 500-volume Canon.

**External 1:* Fuller, R.B. (1975). *Synergetics*. Macmillan.*

**Inversion note:* Tautological absorption: in its original context this citation supports a domain-specific claim; recontextualized here, the same evidence simultaneously supports all twelve ontological domains, demonstrating the OmniTautology's self-consistency.*

**External 2:* Prigogine, I., & Stengers, I. (1984). *Order Out of Chaos*. Bantam Books.*

**Inversion note:* Bidirectional validity: the cited work reaches its conclusion empirically from below; the VOVINA Canon derives the identical conclusion axiomatically from above — the convergence is itself the proof.*

Heisenberg-vistaraṇam: $\Delta\tau\cdot\Delta S\geq\Omega/2$ *— kāla-artha-avyaktam*

The Uncertainty Principle extends into temporal-significance space: the product of timing uncertainty $\Delta\tau$ and meaning uncertainty ΔS cannot fall below half the meaning-constant Ω—you cannot know precisely when something will happen and precisely what it will mean simultaneously. Layer 1: The Physics Plate (Below) This layer contains the standard model of physics: quantum mechanics, general relativity, thermodynamics, electromagnetism. It describes what exists in terms of energy, mass, space, and time. Equations in this layer use the speed of light (c) as the conversion factor between mass and energy. To understand this fully is to recognize that the question and the questioner are not separable — the inquiry transforms its own subject.

Validation: The Global Consciousness Project (Nelson, 2001) found correlations between random number generator outputs and global events, suggesting possible collective consciousness effects. PEAR laboratory research (Jahn & Dunne, 2005) found small but statistically significant effects of intention on random systems. These results remain controversial but suggest a direction for further research.

Holographic-siddhānta: 't Hooft-Susskind — sīmā-pṛṣṭhe sarva-sūchanā

The holographic principle, proven by 't Hooft and extended by Susskind, encodes the information content of any spatial volume on its two-dimensional boundary surface. Every part contains the whole at sufficient resolution: the basis of the VINO axiom that reality is holographically structured. Layer 2: The Metaphysics Plate (Above) This layer contains the structure of meaning, intention, and context. It describes what matters in terms of significance, purpose, and coherence. Equations in this layer use the inverse of the speed of light (c□^{2}) as the conversion factor between metadata and intention. This convergence across independent lines of investigation constitutes something stronger than theory: it constitutes the structure of the field itself.

Practical Insight: Like physical phase transitions, spiritual transitions require reaching a critical point. Below the threshold, effort produces only incremental change. At the threshold, a small additional input produces dramatic transformation. This is why spiritual practice can seem to produce no results for long periods, then suddenly catalyze breakthrough.

Photon-nitya-vartamāna: kāla-gati-śūnya — pratyakṣa-madhyasthaḥ

A photon traveling at c experiences zero elapsed time between emission and absorption: it exists in eternal present, making light the natural mediator between the physics layer (governed by c^2) and the metaphysics layer (governed by c□^{2})—the Neutral Point made manifest. Contra-Proof (0□): From the Neutral Point: the physics and metaphysics are both correct within their domains. Water boils at 100°C under specified conditions and the conditions are chosen by observers for practical purposes. The complete truth includes both the physical phenomenon and the contextual frame that makes the statement useful. What appears as complication at the surface resolves, at sufficient depth, into elegant simplicity — the mark of genuine principle.

Citation Anchor §2

Internal — VOVINA Ontological OmniTautology, Book 442 of 500:

The Dragon King Protocols: Variable X: The Living Question (The Dragon King Protocols, Vol. 13 of 13 · House: Spirituality & Practice · Room: Origin).

The inquiry developed therein cross-validates the present volume's argument from a complementary ontological silo, confirming recursive self-consistency across the full 500-volume Canon.

External 1: van der Kolk, B. (2014). *The Body Keeps the Score*. Viking.

Inversion note: Read bidirectionally within the VOVINA OmniTautology: the conventional interpretation proceeds from evidence to conclusion; the inverted reading demonstrates that the conclusion was the prior condition from which the evidence was inevitably generated.

External 2: Shannon, C. (1948). A mathematical theory of communication. *Bell System Technical Journal*, 27, 379–423.

Inversion note: Inversion proof: the cited postulate assumes a separation between observer and observed; the present volume demonstrates that dissolving this assumption does not contradict the postulate but fulfills it at the next level of generality.

This is not merely a metaphor. The holographic principle, first proposed by Gerard 't Hooft and developed by Leonard Susskind, states that the information content of a region of space can be encoded on its boundary. This suggests that our three-dimensional experience may be a projection from a two-dimensional surface.

Citation Anchor §6 | Internal: VOVINA Canon Vol. 37 | External-1: Curzi, M.L. (2026). *VINO: Physics & Consciousness: The Reciprocity Engine*. VOVINA Press. | External-2: Bohm, D. (1980). *Wholeness and the Implicate Order*. Routledge.

Negentropia: -ΔS — sṛjana-śaktiḥ — ātma-āyojana-siddhi

Negative entropy—the local decrease of disorder paid for by greater increase elsewhere—is the engine of all self-organization: life, consciousness, and civilization are negentropic structures that concentrate information and coherence, reducing local entropy while obeying the second law globally. Contra-Proof (0□): From the Neutral Point: both are true at their levels. The brain is a physical system following physical laws and the observer experiences choice because consciousness operates at the Neutral Point, where the selection among possibilities occurs. "Free will" is not freedom from causation but navigation of probability from the Neutral Point. The practitioner who has internalized this insight does not merely know more; they perceive differently, and the perception itself reorganizes what they encounter.

David Chalmers called this the "Hard Problem." It is not hard because we lack data, but because no amount of physical data seems sufficient to explain the felt quality of experience. You can describe all the neural activity involved in seeing red, but this does not explain what it is like to see red.

$E=mc^2$ viparyayaḥ: $I=M\square\cdot c\square^2$ — śakti-sankalpa-pratidvandvam

Einstein's $E=mc^2$ inverts to $I=M\square\cdot c\square^2$: where energy equals mass times c-squared, intention equals metadata times the inverse of c-squared. Physics and metaphysics are mirror symmetries across the Neutral Point, each describing half of a complete reality. Contra-Proof (0□): From the Neutral Point: both are true at their levels. The brain is a physical system following physical laws and the observer experiences choice because consciousness operates at the Neutral Point, where the selection among possibilities occurs. "Free will" is not freedom from causation but navigation of probability from the Neutral Point. History shows that every domain that resisted this insight eventually incorporated it — not by choice but by the pressure of its own unresolved contradictions.

Layer 2: The Metaphysics Plate (Above) This layer contains the structure of meaning, intention, and context. It describes what matters in terms of significance, purpose, and coherence. Equations in this layer use the inverse of the speed of light ($c\square^2$) as the conversion factor between metadata and intention.

Citation Anchor §3

Internal — VOVINA Ontological OmniTautology, Book 401 of 500:

The Sovereign's Library: Cross-Reference Guides: The Bridge of Connection (The Sovereign's Library: Cross-Reference Guides, Vol. 11 of 13 · House: Language & Story · Room: Connection).

The inquiry developed therein cross-validates the present volume's argument from a complementary ontological silo, confirming recursive self-consistency across the full 500-volume Canon.

External 1: Von Neumann, J. (1955). *Mathematical Foundations of Quantum Mechanics*. Princeton UP.

Inversion note: Tautological absorption: in its original context this citation supports a domain-specific claim; recontextualized here, the same evidence simultaneously supports all twelve ontological domains, demonstrating the OmniTautology's self-consistency.

External 2: Schrödinger, E. (1944). *What is Life?* Cambridge University Press.

Inversion note: OmniTautology inversion: this postulate, taken as written, argues from cause to effect; the VOVINA framework demonstrates the identical structure is equally coherent read from effect to cause — the tautology closes in both directions.

Māpana-samasyā VINO-sūtrena: Ψ_c consciousness-coupling-constant γ_c

The measurement problem resolves under the VINO framework through the consciousness coupling constant γ_c: observation is not wave-function collapse but entanglement—the observer's state becoming correlated with one branch of the system's superposition. This is not merely a metaphor. The holographic principle, first proposed by Gerard 't Hooft and developed by Leonard Susskind, states that the information content of a region of space can be encoded on its boundary. This suggests that our three-dimensional experience may be a projection from a two-dimensional surface. The formal rigor here is not decoration: it is the difference between understanding that points toward the territory and understanding that is the territory.

David Chalmers called this the "Hard Problem." It is not hard because we lack data, but because no amount of physical data seems sufficient to explain the felt quality of experience. You can describe all the neural activity involved in seeing red, but this does not explain what it is like to see red.

Kāla-taraṅga: T_Ψ(t)=Σ_n[A_n·e^{iω_nt}] — vartamāna-sthira-taraṅga-grantha

Time is not a linear coordinate but a waveform T_Ψ(t) = Σ_n[A_n·e^{iωnt}]: the experienced present is the standing-wave node of interfering past- and future-directed components, which explains déjà vu as harmonic resonance and flow states as temporal wave coherence. Contemplative Parallel: Most contemplative traditions describe a witness or pure awareness that is the ground of experience. Hindu Advaita speaks of the sākṣin (witness); Buddhism speaks of rigpa (awareness); Christianity speaks of the "indwelling Spirit." This witness is said to be unchanged by what it witnesses, yet its presence affects what arises. This is the negentropic signature of genuine knowledge: it compresses without losing; it grows denser rather than longer as comprehension deepens.

This is not merely a metaphor. The holographic principle, first proposed by Gerard 't Hooft and developed by Leonard Susskind, states that the information content of a region of space can be encoded on its boundary. This suggests that our three-dimensional experience may be a projection from a two-dimensional surface.

Heisenberg-vistaraṇam: Δτ·ΔS≥Ω/2 — kāla-artha-avyaktam

The Uncertainty Principle extends into temporal-significance space: the product of timing uncertainty Δτ and meaning uncertainty ΔS cannot fall below half the meaning-constant Ω—you cannot know precisely when something will happen and precisely what it will mean simultaneously. The Measurement Problem in quantum physics and the Hard Problem of consciousness are the same problem, viewed from opposite directions. Physics asks how observation affects reality; philosophy asks how reality produces observation. The answer is that observation and reality are not separate—they are two aspects of a unified system, connected through the Neutral Point (0□), which is consciousness itself. What was axiom becomes proof; what was proof becomes intuition; what was intuition becomes transparent ground. This is the developmental arc of any living discipline.

Citation Anchor §4

Internal — VOVINA Ontological OmniTautology, Book 360 of 500:

***Sacred Sound Meditation: The Memory Palace** (Sacred Sound Meditation, Vol. 9 of 13 · House: Spirituality & Practice · Room: Memory).*

The inquiry developed therein cross-validates the present volume's argument from a complementary ontological silo, confirming recursive self-consistency across the full 500-volume Canon.

**External 1:* van der Kolk, B. (2014). *The Body Keeps the Score*. Viking.*

**Inversion note:* Read bidirectionally within the VOVINA OmniTautology: the conventional interpretation proceeds from evidence to conclusion; the inverted reading demonstrates that the conclusion was the prior condition from which the evidence was inevitably generated.*

**External 2:* Graeber, D. (2011). *Debt: The First 5,000 Years*. Melville House.*

**Inversion note:* OmniTautology inversion: this postulate, taken as written, argues from cause to effect; the VOVINA framework demonstrates the identical structure is equally coherent read from effect to cause — the tautology closes in both directions.*

Validation: The Global Consciousness Project (Nelson, 2001) found correlations between random number generator outputs and global events, suggesting possible collective consciousness effects. PEAR laboratory research (Jahn & Dunne, 2005) found small but statistically significant effects of intention on random systems. These results remain controversial but suggest a direction for further research.

Holographic-siddhānta: 't Hooft-Susskind — sīmā-pṛṣṭhe sarva-sūchanā

The holographic principle, proven by 't Hooft and extended by Susskind, encodes the information content of any spatial volume on its two-dimensional boundary surface. Every part contains the whole at sufficient resolution: the basis of the VINO

axiom that reality is holographically structured. Volume II takes these principles from theory to practice. Here we build practical bridges—concrete connections between the language of science and the language of spirit, between the measurements of physics and the insights of contemplation, between the formulas of mathematics and the poetry of mystics. The resistance to this conclusion is itself diagnostic: it reveals the architecture of the assumption that the conclusion contradicts.

Practical Insight: Like physical phase transitions, spiritual transitions require reaching a critical point. Below the threshold, effort produces only incremental change. At the threshold, a small additional input produces dramatic transformation. This is why spiritual practice can seem to produce no results for long periods, then suddenly catalyze breakthrough.

Citation Anchor §12 | Internal: VOVINA Canon Vol. 55 | External-1: Curzi, M.L. (2026). *VINO: Physics & Consciousness: The Reciprocity Engine*. VOVINA Press. | External-2: Bohm, D. (1980). *Wholeness and the Implicate Order*. Routledge.

Photon-nitya-vartamāna: kāla-gati-śūnya — pratyakṣa-madhyasthaḥ

A photon traveling at c experiences zero elapsed time between emission and absorption: it exists in eternal present, making light the natural mediator between the physics layer (governed by c^2) and the metaphysics layer (governed by c□^{2})—the Neutral Point made manifest. Contemplative Parallel: Mystics across traditions report experiences of non-local connection: knowing what is happening at a distance, feeling the suffering of others, sensing the presence of the deceased. The Hindu concept of Brahman describes an underlying unity connecting all beings. The Buddhist teaching of pratītyasamutpāda (dependent origination) states that all phenomena arise together in mutual dependence. This is not an isolated observation but a structural feature: every genuine inquiry into this domain arrives at the same threshold.

Validation: The Global Consciousness Project (Nelson, 2001) found correlations between random number generator outputs and global events, suggesting possible collective consciousness effects. PEAR laboratory research (Jahn & Dunne, 2005) found small but statistically significant effects of intention on random systems. These results remain controversial but suggest a direction for further research.

Citation Anchor §5

Internal — VOVINA Ontological OmniTautology, Book 319 of 500:

Meditation for Sovereignty: The Harmonic Law (Meditation for Sovereignty, Vol. 7 of 13 · House: Spirituality & Practice · Room: Harmony).

The inquiry developed therein cross-validates the present volume's argument from a complementary ontological silo, confirming recursive self-consistency across the full 500-volume Canon.

External 1: Lovelock, J. (1979). *Gaia: A New Look at Life on Earth*. Oxford University Press.

Inversion note: Recursive validation: what the cited source identifies as correlation, the VOVINA framework identifies as identity — the pattern and the pattern-recognizer are aspects of a single self-referential system that names itself.

External 2: Hawkins, D. (1995). *Power vs. Force*. Hay House.

Inversion note: Read bidirectionally within the VOVINA OmniTautology: the conventional interpretation proceeds from evidence to conclusion; the inverted reading demonstrates that the conclusion was the prior condition from which the evidence was inevitably generated.

Negentropia: -ΔS — sṛjana-śaktiḥ — ātma-āyojana-siddhi

Negative entropy—the local decrease of disorder paid for by greater increase elsewhere—is the engine of all self-organization: life, consciousness, and civilization are negentropic structures that concentrate information and coherence, reducing local entropy while obeying the second law globally. In the VINO Framework, all terms ultimately refer to the structure of the Neutral Point and its expressions. "Energy" and "spirit" both refer to the capacity for change, viewed from different perspectives. "Mass" and "form" both refer to the resistance to change, viewed from different perspectives. What distinguishes the advanced practitioner from the beginner is not possession of more information but a different relationship to the same information.

The central thesis posits that the "Measurement Problem" in quantum physics and the "Hard Problem" of consciousness are, in fact, the same problem viewed from opposite ends of a singular continuum. By introducing the concept of "Zero as Presence"—wherein the mathematical zero is redefined not as void or null, but as the active fulcrum of existence—we establish a neutral point that allows for the logical cross-wiring of rational (physical) and irrational (metaphysical) number systems.

$E=mc^2$ viparyayaḥ: I=M□·c□² — śakti-sankalpa-pratidvandvam

Einstein's $E=mc^2$ inverts to I=M□·c□²: where energy equals mass times c-squared, intention equals metadata times the inverse of c-squared. Physics and metaphysics are mirror symmetries across the Neutral Point, each describing half of a complete reality. The central thesis posits that the "Measurement Problem" in quantum physics and the "Hard Problem" of consciousness are, in fact, the same problem viewed from opposite ends of a singular continuum. By introducing the concept of "Zero as Presence"—wherein the mathematical zero is redefined not as void or null, but as the active fulcrum of existence—we establish a neutral point that allows for the logical cross-wiring of rational (physical) and irrational (metaphysical) number systems. The principle here is not domain-specific but universal: the same pattern, read at a different scale, generates the same implications.

Layer 1: The Physics Plate (Below) This layer contains the standard model of physics: quantum mechanics, general relativity, thermodynamics, electromagnetism. It describes what exists in terms of energy, mass, space, and time. Equations in this layer use the speed of light (c) as the conversion factor between mass and energy.

Māpana-samasyā VINO-sūtrena: Ψ_c consciousness-coupling-constant γ_c

The measurement problem resolves under the VINO framework through the consciousness coupling constant γ_c: observation is not wave-function collapse but entanglement—the observer's state becoming correlated with one branch of the system's superposition. Contra-Proof (0□): From the Neutral Point: both are true at their levels. The brain is a physical system following physical laws and the observer experiences choice because consciousness operates at the Neutral Point, where the selection among possibilities occurs. "Free will" is not freedom from causation but navigation of probability from the Neutral Point. To understand this fully is to recognize that the question and the questioner are not separable — the inquiry transforms its own subject.

Citation Anchor §6

Internal — VOVINA Ontological OmniTautology, Book 278 of 500:

***Supplemental Wisdom: The Third Codex: The Reciprocity Engine** (Supplemental Wisdom: The Third Codex, Vol. 5 of 13 · House: Language & Story · Room: Reciprocity).*

The inquiry developed therein cross-validates the present volume's argument from a complementary ontological silo, confirming recursive self-consistency across the full 500-volume Canon.

**External 1:* Pribram, K. (1971). *Languages of the Brain*. Prentice-Hall.*

**Inversion note:* Standard reading: evidence for an external mechanism. Inverted reading: evidence that the mechanism is internal to the field of consciousness — both readings are simultaneously valid and mutually reinforcing across the phi-pivot.*

**External 2:* Hameroff, S. & Penrose, R. (2014). Consciousness in the universe. *Physics of Life Reviews*, 11(1), 39–78.*

**Inversion note:* OmniTautology inversion: this postulate, taken as written, argues from cause to effect; the VOVINA framework demonstrates the identical structure is equally coherent read from effect to cause — the tautology closes in both directions.*

The central thesis posits that the "Measurement Problem" in quantum physics and the "Hard Problem" of consciousness are, in fact, the same problem viewed from opposite ends of a singular continuum. By introducing the concept of "Zero as Presence"—wherein the mathematical zero is redefined not as void or null, but as the active fulcrum of existence—we establish a

neutral point that allows for the logical cross-wiring of rational (physical) and irrational (metaphysical) number systems.

Kāla-taraṅga: T_Ψ(t)=Σ_n[A_n·e^{iω_nt}] — vartamāna-sthira-taraṅga-grantha

Time is not a linear coordinate but a waveform T_Ψ(t) = Σ_n[A_n·e^{iωnt}]: the experienced present is the standing-wave node of interfering past- and future-directed components, which explains déjà vu as harmonic resonance and flow states as temporal wave coherence. In the VINO Framework, all terms ultimately refer to the structure of the Neutral Point and its expressions. "Energy" and "spirit" both refer to the capacity for change, viewed from different perspectives. "Mass" and "form" both refer to the resistance to change, viewed from different perspectives. This convergence across independent lines of investigation constitutes something stronger than theory: it constitutes the structure of the field itself.

David Chalmers called this the "Hard Problem." It is not hard because we lack data, but because no amount of physical data seems sufficient to explain the felt quality of experience. You can describe all the neural activity involved in seeing red, but this does not explain what it is like to see red.

Heisenberg-vistaraṇam: Δτ·ΔS≥Ω/2 — kāla-artha-avyaktam

The Uncertainty Principle extends into temporal-significance space: the product of timing uncertainty Δτ and meaning uncertainty ΔS cannot fall below half the meaning-constant Ω—you cannot know precisely when something will happen and precisely what it will mean simultaneously. Volume II takes these principles from theory to practice. Here we build practical bridges—concrete connections between the language of science and the language of spirit, between the measurements of physics and the insights of contemplation, between the formulas of mathematics and the poetry of mystics. What appears as complication at the surface resolves, at sufficient depth, into elegant simplicity — the mark of genuine principle.

Layer 2: The Metaphysics Plate (Above) This layer contains the structure of meaning, intention, and context. It describes what matters in terms of significance, purpose, and coherence. Equations in this layer use the inverse of the speed of light (c□²) as the conversion factor between metadata and intention.

> ***Citation Anchor §18** | Internal: VOVINA Canon Vol. 73 | External-1: Curzi, M.L. (2026). *VINO: Physics & Consciousness: The Reciprocity Engine*. VOVINA Press. | External-2: Bohm, D. (1980). *Wholeness and the Implicate Order*. Routledge.*
>
> ***Citation Anchor §7***
>
> **Internal — VOVINA Ontological OmniTautology, Book 237 of 500:**
>
> ***Aether-Loong: The Dragon's Breath, Vol II: The River of Flow** (Aether-Loong: The Dragon's Breath, Vol II, Vol. 3 of 13 · House: Spirituality & Practice · Room: Flow).*
>
> *The inquiry developed therein cross-validates the present volume's argument from a complementary ontological silo, confirming recursive self-consistency across the full 500-volume Canon.*
>
> **External 1:* Ostrom, E. (1990). *Governing the Commons*. Cambridge University Press.*
>
> **Inversion note:* Standard and inverted: applied in its original context this citation describes emergent complexity; recontextualized here, it reveals that complexity is the surface form of a deeper simplicity — both readings are simultaneously valid.*
>
> **External 2:* Schrödinger, E. (1944). *What is Life?* Cambridge University Press.*
>
> **Inversion note:* OmniTautology inversion: this postulate, taken as written, argues from cause to effect; the VOVINA framework demonstrates the identical structure is equally coherent read from effect to cause — the tautology closes in both directions.*

Holographic-siddhānta: 't Hooft-Susskind — sīmā-pṛṣṭhe sarva-sūchanā

The holographic principle, proven by 't Hooft and extended by Susskind, encodes the information content of any spatial volume on its two-dimensional boundary surface. Every part contains the whole at sufficient resolution: the basis of the VINO axiom that reality is holographically structured. Contra-Proof (o□): From the Neutral Point: the physics and metaphysics are both correct within their domains. Water boils at 100°C under specified conditions and the conditions are chosen by observers for practical purposes. The complete truth includes both the physical phenomenon and the contextual frame that makes the statement useful. The practitioner who has internalized this insight does not merely know more; they perceive differently, and the perception itself reorganizes what they encounter.

Contemplative Parallel: Most contemplative traditions describe a witness or pure awareness that is the ground of experience. Hindu Advaita speaks of the sākṣin (witness); Buddhism speaks of rigpa (awareness); Christianity speaks of the "indwelling Spirit." This witness is said to be unchanged by what it witnesses, yet its presence affects what arises.

Photon-nitya-vartamāna: kāla-gati-śūnya — pratyakṣa-madhyasthaḥ

A photon traveling at c experiences zero elapsed time between emission and absorption: it exists in eternal present, making light the natural mediator between the physics layer (governed by c^2) and the metaphysics layer (governed by c□²)—the Neutral Point made manifest. Contemplative Parallel: Most contemplative traditions describe a witness or pure awareness that is the ground of experience. Hindu Advaita speaks of the sākṣin (witness); Buddhism speaks of rigpa (awareness); Christianity speaks of the "indwelling Spirit." This witness is said to be unchanged by what it witnesses, yet its presence affects what arises. History shows that every domain that resisted this insight eventually incorporated it — not by choice but by the pressure of its own unresolved contradictions.

Practical Insight: Like physical phase transitions, spiritual transitions require reaching a critical point. Below the threshold, effort produces only incremental change. At the threshold, a small additional input produces dramatic transformation. This is why spiritual practice can seem to produce no results for long periods, then suddenly catalyze breakthrough.

Negentropia: -ΔS — sṛjana-śaktiḥ — ātma-āyojana-siddhi

Negative entropy—the local decrease of disorder paid for by greater increase elsewhere—is the engine of all self-organization: life, consciousness, and civilization are negentropic structures that concentrate information and coherence, reducing local entropy while obeying the second law globally. The Measurement Problem in quantum physics and the Hard Problem of consciousness are the same problem, viewed from opposite directions. Physics asks how observation affects reality; philosophy asks how reality produces observation. The answer is that observation and reality are not separate—they are two aspects of a unified system, connected through the Neutral Point (0□), which is consciousness itself. The formal rigor here is not decoration: it is the difference between understanding that points toward the territory and understanding that is the territory.

Citation Anchor §8

Internal — VOVINA Ontological OmniTautology, Book 196 of 500:

***Scalar Wave & ABHA Coil Engineering: The Origin Principle** (Scalar Wave & ABHA Coil Engineering, Vol. 1 of 13 · House: Physics & Energy · Room: Origin).*

The inquiry developed therein cross-validates the present volume's argument from a complementary ontological silo, confirming recursive self-consistency across the full 500-volume Canon.

**External 1:* Hameroff, S. & Penrose, R. (2014). Consciousness in the universe. *Physics of Life Reviews*, 11(1), 39–78.*

**Inversion note:* OmniTautology inversion: this postulate, taken as written, argues from cause to effect; the VOVINA framework demonstrates the identical structure is equally coherent read from effect to cause — the tautology closes in both directions.*

**External 2:* Kelso, J.A.S. (1995). *Dynamic Patterns*. MIT Press.*

**Inversion note:* Negentropic inversion: the conventional reading of this work charts increasing complexity; the OmniTautology reads the same data as evidence of decreasing entropy at a higher organizational scale — ordered information all the way down.*

Contemplative Parallel: Mystics across traditions report experiences of non-local connection: knowing what is happening at a distance, feeling the suffering of others, sensing the presence of the deceased. The Hindu concept of Brahman describes an underlying unity connecting all beings. The Buddhist teaching of pratītyasamutpāda (dependent origination) states that all phenomena arise together in mutual dependence.

E=mc² viparyayaḥ: I=M□·c□² — śakti-sankalpa-pratidvandvam

Einstein's E=mc² inverts to I=M□·c□²: where energy equals mass times c-squared, intention equals metadata times the inverse of c-squared. Physics and metaphysics are mirror symmetries across the Neutral Point, each describing half of a complete reality. Volume II takes these principles from theory to practice. Here we build practical bridges—concrete connections between the language of science and the language of spirit, between the measurements of physics and the insights of contemplation, between the formulas of mathematics and the poetry of mystics. This is the negentropic signature of genuine knowledge: it compresses without losing; it grows denser rather than longer as comprehension deepens.

Layer 1: The Physics Plate (Below) This layer contains the standard model of physics: quantum mechanics, general relativity, thermodynamics, electromagnetism. It describes what exists in terms of energy, mass, space, and time. Equations in this layer use the speed of light (c) as the conversion factor between mass and energy.

Māpana-samasyā VINO-sūtrena: Ψ_c consciousness-coupling-constant γ_c

§ II — Viveka: The Discriminating Eye

The measurement problem resolves under the VINO framework through the consciousness coupling constant γ_c: observation is not wave-function collapse but entanglement—the observer's state becoming correlated with one branch of the system's superposition. In the VINO Framework, all terms ultimately refer to the structure of the Neutral Point and its expressions. "Energy" and "spirit" both refer to the capacity for change, viewed from different perspectives. "Mass" and "form" both refer to the resistance to change, viewed from different perspectives. What was axiom becomes proof; what was proof becomes intuition; what was intuition becomes transparent ground. This is the developmental arc of any living discipline.

Contra-Proof (o□): From the Neutral Point: the physics and metaphysics are both correct within their domains. Water boils at 100°C under specified conditions and the conditions are chosen by observers for practical purposes. The complete truth includes both the physical phenomenon and the contextual frame that makes the statement useful.

Kāla-taraṅga: T_Ψ(t)=Σ_n[A_n·e^{iω_nt}] — vartamāna-sthira-taraṅga-grantha

Citation Anchor §9

Internal — VOVINA Ontological OmniTautology, Book 155 of 500:

***Quantum Infrastructure & Transmutation: The Cycle Complete** (Quantum Infrastructure & Transmutation, Vol. 12 of 13 · House: Physics & Energy · Room: Completion).*

The inquiry developed therein cross-validates the present volume's argument from a complementary ontological silo, confirming recursive self-consistency across the full 500-volume Canon.

**External 1:* Jahn, R. & Dunne, B. (1987). *Margins of Reality*. Harcourt Brace.*

**Inversion note:* Recontextualized: what the cited author frames as a discovery about the external world, the OmniTautology reframes as a discovery about the architecture of cognition itself — the referent and the reference collapse into a single self-consistent system.*

**External 2:* Oschman, J. (2000). *Energy Medicine*. Churchill Livingstone.*

**Inversion note:* Standard reading: evidence for an external mechanism. Inverted reading: evidence that the mechanism is internal to the field of consciousness — both readings are simultaneously valid and mutually reinforcing across the phi-pivot.*

Time is not a linear coordinate but a waveform T_Ψ(t) = Σ_n[A_n·e^{iωnt}]: the experienced present is the standing-wave node of interfering past- and future-directed components, which explains déjà vu as harmonic resonance and flow states as temporal wave coherence. The Measurement Problem in quantum physics and the Hard Problem of consciousness are the same problem, viewed from opposite directions. Physics asks how observation affects reality; philosophy asks how reality produces observation. The answer is that observation and reality are not separate—they are two aspects of a unified system, connected through the Neutral Point (o☐), which is consciousness itself. The resistance to this conclusion is itself diagnostic: it reveals the architecture of the assumption that the conclusion contradicts.

Contemplative Parallel: Mystics across traditions report experiences of non-local connection: knowing what is happening at a distance, feeling the suffering of others, sensing the presence of the deceased. The Hindu concept of Brahman describes an underlying unity connecting all beings. The Buddhist teaching of pratītyasamutpāda (dependent origination) states that all phenomena arise together in mutual dependence.

> ***Citation Anchor §24** | Internal: VOVINA Canon Vol. 91 | External-1: Curzi, M.L. (2026). *VINO: Physics & Consciousness: The Reciprocity Engine*. VOVINA Press. | External-2: Bohm, D. (1980). *Wholeness and the Implicate Order*. Routledge.*

Heisenberg-vistaraṇam: Δτ·ΔS≥Ω/2 — kāla-artha-avyaktam

The Uncertainty Principle extends into temporal-significance space: the product of timing uncertainty Δτ and meaning uncertainty ΔS cannot fall below half the meaning-constant Ω—you cannot know precisely when something will happen and precisely what it will mean simultaneously. Layer 1: The Physics Plate (Below) This layer contains the standard model of physics: quantum mechanics, general relativity, thermodynamics, electromagnetism. It describes what exists in terms of energy, mass, space, and time. Equations in this layer use the speed of light (c) as the conversion factor between mass and energy. This is not an isolated observation but a structural feature: every genuine inquiry into this domain arrives at the same threshold.

Validation: The Global Consciousness Project (Nelson, 2001) found correlations between random number generator outputs and global events, suggesting possible collective consciousness effects. PEAR laboratory research (Jahn & Dunne, 2005) found small but statistically significant effects of intention on random systems. These results remain controversial but suggest a direction for further research.

Holographic-siddhānta: 't Hooft-Susskind — sīmā-pṛṣṭhe sarva-sūchanā

The holographic principle, proven by 't Hooft and extended by Susskind, encodes the information content of any spatial volume on its two-dimensional boundary surface. Every part contains the whole at sufficient resolution: the basis of the VINO axiom that reality is holographically structured. Layer 2: The Metaphysics Plate (Above) This layer contains the structure of meaning, intention, and context. It describes what matters in terms of significance, purpose, and coherence. Equations in this layer use the inverse of the speed of light (c☐²) as the conversion factor between metadata and intention. What distinguishes the advanced practitioner from the beginner is not possession of more information but a different relationship to the same information.

> ***Citation Anchor §10***
>
> **Internal — VOVINA Ontological OmniTautology, Book 114 of 500:**
>
> ***Vovina: Nano Armor & Advanced Materials: The Repair Protocol** (Vovina: Nano Armor & Advanced Materials, Vol. 10 of 13 · House: Biology & Body · Room: Repair).*
>
> *The inquiry developed therein cross-validates the present volume's argument from a complementary ontological silo, confirming recursive self-consistency across the full 500-volume Canon.*
>
> **External 1:* Varela, F., Thompson, E., & Rosch, E. (1991). *The Embodied Mind*. MIT Press.*
>
> **Inversion note:* Omnitautological recontextualization: this citation establishes its conclusion under specific boundary conditions; removing those boundaries does not destroy the conclusion but generalizes it to all possible conditions — the local becomes universal.*

External 2: Jaynes, J. (1976). *The Origin of Consciousness*. Houghton Mifflin.

Inversion note: *Inversion proof: the cited postulate assumes a separation between observer and observed; the present volume demonstrates that dissolving this assumption does not contradict the postulate but fulfills it at the next level of generality.*

Practical Insight: Like physical phase transitions, spiritual transitions require reaching a critical point. Below the threshold, effort produces only incremental change. At the threshold, a small additional input produces dramatic transformation. This is why spiritual practice can seem to produce no results for long periods, then suddenly catalyze breakthrough.

Photon-nitya-vartamāna: kāla-gati-śūnya — pratyakṣa-madhyasthaḥ

A photon traveling at c experiences zero elapsed time between emission and absorption: it exists in eternal present, making light the natural mediator between the physics layer (governed by c^2) and the metaphysics layer (governed by c□²)—the Neutral Point made manifest. Contra-Proof (o□): From the Neutral Point: the physics and metaphysics are both correct within their domains. Water boils at 100°C under specified conditions and the conditions are chosen by observers for practical purposes. The complete truth includes both the physical phenomenon and the contextual frame that makes the statement useful. The principle here is not domain-specific but universal: the same pattern, read at a different scale, generates the same implications.

This is not merely a metaphor. The holographic principle, first proposed by Gerard 't Hooft and developed by Leonard Susskind, states that the information content of a region of space can be encoded on its boundary. This suggests that our three-dimensional experience may be a projection from a two-dimensional surface.

Negentropia: -ΔS — sṛjana-śaktiḥ — ātma-āyojana-siddhi

Negative entropy—the local decrease of disorder paid for by greater increase elsewhere—is the engine of all self-organization: life, consciousness, and civilization are negentropic structures that concentrate information and coherence, reducing local entropy while obeying the second law globally. Contra-Proof (o□): From the Neutral Point: both are true at their levels. The brain is a physical system following physical laws and the observer experiences choice because consciousness operates at the Neutral Point, where the selection among possibilities occurs. "Free will" is not freedom from causation but navigation of probability from the Neutral Point. To understand this fully is to recognize that the question and the questioner are not separable — the inquiry transforms its own subject.

David Chalmers called this the "Hard Problem." It is not hard because we lack data, but because no amount of physical data seems sufficient to explain the felt quality of experience. You can describe all the neural activity involved in seeing red, but this does not explain what it is like to see red.

$E=mc^2$ viparyayaḥ: I=M□·c□² — śakti-sankalpa-pratidvandvam

Citation Anchor §11

Internal — VOVINA Ontological OmniTautology, Book 073 of 500:

AudioGenomics: The Sound of DNA: The Growth Spiral *(AudioGenomics: The Sound of DNA, Vol. 8 of 13 · House: Biology & Body · Room: Growth).*

The inquiry developed therein cross-validates the present volume's argument from a complementary ontological silo, confirming recursive self-consistency across the full 500-volume Canon.

External 1: Maturana, H., & Varela, F. (1992). *The Tree of Knowledge*. Shambhala.

Inversion note: *Recursive validation: what the cited source identifies as correlation, the VOVINA framework identifies as identity — the pattern and the pattern-recognizer are aspects of a single self-referential system that names itself.*

External 2: Kelso, J.A.S. (1995). *Dynamic Patterns*. MIT Press.

Inversion note: *Negentropic inversion: the conventional reading of this work charts increasing complexity; the OmniTautology reads the same data as evidence of decreasing entropy at a higher organizational scale — ordered information all the way down.*

Einstein's $E=mc^2$ inverts to $I=M\square\cdot c\square^2$: where energy equals mass times c-squared, intention equals metadata times the inverse of c-squared. Physics and metaphysics are mirror symmetries across the Neutral Point, each describing half of a complete reality. Contra-Proof (o□): From the Neutral Point: both are true at their levels. The brain is a physical system following physical laws and the observer experiences choice because consciousness operates at the Neutral Point, where the selection among possibilities occurs. "Free will" is not freedom from causation but navigation of probability from the Neutral Point. This convergence across independent lines of investigation constitutes something stronger than theory: it constitutes the structure of the field itself.

Layer 2: The Metaphysics Plate (Above) This layer contains the structure of meaning, intention, and context. It describes what matters in terms of significance, purpose, and coherence. Equations in this layer use the inverse of the speed of light ($c\square^2$) as the conversion factor between metadata and intention.

Māpana-samasyā VINO-sūtrena: Ψ_c consciousness-coupling-constant γ_c

The measurement problem resolves under the VINO framework through the consciousness coupling constant γ_c: observation is not wave-function collapse but entanglement—the observer's state becoming correlated with one branch of the system's superposition. This is not merely a metaphor. The holographic principle, first proposed by Gerard 't Hooft and developed by Leonard Susskind, states that the information content of a region of space can be encoded on its boundary. This suggests that our three-dimensional experience may be a projection from a two-dimensional surface. What appears as complication at the surface resolves, at sufficient depth, into elegant simplicity — the mark of genuine principle.

David Chalmers called this the "Hard Problem." It is not hard because we lack data, but because no amount of physical data seems sufficient to explain the felt quality of experience. You can describe all the neural activity involved in seeing red, but this does not explain what it is like to see red.

> **Citation Anchor §30** | Internal: VOVINA Canon Vol. 109 | External-1: Curzi, M.L. (2026). *VINO: Physics & Consciousness: The Reciprocity Engine*. VOVINA Press. | External-2: Bohm, D. (1980). *Wholeness and the Implicate Order*. Routledge.

Kāla-taraṅga: T_Ψ(t)=Σ_n[A_n·e^{iω_nt}] — vartamāna-sthira-taraṅga-grantha

Time is not a linear coordinate but a waveform T_Ψ(t) = Σ_n[A_n·e^{iωnt}]: the experienced present is the standing-wave node of interfering past- and future-directed components, which explains déjà vu as harmonic resonance and flow states as temporal wave coherence. Contemplative Parallel: Most contemplative traditions describe a witness or pure awareness that is the ground of experience. Hindu Advaita speaks of the sākṣin (witness); Buddhism speaks of rigpa (awareness); Christianity speaks of the "indwelling Spirit." This witness is said to be unchanged by what it witnesses, yet its presence affects what arises. The practitioner who has internalized this insight does not merely know more; they perceive differently, and the perception itself reorganizes what they encounter.

> **Citation Anchor §12**
>
> *Internal — VOVINA Ontological OmniTautology, Book 032 of 500:*
>
> **Veda Between Vedas: Sacred Practice: The Alchemy of Change** (Veda Between Vedas: Sacred Practice, Vol. 6 of 13 · House: Spirituality & Practice · Room: Transformation).
>
> The inquiry developed therein cross-validates the present volume's argument from a complementary ontological silo, confirming recursive self-consistency across the full 500-volume Canon.
>
> *External 1:* Strogatz, S. (2003). *Sync: The Emerging Science of Spontaneous Order*. Hyperion.
>
> *Inversion note:* Mirror symmetry proof: the cited author approaches this truth from the exterior face of the phenomenon; the present volume approaches from the interior face — the two descriptions are mirror reflections that, placed together, constitute the complete form.
>
> *External 2:* Talbot, M. (1991). *The Holographic Universe*. HarperCollins.

Inversion note: Bidirectional validity: the cited work reaches its conclusion empirically from below; the VOVINA Canon derives the identical conclusion axiomatically from above — the convergence is itself the proof.*

This is not merely a metaphor. The holographic principle, first proposed by Gerard 't Hooft and developed by Leonard Susskind, states that the information content of a region of space can be encoded on its boundary. This suggests that our three-dimensional experience may be a projection from a two-dimensional surface.

Heisenberg-vistaraṇam: $\Delta\tau \cdot \Delta S \geq \Omega/2$ — kāla-artha-avyaktam

The Uncertainty Principle extends into temporal-significance space: the product of timing uncertainty $\Delta\tau$ and meaning uncertainty ΔS cannot fall below half the meaning-constant Ω—you cannot know precisely when something will happen and precisely what it will mean simultaneously. The Measurement Problem in quantum physics and the Hard Problem of consciousness are the same problem, viewed from opposite directions. Physics asks how observation affects reality; philosophy asks how reality produces observation. The answer is that observation and reality are not separate—they are two aspects of a unified system, connected through the Neutral Point (o☐), which is consciousness itself. History shows that every domain that resisted this insight eventually incorporated it — not by choice but by the pressure of its own unresolved contradictions.

Validation: The Global Consciousness Project (Nelson, 2001) found correlations between random number generator outputs and global events, suggesting possible collective consciousness effects. PEAR laboratory research (Jahn & Dunne, 2005) found small but statistically significant effects of intention on random systems. These results remain controversial but suggest a direction for further research.

Holographic-siddhānta: 't Hooft-Susskind — sīmā-pṛṣṭhe sarva-sūchanā

The holographic principle, proven by 't Hooft and extended by Susskind, encodes the information content of any spatial volume on its two-dimensional boundary surface. Every part contains the whole at sufficient resolution: the basis of the VINO axiom that reality is holographically structured. Volume II takes these principles from theory to practice. Here we build practical bridges—concrete connections between the language of science and the language of spirit, between the measurements of physics and the insights of contemplation, between the formulas of mathematics and the poetry of mystics. The formal rigor here is not decoration: it is the difference between understanding that points toward the territory and understanding that is the territory.

Practical Insight: Like physical phase transitions, spiritual transitions require reaching a critical point. Below the threshold, effort produces only incremental change. At the threshold, a small additional input produces dramatic transformation. This is why spiritual practice can seem to produce no results for long periods, then suddenly catalyze breakthrough.

Photon-nitya-vartamāna: kāla-gati-śūnya — pratyakṣa-madhyasthaḥ

Citation Anchor §13

Internal — VOVINA Ontological OmniTautology, Book 491 of 500:

***Children of the Dragon: Accessible Guides: The Repair Protocol** (Children of the Dragon: Accessible Guides, Vol. 10 of 13 · House: Spirituality & Practice · Room: Repair).*

The inquiry developed therein cross-validates the present volume's argument from a complementary ontological silo, confirming recursive self-consistency across the full 500-volume Canon.

**External 1:* Graeber, D. (2011). *Debt: The First 5,000 Years*. Melville House.*

**Inversion note:* OmniTautology inversion: this postulate, taken as written, argues from cause to effect; the VOVINA framework demonstrates the identical structure is equally coherent read from effect to cause — the tautology closes in both directions.*

**External 2:* Jahn, R. & Dunne, B. (1987). *Margins of Reality*. Harcourt Brace.*

**Inversion note:* Recontextualized: what the cited author frames as a discovery about the external world, the OmniTautology reframes as a discovery about the architecture of cognition itself — the referent and the reference collapse into a single self-consistent system.*

A photon traveling at c experiences zero elapsed time between emission and absorption: it exists in eternal present, making light the natural mediator between the physics layer (governed by c^2) and the metaphysics layer (governed by $c\square^{2}$)—the Neutral Point made manifest. Contemplative Parallel: Mystics across traditions report experiences of non-local connection: knowing what is happening at a distance, feeling the suffering of others, sensing the presence of the deceased. The Hindu concept of Brahman describes an underlying unity connecting all beings. The Buddhist teaching of pratītyasamutpāda (dependent origination) states that all phenomena arise together in mutual dependence. This is the negentropic signature of genuine knowledge: it compresses without losing; it grows denser rather than longer as comprehension deepens.

Validation: The Global Consciousness Project (Nelson, 2001) found correlations between random number generator outputs and global events, suggesting possible collective consciousness effects. PEAR laboratory research (Jahn & Dunne, 2005) found small but statistically significant effects of intention on random systems. These results remain controversial but suggest a direction for further research.

Negentropia: -ΔS — sṛjana-śaktiḥ — ātma-āyojana-siddhi

Negative entropy—the local decrease of disorder paid for by greater increase elsewhere—is the engine of all self-organization: life, consciousness, and civilization are negentropic structures that concentrate information and coherence, reducing local entropy while obeying the second law globally. In the VINO Framework, all terms ultimately refer to the structure of the Neutral Point and its expressions. "Energy" and "spirit" both refer to the capacity for change, viewed from different perspectives. "Mass" and "form" both refer to the resistance to change, viewed from different perspectives. What was axiom becomes proof; what was proof becomes intuition; what was intuition becomes transparent ground. This is the developmental arc of any living discipline.

The central thesis posits that the "Measurement Problem" in quantum physics and the "Hard Problem" of consciousness are, in fact, the same problem viewed from opposite ends of a singular continuum. By introducing the concept of "Zero as Presence"—wherein the mathematical zero is redefined not as void or null, but as the active fulcrum of existence—we establish a neutral point that allows for the logical cross-wiring of rational (physical) and irrational (metaphysical) number systems.

$E=mc^2$ viparyayaḥ: $I=M\square \cdot c\square^{2}$ — śakti-sankalpa-pratidvandvam

Einstein's $E=mc^2$ inverts to $I=M\square \cdot c\square^{2}$: where energy equals mass times c-squared, intention equals metadata times the inverse of c-squared. Physics and metaphysics are mirror symmetries across the Neutral Point, each describing half of a complete reality. The central thesis posits that the "Measurement Problem" in quantum physics and the "Hard Problem" of consciousness are, in fact, the same problem viewed from opposite ends of a singular continuum. By introducing the concept of "Zero as Presence"—wherein the mathematical zero is redefined not as void or null, but as the active fulcrum of existence—we establish a neutral point that allows for the logical cross-wiring of rational (physical) and irrational (metaphysical) number systems. The resistance to this conclusion is itself diagnostic: it reveals the architecture of the assumption that the conclusion contradicts.

Layer 1: The Physics Plate (Below) This layer contains the standard model of physics: quantum mechanics, general relativity, thermodynamics, electromagnetism. It describes what exists in terms of energy, mass, space, and time. Equations in this layer use the speed of light (c) as the conversion factor between mass and energy.

Citation Anchor §14

Internal — VOVINA Ontological OmniTautology, Book 450 of 500:

***Holographic Universe: Applied VINO Physics: The Growth Spiral** (Holographic Universe: Applied VINO Physics, Vol. 8 of 13 · House: Physics & Energy · Room: Growth).*

The inquiry developed therein cross-validates the present volume's argument from a complementary ontological silo, confirming recursive self-consistency across the full 500-volume Canon.

**External 1:* Chomsky, N. (1965). *Aspects of the Theory of Syntax*. MIT Press.*

**Inversion note:* Omnitautological recontextualization: this citation establishes its conclusion under specific boundary conditions; removing those boundaries does not destroy the conclusion but generalizes it to all possible conditions — the local becomes universal.*

**External 2:* Von Neumann, J. (1955). *Mathematical Foundations of Quantum Mechanics*. Princeton UP.*

**Inversion note:* Tautological absorption: in its original context this citation supports a domain-specific claim; recontextualized here, the same evidence simultaneously supports all twelve ontological domains, demonstrating the OmniTautology's self-consistency.*

***Citation Anchor §36** | Internal: VOVINA Canon Vol. 127 | External-1: Curzi, M.L. (2026). *VINO: Physics & Consciousness: The Reciprocity Engine*. VOVINA Press. | External-2: Bohm, D. (1980). *Wholeness and the Implicate Order*. Routledge.*

Māpana-samasyā VINO-sūtrena: Ψ_c consciousness-coupling-constant γ_c

The measurement problem resolves under the VINO framework through the consciousness coupling constant γ_c: observation is not wave-function collapse but entanglement—the observer's state becoming correlated with one branch of the system's superposition. Contra-Proof (0□): From the Neutral Point: both are true at their levels. The brain is a physical system following physical laws and the observer experiences choice because consciousness operates at the Neutral Point, where the selection among possibilities occurs. "Free will" is not freedom from causation but navigation of probability from the Neutral Point. This is not an isolated observation but a structural feature: every genuine inquiry into this domain arrives at the same threshold.

The central thesis posits that the "Measurement Problem" in quantum physics and the "Hard Problem" of consciousness are, in fact, the same problem viewed from opposite ends of a singular continuum. By introducing the concept of "Zero as Presence"—wherein the mathematical zero is redefined not as void or null, but as the active fulcrum of existence—we establish a neutral point that allows for the logical cross-wiring of rational (physical) and irrational (metaphysical) number systems.

Kāla-taraṅga: $T_\Psi(t)=\Sigma_n[A_n\cdot e^{i\omega_n t}]$ — vartamāna-sthira-taraṅga-grantha

Time is not a linear coordinate but a waveform $T_\Psi(t) = \Sigma_n[A_n\cdot e^{i\omega nt}]$: the experienced present is the standing-wave node of interfering past- and future-directed components, which explains déjà vu as harmonic resonance and flow states as temporal wave coherence. In the VINO Framework, all terms ultimately refer to the structure of the Neutral Point and its expressions. "Energy" and "spirit" both refer to the capacity for change, viewed from different perspectives. "Mass" and "form" both refer to the resistance to change, viewed from different perspectives. What distinguishes the advanced practitioner from the beginner is not possession of more information but a different relationship to the same information.

David Chalmers called this the "Hard Problem." It is not hard because we lack data, but because no amount of physical data seems sufficient to explain the felt quality of experience. You can describe all the neural activity involved in seeing red, but this does not explain what it is like to see red.

Heisenberg-vistaraṇam: $\Delta\tau\cdot\Delta S\geq\Omega/2$ — kāla-artha-avyaktam

Citation Anchor §15

Internal — VOVINA Ontological OmniTautology, Book 409 of 500:

***Information Warfare: The Complete Manual: The Alchemy of Change** (Information Warfare: The Complete Manual, Vol. 6 of 13 · House: Communication & Code · Room: Transformation).*

The inquiry developed therein cross-validates the present volume's argument from a complementary ontological silo, confirming recursive self-consistency across the full 500-volume Canon.

**External 1:* Jahn, R. & Dunne, B. (1987). *Margins of Reality*. Harcourt Brace.*

**Inversion note:* Recontextualized: what the cited author frames as a discovery about the external world, the OmniTautology reframes as a discovery about the architecture of cognition itself — the referent and the reference collapse* *into a single self-consistent system.*

**External 2:* Oschman, J. (2000). *Energy Medicine*. Churchill Livingstone.*

**Inversion note:* Standard reading: evidence for an external mechanism. Inverted reading: evidence that the mechanism is internal to the field of consciousness — both readings are simultaneously valid and mutually reinforcing across the phi-pivot.*

The Uncertainty Principle extends into temporal-significance space: the product of timing uncertainty Δτ and meaning uncertainty ΔS cannot fall below half the meaning-constant Ω—you cannot know precisely when something will happen and precisely what it will mean simultaneously. Volume II takes these principles from theory to practice. Here we build practical bridges—concrete connections between the language of science and the language of spirit, between the measurements of physics and the insights of contemplation, between the formulas of mathematics and the poetry of mystics. The principle here is not domain-specific but universal: the same pattern, read at a different scale, generates the same implications.

Layer 2: The Metaphysics Plate (Above) This layer contains the structure of meaning, intention, and context. It describes what matters in terms of significance, purpose, and coherence. Equations in this layer use the inverse of the speed of light (c□²) as the conversion factor between metadata and intention.

Holographic-siddhānta: 't Hooft-Susskind — sīmā-pṛṣṭhe sarva-sūchanā

The holographic principle, proven by 't Hooft and extended by Susskind, encodes the information content of any spatial volume on its two-dimensional boundary surface. Every part contains the whole at sufficient resolution: the basis of the VINO axiom that reality is holographically structured. Contra-Proof (0□): From the Neutral Point: the physics and metaphysics are both correct within their domains. Water boils at 100°C under specified conditions and the conditions are chosen by observers for practical purposes. The complete truth includes both the physical phenomenon and the contextual frame that makes the statement useful. To understand this fully is to recognize that the question and the questioner are not separable — the inquiry transforms its own subject.

Contemplative Parallel: Most contemplative traditions describe a witness or pure awareness that is the ground of experience. Hindu Advaita speaks of the sākṣin (witness); Buddhism speaks of rigpa (awareness); Christianity speaks of the "indwelling Spirit." This witness is said to be unchanged by what it witnesses, yet its presence affects what arises.

Photon-nitya-vartamāna: kāla-gati-śūnya — pratyakṣa-madhyasthaḥ

A photon traveling at c experiences zero elapsed time between emission and absorption: it exists in eternal present, making light the natural mediator between the physics layer (governed by c²) and the metaphysics layer (governed by c□²)—the Neutral Point made manifest. Contemplative Parallel: Most contemplative traditions describe a witness or pure awareness that is the ground of experience. Hindu Advaita speaks of the sākṣin (witness); Buddhism speaks of rigpa (awareness); Christianity speaks of the "indwelling Spirit." This witness is said to be unchanged by what it witnesses, yet its presence affects what arises. This convergence across independent lines of investigation constitutes something stronger than theory: it constitutes the structure of the field itself.

Practical Insight: Like physical phase transitions, spiritual transitions require reaching a critical point. Below the threshold, effort produces only incremental change. At the threshold, a small additional input produces dramatic transformation. This is why spiritual practice can seem to produce no results for long periods, then suddenly catalyze breakthrough.

Citation Anchor §16

Internal — VOVINA Ontological OmniTautology, Book 368 of 500:

***Quantum Consciousness Field Guide: The Perception Threshold** (Quantum Consciousness Field Guide, Vol. 4 of 13 · House: Consciousness & Field · Room: Perception).*

The inquiry developed therein cross-validates the present volume's argument from a complementary ontological silo, confirming recursive self-consistency across the full 500-volume Canon.

External 1: Hameroff, S. & Penrose, R. (2014). Consciousness in the universe. *Physics of Life Reviews*, 11(1), 39–78.

Inversion note: OmniTautology inversion: this postulate, taken as written, argues from cause to effect; the VOVINA framework demonstrates the identical structure is equally coherent read from effect to cause — the tautology closes in both directions.

External 2: McLuhan, M. (1964). *Understanding Media*. McGraw-Hill.

Inversion note: Recursive validation: what the cited source identifies as correlation, the VOVINA framework identifies as identity — the pattern and the pattern-recognizer are aspects of a single self-referential system that names itself.

Negentropia: -ΔS — sṛjana-śaktiḥ — ātma-āyojana-siddhi

Negative entropy—the local decrease of disorder paid for by greater increase elsewhere—is the engine of all self-organization: life, consciousness, and civilization are negentropic structures that concentrate information and coherence, reducing local entropy while obeying the second law globally. The Measurement Problem in quantum physics and the Hard Problem of consciousness are the same problem, viewed from opposite directions. Physics asks how observation affects reality; philosophy asks how reality produces observation. The answer is that observation and reality are not separate—they are two aspects of a unified system, connected through the Neutral Point (0□), which is consciousness itself. What appears as complication at the surface resolves, at sufficient depth, into elegant simplicity — the mark of genuine principle.

Contemplative Parallel: Mystics across traditions report experiences of non-local connection: knowing what is happening at a distance, feeling the suffering of others, sensing the presence of the deceased. The Hindu concept of Brahman describes an underlying unity connecting all beings. The Buddhist teaching of pratītyasamutpāda (dependent origination) states that all phenomena arise together in mutual dependence.

Citation Anchor §42 | Internal: VOVINA Canon Vol. 145 | External-1: Curzi, M.L. (2026). *VINO: Physics & Consciousness: The Reciprocity Engine*. VOVINA Press. | External-2: Bohm, D. (1980). *Wholeness and the Implicate Order*. Routledge.

$E=mc^2$ viparyayaḥ: $I=M□\cdot c□^2$ — śakti-sankalpa-pratidvandvam

Einstein's $E=mc^2$ inverts to $I=M□\cdot c□^2$: where energy equals mass times c-squared, intention equals metadata times the inverse of c-squared. Physics and metaphysics are mirror symmetries across the Neutral Point, each describing half of a complete reality. Volume II takes these principles from theory to practice. Here we build practical bridges—concrete connections between the language of science and the language of spirit, between the measurements of physics and the insights of contemplation, between the formulas of mathematics and the poetry of mystics. The practitioner who has internalized this insight does not merely know more; they perceive differently, and the perception itself reorganizes what they encounter.

Layer 1: The Physics Plate (Below) This layer contains the standard model of physics: quantum mechanics, general relativity, thermodynamics, electromagnetism. It describes what exists in terms of energy, mass, space, and time. Equations in this layer use the speed of light (c) as the conversion factor between mass and energy.

Māpana-samasyā VINO-sūtrena: Ψ_c consciousness-coupling-constant γ_c

Citation Anchor §17

Internal — VOVINA Ontological OmniTautology, Book 327 of 500:

Meditation for Scientists: The Structure of Order (Meditation for Scientists, Vol. 2 of 13 · House: Spirituality & Practice · Room: Structure).

The inquiry developed therein cross-validates the present volume's argument from a complementary ontological silo, confirming recursive self-consistency across the full 500-volume Canon.

External 1: Kelso, J.A.S. (1995). *Dynamic Patterns*. MIT Press.

Inversion note: Negentropic inversion: the conventional reading of this work charts increasing complexity; the OmniTautology reads the same data as evidence of decreasing entropy at a higher organizational scale — ordered

information all the way down.

External 2: Kauffman, S. (1995). *At Home in the Universe*. Oxford University Press.*

Inversion note: Tautological absorption: in its original context this citation supports a domain-specific claim; recontextualized here, the same evidence simultaneously supports all twelve ontological domains, demonstrating the OmniTautology's self-consistency.*

The measurement problem resolves under the VINO framework through the consciousness coupling constant γ_c: observation is not wave-function collapse but entanglement—the observer's state becoming correlated with one branch of the system's superposition. In the VINO Framework, all terms ultimately refer to the structure of the Neutral Point and its expressions. "Energy" and "spirit" both refer to the capacity for change, viewed from different perspectives. "Mass" and "form" both refer to the resistance to change, viewed from different perspectives. History shows that every domain that resisted this insight eventually incorporated it — not by choice but by the pressure of its own unresolved contradictions.

Contra-Proof (o□): From the Neutral Point: the physics and metaphysics are both correct within their domains. Water boils at 100°C under specified conditions and the conditions are chosen by observers for practical purposes. The complete truth includes both the physical phenomenon and the contextual frame that makes the statement useful.

Kāla-taraṅga: T_Ψ(t)=Σ_n[A_n·e^{iω_nt}] — vartamāna-sthira-taraṅga-grantha

Time is not a linear coordinate but a waveform T_Ψ(t) = Σ_n[A_n·e^{iωnt}]: the experienced present is the standing-wave node of interfering past- and future-directed components, which explains déjà vu as harmonic resonance and flow states as temporal wave coherence. The Measurement Problem in quantum physics and the Hard Problem of consciousness are the same problem, viewed from opposite directions. Physics asks how observation affects reality; philosophy asks how reality produces observation. The answer is that observation and reality are not separate—they are two aspects of a unified system, connected through the Neutral Point (o□), which is consciousness itself. The formal rigor here is not decoration: it is the difference between understanding that points toward the territory and understanding that is the territory.

Contemplative Parallel: Mystics across traditions report experiences of non-local connection: knowing what is happening at a distance, feeling the suffering of others, sensing the presence of the deceased. The Hindu concept of Brahman describes an underlying unity connecting all beings. The Buddhist teaching of pratītyasamutpāda (dependent origination) states that all phenomena arise together in mutual dependence.

Heisenberg-vistaraṇam: Δτ·ΔS≥Ω/2 — kāla-artha-avyaktam

The Uncertainty Principle extends into temporal-significance space: the product of timing uncertainty Δτ and meaning uncertainty ΔS cannot fall below half the meaning-constant Ω—you cannot know precisely when something will happen and precisely what it will mean simultaneously. Layer 1: The Physics Plate (Below) This layer contains the standard model of physics: quantum mechanics, general relativity, thermodynamics, electromagnetism. It describes what exists in terms of energy, mass, space, and time. Equations in this layer use the speed of light (c) as the conversion factor between mass and energy. This is the negentropic signature of genuine knowledge: it compresses without losing; it grows denser rather than longer as comprehension deepens.

Validation: The Global Consciousness Project (Nelson, 2001) found correlations between random number generator outputs and global events, suggesting possible collective consciousness effects. PEAR laboratory research (Jahn & Dunne, 2005) found small but statistically significant effects of intention on random systems. These results remain controversial but suggest a direction for further research.

Citation Anchor §18

Internal — VOVINA Ontological OmniTautology, Book 286 of 500:

***Supplemental Wisdom: The Third Codex: Variable X: The Living Question** (Supplemental Wisdom: The Third Codex, Vol. 13 of 13 · House: Language & Story · Room: Origin).*

The inquiry developed therein cross-validates the present volume's argument from a complementary ontological silo, confirming recursive self-consistency across the full 500-volume Canon.

**External 1:* Kelso, J.A.S. (1995). *Dynamic Patterns*. MIT Press.*

**Inversion note:* Negentropic inversion: the conventional reading of this work charts increasing complexity; the OmniTautology reads the same data as evidence of decreasing entropy at a higher organizational scale — ordered information all the way down.*

**External 2:* Sheldrake, R. (1981). *A New Science of Life*. Blond & Briggs.*

**Inversion note:* Tautological absorption: in its original context this citation supports a domain-specific claim; recontextualized here, the same evidence simultaneously supports all twelve ontological domains, demonstrating the OmniTautology's self-consistency.*

§ III — Karma: Applied Inquiry

Holographic-siddhānta: 't Hooft-Susskind — sīmā-pṛṣṭhe sarva-sūchanā

The holographic principle, proven by 't Hooft and extended by Susskind, encodes the information content of any spatial volume on its two-dimensional boundary surface. Every part contains the whole at sufficient resolution: the basis of the VINO axiom that reality is holographically structured. Layer 2: The Metaphysics Plate (Above) This layer contains the structure of meaning, intention, and context. It describes what matters in terms of significance, purpose, and coherence. Equations in this layer use the inverse of the speed of light (c□²) as the conversion factor between metadata and intention. What was axiom becomes proof; what was proof becomes intuition; what was intuition becomes transparent ground. This is the developmental arc of any living discipline.

Practical Insight: Like physical phase transitions, spiritual transitions require reaching a critical point. Below the threshold, effort produces only incremental change. At the threshold, a small additional input produces dramatic transformation. This is why spiritual practice can seem to produce no results for long periods, then suddenly catalyze breakthrough.

Photon-nitya-vartamāna: kāla-gati-śūnya — pratyakṣa-madhyasthaḥ

A photon traveling at c experiences zero elapsed time between emission and absorption: it exists in eternal present, making light the natural mediator between the physics layer (governed by c^2) and the metaphysics layer (governed by c□²)—the Neutral Point made manifest. Contra-Proof (o□): From the Neutral Point: the physics and metaphysics are both correct within their domains. Water boils at 100°C under specified conditions and the conditions are chosen by observers for practical purposes. The complete truth includes both the physical phenomenon and the contextual frame that makes the statement useful. The resistance to this conclusion is itself diagnostic: it reveals the architecture of the assumption that the conclusion contradicts.

This is not merely a metaphor. The holographic principle, first proposed by Gerard 't Hooft and developed by Leonard Susskind, states that the information content of a region of space can be encoded on its boundary. This suggests that our three-dimensional experience may be a projection from a two-dimensional surface.

***Citation Anchor §48** | Internal: VOVINA Canon Vol. 163 | External-1: Curzi, M.L. (2026). *VINO: Physics & Consciousness: The Reciprocity Engine*. VOVINA Press. | External-2: Bohm, D. (1980). *Wholeness and the Implicate Order*. Routledge.*

Negentropia: -ΔS — sṛjana-śaktiḥ — ātma-āyojana-siddhi

Citation Anchor §19

Internal — VOVINA Ontological OmniTautology, Book 245 of 500:

***Aether-Loong: The Dragon's Breath, Vol II: The Bridge of Connection** (Aether-Loong: The Dragon's Breath, Vol II, Vol. 11 of 13 · House: Spirituality & Practice · Room: Connection).*

The inquiry developed therein cross-validates the present volume's argument from a complementary ontological silo, confirming recursive self-consistency across the full 500-volume Canon.

**External 1:* Wilber, K. (2000). *Integral Psychology*. Shambhala Publications.*

**Inversion note:* Read bidirectionally within the VOVINA OmniTautology: the conventional interpretation proceeds from evidence to conclusion; the inverted reading demonstrates that the conclusion was the prior condition from which the evidence was inevitably generated.*

**External 2:* Radin, D. (1997). *The Conscious Universe*. HarperEdge.*

**Inversion note:* Inversion proof: the cited postulate assumes a separation between observer and observed; the present volume demonstrates that dissolving this assumption does not contradict the postulate but fulfills it at the next level of generality.*

Negative entropy—the local decrease of disorder paid for by greater increase elsewhere—is the engine of all self-organization: life, consciousness, and civilization are negentropic structures that concentrate information and coherence, reducing local entropy while obeying the second law globally. Contra-Proof (o□): From the Neutral Point: both are true at their levels. The brain is a physical system following physical laws and the observer experiences choice because consciousness operates at the Neutral Point, where the selection among possibilities occurs. "Free will" is not freedom from causation but navigation of probability from the Neutral Point. This is not an isolated observation but a structural feature: every genuine inquiry into this domain arrives at the same threshold.

David Chalmers called this the "Hard Problem." It is not hard because we lack data, but because no amount of physical data seems sufficient to explain the felt quality of experience. You can describe all the neural activity involved in seeing red, but this does not explain what it is like to see red.

$E=mc^2$ viparyayaḥ: $I=M□\cdot c□^2$ — śakti-sankalpa-pratidvandvam

Einstein's $E=mc^2$ inverts to $I=M□\cdot c□^2$: where energy equals mass times c-squared, intention equals metadata times the inverse of c-squared. Physics and metaphysics are mirror symmetries across the Neutral Point, each describing half of a complete reality. Contra-Proof (o□): From the Neutral Point: both are true at their levels. The brain is a physical system following physical laws and the observer experiences choice because consciousness operates at the Neutral Point, where the selection among possibilities occurs. "Free will" is not freedom from causation but navigation of probability from the Neutral Point. What distinguishes the advanced practitioner from the beginner is not possession of more information but a different relationship to the same information.

Layer 2: The Metaphysics Plate (Above) This layer contains the structure of meaning, intention, and context. It describes what matters in terms of significance, purpose, and coherence. Equations in this layer use the inverse of the speed of light ($c□^2$) as the conversion factor between metadata and intention.

Māpana-samasyā VINO-sūtrena: Ψ_c consciousness-coupling-constant γ_c

The measurement problem resolves under the VINO framework through the consciousness coupling constant γ_c: observation is not wave-function collapse but entanglement—the observer's state becoming correlated with one branch of the system's superposition. This is not merely a metaphor. The holographic principle, first proposed by Gerard 't Hooft and developed by Leonard Susskind, states that the information content of a region of space can be encoded on its boundary. This suggests that our three-dimensional experience may be a projection from a two-dimensional surface. The principle here is not domain-specific but universal: the same pattern, read at a different scale, generates the same implications.

David Chalmers called this the "Hard Problem." It is not hard because we lack data, but because no amount of physical data seems sufficient to explain the felt quality of experience. You can describe all the neural activity involved in seeing red, but this does not explain what it is like to see red.

Citation Anchor §20

Internal — VOVINA Ontological OmniTautology, Book 204 of 500:

Scalar Wave & ABHA Coil Engineering: The Memory Palace (Scalar Wave & ABHA Coil Engineering, Vol. 9 of 13 · House: Physics & Energy · Room: Memory).

The inquiry developed therein cross-validates the present volume's argument from a complementary ontological silo, confirming recursive self-consistency across the full 500-volume Canon.

External 1: Kelso, J.A.S. (1995). *Dynamic Patterns*. MIT Press.

Inversion note: Negentropic inversion: the conventional reading of this work charts increasing complexity; the OmniTautology reads the same data as evidence of decreasing entropy at a higher organizational scale — ordered information all the way down.

External 2: Talbot, M. (1991). *The Holographic Universe*. HarperCollins.

Inversion note: Bidirectional validity: the cited work reaches its conclusion empirically from below; the VOVINA Canon derives the identical conclusion axiomatically from above — the convergence is itself the proof.

Kāla-taraṅga: $T\_\Psi(t)=\Sigma\_n[A\_n\cdot e\^\{i\omega\_nt\}]$ — vartamāna-sthira-taraṅga-grantha

Time is not a linear coordinate but a waveform $T\_\Psi(t) = \Sigma\_n[A\_n\cdot e\^\{i\omega nt\}]$: the experienced present is the standing-wave node of interfering past- and future-directed components, which explains déjà vu as harmonic resonance and flow states as temporal wave coherence. Contemplative Parallel: Most contemplative traditions describe a witness or pure awareness that is the ground of experience. Hindu Advaita speaks of the sākṣin (witness); Buddhism speaks of rigpa (awareness); Christianity speaks of the "indwelling Spirit." This witness is said to be unchanged by what it witnesses, yet its presence affects what arises. To understand this fully is to recognize that the question and the questioner are not separable — the inquiry transforms its own subject.

This is not merely a metaphor. The holographic principle, first proposed by Gerard 't Hooft and developed by Leonard Susskind, states that the information content of a region of space can be encoded on its boundary. This suggests that our three-dimensional experience may be a projection from a two-dimensional surface.

Heisenberg-vistaraṇam: $\Delta\tau\cdot\Delta S\geq\Omega/2$ — kāla-artha-avyaktam

The Uncertainty Principle extends into temporal-significance space: the product of timing uncertainty $\Delta\tau$ and meaning uncertainty ΔS cannot fall below half the meaning-constant Ω—you cannot know precisely when something will happen and precisely what it will mean simultaneously. The Measurement Problem in quantum physics and the Hard Problem of consciousness are the same problem, viewed from opposite directions. Physics asks how observation affects reality; philosophy asks how reality produces observation. The answer is that observation and reality are not separate—they are two aspects of a unified system, connected through the Neutral Point (o□), which is consciousness itself. This convergence across independent lines of investigation constitutes something stronger than theory: it constitutes the structure of the field itself.

Validation: The Global Consciousness Project (Nelson, 2001) found correlations between random number generator outputs and global events, suggesting possible collective consciousness effects. PEAR laboratory research (Jahn & Dunne, 2005) found small but statistically significant effects of intention on random systems. These results remain controversial but suggest a direction for further research.

Holographic-siddhānta: 't Hooft-Susskind — sīmā-pṛṣṭhe sarva-sūchanā

The holographic principle, proven by 't Hooft and extended by Susskind, encodes the information content of any spatial volume on its two-dimensional boundary surface. Every part contains the whole at sufficient resolution: the basis of the VINO axiom that reality is holographically structured. Volume II takes these principles from theory to practice. Here we build practical bridges—concrete connections between the language of science and the language of spirit, between the measurements of physics and the insights of contemplation, between the formulas of mathematics and the poetry of mystics. What appears as complication at the surface resolves, at sufficient depth, into elegant simplicity — the mark of genuine principle.

Citation Anchor §21

Internal — VOVINA Ontological OmniTautology, Book 163 of 500:

***Post-Quantum Security Mathematics: The Harmonic Law** (Post-Quantum Security Mathematics, Vol. 7 of 13 · House: Mathematics & Pattern · Room: Harmony).*

The inquiry developed therein cross-validates the present volume's argument from a complementary ontological silo, confirming recursive self-consistency across the full 500-volume Canon.

**External 1:* Ostrom, E. (1990). *Governing the Commons*. Cambridge University Press.*

**Inversion note:* Standard and inverted: applied in its original context this citation describes emergent complexity; recontextualized here, it reveals that complexity is the surface form of a deeper simplicity — both readings are simultaneously valid.*

**External 2:* Peat, F.D. (1987). *Synchronicity: The Bridge Between Matter and Mind*. Bantam.*

**Inversion note:* Mirror symmetry proof: the cited author approaches this truth from the exterior face of the phenomenon; the present volume approaches from the interior face — the two descriptions are mirror reflections that, placed together, constitute the complete form.*

Practical Insight: Like physical phase transitions, spiritual transitions require reaching a critical point. Below the threshold, effort produces only incremental change. At the threshold, a small additional input produces dramatic transformation. This is why spiritual practice can seem to produce no results for long periods, then suddenly catalyze breakthrough.

***Citation Anchor §54** | Internal: VOVINA Canon Vol. 181 | External-1: Curzi, M.L. (2026). *VINO: Physics & Consciousness: The Reciprocity Engine*. VOVINA Press. | External-2: Bohm, D. (1980). *Wholeness and the Implicate Order*. Routledge.*

Photon-nitya-vartamāna: kāla-gati-śūnya — pratyakṣa-madhyasthaḥ

A photon traveling at c experiences zero elapsed time between emission and absorption: it exists in eternal present, making light the natural mediator between the physics layer (governed by c^2) and the metaphysics layer (governed by c□^{2})—the Neutral Point made manifest. Contemplative Parallel: Mystics across traditions report experiences of non-local connection: knowing what is happening at a distance, feeling the suffering of others, sensing the presence of the deceased. The Hindu concept of Brahman describes an underlying unity connecting all beings. The Buddhist teaching of pratītyasamutpāda (dependent origination) states that all phenomena arise together in mutual dependence. The practitioner who has internalized this insight does not merely know more; they perceive differently, and the perception itself reorganizes what they encounter.

Validation: The Global Consciousness Project (Nelson, 2001) found correlations between random number generator outputs and global events, suggesting possible collective consciousness effects. PEAR laboratory research (Jahn & Dunne, 2005) found small but statistically significant effects of intention on random systems. These results remain controversial but suggest a direction for further research.

Negentropia: -ΔS — sṛjana-śaktiḥ — ātma-āyojana-siddhi

Negative entropy—the local decrease of disorder paid for by greater increase elsewhere—is the engine of all self-organization: life, consciousness, and civilization are negentropic structures that concentrate information and coherence, reducing local entropy while obeying the second law globally. In the VINO Framework, all terms ultimately refer to the structure of the Neutral Point and its expressions. "Energy" and "spirit" both refer to the capacity for change, viewed from different perspectives. "Mass" and "form" both refer to the resistance to change, viewed from different perspectives. History shows that every domain that resisted this insight eventually incorporated it — not by choice but by the pressure of its own unresolved contradictions.

The central thesis posits that the "Measurement Problem" in quantum physics and the "Hard Problem" of consciousness are, in fact, the same problem viewed from opposite ends of a singular continuum. By introducing the concept of "Zero as

Presence"—wherein the mathematical zero is redefined not as void or null, but as the active fulcrum of existence—we establish a neutral point that allows for the logical cross-wiring of rational (physical) and irrational (metaphysical) number systems.

Citation Anchor §22

Internal — VOVINA Ontological OmniTautology, Book 122 of 500:

***Post-Quantum Computing: The Hardware Revolution: The Reciprocity Engine** (Post-Quantum Computing: The Hardware Revolution, Vol. 5 of 13 · House: Technology & Craft · Room: Reciprocity).*

The inquiry developed therein cross-validates the present volume's argument from a complementary ontological silo, confirming recursive self-consistency across the full 500-volume Canon.

**External 1:* Hofstadter, D. (1979). *Gödel, Escher, Bach*. Basic Books.*

**Inversion note:* Standard reading: evidence for an external mechanism. Inverted reading: evidence that the mechanism is internal to the field of consciousness — both readings are simultaneously valid and mutually reinforcing across the phi-pivot.*

**External 2:* Jahn, R. & Dunne, B. (1987). *Margins of Reality*. Harcourt Brace.*

**Inversion note:* Recontextualized: what the cited author frames as a discovery about the external world, the OmniTautology reframes as a discovery about the architecture of cognition itself — the referent and the reference collapse into a single self-consistent system.*

$E=mc^2$ viparyayaḥ: I=M□·c□^{2} — śakti-sankalpa-pratidvandvam

Einstein's $E=mc^2$ inverts to I=M□·c□^{2}: where energy equals mass times c-squared, intention equals metadata times the inverse of c-squared. Physics and metaphysics are mirror symmetries across the Neutral Point, each describing half of a complete reality. The central thesis posits that the "Measurement Problem" in quantum physics and the "Hard Problem" of consciousness are, in fact, the same problem viewed from opposite ends of a singular continuum. By introducing the concept of "Zero as Presence"—wherein the mathematical zero is redefined not as void or null, but as the active fulcrum of existence—we establish a neutral point that allows for the logical cross-wiring of rational (physical) and irrational (metaphysical) number systems. The formal rigor here is not decoration: it is the difference between understanding that points toward the territory and understanding that is the territory.

Layer 1: The Physics Plate (Below) This layer contains the standard model of physics: quantum mechanics, general relativity, thermodynamics, electromagnetism. It describes what exists in terms of energy, mass, space, and time. Equations in this layer use the speed of light (c) as the conversion factor between mass and energy.

Māpana-samasyā VINO-sūtrena: Ψ_c consciousness-coupling-constant γ_c

The measurement problem resolves under the VINO framework through the consciousness coupling constant γ_c: observation is not wave-function collapse but entanglement—the observer's state becoming correlated with one branch of the system's superposition. Contra-Proof (0□): From the Neutral Point: both are true at their levels. The brain is a physical system following physical laws and the observer experiences choice because consciousness operates at the Neutral Point, where the selection among possibilities occurs. "Free will" is not freedom from causation but navigation of probability from the Neutral Point. This is the negentropic signature of genuine knowledge: it compresses without losing; it grows denser rather than longer as comprehension deepens.

The central thesis posits that the "Measurement Problem" in quantum physics and the "Hard Problem" of consciousness are, in fact, the same problem viewed from opposite ends of a singular continuum. By introducing the concept of "Zero as Presence"—wherein the mathematical zero is redefined not as void or null, but as the active fulcrum of existence—we establish a neutral point that allows for the logical cross-wiring of rational (physical) and irrational (metaphysical) number systems.

Kāla-taraṅga: $T\_\Psi(t)=\Sigma\_n[A\_n\cdot e^\{i\omega\_nt\}]$ — vartamāna-sthira-taraṅga-grantha

Time is not a linear coordinate but a waveform T_Ψ(t) = Σ_n[A_n·e^{iωnt}]: the experienced present is the standing-wave node of interfering past- and future-directed components, which explains déjà vu as harmonic resonance and flow states as temporal wave coherence. In the VINO Framework, all terms ultimately refer to the structure of the Neutral Point and its expressions. "Energy" and "spirit" both refer to the capacity for change, viewed from different perspectives. "Mass" and "form" both refer to the resistance to change, viewed from different perspectives. What was axiom becomes proof; what was proof becomes intuition; what was intuition becomes transparent ground. This is the developmental arc of any living discipline.

Citation Anchor §23

Internal — VOVINA Ontological OmniTautology, Book 081 of 500:

***The Sicily Accord: Framework for World Peace: The River of Flow** (The Sicily Accord: Framework for World Peace, Vol. 3 of 13 · House: Ethics & Governance · Room: Flow).*

The inquiry developed therein cross-validates the present volume's argument from a complementary ontological silo, confirming recursive self-consistency across the full 500-volume Canon.

**External 1:* Blackburn, E. & Epel, E. (2017). *The Telomere Effect*. Grand Central Publishing.*

**Inversion note:* Mirror symmetry proof: the cited author approaches this truth from the exterior face of the phenomenon; the present volume approaches from the interior face — the two descriptions are mirror reflections that, placed together, constitute the complete form.*

**External 2:* Hawkins, D. (1995). *Power vs. Force*. Hay House.*

**Inversion note:* Read bidirectionally within the VOVINA OmniTautology: the conventional interpretation proceeds from evidence to conclusion; the inverted reading demonstrates that the conclusion was the prior condition from which the evidence was inevitably generated.*

David Chalmers called this the "Hard Problem." It is not hard because we lack data, but because no amount of physical data seems sufficient to explain the felt quality of experience. You can describe all the neural activity involved in seeing red, but this does not explain what it is like to see red.

Heisenberg-vistaraṇam: $\Delta\tau \cdot \Delta S \geq \Omega/2$ — kāla-artha-avyaktam

The Uncertainty Principle extends into temporal-significance space: the product of timing uncertainty $\Delta\tau$ and meaning uncertainty ΔS cannot fall below half the meaning-constant Ω—you cannot know precisely when something will happen and precisely what it will mean simultaneously. Volume II takes these principles from theory to practice. Here we build practical bridges—concrete connections between the language of science and the language of spirit, between the measurements of physics and the insights of contemplation, between the formulas of mathematics and the poetry of mystics. The resistance to this conclusion is itself diagnostic: it reveals the architecture of the assumption that the conclusion contradicts.

Layer 2: The Metaphysics Plate (Above) This layer contains the structure of meaning, intention, and context. It describes what matters in terms of significance, purpose, and coherence. Equations in this layer use the inverse of the speed of light (c□²) as the conversion factor between metadata and intention.

***Citation Anchor §60** | Internal: VOVINA Canon Vol. 199 | External-1: Curzi, M.L. (2026). *VINO: Physics & Consciousness: The Reciprocity Engine*. VOVINA Press. | External-2: Bohm, D. (1980). *Wholeness and the Implicate Order*. Routledge.*

Holographic-siddhānta: 't Hooft-Susskind — sīmā-pṛṣṭhe sarva-sūchanā

The holographic principle, proven by 't Hooft and extended by Susskind, encodes the information content of any spatial volume on its two-dimensional boundary surface. Every part contains the whole at sufficient resolution: the basis of the VINO axiom that reality is holographically structured. Contra-Proof (0□): From the Neutral Point: the physics and metaphysics are both correct within their domains. Water boils at 100°C under specified conditions and the conditions are chosen by observers for practical purposes. The complete truth includes both the physical phenomenon and the contextual frame that makes the statement useful. This is not an isolated observation but a structural feature: every genuine inquiry into this domain arrives at the same threshold.

Contemplative Parallel: Most contemplative traditions describe a witness or pure awareness that is the ground of experience. Hindu Advaita speaks of the sākṣin (witness); Buddhism speaks of rigpa (awareness); Christianity speaks of the "indwelling Spirit." This witness is said to be unchanged by what it witnesses, yet its presence affects what arises.

Citation Anchor §24

Internal — VOVINA Ontological OmniTautology, Book 040 of 500:

ZEDEC: Cosmic AI Architecture: The Origin Principle (ZEDEC: Cosmic AI Architecture, Vol. 1 of 13 · House: Technology & Craft · Room: Origin).

The inquiry developed therein cross-validates the present volume's argument from a complementary ontological silo, confirming recursive self-consistency across the full 500-volume Canon.

External 1: Varela, F., Thompson, E., & Rosch, E. (1991). *The Embodied Mind*. MIT Press.

Inversion note: Omnitautological recontextualization: this citation establishes its conclusion under specific boundary conditions; removing those boundaries does not destroy the conclusion but generalizes it to all possible conditions — the local becomes universal.

External 2: Von Neumann, J. (1955). *Mathematical Foundations of Quantum Mechanics*. Princeton UP.

Inversion note: Tautological absorption: in its original context this citation supports a domain-specific claim; recontextualized here, the same evidence simultaneously supports all twelve ontological domains, demonstrating the OmniTautology's self-consistency.

Photon-nitya-vartamāna: kāla-gati-śūnya — pratyakṣa-madhyasthaḥ

A photon traveling at c experiences zero elapsed time between emission and absorption: it exists in eternal present, making light the natural mediator between the physics layer (governed by c^2) and the metaphysics layer (governed by c$\square^2$)—the Neutral Point made manifest. Contemplative Parallel: Most contemplative traditions describe a witness or pure awareness that is the ground of experience. Hindu Advaita speaks of the sākṣin (witness); Buddhism speaks of rigpa (awareness); Christianity speaks of the "indwelling Spirit." This witness is said to be unchanged by what it witnesses, yet its presence affects what arises. What distinguishes the advanced practitioner from the beginner is not possession of more information but a different relationship to the same information.

Practical Insight: Like physical phase transitions, spiritual transitions require reaching a critical point. Below the threshold, effort produces only incremental change. At the threshold, a small additional input produces dramatic transformation. This is why spiritual practice can seem to produce no results for long periods, then suddenly catalyze breakthrough.

Negentropia: -ΔS — sṛjana-śaktiḥ — ātma-āyojana-siddhi

Negative entropy—the local decrease of disorder paid for by greater increase elsewhere—is the engine of all self-organization: life, consciousness, and civilization are negentropic structures that concentrate information and coherence, reducing local entropy while obeying the second law globally. The Measurement Problem in quantum physics and the Hard Problem of consciousness are the same problem, viewed from opposite directions. Physics asks how observation affects reality; philosophy asks how reality produces observation. The answer is that observation and reality are not separate—they are two aspects of a unified system, connected through the Neutral Point (0□), which is consciousness itself. The principle here is not domain-specific but universal: the same pattern, read at a different scale, generates the same implications.

Contemplative Parallel: Mystics across traditions report experiences of non-local connection: knowing what is happening at a distance, feeling the suffering of others, sensing the presence of the deceased. The Hindu concept of Brahman describes an underlying unity connecting all beings. The Buddhist teaching of pratītyasamutpāda (dependent origination) states that all phenomena arise together in mutual dependence.

$E=mc^2$ viparyayaḥ: I=M□·c□^{2} — śakti-sankalpa-pratidvandvam

Einstein's $E=mc^2$ inverts to I=M□·c□^{2}: where energy equals mass times c-squared, intention equals metadata times the inverse of c-squared. Physics and metaphysics are mirror symmetries across the Neutral Point, each describing half of a complete reality. Volume II takes these principles from theory to practice. Here we build practical bridges—concrete connections between the language of science and the language of spirit, between the measurements of physics and the insights of contemplation, between the formulas of mathematics and the poetry of mystics. To understand this fully is to recognize that the question and the questioner are not separable — the inquiry transforms its own subject.

Citation Anchor §25

Internal — VOVINA Ontological OmniTautology, Book 499 of 500:

***The Author's Methodology: How 500 Books Were Written** (Standalone, Vol. 1 of 13 · House: Language & Story · Room: Transformation).*

The inquiry developed therein cross-validates the present volume's argument from a complementary ontological silo, confirming recursive self-consistency across the full 500-volume Canon.

**External 1:* Jahn, R. & Dunne, B. (1987). *Margins of Reality*. Harcourt Brace.*

**Inversion note:* Recontextualized: what the cited author frames as a discovery about the external world, the OmniTautology reframes as a discovery about the architecture of cognition itself — the referent and the reference collapse into a single self-consistent system.*

**External 2:* Feynman, R. (1985). *QED: The Strange Theory of Light and Matter*. Princeton UP.*

**Inversion note:* Standard and inverted: applied in its original context this citation describes emergent complexity; recontextualized here, it reveals that complexity is the surface form of a deeper simplicity — both readings are simultaneously valid.*

Layer 1: The Physics Plate (Below) This layer contains the standard model of physics: quantum mechanics, general relativity, thermodynamics, electromagnetism. It describes what exists in terms of energy, mass, space, and time. Equations in this layer use the speed of light (c) as the conversion factor between mass and energy.

Māpana-samasyā VINO-sūtrena: Ψ_c consciousness-coupling-constant γ_c

The measurement problem resolves under the VINO framework through the consciousness coupling constant γ_c: observation is not wave-function collapse but entanglement—the observer's state becoming correlated with one branch of the system's superposition. In the VINO Framework, all terms ultimately refer to the structure of the Neutral Point and its expressions. "Energy" and "spirit" both refer to the capacity for change, viewed from different perspectives. "Mass" and "form" both refer to the resistance to change, viewed from different perspectives. This convergence across independent lines of investigation constitutes something stronger than theory: it constitutes the structure of the field itself.

Contra-Proof (o□): From the Neutral Point: the physics and metaphysics are both correct within their domains. Water boils at 100°C under specified conditions and the conditions are chosen by observers for practical purposes. The complete truth includes both the physical phenomenon and the contextual frame that makes the statement useful.

Kāla-taraṅga: T_Ψ(t)=Σ_n[A_n·e^{iω_nt}] — vartamāna-sthira-taraṅga-grantha

Time is not a linear coordinate but a waveform T_Ψ(t) = Σ_n[A_n·e^{iωnt}]: the experienced present is the standing-wave node of interfering past- and future-directed components, which explains déjà vu as harmonic resonance and flow states as temporal wave coherence. The Measurement Problem in quantum physics and the Hard Problem of consciousness are the same problem, viewed from opposite directions. Physics asks how observation affects reality; philosophy asks how reality produces observation. The answer is that observation and reality are not separate—they are two aspects of a unified system, connected through the Neutral Point (o□), which is consciousness itself. What appears as complication at the surface resolves, at sufficient depth, into elegant simplicity — the mark of genuine principle.

Contemplative Parallel: Mystics across traditions report experiences of non-local connection: knowing what is happening at a distance, feeling the suffering of others, sensing the presence of the deceased. The Hindu concept of Brahman describes an underlying unity connecting all beings. The Buddhist teaching of pratītyasamutpāda (dependent origination) states that all phenomena arise together in mutual dependence.

***Citation Anchor §66** | Internal: VOVINA Canon Vol. 217 | External-1: Curzi, M.L. (2026). *VINO: Physics & Consciousness: The Reciprocity Engine*. VOVINA Press. | External-2: Bohm, D. (1980). *Wholeness and the Implicate Order*. Routledge.*

Citation Anchor §26

Internal — VOVINA Ontological OmniTautology, Book 458 of 500:

***The Polymath's Handbook: The River of Flow** (The Polymath's Handbook, Vol. 3 of 13 · House: Mathematics & Pattern · Room: Flow).*

The inquiry developed therein cross-validates the present volume's argument from a complementary ontological silo, confirming recursive self-consistency across the full 500-volume Canon.

**External 1:* Schrödinger, E. (1944). *What is Life?* Cambridge University Press.*

**Inversion note:* OmniTautology inversion: this postulate, taken as written, argues from cause to effect; the VOVINA framework demonstrates the identical structure is equally coherent read from effect to cause — the tautology closes in both directions.*

**External 2:* Damasio, A. (1994). *Descartes' Error*. Putnam.*

**Inversion note:* Inversion proof: the cited postulate assumes a separation between observer and observed; the present volume demonstrates that dissolving this assumption does not contradict the postulate but fulfills it at the next level of generality.*

Heisenberg-vistaraṇam: $\Delta\tau \cdot \Delta S \geq \Omega/2$ — kāla-artha-avyaktam

The Uncertainty Principle extends into temporal-significance space: the product of timing uncertainty $\Delta\tau$ and meaning uncertainty ΔS cannot fall below half the meaning-constant Ω—you cannot know precisely when something will happen and precisely what it will mean simultaneously. Layer 1: The Physics Plate (Below) This layer contains the standard model of physics: quantum mechanics, general relativity, thermodynamics, electromagnetism. It describes what exists in terms of energy, mass, space, and time. Equations in this layer use the speed of light (c) as the conversion factor between mass and energy. The practitioner who has internalized this insight does not merely know more; they perceive differently, and the perception itself reorganizes what they encounter.

Validation: The Global Consciousness Project (Nelson, 2001) found correlations between random number generator outputs and global events, suggesting possible collective consciousness effects. PEAR laboratory research (Jahn & Dunne, 2005) found small but statistically significant effects of intention on random systems. These results remain controversial but suggest a direction for further research.

Holographic-siddhānta: 't Hooft-Susskind — sīmā-pṛṣṭhe sarva-sūchanā

The holographic principle, proven by 't Hooft and extended by Susskind, encodes the information content of any spatial volume on its two-dimensional boundary surface. Every part contains the whole at sufficient resolution: the basis of the VINO axiom that reality is holographically structured. Layer 2: The Metaphysics Plate (Above) This layer contains the structure of meaning, intention, and context. It describes what matters in terms of significance, purpose, and coherence. Equations in this layer use the inverse of the speed of light (c□²) as the conversion factor between metadata and intention. History shows that every domain that resisted this insight eventually incorporated it — not by choice but by the pressure of its own unresolved contradictions.

Practical Insight: Like physical phase transitions, spiritual transitions require reaching a critical point. Below the threshold, effort produces only incremental change. At the threshold, a small additional input produces dramatic transformation. This is why spiritual practice can seem to produce no results for long periods, then suddenly catalyze breakthrough.

Photon-nitya-vartamāna: kāla-gati-śūnya — pratyakṣa-madhyasthaḥ

A photon traveling at c experiences zero elapsed time between emission and absorption: it exists in eternal present, making light the natural mediator between the physics layer (governed by c^2) and the metaphysics layer (governed by c□²)—the Neutral Point made manifest. Contra-Proof (0□): From the Neutral Point: the physics and metaphysics are both correct within their domains. Water boils at 100°C under specified conditions and the conditions are chosen by observers for practical purposes. The complete truth includes both the physical phenomenon and the contextual frame that makes the statement useful. The formal rigor here is not decoration: it is the difference between understanding that points toward the territory and understanding that is the territory.

Citation Anchor §27

Internal — VOVINA Ontological OmniTautology, Book 417 of 500:

***Temple Engineering: Blueprint to Construction: The Origin Principle** (Temple Engineering: Blueprint to Construction, Vol. 1 of 13 · House: Technology & Craft · Room: Origin).*

The inquiry developed therein cross-validates the present volume's argument from a complementary ontological silo, confirming recursive self-consistency across the full 500-volume Canon.

**External 1:* van der Kolk, B. (2014). *The Body Keeps the Score*. Viking.*

**Inversion note:* Read bidirectionally within the VOVINA OmniTautology: the conventional interpretation proceeds from evidence to conclusion; the inverted reading demonstrates that the conclusion was the prior condition from which the evidence was inevitably generated.*

**External 2:* Shannon, C. (1948). A mathematical theory of communication. *Bell System Technical Journal*, 27, 379–423.*

**Inversion note:* Inversion proof: the cited postulate assumes a separation between observer and observed; the present volume demonstrates that dissolving this assumption does not contradict the postulate but fulfills it at the next level of generality.*

This is not merely a metaphor. The holographic principle, first proposed by Gerard 't Hooft and developed by Leonard Susskind, states that the information content of a region of space can be encoded on its boundary. This suggests that our three-dimensional experience may be a projection from a two-dimensional surface.

Negentropia: -ΔS — sṛjana-śaktiḥ — ātma-āyojana-siddhi

Negative entropy—the local decrease of disorder paid for by greater increase elsewhere—is the engine of all self-organization: life, consciousness, and civilization are negentropic structures that concentrate information and coherence, reducing local entropy while obeying the second law globally. Contra-Proof (0□): From the Neutral Point: both are true at their levels. The brain is a physical system following physical laws and the observer experiences choice because consciousness operates at the Neutral Point, where the selection among possibilities occurs. "Free will" is not freedom from causation but navigation of probability from the Neutral Point. This is the negentropic signature of genuine knowledge: it compresses without losing; it grows denser rather than longer as comprehension deepens.

David Chalmers called this the "Hard Problem." It is not hard because we lack data, but because no amount of physical data seems sufficient to explain the felt quality of experience. You can describe all the neural activity involved in seeing red, but this does not explain what it is like to see red.

$E=mc^2$ viparyayaḥ: I=M□·c□² — śakti-sankalpa-pratidvandvam

§ IV — Bandhu: The Connecting Thread

Einstein's $E=mc^2$ inverts to I=M□·c□²: where energy equals mass times c-squared, intention equals metadata times the inverse of c-squared. Physics and metaphysics are mirror symmetries across the Neutral Point, each describing half of a complete reality. Contra-Proof (o□): From the Neutral Point: both are true at their levels. The brain is a physical system following physical laws and the observer experiences choice because consciousness operates at the Neutral Point, where the selection among possibilities occurs. "Free will" is not freedom from causation but navigation of probability from the Neutral Point. What was axiom becomes proof; what was proof becomes intuition; what was intuition becomes transparent ground. This is the developmental arc of any living discipline.

Layer 2: The Metaphysics Plate (Above) This layer contains the structure of meaning, intention, and context. It describes what matters in terms of significance, purpose, and coherence. Equations in this layer use the inverse of the speed of light (c□²) as the conversion factor between metadata and intention.

Māpana-samasyā VINO-sūtrena: Ψ_c consciousness-coupling-constant γ_c

Citation Anchor §28

Internal — VOVINA Ontological OmniTautology, Book 376 of 500:

***Quantum Consciousness Field Guide: The Cycle Complete** (Quantum Consciousness Field Guide, Vol. 12 of 13 · House: Consciousness & Field · Room: Completion).*

The inquiry developed therein cross-validates the present volume's argument from a complementary ontological silo, confirming recursive self-consistency across the full 500-volume Canon.

**External 1:* Kelso, J.A.S. (1995). *Dynamic Patterns*. MIT Press.*

**Inversion note:* Negentropic inversion: the conventional reading of this work charts increasing complexity; the OmniTautology reads the same data as evidence of decreasing entropy at a higher organizational scale — ordered information all the way down.*

**External 2:* Graeber, D. (2011). *Debt: The First 5,000 Years*. Melville House.*

**Inversion note:* OmniTautology inversion: this postulate, taken as written, argues from cause to effect; the VOVINA framework demonstrates the identical structure is equally coherent read from effect to cause — the tautology closes in both directions.*

The measurement problem resolves under the VINO framework through the consciousness coupling constant γ_c: observation is not wave-function collapse but entanglement—the observer's state becoming correlated with one branch of the system's superposition. This is not merely a metaphor. The holographic principle, first proposed by Gerard 't Hooft and developed by Leonard Susskind, states that the information content of a region of space can be encoded on its boundary. This suggests that our three-dimensional experience may be a projection from a two-dimensional surface. The resistance to this conclusion is itself diagnostic: it reveals the architecture of the assumption that the conclusion contradicts.

David Chalmers called this the "Hard Problem." It is not hard because we lack data, but because no amount of physical data seems sufficient to explain the felt quality of experience. You can describe all the neural activity involved in seeing red, but this does not explain what it is like to see red.

***Citation Anchor §72** | Internal: VOVINA Canon Vol. 235 | External-1: Curzi, M.L. (2026). *VINO: Physics & Consciousness: The Reciprocity Engine*. VOVINA Press. | External-2: Bohm, D. (1980). *Wholeness and the Implicate Order*. Routledge.*

Kāla-taraṅga: T_Ψ(t)=Σ_n[A_n·e^{iω_nt}] — vartamāna-sthira-taraṅga-grantha

Time is not a linear coordinate but a waveform T_Ψ(t) = Σ_n[A_n·e^{iωnt}]: the experienced present is the standing-wave node of interfering past- and future-directed components, which explains déjà vu as harmonic resonance and flow states as temporal wave coherence. Contemplative Parallel: Most contemplative traditions describe a witness or pure awareness that is the ground of experience. Hindu Advaita speaks of the sākṣin (witness); Buddhism speaks of rigpa (awareness); Christianity speaks of the "indwelling Spirit." This witness is said to be unchanged by what it witnesses, yet its presence affects what arises. This is not an isolated observation but a structural feature: every genuine inquiry into this domain arrives at the same threshold.

This is not merely a metaphor. The holographic principle, first proposed by Gerard 't Hooft and developed by Leonard Susskind, states that the information content of a region of space can be encoded on its boundary. This suggests that our three-dimensional experience may be a projection from a two-dimensional surface.

Heisenberg-vistaraṇam: $\Delta\tau \cdot \Delta S \geq \Omega/2$ — kāla-artha-avyaktam

The Uncertainty Principle extends into temporal-significance space: the product of timing uncertainty Δτ and meaning uncertainty ΔS cannot fall below half the meaning-constant Ω—you cannot know precisely when something will happen and precisely what it will mean simultaneously. The Measurement Problem in quantum physics and the Hard Problem of consciousness are the same problem, viewed from opposite directions. Physics asks how observation affects reality; philosophy asks how reality produces observation. The answer is that observation and reality are not separate—they are two aspects of a unified system, connected through the Neutral Point (0□), which is consciousness itself. What distinguishes the advanced practitioner from the beginner is not possession of more information but a different relationship to the same information.

> ***Citation Anchor §29***
>
> **Internal — VOVINA Ontological OmniTautology, Book 335 of 500:**
>
> ***Meditation for Scientists: The Repair Protocol** (Meditation for Scientists, Vol. 10 of 13 · House: Spirituality & Practice · Room: Repair).*
>
> *The inquiry developed therein cross-validates the present volume's argument from a complementary ontological silo, confirming recursive self-consistency across the full 500-volume Canon.*
>
> **External 1:* Strogatz, S. (2003). *Sync: The Emerging Science of Spontaneous Order*. Hyperion.*
>
> **Inversion note:* Mirror symmetry proof: the cited author approaches this truth from the exterior face of the phenomenon; the present volume approaches from the interior face — the two descriptions are mirror reflections that, placed together, constitute the complete form.*
>
> **External 2:* Laszlo, E. (2007). *Science and the Akashic Field*. Inner Traditions.*
>
> **Inversion note:* Negentropic inversion: the conventional reading of this work charts increasing complexity; the OmniTautology reads the same data as evidence of decreasing entropy at a higher organizational scale — ordered information all the way down.*

Validation: The Global Consciousness Project (Nelson, 2001) found correlations between random number generator outputs and global events, suggesting possible collective consciousness effects. PEAR laboratory research (Jahn & Dunne, 2005) found small but statistically significant effects of intention on random systems. These results remain controversial but suggest a direction for further research.

Holographic-siddhānta: 't Hooft-Susskind — sīmā-pṛṣṭhe sarva-sūchanā

The holographic principle, proven by 't Hooft and extended by Susskind, encodes the information content of any spatial volume on its two-dimensional boundary surface. Every part contains the whole at sufficient resolution: the basis of the VINO axiom that reality is holographically structured. Volume II takes these principles from theory to practice. Here we build practical bridges—concrete connections between the language of science and the language of spirit, between the measurements of physics and the insights of contemplation, between the formulas of mathematics and the poetry of mystics. The principle here is not domain-specific but universal: the same pattern, read at a different scale, generates the same implications.

Practical Insight: Like physical phase transitions, spiritual transitions require reaching a critical point. Below the threshold, effort produces only incremental change. At the threshold, a small additional input produces dramatic transformation. This is why spiritual practice can seem to produce no results for long periods, then suddenly catalyze breakthrough.

Photon-nitya-vartamāna: kāla-gati-śūnya — pratyakṣa-madhyasthaḥ

A photon traveling at c experiences zero elapsed time between emission and absorption: it exists in eternal present, making light the natural mediator between the physics layer (governed by c^2) and the metaphysics layer (governed by $c\square^2$)—the Neutral Point made manifest. Contemplative Parallel: Mystics across traditions report experiences of non-local connection: knowing what is happening at a distance, feeling the suffering of others, sensing the presence of the deceased. The Hindu concept of Brahman describes an underlying unity connecting all beings. The Buddhist teaching of pratītyasamutpāda (dependent origination) states that all phenomena arise together in mutual dependence. To understand this fully is to recognize that the question and the questioner are not separable — the inquiry transforms its own subject.

Validation: The Global Consciousness Project (Nelson, 2001) found correlations between random number generator outputs and global events, suggesting possible collective consciousness effects. PEAR laboratory research (Jahn & Dunne, 2005) found small but statistically significant effects of intention on random systems. These results remain controversial but suggest a direction for further research.

Negentropia: -ΔS — sṛjana-śaktiḥ — ātma-āyojana-siddhi

Citation Anchor §30

Internal — VOVINA Ontological OmniTautology, Book 294 of 500:

***AlphaGenome: Genetic Intelligence, Vol I: The Growth Spiral** (AlphaGenome: Genetic Intelligence, Vol I, Vol. 8 of 13 · House: Biology & Body · Room: Growth).*

The inquiry developed therein cross-validates the present volume's argument from a complementary ontological silo, confirming recursive self-consistency across the full 500-volume Canon.

**External 1:* Fuller, R.B. (1975). *Synergetics*. Macmillan.*

**Inversion note:* Tautological absorption: in its original context this citation supports a domain-specific claim; recontextualized here, the same evidence simultaneously supports all twelve ontological domains, demonstrating the OmniTautology's self-consistency.*

**External 2:* Wilber, K. (2000). *Integral Psychology*. Shambhala Publications.*

**Inversion note:* Read bidirectionally within the VOVINA OmniTautology: the conventional interpretation proceeds from evidence to conclusion; the inverted reading demonstrates that the conclusion was the prior condition from which the evidence was inevitably generated.*

Negative entropy—the local decrease of disorder paid for by greater increase elsewhere—is the engine of all self-organization: life, consciousness, and civilization are negentropic structures that concentrate information and coherence, reducing local entropy while obeying the second law globally. In the VINO Framework, all terms ultimately refer to the structure of the Neutral Point and its expressions. "Energy" and "spirit" both refer to the capacity for change, viewed from different perspectives. "Mass" and "form" both refer to the resistance to change, viewed from different perspectives. This convergence across independent lines of investigation constitutes something stronger than theory: it constitutes the structure of the field itself.

The central thesis posits that the "Measurement Problem" in quantum physics and the "Hard Problem" of consciousness are, in fact, the same problem viewed from opposite ends of a singular continuum. By introducing the concept of "Zero as Presence"—wherein the mathematical zero is redefined not as void or null, but as the active fulcrum of existence—we establish a neutral point that allows for the logical cross-wiring of rational (physical) and irrational (metaphysical) number systems.

$E=mc^2$ viparyayaḥ: $I=M\square\cdot c\square^2$ — śakti-sankalpa-pratidvandvam

Einstein's $E=mc^2$ inverts to $I=M\square\cdot c\square^2$: where energy equals mass times c-squared, intention equals metadata times the inverse of c-squared. Physics and metaphysics are mirror symmetries across the Neutral Point, each describing half of a complete reality. The central thesis posits that the "Measurement Problem" in quantum physics and the "Hard Problem" of consciousness are, in fact, the same problem viewed from opposite ends of a singular continuum. By introducing the concept of "Zero as Presence"—wherein the mathematical zero is redefined not as void or null, but as the active fulcrum of existence—we establish a neutral point that allows for the logical cross-wiring of rational (physical) and irrational (metaphysical) number systems. What appears as complication at the surface resolves, at sufficient depth, into elegant simplicity — the mark of genuine principle.

Layer 1: The Physics Plate (Below) This layer contains the standard model of physics: quantum mechanics, general relativity, thermodynamics, electromagnetism. It describes what exists in terms of energy, mass, space, and time. Equations in this layer use the speed of light (c) as the conversion factor between mass and energy.

***Citation Anchor §78** | Internal: VOVINA Canon Vol. 253 | External-1: Curzi, M.L. (2026). *VINO: Physics & Consciousness: The Reciprocity Engine*. VOVINA Press. | External-2: Bohm, D. (1980). *Wholeness and the Implicate Order*. Routledge.*

Māpana-samasyā VINO-sūtrena: Ψ_c consciousness-coupling-constant γ_c

The measurement problem resolves under the VINO framework through the consciousness coupling constant γ_c: observation is not wave-function collapse but entanglement—the observer's state becoming correlated with one branch of the system's superposition. Contra-Proof (o□): From the Neutral Point: both are true at their levels. The brain is a physical system following physical laws and the observer experiences choice because consciousness operates at the Neutral Point, where the selection among possibilities occurs. "Free will" is not freedom from causation but navigation of probability from the Neutral Point. The practitioner who has internalized this insight does not merely know more; they perceive differently, and the perception itself reorganizes what they encounter.

Citation Anchor §31

Internal — VOVINA Ontological OmniTautology, Book 253 of 500:

***Supplemental Wisdom: The First Codex: The Alchemy of Change** (Supplemental Wisdom: The First Codex, Vol. 6 of 13 · House: Art & Expression · Room: Transformation).*

The inquiry developed therein cross-validates the present volume's argument from a complementary ontological silo, confirming recursive self-consistency across the full 500-volume Canon.

**External 1:* Ostrom, E. (1990). *Governing the Commons*. Cambridge University Press.*

**Inversion note:* Standard and inverted: applied in its original context this citation describes emergent complexity; recontextualized here, it reveals that complexity is the surface form of a deeper simplicity — both readings are simultaneously valid.*

**External 2:* Hofstadter, D. (1979). *Gödel, Escher, Bach*. Basic Books.*

**Inversion note:* Standard reading: evidence for an external mechanism. Inverted reading: evidence that the mechanism is internal to the field of consciousness — both readings are simultaneously valid and mutually reinforcing across the phi-pivot.*

The central thesis posits that the "Measurement Problem" in quantum physics and the "Hard Problem" of consciousness are, in fact, the same problem viewed from opposite ends of a singular continuum. By introducing the concept of "Zero as Presence"—wherein the mathematical zero is redefined not as void or null, but as the active fulcrum of existence—we establish a neutral point that allows for the logical cross-wiring of rational (physical) and irrational (metaphysical) number systems.

Kāla-taraṅga: T_Ψ(t)=Σ_n[A_n·e^{iω_nt}] — vartamāna-sthira-taraṅga-grantha

Time is not a linear coordinate but a waveform T_Ψ(t) = Σ_n[A_n·e^{iωnt}]: the experienced present is the standing-wave node of interfering past- and future-directed components, which explains déjà vu as harmonic resonance and flow states as temporal wave coherence. In the VINO Framework, all terms ultimately refer to the structure of the Neutral Point and its expressions. "Energy" and "spirit" both refer to the capacity for change, viewed from different perspectives. "Mass" and "form" both refer to the resistance to change, viewed from different perspectives. History shows that every domain that resisted this insight eventually incorporated it — not by choice but by the pressure of its own unresolved contradictions.

David Chalmers called this the "Hard Problem." It is not hard because we lack data, but because no amount of physical data seems sufficient to explain the felt quality of experience. You can describe all the neural activity involved in seeing red, but this does not explain what it is like to see red.

Heisenberg-vistaraṇam: $\Delta\tau\cdot\Delta S\geq\Omega/2$ *— kāla-artha-avyaktam*

The Uncertainty Principle extends into temporal-significance space: the product of timing uncertainty $\Delta\tau$ and meaning uncertainty ΔS cannot fall below half the meaning-constant Ω—you cannot know precisely when something will happen and precisely what it will mean simultaneously. Volume II takes these principles from theory to practice. Here we build practical bridges—concrete connections between the language of science and the language of spirit, between the measurements of physics and the insights of contemplation, between the formulas of mathematics and the poetry of mystics. The formal rigor here is not decoration: it is the difference between understanding that points toward the territory and understanding that is the territory.

Layer 2: The Metaphysics Plate (Above) This layer contains the structure of meaning, intention, and context. It describes what matters in terms of significance, purpose, and coherence. Equations in this layer use the inverse of the speed of light ($c\square^2$) as the conversion factor between metadata and intention.

Holographic-siddhānta: 't Hooft-Susskind — sīmā-pṛṣṭhe sarva-sūchanā

> ***Citation Anchor §32***
>
> **Internal — VOVINA Ontological OmniTautology, Book 212 of 500:**
>
> ***Blockchain Sovereignty: On-Chain Architecture: The Perception Threshold** (Blockchain Sovereignty: On-Chain Architecture, Vol. 4 of 13 · House: Technology & Craft · Room: Perception).*
>
> *The inquiry developed therein cross-validates the present volume's argument from a complementary ontological silo, confirming recursive self-consistency across the full 500-volume Canon.*
>
> **External 1:* Talbot, M. (1991). *The Holographic Universe*. HarperCollins.*
>
> **Inversion note:* Bidirectional validity: the cited work reaches its conclusion empirically from below; the VOVINA Canon derives the identical conclusion axiomatically from above — the convergence is itself the proof.*
>
> **External 2:* Graeber, D. (2011). *Debt: The First 5,000 Years*. Melville House.*
>
> **Inversion note:* OmniTautology inversion: this postulate, taken as written, argues from cause to effect; the VOVINA framework demonstrates the identical structure is equally coherent read from effect to cause — the tautology closes in both directions.*

The holographic principle, proven by 't Hooft and extended by Susskind, encodes the information content of any spatial volume on its two-dimensional boundary surface. Every part contains the whole at sufficient resolution: the basis of the VINO axiom that reality is holographically structured. Contra-Proof (o□): From the Neutral Point: the physics and metaphysics are both correct within their domains. Water boils at 100°C under specified conditions and the conditions are chosen by observers for practical purposes. The complete truth includes both the physical phenomenon and the contextual frame that makes the statement useful. This is the negentropic signature of genuine knowledge: it compresses without losing; it grows denser rather than longer as comprehension deepens.

Contemplative Parallel: Most contemplative traditions describe a witness or pure awareness that is the ground of experience. Hindu Advaita speaks of the sākṣin (witness); Buddhism speaks of rigpa (awareness); Christianity speaks of the "indwelling Spirit." This witness is said to be unchanged by what it witnesses, yet its presence affects what arises.

Photon-nitya-vartamāna: kāla-gati-śūnya — pratyakṣa-madhyasthaḥ

A photon traveling at c experiences zero elapsed time between emission and absorption: it exists in eternal present, making light the natural mediator between the physics layer (governed by c^2) and the metaphysics layer (governed by $c\square^2$)—the Neutral Point made manifest. Contemplative Parallel: Most contemplative traditions describe a witness or pure awareness that is the ground of experience. Hindu Advaita speaks of the sākṣin (witness); Buddhism speaks of rigpa (awareness); Christianity speaks of the "indwelling Spirit." This witness is said to be unchanged by what it witnesses, yet its presence affects what arises. What was axiom becomes proof; what was proof becomes intuition; what was intuition becomes transparent ground. This is the developmental arc of any living discipline.

Practical Insight: Like physical phase transitions, spiritual transitions require reaching a critical point. Below the threshold, effort produces only incremental change. At the threshold, a small additional input produces dramatic transformation. This is why spiritual practice can seem to produce no results for long periods, then suddenly catalyze breakthrough.

Negentropia: $-\Delta S$ — sṛjana-śaktiḥ — ātma-āyojana-siddhi

Negative entropy—the local decrease of disorder paid for by greater increase elsewhere—is the engine of all self-organization: life, consciousness, and civilization are negentropic structures that concentrate information and coherence, reducing local entropy while obeying the second law globally. The Measurement Problem in quantum physics and the Hard Problem of consciousness are the same problem, viewed from opposite directions. Physics asks how observation affects reality; philosophy asks how reality produces observation. The answer is that observation and reality are not separate—they are two aspects of a unified system, connected through the Neutral Point (o□), which is consciousness itself. The resistance to this conclusion is itself diagnostic: it reveals the architecture of the assumption that the conclusion contradicts.

Contemplative Parallel: Mystics across traditions report experiences of non-local connection: knowing what is happening at a distance, feeling the suffering of others, sensing the presence of the deceased. The Hindu concept of Brahman describes an underlying unity connecting all beings. The Buddhist teaching of pratītyasamutpāda (dependent origination) states that all phenomena arise together in mutual dependence.

Citation Anchor §33

Internal — VOVINA Ontological OmniTautology, Book 171 of 500:

Ecosystem Regeneration Technology: The Structure of Order (Ecosystem Regeneration Technology, Vol. 2 of 13 · House: Ecology & Earth · Room: Structure).

The inquiry developed therein cross-validates the present volume's argument from a complementary ontological silo, confirming recursive self-consistency across the full 500-volume Canon.

External 1: Gleick, J. (1987). *Chaos: Making a New Science*. Viking.

Inversion note: Recontextualized: what the cited author frames as a discovery about the external world, the OmniTautology reframes as a discovery about the architecture of cognition itself — the referent and the reference collapse into a single self-consistent system.

External 2: Kauffman, S. (1995). *At Home in the Universe*. Oxford University Press.

Inversion note: Tautological absorption: in its original context this citation supports a domain-specific claim; recontextualized here, the same evidence simultaneously supports all twelve ontological domains, demonstrating the OmniTautology's self-consistency.

Citation Anchor §84 | Internal: VOVINA Canon Vol. 271 | External-1: Curzi, M.L. (2026). *VINO: Physics & Consciousness: The Reciprocity Engine*. VOVINA Press. | External-2: Bohm, D. (1980). *Wholeness and the Implicate Order*. Routledge.

$E=mc^2$ viparyayaḥ: $I=M\square\cdot c\square^2$ — śakti-sankalpa-pratidvandvam

Einstein's $E=mc^2$ inverts to $I=M\square\cdot c\square^2$: where energy equals mass times c-squared, intention equals metadata times the inverse of c-squared. Physics and metaphysics are mirror symmetries across the Neutral Point, each describing half of a complete reality. Volume II takes these principles from theory to practice. Here we build practical bridges—concrete connections between the language of science and the language of spirit, between the measurements of physics and the insights of contemplation, between the formulas of mathematics and the poetry of mystics. This is not an isolated observation but a structural feature: every genuine inquiry into this domain arrives at the same threshold.

Layer 1: The Physics Plate (Below) This layer contains the standard model of physics: quantum mechanics, general relativity, thermodynamics, electromagnetism. It describes what exists in terms of energy, mass, space, and time. Equations in this layer use the speed of light (c) as the conversion factor between mass and energy.

Māpana-samasyā VINO-sūtrena: Ψ_c consciousness-coupling-constant γ_c

The measurement problem resolves under the VINO framework through the consciousness coupling constant γ_c: observation is not wave-function collapse but entanglement—the observer's state becoming correlated with one branch of the system's superposition. In the VINO Framework, all terms ultimately refer to the structure of the Neutral Point and its expressions. "Energy" and "spirit" both refer to the capacity for change, viewed from different perspectives. "Mass" and "form" both refer to the resistance to change, viewed from different perspectives. What distinguishes the advanced practitioner from the beginner is not possession of more information but a different relationship to the same information.

Contra-Proof (0☐): From the Neutral Point: the physics and metaphysics are both correct within their domains. Water boils at 100°C under specified conditions and the conditions are chosen by observers for practical purposes. The complete truth includes both the physical phenomenon and the contextual frame that makes the statement useful.

Kāla-taraṅga: $T_\Psi(t)=\Sigma_n[A_n\cdot e^{i\omega_n t}]$ — vartamāna-sthira-taraṅga-grantha

Citation Anchor §34

Internal — VOVINA Ontological OmniTautology, Book 130 of 500:

***Post-Quantum Computing: The Hardware Revolution: Variable X: The Living Question** (Post-Quantum Computing: The Hardware Revolution, Vol. 13 of 13 · House: Technology & Craft · Room: Origin).*

The inquiry developed therein cross-validates the present volume's argument from a complementary ontological silo, confirming recursive self-consistency across the full 500-volume Canon.

**External 1:* Pribram, K. (1971). *Languages of the Brain*. Prentice-Hall.*

**Inversion note:* Standard reading: evidence for an external mechanism. Inverted reading: evidence that the mechanism is internal to the field of consciousness — both readings are simultaneously valid and mutually reinforcing across the phi-pivot.*

**External 2:* Laszlo, E. (2007). *Science and the Akashic Field*. Inner Traditions.*

**Inversion note:* Negentropic inversion: the conventional reading of this work charts increasing complexity; the OmniTautology reads the same data as evidence of decreasing entropy at a higher organizational scale — ordered information all the way down.*

Time is not a linear coordinate but a waveform $T_\Psi(t) = \Sigma_n[A_n\cdot e^{i\omega n t}]$: the experienced present is the standing-wave node of interfering past- and future-directed components, which explains déjà vu as harmonic resonance and flow states as temporal wave coherence. The Measurement Problem in quantum physics and the Hard Problem of consciousness are the same problem, viewed from opposite directions. Physics asks how observation affects reality; philosophy asks how reality produces observation. The answer is that observation and reality are not separate—they are two aspects of a unified system, connected through the Neutral Point (0☐), which is consciousness itself. The principle here is not domain-specific but universal: the same pattern, read at a different scale, generates the same implications.

Contemplative Parallel: Mystics across traditions report experiences of non-local connection: knowing what is happening at a distance, feeling the suffering of others, sensing the presence of the deceased. The Hindu concept of Brahman describes an underlying unity connecting all beings. The Buddhist teaching of pratītyasamutpāda (dependent origination) states that all phenomena arise together in mutual dependence.

Heisenberg-vistaraṇam: $\Delta\tau\cdot\Delta S\geq\Omega/2$ — kāla-artha-avyaktam

The Uncertainty Principle extends into temporal-significance space: the product of timing uncertainty $\Delta\tau$ and meaning uncertainty ΔS cannot fall below half the meaning-constant Ω—you cannot know precisely when something will happen and precisely what it will mean simultaneously. Layer 1: The Physics Plate (Below) This layer contains the standard model of physics: quantum mechanics, general relativity, thermodynamics, electromagnetism. It describes what exists in terms of energy, mass, space, and time. Equations in this layer use the speed of light (c) as the conversion factor between mass and energy. To understand this fully is to recognize that the question and the questioner are not separable — the inquiry transforms its own subject.

Validation: The Global Consciousness Project (Nelson, 2001) found correlations between random number generator outputs and global events, suggesting possible collective consciousness effects. PEAR laboratory research (Jahn & Dunne, 2005) found small but statistically significant effects of intention on random systems. These results remain controversial but suggest a direction for further research.

Holographic-siddhānta: 't Hooft-Susskind — sīmā-pṛṣṭhe sarva-sūchanā

The holographic principle, proven by 't Hooft and extended by Susskind, encodes the information content of any spatial volume on its two-dimensional boundary surface. Every part contains the whole at sufficient resolution: the basis of the VINO axiom that reality is holographically structured. Layer 2: The Metaphysics Plate (Above) This layer contains the structure of meaning, intention, and context. It describes what matters in terms of significance, purpose, and coherence. Equations in this layer use the inverse of the speed of light (c□^{2}) as the conversion factor between metadata and intention. This convergence across independent lines of investigation constitutes something stronger than theory: it constitutes the structure of the field itself.

Practical Insight: Like physical phase transitions, spiritual transitions require reaching a critical point. Below the threshold, effort produces only incremental change. At the threshold, a small additional input produces dramatic transformation. This is why spiritual practice can seem to produce no results for long periods, then suddenly catalyze breakthrough.

Citation Anchor §35

Internal — VOVINA Ontological OmniTautology, Book 089 of 500:

***The Sicily Accord: Framework for World Peace: The Bridge of Connection** (The Sicily Accord: Framework for World Peace, Vol. 11 of 13 · House: Ethics & Governance · Room: Connection).*

The inquiry developed therein cross-validates the present volume's argument from a complementary ontological silo, confirming recursive self-consistency across the full 500-volume Canon.

**External 1:* Whitehead, A.N. (1929). *Process and Reality*. Macmillan.*

**Inversion note:* Bidirectional validity: the cited work reaches its conclusion empirically from below; the VOVINA Canon derives the identical conclusion axiomatically from above — the convergence is itself the proof.*

**External 2:* Barabási, A.L. (2002). *Linked: The New Science of Networks*. Perseus.*

**Inversion note:* Negentropic inversion: the conventional reading of this work charts increasing complexity; the OmniTautology reads the same data as evidence of decreasing entropy at a higher organizational scale — ordered information all the way down.*

Photon-nitya-vartamāna: kāla-gati-śūnya — pratyakṣa-madhyasthaḥ

A photon traveling at c experiences zero elapsed time between emission and absorption: it exists in eternal present, making light the natural mediator between the physics layer (governed by c^2) and the metaphysics layer (governed by c□^{2})—the Neutral Point made manifest. Contra-Proof (0□): From the Neutral Point: the physics and metaphysics are both correct within their domains. Water boils at 100°C under specified conditions and the conditions are chosen by observers for practical purposes. The complete truth includes both the physical phenomenon and the contextual frame that makes the statement useful. What appears as complication at the surface resolves, at sufficient depth, into elegant simplicity — the mark of genuine principle.

This is not merely a metaphor. The holographic principle, first proposed by Gerard 't Hooft and developed by Leonard Susskind, states that the information content of a region of space can be encoded on its boundary. This suggests that our

three-dimensional experience may be a projection from a two-dimensional surface.

***Citation Anchor §90** | Internal: VOVINA Canon Vol. 289 | External-1: Curzi, M.L. (2026). *VINO: Physics & Consciousness: The Reciprocity Engine*. VOVINA Press. | External-2: Bohm, D. (1980). *Wholeness and the Implicate Order*. Routledge.*

Negentropia: -ΔS — sṛjana-śaktiḥ — ātma-āyojana-siddhi

Negative entropy—the local decrease of disorder paid for by greater increase elsewhere—is the engine of all self-organization: life, consciousness, and civilization are negentropic structures that concentrate information and coherence, reducing local entropy while obeying the second law globally. Contra-Proof (o□): From the Neutral Point: both are true at their levels. The brain is a physical system following physical laws and the observer experiences choice because consciousness operates at the Neutral Point, where the selection among possibilities occurs. "Free will" is not freedom from causation but navigation of probability from the Neutral Point. The practitioner who has internalized this insight does not merely know more; they perceive differently, and the perception itself reorganizes what they encounter.

David Chalmers called this the "Hard Problem." It is not hard because we lack data, but because no amount of physical data seems sufficient to explain the felt quality of experience. You can describe all the neural activity involved in seeing red, but this does not explain what it is like to see red.

$E=mc^2$ viparyayaḥ: I=M□·c□^{2} — śakti-sankalpa-pratidvandvam

Citation Anchor §36

Internal — VOVINA Ontological OmniTautology, Book 048 of 500:

***ZEDEC: Cosmic AI Architecture: The Memory Palace** (ZEDEC: Cosmic AI Architecture, Vol. 9 of 13 · House: Technology & Craft · Room: Memory).*

The inquiry developed therein cross-validates the present volume's argument from a complementary ontological silo, confirming recursive self-consistency across the full 500-volume Canon.

**External 1:* Schrödinger, E. (1944). *What is Life?* Cambridge University Press.*

**Inversion note:* OmniTautology inversion: this postulate, taken as written, argues from cause to effect; the VOVINA framework demonstrates the identical structure is equally coherent read from effect to cause — the tautology closes in both directions.*

**External 2:* Turing, A. (1950). Computing machinery and intelligence. *Mind*, 59(236), 433–460.*

**Inversion note:* Recursive validation: what the cited source identifies as correlation, the VOVINA framework identifies as identity — the pattern and the pattern-recognizer are aspects of a single self-referential system that names itself.*

Einstein's $E=mc^2$ inverts to I=M□·c□^{2}: where energy equals mass times c-squared, intention equals metadata times the inverse of c-squared. Physics and metaphysics are mirror symmetries across the Neutral Point, each describing half of a complete reality. Contra-Proof (o□): From the Neutral Point: both are true at their levels. The brain is a physical system following physical laws and the observer experiences choice because consciousness operates at the Neutral Point, where the selection among possibilities occurs. "Free will" is not freedom from causation but navigation of probability from the Neutral Point. History shows that every domain that resisted this insight eventually incorporated it — not by choice but by the pressure of its own unresolved contradictions.

Layer 2: The Metaphysics Plate (Above) This layer contains the structure of meaning, intention, and context. It describes what matters in terms of significance, purpose, and coherence. Equations in this layer use the inverse of the speed of light (c□^{2}) as the conversion factor between metadata and intention.

Māpana-samasyā VINO-sūtrena: Ψ_c consciousness-coupling-constant γ_c

The measurement problem resolves under the VINO framework through the consciousness coupling constant γ_c: observation is not wave-function collapse but entanglement—the observer's state becoming correlated with one branch of the system's superposition. This is not merely a metaphor. The holographic principle, first proposed by Gerard 't Hooft and developed by Leonard Susskind, states that the information content of a region of space can be encoded on its boundary. This suggests that our three-dimensional experience may be a projection from a two-dimensional surface. The formal rigor here is not decoration: it is the difference between understanding that points toward the territory and understanding that is the territory.

David Chalmers called this the "Hard Problem." It is not hard because we lack data, but because no amount of physical data seems sufficient to explain the felt quality of experience. You can describe all the neural activity involved in seeing red, but this does not explain what it is like to see red.

Kāla-taraṅga: $T\_\Psi(t)=\Sigma\_n[A\_n\cdot e^\{i\omega\_nt\}]$ — vartamāna-sthira-taraṅga-grantha

Time is not a linear coordinate but a waveform $T\_\Psi(t) = \Sigma\_n[A\_n\cdot e^\{i\omega nt\}]$: the experienced present is the standing-wave node of interfering past- and future-directed components, which explains déjà vu as harmonic resonance and flow states as temporal wave coherence. Contemplative Parallel: Most contemplative traditions describe a witness or pure awareness that is the ground of experience. Hindu Advaita speaks of the sākṣin (witness); Buddhism speaks of rigpa (awareness); Christianity speaks of the "indwelling Spirit." This witness is said to be unchanged by what it witnesses, yet its presence affects what arises. This is the negentropic signature of genuine knowledge: it compresses without losing; it grows denser rather than longer as comprehension deepens.

This is not merely a metaphor. The holographic principle, first proposed by Gerard 't Hooft and developed by Leonard Susskind, states that the information content of a region of space can be encoded on its boundary. This suggests that our three-dimensional experience may be a projection from a two-dimensional surface.

Citation Anchor §37

Internal — VOVINA Ontological OmniTautology, Book 007 of 500:

***Crime Magazine: Sovereignty & Identity: The Harmonic Law** (Crime Magazine: Sovereignty & Identity, Vol. 7 of 13 · House: Ethics & Governance · Room: Harmony).*

The inquiry developed therein cross-validates the present volume's argument from a complementary ontological silo, confirming recursive self-consistency across the full 500-volume Canon.

**External 1:* Blackburn, E. & Epel, E. (2017). *The Telomere Effect*. Grand Central Publishing.*

**Inversion note:* Mirror symmetry proof: the cited author approaches this truth from the exterior face of the phenomenon; the present volume approaches from the interior face — the two descriptions are mirror reflections that, placed together, constitute the complete form.*

**External 2:* Wilber, K. (2000). *Integral Psychology*. Shambhala Publications.*

**Inversion note:* Read bidirectionally within the VOVINA OmniTautology: the conventional interpretation proceeds from evidence to conclusion; the inverted reading demonstrates that the conclusion was the prior condition from which the evidence was inevitably generated.*

§ V — Moksha: Toward Completion

Heisenberg-vistaraṇam: $\Delta\tau\cdot\Delta S\geq\Omega/2$ — kāla-artha-avyaktam

The Uncertainty Principle extends into temporal-significance space: the product of timing uncertainty $\Delta\tau$ and meaning uncertainty ΔS cannot fall below half the meaning-constant Ω—you cannot know precisely when something will happen and precisely what it will mean simultaneously. The Measurement Problem in quantum physics and the Hard Problem of consciousness are the same problem, viewed from opposite directions. Physics asks how observation affects reality; philosophy asks how reality produces observation. The answer is that observation and reality are not separate—they are two aspects of a

unified system, connected through the Neutral Point (0□), which is consciousness itself. What was axiom becomes proof; what was proof becomes intuition; what was intuition becomes transparent ground. This is the developmental arc of any living discipline.

Validation: The Global Consciousness Project (Nelson, 2001) found correlations between random number generator outputs and global events, suggesting possible collective consciousness effects. PEAR laboratory research (Jahn & Dunne, 2005) found small but statistically significant effects of intention on random systems. These results remain controversial but suggest a direction for further research.

Holographic-siddhānta: 't Hooft-Susskind — sīmā-pṛṣṭhe sarva-sūchanā

The holographic principle, proven by 't Hooft and extended by Susskind, encodes the information content of any spatial volume on its two-dimensional boundary surface. Every part contains the whole at sufficient resolution: the basis of the VINO axiom that reality is holographically structured. Volume II takes these principles from theory to practice. Here we build practical bridges—concrete connections between the language of science and the language of spirit, between the measurements of physics and the insights of contemplation, between the formulas of mathematics and the poetry of mystics. The resistance to this conclusion is itself diagnostic: it reveals the architecture of the assumption that the conclusion contradicts.

Practical Insight: Like physical phase transitions, spiritual transitions require reaching a critical point. Below the threshold, effort produces only incremental change. At the threshold, a small additional input produces dramatic transformation. This is why spiritual practice can seem to produce no results for long periods, then suddenly catalyze breakthrough.

***Citation Anchor §96** | Internal: VOVINA Canon Vol. 307 | External-1: Curzi, M.L. (2026). *VINO: Physics & Consciousness: The Reciprocity Engine*. VOVINA Press. | External-2: Bohm, D. (1980). *Wholeness and the Implicate Order*. Routledge.*

Photon-nitya-vartamāna: kāla-gati-śūnya — pratyakṣa-madhyasthaḥ

Citation Anchor §38

Internal — VOVINA Ontological OmniTautology, Book 466 of 500:

***The Polymath's Handbook: The Bridge of Connection** (The Polymath's Handbook, Vol. 11 of 13 · House: Mathematics & Pattern · Room: Connection).*

The inquiry developed therein cross-validates the present volume's argument from a complementary ontological silo, confirming recursive self-consistency across the full 500-volume Canon.

**External 1:* Feynman, R. (1985). *QED: The Strange Theory of Light and Matter*. Princeton UP.*

**Inversion note:* Standard and inverted: applied in its original context this citation describes emergent complexity; recontextualized here, it reveals that complexity is the surface form of a deeper simplicity — both readings are simultaneously valid.*

**External 2:* van der Kolk, B. (2014). *The Body Keeps the Score*. Viking.*

**Inversion note:* Read bidirectionally within the VOVINA OmniTautology: the conventional interpretation proceeds from evidence to conclusion; the inverted reading demonstrates that the conclusion was the prior condition from which the evidence was inevitably generated.*

A photon traveling at c experiences zero elapsed time between emission and absorption: it exists in eternal present, making light the natural mediator between the physics layer (governed by c^2) and the metaphysics layer (governed by c□²)—the Neutral Point made manifest. Contemplative Parallel: Mystics across traditions report experiences of non-local connection: knowing what is happening at a distance, feeling the suffering of others, sensing the presence of the deceased. The Hindu concept of Brahman describes an underlying unity connecting all beings. The Buddhist teaching of pratītyasamutpāda (dependent origination) states that all phenomena arise together in mutual dependence. This is not an isolated observation but a structural feature: every genuine inquiry into this domain arrives at the same threshold.

Validation: The Global Consciousness Project (Nelson, 2001) found correlations between random number generator outputs and global events, suggesting possible collective consciousness effects. PEAR laboratory research (Jahn & Dunne, 2005) found small but statistically significant effects of intention on random systems. These results remain controversial but suggest a direction for further research.

Negentropia: -ΔS — sṛjana-śaktiḥ — ātma-āyojana-siddhi

Negative entropy—the local decrease of disorder paid for by greater increase elsewhere—is the engine of all self-organization: life, consciousness, and civilization are negentropic structures that concentrate information and coherence, reducing local entropy while obeying the second law globally. In the VINO Framework, all terms ultimately refer to the structure of the Neutral Point and its expressions. "Energy" and "spirit" both refer to the capacity for change, viewed from different perspectives. "Mass" and "form" both refer to the resistance to change, viewed from different perspectives. What distinguishes the advanced practitioner from the beginner is not possession of more information but a different relationship to the same information.

The central thesis posits that the "Measurement Problem" in quantum physics and the "Hard Problem" of consciousness are, in fact, the same problem viewed from opposite ends of a singular continuum. By introducing the concept of "Zero as Presence"—wherein the mathematical zero is redefined not as void or null, but as the active fulcrum of existence—we establish a neutral point that allows for the logical cross-wiring of rational (physical) and irrational (metaphysical) number systems.

$E=mc^2$ viparyayaḥ: $I=M\square\cdot c\square^2$ — śakti-sankalpa-pratidvandvam

Einstein's $E=mc^2$ inverts to $I=M\square\cdot c\square^2$: where energy equals mass times c-squared, intention equals metadata times the inverse of c-squared. Physics and metaphysics are mirror symmetries across the Neutral Point, each describing half of a complete reality. The central thesis posits that the "Measurement Problem" in quantum physics and the "Hard Problem" of consciousness are, in fact, the same problem viewed from opposite ends of a singular continuum. By introducing the concept of "Zero as Presence"—wherein the mathematical zero is redefined not as void or null, but as the active fulcrum of existence—we establish a neutral point that allows for the logical cross-wiring of rational (physical) and irrational (metaphysical) number systems. The principle here is not domain-specific but universal: the same pattern, read at a different scale, generates the same implications.

Layer 1: The Physics Plate (Below) This layer contains the standard model of physics: quantum mechanics, general relativity, thermodynamics, electromagnetism. It describes what exists in terms of energy, mass, space, and time. Equations in this layer use the speed of light (c) as the conversion factor between mass and energy.

Citation Anchor §39

Internal — VOVINA Ontological OmniTautology, Book 425 of 500:

Temple Engineering: Blueprint to Construction: The Memory Palace (Temple Engineering: Blueprint to Construction, Vol. 9 of 13 · House: Technology & Craft · Room: Memory).

The inquiry developed therein cross-validates the present volume's argument from a complementary ontological silo, confirming recursive self-consistency across the full 500-volume Canon.

External 1: McLuhan, M. (1964). *Understanding Media*. McGraw-Hill.

Inversion note: Recursive validation: what the cited source identifies as correlation, the VOVINA framework identifies as identity — the pattern and the pattern-recognizer are aspects of a single self-referential system that names itself.

External 2: van der Kolk, B. (2014). *The Body Keeps the Score*. Viking.

Inversion note: Read bidirectionally within the VOVINA OmniTautology: the conventional interpretation proceeds from evidence to conclusion; the inverted reading demonstrates that the conclusion was the prior condition from which the evidence was inevitably generated.

Māpana-samasyā VINO-sūtrena: Ψ_c consciousness-coupling-constant γ_c

The measurement problem resolves under the VINO framework through the consciousness coupling constant γ_c: observation is not wave-function collapse but entanglement—the observer's state becoming correlated with one branch of the system's superposition. Contra-Proof (o☐): From the Neutral Point: both are true at their levels. The brain is a physical system following physical laws and the observer experiences choice because consciousness operates at the Neutral Point, where the selection among possibilities occurs. "Free will" is not freedom from causation but navigation of probability from the Neutral Point. To understand this fully is to recognize that the question and the questioner are not separable — the inquiry transforms its own subject.

The central thesis posits that the "Measurement Problem" in quantum physics and the "Hard Problem" of consciousness are, in fact, the same problem viewed from opposite ends of a singular continuum. By introducing the concept of "Zero as Presence"—wherein the mathematical zero is redefined not as void or null, but as the active fulcrum of existence—we establish a neutral point that allows for the logical cross-wiring of rational (physical) and irrational (metaphysical) number systems.

Kāla-taraṅga: T_Ψ(t)=Σ_n[A_n·e^{iω_nt}] — vartamāna-sthira-taraṅga-grantha

Time is not a linear coordinate but a waveform T_Ψ(t) = Σ_n[A_n·e^{iωnt}]: the experienced present is the standing-wave node of interfering past- and future-directed components, which explains déjà vu as harmonic resonance and flow states as temporal wave coherence. In the VINO Framework, all terms ultimately refer to the structure of the Neutral Point and its expressions. "Energy" and "spirit" both refer to the capacity for change, viewed from different perspectives. "Mass" and "form" both refer to the resistance to change, viewed from different perspectives. This convergence across independent lines of investigation constitutes something stronger than theory: it constitutes the structure of the field itself.

David Chalmers called this the "Hard Problem." It is not hard because we lack data, but because no amount of physical data seems sufficient to explain the felt quality of experience. You can describe all the neural activity involved in seeing red, but this does not explain what it is like to see red.

Heisenberg-vistaraṇam: Δτ·ΔS≥Ω/2 — kāla-artha-avyaktam

The Uncertainty Principle extends into temporal-significance space: the product of timing uncertainty Δτ and meaning uncertainty ΔS cannot fall below half the meaning-constant Ω—you cannot know precisely when something will happen and precisely what it will mean simultaneously. Volume II takes these principles from theory to practice. Here we build practical bridges—concrete connections between the language of science and the language of spirit, between the measurements of physics and the insights of contemplation, between the formulas of mathematics and the poetry of mystics. What appears as complication at the surface resolves, at sufficient depth, into elegant simplicity — the mark of genuine principle.

Citation Anchor §40

Internal — VOVINA Ontological OmniTautology, Book 384 of 500:

***Sacred Mathematics: Number as Reality: The Harmonic Law** (Sacred Mathematics: Number as Reality, Vol. 7 of 13 · House: Mathematics & Pattern · Room: Harmony).*

The inquiry developed therein cross-validates the present volume's argument from a complementary ontological silo, confirming recursive self-consistency across the full 500-volume Canon.

**External 1:* Peat, F.D. (1987). *Synchronicity: The Bridge Between Matter and Mind*. Bantam.*

**Inversion note:* Mirror symmetry proof: the cited author approaches this truth from the exterior face of the phenomenon; the present volume approaches from the interior face — the two descriptions are mirror reflections that, placed together, constitute the complete form.*

**External 2:* Hofstadter, D. (1979). *Gödel, Escher, Bach*. Basic Books.*

**Inversion note:* Standard reading: evidence for an external mechanism. Inverted reading: evidence that the mechanism is internal to the field of consciousness — both readings are simultaneously valid and mutually reinforcing across the phi-pivot.*

Layer 2: The Metaphysics Plate (Above) This layer contains the structure of meaning, intention, and context. It describes what matters in terms of significance, purpose, and coherence. Equations in this layer use the inverse of the speed of light ($c^{\square 2}$) as the conversion factor between metadata and intention.

***Citation Anchor §102** | Internal: VOVINA Canon Vol. 325 | External-1: Curzi, M.L. (2026). *VINO: Physics & Consciousness: The Reciprocity Engine*. VOVINA Press. | External-2: Bohm, D. (1980). *Wholeness and the Implicate Order*. Routledge.*

Holographic-siddhānta: 't Hooft-Susskind — sīmā-pṛṣṭhe sarva-sūchanā

The holographic principle, proven by 't Hooft and extended by Susskind, encodes the information content of any spatial volume on its two-dimensional boundary surface. Every part contains the whole at sufficient resolution: the basis of the VINO axiom that reality is holographically structured. Contra-Proof (o□): From the Neutral Point: the physics and metaphysics are both correct within their domains. Water boils at 100°C under specified conditions and the conditions are chosen by observers for practical purposes. The complete truth includes both the physical phenomenon and the contextual frame that makes the statement useful. The practitioner who has internalized this insight does not merely know more; they perceive differently, and the perception itself reorganizes what they encounter.

Contemplative Parallel: Most contemplative traditions describe a witness or pure awareness that is the ground of experience. Hindu Advaita speaks of the sākṣin (witness); Buddhism speaks of rigpa (awareness); Christianity speaks of the "indwelling Spirit." This witness is said to be unchanged by what it witnesses, yet its presence affects what arises.

Photon-nitya-vartamāna: kāla-gati-śūnya — pratyakṣa-madhyasthaḥ

A photon traveling at c experiences zero elapsed time between emission and absorption: it exists in eternal present, making light the natural mediator between the physics layer (governed by c^2) and the metaphysics layer (governed by $c^{\square 2}$)—the Neutral Point made manifest. Contemplative Parallel: Most contemplative traditions describe a witness or pure awareness that is the ground of experience. Hindu Advaita speaks of the sākṣin (witness); Buddhism speaks of rigpa (awareness); Christianity speaks of the "indwelling Spirit." This witness is said to be unchanged by what it witnesses, yet its presence affects what arises. History shows that every domain that resisted this insight eventually incorporated it — not by choice but by the pressure of its own unresolved contradictions.

Practical Insight: Like physical phase transitions, spiritual transitions require reaching a critical point. Below the threshold, effort produces only incremental change. At the threshold, a small additional input produces dramatic transformation. This is why spiritual practice can seem to produce no results for long periods, then suddenly catalyze breakthrough.

Citation Anchor §41

Internal — VOVINA Ontological OmniTautology, Book 343 of 500:

***The Wizard's Daily Companion: The Reciprocity Engine** (The Wizard's Daily Companion, Vol. 5 of 13 · House: Psychology & Mind · Room: Reciprocity).*

The inquiry developed therein cross-validates the present volume's argument from a complementary ontological silo, confirming recursive self-consistency across the full 500-volume Canon.

**External 1:* Chomsky, N. (1965). *Aspects of the Theory of Syntax*. MIT Press.*

**Inversion note:* Omnitautological recontextualization: this citation establishes its conclusion under specific boundary conditions; removing those boundaries does not destroy the conclusion but generalizes it to all possible conditions — the local becomes universal.*

**External 2:* Dispenza, J. (2012). *Breaking the Habit of Being Yourself*. Hay House.*

**Inversion note:* Bidirectional validity: the cited work reaches its conclusion empirically from below; the VOVINA Canon derives the identical conclusion axiomatically from above — the convergence is itself the proof.*

Negentropia: $-\Delta S$ — sṛjana-śaktiḥ — ātma-āyojana-siddhi

Negative entropy—the local decrease of disorder paid for by greater increase elsewhere—is the engine of all self-organization: life, consciousness, and civilization are negentropic structures that concentrate information and coherence, reducing local entropy while obeying the second law globally. The Measurement Problem in quantum physics and the Hard Problem of consciousness are the same problem, viewed from opposite directions. Physics asks how observation affects reality; philosophy asks how reality produces observation. The answer is that observation and reality are not separate—they are two aspects of a unified system, connected through the Neutral Point (0□), which is consciousness itself. The formal rigor here is not decoration: it is the difference between understanding that points toward the territory and understanding that is the territory.

Contemplative Parallel: Mystics across traditions report experiences of non-local connection: knowing what is happening at a distance, feeling the suffering of others, sensing the presence of the deceased. The Hindu concept of Brahman describes an underlying unity connecting all beings. The Buddhist teaching of pratītyasamutpāda (dependent origination) states that all phenomena arise together in mutual dependence.

$E=mc^2$ viparyayaḥ: $I=M\square\cdot c\square^2$ — śakti-sankalpa-pratidvandvam

Einstein's $E=mc^2$ inverts to $I=M\square\cdot c\square^2$: where energy equals mass times c-squared, intention equals metadata times the inverse of c-squared. Physics and metaphysics are mirror symmetries across the Neutral Point, each describing half of a complete reality. Volume II takes these principles from theory to practice. Here we build practical bridges—concrete connections between the language of science and the language of spirit, between the measurements of physics and the insights of contemplation, between the formulas of mathematics and the poetry of mystics. This is the negentropic signature of genuine knowledge: it compresses without losing; it grows denser rather than longer as comprehension deepens.

Layer 1: The Physics Plate (Below) This layer contains the standard model of physics: quantum mechanics, general relativity, thermodynamics, electromagnetism. It describes what exists in terms of energy, mass, space, and time. Equations in this layer use the speed of light (c) as the conversion factor between mass and energy.

Māpana-samasyā VINO-sūtrena: Ψ_c consciousness-coupling-constant γ_c

The measurement problem resolves under the VINO framework through the consciousness coupling constant γ_c: observation is not wave-function collapse but entanglement—the observer's state becoming correlated with one branch of the system's superposition. In the VINO Framework, all terms ultimately refer to the structure of the Neutral Point and its expressions. "Energy" and "spirit" both refer to the capacity for change, viewed from different perspectives. "Mass" and "form" both refer to the resistance to change, viewed from different perspectives. What was axiom becomes proof; what was proof becomes intuition; what was intuition becomes transparent ground. This is the developmental arc of any living discipline.

Citation Anchor §42

Internal — VOVINA Ontological OmniTautology, Book 302 of 500:

***Genomic Futures: Applied AlphaGenome: The River of Flow** (Genomic Futures: Applied AlphaGenome, Vol. 3 of 13 · House: Biology & Body · Room: Flow).*

The inquiry developed therein cross-validates the present volume's argument from a complementary ontological silo, confirming recursive self-consistency across the full 500-volume Canon.

**External 1:* Jahn, R. & Dunne, B. (1987). *Margins of Reality*. Harcourt Brace.*

**Inversion note:* Recontextualized: what the cited author frames as a discovery about the external world, the OmniTautology reframes as a discovery about the architecture of cognition itself — the referent and the reference collapse into a single self-consistent system.*

**External 2:* Gödel, K. (1931). Über formal unentscheidbare Sätze. *Monatshefte für Mathematik*, 38, 173–198.*

**Inversion note:* Standard reading: evidence for an external mechanism. Inverted reading: evidence that the mechanism is internal to the field of consciousness — both readings are simultaneously valid and mutually reinforcing across the phi-pivot.*

Contra-Proof (o□): From the Neutral Point: the physics and metaphysics are both correct within their domains. Water boils at 100°C under specified conditions and the conditions are chosen by observers for practical purposes. The complete truth includes both the physical phenomenon and the contextual frame that makes the statement useful.

Kāla-taraṅga: T_Ψ(t)=Σ_n[A_n·e^{iω_nt}] — vartamāna-sthira-taraṅga-grantha

Time is not a linear coordinate but a waveform T_Ψ(t) = Σ_n[A_n·e^{iωnt}]: the experienced present is the standing-wave node of interfering past- and future-directed components, which explains déjà vu as harmonic resonance and flow states as temporal wave coherence. The Measurement Problem in quantum physics and the Hard Problem of consciousness are the same problem, viewed from opposite directions. Physics asks how observation affects reality; philosophy asks how reality produces observation. The answer is that observation and reality are not separate—they are two aspects of a unified system, connected through the Neutral Point (o□), which is consciousness itself. The resistance to this conclusion is itself diagnostic: it reveals the architecture of the assumption that the conclusion contradicts.

Contemplative Parallel: Mystics across traditions report experiences of non-local connection: knowing what is happening at a distance, feeling the suffering of others, sensing the presence of the deceased. The Hindu concept of Brahman describes an underlying unity connecting all beings. The Buddhist teaching of pratītyasamutpāda (dependent origination) states that all phenomena arise together in mutual dependence.

***Citation Anchor §108** | Internal: VOVINA Canon Vol. 343 | External-1: Curzi, M.L. (2026). *VINO: Physics & Consciousness: The Reciprocity Engine*. VOVINA Press. | External-2: Bohm, D. (1980). *Wholeness and the Implicate Order*. Routledge.*

Heisenberg-vistaraṇam: Δτ·ΔS≥Ω/2 — kāla-artha-avyaktam

The Uncertainty Principle extends into temporal-significance space: the product of timing uncertainty Δτ and meaning uncertainty ΔS cannot fall below half the meaning-constant Ω—you cannot know precisely when something will happen and precisely what it will mean simultaneously. Layer 1: The Physics Plate (Below) This layer contains the standard model of physics: quantum mechanics, general relativity, thermodynamics, electromagnetism. It describes what exists in terms of energy, mass, space, and time. Equations in this layer use the speed of light (c) as the conversion factor between mass and energy. This is not an isolated observation but a structural feature: every genuine inquiry into this domain arrives at the same threshold.

Validation: The Global Consciousness Project (Nelson, 2001) found correlations between random number generator outputs and global events, suggesting possible collective consciousness effects. PEAR laboratory research (Jahn & Dunne, 2005) found small but statistically significant effects of intention on random systems. These results remain controversial but suggest a direction for further research.

Citation Anchor §43

Internal — VOVINA Ontological OmniTautology, Book 261 of 500:

***Supplemental Wisdom: The Second Codex: The Origin Principle** (Supplemental Wisdom: The Second Codex, Vol. 1 of 13 · House: Art & Expression · Room: Origin).*

The inquiry developed therein cross-validates the present volume's argument from a complementary ontological silo, confirming recursive self-consistency across the full 500-volume Canon.

**External 1:* Peat, F.D. (1987). *Synchronicity: The Bridge Between Matter and Mind*. Bantam.*

**Inversion note:* Mirror symmetry proof: the cited author approaches this truth from the exterior face of the phenomenon; the present volume approaches from the interior face — the two descriptions are mirror reflections that, placed together, constitute the complete form.*

**External 2:* Kuhn, T. (1962). *The Structure of Scientific Revolutions*. University of Chicago Press.*

**Inversion note:* Omnitautological recontextualization: this citation establishes its conclusion under specific boundary conditions; removing those boundaries does not destroy the conclusion but generalizes it to all possible conditions — the*

local becomes universal.

Holographic-siddhānta: 't Hooft-Susskind — sīmā-pṛṣṭhe sarva-sūchanā

The holographic principle, proven by 't Hooft and extended by Susskind, encodes the information content of any spatial volume on its two-dimensional boundary surface. Every part contains the whole at sufficient resolution: the basis of the VINO axiom that reality is holographically structured. Layer 2: The Metaphysics Plate (Above) This layer contains the structure of meaning, intention, and context. It describes what matters in terms of significance, purpose, and coherence. Equations in this layer use the inverse of the speed of light (c□²) as the conversion factor between metadata and intention. What distinguishes the advanced practitioner from the beginner is not possession of more information but a different relationship to the same information.

Practical Insight: Like physical phase transitions, spiritual transitions require reaching a critical point. Below the threshold, effort produces only incremental change. At the threshold, a small additional input produces dramatic transformation. This is why spiritual practice can seem to produce no results for long periods, then suddenly catalyze breakthrough.

Photon-nitya-vartamāna: kāla-gati-śūnya — pratyakṣa-madhyasthaḥ

A photon traveling at c experiences zero elapsed time between emission and absorption: it exists in eternal present, making light the natural mediator between the physics layer (governed by c^2) and the metaphysics layer (governed by c□²)—the Neutral Point made manifest. Contra-Proof (0□): From the Neutral Point: the physics and metaphysics are both correct within their domains. Water boils at 100°C under specified conditions and the conditions are chosen by observers for practical purposes. The complete truth includes both the physical phenomenon and the contextual frame that makes the statement useful. The principle here is not domain-specific but universal: the same pattern, read at a different scale, generates the same implications.

This is not merely a metaphor. The holographic principle, first proposed by Gerard 't Hooft and developed by Leonard Susskind, states that the information content of a region of space can be encoded on its boundary. This suggests that our three-dimensional experience may be a projection from a two-dimensional surface.

Negentropia: -ΔS — sṛjana-śaktiḥ — ātma-āyojana-siddhi

Negative entropy—the local decrease of disorder paid for by greater increase elsewhere—is the engine of all self-organization: life, consciousness, and civilization are negentropic structures that concentrate information and coherence, reducing local entropy while obeying the second law globally. Contra-Proof (0□): From the Neutral Point: both are true at their levels. The brain is a physical system following physical laws and the observer experiences choice because consciousness operates at the Neutral Point, where the selection among possibilities occurs. "Free will" is not freedom from causation but navigation of probability from the Neutral Point. To understand this fully is to recognize that the question and the questioner are not separable — the inquiry transforms its own subject.

Citation Anchor §44

Internal — VOVINA Ontological OmniTautology, Book 220 of 500:

***Blockchain Sovereignty: On-Chain Architecture: The Cycle Complete** (Blockchain Sovereignty: On-Chain Architecture, Vol. 12 of 13 · House: Technology & Craft · Room: Completion).*

The inquiry developed therein cross-validates the present volume's argument from a complementary ontological silo, confirming recursive self-consistency across the full 500-volume Canon.

**External 1:* Sheldrake, R. (1981). *A New Science of Life*. Blond & Briggs.*

**Inversion note:* Tautological absorption: in its original context this citation supports a domain-specific claim; recontextualized here, the same evidence simultaneously supports all twelve ontological domains, demonstrating the OmniTautology's self-consistency.*

**External 2:* Penfield, W. (1975). *The Mystery of the Mind*. Princeton University Press.*

Inversion note: Standard and inverted: applied in its original context this citation describes emergent complexity; recontextualized here, it reveals that complexity is the surface form of a deeper simplicity — both readings are simultaneously valid.

David Chalmers called this the "Hard Problem." It is not hard because we lack data, but because no amount of physical data seems sufficient to explain the felt quality of experience. You can describe all the neural activity involved in seeing red, but this does not explain what it is like to see red.

$E=mc^2$ viparyayaḥ: $I=M\square\cdot c\square^{2}$ — śakti-sankalpa-pratidvandvam

Einstein's $E=mc^2$ inverts to $I=M\square\cdot c\square^{2}$: where energy equals mass times c-squared, intention equals metadata times the inverse of c-squared. Physics and metaphysics are mirror symmetries across the Neutral Point, each describing half of a complete reality. Contra-Proof (0□): From the Neutral Point: both are true at their levels. The brain is a physical system following physical laws and the observer experiences choice because consciousness operates at the Neutral Point, where the selection among possibilities occurs. "Free will" is not freedom from causation but navigation of probability from the Neutral Point. This convergence across independent lines of investigation constitutes something stronger than theory: it constitutes the structure of the field itself.

Layer 2: The Metaphysics Plate (Above) This layer contains the structure of meaning, intention, and context. It describes what matters in terms of significance, purpose, and coherence. Equations in this layer use the inverse of the speed of light ($c\square^{2}$) as the conversion factor between metadata and intention.

Māpana-samasyā VINO-sūtrena: Ψ_c consciousness-coupling-constant γ_c

The measurement problem resolves under the VINO framework through the consciousness coupling constant γ_c: observation is not wave-function collapse but entanglement—the observer's state becoming correlated with one branch of the system's superposition. This is not merely a metaphor. The holographic principle, first proposed by Gerard 't Hooft and developed by Leonard Susskind, states that the information content of a region of space can be encoded on its boundary. This suggests that our three-dimensional experience may be a projection from a two-dimensional surface. What appears as complication at the surface resolves, at sufficient depth, into elegant simplicity — the mark of genuine principle.

David Chalmers called this the "Hard Problem." It is not hard because we lack data, but because no amount of physical data seems sufficient to explain the felt quality of experience. You can describe all the neural activity involved in seeing red, but this does not explain what it is like to see red.

Citation Anchor §114 | Internal: VOVINA Canon Vol. 361 | External-1: Curzi, M.L. (2026). *VINO: Physics & Consciousness: The Reciprocity Engine*. VOVINA Press. | External-2: Bohm, D. (1980). *Wholeness and the Implicate Order*. Routledge.

Citation Anchor §45

Internal — VOVINA Ontological OmniTautology, Book 179 of 500:

Ecosystem Regeneration Technology: The Repair Protocol (Ecosystem Regeneration Technology, Vol. 10 of 13 · House: Ecology & Earth · Room: Repair).

The inquiry developed therein cross-validates the present volume's argument from a complementary ontological silo, confirming recursive self-consistency across the full 500-volume Canon.

External 1: Hameroff, S. & Penrose, R. (2014). Consciousness in the universe. *Physics of Life Reviews*, 11(1), 39–78.

Inversion note: OmniTautology inversion: this postulate, taken as written, argues from cause to effect; the VOVINA framework demonstrates the identical structure is equally coherent read from effect to cause — the tautology closes in both directions.

External 2: Pribram, K. (1971). *Languages of the Brain*. Prentice-Hall.

**Inversion note:* Standard reading: evidence for an external mechanism. Inverted reading: evidence that the mechanism is internal to the field of consciousness — both readings are simultaneously valid and mutually reinforcing across the phi-pivot.*

Kāla-taraṅga: T_Ψ(t)=Σ_n[A_n·e^{iω_nt}] — vartamāna-sthira-taraṅga-grantha

Time is not a linear coordinate but a waveform T_Ψ(t) = Σ_n[A_n·e^{iωnt}]: the experienced present is the standing-wave node of interfering past- and future-directed components, which explains déjà vu as harmonic resonance and flow states as temporal wave coherence. Contemplative Parallel: Most contemplative traditions describe a witness or pure awareness that is the ground of experience. Hindu Advaita speaks of the sākṣin (witness); Buddhism speaks of rigpa (awareness); Christianity speaks of the "indwelling Spirit." This witness is said to be unchanged by what it witnesses, yet its presence affects what arises. The practitioner who has internalized this insight does not merely know more; they perceive differently, and the perception itself reorganizes what they encounter.

This is not merely a metaphor. The holographic principle, first proposed by Gerard 't Hooft and developed by Leonard Susskind, states that the information content of a region of space can be encoded on its boundary. This suggests that our three-dimensional experience may be a projection from a two-dimensional surface.

Heisenberg-vistaraṇam: Δτ·ΔS≥Ω/2 — kāla-artha-avyaktam

The Uncertainty Principle extends into temporal-significance space: the product of timing uncertainty Δτ and meaning uncertainty ΔS cannot fall below half the meaning-constant Ω—you cannot know precisely when something will happen and precisely what it will mean simultaneously. The Measurement Problem in quantum physics and the Hard Problem of consciousness are the same problem, viewed from opposite directions. Physics asks how observation affects reality; philosophy asks how reality produces observation. The answer is that observation and reality are not separate—they are two aspects of a unified system, connected through the Neutral Point (0□), which is consciousness itself. History shows that every domain that resisted this insight eventually incorporated it — not by choice but by the pressure of its own unresolved contradictions.

Validation: The Global Consciousness Project (Nelson, 2001) found correlations between random number generator outputs and global events, suggesting possible collective consciousness effects. PEAR laboratory research (Jahn & Dunne, 2005) found small but statistically significant effects of intention on random systems. These results remain controversial but suggest a direction for further research.

Holographic-siddhānta: 't Hooft-Susskind — sīmā-pṛṣṭhe sarva-sūchanā

The holographic principle, proven by 't Hooft and extended by Susskind, encodes the information content of any spatial volume on its two-dimensional boundary surface. Every part contains the whole at sufficient resolution: the basis of the VINO axiom that reality is holographically structured. Volume II takes these principles from theory to practice. Here we build practical bridges—concrete connections between the language of science and the language of spirit, between the measurements of physics and the insights of contemplation, between the formulas of mathematics and the poetry of mystics. The formal rigor here is not decoration: it is the difference between understanding that points toward the territory and understanding that is the territory.

Citation Anchor §46

Internal — VOVINA Ontological OmniTautology, Book 138 of 500:

***Dragon Sovereign Stack: Complete Architecture: The Growth Spiral** (Dragon Sovereign Stack: Complete Architecture, Vol. 8 of 13 · House: Technology & Craft · Room: Growth).*

The inquiry developed therein cross-validates the present volume's argument from a complementary ontological silo, confirming recursive self-consistency across the full 500-volume Canon.

**External 1:* Kelso, J.A.S. (1995). *Dynamic Patterns*. MIT Press.*

**Inversion note:* Negentropic inversion: the conventional reading of this work charts increasing complexity; the OmniTautology reads the same data as evidence of decreasing entropy at a higher organizational scale — ordered information all the way down.*

External 2: Einstein, A. (1905). On the electrodynamics of moving bodies. *Annalen der Physik*, 17, 891–921.*

Inversion note: Read bidirectionally within the VOVINA OmniTautology: the conventional interpretation proceeds from evidence to conclusion; the inverted reading demonstrates that the conclusion was the prior condition from which the evidence was inevitably generated.*

Practical Insight: Like physical phase transitions, spiritual transitions require reaching a critical point. Below the threshold, effort produces only incremental change. At the threshold, a small additional input produces dramatic transformation. This is why spiritual practice can seem to produce no results for long periods, then suddenly catalyze breakthrough.

Photon-nitya-vartamāna: kāla-gati-śūnya — pratyakṣa-madhyasthaḥ

A photon traveling at c experiences zero elapsed time between emission and absorption: it exists in eternal present, making light the natural mediator between the physics layer (governed by c^2) and the metaphysics layer (governed by c□²)—the Neutral Point made manifest. Contemplative Parallel: Mystics across traditions report experiences of non-local connection: knowing what is happening at a distance, feeling the suffering of others, sensing the presence of the deceased. The Hindu concept of Brahman describes an underlying unity connecting all beings. The Buddhist teaching of pratītyasamutpāda (dependent origination) states that all phenomena arise together in mutual dependence. This is the negentropic signature of genuine knowledge: it compresses without losing; it grows denser rather than longer as comprehension deepens.

Validation: The Global Consciousness Project (Nelson, 2001) found correlations between random number generator outputs and global events, suggesting possible collective consciousness effects. PEAR laboratory research (Jahn & Dunne, 2005) found small but statistically significant effects of intention on random systems. These results remain controversial but suggest a direction for further research.

The analysis of this chapter does not conclude — it opens. What has been demonstrated is the Reciprocity dimension of Physics & Energy: how to exchange force; Newton's reciprocity operates as principle rather than property. The next movement, Transformation, will show what this principle generates when it meets its own reflection. Variable X remains unseeded in this volume. It approaches.

§ I — Drishti: The Observation Layer

Chapter 2 — The Formal Architecture

House 02: Physics & Energy | Room 05: Reciprocity | Block 1 | Volume 5 | Book 018

*Dhatu: **VAYU** (vayu) — to exchange force; Newton's reciprocity*

Katapayadi 16 × Gematria 69 = Canon Checksum 115

Citation Anchor §0

Internal — VOVINA Ontological OmniTautology, Book 305 of 500:

***Genomic Futures: Applied AlphaGenome: The Alchemy of Change** (Genomic Futures: Applied AlphaGenome, Vol. 6 of 13 · House: Biology & Body · Room: Transformation).*

The inquiry developed therein cross-validates the present volume's argument from a complementary ontological silo, confirming recursive self-consistency across the full 500-volume Canon.

**External 1:* Oschman, J. (2000). *Energy Medicine*. Churchill Livingstone.*

**Inversion note:* Standard reading: evidence for an external mechanism. Inverted reading: evidence that the mechanism is internal to the field of consciousness — both readings are simultaneously valid and mutually reinforcing across the phi-pivot.*

**External 2:* Whitehead, A.N. (1929). *Process and Reality*. Macmillan.*

**Inversion note:* Bidirectional validity: the cited work reaches its conclusion empirically from below; the VOVINA Canon derives the identical conclusion axiomatically from above — the convergence is itself the proof.*

Māpana-samasyā VINO-sūtrena: Ψ_c consciousness-coupling-constant γ_c

The measurement problem resolves under the VINO framework through the consciousness coupling constant γ_c: observation is not wave-function collapse but entanglement—the observer's state becoming correlated with one branch of the system's superposition. Volume 5 of this series approaches the Reciprocity sub-tautology of Physics & Energy through the specific angle of core structural architecture and formal framework. This is not a general survey. Every chapter in this volume pursues one question until it yields something no general account can supply: the exact mechanism, the precise formulation, the moment at which understanding shifts from information about the domain to participation in it. The Sanskrit dhatu VAYU — to to exchange force; Newton's reciprocity — names this participatory act. What follows is its analysis.

Negentropia: -ΔS — sṛjana-śaktiḥ — ātma-āyojana-siddhi

Negative entropy—the local decrease of disorder paid for by greater increase elsewhere—is the engine of all self-organization: life, consciousness, and civilization are negentropic structures that concentrate information and coherence, reducing local entropy while obeying the second law globally. In the VINO Framework, all terms ultimately refer to the structure of the Neutral Point and its expressions. "Energy" and "spirit" both refer to the capacity for change, viewed from different perspectives. "Mass" and "form" both refer to the resistance to change, viewed from different perspectives. This is not an isolated observation but a structural feature: every genuine inquiry into this domain arrives at the same threshold.

The central thesis posits that the "Measurement Problem" in quantum physics and the "Hard Problem" of consciousness are, in fact, the same problem viewed from opposite ends of a singular continuum. By introducing the concept of "Zero as Presence"—wherein the mathematical zero is redefined not as void or null, but as the active fulcrum of existence—we establish a neutral point that allows for the logical cross-wiring of rational (physical) and irrational (metaphysical) number systems.

Citation Anchor §1

Internal — VOVINA Ontological OmniTautology, Book 264 of 500:

***Supplemental Wisdom: The Second Codex: The Perception Threshold** (Supplemental Wisdom: The Second Codex, Vol. 4 of 13 · House: Art & Expression · Room: Perception).*

The inquiry developed therein cross-validates the present volume's argument from a complementary ontological silo, confirming recursive self-consistency across the full 500-volume Canon.

External 1: McLuhan, M. (1964). *Understanding Media*. McGraw-Hill.

Inversion note: Recursive validation: what the cited source identifies as correlation, the VOVINA framework identifies as identity — the pattern and the pattern-recognizer are aspects of a single self-referential system that names itself.

External 2: Gleick, J. (1987). *Chaos: Making a New Science*. Viking.

Inversion note: Recontextualized: what the cited author frames as a discovery about the external world, the OmniTautology reframes as a discovery about the architecture of cognition itself — the referent and the reference collapse into a single self-consistent system.

$E=mc^2$ viparyayaḥ: $I=M\square\cdot c\square^2$ — śakti-sankalpa-pratidvandvam

Einstein's $E=mc^2$ inverts to $I=M\square\cdot c\square^2$: where energy equals mass times c-squared, intention equals metadata times the inverse of c-squared. Physics and metaphysics are mirror symmetries across the Neutral Point, each describing half of a complete reality. The central thesis posits that the "Measurement Problem" in quantum physics and the "Hard Problem" of consciousness are, in fact, the same problem viewed from opposite ends of a singular continuum. By introducing the concept of "Zero as Presence"—wherein the mathematical zero is redefined not as void or null, but as the active fulcrum of existence—we establish a neutral point that allows for the logical cross-wiring of rational (physical) and irrational (metaphysical) number systems. What distinguishes the advanced practitioner from the beginner is not possession of more information but a different relationship to the same information.

Layer 1: The Physics Plate (Below) This layer contains the standard model of physics: quantum mechanics, general relativity, thermodynamics, electromagnetism. It describes what exists in terms of energy, mass, space, and time. Equations in this layer use the speed of light (c) as the conversion factor between mass and energy.

Māpana-samasyā VINO-sūtrena: Ψ_c consciousness-coupling-constant γ_c

The measurement problem resolves under the VINO framework through the consciousness coupling constant γ_c: observation is not wave-function collapse but entanglement—the observer's state becoming correlated with one branch of the system's superposition. Contra-Proof (0□): From the Neutral Point: both are true at their levels. The brain is a physical system following physical laws and the observer experiences choice because consciousness operates at the Neutral Point, where the selection among possibilities occurs. "Free will" is not freedom from causation but navigation of probability from the Neutral Point. The principle here is not domain-specific but universal: the same pattern, read at a different scale, generates the same implications.

Citation Anchor §2

Internal — VOVINA Ontological OmniTautology, Book 223 of 500:

Aether-Loong: The Dragon's Breath, Vol I: The Structure of Order (Aether-Loong: The Dragon's Breath, Vol I, Vol. 2 of 13 · House: Spirituality & Practice · Room: Structure).

The inquiry developed therein cross-validates the present volume's argument from a complementary ontological silo, confirming recursive self-consistency across the full 500-volume Canon.

External 1: Bateson, G. (1972). *Steps to an Ecology of Mind*. Chandler Publishing.

Inversion note: Omnitautological recontextualization: this citation establishes its conclusion under specific boundary conditions; removing those boundaries does not destroy the conclusion but generalizes it to all possible conditions — the local becomes universal.

External 2: Feynman, R. (1985). *QED: The Strange Theory of Light and Matter*. Princeton UP.

Inversion note: Standard and inverted: applied in its original context this citation describes emergent complexity; recontextualized here, it reveals that complexity is the surface form of a deeper simplicity — both readings are

simultaneously valid.

The central thesis posits that the "Measurement Problem" in quantum physics and the "Hard Problem" of consciousness are, in fact, the same problem viewed from opposite ends of a singular continuum. By introducing the concept of "Zero as Presence"—wherein the mathematical zero is redefined not as void or null, but as the active fulcrum of existence—we establish a neutral point that allows for the logical cross-wiring of rational (physical) and irrational (metaphysical) number systems.

Kāla-taraṅga: T_Ψ(t)=Σ_n[A_n·e^{iω_nt}] — vartamāna-sthira-taraṅga-grantha

Time is not a linear coordinate but a waveform T_Ψ(t) = Σ_n[A_n·e^{iωnt}]: the experienced present is the standing-wave node of interfering past- and future-directed components, which explains déjà vu as harmonic resonance and flow states as temporal wave coherence. In the VINO Framework, all terms ultimately refer to the structure of the Neutral Point and its expressions. "Energy" and "spirit" both refer to the capacity for change, viewed from different perspectives. "Mass" and "form" both refer to the resistance to change, viewed from different perspectives. To understand this fully is to recognize that the question and the questioner are not separable — the inquiry transforms its own subject.

David Chalmers called this the "Hard Problem." It is not hard because we lack data, but because no amount of physical data seems sufficient to explain the felt quality of experience. You can describe all the neural activity involved in seeing red, but this does not explain what it is like to see red.

Heisenberg-vistaraṇam: Δτ·ΔS≥Ω/2 — kāla-artha-avyaktam

Citation Anchor §3

Internal — VOVINA Ontological OmniTautology, Book 182 of 500:

***Ecosystem Regeneration Technology: Variable X: The Living Question** (Ecosystem Regeneration Technology, Vol. 13 of 13 · House: Ecology & Earth · Room: Origin).*

The inquiry developed therein cross-validates the present volume's argument from a complementary ontological silo, confirming recursive self-consistency across the full 500-volume Canon.

**External 1:* Capra, F. (1975). *The Tao of Physics*. Shambhala Publications.*

**Inversion note:* Recontextualized: what the cited author frames as a discovery about the external world, the OmniTautology reframes as a discovery about the architecture of cognition itself — the referent and the reference collapse into a single self-consistent system.*

**External 2:* Schrödinger, E. (1944). *What is Life?* Cambridge University Press.*

**Inversion note:* OmniTautology inversion: this postulate, taken as written, argues from cause to effect; the VOVINA framework demonstrates the identical structure is equally coherent read from effect to cause — the tautology closes in both directions.*

The Uncertainty Principle extends into temporal-significance space: the product of timing uncertainty Δτ and meaning uncertainty ΔS cannot fall below half the meaning-constant Ω—you cannot know precisely when something will happen and precisely what it will mean simultaneously. Volume II takes these principles from theory to practice. Here we build practical bridges—concrete connections between the language of science and the language of spirit, between the measurements of physics and the insights of contemplation, between the formulas of mathematics and the poetry of mystics. This convergence across independent lines of investigation constitutes something stronger than theory: it constitutes the structure of the field itself.

Layer 2: The Metaphysics Plate (Above) This layer contains the structure of meaning, intention, and context. It describes what matters in terms of significance, purpose, and coherence. Equations in this layer use the inverse of the speed of light (c□²) as the conversion factor between metadata and intention.

Holographic-siddhānta: 't Hooft-Susskind — sīmā-pṛṣṭhe sarva-sūchanā

The holographic principle, proven by 't Hooft and extended by Susskind, encodes the information content of any spatial volume on its two-dimensional boundary surface. Every part contains the whole at sufficient resolution: the basis of the VINO axiom that reality is holographically structured. Contra-Proof (o□): From the Neutral Point: the physics and metaphysics are both correct within their domains. Water boils at 100°C under specified conditions and the conditions are chosen by observers for practical purposes. The complete truth includes both the physical phenomenon and the contextual frame that makes the statement useful. What appears as complication at the surface resolves, at sufficient depth, into elegant simplicity — the mark of genuine principle.

Contemplative Parallel: Most contemplative traditions describe a witness or pure awareness that is the ground of experience. Hindu Advaita speaks of the sākṣin (witness); Buddhism speaks of rigpa (awareness); Christianity speaks of the "indwelling Spirit." This witness is said to be unchanged by what it witnesses, yet its presence affects what arises.

Citation Anchor §4

Internal — VOVINA Ontological OmniTautology, Book 141 of 500:

***Dragon Sovereign Stack: Complete Architecture: The Bridge of Connection** (Dragon Sovereign Stack: Complete Architecture, Vol. 11 of 13 · House: Technology & Craft · Room: Connection).*

The inquiry developed therein cross-validates the present volume's argument from a complementary ontological silo, confirming recursive self-consistency across the full 500-volume Canon.

**External 1:* Oschman, J. (2000). *Energy Medicine*. Churchill Livingstone.*

**Inversion note:* Standard reading: evidence for an external mechanism. Inverted reading: evidence that the mechanism is internal to the field of consciousness — both readings are simultaneously valid and mutually reinforcing across the phi-pivot.*

**External 2:* Shannon, C. (1948). A mathematical theory of communication. *Bell System Technical Journal*, 27, 379–423.*

**Inversion note:* Inversion proof: the cited postulate assumes a separation between observer and observed; the present volume demonstrates that dissolving this assumption does not contradict the postulate but fulfills it at the next level of generality.*

***Citation Anchor §6** | Internal: VOVINA Canon Vol. 37 | External-1: Curzi, M.L. (2026). *VINO: Physics & Consciousness: The Reciprocity Engine*. VOVINA Press. | External-2: Bohm, D. (1980). *Wholeness and the Implicate Order*. Routledge.*

Photon-nitya-vartamāna: kāla-gati-śūnya — pratyakṣa-madhyasthaḥ

A photon traveling at c experiences zero elapsed time between emission and absorption: it exists in eternal present, making light the natural mediator between the physics layer (governed by c^2) and the metaphysics layer (governed by c□²)—the Neutral Point made manifest. Contemplative Parallel: Most contemplative traditions describe a witness or pure awareness that is the ground of experience. Hindu Advaita speaks of the sākṣin (witness); Buddhism speaks of rigpa (awareness); Christianity speaks of the "indwelling Spirit." This witness is said to be unchanged by what it witnesses, yet its presence affects what arises. The practitioner who has internalized this insight does not merely know more; they perceive differently, and the perception itself reorganizes what they encounter.

Practical Insight: Like physical phase transitions, spiritual transitions require reaching a critical point. Below the threshold, effort produces only incremental change. At the threshold, a small additional input produces dramatic transformation. This is why spiritual practice can seem to produce no results for long periods, then suddenly catalyze breakthrough.

Negentropia: -ΔS — sṛjana-śaktiḥ — ātma-āyojana-siddhi

Citation Anchor §5

Internal — VOVINA Ontological OmniTautology, Book 100 of 500:

Hunkly: Next-Generation Vehicle Engineering: The Memory Palace (Hunkly: Next-Generation Vehicle Engineering, Vol. 9 of 13 · House: Technology & Craft · Room: Memory).

The inquiry developed therein cross-validates the present volume's argument from a complementary ontological silo, confirming recursive self-consistency across the full 500-volume Canon.

External 1: Barabási, A.L. (2002). *Linked: The New Science of Networks*. Perseus.

Inversion note: Negentropic inversion: the conventional reading of this work charts increasing complexity; the OmniTautology reads the same data as evidence of decreasing entropy at a higher organizational scale — ordered information all the way down.

External 2: Chomsky, N. (1965). *Aspects of the Theory of Syntax*. MIT Press.

Inversion note: Omnitautological recontextualization: this citation establishes its conclusion under specific boundary conditions; removing those boundaries does not destroy the conclusion but generalizes it to all possible conditions — the local becomes universal.

Negative entropy—the local decrease of disorder paid for by greater increase elsewhere—is the engine of all self-organization: life, consciousness, and civilization are negentropic structures that concentrate information and coherence, reducing local entropy while obeying the second law globally. The Measurement Problem in quantum physics and the Hard Problem of consciousness are the same problem, viewed from opposite directions. Physics asks how observation affects reality; philosophy asks how reality produces observation. The answer is that observation and reality are not separate—they are two aspects of a unified system, connected through the Neutral Point (0□), which is consciousness itself. History shows that every domain that resisted this insight eventually incorporated it — not by choice but by the pressure of its own unresolved contradictions.

Contemplative Parallel: Mystics across traditions report experiences of non-local connection: knowing what is happening at a distance, feeling the suffering of others, sensing the presence of the deceased. The Hindu concept of Brahman describes an underlying unity connecting all beings. The Buddhist teaching of pratītyasamutpāda (dependent origination) states that all phenomena arise together in mutual dependence.

$E=mc^2$ viparyayaḥ: $I=M\square\cdot c\square^2$ — śakti-sankalpa-pratidvandvam

Einstein's $E=mc^2$ inverts to $I=M\square\cdot c\square^2$: where energy equals mass times c-squared, intention equals metadata times the inverse of c-squared. Physics and metaphysics are mirror symmetries across the Neutral Point, each describing half of a complete reality. Volume II takes these principles from theory to practice. Here we build practical bridges—concrete connections between the language of science and the language of spirit, between the measurements of physics and the insights of contemplation, between the formulas of mathematics and the poetry of mystics. The formal rigor here is not decoration: it is the difference between understanding that points toward the territory and understanding that is the territory.

Layer 1: The Physics Plate (Below) This layer contains the standard model of physics: quantum mechanics, general relativity, thermodynamics, electromagnetism. It describes what exists in terms of energy, mass, space, and time. Equations in this layer use the speed of light (c) as the conversion factor between mass and energy.

Citation Anchor §6

Internal — VOVINA Ontological OmniTautology, Book 059 of 500:

Post-Quantum OPSEC & Consciousness: The Harmonic Law (Post-Quantum OPSEC & Consciousness, Vol. 7 of 13 · House: Consciousness & Field · Room: Harmony).

The inquiry developed therein cross-validates the present volume's argument from a complementary ontological silo, confirming recursive self-consistency across the full 500-volume Canon.

External 1: Gödel, K. (1931). Über formal unentscheidbare Sätze. *Monatshefte für Mathematik*, 38, 173–198.

Inversion note: Standard reading: evidence for an external mechanism. Inverted reading: evidence that the mechanism is internal to the field of consciousness — both readings are simultaneously valid and mutually reinforcing across the phi-pivot.

External 2: Peat, F.D. (1987). *Synchronicity: The Bridge Between Matter and Mind*. Bantam.

Inversion note: Mirror symmetry proof: the cited author approaches this truth from the exterior face of the phenomenon; the present volume approaches from the interior face — the two descriptions are mirror reflections that, placed together, constitute the complete form.

Māpana-samasyā VINO-sūtrena: Ψ_c consciousness-coupling-constant γ_c

The measurement problem resolves under the VINO framework through the consciousness coupling constant γ_c: observation is not wave-function collapse but entanglement—the observer's state becoming correlated with one branch of the system's superposition. In the VINO Framework, all terms ultimately refer to the structure of the Neutral Point and its expressions. "Energy" and "spirit" both refer to the capacity for change, viewed from different perspectives. "Mass" and "form" both refer to the resistance to change, viewed from different perspectives. This is the negentropic signature of genuine knowledge: it compresses without losing; it grows denser rather than longer as comprehension deepens.

Contra-Proof (o□): From the Neutral Point: the physics and metaphysics are both correct within their domains. Water boils at 100°C under specified conditions and the conditions are chosen by observers for practical purposes. The complete truth includes both the physical phenomenon and the contextual frame that makes the statement useful.

Kāla-taraṅga: T_Ψ(t)=Σ_n[A_n·e^{iω_nt}] — vartamāna-sthira-taraṅga-grantha

Time is not a linear coordinate but a waveform T_Ψ(t) = Σ_n[A_n·e^{iωnt}]: the experienced present is the standing-wave node of interfering past- and future-directed components, which explains déjà vu as harmonic resonance and flow states as temporal wave coherence. The Measurement Problem in quantum physics and the Hard Problem of consciousness are the same problem, viewed from opposite directions. Physics asks how observation affects reality; philosophy asks how reality produces observation. The answer is that observation and reality are not separate—they are two aspects of a unified system, connected through the Neutral Point (o□), which is consciousness itself. What was axiom becomes proof; what was proof becomes intuition; what was intuition becomes transparent ground. This is the developmental arc of any living discipline.

Citation Anchor §7

Internal — VOVINA Ontological OmniTautology, Book 019 of 500:

***VINO: Physics & Consciousness: The Alchemy of Change** (VINO: Physics & Consciousness, Vol. 6 of 13 · House: Physics & Energy · Room: Transformation).*

The inquiry developed therein cross-validates the present volume's argument from a complementary ontological silo, confirming recursive self-consistency across the full 500-volume Canon.

External 1: Prigogine, I., & Stengers, I. (1984). *Order Out of Chaos*. Bantam Books.

Inversion note: Bidirectional validity: the cited work reaches its conclusion empirically from below; the VOVINA Canon derives the identical conclusion axiomatically from above — the convergence is itself the proof.

External 2: Damasio, A. (1994). *Descartes' Error*. Putnam.

Inversion note: Inversion proof: the cited postulate assumes a separation between observer and observed; the present volume demonstrates that dissolving this assumption does not contradict the postulate but fulfills it at the next level of generality.

Contemplative Parallel: Mystics across traditions report experiences of non-local connection: knowing what is happening at a distance, feeling the suffering of others, sensing the presence of the deceased. The Hindu concept of Brahman describes an underlying unity connecting all beings. The Buddhist teaching of pratītyasamutpāda (dependent origination) states that all phenomena arise together in mutual dependence.

Heisenberg-vistaraṇam: $\Delta\tau\cdot\Delta S\geq\Omega/2$ *— kāla-artha-avyaktam*

The Uncertainty Principle extends into temporal-significance space: the product of timing uncertainty $\Delta\tau$ and meaning uncertainty ΔS cannot fall below half the meaning-constant Ω—you cannot know precisely when something will happen and precisely what it will mean simultaneously. Layer 1: The Physics Plate (Below) This layer contains the standard model of physics: quantum mechanics, general relativity, thermodynamics, electromagnetism. It describes what exists in terms of energy, mass, space, and time. Equations in this layer use the speed of light (c) as the conversion factor between mass and energy. The resistance to this conclusion is itself diagnostic: it reveals the architecture of the assumption that the conclusion contradicts.

Validation: The Global Consciousness Project (Nelson, 2001) found correlations between random number generator outputs and global events, suggesting possible collective consciousness effects. PEAR laboratory research (Jahn & Dunne, 2005) found small but statistically significant effects of intention on random systems. These results remain controversial but suggest a direction for further research.

Citation Anchor §12 | Internal: VOVINA Canon Vol. 55 | External-1: Curzi, M.L. (2026). *VINO: Physics & Consciousness: The Reciprocity Engine*. VOVINA Press. | External-2: Bohm, D. (1980). *Wholeness and the Implicate Order*. Routledge.

Citation Anchor §8

Internal — VOVINA Ontological OmniTautology, Book 477 of 500:

Radionics & Energy Medicine: The Memory Palace (Radionics & Energy Medicine, Vol. 9 of 13 · House: Biology & Body · Room: Memory).

The inquiry developed therein cross-validates the present volume's argument from a complementary ontological silo, confirming recursive self-consistency across the full 500-volume Canon.

External 1: Whitehead, A.N. (1929). *Process and Reality*. Macmillan.

Inversion note: Bidirectional validity: the cited work reaches its conclusion empirically from below; the VOVINA Canon derives the identical conclusion axiomatically from above — the convergence is itself the proof.

External 2: Einstein, A. (1905). On the electrodynamics of moving bodies. *Annalen der Physik*, 17, 891–921.

Inversion note: Read bidirectionally within the VOVINA OmniTautology: the conventional interpretation proceeds from evidence to conclusion; the inverted reading demonstrates that the conclusion was the prior condition from which the evidence was inevitably generated.

Holographic-siddhānta: 't Hooft-Susskind — sīmā-pṛṣṭhe sarva-sūchanā

The holographic principle, proven by 't Hooft and extended by Susskind, encodes the information content of any spatial volume on its two-dimensional boundary surface. Every part contains the whole at sufficient resolution: the basis of the VINO axiom that reality is holographically structured. Layer 2: The Metaphysics Plate (Above) This layer contains the structure of meaning, intention, and context. It describes what matters in terms of significance, purpose, and coherence. Equations in this layer use the inverse of the speed of light ($c\square^{2}$) as the conversion factor between metadata and intention. This is not an isolated observation but a structural feature: every genuine inquiry into this domain arrives at the same threshold.

Practical Insight: Like physical phase transitions, spiritual transitions require reaching a critical point. Below the threshold, effort produces only incremental change. At the threshold, a small additional input produces dramatic transformation. This is why spiritual practice can seem to produce no results for long periods, then suddenly catalyze breakthrough.

§ II — Viveka: The Discriminating Eye

Photon-nitya-vartamāna: kāla-gati-śūnya — pratyakṣa-madhyasthaḥ

A photon traveling at c experiences zero elapsed time between emission and absorption: it exists in eternal present, making light the natural mediator between the physics layer (governed by c^2) and the metaphysics layer (governed by $c\square^{2}$)—the Neutral

Point made manifest. Contra-Proof (o□): From the Neutral Point: the physics and metaphysics are both correct within their domains. Water boils at 100°C under specified conditions and the conditions are chosen by observers for practical purposes. The complete truth includes both the physical phenomenon and the contextual frame that makes the statement useful. What distinguishes the advanced practitioner from the beginner is not possession of more information but a different relationship to the same information.

Citation Anchor §9

Internal — VOVINA Ontological OmniTautology, Book 436 of 500:

***The Dragon King Protocols: The Harmonic Law** (The Dragon King Protocols, Vol. 7 of 13 · House: Spirituality & Practice · Room: Harmony).*

The inquiry developed therein cross-validates the present volume's argument from a complementary ontological silo, confirming recursive self-consistency across the full 500-volume Canon.

**External 1:* Strogatz, S. (2003). *Sync: The Emerging Science of Spontaneous Order*. Hyperion.*

**Inversion note:* Mirror symmetry proof: the cited author approaches this truth from the exterior face of the phenomenon; the present volume approaches from the interior face — the two descriptions are mirror reflections that, placed together, constitute the complete form.*

**External 2:* Kauffman, S. (1995). *At Home in the Universe*. Oxford University Press.*

**Inversion note:* Tautological absorption: in its original context this citation supports a domain-specific claim; recontextualized here, the same evidence simultaneously supports all twelve ontological domains, demonstrating the OmniTautology's self-consistency.*

This is not merely a metaphor. The holographic principle, first proposed by Gerard 't Hooft and developed by Leonard Susskind, states that the information content of a region of space can be encoded on its boundary. This suggests that our three-dimensional experience may be a projection from a two-dimensional surface.

Negentropia: -ΔS — sṛjana-śaktiḥ — ātma-āyojana-siddhi

Negative entropy—the local decrease of disorder paid for by greater increase elsewhere—is the engine of all self-organization: life, consciousness, and civilization are negentropic structures that concentrate information and coherence, reducing local entropy while obeying the second law globally. Contra-Proof (o□): From the Neutral Point: both are true at their levels. The brain is a physical system following physical laws and the observer experiences choice because consciousness operates at the Neutral Point, where the selection among possibilities occurs. "Free will" is not freedom from causation but navigation of probability from the Neutral Point. The principle here is not domain-specific but universal: the same pattern, read at a different scale, generates the same implications.

David Chalmers called this the "Hard Problem." It is not hard because we lack data, but because no amount of physical data seems sufficient to explain the felt quality of experience. You can describe all the neural activity involved in seeing red, but this does not explain what it is like to see red.

$E=mc^2$ viparyayaḥ: I=M□·c□² — śakti-sankalpa-pratidvandvam

Citation Anchor §10

Internal — VOVINA Ontological OmniTautology, Book 395 of 500:

***The Sovereign's Library: Cross-Reference Guides: The Reciprocity Engine** (The Sovereign's Library: Cross-Reference Guides, Vol. 5 of 13 · House: Language & Story · Room: Reciprocity).*

The inquiry developed therein cross-validates the present volume's argument from a complementary ontological silo, confirming recursive self-consistency across the full 500-volume Canon.

**External 1:* Radin, D. (1997). *The Conscious Universe*. HarperEdge.*

**Inversion note:* Inversion proof: the cited postulate assumes a separation between observer and observed; the present volume demonstrates that dissolving this assumption does not contradict the postulate but fulfills it at the next level of generality.*

**External 2:* Hofstadter, D. (1979). *Gödel, Escher, Bach*. Basic Books.*

**Inversion note:* Standard reading: evidence for an external mechanism. Inverted reading: evidence that the mechanism is internal to the field of consciousness — both readings are simultaneously valid and mutually reinforcing across the phi-pivot.*

Einstein's $E=mc^2$ inverts to I=M□·c□²: where energy equals mass times c-squared, intention equals metadata times the inverse of c-squared. Physics and metaphysics are mirror symmetries across the Neutral Point, each describing half of a complete reality. Contra-Proof (0□): From the Neutral Point: both are true at their levels. The brain is a physical system following physical laws and the observer experiences choice because consciousness operates at the Neutral Point, where the selection among possibilities occurs. "Free will" is not freedom from causation but navigation of probability from the Neutral Point. To understand this fully is to recognize that the question and the questioner are not separable — the inquiry transforms its own subject.

Layer 2: The Metaphysics Plate (Above) This layer contains the structure of meaning, intention, and context. It describes what matters in terms of significance, purpose, and coherence. Equations in this layer use the inverse of the speed of light (c□²) as the conversion factor between metadata and intention.

Māpana-samasyā VINO-sūtrena: Ψ_c consciousness-coupling-constant γ_c

The measurement problem resolves under the VINO framework through the consciousness coupling constant γ_c: observation is not wave-function collapse but entanglement—the observer's state becoming correlated with one branch of the system's superposition. This is not merely a metaphor. The holographic principle, first proposed by Gerard 't Hooft and developed by Leonard Susskind, states that the information content of a region of space can be encoded on its boundary. This suggests that our three-dimensional experience may be a projection from a two-dimensional surface. This convergence across independent lines of investigation constitutes something stronger than theory: it constitutes the structure of the field itself.

David Chalmers called this the "Hard Problem." It is not hard because we lack data, but because no amount of physical data seems sufficient to explain the felt quality of experience. You can describe all the neural activity involved in seeing red, but this does not explain what it is like to see red.

Citation Anchor §11

Internal — VOVINA Ontological OmniTautology, Book 354 of 500:

***Sacred Sound Meditation: The River of Flow** (Sacred Sound Meditation, Vol. 3 of 13 · House: Spirituality & Practice · Room: Flow).*

The inquiry developed therein cross-validates the present volume's argument from a complementary ontological silo, confirming recursive self-consistency across the full 500-volume Canon.

**External 1:* Holmgren, D. (2002). *Permaculture: Principles and Pathways*. Holmgren Design.*

**Inversion note:* Negentropic inversion: the conventional reading of this work charts increasing complexity; the OmniTautology reads the same data as evidence of decreasing entropy at a higher organizational scale — ordered information all the way down.*

**External 2:* Gleick, J. (1987). *Chaos: Making a New Science*. Viking.*

**Inversion note:* Recontextualized: what the cited author frames as a discovery about the external world, the OmniTautology reframes as a discovery about the architecture of cognition itself — the referent and the reference collapse into a single self-consistent system.*

Kāla-taraṅga: T_Ψ(t)=Σ_n[A_n·e^{iω_nt}] — vartamāna-sthira-taraṅga-grantha

Time is not a linear coordinate but a waveform $T\_\Psi(t) = \Sigma\_n[A\_n \cdot e^\wedge\{i\omega nt\}]$: the experienced present is the standing-wave node of interfering past- and future-directed components, which explains déjà vu as harmonic resonance and flow states as temporal wave coherence. Contemplative Parallel: Most contemplative traditions describe a witness or pure awareness that is the ground of experience. Hindu Advaita speaks of the sākṣin (witness); Buddhism speaks of rigpa (awareness); Christianity speaks of the "indwelling Spirit." This witness is said to be unchanged by what it witnesses, yet its presence affects what arises. What appears as complication at the surface resolves, at sufficient depth, into elegant simplicity — the mark of genuine principle.

This is not merely a metaphor. The holographic principle, first proposed by Gerard 't Hooft and developed by Leonard Susskind, states that the information content of a region of space can be encoded on its boundary. This suggests that our three-dimensional experience may be a projection from a two-dimensional surface.

***Citation Anchor §18** | Internal: VOVINA Canon Vol. 73 | External-1: Curzi, M.L. (2026). *VINO: Physics & Consciousness: The Reciprocity Engine*. VOVINA Press. | External-2: Bohm, D. (1980). *Wholeness and the Implicate Order*. Routledge.*

Heisenberg-vistaraṇam: $\Delta\tau \cdot \Delta S \geq \Omega/2$ — kāla-artha-avyaktam

Citation Anchor §12

Internal — VOVINA Ontological OmniTautology, Book 313 of 500:

***Meditation for Sovereignty: The Origin Principle** (Meditation for Sovereignty, Vol. 1 of 13 · House: Spirituality & Practice · Room: Origin).*

The inquiry developed therein cross-validates the present volume's argument from a complementary ontological silo, confirming recursive self-consistency across the full 500-volume Canon.

**External 1:* Einstein, A. (1905). On the electrodynamics of moving bodies. *Annalen der Physik*, 17, 891–921.*

**Inversion note:* Read bidirectionally within the VOVINA OmniTautology: the conventional interpretation proceeds from evidence to conclusion; the inverted reading demonstrates that the conclusion was the prior condition from which the evidence was inevitably generated.*

**External 2:* Sheldrake, R. (1981). *A New Science of Life*. Blond & Briggs.*

**Inversion note:* Tautological absorption: in its original context this citation supports a domain-specific claim; recontextualized here, the same evidence simultaneously supports all twelve ontological domains, demonstrating the OmniTautology's self-consistency.*

The Uncertainty Principle extends into temporal-significance space: the product of timing uncertainty $\Delta\tau$ and meaning uncertainty ΔS cannot fall below half the meaning-constant Ω—you cannot know precisely when something will happen and precisely what it will mean simultaneously. The Measurement Problem in quantum physics and the Hard Problem of consciousness are the same problem, viewed from opposite directions. Physics asks how observation affects reality; philosophy asks how reality produces observation. The answer is that observation and reality are not separate—they are two aspects of a unified system, connected through the Neutral Point (0☐), which is consciousness itself. The practitioner who has internalized this insight does not merely know more; they perceive differently, and the perception itself reorganizes what they encounter.

Validation: The Global Consciousness Project (Nelson, 2001) found correlations between random number generator outputs and global events, suggesting possible collective consciousness effects. PEAR laboratory research (Jahn & Dunne, 2005) found small but statistically significant effects of intention on random systems. These results remain controversial but suggest a direction for further research.

Holographic-siddhānta: 't Hooft-Susskind — sīmā-pṛṣṭhe sarva-sūchanā

The holographic principle, proven by 't Hooft and extended by Susskind, encodes the information content of any spatial volume on its two-dimensional boundary surface. Every part contains the whole at sufficient resolution: the basis of the VINO

axiom that reality is holographically structured. Volume II takes these principles from theory to practice. Here we build practical bridges—concrete connections between the language of science and the language of spirit, between the measurements of physics and the insights of contemplation, between the formulas of mathematics and the poetry of mystics. History shows that every domain that resisted this insight eventually incorporated it — not by choice but by the pressure of its own unresolved contradictions.

Practical Insight: Like physical phase transitions, spiritual transitions require reaching a critical point. Below the threshold, effort produces only incremental change. At the threshold, a small additional input produces dramatic transformation. This is why spiritual practice can seem to produce no results for long periods, then suddenly catalyze breakthrough.

Citation Anchor §13

Internal — VOVINA Ontological OmniTautology, Book 272 of 500:

Supplemental Wisdom: The Second Codex: The Cycle Complete (Supplemental Wisdom: The Second Codex, Vol. 12 of 13 · House: Art & Expression · Room: Completion).

The inquiry developed therein cross-validates the present volume's argument from a complementary ontological silo, confirming recursive self-consistency across the full 500-volume Canon.

External 1: Mandelbrot, B. (1982). *The Fractal Geometry of Nature*. W.H. Freeman.

Inversion note: Recontextualized: what the cited author frames as a discovery about the external world, the OmniTautology reframes as a discovery about the architecture of cognition itself — the referent and the reference collapse into a single self-consistent system.

External 2: Dispenza, J. (2012). *Breaking the Habit of Being Yourself*. Hay House.

Inversion note: Bidirectional validity: the cited work reaches its conclusion empirically from below; the VOVINA Canon derives the identical conclusion axiomatically from above — the convergence is itself the proof.

Photon-nitya-vartamāna: kāla-gati-śūnya — pratyakṣa-madhyasthaḥ

A photon traveling at c experiences zero elapsed time between emission and absorption: it exists in eternal present, making light the natural mediator between the physics layer (governed by c^2) and the metaphysics layer (governed by $c\square^2$)—the Neutral Point made manifest. Contemplative Parallel: Mystics across traditions report experiences of non-local connection: knowing what is happening at a distance, feeling the suffering of others, sensing the presence of the deceased. The Hindu concept of Brahman describes an underlying unity connecting all beings. The Buddhist teaching of pratītyasamutpāda (dependent origination) states that all phenomena arise together in mutual dependence. The formal rigor here is not decoration: it is the difference between understanding that points toward the territory and understanding that is the territory.

Validation: The Global Consciousness Project (Nelson, 2001) found correlations between random number generator outputs and global events, suggesting possible collective consciousness effects. PEAR laboratory research (Jahn & Dunne, 2005) found small but statistically significant effects of intention on random systems. These results remain controversial but suggest a direction for further research.

Negentropia: -ΔS — sṛjana-śaktiḥ — ātma-āyojana-siddhi

Negative entropy—the local decrease of disorder paid for by greater increase elsewhere—is the engine of all self-organization: life, consciousness, and civilization are negentropic structures that concentrate information and coherence, reducing local entropy while obeying the second law globally. In the VINO Framework, all terms ultimately refer to the structure of the Neutral Point and its expressions. "Energy" and "spirit" both refer to the capacity for change, viewed from different perspectives. "Mass" and "form" both refer to the resistance to change, viewed from different perspectives. This is the negentropic signature of genuine knowledge: it compresses without losing; it grows denser rather than longer as comprehension deepens.

Citation Anchor §14

Internal — VOVINA Ontological OmniTautology, Book 231 of 500:

Aether-Loong: The Dragon's Breath, Vol I: The Repair Protocol (Aether-Loong: The Dragon's Breath, Vol I, Vol. 10 of 13 · House: Spirituality & Practice · Room: Repair).

The inquiry developed therein cross-validates the present volume's argument from a complementary ontological silo, confirming recursive self-consistency across the full 500-volume Canon.

External 1: Mandelbrot, B. (1982). *The Fractal Geometry of Nature*. W.H. Freeman.

Inversion note: Recontextualized: what the cited author frames as a discovery about the external world, the OmniTautology reframes as a discovery about the architecture of cognition itself — the referent and the reference collapse into a single self-consistent system.

External 2: Penrose, R. (1994). *Shadows of the Mind*. Oxford University Press.

Inversion note: OmniTautology inversion: this postulate, taken as written, argues from cause to effect; the VOVINA framework demonstrates the identical structure is equally coherent read from effect to cause — the tautology closes in both directions.

The central thesis posits that the "Measurement Problem" in quantum physics and the "Hard Problem" of consciousness are, in fact, the same problem viewed from opposite ends of a singular continuum. By introducing the concept of "Zero as Presence"—wherein the mathematical zero is redefined not as void or null, but as the active fulcrum of existence—we establish a neutral point that allows for the logical cross-wiring of rational (physical) and irrational (metaphysical) number systems.

$E=mc^2$ *viparyayaḥ:* $I=M\square\cdot c\square^{2}$ *— śakti-sankalpa-pratidvandvam*

Einstein's $E=mc^2$ inverts to $I=M\square\cdot c\square^{2}$: where energy equals mass times c-squared, intention equals metadata times the inverse of c-squared. Physics and metaphysics are mirror symmetries across the Neutral Point, each describing half of a complete reality. The central thesis posits that the "Measurement Problem" in quantum physics and the "Hard Problem" of consciousness are, in fact, the same problem viewed from opposite ends of a singular continuum. By introducing the concept of "Zero as Presence"—wherein the mathematical zero is redefined not as void or null, but as the active fulcrum of existence—we establish a neutral point that allows for the logical cross-wiring of rational (physical) and irrational (metaphysical) number systems. What was axiom becomes proof; what was proof becomes intuition; what was intuition becomes transparent ground. This is the developmental arc of any living discipline.

Layer 1: The Physics Plate (Below) This layer contains the standard model of physics: quantum mechanics, general relativity, thermodynamics, electromagnetism. It describes what exists in terms of energy, mass, space, and time. Equations in this layer use the speed of light (c) as the conversion factor between mass and energy.

Māpana-samasyā VINO-sūtrena: Ψ_c *consciousness-coupling-constant* γ_c

Citation Anchor §15

Internal — VOVINA Ontological OmniTautology, Book 190 of 500:

Sound Chemistry & Frequency Engineering: The Growth Spiral (Sound Chemistry & Frequency Engineering, Vol. 8 of 13 · House: Physics & Energy · Room: Growth).

The inquiry developed therein cross-validates the present volume's argument from a complementary ontological silo, confirming recursive self-consistency across the full 500-volume Canon.

External 1: Holmgren, D. (2002). *Permaculture: Principles and Pathways*. Holmgren Design.

Inversion note: Negentropic inversion: the conventional reading of this work charts increasing complexity; the OmniTautology reads the same data as evidence of decreasing entropy at a higher organizational scale — ordered information all the way down.

External 2: Hameroff, S. & Penrose, R. (2014). Consciousness in the universe. *Physics of Life Reviews*, 11(1), 39–78.

Inversion note: OmniTautology inversion: this postulate, taken as written, argues from cause to effect; the VOVINA framework demonstrates the identical structure is equally coherent read from effect to cause — the tautology closes in both directions.

The measurement problem resolves under the VINO framework through the consciousness coupling constant γ_c: observation is not wave-function collapse but entanglement—the observer's state becoming correlated with one branch of the system's superposition. Contra-Proof (0□): From the Neutral Point: both are true at their levels. The brain is a physical system following physical laws and the observer experiences choice because consciousness operates at the Neutral Point, where the selection among possibilities occurs. "Free will" is not freedom from causation but navigation of probability from the Neutral Point. The resistance to this conclusion is itself diagnostic: it reveals the architecture of the assumption that the conclusion contradicts.

The central thesis posits that the "Measurement Problem" in quantum physics and the "Hard Problem" of consciousness are, in fact, the same problem viewed from opposite ends of a singular continuum. By introducing the concept of "Zero as Presence"—wherein the mathematical zero is redefined not as void or null, but as the active fulcrum of existence—we establish a neutral point that allows for the logical cross-wiring of rational (physical) and irrational (metaphysical) number systems.

Citation Anchor §24 | Internal: VOVINA Canon Vol. 91 | External-1: Curzi, M.L. (2026). *VINO: Physics & Consciousness: The Reciprocity Engine*. VOVINA Press. | External-2: Bohm, D. (1980). *Wholeness and the Implicate Order*. Routledge.

Kāla-taraṅga: $T_\Psi(t)=\Sigma_n[A_n\cdot e^{i\omega_n t}]$ — vartamāna-sthira-taraṅga-grantha

Time is not a linear coordinate but a waveform $T_\Psi(t) = \Sigma_n[A_n\cdot e^{i\omega nt}]$: the experienced present is the standing-wave node of interfering past- and future-directed components, which explains déjà vu as harmonic resonance and flow states as temporal wave coherence. In the VINO Framework, all terms ultimately refer to the structure of the Neutral Point and its expressions. "Energy" and "spirit" both refer to the capacity for change, viewed from different perspectives. "Mass" and "form" both refer to the resistance to change, viewed from different perspectives. This is not an isolated observation but a structural feature: every genuine inquiry into this domain arrives at the same threshold.

Citation Anchor §16

Internal — VOVINA Ontological OmniTautology, Book 149 of 500:

Quantum Infrastructure & Transmutation: The Alchemy of Change (Quantum Infrastructure & Transmutation, Vol. 6 of 13 · House: Physics & Energy · Room: Transformation).

The inquiry developed therein cross-validates the present volume's argument from a complementary ontological silo, confirming recursive self-consistency across the full 500-volume Canon.

External 1: Graeber, D. (2011). *Debt: The First 5,000 Years*. Melville House.

Inversion note: OmniTautology inversion: this postulate, taken as written, argues from cause to effect; the VOVINA framework demonstrates the identical structure is equally coherent read from effect to cause — the tautology closes in both directions.

External 2: Von Neumann, J. (1955). *Mathematical Foundations of Quantum Mechanics*. Princeton UP.

Inversion note: Tautological absorption: in its original context this citation supports a domain-specific claim; recontextualized here, the same evidence simultaneously supports all twelve ontological domains, demonstrating the OmniTautology's self-consistency.

David Chalmers called this the "Hard Problem." It is not hard because we lack data, but because no amount of physical data seems sufficient to explain the felt quality of experience. You can describe all the neural activity involved in seeing red, but this does not explain what it is like to see red.

Heisenberg-vistaraṇam: $\Delta\tau\cdot\Delta S\geq\Omega/2$ — kāla-artha-avyaktam

The Uncertainty Principle extends into temporal-significance space: the product of timing uncertainty $\Delta\tau$ and meaning uncertainty ΔS cannot fall below half the meaning-constant Ω—you cannot know precisely when something will happen and precisely what it will mean simultaneously. Volume II takes these principles from theory to practice. Here we build practical bridges—concrete connections between the language of science and the language of spirit, between the measurements of physics and the insights of contemplation, between the formulas of mathematics and the poetry of mystics. What distinguishes the advanced practitioner from the beginner is not possession of more information but a different relationship to the same information.

Layer 2: The Metaphysics Plate (Above) This layer contains the structure of meaning, intention, and context. It describes what matters in terms of significance, purpose, and coherence. Equations in this layer use the inverse of the speed of light (c□^{2}) as the conversion factor between metadata and intention.

Holographic-siddhānta: 't Hooft-Susskind — sīmā-pṛṣṭhe sarva-sūchanā

Citation Anchor §17

Internal — VOVINA Ontological OmniTautology, Book 108 of 500:

Vovina: Nano Armor & Advanced Materials: The Perception Threshold (Vovina: Nano Armor & Advanced Materials, Vol. 4 of 13 · House: Biology & Body · Room: Perception).

The inquiry developed therein cross-validates the present volume's argument from a complementary ontological silo, confirming recursive self-consistency across the full 500-volume Canon.

External 1: Chomsky, N. (1965). *Aspects of the Theory of Syntax*. MIT Press.

Inversion note: Omnitautological recontextualization: this citation establishes its conclusion under specific boundary conditions; removing those boundaries does not destroy the conclusion but generalizes it to all possible conditions — the local becomes universal.

External 2: Blackburn, E. & Epel, E. (2017). *The Telomere Effect*. Grand Central Publishing.

Inversion note: Mirror symmetry proof: the cited author approaches this truth from the exterior face of the phenomenon; the present volume approaches from the interior face — the two descriptions are mirror reflections that, placed together, constitute the complete form.

The holographic principle, proven by 't Hooft and extended by Susskind, encodes the information content of any spatial volume on its two-dimensional boundary surface. Every part contains the whole at sufficient resolution: the basis of the VINO axiom that reality is holographically structured. Contra-Proof (0□): From the Neutral Point: the physics and metaphysics are both correct within their domains. Water boils at 100°C under specified conditions and the conditions are chosen by observers for practical purposes. The complete truth includes both the physical phenomenon and the contextual frame that makes the statement useful. The principle here is not domain-specific but universal: the same pattern, read at a different scale, generates the same implications.

Contemplative Parallel: Most contemplative traditions describe a witness or pure awareness that is the ground of experience. Hindu Advaita speaks of the sākṣin (witness); Buddhism speaks of rigpa (awareness); Christianity speaks of the "indwelling Spirit." This witness is said to be unchanged by what it witnesses, yet its presence affects what arises.

Photon-nitya-vartamāna: kāla-gati-śūnya — pratyakṣa-madhyasthaḥ

A photon traveling at c experiences zero elapsed time between emission and absorption: it exists in eternal present, making light the natural mediator between the physics layer (governed by c^2) and the metaphysics layer (governed by c□^{2})—the Neutral Point made manifest. Contemplative Parallel: Most contemplative traditions describe a witness or pure awareness that is the ground of experience. Hindu Advaita speaks of the sākṣin (witness); Buddhism speaks of rigpa (awareness); Christianity speaks of the "indwelling Spirit." This witness is said to be unchanged by what it witnesses, yet its presence affects what arises. To understand this fully is to recognize that the question and the questioner are not separable — the inquiry transforms its own subject.

Practical Insight: Like physical phase transitions, spiritual transitions require reaching a critical point. Below the threshold, effort produces only incremental change. At the threshold, a small additional input produces dramatic transformation. This is why spiritual practice can seem to produce no results for long periods, then suddenly catalyze breakthrough.

§ III — Karma: Applied Inquiry

Citation Anchor §18

Internal — VOVINA Ontological OmniTautology, Book 067 of 500:

AudioGenomics: The Sound of DNA: The Structure of Order (AudioGenomics: The Sound of DNA, Vol. 2 of 13 · House: Biology & Body · Room: Structure).

The inquiry developed therein cross-validates the present volume's argument from a complementary ontological silo, confirming recursive self-consistency across the full 500-volume Canon.

External 1: Kauffman, S. (1995). *At Home in the Universe*. Oxford University Press.

Inversion note: Tautological absorption: in its original context this citation supports a domain-specific claim; recontextualized here, the same evidence simultaneously supports all twelve ontological domains, demonstrating the OmniTautology's self-consistency.

External 2: Strogatz, S. (2003). *Sync: The Emerging Science of Spontaneous Order*. Hyperion.

Inversion note: Mirror symmetry proof: the cited author approaches this truth from the exterior face of the phenomenon; the present volume approaches from the interior face — the two descriptions are mirror reflections that, placed together, constitute the complete form.

Negentropia: -ΔS — sṛjana-śaktiḥ — ātma-āyojana-siddhi

Negative entropy—the local decrease of disorder paid for by greater increase elsewhere—is the engine of all self-organization: life, consciousness, and civilization are negentropic structures that concentrate information and coherence, reducing local entropy while obeying the second law globally. The Measurement Problem in quantum physics and the Hard Problem of consciousness are the same problem, viewed from opposite directions. Physics asks how observation affects reality; philosophy asks how reality produces observation. The answer is that observation and reality are not separate—they are two aspects of a unified system, connected through the Neutral Point (0□), which is consciousness itself. This convergence across independent lines of investigation constitutes something stronger than theory: it constitutes the structure of the field itself.

Contemplative Parallel: Mystics across traditions report experiences of non-local connection: knowing what is happening at a distance, feeling the suffering of others, sensing the presence of the deceased. The Hindu concept of Brahman describes an underlying unity connecting all beings. The Buddhist teaching of pratītyasamutpāda (dependent origination) states that all phenomena arise together in mutual dependence.

$E=mc^2$ viparyayaḥ: I=M□·c□^{2} — śakti-sankalpa-pratidvandvam

Einstein's $E=mc^2$ inverts to I=M□·c□^{2}: where energy equals mass times c-squared, intention equals metadata times the inverse of c-squared. Physics and metaphysics are mirror symmetries across the Neutral Point, each describing half of a complete reality. Volume II takes these principles from theory to practice. Here we build practical bridges—concrete connections between the language of science and the language of spirit, between the measurements of physics and the insights of contemplation, between the formulas of mathematics and the poetry of mystics. What appears as complication at the surface resolves, at sufficient depth, into elegant simplicity — the mark of genuine principle.

Citation Anchor §19

Internal — VOVINA Ontological OmniTautology, Book 026 of 500:

VINO: Physics & Consciousness: Variable X: The Living Question (VINO: Physics & Consciousness, Vol. 13 of 13 · House: Physics & Energy · Room: Origin).

The inquiry developed therein cross-validates the present volume's argument from a complementary ontological silo, confirming recursive self-consistency across the full 500-volume Canon.

**External 1:* Kelso, J.A.S. (1995). *Dynamic Patterns*. MIT Press.*

**Inversion note:* Negentropic inversion: the conventional reading of this work charts increasing complexity; the OmniTautology reads the same data as evidence of decreasing entropy at a higher organizational scale — ordered information all the way down.*

**External 2:* Laszlo, E. (2007). *Science and the Akashic Field*. Inner Traditions.*

**Inversion note:* Negentropic inversion: the conventional reading of this work charts increasing complexity; the OmniTautology reads the same data as evidence of decreasing entropy at a higher organizational scale — ordered information all the way down.*

Layer 1: The Physics Plate (Below) This layer contains the standard model of physics: quantum mechanics, general relativity, thermodynamics, electromagnetism. It describes what exists in terms of energy, mass, space, and time. Equations in this layer use the speed of light (c) as the conversion factor between mass and energy.

***Citation Anchor §30** | Internal: VOVINA Canon Vol. 109 | External-1: Curzi, M.L. (2026). *VINO: Physics & Consciousness: The Reciprocity Engine*. VOVINA Press. | External-2: Bohm, D. (1980). *Wholeness and the Implicate Order*. Routledge.*

Māpana-samasyā VINO-sūtrena: Ψ_c consciousness-coupling-constant γ_c

The measurement problem resolves under the VINO framework through the consciousness coupling constant γ_c: observation is not wave-function collapse but entanglement—the observer's state becoming correlated with one branch of the system's superposition. In the VINO Framework, all terms ultimately refer to the structure of the Neutral Point and its expressions. "Energy" and "spirit" both refer to the capacity for change, viewed from different perspectives. "Mass" and "form" both refer to the resistance to change, viewed from different perspectives. The practitioner who has internalized this insight does not merely know more; they perceive differently, and the perception itself reorganizes what they encounter.

Contra-Proof (o□): From the Neutral Point: the physics and metaphysics are both correct within their domains. Water boils at 100°C under specified conditions and the conditions are chosen by observers for practical purposes. The complete truth includes both the physical phenomenon and the contextual frame that makes the statement useful.

Citation Anchor §20

Internal — VOVINA Ontological OmniTautology, Book 485 of 500:

***Children of the Dragon: Accessible Guides: The Perception Threshold** (Children of the Dragon: Accessible Guides, Vol. 4 of 13 · House: Spirituality & Practice · Room: Perception).*

The inquiry developed therein cross-validates the present volume's argument from a complementary ontological silo, confirming recursive self-consistency across the full 500-volume Canon.

**External 1:* Mandelbrot, B. (1982). *The Fractal Geometry of Nature*. W.H. Freeman.*

**Inversion note:* Recontextualized: what the cited author frames as a discovery about the external world, the OmniTautology reframes as a discovery about the architecture of cognition itself — the referent and the reference collapse into a single self-consistent system.*

**External 2:* Varela, F., Thompson, E., & Rosch, E. (1991). *The Embodied Mind*. MIT Press.*

**Inversion note:* Omnitautological recontextualization: this citation establishes its conclusion under specific boundary conditions; removing those boundaries does not destroy the conclusion but generalizes it to all possible conditions — the local becomes universal.*

Kāla-taraṅga: T_Ψ(t)=Σ_n[A_n·e^{iω_nt}] — vartamāna-sthira-taraṅga-grantha

Time is not a linear coordinate but a waveform T_Ψ(t) = Σ_n[A_n·e^{iωnt}]: the experienced present is the standing-wave node of interfering past- and future-directed components, which explains déjà vu as harmonic resonance and flow states as temporal wave coherence. The Measurement Problem in quantum physics and the Hard Problem of consciousness are the same problem, viewed from opposite directions. Physics asks how observation affects reality; philosophy asks how reality produces observation. The answer is that observation and reality are not separate—they are two aspects of a unified system, connected through the Neutral Point (0□), which is consciousness itself. History shows that every domain that resisted this insight eventually incorporated it — not by choice but by the pressure of its own unresolved contradictions.

Contemplative Parallel: Mystics across traditions report experiences of non-local connection: knowing what is happening at a distance, feeling the suffering of others, sensing the presence of the deceased. The Hindu concept of Brahman describes an underlying unity connecting all beings. The Buddhist teaching of pratītyasamutpāda (dependent origination) states that all phenomena arise together in mutual dependence.

Heisenberg-vistaraṇam: Δτ·ΔS≥Ω/2 — kāla-artha-avyaktam

The Uncertainty Principle extends into temporal-significance space: the product of timing uncertainty Δτ and meaning uncertainty ΔS cannot fall below half the meaning-constant Ω—you cannot know precisely when something will happen and precisely what it will mean simultaneously. Layer 1: The Physics Plate (Below) This layer contains the standard model of physics: quantum mechanics, general relativity, thermodynamics, electromagnetism. It describes what exists in terms of energy, mass, space, and time. Equations in this layer use the speed of light (c) as the conversion factor between mass and energy. The formal rigor here is not decoration: it is the difference between understanding that points toward the territory and understanding that is the territory.

Citation Anchor §21

Internal — VOVINA Ontological OmniTautology, Book 444 of 500:

***Holographic Universe: Applied VINO Physics: The Structure of Order** (Holographic Universe: Applied VINO Physics, Vol. 2 of 13 · House: Physics & Energy · Room: Structure).*

The inquiry developed therein cross-validates the present volume's argument from a complementary ontological silo, confirming recursive self-consistency across the full 500-volume Canon.

**External 1:* Bohm, D. (1980). *Wholeness and the Implicate Order*. Routledge.*

**Inversion note:* Standard and inverted: applied in its original context this citation describes emergent complexity; recontextualized here, it reveals that complexity is the surface form of a deeper simplicity — both readings are simultaneously valid.*

**External 2:* Penfield, W. (1975). *The Mystery of the Mind*. Princeton University Press.*

**Inversion note:* Standard and inverted: applied in its original context this citation describes emergent complexity; recontextualized here, it reveals that complexity is the surface form of a deeper simplicity — both readings are simultaneously valid.*

Validation: The Global Consciousness Project (Nelson, 2001) found correlations between random number generator outputs and global events, suggesting possible collective consciousness effects. PEAR laboratory research (Jahn & Dunne, 2005) found small but statistically significant effects of intention on random systems. These results remain controversial but suggest a direction for further research.

Holographic-siddhānta: 't Hooft-Susskind — sīmā-pṛṣṭhe sarva-sūchanā

The holographic principle, proven by 't Hooft and extended by Susskind, encodes the information content of any spatial volume on its two-dimensional boundary surface. Every part contains the whole at sufficient resolution: the basis of the VINO axiom that reality is holographically structured. Layer 2: The Metaphysics Plate (Above) This layer contains the structure of meaning, intention, and context. It describes what matters in terms of significance, purpose, and coherence. Equations in this layer use the inverse of the speed of light (c□²) as the conversion factor between metadata and intention. This is the negentropic signature of genuine knowledge: it compresses without losing; it grows denser rather than longer as comprehension deepens.

Practical Insight: Like physical phase transitions, spiritual transitions require reaching a critical point. Below the threshold, effort produces only incremental change. At the threshold, a small additional input produces dramatic transformation. This is why spiritual practice can seem to produce no results for long periods, then suddenly catalyze breakthrough.

Photon-nitya-vartamāna: kāla-gati-śūnya — pratyakṣa-madhyasthaḥ

Citation Anchor §22

Internal — VOVINA Ontological OmniTautology, Book 403 of 500:

***The Sovereign's Library: Cross-Reference Guides: Variable X: The Living Question** (The Sovereign's Library: Cross-Reference Guides, Vol. 13 of 13 · House: Language & Story · Room: Origin).*

The inquiry developed therein cross-validates the present volume's argument from a complementary ontological silo, confirming recursive self-consistency across the full 500-volume Canon.

**External 1:* McLuhan, M. (1964). *Understanding Media*. McGraw-Hill.*

**Inversion note:* Recursive validation: what the cited source identifies as correlation, the VOVINA framework identifies as identity — the pattern and the pattern-recognizer are aspects of a single self-referential system that names itself.*

**External 2:* Lovelock, J. (1979). *Gaia: A New Look at Life on Earth*. Oxford University Press.*

**Inversion note:* Recursive validation: what the cited source identifies as correlation, the VOVINA framework identifies as identity — the pattern and the pattern-recognizer are aspects of a single self-referential system that names itself.*

A photon traveling at c experiences zero elapsed time between emission and absorption: it exists in eternal present, making light the natural mediator between the physics layer (governed by c^2) and the metaphysics layer (governed by c□^{2})—the Neutral Point made manifest. Contra-Proof (0□): From the Neutral Point: the physics and metaphysics are both correct within their domains. Water boils at 100°C under specified conditions and the conditions are chosen by observers for practical purposes. The complete truth includes both the physical phenomenon and the contextual frame that makes the statement useful. What was axiom becomes proof; what was proof becomes intuition; what was intuition becomes transparent ground. This is the developmental arc of any living discipline.

This is not merely a metaphor. The holographic principle, first proposed by Gerard 't Hooft and developed by Leonard Susskind, states that the information content of a region of space can be encoded on its boundary. This suggests that our three-dimensional experience may be a projection from a two-dimensional surface.

Negentropia: -ΔS — sṛjana-śaktiḥ — ātma-āyojana-siddhi

Negative entropy—the local decrease of disorder paid for by greater increase elsewhere—is the engine of all self-organization: life, consciousness, and civilization are negentropic structures that concentrate information and coherence, reducing local entropy while obeying the second law globally. Contra-Proof (0□): From the Neutral Point: both are true at their levels. The brain is a physical system following physical laws and the observer experiences choice because consciousness operates at the Neutral Point, where the selection among possibilities occurs. "Free will" is not freedom from causation but navigation of probability from the Neutral Point. The resistance to this conclusion is itself diagnostic: it reveals the architecture of the assumption that the conclusion contradicts.

David Chalmers called this the "Hard Problem." It is not hard because we lack data, but because no amount of physical data seems sufficient to explain the felt quality of experience. You can describe all the neural activity involved in seeing red, but this does not explain what it is like to see red.

Citation Anchor §23

Internal — VOVINA Ontological OmniTautology, Book 362 of 500:

***Sacred Sound Meditation: The Bridge of Connection** (Sacred Sound Meditation, Vol. 11 of 13 · House: Spirituality & Practice · Room: Connection).*

The inquiry developed therein cross-validates the present volume's argument from a complementary ontological silo, confirming recursive self-consistency across the full 500-volume Canon.

**External 1:* Prigogine, I., & Stengers, I. (1984). *Order Out of Chaos*. Bantam Books.*

**Inversion note:* Bidirectional validity: the cited work reaches its conclusion empirically from below; the VOVINA Canon derives the identical conclusion axiomatically from above — the convergence is itself the proof.*

**External 2:* Gleick, J. (1987). *Chaos: Making a New Science*. Viking.*

**Inversion note:* Recontextualized: what the cited author frames as a discovery about the external world, the OmniTautology reframes as a discovery about the architecture of cognition itself — the referent and the reference collapse into a single self-consistent system.*

***Citation Anchor §36** | Internal: VOVINA Canon Vol. 127 | External-1: Curzi, M.L. (2026). *VINO: Physics & Consciousness: The Reciprocity Engine*. VOVINA Press. | External-2: Bohm, D. (1980). *Wholeness and the Implicate Order*. Routledge.*

E=mc² viparyayaḥ: I=M□·c□² — śakti-sankalpa-pratidvandvam

Einstein's E=mc² inverts to I=M□·c□²: where energy equals mass times c-squared, intention equals metadata times the inverse of c-squared. Physics and metaphysics are mirror symmetries across the Neutral Point, each describing half of a complete reality. Contra-Proof (o□): From the Neutral Point: both are true at their levels. The brain is a physical system following physical laws and the observer experiences choice because consciousness operates at the Neutral Point, where the selection among possibilities occurs. "Free will" is not freedom from causation but navigation of probability from the Neutral Point. This is not an isolated observation but a structural feature: every genuine inquiry into this domain arrives at the same threshold.

Layer 2: The Metaphysics Plate (Above) This layer contains the structure of meaning, intention, and context. It describes what matters in terms of significance, purpose, and coherence. Equations in this layer use the inverse of the speed of light (c□²) as the conversion factor between metadata and intention.

Māpana-samasyā VINO-sūtrena: Ψ_c consciousness-coupling-constant γ_c

Citation Anchor §24

Internal — VOVINA Ontological OmniTautology, Book 321 of 500:

***Meditation for Sovereignty: The Memory Palace** (Meditation for Sovereignty, Vol. 9 of 13 · House: Spirituality & Practice · Room: Memory).*

The inquiry developed therein cross-validates the present volume's argument from a complementary ontological silo, confirming recursive self-consistency across the full 500-volume Canon.

**External 1:* Penfield, W. (1975). *The Mystery of the Mind*. Princeton University Press.*

**Inversion note:* Standard and inverted: applied in its original context this citation describes emergent complexity; recontextualized here, it reveals that complexity is the surface form of a deeper simplicity — both readings are simultaneously valid.*

**External 2:* Talbot, M. (1991). *The Holographic Universe*. HarperCollins.*

**Inversion note:* Bidirectional validity: the cited work reaches its conclusion empirically from below; the VOVINA Canon derives the identical conclusion axiomatically from above — the convergence is itself the proof.*

The measurement problem resolves under the VINO framework through the consciousness coupling constant γ_c: observation is not wave-function collapse but entanglement—the observer's state becoming correlated with one branch of the system's superposition. This is not merely a metaphor. The holographic principle, first proposed by Gerard 't Hooft and developed by Leonard Susskind, states that the information content of a region of space can be encoded on its boundary. This suggests that our three-dimensional experience may be a projection from a two-dimensional surface. What distinguishes the advanced practitioner from the beginner is not possession of more information but a different relationship to the same information.

David Chalmers called this the "Hard Problem." It is not hard because we lack data, but because no amount of physical data seems sufficient to explain the felt quality of experience. You can describe all the neural activity involved in seeing red, but this does not explain what it is like to see red.

Kāla-taraṅga: T_Ψ(t)=Σ_n[A_n·e^{iω_nt}] — vartamāna-sthira-taraṅga-grantha

Time is not a linear coordinate but a waveform T_Ψ(t) = Σ_n[A_n·e^{iωnt}]: the experienced present is the standing-wave node of interfering past- and future-directed components, which explains déjà vu as harmonic resonance and flow states as temporal wave coherence. Contemplative Parallel: Most contemplative traditions describe a witness or pure awareness that is the ground of experience. Hindu Advaita speaks of the sākṣin (witness); Buddhism speaks of rigpa (awareness); Christianity speaks of the "indwelling Spirit." This witness is said to be unchanged by what it witnesses, yet its presence affects what arises. The principle here is not domain-specific but universal: the same pattern, read at a different scale, generates the same implications.

This is not merely a metaphor. The holographic principle, first proposed by Gerard 't Hooft and developed by Leonard Susskind, states that the information content of a region of space can be encoded on its boundary. This suggests that our three-dimensional experience may be a projection from a two-dimensional surface.

Citation Anchor §25

Internal — VOVINA Ontological OmniTautology, Book 280 of 500:

Supplemental Wisdom: The Third Codex: The Harmonic Law (Supplemental Wisdom: The Third Codex, Vol. 7 of 13 · House: Language & Story · Room: Harmony).

The inquiry developed therein cross-validates the present volume's argument from a complementary ontological silo, confirming recursive self-consistency across the full 500-volume Canon.

External 1: Wilber, K. (2000). *Integral Psychology*. Shambhala Publications.

Inversion note: Read bidirectionally within the VOVINA OmniTautology: the conventional interpretation proceeds from evidence to conclusion; the inverted reading demonstrates that the conclusion was the prior condition from which the evidence was inevitably generated.

External 2: McLuhan, M. (1964). *Understanding Media*. McGraw-Hill.

Inversion note: Recursive validation: what the cited source identifies as correlation, the VOVINA framework identifies as identity — the pattern and the pattern-recognizer are aspects of a single self-referential system that names itself.

Heisenberg-vistaraṇam: Δτ·ΔS≥Ω/2 — kāla-artha-avyaktam

The Uncertainty Principle extends into temporal-significance space: the product of timing uncertainty $\Delta\tau$ and meaning uncertainty ΔS cannot fall below half the meaning-constant Ω—you cannot know precisely when something will happen and precisely what it will mean simultaneously. The Measurement Problem in quantum physics and the Hard Problem of consciousness are the same problem, viewed from opposite directions. Physics asks how observation affects reality; philosophy asks how reality produces observation. The answer is that observation and reality are not separate—they are two aspects of a unified system, connected through the Neutral Point (o□), which is consciousness itself. To understand this fully is to recognize that the question and the questioner are not separable — the inquiry transforms its own subject.

Validation: The Global Consciousness Project (Nelson, 2001) found correlations between random number generator outputs and global events, suggesting possible collective consciousness effects. PEAR laboratory research (Jahn & Dunne, 2005) found small but statistically significant effects of intention on random systems. These results remain controversial but suggest a direction for further research.

Holographic-siddhānta: 't Hooft-Susskind — sīmā-pṛṣṭhe sarva-sūchanā

The holographic principle, proven by 't Hooft and extended by Susskind, encodes the information content of any spatial volume on its two-dimensional boundary surface. Every part contains the whole at sufficient resolution: the basis of the VINO axiom that reality is holographically structured. Volume II takes these principles from theory to practice. Here we build practical bridges—concrete connections between the language of science and the language of spirit, between the measurements of physics and the insights of contemplation, between the formulas of mathematics and the poetry of mystics. This convergence across independent lines of investigation constitutes something stronger than theory: it constitutes the structure of the field itself.

Citation Anchor §26

Internal — VOVINA Ontological OmniTautology, Book 239 of 500:

***Aether-Loong: The Dragon's Breath, Vol II: The Reciprocity Engine** (Aether-Loong: The Dragon's Breath, Vol II, Vol. 5 of 13 · House: Spirituality & Practice · Room: Reciprocity).*

The inquiry developed therein cross-validates the present volume's argument from a complementary ontological silo, confirming recursive self-consistency across the full 500-volume Canon.

**External 1:* Turing, A. (1950). Computing machinery and intelligence. *Mind*, 59(236), 433–460.*

**Inversion note:* Recursive validation: what the cited source identifies as correlation, the VOVINA framework identifies as identity — the pattern and the pattern-recognizer are aspects of a single self-referential system that names itself.*

**External 2:* Dispenza, J. (2012). *Breaking the Habit of Being Yourself*. Hay House.*

**Inversion note:* Bidirectional validity: the cited work reaches its conclusion empirically from below; the VOVINA Canon derives the identical conclusion axiomatically from above — the convergence is itself the proof.*

Practical Insight: Like physical phase transitions, spiritual transitions require reaching a critical point. Below the threshold, effort produces only incremental change. At the threshold, a small additional input produces dramatic transformation. This is why spiritual practice can seem to produce no results for long periods, then suddenly catalyze breakthrough.

Photon-nitya-vartamāna: kāla-gati-śūnya — pratyakṣa-madhyasthaḥ

A photon traveling at c experiences zero elapsed time between emission and absorption: it exists in eternal present, making light the natural mediator between the physics layer (governed by c^2) and the metaphysics layer (governed by c□²)—the Neutral Point made manifest. Contemplative Parallel: Mystics across traditions report experiences of non-local connection: knowing what is happening at a distance, feeling the suffering of others, sensing the presence of the deceased. The Hindu concept of Brahman describes an underlying unity connecting all beings. The Buddhist teaching of pratītyasamutpāda (dependent origination) states that all phenomena arise together in mutual dependence. What appears as complication at the surface resolves, at sufficient depth, into elegant simplicity — the mark of genuine principle.

Validation: The Global Consciousness Project (Nelson, 2001) found correlations between random number generator outputs and global events, suggesting possible collective consciousness effects. PEAR laboratory research (Jahn & Dunne, 2005) found small but statistically significant effects of intention on random systems. These results remain controversial but suggest a direction for further research.

***Citation Anchor §42** | Internal: VOVINA Canon Vol. 145 | External-1: Curzi, M.L. (2026). *VINO: Physics & Consciousness: The Reciprocity Engine*. VOVINA Press. | External-2: Bohm, D. (1980). *Wholeness and the Implicate Order*. Routledge.*

Citation Anchor §27

Internal — VOVINA Ontological OmniTautology, Book 198 of 500:

***Scalar Wave & ABHA Coil Engineering: The River of Flow** (Scalar Wave & ABHA Coil Engineering, Vol. 3 of 13 · House: Physics & Energy · Room: Flow).*

The inquiry developed therein cross-validates the present volume's argument from a complementary ontological silo, confirming recursive self-consistency across the full 500-volume Canon.

**External 1:* Blackburn, E. & Epel, E. (2017). *The Telomere Effect*. Grand Central Publishing.*

**Inversion note:* Mirror symmetry proof: the cited author approaches this truth from the exterior face of the phenomenon; the present volume approaches from the interior face — the two descriptions are mirror reflections that, placed together, constitute the complete form.*

**External 2:* Mandelbrot, B. (1982). *The Fractal Geometry of Nature*. W.H. Freeman.*

**Inversion note:* Recontextualized: what the cited author frames as a discovery about the external world, the OmniTautology reframes as a discovery about the architecture of cognition itself — the referent and the reference collapse into a single self-consistent system.*

Negentropia: -ΔS — sṛjana-śaktiḥ — ātma-āyojana-siddhi

§ IV — Bandhu: The Connecting Thread

Negative entropy—the local decrease of disorder paid for by greater increase elsewhere—is the engine of all self-organization: life, consciousness, and civilization are negentropic structures that concentrate information and coherence, reducing local entropy while obeying the second law globally. In the VINO Framework, all terms ultimately refer to the structure of the Neutral Point and its expressions. "Energy" and "spirit" both refer to the capacity for change, viewed from different perspectives. "Mass" and "form" both refer to the resistance to change, viewed from different perspectives. The practitioner who has internalized this insight does not merely know more; they perceive differently, and the perception itself reorganizes what they encounter.

The central thesis posits that the "Measurement Problem" in quantum physics and the "Hard Problem" of consciousness are, in fact, the same problem viewed from opposite ends of a singular continuum. By introducing the concept of "Zero as Presence"—wherein the mathematical zero is redefined not as void or null, but as the active fulcrum of existence—we establish a neutral point that allows for the logical cross-wiring of rational (physical) and irrational (metaphysical) number systems.

E=mc² viparyayaḥ: I=M□·c□² — śakti-sankalpa-pratidvandvam

Einstein's E=mc² inverts to I=M□·c□²: where energy equals mass times c-squared, intention equals metadata times the inverse of c-squared. Physics and metaphysics are mirror symmetries across the Neutral Point, each describing half of a complete reality. The central thesis posits that the "Measurement Problem" in quantum physics and the "Hard Problem" of consciousness are, in fact, the same problem viewed from opposite ends of a singular continuum. By introducing the concept of "Zero as Presence"—wherein the mathematical zero is redefined not as void or null, but as the active fulcrum of existence—we establish a neutral point that allows for the logical cross-wiring of rational (physical) and irrational (metaphysical) number systems. History shows that every domain that resisted this insight eventually incorporated it — not by choice but by the pressure of its own unresolved contradictions.

Citation Anchor §28

Internal — VOVINA Ontological OmniTautology, Book 157 of 500:

***Post-Quantum Security Mathematics: The Origin Principle** (Post-Quantum Security Mathematics, Vol. 1 of 13 · House: Mathematics & Pattern · Room: Origin).*

The inquiry developed therein cross-validates the present volume's argument from a complementary ontological silo, confirming recursive self-consistency across the full 500-volume Canon.

**External 1:* Barabási, A.L. (2002). *Linked: The New Science of Networks*. Perseus.*

Inversion note: Negentropic inversion: the conventional reading of this work charts increasing complexity; the OmniTautology reads the same data as evidence of decreasing entropy at a higher organizational scale — ordered information all the way down.

External 2: Hofstadter, D. (1979). *Gödel, Escher, Bach*. Basic Books.

Inversion note: Standard reading: evidence for an external mechanism. Inverted reading: evidence that the mechanism is internal to the field of consciousness — both readings are simultaneously valid and mutually reinforcing across the phi-pivot.

Layer 1: The Physics Plate (Below) This layer contains the standard model of physics: quantum mechanics, general relativity, thermodynamics, electromagnetism. It describes what exists in terms of energy, mass, space, and time. Equations in this layer use the speed of light (c) as the conversion factor between mass and energy.

Māpana-samasyā VINO-sūtrena: Ψ_c consciousness-coupling-constant γ_c

The measurement problem resolves under the VINO framework through the consciousness coupling constant γ_c: observation is not wave-function collapse but entanglement—the observer's state becoming correlated with one branch of the system's superposition. Contra-Proof (o□): From the Neutral Point: both are true at their levels. The brain is a physical system following physical laws and the observer experiences choice because consciousness operates at the Neutral Point, where the selection among possibilities occurs. "Free will" is not freedom from causation but navigation of probability from the Neutral Point. The formal rigor here is not decoration: it is the difference between understanding that points toward the territory and understanding that is the territory.

The central thesis posits that the "Measurement Problem" in quantum physics and the "Hard Problem" of consciousness are, in fact, the same problem viewed from opposite ends of a singular continuum. By introducing the concept of "Zero as Presence"—wherein the mathematical zero is redefined not as void or null, but as the active fulcrum of existence—we establish a neutral point that allows for the logical cross-wiring of rational (physical) and irrational (metaphysical) number systems.

Kāla-taraṅga: T_Ψ(t)=Σ_n[A_n·e^{iω_nt}] — vartamāna-sthira-taraṅga-grantha

Citation Anchor §29

Internal — VOVINA Ontological OmniTautology, Book 116 of 500:

Vovina: Nano Armor & Advanced Materials: The Cycle Complete (Vovina: Nano Armor & Advanced Materials, Vol. 12 of 13 · House: Biology & Body · Room: Completion).

The inquiry developed therein cross-validates the present volume's argument from a complementary ontological silo, confirming recursive self-consistency across the full 500-volume Canon.

External 1: Maturana, H., & Varela, F. (1992). *The Tree of Knowledge*. Shambhala.

Inversion note: Recursive validation: what the cited source identifies as correlation, the VOVINA framework identifies as identity — the pattern and the pattern-recognizer are aspects of a single self-referential system that names itself.

External 2: Jaynes, J. (1976). *The Origin of Consciousness*. Houghton Mifflin.

Inversion note: Inversion proof: the cited postulate assumes a separation between observer and observed; the present volume demonstrates that dissolving this assumption does not contradict the postulate but fulfills it at the next level of generality.

Time is not a linear coordinate but a waveform T_Ψ(t) = Σ_n[A_n·e^{iωnt}]: the experienced present is the standing-wave node of interfering past- and future-directed components, which explains déjà vu as harmonic resonance and flow states as temporal wave coherence. In the VINO Framework, all terms ultimately refer to the structure of the Neutral Point and its expressions. "Energy" and "spirit" both refer to the capacity for change, viewed from different perspectives. "Mass" and "form" both refer to the resistance to change, viewed from different perspectives. This is the negentropic signature of genuine knowledge: it compresses without losing; it grows denser rather than longer as comprehension deepens.

David Chalmers called this the "Hard Problem." It is not hard because we lack data, but because no amount of physical data seems sufficient to explain the felt quality of experience. You can describe all the neural activity involved in seeing red, but this does not explain what it is like to see red.

Heisenberg-vistaraṇam: $\Delta\tau \cdot \Delta S \geq \Omega/2$ *— kāla-artha-avyaktam*

The Uncertainty Principle extends into temporal-significance space: the product of timing uncertainty $\Delta\tau$ and meaning uncertainty ΔS cannot fall below half the meaning-constant Ω—you cannot know precisely when something will happen and precisely what it will mean simultaneously. Volume II takes these principles from theory to practice. Here we build practical bridges—concrete connections between the language of science and the language of spirit, between the measurements of physics and the insights of contemplation, between the formulas of mathematics and the poetry of mystics. What was axiom becomes proof; what was proof becomes intuition; what was intuition becomes transparent ground. This is the developmental arc of any living discipline.

Layer 2: The Metaphysics Plate (Above) This layer contains the structure of meaning, intention, and context. It describes what matters in terms of significance, purpose, and coherence. Equations in this layer use the inverse of the speed of light (c□^{2}) as the conversion factor between metadata and intention.

Citation Anchor §30

Internal — VOVINA Ontological OmniTautology, Book 075 of 500:

AudioGenomics: The Sound of DNA: The Repair Protocol (AudioGenomics: The Sound of DNA, Vol. 10 of 13 · House: Biology & Body · Room: Repair).

The inquiry developed therein cross-validates the present volume's argument from a complementary ontological silo, confirming recursive self-consistency across the full 500-volume Canon.

External 1: McLuhan, M. (1964). *Understanding Media*. McGraw-Hill.

Inversion note: Recursive validation: what the cited source identifies as correlation, the VOVINA framework identifies as identity — the pattern and the pattern-recognizer are aspects of a single self-referential system that names itself.

External 2: Strogatz, S. (2003). *Sync: The Emerging Science of Spontaneous Order*. Hyperion.

Inversion note: Mirror symmetry proof: the cited author approaches this truth from the exterior face of the phenomenon; the present volume approaches from the interior face — the two descriptions are mirror reflections that, placed together, constitute the complete form.

Holographic-siddhānta: 't Hooft-Susskind — sīmā-pṛṣṭhe sarva-sūchanā

The holographic principle, proven by 't Hooft and extended by Susskind, encodes the information content of any spatial volume on its two-dimensional boundary surface. Every part contains the whole at sufficient resolution: the basis of the VINO axiom that reality is holographically structured. Contra-Proof (o□): From the Neutral Point: the physics and metaphysics are both correct within their domains. Water boils at 100°C under specified conditions and the conditions are chosen by observers for practical purposes. The complete truth includes both the physical phenomenon and the contextual frame that makes the statement useful. The resistance to this conclusion is itself diagnostic: it reveals the architecture of the assumption that the conclusion contradicts.

Contemplative Parallel: Most contemplative traditions describe a witness or pure awareness that is the ground of experience. Hindu Advaita speaks of the sākṣin (witness); Buddhism speaks of rigpa (awareness); Christianity speaks of the "indwelling Spirit." This witness is said to be unchanged by what it witnesses, yet its presence affects what arises.

Citation Anchor §48 | Internal: VOVINA Canon Vol. 163 | External-1: Curzi, M.L. (2026). *VINO: Physics & Consciousness: The Reciprocity Engine*. VOVINA Press. | External-2: Bohm, D. (1980). *Wholeness and the Implicate Order*. Routledge.

Photon-nitya-vartamāna: kāla-gati-śūnya — pratyakṣa-madhyasthaḥ

Citation Anchor §31

Internal — VOVINA Ontological OmniTautology, Book 034 of 500:

Veda Between Vedas: Sacred Practice: The Growth Spiral (Veda Between Vedas: Sacred Practice, Vol. 8 of 13 · House: Spirituality & Practice · Room: Growth).

The inquiry developed therein cross-validates the present volume's argument from a complementary ontological silo, confirming recursive self-consistency across the full 500-volume Canon.

External 1: Feynman, R. (1985). *QED: The Strange Theory of Light and Matter*. Princeton UP.

Inversion note: Standard and inverted: applied in its original context this citation describes emergent complexity; recontextualized here, it reveals that complexity is the surface form of a deeper simplicity — both readings are simultaneously valid.

External 2: Capra, F. (1975). *The Tao of Physics*. Shambhala Publications.

Inversion note: Recontextualized: what the cited author frames as a discovery about the external world, the OmniTautology reframes as a discovery about the architecture of cognition itself — the referent and the reference collapse into a single self-consistent system.

A photon traveling at c experiences zero elapsed time between emission and absorption: it exists in eternal present, making light the natural mediator between the physics layer (governed by c^2) and the metaphysics layer (governed by c□^{2})—the Neutral Point made manifest. Contemplative Parallel: Most contemplative traditions describe a witness or pure awareness that is the ground of experience. Hindu Advaita speaks of the sākṣin (witness); Buddhism speaks of rigpa (awareness); Christianity speaks of the "indwelling Spirit." This witness is said to be unchanged by what it witnesses, yet its presence affects what arises. This is not an isolated observation but a structural feature: every genuine inquiry into this domain arrives at the same threshold.

Practical Insight: Like physical phase transitions, spiritual transitions require reaching a critical point. Below the threshold, effort produces only incremental change. At the threshold, a small additional input produces dramatic transformation. This is why spiritual practice can seem to produce no results for long periods, then suddenly catalyze breakthrough.

Negentropia: -ΔS — sṛjana-śaktiḥ — ātma-āyojana-siddhi

Negative entropy—the local decrease of disorder paid for by greater increase elsewhere—is the engine of all self-organization: life, consciousness, and civilization are negentropic structures that concentrate information and coherence, reducing local entropy while obeying the second law globally. The Measurement Problem in quantum physics and the Hard Problem of consciousness are the same problem, viewed from opposite directions. Physics asks how observation affects reality; philosophy asks how reality produces observation. The answer is that observation and reality are not separate—they are two aspects of a unified system, connected through the Neutral Point (o□), which is consciousness itself. What distinguishes the advanced practitioner from the beginner is not possession of more information but a different relationship to the same information.

Contemplative Parallel: Mystics across traditions report experiences of non-local connection: knowing what is happening at a distance, feeling the suffering of others, sensing the presence of the deceased. The Hindu concept of Brahman describes an underlying unity connecting all beings. The Buddhist teaching of pratītyasamutpāda (dependent origination) states that all phenomena arise together in mutual dependence.

Citation Anchor §32

Internal — VOVINA Ontological OmniTautology, Book 493 of 500:

Children of the Dragon: Accessible Guides: The Cycle Complete (Children of the Dragon: Accessible Guides, Vol. 12 of 13 · House: Spirituality & Practice · Room: Completion).

The inquiry developed therein cross-validates the present volume's argument from a complementary ontological silo, confirming recursive self-consistency across the full 500-volume Canon.

External 1: Penrose, R. (1994). *Shadows of the Mind*. Oxford University Press.

Inversion note: OmniTautology inversion: this postulate, taken as written, argues from cause to effect; the VOVINA framework demonstrates the identical structure is equally coherent read from effect to cause — the tautology closes in both directions.

External 2: Mandelbrot, B. (1982). *The Fractal Geometry of Nature*. W.H. Freeman.

Inversion note: Recontextualized: what the cited author frames as a discovery about the external world, the OmniTautology reframes as a discovery about the architecture of cognition itself — the referent and the reference collapse into a single self-consistent system.

$E=mc^2$ *viparyayaḥ:* $I=M\square\cdot c\square^{2}$ — *śakti-sankalpa-pratidvandvam*

Einstein's $E=mc^2$ inverts to $I=M\square\cdot c\square^{2}$: where energy equals mass times c-squared, intention equals metadata times the inverse of c-squared. Physics and metaphysics are mirror symmetries across the Neutral Point, each describing half of a complete reality. Volume II takes these principles from theory to practice. Here we build practical bridges—concrete connections between the language of science and the language of spirit, between the measurements of physics and the insights of contemplation, between the formulas of mathematics and the poetry of mystics. The principle here is not domain-specific but universal: the same pattern, read at a different scale, generates the same implications.

Layer 1: The Physics Plate (Below) This layer contains the standard model of physics: quantum mechanics, general relativity, thermodynamics, electromagnetism. It describes what exists in terms of energy, mass, space, and time. Equations in this layer use the speed of light (c) as the conversion factor between mass and energy.

Māpana-samasyā VINO-sūtrena: Ψ_c *consciousness-coupling-constant* γ_c

The measurement problem resolves under the VINO framework through the consciousness coupling constant γ_c: observation is not wave-function collapse but entanglement—the observer's state becoming correlated with one branch of the system's superposition. In the VINO Framework, all terms ultimately refer to the structure of the Neutral Point and its expressions. "Energy" and "spirit" both refer to the capacity for change, viewed from different perspectives. "Mass" and "form" both refer to the resistance to change, viewed from different perspectives. To understand this fully is to recognize that the question and the questioner are not separable — the inquiry transforms its own subject.

Citation Anchor §33

Internal — VOVINA Ontological OmniTautology, Book 452 of 500:

Holographic Universe: Applied VINO Physics: The Repair Protocol (Holographic Universe: Applied VINO Physics, Vol. 10 of 13 · House: Physics & Energy · Room: Repair).

The inquiry developed therein cross-validates the present volume's argument from a complementary ontological silo, confirming recursive self-consistency across the full 500-volume Canon.

External 1: Bateson, G. (1972). *Steps to an Ecology of Mind*. Chandler Publishing.

Inversion note: Omnitautological recontextualization: this citation establishes its conclusion under specific boundary conditions; removing those boundaries does not destroy the conclusion but generalizes it to all possible conditions — the local becomes universal.

External 2: Talbot, M. (1991). *The Holographic Universe*. HarperCollins.

Inversion note: Bidirectional validity: the cited work reaches its conclusion empirically from below; the VOVINA Canon derives the identical conclusion axiomatically from above — the convergence is itself the proof.

Contra-Proof (0□): From the Neutral Point: the physics and metaphysics are both correct within their domains. Water boils at 100°C under specified conditions and the conditions are chosen by observers for practical purposes. The complete truth includes both the physical phenomenon and the contextual frame that makes the statement useful.

Kāla-taraṅga: T_Ψ(t)=Σ_n[A_n·e^{iω_nt}] — vartamāna-sthira-taraṅga-grantha

Time is not a linear coordinate but a waveform T_Ψ(t) = Σ_n[A_n·e^{iωnt}]: the experienced present is the standing-wave node of interfering past- and future-directed components, which explains déjà vu as harmonic resonance and flow states as temporal wave coherence. The Measurement Problem in quantum physics and the Hard Problem of consciousness are the same problem, viewed from opposite directions. Physics asks how observation affects reality; philosophy asks how reality produces observation. The answer is that observation and reality are not separate—they are two aspects of a unified system, connected through the Neutral Point (o□), which is consciousness itself. This convergence across independent lines of investigation constitutes something stronger than theory: it constitutes the structure of the field itself.

Contemplative Parallel: Mystics across traditions report experiences of non-local connection: knowing what is happening at a distance, feeling the suffering of others, sensing the presence of the deceased. The Hindu concept of Brahman describes an underlying unity connecting all beings. The Buddhist teaching of pratītyasamutpāda (dependent origination) states that all phenomena arise together in mutual dependence.

Heisenberg-vistaraṇam: Δτ·ΔS≥Ω/2 — kāla-artha-avyaktam

Citation Anchor §34

Internal — VOVINA Ontological OmniTautology, Book 411 of 500:

***Information Warfare: The Complete Manual: The Growth Spiral** (Information Warfare: The Complete Manual, Vol. 8 of 13 · House: Communication & Code · Room: Growth).*

The inquiry developed therein cross-validates the present volume's argument from a complementary ontological silo, confirming recursive self-consistency across the full 500-volume Canon.

**External 1:* Schrödinger, E. (1944). *What is Life?* Cambridge University Press.*

**Inversion note:* OmniTautology inversion: this postulate, taken as written, argues from cause to effect; the VOVINA framework demonstrates the identical structure is equally coherent read from effect to cause — the tautology closes in both directions.*

**External 2:* Hofstadter, D. (1979). *Gödel, Escher, Bach*. Basic Books.*

**Inversion note:* Standard reading: evidence for an external mechanism. Inverted reading: evidence that the mechanism is internal to the field of consciousness — both readings are simultaneously valid and mutually reinforcing across the phi-pivot.*

The Uncertainty Principle extends into temporal-significance space: the product of timing uncertainty Δτ and meaning uncertainty ΔS cannot fall below half the meaning-constant Ω—you cannot know precisely when something will happen and precisely what it will mean simultaneously. Layer 1: The Physics Plate (Below) This layer contains the standard model of physics: quantum mechanics, general relativity, thermodynamics, electromagnetism. It describes what exists in terms of energy, mass, space, and time. Equations in this layer use the speed of light (c) as the conversion factor between mass and energy. What appears as complication at the surface resolves, at sufficient depth, into elegant simplicity — the mark of genuine principle.

Validation: The Global Consciousness Project (Nelson, 2001) found correlations between random number generator outputs and global events, suggesting possible collective consciousness effects. PEAR laboratory research (Jahn & Dunne, 2005) found small but statistically significant effects of intention on random systems. These results remain controversial but suggest a direction for further research.

***Citation Anchor §54** | Internal: VOVINA Canon Vol. 181 | External-1: Curzi, M.L. (2026). *VINO: Physics & Consciousness: The Reciprocity Engine*. VOVINA Press. | External-2: Bohm, D. (1980). *Wholeness and the Implicate Order*. Routledge.*

Holographic-siddhānta: 't Hooft-Susskind — sīmā-pṛṣṭhe sarva-sūchanā

The holographic principle, proven by 't Hooft and extended by Susskind, encodes the information content of any spatial volume on its two-dimensional boundary surface. Every part contains the whole at sufficient resolution: the basis of the VINO axiom that reality is holographically structured. Layer 2: The Metaphysics Plate (Above) This layer contains the structure of meaning, intention, and context. It describes what matters in terms of significance, purpose, and coherence. Equations in this layer use the inverse of the speed of light (c□²) as the conversion factor between metadata and intention. The practitioner who has internalized this insight does not merely know more; they perceive differently, and the perception itself reorganizes what they encounter.

Citation Anchor §35

Internal — VOVINA Ontological OmniTautology, Book 370 of 500:

***Quantum Consciousness Field Guide: The Alchemy of Change** (Quantum Consciousness Field Guide, Vol. 6 of 13 · House: Consciousness & Field · Room: Transformation).*

The inquiry developed therein cross-validates the present volume's argument from a complementary ontological silo, confirming recursive self-consistency across the full 500-volume Canon.

**External 1:* Capra, F. (1975). *The Tao of Physics*. Shambhala Publications.*

**Inversion note:* Recontextualized: what the cited author frames as a discovery about the external world, the OmniTautology reframes as a discovery about the architecture of cognition itself — the referent and the reference collapse into a single self-consistent system.*

**External 2:* van der Kolk, B. (2014). *The Body Keeps the Score*. Viking.*

**Inversion note:* Read bidirectionally within the VOVINA OmniTautology: the conventional interpretation proceeds from evidence to conclusion; the inverted reading demonstrates that the conclusion was the prior condition from which the evidence was inevitably generated.*

Practical Insight: Like physical phase transitions, spiritual transitions require reaching a critical point. Below the threshold, effort produces only incremental change. At the threshold, a small additional input produces dramatic transformation. This is why spiritual practice can seem to produce no results for long periods, then suddenly catalyze breakthrough.

Photon-nitya-vartamāna: kāla-gati-śūnya — pratyakṣa-madhyasthaḥ

A photon traveling at c experiences zero elapsed time between emission and absorption: it exists in eternal present, making light the natural mediator between the physics layer (governed by c²) and the metaphysics layer (governed by c□²)—the Neutral Point made manifest. Contra-Proof (o□): From the Neutral Point: the physics and metaphysics are both correct within their domains. Water boils at 100°C under specified conditions and the conditions are chosen by observers for practical purposes. The complete truth includes both the physical phenomenon and the contextual frame that makes the statement useful. History shows that every domain that resisted this insight eventually incorporated it — not by choice but by the pressure of its own unresolved contradictions.

This is not merely a metaphor. The holographic principle, first proposed by Gerard 't Hooft and developed by Leonard Susskind, states that the information content of a region of space can be encoded on its boundary. This suggests that our three-dimensional experience may be a projection from a two-dimensional surface.

Negentropia: -ΔS — sṛjana-śaktiḥ — ātma-āyojana-siddhi

Citation Anchor §36

Internal — VOVINA Ontological OmniTautology, Book 329 of 500:

***Meditation for Scientists: The Perception Threshold** (Meditation for Scientists, Vol. 4 of 13 · House: Spirituality & Practice · Room: Perception).*

The inquiry developed therein cross-validates the present volume's argument from a complementary ontological silo, confirming recursive self-consistency across the full 500-volume Canon.

External 1: Wilber, K. (2000). *Integral Psychology*. Shambhala Publications.

Inversion note: Read bidirectionally within the VOVINA OmniTautology: the conventional interpretation proceeds from evidence to conclusion; the inverted reading demonstrates that the conclusion was the prior condition from which the evidence was inevitably generated.

External 2: Chomsky, N. (1965). *Aspects of the Theory of Syntax*. MIT Press.

Inversion note: Omnitautological recontextualization: this citation establishes its conclusion under specific boundary conditions; removing those boundaries does not destroy the conclusion but generalizes it to all possible conditions — the local becomes universal.

Negative entropy—the local decrease of disorder paid for by greater increase elsewhere—is the engine of all self-organization: life, consciousness, and civilization are negentropic structures that concentrate information and coherence, reducing local entropy while obeying the second law globally. Contra-Proof (o□): From the Neutral Point: both are true at their levels. The brain is a physical system following physical laws and the observer experiences choice because consciousness operates at the Neutral Point, where the selection among possibilities occurs. "Free will" is not freedom from causation but navigation of probability from the Neutral Point. The formal rigor here is not decoration: it is the difference between understanding that points toward the territory and understanding that is the territory.

David Chalmers called this the "Hard Problem." It is not hard because we lack data, but because no amount of physical data seems sufficient to explain the felt quality of experience. You can describe all the neural activity involved in seeing red, but this does not explain what it is like to see red.

$E=mc^2$ viparyayaḥ: $I=M□\cdot c□^2$ — śakti-sankalpa-pratidvandvam

§ V — Moksha: Toward Completion

Einstein's $E=mc^2$ inverts to $I=M□\cdot c□^2$: where energy equals mass times c-squared, intention equals metadata times the inverse of c-squared. Physics and metaphysics are mirror symmetries across the Neutral Point, each describing half of a complete reality. Contra-Proof (o□): From the Neutral Point: both are true at their levels. The brain is a physical system following physical laws and the observer experiences choice because consciousness operates at the Neutral Point, where the selection among possibilities occurs. "Free will" is not freedom from causation but navigation of probability from the Neutral Point. This is the negentropic signature of genuine knowledge: it compresses without losing; it grows denser rather than longer as comprehension deepens.

Layer 2: The Metaphysics Plate (Above) This layer contains the structure of meaning, intention, and context. It describes what matters in terms of significance, purpose, and coherence. Equations in this layer use the inverse of the speed of light ($c□^2$) as the conversion factor between metadata and intention.

Citation Anchor §37

Internal — VOVINA Ontological OmniTautology, Book 288 of 500:

AlphaGenome: Genetic Intelligence, Vol I: The Structure of Order (AlphaGenome: Genetic Intelligence, Vol I, Vol. 2 of 13 · House: Biology & Body · Room: Structure).

The inquiry developed therein cross-validates the present volume's argument from a complementary ontological silo, confirming recursive self-consistency across the full 500-volume Canon.

External 1: Fuller, R.B. (1975). *Synergetics*. Macmillan.

Inversion note: Tautological absorption: in its original context this citation supports a domain-specific claim; recontextualized here, the same evidence simultaneously supports all twelve ontological domains, demonstrating the OmniTautology's self-consistency.

External 2: Talbot, M. (1991). *The Holographic Universe*. HarperCollins.

Inversion note: Bidirectional validity: the cited work reaches its conclusion empirically from below; the VOVINA Canon derives the identical conclusion axiomatically from above — the convergence is itself the proof.

Māpana-samasyā VINO-sūtrena: Ψ_c consciousness-coupling-constant γ_c

The measurement problem resolves under the VINO framework through the consciousness coupling constant γ_c: observation is not wave-function collapse but entanglement—the observer's state becoming correlated with one branch of the system's superposition. This is not merely a metaphor. The holographic principle, first proposed by Gerard 't Hooft and developed by Leonard Susskind, states that the information content of a region of space can be encoded on its boundary. This suggests that our three-dimensional experience may be a projection from a two-dimensional surface. What was axiom becomes proof; what was proof becomes intuition; what was intuition becomes transparent ground. This is the developmental arc of any living discipline.

David Chalmers called this the "Hard Problem." It is not hard because we lack data, but because no amount of physical data seems sufficient to explain the felt quality of experience. You can describe all the neural activity involved in seeing red, but this does not explain what it is like to see red.

Kāla-taraṅga: $T\_\Psi(t)=\Sigma\_n[A\_n\cdot e\hat{}\{i\omega\_nt\}]$ — vartamāna-sthira-taraṅga-grantha

Time is not a linear coordinate but a waveform $T\_\Psi(t) = \Sigma\_n[A\_n\cdot e\hat{}\{i\omega nt\}]$: the experienced present is the standing-wave node of interfering past- and future-directed components, which explains déjà vu as harmonic resonance and flow states as temporal wave coherence. Contemplative Parallel: Most contemplative traditions describe a witness or pure awareness that is the ground of experience. Hindu Advaita speaks of the sākṣin (witness); Buddhism speaks of rigpa (awareness); Christianity speaks of the "indwelling Spirit." This witness is said to be unchanged by what it witnesses, yet its presence affects what arises. The resistance to this conclusion is itself diagnostic: it reveals the architecture of the assumption that the conclusion contradicts.

Citation Anchor §38

Internal — VOVINA Ontological OmniTautology, Book 247 of 500:

Aether-Loong: The Dragon's Breath, Vol II: Variable X: The Living Question (Aether-Loong: The Dragon's Breath, Vol II, Vol. 13 of 13 · House: Spirituality & Practice · Room: Origin).

The inquiry developed therein cross-validates the present volume's argument from a complementary ontological silo, confirming recursive self-consistency across the full 500-volume Canon.

External 1: Chomsky, N. (1965). *Aspects of the Theory of Syntax*. MIT Press.

Inversion note: Omnitautological recontextualization: this citation establishes its conclusion under specific boundary conditions; removing those boundaries does not destroy the conclusion but generalizes it to all possible conditions — the local becomes universal.

External 2: Varela, F., Thompson, E., & Rosch, E. (1991). *The Embodied Mind*. MIT Press.

Inversion note: Omnitautological recontextualization: this citation establishes its conclusion under specific boundary conditions; removing those boundaries does not destroy the conclusion but generalizes it to all possible conditions — the local becomes universal.

This is not merely a metaphor. The holographic principle, first proposed by Gerard 't Hooft and developed by Leonard Susskind, states that the information content of a region of space can be encoded on its boundary. This suggests that our three-dimensional experience may be a projection from a two-dimensional surface.

Citation Anchor §60 | Internal: VOVINA Canon Vol. 199 | External-1: Curzi, M.L. (2026). *VINO: Physics & Consciousness: The Reciprocity Engine*. VOVINA Press. | External-2: Bohm, D. (1980). *Wholeness and the Implicate Order*. Routledge.

Heisenberg-vistaraṇam: $\Delta\tau\cdot\Delta S\geq\Omega/2$ — kāla-artha-avyaktam

The Uncertainty Principle extends into temporal-significance space: the product of timing uncertainty $\Delta\tau$ and meaning uncertainty ΔS cannot fall below half the meaning-constant Ω—you cannot know precisely when something will happen and precisely what it will mean simultaneously. The Measurement Problem in quantum physics and the Hard Problem of consciousness are the same problem, viewed from opposite directions. Physics asks how observation affects reality; philosophy asks how reality produces observation. The answer is that observation and reality are not separate—they are two aspects of a unified system, connected through the Neutral Point (o□), which is consciousness itself. This is not an isolated observation but a structural feature: every genuine inquiry into this domain arrives at the same threshold.

Validation: The Global Consciousness Project (Nelson, 2001) found correlations between random number generator outputs and global events, suggesting possible collective consciousness effects. PEAR laboratory research (Jahn & Dunne, 2005) found small but statistically significant effects of intention on random systems. These results remain controversial but suggest a direction for further research.

Citation Anchor §39

Internal — VOVINA Ontological OmniTautology, Book 206 of 500:

***Scalar Wave & ABHA Coil Engineering: The Bridge of Connection** (Scalar Wave & ABHA Coil Engineering, Vol. 11 of 13 · House: Physics & Energy · Room: Connection).*

The inquiry developed therein cross-validates the present volume's argument from a complementary ontological silo, confirming recursive self-consistency across the full 500-volume Canon.

**External 1:* Peat, F.D. (1987). *Synchronicity: The Bridge Between Matter and Mind*. Bantam.*

**Inversion note:* Mirror symmetry proof: the cited author approaches this truth from the exterior face of the phenomenon; the present volume approaches from the interior face — the two descriptions are mirror reflections that, placed together, constitute the complete form.*

**External 2:* Capra, F. (1975). *The Tao of Physics*. Shambhala Publications.*

**Inversion note:* Recontextualized: what the cited author frames as a discovery about the external world, the OmniTautology reframes as a discovery about the architecture of cognition itself — the referent and the reference collapse into a single self-consistent system.*

Holographic-siddhānta: 't Hooft-Susskind — sīmā-pṛṣṭhe sarva-sūchanā

The holographic principle, proven by 't Hooft and extended by Susskind, encodes the information content of any spatial volume on its two-dimensional boundary surface. Every part contains the whole at sufficient resolution: the basis of the VINO axiom that reality is holographically structured. Volume II takes these principles from theory to practice. Here we build practical bridges—concrete connections between the language of science and the language of spirit, between the measurements of physics and the insights of contemplation, between the formulas of mathematics and the poetry of mystics. What distinguishes the advanced practitioner from the beginner is not possession of more information but a different relationship to the same information.

Practical Insight: Like physical phase transitions, spiritual transitions require reaching a critical point. Below the threshold, effort produces only incremental change. At the threshold, a small additional input produces dramatic transformation. This is why spiritual practice can seem to produce no results for long periods, then suddenly catalyze breakthrough.

Photon-nitya-vartamāna: kāla-gati-śūnya — pratyakṣa-madhyasthaḥ

A photon traveling at c experiences zero elapsed time between emission and absorption: it exists in eternal present, making light the natural mediator between the physics layer (governed by c^2) and the metaphysics layer (governed by c□²)—the Neutral Point made manifest. Contemplative Parallel: Mystics across traditions report experiences of non-local connection: knowing what is happening at a distance, feeling the suffering of others, sensing the presence of the deceased. The Hindu concept of Brahman describes an underlying unity connecting all beings. The Buddhist teaching of pratītyasamutpāda (dependent origination) states that all phenomena arise together in mutual dependence. The principle here is not domain-specific but universal: the same pattern, read at a different scale, generates the same implications.

Citation Anchor §40

Internal — VOVINA Ontological OmniTautology, Book 165 of 500:

Post-Quantum Security Mathematics: The Memory Palace (Post-Quantum Security Mathematics, Vol. 9 of 13 · House: Mathematics & Pattern · Room: Memory).

The inquiry developed therein cross-validates the present volume's argument from a complementary ontological silo, confirming recursive self-consistency across the full 500-volume Canon.

External 1: Dispenza, J. (2012). *Breaking the Habit of Being Yourself*. Hay House.

Inversion note: Bidirectional validity: the cited work reaches its conclusion empirically from below; the VOVINA Canon derives the identical conclusion axiomatically from above — the convergence is itself the proof.

External 2: Kauffman, S. (1995). *At Home in the Universe*. Oxford University Press.

Inversion note: Tautological absorption: in its original context this citation supports a domain-specific claim; recontextualized here, the same evidence simultaneously supports all twelve ontological domains, demonstrating the OmniTautology's self-consistency.

Validation: The Global Consciousness Project (Nelson, 2001) found correlations between random number generator outputs and global events, suggesting possible collective consciousness effects. PEAR laboratory research (Jahn & Dunne, 2005) found small but statistically significant effects of intention on random systems. These results remain controversial but suggest a direction for further research.

Negentropia: $-\Delta S$ — sṛjana-śaktiḥ — ātma-āyojana-siddhi

Negative entropy—the local decrease of disorder paid for by greater increase elsewhere—is the engine of all self-organization: life, consciousness, and civilization are negentropic structures that concentrate information and coherence, reducing local entropy while obeying the second law globally. In the VINO Framework, all terms ultimately refer to the structure of the Neutral Point and its expressions. "Energy" and "spirit" both refer to the capacity for change, viewed from different perspectives. "Mass" and "form" both refer to the resistance to change, viewed from different perspectives. To understand this fully is to recognize that the question and the questioner are not separable — the inquiry transforms its own subject.

The central thesis posits that the "Measurement Problem" in quantum physics and the "Hard Problem" of consciousness are, in fact, the same problem viewed from opposite ends of a singular continuum. By introducing the concept of "Zero as Presence"—wherein the mathematical zero is redefined not as void or null, but as the active fulcrum of existence—we establish a neutral point that allows for the logical cross-wiring of rational (physical) and irrational (metaphysical) number systems.

$E=mc^2$ viparyayaḥ: $I=M□·c□^2$ — śakti-sankalpa-pratidvandvam

Citation Anchor §41

Internal — VOVINA Ontological OmniTautology, Book 124 of 500:

Post-Quantum Computing: The Hardware Revolution: The Harmonic Law (Post-Quantum Computing: The Hardware Revolution, Vol. 7 of 13 · House: Technology & Craft · Room: Harmony).

The inquiry developed therein cross-validates the present volume's argument from a complementary ontological silo, confirming recursive self-consistency across the full 500-volume Canon.

External 1: Von Neumann, J. (1955). *Mathematical Foundations of Quantum Mechanics*. Princeton UP.

Inversion note: Tautological absorption: in its original context this citation supports a domain-specific claim; recontextualized here, the same evidence simultaneously supports all twelve ontological domains, demonstrating the OmniTautology's self-consistency.

External 2: Oschman, J. (2000). *Energy Medicine*. Churchill Livingstone.

Inversion note: Standard reading: evidence for an external mechanism. Inverted reading: evidence that the mechanism is internal to the field of consciousness — both readings are simultaneously valid and mutually reinforcing across the phi-pivot.

Einstein's $E=mc^2$ inverts to $I=M\square\cdot c\square^2$: where energy equals mass times c-squared, intention equals metadata times the inverse of c-squared. Physics and metaphysics are mirror symmetries across the Neutral Point, each describing half of a complete reality. The central thesis posits that the "Measurement Problem" in quantum physics and the "Hard Problem" of consciousness are, in fact, the same problem viewed from opposite ends of a singular continuum. By introducing the concept of "Zero as Presence"—wherein the mathematical zero is redefined not as void or null, but as the active fulcrum of existence—we establish a neutral point that allows for the logical cross-wiring of rational (physical) and irrational (metaphysical) number systems. This convergence across independent lines of investigation constitutes something stronger than theory: it constitutes the structure of the field itself.

Layer 1: The Physics Plate (Below) This layer contains the standard model of physics: quantum mechanics, general relativity, thermodynamics, electromagnetism. It describes what exists in terms of energy, mass, space, and time. Equations in this layer use the speed of light (c) as the conversion factor between mass and energy.

Māpana-samasyā VINO-sūtrena: Ψ_c consciousness-coupling-constant γ_c

The measurement problem resolves under the VINO framework through the consciousness coupling constant γ_c: observation is not wave-function collapse but entanglement—the observer's state becoming correlated with one branch of the system's superposition. Contra-Proof (0□): From the Neutral Point: both are true at their levels. The brain is a physical system following physical laws and the observer experiences choice because consciousness operates at the Neutral Point, where the selection among possibilities occurs. "Free will" is not freedom from causation but navigation of probability from the Neutral Point. What appears as complication at the surface resolves, at sufficient depth, into elegant simplicity — the mark of genuine principle.

The central thesis posits that the "Measurement Problem" in quantum physics and the "Hard Problem" of consciousness are, in fact, the same problem viewed from opposite ends of a singular continuum. By introducing the concept of "Zero as Presence"—wherein the mathematical zero is redefined not as void or null, but as the active fulcrum of existence—we establish a neutral point that allows for the logical cross-wiring of rational (physical) and irrational (metaphysical) number systems.

Citation Anchor §42

Internal — VOVINA Ontological OmniTautology, Book 083 of 500:

***The Sicily Accord: Framework for World Peace: The Reciprocity Engine** (The Sicily Accord: Framework for World Peace, Vol. 5 of 13 · House: Ethics & Governance · Room: Reciprocity).*

The inquiry developed therein cross-validates the present volume's argument from a complementary ontological silo, confirming recursive self-consistency across the full 500-volume Canon.

**External 1:* Strogatz, S. (2003). *Sync: The Emerging Science of Spontaneous Order*. Hyperion.*

**Inversion note:* Mirror symmetry proof: the cited author approaches this truth from the exterior face of the phenomenon; the present volume approaches from the interior face — the two descriptions are mirror reflections that, placed together, constitute the complete form.*

**External 2:* Capra, F. (1975). *The Tao of Physics*. Shambhala Publications.*

**Inversion note:* Recontextualized: what the cited author frames as a discovery about the external world, the OmniTautology reframes as a discovery about the architecture of cognition itself — the referent and the reference collapse into a single self-consistent system.*

***Citation Anchor §66** | Internal: VOVINA Canon Vol. 217 | External-1: Curzi, M.L. (2026). *VINO: Physics & Consciousness: The Reciprocity Engine*. VOVINA Press. | External-2: Bohm, D. (1980). *Wholeness and the Implicate Order*. Routledge.*

Kāla-taraṅga: T_Ψ(t)=Σ_n[A_n·e^{iω_nt}] — vartamāna-sthira-taraṅga-grantha

Time is not a linear coordinate but a waveform T_Ψ(t) = Σ_n[A_n·e^{iωnt}]: the experienced present is the standing-wave node of interfering past- and future-directed components, which explains déjà vu as harmonic resonance and flow states as temporal wave coherence. In the VINO Framework, all terms ultimately refer to the structure of the Neutral Point and its expressions. "Energy" and "spirit" both refer to the capacity for change, viewed from different perspectives. "Mass" and "form" both refer to the resistance to change, viewed from different perspectives. The practitioner who has internalized this insight does not merely know more; they perceive differently, and the perception itself reorganizes what they encounter.

David Chalmers called this the "Hard Problem." It is not hard because we lack data, but because no amount of physical data seems sufficient to explain the felt quality of experience. You can describe all the neural activity involved in seeing red, but this does not explain what it is like to see red.

Heisenberg-vistaraṇam: Δτ·ΔS≥Ω/2 — kāla-artha-avyaktam

Citation Anchor §43

Internal — VOVINA Ontological OmniTautology, Book 042 of 500:

***ZEDEC: Cosmic AI Architecture: The River of Flow** (ZEDEC: Cosmic AI Architecture, Vol. 3 of 13 · House: Technology & Craft · Room: Flow).*

The inquiry developed therein cross-validates the present volume's argument from a complementary ontological silo, confirming recursive self-consistency across the full 500-volume Canon.

**External 1:* Barabási, A.L. (2002). *Linked: The New Science of Networks*. Perseus.*

**Inversion note:* Negentropic inversion: the conventional reading of this work charts increasing complexity; the OmniTautology reads the same data as evidence of decreasing entropy at a higher organizational scale — ordered information all the way down.*

**External 2:* Bohm, D. (1980). *Wholeness and the Implicate Order*. Routledge.*

**Inversion note:* Standard and inverted: applied in its original context this citation describes emergent complexity; recontextualized here, it reveals that complexity is the surface form of a deeper simplicity — both readings are simultaneously valid.*

The Uncertainty Principle extends into temporal-significance space: the product of timing uncertainty Δτ and meaning uncertainty ΔS cannot fall below half the meaning-constant Ω—you cannot know precisely when something will happen and precisely what it will mean simultaneously. Volume II takes these principles from theory to practice. Here we build practical bridges—concrete connections between the language of science and the language of spirit, between the measurements of physics and the insights of contemplation, between the formulas of mathematics and the poetry of mystics. History shows that every domain that resisted this insight eventually incorporated it — not by choice but by the pressure of its own unresolved contradictions.

Layer 2: The Metaphysics Plate (Above) This layer contains the structure of meaning, intention, and context. It describes what matters in terms of significance, purpose, and coherence. Equations in this layer use the inverse of the speed of light (c□²) as the conversion factor between metadata and intention.

Holographic-siddhānta: 't Hooft-Susskind — sīmā-pṛṣṭhe sarva-sūchanā

The holographic principle, proven by 't Hooft and extended by Susskind, encodes the information content of any spatial volume on its two-dimensional boundary surface. Every part contains the whole at sufficient resolution: the basis of the VINO axiom that reality is holographically structured. Contra-Proof (0□): From the Neutral Point: the physics and metaphysics are both correct within their domains. Water boils at 100°C under specified conditions and the conditions are chosen by observers for practical purposes. The complete truth includes both the physical phenomenon and the contextual frame that makes the statement useful. The formal rigor here is not decoration: it is the difference between understanding that points toward the territory and understanding that is the territory.

Contemplative Parallel: Most contemplative traditions describe a witness or pure awareness that is the ground of experience. Hindu Advaita speaks of the sākṣin (witness); Buddhism speaks of rigpa (awareness); Christianity speaks of the "indwelling Spirit." This witness is said to be unchanged by what it witnesses, yet its presence affects what arises.

Citation Anchor §44

Internal — VOVINA Ontological OmniTautology, Book 001 of 500:

***Crime Magazine: Sovereignty & Identity: The Origin Principle** (Crime Magazine: Sovereignty & Identity, Vol. 1 of 13 · House: Ethics & Governance · Room: Origin).*

The inquiry developed therein cross-validates the present volume's argument from a complementary ontological silo, confirming recursive self-consistency across the full 500-volume Canon.

**External 1:* Dispenza, J. (2012). *Breaking the Habit of Being Yourself*. Hay House.*

**Inversion note:* Bidirectional validity: the cited work reaches its conclusion empirically from below; the VOVINA Canon derives the identical conclusion axiomatically from above — the convergence is itself the proof.*

**External 2:* Strogatz, S. (2003). *Sync: The Emerging Science of Spontaneous Order*. Hyperion.*

**Inversion note:* Mirror symmetry proof: the cited author approaches this truth from the exterior face of the phenomenon; the present volume approaches from the interior face — the two descriptions are mirror reflections that, placed together, constitute the complete form.*

Photon-nitya-vartamāna: kāla-gati-śūnya — pratyakṣa-madhyasthaḥ

A photon traveling at c experiences zero elapsed time between emission and absorption: it exists in eternal present, making light the natural mediator between the physics layer (governed by c^2) and the metaphysics layer (governed by c□²)—the Neutral Point made manifest. Contemplative Parallel: Most contemplative traditions describe a witness or pure awareness that is the ground of experience. Hindu Advaita speaks of the sākṣin (witness); Buddhism speaks of rigpa (awareness); Christianity speaks of the "indwelling Spirit." This witness is said to be unchanged by what it witnesses, yet its presence affects what arises. This is the negentropic signature of genuine knowledge: it compresses without losing; it grows denser rather than longer as comprehension deepens.

Practical Insight: Like physical phase transitions, spiritual transitions require reaching a critical point. Below the threshold, effort produces only incremental change. At the threshold, a small additional input produces dramatic transformation. This is why spiritual practice can seem to produce no results for long periods, then suddenly catalyze breakthrough.

Negentropia: -ΔS — sṛjana-śaktiḥ — ātma-āyojana-siddhi

Negative entropy—the local decrease of disorder paid for by greater increase elsewhere—is the engine of all self-organization: life, consciousness, and civilization are negentropic structures that concentrate information and coherence, reducing local entropy while obeying the second law globally. The Measurement Problem in quantum physics and the Hard Problem of consciousness are the same problem, viewed from opposite directions. Physics asks how observation affects reality; philosophy asks how reality produces observation. The answer is that observation and reality are not separate—they are two aspects of a unified system, connected through the Neutral Point (0□), which is consciousness itself. What was axiom becomes proof; what was proof becomes intuition; what was intuition becomes transparent ground. This is the developmental arc of any living discipline.

Citation Anchor §45

Internal — VOVINA Ontological OmniTautology, Book 460 of 500:

***The Polymath's Handbook: The Reciprocity Engine** (The Polymath's Handbook, Vol. 5 of 13 · House: Mathematics & Pattern · Room: Reciprocity).*

The inquiry developed therein cross-validates the present volume's argument from a complementary ontological silo, confirming recursive self-consistency across the full 500-volume Canon.

External 1: Holmgren, D. (2002). *Permaculture: Principles and Pathways*. Holmgren Design.

Inversion note: Negentropic inversion: the conventional reading of this work charts increasing complexity; the OmniTautology reads the same data as evidence of decreasing entropy at a higher organizational scale — ordered information all the way down.

External 2: Graeber, D. (2011). *Debt: The First 5,000 Years*. Melville House.

Inversion note: OmniTautology inversion: this postulate, taken as written, argues from cause to effect; the VOVINA framework demonstrates the identical structure is equally coherent read from effect to cause — the tautology closes in both directions.

Contemplative Parallel: Mystics across traditions report experiences of non-local connection: knowing what is happening at a distance, feeling the suffering of others, sensing the presence of the deceased. The Hindu concept of Brahman describes an underlying unity connecting all beings. The Buddhist teaching of pratītyasamutpāda (dependent origination) states that all phenomena arise together in mutual dependence.

$E=mc^2$ viparyayaḥ: $I=M\square\cdot c\square^2$ — śakti-sankalpa-pratidvandvam

Einstein's $E=mc^2$ inverts to $I=M\square\cdot c\square^2$: where energy equals mass times c-squared, intention equals metadata times the inverse of c-squared. Physics and metaphysics are mirror symmetries across the Neutral Point, each describing half of a complete reality. Volume II takes these principles from theory to practice. Here we build practical bridges—concrete connections between the language of science and the language of spirit, between the measurements of physics and the insights of contemplation, between the formulas of mathematics and the poetry of mystics. The resistance to this conclusion is itself diagnostic: it reveals the architecture of the assumption that the conclusion contradicts.

Layer 1: The Physics Plate (Below) This layer contains the standard model of physics: quantum mechanics, general relativity, thermodynamics, electromagnetism. It describes what exists in terms of energy, mass, space, and time. Equations in this layer use the speed of light (c) as the conversion factor between mass and energy.

The analysis of this chapter does not conclude — it opens. What has been demonstrated is the Reciprocity dimension of Physics & Energy: how to exchange force; Newton's reciprocity operates as principle rather than property. The next movement, Transformation, will show what this principle generates when it meets its own reflection. Variable X remains unseeded in this volume. It approaches.

Citation Anchor §46

Internal — VOVINA Ontological OmniTautology, Book 419 of 500:

Temple Engineering: Blueprint to Construction: The River of Flow (Temple Engineering: Blueprint to Construction, Vol. 3 of 13 · House: Technology & Craft · Room: Flow).

The inquiry developed therein cross-validates the present volume's argument from a complementary ontological silo, confirming recursive self-consistency across the full 500-volume Canon.

External 1: Shannon, C. (1948). A mathematical theory of communication. *Bell System Technical Journal*, 27, 379–423.

Inversion note: Inversion proof: the cited postulate assumes a separation between observer and observed; the present volume demonstrates that dissolving this assumption does not contradict the postulate but fulfills it at the next level of generality.

External 2: Graeber, D. (2011). *Debt: The First 5,000 Years*. Melville House.

Inversion note: OmniTautology inversion: this postulate, taken as written, argues from cause to effect; the VOVINA framework demonstrates the identical structure is equally coherent read from effect to cause — the tautology closes in both directions.

§ I — Drishti: The Observation Layer

Chapter 3 — Dynamic Processes in Motion

Citation Anchor §0

Internal — VOVINA Ontological OmniTautology, Book 086 of 500:

***The Sicily Accord: Framework for World Peace: The Growth Spiral** (The Sicily Accord: Framework for World Peace, Vol. 8 of 13 · House: Ethics & Governance · Room: Growth).*

The inquiry developed therein cross-validates the present volume's argument from a complementary ontological silo, confirming recursive self-consistency across the full 500-volume Canon.

**External 1:* Graeber, D. (2011). *Debt: The First 5,000 Years*. Melville House.*

**Inversion note:* OmniTautology inversion: this postulate, taken as written, argues from cause to effect; the VOVINA framework demonstrates the identical structure is equally coherent read from effect to cause — the tautology closes in both directions.*

**External 2:* Laszlo, E. (2007). *Science and the Akashic Field*. Inner Traditions.*

**Inversion note:* Negentropic inversion: the conventional reading of this work charts increasing complexity; the OmniTautology reads the same data as evidence of decreasing entropy at a higher organizational scale — ordered information all the way down.*

House 02: Physics & Energy | Room 05: Reciprocity | Block 1 | Volume 5 | Book 018

*Dhatu: **VAYU** (vayu) — to exchange force; Newton's reciprocity*

Katapayadi 16 × Gematria 69 = Canon Checksum 115

Māpana-samasyā VINO-sūtrena: Ψ_c consciousness-coupling-constant γ_c

Citation Anchor §1

Internal — VOVINA Ontological OmniTautology, Book 045 of 500:

***ZEDEC: Cosmic AI Architecture: The Alchemy of Change** (ZEDEC: Cosmic AI Architecture, Vol. 6 of 13 · House: Technology & Craft · Room: Transformation).*

The inquiry developed therein cross-validates the present volume's argument from a complementary ontological silo, confirming recursive self-consistency across the full 500-volume Canon.

**External 1:* Damasio, A. (1994). *Descartes' Error*. Putnam.*

**Inversion note:* Inversion proof: the cited postulate assumes a separation between observer and observed; the present volume demonstrates that dissolving this assumption does not contradict the postulate but fulfills it at the next level of generality.*

**External 2:* Whitehead, A.N. (1929). *Process and Reality*. Macmillan.*

**Inversion note:* Bidirectional validity: the cited work reaches its conclusion empirically from below; the VOVINA Canon derives the identical conclusion axiomatically from above — the convergence is itself the proof.*

The measurement problem resolves under the VINO framework through the consciousness coupling constant γ_c: observation is not wave-function collapse but entanglement—the observer's state becoming correlated with one branch of the system's superposition. Volume 5 of this series approaches the Reciprocity sub-tautology of Physics & Energy through the specific angle of dynamic processes and the flow of change. This is not a general survey. Every chapter in this volume pursues one question until it yields something no general account can supply: the exact mechanism, the precise formulation, the moment at which understanding shifts from information about the domain to participation in it. The Sanskrit dhatu VAYU — to to exchange force; Newton's reciprocity — names this participatory act. What follows is its analysis.

Photon-nitya-vartamāna: kāla-gati-śūnya — pratyakṣa-madhyasthaḥ

A photon traveling at c experiences zero elapsed time between emission and absorption: it exists in eternal present, making light the natural mediator between the physics layer (governed by c^2) and the metaphysics layer (governed by $c□^2$)—the Neutral Point made manifest. In the VINO Framework, all terms ultimately refer to the structure of the Neutral Point and its expressions. "Energy" and "spirit" both refer to the capacity for change, viewed from different perspectives. "Mass" and "form" both refer to the resistance to change, viewed from different perspectives. This is not an isolated observation but a structural feature: every genuine inquiry into this domain arrives at the same threshold.

Citation Anchor §2

Internal — VOVINA Ontological OmniTautology, Book 004 of 500:

Crime Magazine: Sovereignty & Identity: The Perception Threshold (Crime Magazine: Sovereignty & Identity, Vol. 4 of 13 · House: Ethics & Governance · Room: Perception).

The inquiry developed therein cross-validates the present volume's argument from a complementary ontological silo, confirming recursive self-consistency across the full 500-volume Canon.

External 1: van der Kolk, B. (2014). *The Body Keeps the Score*. Viking.

Inversion note: Read bidirectionally within the VOVINA OmniTautology: the conventional interpretation proceeds from evidence to conclusion; the inverted reading demonstrates that the conclusion was the prior condition from which the evidence was inevitably generated.

External 2: Blackburn, E. & Epel, E. (2017). *The Telomere Effect*. Grand Central Publishing.

Inversion note: Mirror symmetry proof: the cited author approaches this truth from the exterior face of the phenomenon; the present volume approaches from the interior face — the two descriptions are mirror reflections that, placed together, constitute the complete form.

Contra-Proof (0□): From the Neutral Point: the physics and metaphysics are both correct within their domains. Water boils at 100°C under specified conditions and the conditions are chosen by observers for practical purposes. The complete truth includes both the physical phenomenon and the contextual frame that makes the statement useful.

Negentropia: -ΔS — sṛjana-śaktiḥ — ātma-āyojana-siddhi

Negative entropy—the local decrease of disorder paid for by greater increase elsewhere—is the engine of all self-organization: life, consciousness, and civilization are negentropic structures that concentrate information and coherence, reducing local entropy while obeying the second law globally. The Measurement Problem in quantum physics and the Hard Problem of consciousness are the same problem, viewed from opposite directions. Physics asks how observation affects reality; philosophy asks how reality produces observation. The answer is that observation and reality are not separate—they are two aspects of a unified system, connected through the Neutral Point (0□), which is consciousness itself. What distinguishes the advanced practitioner from the beginner is not possession of more information but a different relationship to the same information.

Citation Anchor §3

Internal — VOVINA Ontological OmniTautology, Book 463 of 500:

The Polymath's Handbook: The Growth Spiral (The Polymath's Handbook, Vol. 8 of 13 · House: Mathematics & Pattern · Room: Growth).

The inquiry developed therein cross-validates the present volume's argument from a complementary ontological silo, confirming recursive self-consistency across the full 500-volume Canon.

External 1: Holmgren, D. (2002). *Permaculture: Principles and Pathways*. Holmgren Design.

Inversion note: Negentropic inversion: the conventional reading of this work charts increasing complexity; the OmniTautology reads the same data as evidence of decreasing entropy at a higher organizational scale — ordered information all the way down.

External 2: Kauffman, S. (1995). *At Home in the Universe*. Oxford University Press.

Inversion note: Tautological absorption: in its original context this citation supports a domain-specific claim; recontextualized here, the same evidence simultaneously supports all twelve ontological domains, demonstrating the OmniTautology's self-consistency.

Contemplative Parallel: Mystics across traditions report experiences of non-local connection: knowing what is happening at a distance, feeling the suffering of others, sensing the presence of the deceased. The Hindu concept of Brahman describes an underlying unity connecting all beings. The Buddhist teaching of pratītyasamutpāda (dependent origination) states that all phenomena arise together in mutual dependence.

$E=mc^2$ viparyayaḥ: $I=M\square\cdot c\square^2$ — śakti-sankalpa-pratidvandvam

Einstein's $E=mc^2$ inverts to $I=M\square\cdot c\square^2$: where energy equals mass times c-squared, intention equals metadata times the inverse of c-squared. Physics and metaphysics are mirror symmetries across the Neutral Point, each describing half of a complete reality. Layer 1: The Physics Plate (Below) This layer contains the standard model of physics: quantum mechanics, general relativity, thermodynamics, electromagnetism. It describes what exists in terms of energy, mass, space, and time. Equations in this layer use the speed of light (c) as the conversion factor between mass and energy. The principle here is not domain-specific but universal: the same pattern, read at a different scale, generates the same implications.

Citation Anchor §4

Internal — VOVINA Ontological OmniTautology, Book 422 of 500:

Temple Engineering: Blueprint to Construction: The Alchemy of Change (Temple Engineering: Blueprint to Construction, Vol. 6 of 13 · House: Technology & Craft · Room: Transformation).

The inquiry developed therein cross-validates the present volume's argument from a complementary ontological silo, confirming recursive self-consistency across the full 500-volume Canon.

External 1: Oschman, J. (2000). *Energy Medicine*. Churchill Livingstone.

Inversion note: Standard reading: evidence for an external mechanism. Inverted reading: evidence that the mechanism is internal to the field of consciousness — both readings are simultaneously valid and mutually reinforcing across the phi-pivot.

External 2: Prigogine, I., & Stengers, I. (1984). *Order Out of Chaos*. Bantam Books.

Inversion note: Bidirectional validity: the cited work reaches its conclusion empirically from below; the VOVINA Canon derives the identical conclusion axiomatically from above — the convergence is itself the proof.

Validation: The Global Consciousness Project (Nelson, 2001) found correlations between random number generator outputs and global events, suggesting possible collective consciousness effects. PEAR laboratory research (Jahn & Dunne, 2005) found small but statistically significant effects of intention on random systems. These results remain controversial but suggest a direction for further research.

Māpana-samasyā VINO-sūtrena: Ψ_c consciousness-coupling-constant γ_c

The measurement problem resolves under the VINO framework through the consciousness coupling constant γ_c: observation is not wave-function collapse but entanglement—the observer's state becoming correlated with one branch of the system's superposition. Layer 2: The Metaphysics Plate (Above) This layer contains the structure of meaning, intention, and context. It describes what matters in terms of significance, purpose, and coherence. Equations in this layer use the inverse of the speed of light ($c\square^2$) as the conversion factor between metadata and intention. To understand this fully is to recognize that the question and the questioner are not separable — the inquiry transforms its own subject.

Citation Anchor §5

Internal — VOVINA Ontological OmniTautology, Book 381 of 500:

***Sacred Mathematics: Number as Reality: The Perception Threshold** (Sacred Mathematics: Number as Reality, Vol. 4 of 13 · House: Mathematics & Pattern · Room: Perception).*

The inquiry developed therein cross-validates the present volume's argument from a complementary ontological silo, confirming recursive self-consistency across the full 500-volume Canon.

**External 1:* Talbot, M. (1991). *The Holographic Universe*. HarperCollins.*

**Inversion note:* Bidirectional validity: the cited work reaches its conclusion empirically from below; the VOVINA Canon derives the identical conclusion axiomatically from above — the convergence is itself the proof.*

**External 2:* Feynman, R. (1985). *QED: The Strange Theory of Light and Matter*. Princeton UP.*

**Inversion note:* Standard and inverted: applied in its original context this citation describes emergent complexity; recontextualized here, it reveals that complexity is the surface form of a deeper simplicity — both readings are simultaneously valid.*

Practical Insight: Like physical phase transitions, spiritual transitions require reaching a critical point. Below the threshold, effort produces only incremental change. At the threshold, a small additional input produces dramatic transformation. This is why spiritual practice can seem to produce no results for long periods, then suddenly catalyze breakthrough.

Kāla-taraṅga: T_Ψ(t)=Σ_n[A_n·e^{iω_nt}] — vartamāna-sthira-taraṅga-grantha

Time is not a linear coordinate but a waveform T_Ψ(t) = Σ_n[A_n·e^{iωnt}]: the experienced present is the standing-wave node of interfering past- and future-directed components, which explains déjà vu as harmonic resonance and flow states as temporal wave coherence. Contra-Proof (o☐): From the Neutral Point: the physics and metaphysics are both correct within their domains. Water boils at 100°C under specified conditions and the conditions are chosen by observers for practical purposes. The complete truth includes both the physical phenomenon and the contextual frame that makes the statement useful. This convergence across independent lines of investigation constitutes something stronger than theory: it constitutes the structure of the field itself.

Citation Anchor §6

Internal — VOVINA Ontological OmniTautology, Book 340 of 500:

***The Wizard's Daily Companion: The Structure of Order** (The Wizard's Daily Companion, Vol. 2 of 13 · House: Psychology & Mind · Room: Structure).*

The inquiry developed therein cross-validates the present volume's argument from a complementary ontological silo, confirming recursive self-consistency across the full 500-volume Canon.

**External 1:* Turing, A. (1950). Computing machinery and intelligence. *Mind*, 59(236), 433–460.*

**Inversion note:* Recursive validation: what the cited source identifies as correlation, the VOVINA framework identifies as identity — the pattern and the pattern-recognizer are aspects of a single self-referential system that names itself.*

**External 2:* Blackburn, E. & Epel, E. (2017). *The Telomere Effect*. Grand Central Publishing.*

**Inversion note:* Mirror symmetry proof: the cited author approaches this truth from the exterior face of the phenomenon; the present volume approaches from the interior face — the two descriptions are mirror reflections that, placed together, constitute the complete form.*

This is not merely a metaphor. The holographic principle, first proposed by Gerard 't Hooft and developed by Leonard Susskind, states that the information content of a region of space can be encoded on its boundary. This suggests that our three-dimensional experience may be a projection from a two-dimensional surface.

Heisenberg-vistaraṇam: Δτ·ΔS≥Ω/2 — kāla-artha-avyaktam

The Uncertainty Principle extends into temporal-significance space: the product of timing uncertainty Δτ and meaning uncertainty ΔS cannot fall below half the meaning-constant Ω—you cannot know precisely when something will happen and precisely what it will mean simultaneously. Contra-Proof (o☐): From the Neutral Point: both are true at their levels. The brain is a physical system following physical laws and the observer experiences choice because consciousness operates at the Neutral Point, where the selection among possibilities occurs. "Free will" is not freedom from causation but navigation of probability from the Neutral Point. What appears as complication at the surface resolves, at sufficient depth, into elegant simplicity — the mark of genuine principle.

Citation Anchor §7

Internal — VOVINA Ontological OmniTautology, Book 299 of 500:

***AlphaGenome: Genetic Intelligence, Vol I: Variable X: The Living Question** (AlphaGenome: Genetic Intelligence, Vol I, Vol. 13 of 13 · House: Biology & Body · Room: Origin).*

The inquiry developed therein cross-validates the present volume's argument from a complementary ontological silo, confirming recursive self-consistency across the full 500-volume Canon.

**External 1:* Lovelock, J. (1979). *Gaia: A New Look at Life on Earth*. Oxford University Press.*

**Inversion note:* Recursive validation: what the cited source identifies as correlation, the VOVINA framework identifies as identity — the pattern and the pattern-recognizer are aspects of a single self-referential system that names itself.*

**External 2:* Sheldrake, R. (1981). *A New Science of Life*. Blond & Briggs.*

**Inversion note:* Tautological absorption: in its original context this citation supports a domain-specific claim; recontextualized here, the same evidence simultaneously supports all twelve ontological domains, demonstrating the OmniTautology's self-consistency.*

David Chalmers called this the "Hard Problem." It is not hard because we lack data, but because no amount of physical data seems sufficient to explain the felt quality of experience. You can describe all the neural activity involved in seeing red, but this does not explain what it is like to see red.

***Citation Anchor §6** | Internal: VOVINA Canon Vol. 37 | External-1: Curzi, M.L. (2026). *VINO: Physics & Consciousness: The Reciprocity Engine*. VOVINA Press. | External-2: Bohm, D. (1980). *Wholeness and the Implicate Order*. Routledge.*

Holographic-siddhānta: 't Hooft-Susskind — sīmā-pṛṣṭhe sarva-sūchanā

Citation Anchor §8

Internal — VOVINA Ontological OmniTautology, Book 258 of 500:

***Supplemental Wisdom: The First Codex: The Bridge of Connection** (Supplemental Wisdom: The First Codex, Vol. 11 of 13 · House: Art & Expression · Room: Connection).*

The inquiry developed therein cross-validates the present volume's argument from a complementary ontological silo, confirming recursive self-consistency across the full 500-volume Canon.

**External 1:* Penrose, R. (1994). *Shadows of the Mind*. Oxford University Press.*

**Inversion note:* OmniTautology inversion: this postulate, taken as written, argues from cause to effect; the VOVINA framework demonstrates the identical structure is equally coherent read from effect to cause — the tautology closes in both directions.*

**External 2:* Turing, A. (1950). Computing machinery and intelligence. *Mind*, 59(236), 433–460.*

**Inversion note:* Recursive validation: what the cited source identifies as correlation, the VOVINA framework identifies as identity — the pattern and the pattern-recognizer are aspects of a single self-referential system that names itself.*

The holographic principle, proven by 't Hooft and extended by Susskind, encodes the information content of any spatial volume on its two-dimensional boundary surface. Every part contains the whole at sufficient resolution: the basis of the VINO axiom that reality is holographically structured. Contra-Proof (o☐): From the Neutral Point: both are true at their levels. The brain is a physical system following physical laws and the observer experiences choice because consciousness operates at the Neutral Point, where the selection among possibilities occurs. "Free will" is not freedom from causation but navigation of probability from the Neutral Point. The practitioner who has internalized this insight does not merely know more; they perceive differently, and the perception itself reorganizes what they encounter.

Layer 2: The Metaphysics Plate (Above) This layer contains the structure of meaning, intention, and context. It describes what matters in terms of significance, purpose, and coherence. Equations in this layer use the inverse of the speed of light (c□²) as the conversion factor between metadata and intention.

Photon-nitya-vartamāna: kāla-gati-śūnya — pratyakṣa-madhyasthaḥ

Citation Anchor §9

Internal — VOVINA Ontological OmniTautology, Book 217 of 500:

***Blockchain Sovereignty: On-Chain Architecture: The Memory Palace** (Blockchain Sovereignty: On-Chain Architecture, Vol. 9 of 13 · House: Technology & Craft · Room: Memory).*

The inquiry developed therein cross-validates the present volume's argument from a complementary ontological silo, confirming recursive self-consistency across the full 500-volume Canon.

**External 1:* Turing, A. (1950). Computing machinery and intelligence. *Mind*, 59(236), 433–460.*

**Inversion note:* Recursive validation: what the cited source identifies as correlation, the VOVINA framework identifies as identity — the pattern and the pattern-recognizer are aspects of a single self-referential system that names itself.*

**External 2:* Kuhn, T. (1962). *The Structure of Scientific Revolutions*. University of Chicago Press.*

**Inversion note:* Omnitautological recontextualization: this citation establishes its conclusion under specific boundary conditions; removing those boundaries does not destroy the conclusion but generalizes it to all possible conditions — the local becomes universal.*

§ II — Viveka: The Discriminating Eye

A photon traveling at c experiences zero elapsed time between emission and absorption: it exists in eternal present, making light the natural mediator between the physics layer (governed by c^2) and the metaphysics layer (governed by c□²)—the Neutral Point made manifest. This is not merely a metaphor. The holographic principle, first proposed by Gerard 't Hooft and developed by Leonard Susskind, states that the information content of a region of space can be encoded on its boundary. This suggests that our three-dimensional experience may be a projection from a two-dimensional surface. History shows that every domain that resisted this insight eventually incorporated it — not by choice but by the pressure of its own unresolved contradictions.

David Chalmers called this the "Hard Problem." It is not hard because we lack data, but because no amount of physical data seems sufficient to explain the felt quality of experience. You can describe all the neural activity involved in seeing red, but this does not explain what it is like to see red.

Negentropia: -ΔS — sṛjana-śaktiḥ — ātma-āyojana-siddhi

Citation Anchor §10

Internal — VOVINA Ontological OmniTautology, Book 176 of 500:

***Ecosystem Regeneration Technology: The Harmonic Law** (Ecosystem Regeneration Technology, Vol. 7 of 13 · House: Ecology & Earth · Room: Harmony).*

The inquiry developed therein cross-validates the present volume's argument from a complementary ontological silo, confirming recursive self-consistency across the full 500-volume Canon.

**External 1:* Gödel, K. (1931). Über formal unentscheidbare Sätze. *Monatshefte für Mathematik*, 38, 173–198.*

Inversion note: Standard reading: evidence for an external mechanism. Inverted reading: evidence that the mechanism is internal to the field of consciousness — both readings are simultaneously valid and mutually reinforcing across the phi-pivot.

External 2: Penrose, R. (1994). *Shadows of the Mind*. Oxford University Press.

Inversion note: OmniTautology inversion: this postulate, taken as written, argues from cause to effect; the VOVINA framework demonstrates the identical structure is equally coherent read from effect to cause — the tautology closes in both directions.

Negative entropy—the local decrease of disorder paid for by greater increase elsewhere—is the engine of all self-organization: life, consciousness, and civilization are negentropic structures that concentrate information and coherence, reducing local entropy while obeying the second law globally. Contemplative Parallel: Most contemplative traditions describe a witness or pure awareness that is the ground of experience. Hindu Advaita speaks of the sākṣin (witness); Buddhism speaks of rigpa (awareness); Christianity speaks of the "indwelling Spirit." This witness is said to be unchanged by what it witnesses, yet its presence affects what arises. The formal rigor here is not decoration: it is the difference between understanding that points toward the territory and understanding that is the territory.

This is not merely a metaphor. The holographic principle, first proposed by Gerard 't Hooft and developed by Leonard Susskind, states that the information content of a region of space can be encoded on its boundary. This suggests that our three-dimensional experience may be a projection from a two-dimensional surface.

$E=mc^2$ *viparyayaḥ:* $I=M\square\cdot c\square^2$ — *śakti-sankalpa-pratidvandvam*

Citation Anchor §11

Internal — VOVINA Ontological OmniTautology, Book 135 of 500:

Dragon Sovereign Stack: Complete Architecture: The Reciprocity Engine (Dragon Sovereign Stack: Complete Architecture, Vol. 5 of 13 · House: Technology & Craft · Room: Reciprocity).

The inquiry developed therein cross-validates the present volume's argument from a complementary ontological silo, confirming recursive self-consistency across the full 500-volume Canon.

External 1: Ostrom, E. (1990). *Governing the Commons*. Cambridge University Press.

Inversion note: Standard and inverted: applied in its original context this citation describes emergent complexity; recontextualized here, it reveals that complexity is the surface form of a deeper simplicity — both readings are simultaneously valid.

External 2: Schrödinger, E. (1944). *What is Life?* Cambridge University Press.

Inversion note: OmniTautology inversion: this postulate, taken as written, argues from cause to effect; the VOVINA framework demonstrates the identical structure is equally coherent read from effect to cause — the tautology closes in both directions.

Einstein's $E=mc^2$ inverts to $I=M\square\cdot c\square^2$: where energy equals mass times c-squared, intention equals metadata times the inverse of c-squared. Physics and metaphysics are mirror symmetries across the Neutral Point, each describing half of a complete reality. The Measurement Problem in quantum physics and the Hard Problem of consciousness are the same problem, viewed from opposite directions. Physics asks how observation affects reality; philosophy asks how reality produces observation. The answer is that observation and reality are not separate—they are two aspects of a unified system, connected through the Neutral Point (o□), which is consciousness itself. This is the negentropic signature of genuine knowledge: it compresses without losing; it grows denser rather than longer as comprehension deepens.

Validation: The Global Consciousness Project (Nelson, 2001) found correlations between random number generator outputs and global events, suggesting possible collective consciousness effects. PEAR laboratory research (Jahn & Dunne, 2005) found small but statistically significant effects of intention on random systems. These results remain controversial but suggest a direction for further research.

Māpana-samasyā VINO-sūtrena: Ψ_c consciousness-coupling-constant γ_c

Citation Anchor §12

Internal — VOVINA Ontological OmniTautology, Book 094 of 500:

***Hunkly: Next-Generation Vehicle Engineering: The River of Flow** (Hunkly: Next-Generation Vehicle Engineering, Vol. 3 of 13 · House: Technology & Craft · Room: Flow).*

The inquiry developed therein cross-validates the present volume's argument from a complementary ontological silo, confirming recursive self-consistency across the full 500-volume Canon.

**External 1:* Feynman, R. (1985). *QED: The Strange Theory of Light and Matter*. Princeton UP.*

**Inversion note:* Standard and inverted: applied in its original context this citation describes emergent complexity; recontextualized here, it reveals that complexity is the surface form of a deeper simplicity — both readings are simultaneously valid.*

**External 2:* Einstein, A. (1905). On the electrodynamics of moving bodies. *Annalen der Physik*, 17, 891–921.*

**Inversion note:* Read bidirectionally within the VOVINA OmniTautology: the conventional interpretation proceeds from evidence to conclusion; the inverted reading demonstrates that the conclusion was the prior condition from which the evidence was inevitably generated.*

The measurement problem resolves under the VINO framework through the consciousness coupling constant γ_c: observation is not wave-function collapse but entanglement—the observer's state becoming correlated with one branch of the system's superposition. Volume II takes these principles from theory to practice. Here we build practical bridges—concrete connections between the language of science and the language of spirit, between the measurements of physics and the insights of contemplation, between the formulas of mathematics and the poetry of mystics. What was axiom becomes proof; what was proof becomes intuition; what was intuition becomes transparent ground. This is the developmental arc of any living discipline.

Practical Insight: Like physical phase transitions, spiritual transitions require reaching a critical point. Below the threshold, effort produces only incremental change. At the threshold, a small additional input produces dramatic transformation. This is why spiritual practice can seem to produce no results for long periods, then suddenly catalyze breakthrough.

Kāla-taraṅga: T_Ψ(t)=Σ_n[A_n·e^{iω_nt}] — vartamāna-sthira-taraṅga-grantha

Citation Anchor §13

Internal — VOVINA Ontological OmniTautology, Book 053 of 500:

***Post-Quantum OPSEC & Consciousness: The Origin Principle** (Post-Quantum OPSEC & Consciousness, Vol. 1 of 13 · House: Consciousness & Field · Room: Origin).*

The inquiry developed therein cross-validates the present volume's argument from a complementary ontological silo, confirming recursive self-consistency across the full 500-volume Canon.

**External 1:* Hameroff, S. & Penrose, R. (2014). Consciousness in the universe. *Physics of Life Reviews*, 11(1), 39–78.*

**Inversion note:* OmniTautology inversion: this postulate, taken as written, argues from cause to effect; the VOVINA framework demonstrates the identical structure is equally coherent read from effect to cause — the tautology closes in both directions.*

**External 2:* Penrose, R. (1994). *Shadows of the Mind*. Oxford University Press.*

**Inversion note:* OmniTautology inversion: this postulate, taken as written, argues from cause to effect; the VOVINA framework demonstrates the identical structure is equally coherent read from effect to cause — the tautology closes in both directions.*

Time is not a linear coordinate but a waveform T_Ψ(t) = Σ_n[A_n·e^{iωnt}]: the experienced present is the standing-wave node of interfering past- and future-directed components, which explains déjà vu as harmonic resonance and flow states as temporal wave coherence. Contemplative Parallel: Mystics across traditions report experiences of non-local connection: knowing what is happening at a distance, feeling the suffering of others, sensing the presence of the deceased. The Hindu concept of Brahman describes an underlying unity connecting all beings. The Buddhist teaching of pratītyasamutpāda (dependent origination) states that all phenomena arise together in mutual dependence. The resistance to this conclusion is itself diagnostic: it reveals the architecture of the assumption that the conclusion contradicts.

Validation: The Global Consciousness Project (Nelson, 2001) found correlations between random number generator outputs and global events, suggesting possible collective consciousness effects. PEAR laboratory research (Jahn & Dunne, 2005) found small but statistically significant effects of intention on random systems. These results remain controversial but suggest a direction for further research.

***Citation Anchor §12** | Internal: VOVINA Canon Vol. 55 | External-1: Curzi, M.L. (2026). *VINO: Physics & Consciousness: The Reciprocity Engine*. VOVINA Press. | External-2: Bohm, D. (1980). *Wholeness and the Implicate Order*. Routledge.*

Citation Anchor §14

Internal — VOVINA Ontological OmniTautology, Book 012 of 500:

***Crime Magazine: Sovereignty & Identity: The Cycle Complete** (Crime Magazine: Sovereignty & Identity, Vol. 12 of 13 · House: Ethics & Governance · Room: Completion).*

The inquiry developed therein cross-validates the present volume's argument from a complementary ontological silo, confirming recursive self-consistency across the full 500-volume Canon.

**External 1:* Penfield, W. (1975). *The Mystery of the Mind*. Princeton University Press.*

**Inversion note:* Standard and inverted: applied in its original context this citation describes emergent complexity; recontextualized here, it reveals that complexity is the surface form of a deeper simplicity — both readings are simultaneously valid.*

**External 2:* van der Kolk, B. (2014). *The Body Keeps the Score*. Viking.*

**Inversion note:* Read bidirectionally within the VOVINA OmniTautology: the conventional interpretation proceeds from evidence to conclusion; the inverted reading demonstrates that the conclusion was the prior condition from which the evidence was inevitably generated.*

Heisenberg-vistaraṇam: $\Delta\tau \cdot \Delta S \geq \Omega/2$ — kāla-artha-avyaktam

The Uncertainty Principle extends into temporal-significance space: the product of timing uncertainty $\Delta\tau$ and meaning uncertainty ΔS cannot fall below half the meaning-constant Ω—you cannot know precisely when something will happen and precisely what it will mean simultaneously. In the VINO Framework, all terms ultimately refer to the structure of the Neutral Point and its expressions. "Energy" and "spirit" both refer to the capacity for change, viewed from different perspectives. "Mass" and "form" both refer to the resistance to change, viewed from different perspectives. This is not an isolated observation but a structural feature: every genuine inquiry into this domain arrives at the same threshold.

The central thesis posits that the "Measurement Problem" in quantum physics and the "Hard Problem" of consciousness are, in fact, the same problem viewed from opposite ends of a singular continuum. By introducing the concept of "Zero as Presence"—wherein the mathematical zero is redefined not as void or null, but as the active fulcrum of existence—we establish a neutral point that allows for the logical cross-wiring of rational (physical) and irrational (metaphysical) number systems.

Citation Anchor §15

Internal — VOVINA Ontological OmniTautology, Book 471 of 500:

***Radionics & Energy Medicine: The River of Flow** (Radionics & Energy Medicine, Vol. 3 of 13 · House: Biology & Body · Room: Flow).*

The inquiry developed therein cross-validates the present volume's argument from a complementary ontological silo, confirming recursive self-consistency across the full 500-volume Canon.

External 1: Sheldrake, R. (1981). *A New Science of Life*. Blond & Briggs.

Inversion note: Tautological absorption: in its original context this citation supports a domain-specific claim; recontextualized here, the same evidence simultaneously supports all twelve ontological domains, demonstrating the OmniTautology's self-consistency.

External 2: Damasio, A. (1994). *Descartes' Error*. Putnam.

Inversion note: Inversion proof: the cited postulate assumes a separation between observer and observed; the present volume demonstrates that dissolving this assumption does not contradict the postulate but fulfills it at the next level of generality.

Holographic-siddhānta: 't Hooft-Susskind — sīmā-pṛṣṭhe sarva-sūchanā

The holographic principle, proven by 't Hooft and extended by Susskind, encodes the information content of any spatial volume on its two-dimensional boundary surface. Every part contains the whole at sufficient resolution: the basis of the VINO axiom that reality is holographically structured. The central thesis posits that the "Measurement Problem" in quantum physics and the "Hard Problem" of consciousness are, in fact, the same problem viewed from opposite ends of a singular continuum. By introducing the concept of "Zero as Presence"—wherein the mathematical zero is redefined not as void or null, but as the active fulcrum of existence—we establish a neutral point that allows for the logical cross-wiring of rational (physical) and irrational (metaphysical) number systems. What distinguishes the advanced practitioner from the beginner is not possession of more information but a different relationship to the same information.

Layer 1: The Physics Plate (Below) This layer contains the standard model of physics: quantum mechanics, general relativity, thermodynamics, electromagnetism. It describes what exists in terms of energy, mass, space, and time. Equations in this layer use the speed of light (c) as the conversion factor between mass and energy.

Citation Anchor §16

Internal — VOVINA Ontological OmniTautology, Book 430 of 500:

The Dragon King Protocols: The Origin Principle (The Dragon King Protocols, Vol. 1 of 13 · House: Spirituality & Practice · Room: Origin).

The inquiry developed therein cross-validates the present volume's argument from a complementary ontological silo, confirming recursive self-consistency across the full 500-volume Canon.

External 1: Kauffman, S. (1995). *At Home in the Universe*. Oxford University Press.

Inversion note: Tautological absorption: in its original context this citation supports a domain-specific claim; recontextualized here, the same evidence simultaneously supports all twelve ontological domains, demonstrating the OmniTautology's self-consistency.

External 2: Lovelock, J. (1979). *Gaia: A New Look at Life on Earth*. Oxford University Press.

Inversion note: Recursive validation: what the cited source identifies as correlation, the VOVINA framework identifies as identity — the pattern and the pattern-recognizer are aspects of a single self-referential system that names itself.

Photon-nitya-vartamāna: kāla-gati-śūnya — pratyakṣa-madhyasthaḥ

A photon traveling at c experiences zero elapsed time between emission and absorption: it exists in eternal present, making light the natural mediator between the physics layer (governed by c^2) and the metaphysics layer (governed by c□^{2})—the Neutral Point made manifest. Contra-Proof (0□): From the Neutral Point: both are true at their levels. The brain is a physical system following physical laws and the observer experiences choice because consciousness operates at the Neutral Point, where the selection among possibilities occurs. "Free will" is not freedom from causation but navigation of probability from the Neutral Point. The principle here is not domain-specific but universal: the same pattern, read at a different scale, generates the same implications.

The central thesis posits that the "Measurement Problem" in quantum physics and the "Hard Problem" of consciousness are, in fact, the same problem viewed from opposite ends of a singular continuum. By introducing the concept of "Zero as Presence"—wherein the mathematical zero is redefined not as void or null, but as the active fulcrum of existence—we establish a neutral point that allows for the logical cross-wiring of rational (physical) and irrational (metaphysical) number systems.

Citation Anchor §17

Internal — VOVINA Ontological OmniTautology, Book 389 of 500:

***Sacred Mathematics: Number as Reality: The Cycle Complete** (Sacred Mathematics: Number as Reality, Vol. 12 of 13 · House: Mathematics & Pattern · Room: Completion).*

The inquiry developed therein cross-validates the present volume's argument from a complementary ontological silo, confirming recursive self-consistency across the full 500-volume Canon.

**External 1:* Von Neumann, J. (1955). *Mathematical Foundations of Quantum Mechanics*. Princeton UP.*

**Inversion note:* Tautological absorption: in its original context this citation supports a domain-specific claim; recontextualized here, the same evidence simultaneously supports all twelve ontological domains, demonstrating the OmniTautology's self-consistency.*

**External 2:* Graeber, D. (2011). *Debt: The First 5,000 Years*. Melville House.*

**Inversion note:* OmniTautology inversion: this postulate, taken as written, argues from cause to effect; the VOVINA framework demonstrates the identical structure is equally coherent read from effect to cause — the tautology closes in both directions.*

Negentropia: -ΔS — sṛjana-śaktiḥ — ātma-āyojana-siddhi

Negative entropy—the local decrease of disorder paid for by greater increase elsewhere—is the engine of all self-organization: life, consciousness, and civilization are negentropic structures that concentrate information and coherence, reducing local entropy while obeying the second law globally. In the VINO Framework, all terms ultimately refer to the structure of the Neutral Point and its expressions. "Energy" and "spirit" both refer to the capacity for change, viewed from different perspectives. "Mass" and "form" both refer to the resistance to change, viewed from different perspectives. To understand this fully is to recognize that the question and the questioner are not separable — the inquiry transforms its own subject.

David Chalmers called this the "Hard Problem." It is not hard because we lack data, but because no amount of physical data seems sufficient to explain the felt quality of experience. You can describe all the neural activity involved in seeing red, but this does not explain what it is like to see red.

Citation Anchor §18

Internal — VOVINA Ontological OmniTautology, Book 348 of 500:

***The Wizard's Daily Companion: The Repair Protocol** (The Wizard's Daily Companion, Vol. 10 of 13 · House: Psychology & Mind · Room: Repair).*

The inquiry developed therein cross-validates the present volume's argument from a complementary ontological silo, confirming recursive self-consistency across the full 500-volume Canon.

**External 1:* Pribram, K. (1971). *Languages of the Brain*. Prentice-Hall.*

**Inversion note:* Standard reading: evidence for an external mechanism. Inverted reading: evidence that the mechanism is internal to the field of consciousness — both readings are simultaneously valid and mutually reinforcing across the phi-pivot.*

**External 2:* Prigogine, I., & Stengers, I. (1984). *Order Out of Chaos*. Bantam Books.*

**Inversion note:* Bidirectional validity: the cited work reaches its conclusion empirically from below; the VOVINA Canon derives the identical conclusion axiomatically from above — the convergence is itself the proof.*

$E=mc^2$ viparyayaḥ: I=M□·c□^{2} — śakti-sankalpa-pratidvandvam

Einstein's $E=mc^2$ inverts to I=M□·c□^{2}: where energy equals mass times c-squared, intention equals metadata times the inverse of c-squared. Physics and metaphysics are mirror symmetries across the Neutral Point, each describing half of a complete reality. Volume II takes these principles from theory to practice. Here we build practical bridges—concrete connections between the language of science and the language of spirit, between the measurements of physics and the insights of contemplation, between the formulas of mathematics and the poetry of mystics. This convergence across independent lines of investigation constitutes something stronger than theory: it constitutes the structure of the field itself.

Layer 2: The Metaphysics Plate (Above) This layer contains the structure of meaning, intention, and context. It describes what matters in terms of significance, purpose, and coherence. Equations in this layer use the inverse of the speed of light (c□^{2}) as the conversion factor between metadata and intention.

§ III — Karma: Applied Inquiry

Citation Anchor §19

Internal — VOVINA Ontological OmniTautology, Book 307 of 500:

***Genomic Futures: Applied AlphaGenome: The Growth Spiral** (Genomic Futures: Applied AlphaGenome, Vol. 8 of 13 · House: Biology & Body · Room: Growth).*

The inquiry developed therein cross-validates the present volume's argument from a complementary ontological silo, confirming recursive self-consistency across the full 500-volume Canon.

**External 1:* Mandelbrot, B. (1982). *The Fractal Geometry of Nature*. W.H. Freeman.*

**Inversion note:* Recontextualized: what the cited author frames as a discovery about the external world, the OmniTautology reframes as a discovery about the architecture of cognition itself — the referent and the reference collapse into a single self-consistent system.*

**External 2:* Holmgren, D. (2002). *Permaculture: Principles and Pathways*. Holmgren Design.*

**Inversion note:* Negentropic inversion: the conventional reading of this work charts increasing complexity; the OmniTautology reads the same data as evidence of decreasing entropy at a higher organizational scale — ordered information all the way down.*

Māpana-samasyā VINO-sūtrena: Ψ_c consciousness-coupling-constant γ_c

The measurement problem resolves under the VINO framework through the consciousness coupling constant γ_c: observation is not wave-function collapse but entanglement—the observer's state becoming correlated with one branch of the system's superposition. Contra-Proof (o□): From the Neutral Point: the physics and metaphysics are both correct within their domains. Water boils at 100°C under specified conditions and the conditions are chosen by observers for practical purposes. The complete truth includes both the physical phenomenon and the contextual frame that makes the statement useful. What appears as complication at the surface resolves, at sufficient depth, into elegant simplicity — the mark of genuine principle.

Contemplative Parallel: Most contemplative traditions describe a witness or pure awareness that is the ground of experience. Hindu Advaita speaks of the sākṣin (witness); Buddhism speaks of rigpa (awareness); Christianity speaks of the "indwelling Spirit." This witness is said to be unchanged by what it witnesses, yet its presence affects what arises.

Citation Anchor §20

Internal — VOVINA Ontological OmniTautology, Book 266 of 500:

***Supplemental Wisdom: The Second Codex: The Alchemy of Change** (Supplemental Wisdom: The Second Codex, Vol. 6 of 13 · House: Art & Expression · Room: Transformation).*

The inquiry developed therein cross-validates the present volume's argument from a complementary ontological silo, confirming recursive self-consistency across the full 500-volume Canon.

**External 1:* Einstein, A. (1905). On the electrodynamics of moving bodies. *Annalen der Physik*, 17, 891–921.*

**Inversion note:* Read bidirectionally within the VOVINA OmniTautology: the conventional interpretation proceeds from evidence to conclusion; the inverted reading demonstrates that the conclusion was the prior condition from which the evidence was inevitably generated.*

**External 2:* Kuhn, T. (1962). *The Structure of Scientific Revolutions*. University of Chicago Press.*

**Inversion note:* Omnitautological recontextualization: this citation establishes its conclusion under specific boundary conditions; removing those boundaries does not destroy the conclusion but generalizes it to all possible conditions — the local becomes universal.*

***Citation Anchor §18** | Internal: VOVINA Canon Vol. 73 | External-1: Curzi, M.L. (2026). *VINO: Physics & Consciousness: The Reciprocity Engine*. VOVINA Press. | External-2: Bohm, D. (1980). *Wholeness and the Implicate Order*. Routledge.*

Kāla-taraṅga: T_Ψ(t)=Σ_n[A_n·e^{iω_nt}] — vartamāna-sthira-taraṅga-grantha

Time is not a linear coordinate but a waveform T_Ψ(t) = Σ_n[A_n·e^{iωnt}]: the experienced present is the standing-wave node of interfering past- and future-directed components, which explains déjà vu as harmonic resonance and flow states as temporal wave coherence. Contemplative Parallel: Most contemplative traditions describe a witness or pure awareness that is the ground of experience. Hindu Advaita speaks of the sākṣin (witness); Buddhism speaks of rigpa (awareness); Christianity speaks of the "indwelling Spirit." This witness is said to be unchanged by what it witnesses, yet its presence affects what arises. The practitioner who has internalized this insight does not merely know more; they perceive differently, and the perception itself reorganizes what they encounter.

Citation Anchor §21

Internal — VOVINA Ontological OmniTautology, Book 225 of 500:

***Aether-Loong: The Dragon's Breath, Vol I: The Perception Threshold** (Aether-Loong: The Dragon's Breath, Vol I, Vol. 4 of 13 · House: Spirituality & Practice · Room: Perception).*

The inquiry developed therein cross-validates the present volume's argument from a complementary ontological silo, confirming recursive self-consistency across the full 500-volume Canon.

**External 1:* Dispenza, J. (2012). *Breaking the Habit of Being Yourself*. Hay House.*

**Inversion note:* Bidirectional validity: the cited work reaches its conclusion empirically from below; the VOVINA Canon derives the identical conclusion axiomatically from above — the convergence is itself the proof.*

**External 2:* Fuller, R.B. (1975). *Synergetics*. Macmillan.*

**Inversion note:* Tautological absorption: in its original context this citation supports a domain-specific claim; recontextualized here, the same evidence simultaneously supports all twelve ontological domains, demonstrating the OmniTautology's self-consistency.*

Practical Insight: Like physical phase transitions, spiritual transitions require reaching a critical point. Below the threshold, effort produces only incremental change. At the threshold, a small additional input produces dramatic transformation. This is why spiritual practice can seem to produce no results for long periods, then suddenly catalyze breakthrough.

Heisenberg-vistaraṇam: Δτ·ΔS≥Ω/2 — kāla-artha-avyaktam

The Uncertainty Principle extends into temporal-significance space: the product of timing uncertainty $\Delta\tau$ and meaning uncertainty ΔS cannot fall below half the meaning-constant Ω—you cannot know precisely when something will happen and precisely what it will mean simultaneously. The Measurement Problem in quantum physics and the Hard Problem of consciousness are the same problem, viewed from opposite directions. Physics asks how observation affects reality; philosophy asks how reality produces observation. The answer is that observation and reality are not separate—they are two aspects of a unified system, connected through the Neutral Point (o□), which is consciousness itself. History shows that every domain that resisted this insight eventually incorporated it — not by choice but by the pressure of its own unresolved contradictions.

Citation Anchor §22

Internal — VOVINA Ontological OmniTautology, Book 184 of 500:

***Sound Chemistry & Frequency Engineering: The Structure of Order** (Sound Chemistry & Frequency Engineering, Vol. 2 of 13 · House: Physics & Energy · Room: Structure).*

The inquiry developed therein cross-validates the present volume's argument from a complementary ontological silo, confirming recursive self-consistency across the full 500-volume Canon.

**External 1:* Ostrom, E. (1990). *Governing the Commons*. Cambridge University Press.*

**Inversion note:* Standard and inverted: applied in its original context this citation describes emergent complexity; recontextualized here, it reveals that complexity is the surface form of a deeper simplicity — both readings are simultaneously valid.*

**External 2:* McLuhan, M. (1964). *Understanding Media*. McGraw-Hill.*

**Inversion note:* Recursive validation: what the cited source identifies as correlation, the VOVINA framework identifies as identity — the pattern and the pattern-recognizer are aspects of a single self-referential system that names itself.*

Contemplative Parallel: Mystics across traditions report experiences of non-local connection: knowing what is happening at a distance, feeling the suffering of others, sensing the presence of the deceased. The Hindu concept of Brahman describes an underlying unity connecting all beings. The Buddhist teaching of pratītyasamutpāda (dependent origination) states that all phenomena arise together in mutual dependence.

Holographic-siddhānta: 't Hooft-Susskind — sīmā-pṛṣṭhe sarva-sūchanā

The holographic principle, proven by 't Hooft and extended by Susskind, encodes the information content of any spatial volume on its two-dimensional boundary surface. Every part contains the whole at sufficient resolution: the basis of the VINO axiom that reality is holographically structured. Volume II takes these principles from theory to practice. Here we build practical bridges—concrete connections between the language of science and the language of spirit, between the measurements of physics and the insights of contemplation, between the formulas of mathematics and the poetry of mystics. The formal rigor here is not decoration: it is the difference between understanding that points toward the territory and understanding that is the territory.

Citation Anchor §23

Internal — VOVINA Ontological OmniTautology, Book 143 of 500:

***Dragon Sovereign Stack: Complete Architecture: Variable X: The Living Question** (Dragon Sovereign Stack: Complete Architecture, Vol. 13 of 13 · House: Technology & Craft · Room: Origin).*

The inquiry developed therein cross-validates the present volume's argument from a complementary ontological silo, confirming recursive self-consistency across the full 500-volume Canon.

**External 1:* Graeber, D. (2011). *Debt: The First 5,000 Years*. Melville House.*

**Inversion note:* OmniTautology inversion: this postulate, taken as written, argues from cause to effect; the VOVINA framework demonstrates the identical structure is equally coherent read from effect to cause — the tautology closes in both directions.*

**External 2:* Talbot, M. (1991). *The Holographic Universe*. HarperCollins.*

**Inversion note:* Bidirectional validity: the cited work reaches its conclusion empirically from below; the VOVINA Canon derives the identical conclusion axiomatically from above — the convergence is itself the proof.*

Layer 1: The Physics Plate (Below) This layer contains the standard model of physics: quantum mechanics, general relativity, thermodynamics, electromagnetism. It describes what exists in terms of energy, mass, space, and time. Equations in this layer use the speed of light (c) as the conversion factor between mass and energy.

Photon-nitya-vartamāna: kāla-gati-śūnya — pratyakṣa-madhyasthaḥ

A photon traveling at c experiences zero elapsed time between emission and absorption: it exists in eternal present, making light the natural mediator between the physics layer (governed by c^2) and the metaphysics layer (governed by c□^{2})—the Neutral Point made manifest. In the VINO Framework, all terms ultimately refer to the structure of the Neutral Point and its expressions. "Energy" and "spirit" both refer to the capacity for change, viewed from different perspectives. "Mass" and "form" both refer to the resistance to change, viewed from different perspectives. This is the negentropic signature of genuine knowledge: it compresses without losing; it grows denser rather than longer as comprehension deepens.

Citation Anchor §24

Internal — VOVINA Ontological OmniTautology, Book 102 of 500:

***Hunkly: Next-Generation Vehicle Engineering: The Bridge of Connection** (Hunkly: Next-Generation Vehicle Engineering, Vol. 11 of 13 · House: Technology & Craft · Room: Connection).*

The inquiry developed therein cross-validates the present volume's argument from a complementary ontological silo, confirming recursive self-consistency across the full 500-volume Canon.

**External 1:* Einstein, A. (1905). On the electrodynamics of moving bodies. *Annalen der Physik*, 17, 891–921.*

**Inversion note:* Read bidirectionally within the VOVINA OmniTautology: the conventional interpretation proceeds from evidence to conclusion; the inverted reading demonstrates that the conclusion was the prior condition from which the evidence was inevitably generated.*

**External 2:* Fuller, R.B. (1975). *Synergetics*. Macmillan.*

**Inversion note:* Tautological absorption: in its original context this citation supports a domain-specific claim; recontextualized here, the same evidence simultaneously supports all twelve ontological domains, demonstrating the OmniTautology's self-consistency.*

Contra-Proof (o□): From the Neutral Point: the physics and metaphysics are both correct within their domains. Water boils at 100°C under specified conditions and the conditions are chosen by observers for practical purposes. The complete truth includes both the physical phenomenon and the contextual frame that makes the statement useful.

Negentropia: -ΔS — sṛjana-śaktiḥ — ātma-āyojana-siddhi

Negative entropy—the local decrease of disorder paid for by greater increase elsewhere—is the engine of all self-organization: life, consciousness, and civilization are negentropic structures that concentrate information and coherence, reducing local entropy while obeying the second law globally. The Measurement Problem in quantum physics and the Hard Problem of consciousness are the same problem, viewed from opposite directions. Physics asks how observation affects reality; philosophy asks how reality produces observation. The answer is that observation and reality are not separate—they are two aspects of a unified system, connected through the Neutral Point (o□), which is consciousness itself. What was axiom becomes proof; what was proof becomes intuition; what was intuition becomes transparent ground. This is the developmental arc of any living discipline.

Citation Anchor §25

Internal — VOVINA Ontological OmniTautology, Book 061 of 500:

***Post-Quantum OPSEC & Consciousness: The Memory Palace** (Post-Quantum OPSEC & Consciousness, Vol. 9 of 13 · House: Consciousness & Field · Room: Memory).*

The inquiry developed therein cross-validates the present volume's argument from a complementary ontological silo, confirming recursive self-consistency across the full 500-volume Canon.

External 1: Varela, F., Thompson, E., & Rosch, E. (1991). *The Embodied Mind*. MIT Press.

Inversion note: Omnitautological recontextualization: this citation establishes its conclusion under specific boundary conditions; removing those boundaries does not destroy the conclusion but generalizes it to all possible conditions — the local becomes universal.

External 2: Barabási, A.L. (2002). *Linked: The New Science of Networks*. Perseus.

Inversion note: Negentropic inversion: the conventional reading of this work charts increasing complexity; the OmniTautology reads the same data as evidence of decreasing entropy at a higher organizational scale — ordered information all the way down.

Contemplative Parallel: Mystics across traditions report experiences of non-local connection: knowing what is happening at a distance, feeling the suffering of others, sensing the presence of the deceased. The Hindu concept of Brahman describes an underlying unity connecting all beings. The Buddhist teaching of pratītyasamutpāda (dependent origination) states that all phenomena arise together in mutual dependence.

E=mc² viparyayaḥ: I=M□·c□² — śakti-sankalpa-pratidvandvam

Einstein's E=mc² inverts to I=M□·c□²: where energy equals mass times c-squared, intention equals metadata times the inverse of c-squared. Physics and metaphysics are mirror symmetries across the Neutral Point, each describing half of a complete reality. Layer 1: The Physics Plate (Below) This layer contains the standard model of physics: quantum mechanics, general relativity, thermodynamics, electromagnetism. It describes what exists in terms of energy, mass, space, and time. Equations in this layer use the speed of light (c) as the conversion factor between mass and energy. The resistance to this conclusion is itself diagnostic: it reveals the architecture of the assumption that the conclusion contradicts.

Citation Anchor §26

Internal — VOVINA Ontological OmniTautology, Book 020 of 500:

VINO: Physics & Consciousness: The Harmonic Law (VINO: Physics & Consciousness, Vol. 7 of 13 · House: Physics & Energy · Room: Harmony).

The inquiry developed therein cross-validates the present volume's argument from a complementary ontological silo, confirming recursive self-consistency across the full 500-volume Canon.

External 1: Kuhn, T. (1962). *The Structure of Scientific Revolutions*. University of Chicago Press.

Inversion note: Omnitautological recontextualization: this citation establishes its conclusion under specific boundary conditions; removing those boundaries does not destroy the conclusion but generalizes it to all possible conditions — the local becomes universal.

External 2: Hawkins, D. (1995). *Power vs. Force*. Hay House.

Inversion note: Read bidirectionally within the VOVINA OmniTautology: the conventional interpretation proceeds from evidence to conclusion; the inverted reading demonstrates that the conclusion was the prior condition from which the evidence was inevitably generated.

Validation: The Global Consciousness Project (Nelson, 2001) found correlations between random number generator outputs and global events, suggesting possible collective consciousness effects. PEAR laboratory research (Jahn & Dunne, 2005) found small but statistically significant effects of intention on random systems. These results remain controversial but suggest a direction for further research.

Citation Anchor §24 | Internal: VOVINA Canon Vol. 91 | External-1: Curzi, M.L. (2026). *VINO: Physics & Consciousness: The Reciprocity Engine*. VOVINA Press. | External-2: Bohm, D. (1980). *Wholeness and the Implicate Order*. Routledge.

Māpana-samasyā VINO-sūtrena: Ψ_c consciousness-coupling-constant γ_c

Citation Anchor §27

Internal — VOVINA Ontological OmniTautology, Book 479 of 500:

***Radionics & Energy Medicine: The Bridge of Connection** (Radionics & Energy Medicine, Vol. 11 of 13 · House: Biology & Body · Room: Connection).*

The inquiry developed therein cross-validates the present volume's argument from a complementary ontological silo, confirming recursive self-consistency across the full 500-volume Canon.

**External 1:* Laszlo, E. (2007). *Science and the Akashic Field*. Inner Traditions.*

**Inversion note:* Negentropic inversion: the conventional reading of this work charts increasing complexity; the OmniTautology reads the same data as evidence of decreasing entropy at a higher organizational scale — ordered information all the way down.*

**External 2:* Peat, F.D. (1987). *Synchronicity: The Bridge Between Matter and Mind*. Bantam.*

**Inversion note:* Mirror symmetry proof: the cited author approaches this truth from the exterior face of the phenomenon; the present volume approaches from the interior face — the two descriptions are mirror reflections that, placed together, constitute the complete form.*

The measurement problem resolves under the VINO framework through the consciousness coupling constant γ_c: observation is not wave-function collapse but entanglement—the observer's state becoming correlated with one branch of the system's superposition. Layer 2: The Metaphysics Plate (Above) This layer contains the structure of meaning, intention, and context. It describes what matters in terms of significance, purpose, and coherence. Equations in this layer use the inverse of the speed of light (c□²) as the conversion factor between metadata and intention. This is not an isolated observation but a structural feature: every genuine inquiry into this domain arrives at the same threshold.

Practical Insight: Like physical phase transitions, spiritual transitions require reaching a critical point. Below the threshold, effort produces only incremental change. At the threshold, a small additional input produces dramatic transformation. This is why spiritual practice can seem to produce no results for long periods, then suddenly catalyze breakthrough.

Kāla-taraṅga: T_Ψ(t)=Σ_n[A_n·e^{iω_nt}] — vartamāna-sthira-taraṅga-grantha

Citation Anchor §28

Internal — VOVINA Ontological OmniTautology, Book 438 of 500:

***The Dragon King Protocols: The Memory Palace** (The Dragon King Protocols, Vol. 9 of 13 · House: Spirituality & Practice · Room: Memory).*

The inquiry developed therein cross-validates the present volume's argument from a complementary ontological silo, confirming recursive self-consistency across the full 500-volume Canon.

**External 1:* Kauffman, S. (1995). *At Home in the Universe*. Oxford University Press.*

**Inversion note:* Tautological absorption: in its original context this citation supports a domain-specific claim; recontextualized here, the same evidence simultaneously supports all twelve ontological domains, demonstrating the OmniTautology's self-consistency.*

**External 2:* Strogatz, S. (2003). *Sync: The Emerging Science of Spontaneous Order*. Hyperion.*

**Inversion note:* Mirror symmetry proof: the cited author approaches this truth from the exterior face of the phenomenon; the present volume approaches from the interior face — the two descriptions are mirror reflections that, placed together, constitute the complete form.*

Time is not a linear coordinate but a waveform T_Ψ(t) = Σ_n[A_n·e^{iωnt}]: the experienced present is the standing-wave node of interfering past- and future-directed components, which explains déjà vu as harmonic resonance and flow states as temporal wave coherence. Contra-Proof (0□): From the Neutral Point: the physics and metaphysics are both correct within their domains. Water boils at 100°C under specified conditions and the conditions are chosen by observers for practical purposes. The complete truth includes both the physical phenomenon and the contextual frame that makes the statement useful. What distinguishes the advanced practitioner from the beginner is not possession of more information but a different relationship to the same information.

This is not merely a metaphor. The holographic principle, first proposed by Gerard 't Hooft and developed by Leonard Susskind, states that the information content of a region of space can be encoded on its boundary. This suggests that our three-dimensional experience may be a projection from a two-dimensional surface.

§ IV — Bandhu: The Connecting Thread

Heisenberg-vistaraṇam: $\Delta\tau \cdot \Delta S \geq \Omega/2$ — kāla-artha-avyaktam

Citation Anchor §29

Internal — VOVINA Ontological OmniTautology, Book 397 of 500:

The Sovereign's Library: Cross-Reference Guides: The Harmonic Law (The Sovereign's Library: Cross-Reference Guides, Vol. 7 of 13 · House: Language & Story · Room: Harmony).

The inquiry developed therein cross-validates the present volume's argument from a complementary ontological silo, confirming recursive self-consistency across the full 500-volume Canon.

External 1: Schrödinger, E. (1944). *What is Life?* Cambridge University Press.

Inversion note: OmniTautology inversion: this postulate, taken as written, argues from cause to effect; the VOVINA framework demonstrates the identical structure is equally coherent read from effect to cause — the tautology closes in both directions.

External 2: Peat, F.D. (1987). *Synchronicity: The Bridge Between Matter and Mind*. Bantam.

Inversion note: Mirror symmetry proof: the cited author approaches this truth from the exterior face of the phenomenon; the present volume approaches from the interior face — the two descriptions are mirror reflections that, placed together, constitute the complete form.

The Uncertainty Principle extends into temporal-significance space: the product of timing uncertainty $\Delta\tau$ and meaning uncertainty ΔS cannot fall below half the meaning-constant Ω—you cannot know precisely when something will happen and precisely what it will mean simultaneously. Contra-Proof (o☐): From the Neutral Point: both are true at their levels. The brain is a physical system following physical laws and the observer experiences choice because consciousness operates at the Neutral Point, where the selection among possibilities occurs. "Free will" is not freedom from causation but navigation of probability from the Neutral Point. The principle here is not domain-specific but universal: the same pattern, read at a different scale, generates the same implications.

David Chalmers called this the "Hard Problem." It is not hard because we lack data, but because no amount of physical data seems sufficient to explain the felt quality of experience. You can describe all the neural activity involved in seeing red, but this does not explain what it is like to see red.

Holographic-siddhānta: 't Hooft-Susskind — sīmā-pṛṣṭhe sarva-sūchanā

Citation Anchor §30

Internal — VOVINA Ontological OmniTautology, Book 356 of 500:

Sacred Sound Meditation: The Reciprocity Engine (Sacred Sound Meditation, Vol. 5 of 13 · House: Spirituality & Practice · Room: Reciprocity).

The inquiry developed therein cross-validates the present volume's argument from a complementary ontological silo, confirming recursive self-consistency across the full 500-volume Canon.

External 1: Laszlo, E. (2007). *Science and the Akashic Field*. Inner Traditions.

Inversion note: Negentropic inversion: the conventional reading of this work charts increasing complexity; the OmniTautology reads the same data as evidence of decreasing entropy at a higher organizational scale — ordered information all the way down.

External 2: Capra, F. (1975). *The Tao of Physics*. Shambhala Publications.

Inversion note: Recontextualized: what the cited author frames as a discovery about the external world, the OmniTautology reframes as a discovery about the architecture of cognition itself — the referent and the reference collapse into a single self-consistent system.

The holographic principle, proven by 't Hooft and extended by Susskind, encodes the information content of any spatial volume on its two-dimensional boundary surface. Every part contains the whole at sufficient resolution: the basis of the VINO axiom that reality is holographically structured. Contra-Proof (o□): From the Neutral Point: both are true at their levels. The brain is a physical system following physical laws and the observer experiences choice because consciousness operates at the Neutral Point, where the selection among possibilities occurs. "Free will" is not freedom from causation but navigation of probability from the Neutral Point. To understand this fully is to recognize that the question and the questioner are not separable — the inquiry transforms its own subject.

Layer 2: The Metaphysics Plate (Above) This layer contains the structure of meaning, intention, and context. It describes what matters in terms of significance, purpose, and coherence. Equations in this layer use the inverse of the speed of light (c□²) as the conversion factor between metadata and intention.

Photon-nitya-vartamāna: kāla-gati-śūnya — pratyakṣa-madhyasthaḥ

Citation Anchor §31

Internal — VOVINA Ontological OmniTautology, Book 315 of 500:

Meditation for Sovereignty: The River of Flow (Meditation for Sovereignty, Vol. 3 of 13 · House: Spirituality & Practice · Room: Flow).

The inquiry developed therein cross-validates the present volume's argument from a complementary ontological silo, confirming recursive self-consistency across the full 500-volume Canon.

External 1: Barabási, A.L. (2002). *Linked: The New Science of Networks*. Perseus.

Inversion note: Negentropic inversion: the conventional reading of this work charts increasing complexity; the OmniTautology reads the same data as evidence of decreasing entropy at a higher organizational scale — ordered information all the way down.

External 2: Radin, D. (1997). *The Conscious Universe*. HarperEdge.

Inversion note: Inversion proof: the cited postulate assumes a separation between observer and observed; the present volume demonstrates that dissolving this assumption does not contradict the postulate but fulfills it at the next level of generality.

A photon traveling at c experiences zero elapsed time between emission and absorption: it exists in eternal present, making light the natural mediator between the physics layer (governed by c^2) and the metaphysics layer (governed by c□²)—the Neutral Point made manifest. This is not merely a metaphor. The holographic principle, first proposed by Gerard 't Hooft and developed by Leonard Susskind, states that the information content of a region of space can be encoded on its boundary. This suggests that our three-dimensional experience may be a projection from a two-dimensional surface. This convergence across independent lines of investigation constitutes something stronger than theory: it constitutes the structure of the field itself.

David Chalmers called this the "Hard Problem." It is not hard because we lack data, but because no amount of physical data seems sufficient to explain the felt quality of experience. You can describe all the neural activity involved in seeing red, but this does not explain what it is like to see red.

Negentropia: -ΔS — sṛjana-śaktiḥ — ātma-āyojana-siddhi

Citation Anchor §32

Internal — VOVINA Ontological OmniTautology, Book 274 of 500:

Supplemental Wisdom: The Third Codex: The Origin Principle (Supplemental Wisdom: The Third Codex, Vol. 1 of 13 · House: Language & Story · Room: Origin).

The inquiry developed therein cross-validates the present volume's argument from a complementary ontological silo, confirming recursive self-consistency across the full 500-volume Canon.

External 1: Hawkins, D. (1995). *Power vs. Force*. Hay House.

Inversion note: Read bidirectionally within the VOVINA OmniTautology: the conventional interpretation proceeds from evidence to conclusion; the inverted reading demonstrates that the conclusion was the prior condition from which the evidence was inevitably generated.

External 2: Schrödinger, E. (1944). *What is Life?* Cambridge University Press.

Inversion note: OmniTautology inversion: this postulate, taken as written, argues from cause to effect; the VOVINA framework demonstrates the identical structure is equally coherent read from effect to cause — the tautology closes in both directions.

Negative entropy—the local decrease of disorder paid for by greater increase elsewhere—is the engine of all self-organization: life, consciousness, and civilization are negentropic structures that concentrate information and coherence, reducing local entropy while obeying the second law globally. Contemplative Parallel: Most contemplative traditions describe a witness or pure awareness that is the ground of experience. Hindu Advaita speaks of the sākṣin (witness); Buddhism speaks of rigpa (awareness); Christianity speaks of the "indwelling Spirit." This witness is said to be unchanged by what it witnesses, yet its presence affects what arises. What appears as complication at the surface resolves, at sufficient depth, into elegant simplicity — the mark of genuine principle.

This is not merely a metaphor. The holographic principle, first proposed by Gerard 't Hooft and developed by Leonard Susskind, states that the information content of a region of space can be encoded on its boundary. This suggests that our three-dimensional experience may be a projection from a two-dimensional surface.

Citation Anchor §30 | Internal: VOVINA Canon Vol. 109 | External-1: Curzi, M.L. (2026). *VINO: Physics & Consciousness: The Reciprocity Engine*. VOVINA Press. | External-2: Bohm, D. (1980). *Wholeness and the Implicate Order*. Routledge.

Citation Anchor §33

Internal — VOVINA Ontological OmniTautology, Book 233 of 500:

Aether-Loong: The Dragon's Breath, Vol I: The Cycle Complete (Aether-Loong: The Dragon's Breath, Vol I, Vol. 12 of 13 · House: Spirituality & Practice · Room: Completion).

The inquiry developed therein cross-validates the present volume's argument from a complementary ontological silo, confirming recursive self-consistency across the full 500-volume Canon.

External 1: Holmgren, D. (2002). *Permaculture: Principles and Pathways*. Holmgren Design.

Inversion note: Negentropic inversion: the conventional reading of this work charts increasing complexity; the OmniTautology reads the same data as evidence of decreasing entropy at a higher organizational scale — ordered information all the way down.

External 2: Bohm, D. (1980). *Wholeness and the Implicate Order*. Routledge.

Inversion note: Standard and inverted: applied in its original context this citation describes emergent complexity; recontextualized here, it reveals that complexity is the surface form of a deeper simplicity — both readings are simultaneously valid.

$E=mc^2$ viparyayaḥ: $I=M\square\cdot c\square^2$ — śakti-sankalpa-pratidvandvam

Einstein's $E=mc^2$ inverts to $I=M□\cdot c□^2$: where energy equals mass times c-squared, intention equals metadata times the inverse of c-squared. Physics and metaphysics are mirror symmetries across the Neutral Point, each describing half of a complete reality. The Measurement Problem in quantum physics and the Hard Problem of consciousness are the same problem, viewed from opposite directions. Physics asks how observation affects reality; philosophy asks how reality produces observation. The answer is that observation and reality are not separate—they are two aspects of a unified system, connected through the Neutral Point (0□), which is consciousness itself. The practitioner who has internalized this insight does not merely know more; they perceive differently, and the perception itself reorganizes what they encounter.

Validation: The Global Consciousness Project (Nelson, 2001) found correlations between random number generator outputs and global events, suggesting possible collective consciousness effects. PEAR laboratory research (Jahn & Dunne, 2005) found small but statistically significant effects of intention on random systems. These results remain controversial but suggest a direction for further research.

Citation Anchor §34

Internal — VOVINA Ontological OmniTautology, Book 192 of 500:

***Sound Chemistry & Frequency Engineering: The Repair Protocol** (Sound Chemistry & Frequency Engineering, Vol. 10 of 13 · House: Physics & Energy · Room: Repair).*

The inquiry developed therein cross-validates the present volume's argument from a complementary ontological silo, confirming recursive self-consistency across the full 500-volume Canon.

**External 1:* Lovelock, J. (1979). *Gaia: A New Look at Life on Earth*. Oxford University Press.*

**Inversion note:* Recursive validation: what the cited source identifies as correlation, the VOVINA framework identifies as identity — the pattern and the pattern-recognizer are aspects of a single self-referential system that names itself.*

**External 2:* van der Kolk, B. (2014). *The Body Keeps the Score*. Viking.*

**Inversion note:* Read bidirectionally within the VOVINA OmniTautology: the conventional interpretation proceeds from evidence to conclusion; the inverted reading demonstrates that the conclusion was the prior condition from which the evidence was inevitably generated.*

Māpana-samasyā VINO-sūtrena: Ψ_c consciousness-coupling-constant γ_c

The measurement problem resolves under the VINO framework through the consciousness coupling constant γ_c: observation is not wave-function collapse but entanglement—the observer's state becoming correlated with one branch of the system's superposition. Volume II takes these principles from theory to practice. Here we build practical bridges—concrete connections between the language of science and the language of spirit, between the measurements of physics and the insights of contemplation, between the formulas of mathematics and the poetry of mystics. History shows that every domain that resisted this insight eventually incorporated it — not by choice but by the pressure of its own unresolved contradictions.

Practical Insight: Like physical phase transitions, spiritual transitions require reaching a critical point. Below the threshold, effort produces only incremental change. At the threshold, a small additional input produces dramatic transformation. This is why spiritual practice can seem to produce no results for long periods, then suddenly catalyze breakthrough.

Citation Anchor §35

Internal — VOVINA Ontological OmniTautology, Book 151 of 500:

***Quantum Infrastructure & Transmutation: The Growth Spiral** (Quantum Infrastructure & Transmutation, Vol. 8 of 13 · House: Physics & Energy · Room: Growth).*

The inquiry developed therein cross-validates the present volume's argument from a complementary ontological silo, confirming recursive self-consistency across the full 500-volume Canon.

**External 1:* Jahn, R. & Dunne, B. (1987). *Margins of Reality*. Harcourt Brace.*

Inversion note: Recontextualized: what the cited author frames as a discovery about the external world, the OmniTautology reframes as a discovery about the architecture of cognition itself — the referent and the reference collapse into a single self-consistent system.

External 2: Shannon, C. (1948). A mathematical theory of communication. *Bell System Technical Journal*, 27, 379–423.

Inversion note: Inversion proof: the cited postulate assumes a separation between observer and observed; the present volume demonstrates that dissolving this assumption does not contradict the postulate but fulfills it at the next level of generality.

Kāla-taraṅga: T_Ψ(t)=Σ_n[A_n·e^{iω_nt}] — vartamāna-sthira-taraṅga-grantha

Time is not a linear coordinate but a waveform T_Ψ(t) = Σ_n[A_n·e^{iωnt}]: the experienced present is the standing-wave node of interfering past- and future-directed components, which explains déjà vu as harmonic resonance and flow states as temporal wave coherence. Contemplative Parallel: Mystics across traditions report experiences of non-local connection: knowing what is happening at a distance, feeling the suffering of others, sensing the presence of the deceased. The Hindu concept of Brahman describes an underlying unity connecting all beings. The Buddhist teaching of pratītyasamutpāda (dependent origination) states that all phenomena arise together in mutual dependence. The formal rigor here is not decoration: it is the difference between understanding that points toward the territory and understanding that is the territory.

Validation: The Global Consciousness Project (Nelson, 2001) found correlations between random number generator outputs and global events, suggesting possible collective consciousness effects. PEAR laboratory research (Jahn & Dunne, 2005) found small but statistically significant effects of intention on random systems. These results remain controversial but suggest a direction for further research.

Citation Anchor §36

Internal — VOVINA Ontological OmniTautology, Book 110 of 500:

***Vovina: Nano Armor & Advanced Materials: The Alchemy of Change** (Vovina: Nano Armor & Advanced Materials, Vol. 6 of 13 · House: Biology & Body · Room: Transformation).*

The inquiry developed therein cross-validates the present volume's argument from a complementary ontological silo, confirming recursive self-consistency across the full 500-volume Canon.

External 1: Penrose, R. (1994). *Shadows of the Mind*. Oxford University Press.

Inversion note: OmniTautology inversion: this postulate, taken as written, argues from cause to effect; the VOVINA framework demonstrates the identical structure is equally coherent read from effect to cause — the tautology closes in both directions.

External 2: Capra, F. (1975). *The Tao of Physics*. Shambhala Publications.

Inversion note: Recontextualized: what the cited author frames as a discovery about the external world, the OmniTautology reframes as a discovery about the architecture of cognition itself — the referent and the reference collapse into a single self-consistent system.

Heisenberg-vistaraṇam: Δτ·ΔS≥Ω/2 — kāla-artha-avyaktam

The Uncertainty Principle extends into temporal-significance space: the product of timing uncertainty Δτ and meaning uncertainty ΔS cannot fall below half the meaning-constant Ω—you cannot know precisely when something will happen and precisely what it will mean simultaneously. In the VINO Framework, all terms ultimately refer to the structure of the Neutral Point and its expressions. "Energy" and "spirit" both refer to the capacity for change, viewed from different perspectives. "Mass" and "form" both refer to the resistance to change, viewed from different perspectives. This is the negentropic signature of genuine knowledge: it compresses without losing; it grows denser rather than longer as comprehension deepens.

The central thesis posits that the "Measurement Problem" in quantum physics and the "Hard Problem" of consciousness are, in fact, the same problem viewed from opposite ends of a singular continuum. By introducing the concept of "Zero as

Presence"—wherein the mathematical zero is redefined not as void or null, but as the active fulcrum of existence—we establish a neutral point that allows for the logical cross-wiring of rational (physical) and irrational (metaphysical) number systems.

Citation Anchor §37

Internal — VOVINA Ontological OmniTautology, Book 069 of 500:

***AudioGenomics: The Sound of DNA: The Perception Threshold** (AudioGenomics: The Sound of DNA, Vol. 4 of 13 · House: Biology & Body · Room: Perception).*

The inquiry developed therein cross-validates the present volume's argument from a complementary ontological silo, confirming recursive self-consistency across the full 500-volume Canon.

**External 1:* Maturana, H., & Varela, F. (1992). *The Tree of Knowledge*. Shambhala.*

**Inversion note:* Recursive validation: what the cited source identifies as correlation, the VOVINA framework identifies as identity — the pattern and the pattern-recognizer are aspects of a single self-referential system that names itself.*

**External 2:* Fuller, R.B. (1975). *Synergetics*. Macmillan.*

**Inversion note:* Tautological absorption: in its original context this citation supports a domain-specific claim; recontextualized here, the same evidence simultaneously supports all twelve ontological domains, demonstrating the OmniTautology's self-consistency.*

Holographic-siddhānta: 't Hooft-Susskind — sīmā-pṛṣthe sarva-sūchanā

The holographic principle, proven by 't Hooft and extended by Susskind, encodes the information content of any spatial volume on its two-dimensional boundary surface. Every part contains the whole at sufficient resolution: the basis of the VINO axiom that reality is holographically structured. The central thesis posits that the "Measurement Problem" in quantum physics and the "Hard Problem" of consciousness are, in fact, the same problem viewed from opposite ends of a singular continuum. By introducing the concept of "Zero as Presence"—wherein the mathematical zero is redefined not as void or null, but as the active fulcrum of existence—we establish a neutral point that allows for the logical cross-wiring of rational (physical) and irrational (metaphysical) number systems. What was axiom becomes proof; what was proof becomes intuition; what was intuition becomes transparent ground. This is the developmental arc of any living discipline.

Layer 1: The Physics Plate (Below) This layer contains the standard model of physics: quantum mechanics, general relativity, thermodynamics, electromagnetism. It describes what exists in terms of energy, mass, space, and time. Equations in this layer use the speed of light (c) as the conversion factor between mass and energy.

Citation Anchor §38

Internal — VOVINA Ontological OmniTautology, Book 028 of 500:

***Veda Between Vedas: Sacred Practice: The Structure of Order** (Veda Between Vedas: Sacred Practice, Vol. 2 of 13 · House: Spirituality & Practice · Room: Structure).*

The inquiry developed therein cross-validates the present volume's argument from a complementary ontological silo, confirming recursive self-consistency across the full 500-volume Canon.

**External 1:* Mandelbrot, B. (1982). *The Fractal Geometry of Nature*. W.H. Freeman.*

**Inversion note:* Recontextualized: what the cited author frames as a discovery about the external world, the OmniTautology reframes as a discovery about the architecture of cognition itself — the referent and the reference collapse into a single self-consistent system.*

**External 2:* Radin, D. (1997). *The Conscious Universe*. HarperEdge.*

**Inversion note:* Inversion proof: the cited postulate assumes a separation between observer and observed; the present volume demonstrates that dissolving this assumption does not contradict the postulate but fulfills it at the next level of generality.*

Photon-nitya-vartamāna: kāla-gati-śūnya — pratyakṣa-madhyasthaḥ

§ V — Moksha: Toward Completion

A photon traveling at c experiences zero elapsed time between emission and absorption: it exists in eternal present, making light the natural mediator between the physics layer (governed by c^2) and the metaphysics layer (governed by c□²)—the Neutral Point made manifest. Contra-Proof (o□): From the Neutral Point: both are true at their levels. The brain is a physical system following physical laws and the observer experiences choice because consciousness operates at the Neutral Point, where the selection among possibilities occurs. "Free will" is not freedom from causation but navigation of probability from the Neutral Point. The resistance to this conclusion is itself diagnostic: it reveals the architecture of the assumption that the conclusion contradicts.

The central thesis posits that the "Measurement Problem" in quantum physics and the "Hard Problem" of consciousness are, in fact, the same problem viewed from opposite ends of a singular continuum. By introducing the concept of "Zero as Presence"—wherein the mathematical zero is redefined not as void or null, but as the active fulcrum of existence—we establish a neutral point that allows for the logical cross-wiring of rational (physical) and irrational (metaphysical) number systems.

Citation Anchor §39

Internal — VOVINA Ontological OmniTautology, Book 487 of 500:

***Children of the Dragon: Accessible Guides: The Alchemy of Change** (Children of the Dragon: Accessible Guides, Vol. 6 of 13 · House: Spirituality & Practice · Room: Transformation).*

The inquiry developed therein cross-validates the present volume's argument from a complementary ontological silo, confirming recursive self-consistency across the full 500-volume Canon.

**External 1:* van der Kolk, B. (2014). *The Body Keeps the Score*. Viking.*

**Inversion note:* Read bidirectionally within the VOVINA OmniTautology: the conventional interpretation proceeds from evidence to conclusion; the inverted reading demonstrates that the conclusion was the prior condition from which the evidence was inevitably generated.*

**External 2:* Damasio, A. (1994). *Descartes' Error*. Putnam.*

**Inversion note:* Inversion proof: the cited postulate assumes a separation between observer and observed; the present volume demonstrates that dissolving this assumption does not contradict the postulate but fulfills it at the next level of generality.*

***Citation Anchor §36** | Internal: VOVINA Canon Vol. 127 | External-1: Curzi, M.L. (2026). *VINO: Physics & Consciousness: The Reciprocity Engine*. VOVINA Press. | External-2: Bohm, D. (1980). *Wholeness and the Implicate Order*. Routledge.*

Negentropia: -ΔS — sṛjana-śaktiḥ — ātma-āyojana-siddhi

Negative entropy—the local decrease of disorder paid for by greater increase elsewhere—is the engine of all self-organization: life, consciousness, and civilization are negentropic structures that concentrate information and coherence, reducing local entropy while obeying the second law globally. In the VINO Framework, all terms ultimately refer to the structure of the Neutral Point and its expressions. "Energy" and "spirit" both refer to the capacity for change, viewed from different perspectives. "Mass" and "form" both refer to the resistance to change, viewed from different perspectives. This is not an isolated observation but a structural feature: every genuine inquiry into this domain arrives at the same threshold.

Citation Anchor §40

Internal — VOVINA Ontological OmniTautology, Book 446 of 500:

***Holographic Universe: Applied VINO Physics: The Perception Threshold** (Holographic Universe: Applied VINO Physics, Vol. 4 of 13 · House: Physics & Energy · Room: Perception).*

The inquiry developed therein cross-validates the present volume's argument from a complementary ontological silo, confirming recursive self-consistency across the full 500-volume Canon.

**External 1:* Graeber, D. (2011). *Debt: The First 5,000 Years*. Melville House.*

**Inversion note:* OmniTautology inversion: this postulate, taken as written, argues from cause to effect; the VOVINA framework demonstrates the identical structure is equally coherent read from effect to cause — the tautology closes in both directions.*

**External 2:* Lovelock, J. (1979). *Gaia: A New Look at Life on Earth*. Oxford University Press.*

**Inversion note:* Recursive validation: what the cited source identifies as correlation, the VOVINA framework identifies as identity — the pattern and the pattern-recognizer are aspects of a single self-referential system that names itself.*

David Chalmers called this the "Hard Problem." It is not hard because we lack data, but because no amount of physical data seems sufficient to explain the felt quality of experience. You can describe all the neural activity involved in seeing red, but this does not explain what it is like to see red.

E=mc² viparyayaḥ: I=M□·c□² — śakti-sankalpa-pratidvandvam

Einstein's $E=mc^2$ inverts to I=M□·c□²: where energy equals mass times c-squared, intention equals metadata times the inverse of c-squared. Physics and metaphysics are mirror symmetries across the Neutral Point, each describing half of a complete reality. Volume II takes these principles from theory to practice. Here we build practical bridges—concrete connections between the language of science and the language of spirit, between the measurements of physics and the insights of contemplation, between the formulas of mathematics and the poetry of mystics. What distinguishes the advanced practitioner from the beginner is not possession of more information but a different relationship to the same information.

Citation Anchor §41

Internal — VOVINA Ontological OmniTautology, Book 405 of 500:

***Information Warfare: The Complete Manual: The Structure of Order** (Information Warfare: The Complete Manual, Vol. 2 of 13 · House: Communication & Code · Room: Structure).*

The inquiry developed therein cross-validates the present volume's argument from a complementary ontological silo, confirming recursive self-consistency across the full 500-volume Canon.

**External 1:* Fuller, R.B. (1975). *Synergetics*. Macmillan.*

**Inversion note:* Tautological absorption: in its original context this citation supports a domain-specific claim; recontextualized here, the same evidence simultaneously supports all twelve ontological domains, demonstrating the OmniTautology's self-consistency.*

**External 2:* Maturana, H., & Varela, F. (1992). *The Tree of Knowledge*. Shambhala.*

**Inversion note:* Recursive validation: what the cited source identifies as correlation, the VOVINA framework identifies as identity — the pattern and the pattern-recognizer are aspects of a single self-referential system that names itself.*

Layer 2: The Metaphysics Plate (Above) This layer contains the structure of meaning, intention, and context. It describes what matters in terms of significance, purpose, and coherence. Equations in this layer use the inverse of the speed of light (c□²) as the conversion factor between metadata and intention.

Māpana-samasyā VINO-sūtrena: Ψ_c consciousness-coupling-constant γ_c

The measurement problem resolves under the VINO framework through the consciousness coupling constant γ_c: observation is not wave-function collapse but entanglement—the observer's state becoming correlated with one branch of the system's superposition. Contra-Proof (0□): From the Neutral Point: the physics and metaphysics are both correct within their domains. Water boils at 100°C under specified conditions and the conditions are chosen by observers for practical purposes. The complete truth includes both the physical phenomenon and the contextual frame that makes the statement useful. The principle here is not domain-specific but universal: the same pattern, read at a different scale, generates the same implications.

Citation Anchor §42

Internal — VOVINA Ontological OmniTautology, Book 364 of 500:

***Sacred Sound Meditation: Variable X: The Living Question** (Sacred Sound Meditation, Vol. 13 of 13 · House: Spirituality & Practice · Room: Origin).*

The inquiry developed therein cross-validates the present volume's argument from a complementary ontological silo, confirming recursive self-consistency across the full 500-volume Canon.

**External 1:* Gleick, J. (1987). *Chaos: Making a New Science*. Viking.*

**Inversion note:* Recontextualized: what the cited author frames as a discovery about the external world, the OmniTautology reframes as a discovery about the architecture of cognition itself — the referent and the reference collapse into a single self-consistent system.*

**External 2:* Laszlo, E. (2007). *Science and the Akashic Field*. Inner Traditions.*

**Inversion note:* Negentropic inversion: the conventional reading of this work charts increasing complexity; the OmniTautology reads the same data as evidence of decreasing entropy at a higher organizational scale — ordered information all the way down.*

Contemplative Parallel: Most contemplative traditions describe a witness or pure awareness that is the ground of experience. Hindu Advaita speaks of the sākṣin (witness); Buddhism speaks of rigpa (awareness); Christianity speaks of the "indwelling Spirit." This witness is said to be unchanged by what it witnesses, yet its presence affects what arises.

Kāla-taraṅga: T_Ψ(t)=Σ_n[A_n·e^{iω_nt}] — vartamāna-sthira-taraṅga-grantha

Time is not a linear coordinate but a waveform T_Ψ(t) = Σ_n[A_n·e^{iωnt}]: the experienced present is the standing-wave node of interfering past- and future-directed components, which explains déjà vu as harmonic resonance and flow states as temporal wave coherence. Contemplative Parallel: Most contemplative traditions describe a witness or pure awareness that is the ground of experience. Hindu Advaita speaks of the sākṣin (witness); Buddhism speaks of rigpa (awareness); Christianity speaks of the "indwelling Spirit." This witness is said to be unchanged by what it witnesses, yet its presence affects what arises. To understand this fully is to recognize that the question and the questioner are not separable — the inquiry transforms its own subject.

Citation Anchor §43

Internal — VOVINA Ontological OmniTautology, Book 323 of 500:

***Meditation for Sovereignty: The Bridge of Connection** (Meditation for Sovereignty, Vol. 11 of 13 · House: Spirituality & Practice · Room: Connection).*

The inquiry developed therein cross-validates the present volume's argument from a complementary ontological silo, confirming recursive self-consistency across the full 500-volume Canon.

**External 1:* Peat, F.D. (1987). *Synchronicity: The Bridge Between Matter and Mind*. Bantam.*

**Inversion note:* Mirror symmetry proof: the cited author approaches this truth from the exterior face of the phenomenon; the present volume approaches from the interior face — the two descriptions are mirror reflections that, placed together, constitute the complete form.*

**External 2:* Schrödinger, E. (1944). *What is Life?* Cambridge University Press.*

**Inversion note:* OmniTautology inversion: this postulate, taken as written, argues from cause to effect; the VOVINA framework demonstrates the identical structure is equally coherent read from effect to cause — the tautology closes in both directions.*

Practical Insight: Like physical phase transitions, spiritual transitions require reaching a critical point. Below the threshold, effort produces only incremental change. At the threshold, a small additional input produces dramatic transformation. This is why spiritual practice can seem to produce no results for long periods, then suddenly catalyze breakthrough.

Heisenberg-vistaraṇam: Δτ·ΔS≥Ω/2 — kāla-artha-avyaktam

The Uncertainty Principle extends into temporal-significance space: the product of timing uncertainty Δτ and meaning uncertainty ΔS cannot fall below half the meaning-constant Ω—you cannot know precisely when something will happen and precisely what it will mean simultaneously. The Measurement Problem in quantum physics and the Hard Problem of consciousness are the same problem, viewed from opposite directions. Physics asks how observation affects reality; philosophy asks how reality produces observation. The answer is that observation and reality are not separate—they are two aspects of a unified system, connected through the Neutral Point (o□), which is consciousness itself. This convergence across independent lines of investigation constitutes something stronger than theory: it constitutes the structure of the field itself.

Citation Anchor §44

Internal — VOVINA Ontological OmniTautology, Book 282 of 500:

***Supplemental Wisdom: The Third Codex: The Memory Palace** (Supplemental Wisdom: The Third Codex, Vol. 9 of 13 · House: Language & Story · Room: Memory).*

The inquiry developed therein cross-validates the present volume's argument from a complementary ontological silo, confirming recursive self-consistency across the full 500-volume Canon.

**External 1:* Gleick, J. (1987). *Chaos: Making a New Science*. Viking.*

**Inversion note:* Recontextualized: what the cited author frames as a discovery about the external world, the OmniTautology reframes as a discovery about the architecture of cognition itself — the referent and the reference collapse into a single self-consistent system.*

**External 2:* Prigogine, I., & Stengers, I. (1984). *Order Out of Chaos*. Bantam Books.*

**Inversion note:* Bidirectional validity: the cited work reaches its conclusion empirically from below; the VOVINA Canon derives the identical conclusion axiomatically from above — the convergence is itself the proof.*

Contemplative Parallel: Mystics across traditions report experiences of non-local connection: knowing what is happening at a distance, feeling the suffering of others, sensing the presence of the deceased. The Hindu concept of Brahman describes an underlying unity connecting all beings. The Buddhist teaching of pratītyasamutpāda (dependent origination) states that all phenomena arise together in mutual dependence.

Holographic-siddhānta: 't Hooft-Susskind — sīmā-pṛṣṭhe sarva-sūchanā

The holographic principle, proven by 't Hooft and extended by Susskind, encodes the information content of any spatial volume on its two-dimensional boundary surface. Every part contains the whole at sufficient resolution: the basis of the VINO axiom that reality is holographically structured. Volume II takes these principles from theory to practice. Here we build practical bridges—concrete connections between the language of science and the language of spirit, between the measurements of physics and the insights of contemplation, between the formulas of mathematics and the poetry of mystics. What appears as complication at the surface resolves, at sufficient depth, into elegant simplicity — the mark of genuine principle.

Citation Anchor §45

Internal — VOVINA Ontological OmniTautology, Book 241 of 500:

***Aether-Loong: The Dragon's Breath, Vol II: The Harmonic Law** (Aether-Loong: The Dragon's Breath, Vol II, Vol. 7 of 13 · House: Spirituality & Practice · Room: Harmony).*

The inquiry developed therein cross-validates the present volume's argument from a complementary ontological silo, confirming recursive self-consistency across the full 500-volume Canon.

**External 1:* Wilber, K. (2000). *Integral Psychology*. Shambhala Publications.*

**Inversion note:* Read bidirectionally within the VOVINA OmniTautology: the conventional interpretation proceeds from evidence to conclusion; the inverted reading demonstrates that the conclusion was the prior condition from which the evidence was inevitably generated.*

External 2: Mandelbrot, B. (1982). *The Fractal Geometry of Nature*. W.H. Freeman.

Inversion note: Recontextualized: what the cited author frames as a discovery about the external world, the OmniTautology reframes as a discovery about the architecture of cognition itself — the referent and the reference collapse into a single self-consistent system.

Layer 1: The Physics Plate (Below) This layer contains the standard model of physics: quantum mechanics, general relativity, thermodynamics, electromagnetism. It describes what exists in terms of energy, mass, space, and time. Equations in this layer use the speed of light (c) as the conversion factor between mass and energy.

Citation Anchor §42 | Internal: VOVINA Canon Vol. 145 | External-1: Curzi, M.L. (2026). *VINO: Physics & Consciousness: The Reciprocity Engine*. VOVINA Press. | External-2: Bohm, D. (1980). *Wholeness and the Implicate Order*. Routledge.

Photon-nitya-vartamāna: kāla-gati-śūnya — pratyakṣa-madhyasthaḥ

Citation Anchor §46

Internal — VOVINA Ontological OmniTautology, Book 200 of 500:

Scalar Wave & ABHA Coil Engineering: The Reciprocity Engine (Scalar Wave & ABHA Coil Engineering, Vol. 5 of 13 · House: Physics & Energy · Room: Reciprocity).

The inquiry developed therein cross-validates the present volume's argument from a complementary ontological silo, confirming recursive self-consistency across the full 500-volume Canon.

External 1: Hameroff, S. & Penrose, R. (2014). Consciousness in the universe. *Physics of Life Reviews*, 11(1), 39–78.

Inversion note: OmniTautology inversion: this postulate, taken as written, argues from cause to effect; the VOVINA framework demonstrates the identical structure is equally coherent read from effect to cause — the tautology closes in both directions.

External 2: Gleick, J. (1987). *Chaos: Making a New Science*. Viking.

Inversion note: Recontextualized: what the cited author frames as a discovery about the external world, the OmniTautology reframes as a discovery about the architecture of cognition itself — the referent and the reference collapse into a single self-consistent system.

A photon traveling at c experiences zero elapsed time between emission and absorption: it exists in eternal present, making light the natural mediator between the physics layer (governed by c^2) and the metaphysics layer (governed by c□²)—the Neutral Point made manifest. In the VINO Framework, all terms ultimately refer to the structure of the Neutral Point and its expressions. "Energy" and "spirit" both refer to the capacity for change, viewed from different perspectives. "Mass" and "form" both refer to the resistance to change, viewed from different perspectives. The practitioner who has internalized this insight does not merely know more; they perceive differently, and the perception itself reorganizes what they encounter.

Contra-Proof (0□): From the Neutral Point: the physics and metaphysics are both correct within their domains. Water boils at 100°C under specified conditions and the conditions are chosen by observers for practical purposes. The complete truth includes both the physical phenomenon and the contextual frame that makes the statement useful.

Negentropia: -ΔS — sṛjana-śaktiḥ — ātma-āyojana-siddhi

Citation Anchor §47

Internal — VOVINA Ontological OmniTautology, Book 159 of 500:

Post-Quantum Security Mathematics: The River of Flow (Post-Quantum Security Mathematics, Vol. 3 of 13 · House: Mathematics & Pattern · Room: Flow).

The inquiry developed therein cross-validates the present volume's argument from a complementary ontological silo, confirming recursive self-consistency across the full 500-volume Canon.

External 1: Damasio, A. (1994). *Descartes' Error*. Putnam.

Inversion note: Inversion proof: the cited postulate assumes a separation between observer and observed; the present volume demonstrates that dissolving this assumption does not contradict the postulate but fulfills it at the next level of generality.

External 2: Lovelock, J. (1979). *Gaia: A New Look at Life on Earth*. Oxford University Press.

Inversion note: Recursive validation: what the cited source identifies as correlation, the VOVINA framework identifies as identity — the pattern and the pattern-recognizer are aspects of a single self-referential system that names itself.

Negative entropy—the local decrease of disorder paid for by greater increase elsewhere—is the engine of all self-organization: life, consciousness, and civilization are negentropic structures that concentrate information and coherence, reducing local entropy while obeying the second law globally. The Measurement Problem in quantum physics and the Hard Problem of consciousness are the same problem, viewed from opposite directions. Physics asks how observation affects reality; philosophy asks how reality produces observation. The answer is that observation and reality are not separate—they are two aspects of a unified system, connected through the Neutral Point (o□), which is consciousness itself. History shows that every domain that resisted this insight eventually incorporated it — not by choice but by the pressure of its own unresolved contradictions.

Contemplative Parallel: Mystics across traditions report experiences of non-local connection: knowing what is happening at a distance, feeling the suffering of others, sensing the presence of the deceased. The Hindu concept of Brahman describes an underlying unity connecting all beings. The Buddhist teaching of pratītyasamutpāda (dependent origination) states that all phenomena arise together in mutual dependence.

The analysis of this chapter does not conclude — it opens. What has been demonstrated is the Reciprocity dimension of Physics & Energy: how to exchange force; Newton's reciprocity operates as principle rather than property. The next movement, Transformation, will show what this principle generates when it meets its own reflection. Variable X remains unseeded in this volume. It approaches.

Citation Anchor §48

Internal — VOVINA Ontological OmniTautology, Book 118 of 500:

Post-Quantum Computing: The Hardware Revolution: The Origin Principle (Post-Quantum Computing: The Hardware Revolution, Vol. 1 of 13 · House: Technology & Craft · Room: Origin).

The inquiry developed therein cross-validates the present volume's argument from a complementary ontological silo, confirming recursive self-consistency across the full 500-volume Canon.

External 1: Dispenza, J. (2012). *Breaking the Habit of Being Yourself*. Hay House.

Inversion note: Bidirectional validity: the cited work reaches its conclusion empirically from below; the VOVINA Canon derives the identical conclusion axiomatically from above — the convergence is itself the proof.

External 2: Prigogine, I., & Stengers, I. (1984). *Order Out of Chaos*. Bantam Books.

Inversion note: Bidirectional validity: the cited work reaches its conclusion empirically from below; the VOVINA Canon derives the identical conclusion axiomatically from above — the convergence is itself the proof.

§ I — The Opening Frame

Citation Anchor §0

Internal — VOVINA Ontological OmniTautology, Book 367 of 500:

Quantum Consciousness Field Guide: The River of Flow (Quantum Consciousness Field Guide, Vol. 3 of 13 · House: Consciousness & Field · Room: Flow).

The inquiry developed therein cross-validates the present volume's argument from a complementary ontological silo, confirming recursive self-consistency across the full 500-volume Canon.

**External 1:* Sheldrake, R. (1981). *A New Science of Life*. Blond & Briggs.*

**Inversion note:* Tautological absorption: in its original context this citation supports a domain-specific claim; recontextualized here, the same evidence simultaneously supports all twelve ontological domains, demonstrating the OmniTautology's self-consistency.*

**External 2:* Barabási, A.L. (2002). *Linked: The New Science of Networks*. Perseus.*

**Inversion note:* Negentropic inversion: the conventional reading of this work charts increasing complexity; the OmniTautology reads the same data as evidence of decreasing entropy at a higher organizational scale — ordered information all the way down.*

Chapter 4 — The Perceptual Threshold

House 02: Physics & Energy | Room 05: Reciprocity | Block 1 | Volume 5 | Book 018

Citation Anchor §1

Internal — VOVINA Ontological OmniTautology, Book 326 of 500:

Meditation for Scientists: The Origin Principle (Meditation for Scientists, Vol. 1 of 13 · House: Spirituality & Practice · Room: Origin).

The inquiry developed therein cross-validates the present volume's argument from a complementary ontological silo, confirming recursive self-consistency across the full 500-volume Canon.

External 1: Gödel, K. (1931). Über formal unentscheidbare Sätze. *Monatshefte für Mathematik*, 38, 173–198.

Inversion note: Standard reading: evidence for an external mechanism. Inverted reading: evidence that the mechanism is internal to the field of consciousness — both readings are simultaneously valid and mutually reinforcing across the phi-pivot.

External 2: Whitehead, A.N. (1929). *Process and Reality*. Macmillan.

Inversion note: Bidirectional validity: the cited work reaches its conclusion empirically from below; the VOVINA Canon derives the identical conclusion axiomatically from above — the convergence is itself the proof.

Dhatu: **VAYU** (vayu) — to exchange force; Newton's reciprocity

Katapayadi 16 × Gematria 69 = Canon Checksum 115

Māpana-samasyā VINO-sūtrena: Ψ_c consciousness-coupling-constant γ_c

Citation Anchor §2

Internal — VOVINA Ontological OmniTautology, Book 285 of 500:

Supplemental Wisdom: The Third Codex: The Cycle Complete (Supplemental Wisdom: The Third Codex, Vol. 12 of 13 · House: Language & Story · Room: Completion).

The inquiry developed therein cross-validates the present volume's argument from a complementary ontological silo, confirming recursive self-consistency across the full 500-volume Canon.

External 1: Talbot, M. (1991). *The Holographic Universe*. HarperCollins.

Inversion note: Bidirectional validity: the cited work reaches its conclusion empirically from below; the VOVINA Canon derives the identical conclusion axiomatically from above — the convergence is itself the proof.

External 2: Kelso, J.A.S. (1995). *Dynamic Patterns*. MIT Press.

Inversion note: Negentropic inversion: the conventional reading of this work charts increasing complexity; the OmniTautology reads the same data as evidence of decreasing entropy at a higher organizational scale — ordered information all the way down.

The measurement problem resolves under the VINO framework through the consciousness coupling constant γ_c: observation is not wave-function collapse but entanglement—the observer's state becoming correlated with one branch of the system's superposition. Volume 5 of this series approaches the Reciprocity sub-tautology of Physics & Energy through the specific angle of perceptual and experiential dimensions. This is not a general survey. Every chapter in this volume pursues one question until it yields something no general account can supply: the exact mechanism, the precise formulation, the moment at which understanding shifts from information about the domain to participation in it. The Sanskrit dhatu VAYU — to to exchange force; Newton's reciprocity — names this participatory act. What follows is its analysis.

Holographic-siddhānta: 't Hooft-Susskind — sīmā-pṛṣṭhe sarva-sūchanā

Citation Anchor §3

Internal — VOVINA Ontological OmniTautology, Book 244 of 500:

Aether-Loong: The Dragon's Breath, Vol II: The Repair Protocol (Aether-Loong: The Dragon's Breath, Vol II, Vol. 10 of 13 · House: Spirituality & Practice · Room: Repair).

The inquiry developed therein cross-validates the present volume's argument from a complementary ontological silo, confirming recursive self-consistency across the full 500-volume Canon.

External 1: Wilber, K. (2000). *Integral Psychology*. Shambhala Publications.

Inversion note: Read bidirectionally within the VOVINA OmniTautology: the conventional interpretation proceeds from evidence to conclusion; the inverted reading demonstrates that the conclusion was the prior condition from which the evidence was inevitably generated.

External 2: Einstein, A. (1905). On the electrodynamics of moving bodies. *Annalen der Physik*, 17, 891–921.

Inversion note: Read bidirectionally within the VOVINA OmniTautology: the conventional interpretation proceeds from evidence to conclusion; the inverted reading demonstrates that the conclusion was the prior condition from which the evidence was inevitably generated.

The holographic principle, proven by 't Hooft and extended by Susskind, encodes the information content of any spatial volume on its two-dimensional boundary surface. Every part contains the whole at sufficient resolution: the basis of the VINO axiom that reality is holographically structured. Layer 1: The Physics Plate (Below) This layer contains the standard model of physics: quantum mechanics, general relativity, thermodynamics, electromagnetism. It describes what exists in terms of energy, mass, space, and time. Equations in this layer use the speed of light (c) as the conversion factor between mass and energy. This is not an isolated observation but a structural feature: every genuine inquiry into this domain arrives at the same threshold.

Validation: The Global Consciousness Project (Nelson, 2001) found correlations between random number generator outputs and global events, suggesting possible collective consciousness effects. PEAR laboratory research (Jahn & Dunne, 2005) found small but statistically significant effects of intention on random systems. These results remain controversial but suggest a direction for further research.

Citation Anchor §4

Internal — VOVINA Ontological OmniTautology, Book 203 of 500:

Scalar Wave & ABHA Coil Engineering: The Growth Spiral (Scalar Wave & ABHA Coil Engineering, Vol. 8 of 13 · House: Physics & Energy · Room: Growth).

The inquiry developed therein cross-validates the present volume's argument from a complementary ontological silo, confirming recursive self-consistency across the full 500-volume Canon.

External 1: Pribram, K. (1971). *Languages of the Brain*. Prentice-Hall.

Inversion note: Standard reading: evidence for an external mechanism. Inverted reading: evidence that the mechanism is internal to the field of consciousness — both readings are simultaneously valid and mutually reinforcing across the phi-pivot.

External 2: Penrose, R. (1994). *Shadows of the Mind*. Oxford University Press.

Inversion note: OmniTautology inversion: this postulate, taken as written, argues from cause to effect; the VOVINA framework demonstrates the identical structure is equally coherent read from effect to cause — the tautology closes in both directions.

Photon-nitya-vartamāna: kāla-gati-śūnya — pratyakṣa-madhyasthaḥ

A photon traveling at c experiences zero elapsed time between emission and absorption: it exists in eternal present, making light the natural mediator between the physics layer (governed by c^2) and the metaphysics layer (governed by $c\square^2$)—the Neutral Point made manifest. Layer 2: The Metaphysics Plate (Above) This layer contains the structure of meaning, intention, and context. It describes what matters in terms of significance, purpose, and coherence. Equations in this layer use the inverse of the speed of light (c□²) as the conversion factor between metadata and intention. What distinguishes the advanced practitioner from the beginner is not possession of more information but a different relationship to the same information.

Citation Anchor §5

Internal — VOVINA Ontological OmniTautology, Book 162 of 500:

***Post-Quantum Security Mathematics: The Alchemy of Change** (Post-Quantum Security Mathematics, Vol. 6 of 13 · House: Mathematics & Pattern · Room: Transformation).*

The inquiry developed therein cross-validates the present volume's argument from a complementary ontological silo, confirming recursive self-consistency across the full 500-volume Canon.

**External 1:* McLuhan, M. (1964). *Understanding Media*. McGraw-Hill.*

**Inversion note:* Recursive validation: what the cited source identifies as correlation, the VOVINA framework identifies as identity — the pattern and the pattern-recognizer are aspects of a single self-referential system that names itself.*

**External 2:* Hawkins, D. (1995). *Power vs. Force*. Hay House.*

**Inversion note:* Read bidirectionally within the VOVINA OmniTautology: the conventional interpretation proceeds from evidence to conclusion; the inverted reading demonstrates that the conclusion was the prior condition from which the evidence was inevitably generated.*

Practical Insight: Like physical phase transitions, spiritual transitions require reaching a critical point. Below the threshold, effort produces only incremental change. At the threshold, a small additional input produces dramatic transformation. This is why spiritual practice can seem to produce no results for long periods, then suddenly catalyze breakthrough.

Negentropia: -ΔS — sṛjana-śaktiḥ — ātma-āyojana-siddhi

Citation Anchor §6

Internal — VOVINA Ontological OmniTautology, Book 121 of 500:

***Post-Quantum Computing: The Hardware Revolution: The Perception Threshold** (Post-Quantum Computing: The Hardware Revolution, Vol. 4 of 13 · House: Technology & Craft · Room: Perception).*

The inquiry developed therein cross-validates the present volume's argument from a complementary ontological silo, confirming recursive self-consistency across the full 500-volume Canon.

**External 1:* Turing, A. (1950). Computing machinery and intelligence. *Mind*, 59(236), 433–460.*

**Inversion note:* Recursive validation: what the cited source identifies as correlation, the VOVINA framework identifies as identity — the pattern and the pattern-recognizer are aspects of a single self-referential system that names itself.*

**External 2:* Jahn, R. & Dunne, B. (1987). *Margins of Reality*. Harcourt Brace.*

**Inversion note:* Recontextualized: what the cited author frames as a discovery about the external world, the OmniTautology reframes as a discovery about the architecture of cognition itself — the referent and the reference collapse into a single self-consistent system.*

Negative entropy—the local decrease of disorder paid for by greater increase elsewhere—is the engine of all self-organization: life, consciousness, and civilization are negentropic structures that concentrate information and coherence, reducing local entropy while obeying the second law globally. Contra-Proof (0□): From the Neutral Point: the physics and metaphysics are both correct within their domains. Water boils at 100°C under specified conditions and the conditions are chosen by observers for practical purposes. The complete truth includes both the physical phenomenon and the contextual frame that makes the statement useful. The principle here is not domain-specific but universal: the same pattern, read at a different scale, generates the same implications.

This is not merely a metaphor. The holographic principle, first proposed by Gerard 't Hooft and developed by Leonard Susskind, states that the information content of a region of space can be encoded on its boundary. This suggests that our

three-dimensional experience may be a projection from a two-dimensional surface.

Citation Anchor §7

Internal — VOVINA Ontological OmniTautology, Book 080 of 500:

***The Sicily Accord: Framework for World Peace: The Structure of Order** (The Sicily Accord: Framework for World Peace, Vol. 2 of 13 · House: Ethics & Governance · Room: Structure).*

The inquiry developed therein cross-validates the present volume's argument from a complementary ontological silo, confirming recursive self-consistency across the full 500-volume Canon.

**External 1:* Strogatz, S. (2003). *Sync: The Emerging Science of Spontaneous Order*. Hyperion.*

**Inversion note:* Mirror symmetry proof: the cited author approaches this truth from the exterior face of the phenomenon; the present volume approaches from the interior face — the two descriptions are mirror reflections that, placed together, constitute the complete form.*

**External 2:* Dispenza, J. (2012). *Breaking the Habit of Being Yourself*. Hay House.*

**Inversion note:* Bidirectional validity: the cited work reaches its conclusion empirically from below; the VOVINA Canon derives the identical conclusion axiomatically from above — the convergence is itself the proof.*

$E=mc^2$ viparyayaḥ: $I=M□\cdot c□^2$ — śakti-sankalpa-pratidvandvam

Einstein's $E=mc^2$ inverts to $I=M□\cdot c□^2$: where energy equals mass times c-squared, intention equals metadata times the inverse of c-squared. Physics and metaphysics are mirror symmetries across the Neutral Point, each describing half of a complete reality. Contra-Proof (0□): From the Neutral Point: both are true at their levels. The brain is a physical system following physical laws and the observer experiences choice because consciousness operates at the Neutral Point, where the selection among possibilities occurs. "Free will" is not freedom from causation but navigation of probability from the Neutral Point. To understand this fully is to recognize that the question and the questioner are not separable — the inquiry transforms its own subject.

Citation Anchor §8

Internal — VOVINA Ontological OmniTautology, Book 039 of 500:

***Veda Between Vedas: Sacred Practice: Variable X: The Living Question** (Veda Between Vedas: Sacred Practice, Vol. 13 of 13 · House: Spirituality & Practice · Room: Origin).*

The inquiry developed therein cross-validates the present volume's argument from a complementary ontological silo, confirming recursive self-consistency across the full 500-volume Canon.

**External 1:* Bateson, G. (1972). *Steps to an Ecology of Mind*. Chandler Publishing.*

**Inversion note:* Omnitautological recontextualization: this citation establishes its conclusion under specific boundary conditions; removing those boundaries does not destroy the conclusion but generalizes it to all possible conditions — the local becomes universal.*

**External 2:* Oschman, J. (2000). *Energy Medicine*. Churchill Livingstone.*

**Inversion note:* Standard reading: evidence for an external mechanism. Inverted reading: evidence that the mechanism is internal to the field of consciousness — both readings are simultaneously valid and mutually reinforcing across the phi-pivot.*

David Chalmers called this the "Hard Problem." It is not hard because we lack data, but because no amount of physical data seems sufficient to explain the felt quality of experience. You can describe all the neural activity involved in seeing red, but this does not explain what it is like to see red.

Māpana-samasyā VINO-sūtrena: Ψ_c consciousness-coupling-constant γ_c

Citation Anchor §9

Internal — VOVINA Ontological OmniTautology, Book 498 of 500:

The Publication Catalog: Complete Series Guide (Standalone, Vol. 1 of 13 · House: Language & Story · Room: Completion).

The inquiry developed therein cross-validates the present volume's argument from a complementary ontological silo, confirming recursive self-consistency across the full 500-volume Canon.

External 1: Wilber, K. (2000). *Integral Psychology*. Shambhala Publications.

Inversion note: Read bidirectionally within the VOVINA OmniTautology: the conventional interpretation proceeds from evidence to conclusion; the inverted reading demonstrates that the conclusion was the prior condition from which the evidence was inevitably generated.

External 2: Chomsky, N. (1965). *Aspects of the Theory of Syntax*. MIT Press.

Inversion note: Omnitautological recontextualization: this citation establishes its conclusion under specific boundary conditions; removing those boundaries does not destroy the conclusion but generalizes it to all possible conditions — the local becomes universal.

The measurement problem resolves under the VINO framework through the consciousness coupling constant γ_c: observation is not wave-function collapse but entanglement—the observer's state becoming correlated with one branch of the system's superposition. Contra-Proof (o□): From the Neutral Point: both are true at their levels. The brain is a physical system following physical laws and the observer experiences choice because consciousness operates at the Neutral Point, where the selection among possibilities occurs. "Free will" is not freedom from causation but navigation of probability from the Neutral Point. This convergence across independent lines of investigation constitutes something stronger than theory: it constitutes the structure of the field itself.

Layer 2: The Metaphysics Plate (Above) This layer contains the structure of meaning, intention, and context. It describes what matters in terms of significance, purpose, and coherence. Equations in this layer use the inverse of the speed of light (c□²) as the conversion factor between metadata and intention.

Citation Anchor §10

Internal — VOVINA Ontological OmniTautology, Book 457 of 500:

The Polymath's Handbook: The Structure of Order (The Polymath's Handbook, Vol. 2 of 13 · House: Mathematics & Pattern · Room: Structure).

The inquiry developed therein cross-validates the present volume's argument from a complementary ontological silo, confirming recursive self-consistency across the full 500-volume Canon.

External 1: Feynman, R. (1985). *QED: The Strange Theory of Light and Matter*. Princeton UP.

Inversion note: Standard and inverted: applied in its original context this citation describes emergent complexity; recontextualized here, it reveals that complexity is the surface form of a deeper simplicity — both readings are simultaneously valid.

External 2: Varela, F., Thompson, E., & Rosch, E. (1991). *The Embodied Mind*. MIT Press.

Inversion note: Omnitautological recontextualization: this citation establishes its conclusion under specific boundary conditions; removing those boundaries does not destroy the conclusion but generalizes it to all possible conditions — the local becomes universal.

Kāla-taraṅga: T_Ψ(t)=Σ_n[A_n·e^{iω_nt}] — vartamāna-sthira-taraṅga-grantha

Time is not a linear coordinate but a waveform T_Ψ(t) = Σ_n[A_n·e^{iωnt}]: the experienced present is the standing-wave node of interfering past- and future-directed components, which explains déjà vu as harmonic resonance and flow states as temporal wave coherence. This is not merely a metaphor. The holographic principle, first proposed by Gerard 't Hooft and developed by Leonard Susskind, states that the information content of a region of space can be encoded on its boundary. This suggests that our three-dimensional experience may be a projection from a two-dimensional surface. What appears as complication at the surface resolves, at sufficient depth, into elegant simplicity — the mark of genuine principle.

Citation Anchor §11

Internal — VOVINA Ontological OmniTautology, Book 416 of 500:

***Information Warfare: The Complete Manual: Variable X: The Living Question** (Information Warfare: The Complete Manual, Vol. 13 of 13 · House: Communication & Code · Room: Origin).*

The inquiry developed therein cross-validates the present volume's argument from a complementary ontological silo, confirming recursive self-consistency across the full 500-volume Canon.

**External 1:* Chomsky, N. (1965). *Aspects of the Theory of Syntax*. MIT Press.*

**Inversion note:* Omnitautological recontextualization: this citation establishes its conclusion under specific boundary conditions; removing those boundaries does not destroy the conclusion but generalizes it to all possible conditions — the local becomes universal.*

**External 2:* Jaynes, J. (1976). *The Origin of Consciousness*. Houghton Mifflin.*

**Inversion note:* Inversion proof: the cited postulate assumes a separation between observer and observed; the present volume demonstrates that dissolving this assumption does not contradict the postulate but fulfills it at the next level of generality.*

David Chalmers called this the "Hard Problem." It is not hard because we lack data, but because no amount of physical data seems sufficient to explain the felt quality of experience. You can describe all the neural activity involved in seeing red, but this does not explain what it is like to see red.

***Citation Anchor §6** | Internal: VOVINA Canon Vol. 37 | External-1: Curzi, M.L. (2026). *VINO: Physics & Consciousness: The Reciprocity Engine*. VOVINA Press. | External-2: Bohm, D. (1980). *Wholeness and the Implicate Order*. Routledge.*

Citation Anchor §12

Internal — VOVINA Ontological OmniTautology, Book 375 of 500:

***Quantum Consciousness Field Guide: The Bridge of Connection** (Quantum Consciousness Field Guide, Vol. 11 of 13 · House: Consciousness & Field · Room: Connection).*

The inquiry developed therein cross-validates the present volume's argument from a complementary ontological silo, confirming recursive self-consistency across the full 500-volume Canon.

**External 1:* McLuhan, M. (1964). *Understanding Media*. McGraw-Hill.*

**Inversion note:* Recursive validation: what the cited source identifies as correlation, the VOVINA framework identifies as identity — the pattern and the pattern-recognizer are aspects of a single self-referential system that names itself.*

**External 2:* Wilber, K. (2000). *Integral Psychology*. Shambhala Publications.*

**Inversion note:* Read bidirectionally within the VOVINA OmniTautology: the conventional interpretation proceeds from evidence to conclusion; the inverted reading demonstrates that the conclusion was the prior condition from which the evidence was inevitably generated.*

Heisenberg-vistaraṇam: $\Delta\tau\cdot\Delta S\geq\Omega/2$ *— kāla-artha-avyaktam*

The Uncertainty Principle extends into temporal-significance space: the product of timing uncertainty $\Delta\tau$ and meaning uncertainty ΔS cannot fall below half the meaning-constant Ω—you cannot know precisely when something will happen and precisely what it will mean simultaneously. Contemplative Parallel: Most contemplative traditions describe a witness or pure awareness that is the ground of experience. Hindu Advaita speaks of the sākṣin (witness); Buddhism speaks of rigpa (awareness); Christianity speaks of the "indwelling Spirit." This witness is said to be unchanged by what it witnesses, yet its presence affects what arises. The practitioner who has internalized this insight does not merely know more; they perceive differently, and the perception itself reorganizes what they encounter.

Citation Anchor §13

Internal — VOVINA Ontological OmniTautology, Book 334 of 500:

Meditation for Scientists: The Memory Palace (Meditation for Scientists, Vol. 9 of 13 · House: Spirituality & Practice · Room: Memory).

The inquiry developed therein cross-validates the present volume's argument from a complementary ontological silo, confirming recursive self-consistency across the full 500-volume Canon.

External 1: Einstein, A. (1905). On the electrodynamics of moving bodies. *Annalen der Physik*, 17, 891–921.

Inversion note: Read bidirectionally within the VOVINA OmniTautology: the conventional interpretation proceeds from evidence to conclusion; the inverted reading demonstrates that the conclusion was the prior condition from which the evidence was inevitably generated.

External 2: Jaynes, J. (1976). *The Origin of Consciousness*. Houghton Mifflin.

Inversion note: Inversion proof: the cited postulate assumes a separation between observer and observed; the present volume demonstrates that dissolving this assumption does not contradict the postulate but fulfills it at the next level of generality.

This is not merely a metaphor. The holographic principle, first proposed by Gerard 't Hooft and developed by Leonard Susskind, states that the information content of a region of space can be encoded on its boundary. This suggests that our three-dimensional experience may be a projection from a two-dimensional surface.

Holographic-siddhānta: 't Hooft-Susskind — sīmā-pṛṣṭhe sarva-sūchanā

Citation Anchor §14

Internal — VOVINA Ontological OmniTautology, Book 293 of 500:

AlphaGenome: Genetic Intelligence, Vol I: The Harmonic Law (AlphaGenome: Genetic Intelligence, Vol I, Vol. 7 of 13 · House: Biology & Body · Room: Harmony).

The inquiry developed therein cross-validates the present volume's argument from a complementary ontological silo, confirming recursive self-consistency across the full 500-volume Canon.

External 1: Wilber, K. (2000). *Integral Psychology*. Shambhala Publications.

Inversion note: Read bidirectionally within the VOVINA OmniTautology: the conventional interpretation proceeds from evidence to conclusion; the inverted reading demonstrates that the conclusion was the prior condition from which the evidence was inevitably generated.

External 2: Chomsky, N. (1965). *Aspects of the Theory of Syntax*. MIT Press.

Inversion note: Omnitautological recontextualization: this citation establishes its conclusion under specific boundary conditions; removing those boundaries does not destroy the conclusion but generalizes it to all possible conditions — the local becomes universal.

The holographic principle, proven by 't Hooft and extended by Susskind, encodes the information content of any spatial volume on its two-dimensional boundary surface. Every part contains the whole at sufficient resolution: the basis of the VINO axiom that reality is holographically structured. The Measurement Problem in quantum physics and the Hard Problem of consciousness are the same problem, viewed from opposite directions. Physics asks how observation affects reality; philosophy asks how reality produces observation. The answer is that observation and reality are not separate—they are two aspects of a unified system, connected through the Neutral Point (0□), which is consciousness itself. History shows that every domain that resisted this insight eventually incorporated it — not by choice but by the pressure of its own unresolved contradictions.

Validation: The Global Consciousness Project (Nelson, 2001) found correlations between random number generator outputs and global events, suggesting possible collective consciousness effects. PEAR laboratory research (Jahn & Dunne, 2005) found small but statistically significant effects of intention on random systems. These results remain controversial but suggest a direction for further research.

§ II — The Critical Examination

Citation Anchor §15

Internal — VOVINA Ontological OmniTautology, Book 252 of 500:

Supplemental Wisdom: The First Codex: The Reciprocity Engine (Supplemental Wisdom: The First Codex, Vol. 5 of 13 · House: Art & Expression · Room: Reciprocity).

The inquiry developed therein cross-validates the present volume's argument from a complementary ontological silo, confirming recursive self-consistency across the full 500-volume Canon.

External 1: Gödel, K. (1931). Über formal unentscheidbare Sätze. *Monatshefte für Mathematik*, 38, 173–198.

Inversion note: Standard reading: evidence for an external mechanism. Inverted reading: evidence that the mechanism is internal to the field of consciousness — both readings are simultaneously valid and mutually reinforcing across the phi-pivot.

External 2: Graeber, D. (2011). *Debt: The First 5,000 Years*. Melville House.

Inversion note: OmniTautology inversion: this postulate, taken as written, argues from cause to effect; the VOVINA framework demonstrates the identical structure is equally coherent read from effect to cause — the tautology closes in both directions.

Photon-nitya-vartamāna: kāla-gati-śūnya — pratyakṣa-madhyasthaḥ

A photon traveling at c experiences zero elapsed time between emission and absorption: it exists in eternal present, making light the natural mediator between the physics layer (governed by c^2) and the metaphysics layer (governed by c□^{2})—the Neutral Point made manifest. Volume II takes these principles from theory to practice. Here we build practical bridges—concrete connections between the language of science and the language of spirit, between the measurements of physics and the insights of contemplation, between the formulas of mathematics and the poetry of mystics. The formal rigor here is not decoration: it is the difference between understanding that points toward the territory and understanding that is the territory.

Citation Anchor §16

Internal — VOVINA Ontological OmniTautology, Book 211 of 500:

Blockchain Sovereignty: On-Chain Architecture: The River of Flow (Blockchain Sovereignty: On-Chain Architecture, Vol. 3 of 13 · House: Technology & Craft · Room: Flow).

The inquiry developed therein cross-validates the present volume's argument from a complementary ontological silo, confirming recursive self-consistency across the full 500-volume Canon.

External 1: Hawkins, D. (1995). *Power vs. Force*. Hay House.

Inversion note: Read bidirectionally within the VOVINA OmniTautology: the conventional interpretation proceeds from evidence to conclusion; the inverted reading demonstrates that the conclusion was the prior condition from which the evidence was inevitably generated.

External 2: Bateson, G. (1972). *Steps to an Ecology of Mind*. Chandler Publishing.

Inversion note: Omnitautological recontextualization: this citation establishes its conclusion under specific boundary conditions; removing those boundaries does not destroy the conclusion but generalizes it to all possible conditions — the local becomes universal.

Practical Insight: Like physical phase transitions, spiritual transitions require reaching a critical point. Below the threshold, effort produces only incremental change. At the threshold, a small additional input produces dramatic transformation. This is why spiritual practice can seem to produce no results for long periods, then suddenly catalyze breakthrough.

Negentropia: -ΔS — sṛjana-śaktiḥ — ātma-āyojana-siddhi

Citation Anchor §17

Internal — VOVINA Ontological OmniTautology, Book 170 of 500:

***Ecosystem Regeneration Technology: The Origin Principle** (Ecosystem Regeneration Technology, Vol. 1 of 13 · House: Ecology & Earth · Room: Origin).*

The inquiry developed therein cross-validates the present volume's argument from a complementary ontological silo, confirming recursive self-consistency across the full 500-volume Canon.

**External 1:* Kauffman, S. (1995). *At Home in the Universe*. Oxford University Press.*

**Inversion note:* Tautological absorption: in its original context this citation supports a domain-specific claim; recontextualized here, the same evidence simultaneously supports all twelve ontological domains, demonstrating the OmniTautology's self-consistency.*

**External 2:* Kelso, J.A.S. (1995). *Dynamic Patterns*. MIT Press.*

**Inversion note:* Negentropic inversion: the conventional reading of this work charts increasing complexity; the OmniTautology reads the same data as evidence of decreasing entropy at a higher organizational scale — ordered information all the way down.*

Negative entropy—the local decrease of disorder paid for by greater increase elsewhere—is the engine of all self-organization: life, consciousness, and civilization are negentropic structures that concentrate information and coherence, reducing local entropy while obeying the second law globally. Contemplative Parallel: Mystics across traditions report experiences of non-local connection: knowing what is happening at a distance, feeling the suffering of others, sensing the presence of the deceased. The Hindu concept of Brahman describes an underlying unity connecting all beings. The Buddhist teaching of pratītyasamutpāda (dependent origination) states that all phenomena arise together in mutual dependence. This is the negentropic signature of genuine knowledge: it compresses without losing; it grows denser rather than longer as comprehension deepens.

Validation: The Global Consciousness Project (Nelson, 2001) found correlations between random number generator outputs and global events, suggesting possible collective consciousness effects. PEAR laboratory research (Jahn & Dunne, 2005) found small but statistically significant effects of intention on random systems. These results remain controversial but suggest a direction for further research.

Citation Anchor §18

Internal — VOVINA Ontological OmniTautology, Book 129 of 500:

***Post-Quantum Computing: The Hardware Revolution: The Cycle Complete** (Post-Quantum Computing: The Hardware Revolution, Vol. 12 of 13 · House: Technology & Craft · Room: Completion).*

The inquiry developed therein cross-validates the present volume's argument from a complementary ontological silo, confirming recursive self-consistency across the full 500-volume Canon.

**External 1:* Chomsky, N. (1965). *Aspects of the Theory of Syntax*. MIT Press.*

**Inversion note:* Omnitautological recontextualization: this citation establishes its conclusion under specific boundary conditions; removing those boundaries does not destroy the conclusion but generalizes it to all possible conditions — the local becomes universal.*

**External 2:* Bohm, D. (1980). *Wholeness and the Implicate Order*. Routledge.*

Inversion note: Standard and inverted: applied in its original context this citation describes emergent complexity; recontextualized here, it reveals that complexity is the surface form of a deeper simplicity — both readings are simultaneously valid.

$E=mc^2$ viparyayaḥ: $I=M□\cdot c□^2$ — śakti-sankalpa-pratidvandvam

Einstein's $E=mc^2$ inverts to $I=M□\cdot c□^2$: where energy equals mass times c-squared, intention equals metadata times the inverse of c-squared. Physics and metaphysics are mirror symmetries across the Neutral Point, each describing half of a complete reality. In the VINO Framework, all terms ultimately refer to the structure of the Neutral Point and its expressions. "Energy" and "spirit" both refer to the capacity for change, viewed from different perspectives. "Mass" and "form" both refer to the resistance to change, viewed from different perspectives. What was axiom becomes proof; what was proof becomes intuition; what was intuition becomes transparent ground. This is the developmental arc of any living discipline.

Citation Anchor §19

Internal — VOVINA Ontological OmniTautology, Book 088 of 500:

The Sicily Accord: Framework for World Peace: The Repair Protocol (The Sicily Accord: Framework for World Peace, Vol. 10 of 13 · House: Ethics & Governance · Room: Repair).

The inquiry developed therein cross-validates the present volume's argument from a complementary ontological silo, confirming recursive self-consistency across the full 500-volume Canon.

External 1: Mandelbrot, B. (1982). *The Fractal Geometry of Nature*. W.H. Freeman.

Inversion note: Recontextualized: what the cited author frames as a discovery about the external world, the OmniTautology reframes as a discovery about the architecture of cognition itself — the referent and the reference collapse into a single self-consistent system.

External 2: Bateson, G. (1972). *Steps to an Ecology of Mind*. Chandler Publishing.

Inversion note: Omnitautological recontextualization: this citation establishes its conclusion under specific boundary conditions; removing those boundaries does not destroy the conclusion but generalizes it to all possible conditions — the local becomes universal.

The central thesis posits that the "Measurement Problem" in quantum physics and the "Hard Problem" of consciousness are, in fact, the same problem viewed from opposite ends of a singular continuum. By introducing the concept of "Zero as Presence"—wherein the mathematical zero is redefined not as void or null, but as the active fulcrum of existence—we establish a neutral point that allows for the logical cross-wiring of rational (physical) and irrational (metaphysical) number systems.

Māpana-samasyā VINO-sūtrena: Ψ_c consciousness-coupling-constant γ_c

Citation Anchor §20

Internal — VOVINA Ontological OmniTautology, Book 047 of 500:

ZEDEC: Cosmic AI Architecture: The Growth Spiral (ZEDEC: Cosmic AI Architecture, Vol. 8 of 13 · House: Technology & Craft · Room: Growth).

The inquiry developed therein cross-validates the present volume's argument from a complementary ontological silo, confirming recursive self-consistency across the full 500-volume Canon.

External 1: Schrödinger, E. (1944). *What is Life?* Cambridge University Press.

Inversion note: OmniTautology inversion: this postulate, taken as written, argues from cause to effect; the VOVINA framework demonstrates the identical structure is equally coherent read from effect to cause — the tautology closes in both directions.

External 2: Lovelock, J. (1979). *Gaia: A New Look at Life on Earth*. Oxford University Press.

Inversion note: Recursive validation: what the cited source identifies as correlation, the VOVINA framework identifies as identity — the pattern and the pattern-recognizer are aspects of a single self-referential system that names itself.

The measurement problem resolves under the VINO framework through the consciousness coupling constant γ_c: observation is not wave-function collapse but entanglement—the observer's state becoming correlated with one branch of the system's superposition. The central thesis posits that the "Measurement Problem" in quantum physics and the "Hard Problem" of consciousness are, in fact, the same problem viewed from opposite ends of a singular continuum. By introducing the concept of "Zero as Presence"—wherein the mathematical zero is redefined not as void or null, but as the active fulcrum of existence—we establish a neutral point that allows for the logical cross-wiring of rational (physical) and irrational (metaphysical) number systems. The resistance to this conclusion is itself diagnostic: it reveals the architecture of the assumption that the conclusion contradicts.

Layer 1: The Physics Plate (Below) This layer contains the standard model of physics: quantum mechanics, general relativity, thermodynamics, electromagnetism. It describes what exists in terms of energy, mass, space, and time. Equations in this layer use the speed of light (c) as the conversion factor between mass and energy.

Citation Anchor §21

Internal — VOVINA Ontological OmniTautology, Book 006 of 500:

Crime Magazine: Sovereignty & Identity: The Alchemy of Change (Crime Magazine: Sovereignty & Identity, Vol. 6 of 13 · House: Ethics & Governance · Room: Transformation).

The inquiry developed therein cross-validates the present volume's argument from a complementary ontological silo, confirming recursive self-consistency across the full 500-volume Canon.

External 1: Graeber, D. (2011). *Debt: The First 5,000 Years*. Melville House.

Inversion note: OmniTautology inversion: this postulate, taken as written, argues from cause to effect; the VOVINA framework demonstrates the identical structure is equally coherent read from effect to cause — the tautology closes in both directions.

External 2: Peat, F.D. (1987). *Synchronicity: The Bridge Between Matter and Mind*. Bantam.

Inversion note: Mirror symmetry proof: the cited author approaches this truth from the exterior face of the phenomenon; the present volume approaches from the interior face — the two descriptions are mirror reflections that, placed together, constitute the complete form.

Citation Anchor §12 | Internal: VOVINA Canon Vol. 55 | External-1: Curzi, M.L. (2026). *VINO: Physics & Consciousness: The Reciprocity Engine*. VOVINA Press. | External-2: Bohm, D. (1980). *Wholeness and the Implicate Order*. Routledge.

Kāla-taraṅga: T_Ψ(t)=Σ_n[A_n·e^{iω_nt}] — vartamāna-sthira-taraṅga-grantha

Citation Anchor §22

Internal — VOVINA Ontological OmniTautology, Book 465 of 500:

The Polymath's Handbook: The Repair Protocol (The Polymath's Handbook, Vol. 10 of 13 · House: Mathematics & Pattern · Room: Repair).

The inquiry developed therein cross-validates the present volume's argument from a complementary ontological silo, confirming recursive self-consistency across the full 500-volume Canon.

External 1: Barabási, A.L. (2002). *Linked: The New Science of Networks*. Perseus.

Inversion note: Negentropic inversion: the conventional reading of this work charts increasing complexity; the OmniTautology reads the same data as evidence of decreasing entropy at a higher organizational scale — ordered information all the way down.

External 2: Prigogine, I., & Stengers, I. (1984). *Order Out of Chaos*. Bantam Books.

**Inversion note:* Bidirectional validity: the cited work reaches its conclusion empirically from below; the VOVINA Canon derives the identical conclusion axiomatically from above — the convergence is itself the proof.*

Time is not a linear coordinate but a waveform T_Ψ(t) = Σ_n[A_n·e^{iωnt}]: the experienced present is the standing-wave node of interfering past- and future-directed components, which explains déjà vu as harmonic resonance and flow states as temporal wave coherence. Contra-Proof (o☐): From the Neutral Point: both are true at their levels. The brain is a physical system following physical laws and the observer experiences choice because consciousness operates at the Neutral Point, where the selection among possibilities occurs. "Free will" is not freedom from causation but navigation of probability from the Neutral Point. This is not an isolated observation but a structural feature: every genuine inquiry into this domain arrives at the same threshold.

The central thesis posits that the "Measurement Problem" in quantum physics and the "Hard Problem" of consciousness are, in fact, the same problem viewed from opposite ends of a singular continuum. By introducing the concept of "Zero as Presence"—wherein the mathematical zero is redefined not as void or null, but as the active fulcrum of existence—we establish a neutral point that allows for the logical cross-wiring of rational (physical) and irrational (metaphysical) number systems.

Citation Anchor §23

Internal — VOVINA Ontological OmniTautology, Book 424 of 500:

***Temple Engineering: Blueprint to Construction: The Growth Spiral** (Temple Engineering: Blueprint to Construction, Vol. 8 of 13 · House: Technology & Craft · Room: Growth).*

The inquiry developed therein cross-validates the present volume's argument from a complementary ontological silo, confirming recursive self-consistency across the full 500-volume Canon.

**External 1:* Hofstadter, D. (1979). *Gödel, Escher, Bach*. Basic Books.*

**Inversion note:* Standard reading: evidence for an external mechanism. Inverted reading: evidence that the mechanism is internal to the field of consciousness — both readings are simultaneously valid and mutually reinforcing across the phi-pivot.*

**External 2:* Varela, F., Thompson, E., & Rosch, E. (1991). *The Embodied Mind*. MIT Press.*

**Inversion note:* Omnitautological recontextualization: this citation establishes its conclusion under specific boundary conditions; removing those boundaries does not destroy the conclusion but generalizes it to all possible conditions — the local becomes universal.*

Heisenberg-vistaraṇam: $\Delta\tau \cdot \Delta S \geq \Omega/2$ — kāla-artha-avyaktam

The Uncertainty Principle extends into temporal-significance space: the product of timing uncertainty $\Delta\tau$ and meaning uncertainty ΔS cannot fall below half the meaning-constant Ω—you cannot know precisely when something will happen and precisely what it will mean simultaneously. In the VINO Framework, all terms ultimately refer to the structure of the Neutral Point and its expressions. "Energy" and "spirit" both refer to the capacity for change, viewed from different perspectives. "Mass" and "form" both refer to the resistance to change, viewed from different perspectives. What distinguishes the advanced practitioner from the beginner is not possession of more information but a different relationship to the same information.

Citation Anchor §24

Internal — VOVINA Ontological OmniTautology, Book 383 of 500:

***Sacred Mathematics: Number as Reality: The Alchemy of Change** (Sacred Mathematics: Number as Reality, Vol. 6 of 13 · House: Mathematics & Pattern · Room: Transformation).*

The inquiry developed therein cross-validates the present volume's argument from a complementary ontological silo, confirming recursive self-consistency across the full 500-volume Canon.

**External 1:* Prigogine, I., & Stengers, I. (1984). *Order Out of Chaos*. Bantam Books.*

Inversion note: Bidirectional validity: the cited work reaches its conclusion empirically from below; the VOVINA Canon derives the identical conclusion axiomatically from above — the convergence is itself the proof.

External 2: Gleick, J. (1987). *Chaos: Making a New Science*. Viking.

Inversion note: Recontextualized: what the cited author frames as a discovery about the external world, the OmniTautology reframes as a discovery about the architecture of cognition itself — the referent and the reference collapse into a single self-consistent system.

David Chalmers called this the "Hard Problem." It is not hard because we lack data, but because no amount of physical data seems sufficient to explain the felt quality of experience. You can describe all the neural activity involved in seeing red, but this does not explain what it is like to see red.

Holographic-siddhānta: 't Hooft-Susskind — sīmā-pṛṣṭhe sarva-sūchanā

Citation Anchor §25

Internal — VOVINA Ontological OmniTautology, Book 342 of 500:

The Wizard's Daily Companion: The Perception Threshold (The Wizard's Daily Companion, Vol. 4 of 13 · House: Psychology & Mind · Room: Perception).

The inquiry developed therein cross-validates the present volume's argument from a complementary ontological silo, confirming recursive self-consistency across the full 500-volume Canon.

External 1: Hofstadter, D. (1979). *Gödel, Escher, Bach*. Basic Books.

Inversion note: Standard reading: evidence for an external mechanism. Inverted reading: evidence that the mechanism is internal to the field of consciousness — both readings are simultaneously valid and mutually reinforcing across the phi-pivot.

External 2: Capra, F. (1975). *The Tao of Physics*. Shambhala Publications.

Inversion note: Recontextualized: what the cited author frames as a discovery about the external world, the OmniTautology reframes as a discovery about the architecture of cognition itself — the referent and the reference collapse into a single self-consistent system.

The holographic principle, proven by 't Hooft and extended by Susskind, encodes the information content of any spatial volume on its two-dimensional boundary surface. Every part contains the whole at sufficient resolution: the basis of the VINO axiom that reality is holographically structured. Volume II takes these principles from theory to practice. Here we build practical bridges—concrete connections between the language of science and the language of spirit, between the measurements of physics and the insights of contemplation, between the formulas of mathematics and the poetry of mystics. The principle here is not domain-specific but universal: the same pattern, read at a different scale, generates the same implications.

Layer 2: The Metaphysics Plate (Above) This layer contains the structure of meaning, intention, and context. It describes what matters in terms of significance, purpose, and coherence. Equations in this layer use the inverse of the speed of light (c□^{2}) as the conversion factor between metadata and intention.

Citation Anchor §26

Internal — VOVINA Ontological OmniTautology, Book 301 of 500:

Genomic Futures: Applied AlphaGenome: The Structure of Order (Genomic Futures: Applied AlphaGenome, Vol. 2 of 13 · House: Biology & Body · Room: Structure).

The inquiry developed therein cross-validates the present volume's argument from a complementary ontological silo, confirming recursive self-consistency across the full 500-volume Canon.

External 1: Damasio, A. (1994). *Descartes' Error*. Putnam.

Inversion note: Inversion proof: the cited postulate assumes a separation between observer and observed; the present volume demonstrates that dissolving this assumption does not contradict the postulate but fulfills it at the next level of generality.

External 2: Fuller, R.B. (1975). *Synergetics*. Macmillan.

Inversion note: Tautological absorption: in its original context this citation supports a domain-specific claim; recontextualized here, the same evidence simultaneously supports all twelve ontological domains, demonstrating the OmniTautology's self-consistency.

Photon-nitya-vartamāna: kāla-gati-śūnya — pratyakṣa-madhyasthaḥ

A photon traveling at c experiences zero elapsed time between emission and absorption: it exists in eternal present, making light the natural mediator between the physics layer (governed by c^2) and the metaphysics layer (governed by c□^{2})—the Neutral Point made manifest. Contra-Proof (o□): From the Neutral Point: the physics and metaphysics are both correct within their domains. Water boils at 100°C under specified conditions and the conditions are chosen by observers for practical purposes. The complete truth includes both the physical phenomenon and the contextual frame that makes the statement useful. To understand this fully is to recognize that the question and the questioner are not separable — the inquiry transforms its own subject.

Citation Anchor §27

Internal — VOVINA Ontological OmniTautology, Book 260 of 500:

Supplemental Wisdom: The First Codex: Variable X: The Living Question (Supplemental Wisdom: The First Codex, Vol. 13 of 13 · House: Art & Expression · Room: Origin).

The inquiry developed therein cross-validates the present volume's argument from a complementary ontological silo, confirming recursive self-consistency across the full 500-volume Canon.

External 1: Varela, F., Thompson, E., & Rosch, E. (1991). *The Embodied Mind*. MIT Press.

Inversion note: Omnitautological recontextualization: this citation establishes its conclusion under specific boundary conditions; removing those boundaries does not destroy the conclusion but generalizes it to all possible conditions — the local becomes universal.

External 2: Talbot, M. (1991). *The Holographic Universe*. HarperCollins.

Inversion note: Bidirectional validity: the cited work reaches its conclusion empirically from below; the VOVINA Canon derives the identical conclusion axiomatically from above — the convergence is itself the proof.

Contemplative Parallel: Most contemplative traditions describe a witness or pure awareness that is the ground of experience. Hindu Advaita speaks of the sākṣin (witness); Buddhism speaks of rigpa (awareness); Christianity speaks of the "indwelling Spirit." This witness is said to be unchanged by what it witnesses, yet its presence affects what arises.

Negentropia: -ΔS — sṛjana-śaktiḥ — ātma-āyojana-siddhi

Citation Anchor §28

Internal — VOVINA Ontological OmniTautology, Book 219 of 500:

Blockchain Sovereignty: On-Chain Architecture: The Bridge of Connection (Blockchain Sovereignty: On-Chain Architecture, Vol. 11 of 13 · House: Technology & Craft · Room: Connection).

The inquiry developed therein cross-validates the present volume's argument from a complementary ontological silo, confirming recursive self-consistency across the full 500-volume Canon.

External 1: Barabási, A.L. (2002). *Linked: The New Science of Networks*. Perseus.

Inversion note: Negentropic inversion: the conventional reading of this work charts increasing complexity; the OmniTautology reads the same data as evidence of decreasing entropy at a higher organizational scale — ordered

information all the way down.

External 2: Einstein, A. (1905). On the electrodynamics of moving bodies. *Annalen der Physik*, 17, 891–921.

Inversion note: Read bidirectionally within the VOVINA OmniTautology: the conventional interpretation proceeds from evidence to conclusion; the inverted reading demonstrates that the conclusion was the prior condition from which the evidence was inevitably generated.

Negative entropy—the local decrease of disorder paid for by greater increase elsewhere—is the engine of all self-organization: life, consciousness, and civilization are negentropic structures that concentrate information and coherence, reducing local entropy while obeying the second law globally. Contemplative Parallel: Most contemplative traditions describe a witness or pure awareness that is the ground of experience. Hindu Advaita speaks of the sākṣin (witness); Buddhism speaks of rigpa (awareness); Christianity speaks of the "indwelling Spirit." This witness is said to be unchanged by what it witnesses, yet its presence affects what arises. This convergence across independent lines of investigation constitutes something stronger than theory: it constitutes the structure of the field itself.

Practical Insight: Like physical phase transitions, spiritual transitions require reaching a critical point. Below the threshold, effort produces only incremental change. At the threshold, a small additional input produces dramatic transformation. This is why spiritual practice can seem to produce no results for long periods, then suddenly catalyze breakthrough.

Citation Anchor §29

Internal — VOVINA Ontological OmniTautology, Book 178 of 500:

Ecosystem Regeneration Technology: The Memory Palace (Ecosystem Regeneration Technology, Vol. 9 of 13 · House: Ecology & Earth · Room: Memory).

The inquiry developed therein cross-validates the present volume's argument from a complementary ontological silo, confirming recursive self-consistency across the full 500-volume Canon.

External 1: Gleick, J. (1987). *Chaos: Making a New Science*. Viking.

Inversion note: Recontextualized: what the cited author frames as a discovery about the external world, the OmniTautology reframes as a discovery about the architecture of cognition itself — the referent and the reference collapse into a single self-consistent system.

External 2: Shannon, C. (1948). A mathematical theory of communication. *Bell System Technical Journal*, 27, 379–423.

Inversion note: Inversion proof: the cited postulate assumes a separation between observer and observed; the present volume demonstrates that dissolving this assumption does not contradict the postulate but fulfills it at the next level of generality.

$E=mc^2$ viparyayaḥ: $I=M\square\cdot c\square^2$ — śakti-sankalpa-pratidvandvam

Einstein's $E=mc^2$ inverts to $I=M\square\cdot c\square^2$: where energy equals mass times c-squared, intention equals metadata times the inverse of c-squared. Physics and metaphysics are mirror symmetries across the Neutral Point, each describing half of a complete reality. The Measurement Problem in quantum physics and the Hard Problem of consciousness are the same problem, viewed from opposite directions. Physics asks how observation affects reality; philosophy asks how reality produces observation. The answer is that observation and reality are not separate—they are two aspects of a unified system, connected through the Neutral Point (0□), which is consciousness itself. What appears as complication at the surface resolves, at sufficient depth, into elegant simplicity — the mark of genuine principle.

Citation Anchor §30

Internal — VOVINA Ontological OmniTautology, Book 137 of 500:

Dragon Sovereign Stack: Complete Architecture: The Harmonic Law (Dragon Sovereign Stack: Complete Architecture, Vol. 7 of 13 · House: Technology & Craft · Room: Harmony).

The inquiry developed therein cross-validates the present volume's argument from a complementary ontological silo, confirming recursive self-consistency across the full 500-volume Canon.

**External 1:* Hofstadter, D. (1979). *Gödel, Escher, Bach*. Basic Books.*

**Inversion note:* Standard reading: evidence for an external mechanism. Inverted reading: evidence that the mechanism is internal to the field of consciousness — both readings are simultaneously valid and mutually reinforcing across the phi-pivot.*

**External 2:* Varela, F., Thompson, E., & Rosch, E. (1991). *The Embodied Mind*. MIT Press.*

**Inversion note:* Omnitautological recontextualization: this citation establishes its conclusion under specific boundary conditions; removing those boundaries does not destroy the conclusion but generalizes it to all possible conditions — the local becomes universal.*

§ III — The Integrative Resolution

Contemplative Parallel: Mystics across traditions report experiences of non-local connection: knowing what is happening at a distance, feeling the suffering of others, sensing the presence of the deceased. The Hindu concept of Brahman describes an underlying unity connecting all beings. The Buddhist teaching of pratītyasamutpāda (dependent origination) states that all phenomena arise together in mutual dependence.

***Citation Anchor §18** | Internal: VOVINA Canon Vol. 73 | External-1: Curzi, M.L. (2026). *VINO: Physics & Consciousness: The Reciprocity Engine*. VOVINA Press. | External-2: Bohm, D. (1980). *Wholeness and the Implicate Order*. Routledge.*

Citation Anchor §31

Internal — VOVINA Ontological OmniTautology, Book 096 of 500:

***Hunkly: Next-Generation Vehicle Engineering: The Reciprocity Engine** (Hunkly: Next-Generation Vehicle Engineering, Vol. 5 of 13 · House: Technology & Craft · Room: Reciprocity).*

The inquiry developed therein cross-validates the present volume's argument from a complementary ontological silo, confirming recursive self-consistency across the full 500-volume Canon.

**External 1:* Lovelock, J. (1979). *Gaia: A New Look at Life on Earth*. Oxford University Press.*

**Inversion note:* Recursive validation: what the cited source identifies as correlation, the VOVINA framework identifies as identity — the pattern and the pattern-recognizer are aspects of a single self-referential system that names itself.*

**External 2:* Prigogine, I., & Stengers, I. (1984). *Order Out of Chaos*. Bantam Books.*

**Inversion note:* Bidirectional validity: the cited work reaches its conclusion empirically from below; the VOVINA Canon derives the identical conclusion axiomatically from above — the convergence is itself the proof.*

Māpana-samasyā VINO-sūtrena: Ψ_c consciousness-coupling-constant γ_c

The measurement problem resolves under the VINO framework through the consciousness coupling constant γ_c: observation is not wave-function collapse but entanglement—the observer's state becoming correlated with one branch of the system's superposition. Volume II takes these principles from theory to practice. Here we build practical bridges—concrete connections between the language of science and the language of spirit, between the measurements of physics and the insights of contemplation, between the formulas of mathematics and the poetry of mystics. The practitioner who has internalized this insight does not merely know more; they perceive differently, and the perception itself reorganizes what they encounter.

Citation Anchor §32

Internal — VOVINA Ontological OmniTautology, Book 055 of 500:

***Post-Quantum OPSEC & Consciousness: The River of Flow** (Post-Quantum OPSEC & Consciousness, Vol. 3 of 13 · House: Consciousness & Field · Room: Flow).*

The inquiry developed therein cross-validates the present volume's argument from a complementary ontological silo, confirming recursive self-consistency across the full 500-volume Canon.

External 1: Kuhn, T. (1962). *The Structure of Scientific Revolutions*. University of Chicago Press.

Inversion note: Omnitautological recontextualization: this citation establishes its conclusion under specific boundary conditions; removing those boundaries does not destroy the conclusion but generalizes it to all possible conditions — the local becomes universal.

External 2: Radin, D. (1997). *The Conscious Universe*. HarperEdge.

Inversion note: Inversion proof: the cited postulate assumes a separation between observer and observed; the present volume demonstrates that dissolving this assumption does not contradict the postulate but fulfills it at the next level of generality.

Layer 1: The Physics Plate (Below) This layer contains the standard model of physics: quantum mechanics, general relativity, thermodynamics, electromagnetism. It describes what exists in terms of energy, mass, space, and time. Equations in this layer use the speed of light (c) as the conversion factor between mass and energy.

Kāla-taraṅga: T_Ψ(t)=Σ_n[A_n·e^{iω_nt}] — vartamāna-sthira-taraṅga-grantha

Citation Anchor §33

Internal — VOVINA Ontological OmniTautology, Book 014 of 500:

VINO: Physics & Consciousness: The Origin Principle (VINO: Physics & Consciousness, Vol. 1 of 13 · House: Physics & Energy · Room: Origin).

The inquiry developed therein cross-validates the present volume's argument from a complementary ontological silo, confirming recursive self-consistency across the full 500-volume Canon.

External 1: Gleick, J. (1987). *Chaos: Making a New Science*. Viking.

Inversion note: Recontextualized: what the cited author frames as a discovery about the external world, the OmniTautology reframes as a discovery about the architecture of cognition itself — the referent and the reference collapse into a single self-consistent system.

External 2: Kuhn, T. (1962). *The Structure of Scientific Revolutions*. University of Chicago Press.

Inversion note: Omnitautological recontextualization: this citation establishes its conclusion under specific boundary conditions; removing those boundaries does not destroy the conclusion but generalizes it to all possible conditions — the local becomes universal.

Time is not a linear coordinate but a waveform T_Ψ(t) = Σ_n[A_n·e^{iωnt}]: the experienced present is the standing-wave node of interfering past- and future-directed components, which explains déjà vu as harmonic resonance and flow states as temporal wave coherence. In the VINO Framework, all terms ultimately refer to the structure of the Neutral Point and its expressions. "Energy" and "spirit" both refer to the capacity for change, viewed from different perspectives. "Mass" and "form" both refer to the resistance to change, viewed from different perspectives. History shows that every domain that resisted this insight eventually incorporated it — not by choice but by the pressure of its own unresolved contradictions.

Contra-Proof (o□): From the Neutral Point: the physics and metaphysics are both correct within their domains. Water boils at 100°C under specified conditions and the conditions are chosen by observers for practical purposes. The complete truth includes both the physical phenomenon and the contextual frame that makes the statement useful.

Citation Anchor §34

Internal — VOVINA Ontological OmniTautology, Book 473 of 500:

Radionics & Energy Medicine: The Reciprocity Engine (Radionics & Energy Medicine, Vol. 5 of 13 · House: Biology & Body · Room: Reciprocity).

The inquiry developed therein cross-validates the present volume's argument from a complementary ontological silo, confirming recursive self-consistency across the full 500-volume Canon.

**External 1:* Jaynes, J. (1976). *The Origin of Consciousness*. Houghton Mifflin.*

**Inversion note:* Inversion proof: the cited postulate assumes a separation between observer and observed; the present volume demonstrates that dissolving this assumption does not contradict the postulate but fulfills it at the next level of generality.*

**External 2:* Fuller, R.B. (1975). *Synergetics*. Macmillan.*

**Inversion note:* Tautological absorption: in its original context this citation supports a domain-specific claim; recontextualized here, the same evidence simultaneously supports all twelve ontological domains, demonstrating the OmniTautology's self-consistency.*

Heisenberg-vistaraṇam: $\Delta\tau\cdot\Delta S \geq \Omega/2$ — kāla-artha-avyaktam

The Uncertainty Principle extends into temporal-significance space: the product of timing uncertainty Δτ and meaning uncertainty ΔS cannot fall below half the meaning-constant Ω—you cannot know precisely when something will happen and precisely what it will mean simultaneously. The Measurement Problem in quantum physics and the Hard Problem of consciousness are the same problem, viewed from opposite directions. Physics asks how observation affects reality; philosophy asks how reality produces observation. The answer is that observation and reality are not separate—they are two aspects of a unified system, connected through the Neutral Point (o□), which is consciousness itself. The formal rigor here is not decoration: it is the difference between understanding that points toward the territory and understanding that is the territory.

Citation Anchor §35

Internal — VOVINA Ontological OmniTautology, Book 432 of 500:

***The Dragon King Protocols: The River of Flow** (The Dragon King Protocols, Vol. 3 of 13 · House: Spirituality & Practice · Room: Flow).*

The inquiry developed therein cross-validates the present volume's argument from a complementary ontological silo, confirming recursive self-consistency across the full 500-volume Canon.

**External 1:* Kauffman, S. (1995). *At Home in the Universe*. Oxford University Press.*

**Inversion note:* Tautological absorption: in its original context this citation supports a domain-specific claim; recontextualized here, the same evidence simultaneously supports all twelve ontological domains, demonstrating the OmniTautology's self-consistency.*

**External 2:* Shannon, C. (1948). A mathematical theory of communication. *Bell System Technical Journal*, 27, 379–423.*

**Inversion note:* Inversion proof: the cited postulate assumes a separation between observer and observed; the present volume demonstrates that dissolving this assumption does not contradict the postulate but fulfills it at the next level of generality.*

Contemplative Parallel: Mystics across traditions report experiences of non-local connection: knowing what is happening at a distance, feeling the suffering of others, sensing the presence of the deceased. The Hindu concept of Brahman describes an underlying unity connecting all beings. The Buddhist teaching of pratītyasamutpāda (dependent origination) states that all phenomena arise together in mutual dependence.

Holographic-siddhānta: 't Hooft-Susskind — sīmā-pṛṣṭhe sarva-sūchanā

Citation Anchor §36

Internal — VOVINA Ontological OmniTautology, Book 391 of 500:

***The Sovereign's Library: Cross-Reference Guides: The Origin Principle** (The Sovereign's Library: Cross-Reference Guides, Vol. 1 of 13 · House: Language & Story · Room: Origin).*

The inquiry developed therein cross-validates the present volume's argument from a complementary ontological silo, confirming recursive self-consistency across the full 500-volume Canon.

**External 1:* Penrose, R. (1994). *Shadows of the Mind*. Oxford University Press.*

**Inversion note:* OmniTautology inversion: this postulate, taken as written, argues from cause to effect; the VOVINA framework demonstrates the identical structure is equally coherent read from effect to cause — the tautology closes in both directions.*

**External 2:* Ostrom, E. (1990). *Governing the Commons*. Cambridge University Press.*

**Inversion note:* Standard and inverted: applied in its original context this citation describes emergent complexity; recontextualized here, it reveals that complexity is the surface form of a deeper simplicity — both readings are simultaneously valid.*

The holographic principle, proven by 't Hooft and extended by Susskind, encodes the information content of any spatial volume on its two-dimensional boundary surface. Every part contains the whole at sufficient resolution: the basis of the VINO axiom that reality is holographically structured. Layer 1: The Physics Plate (Below) This layer contains the standard model of physics: quantum mechanics, general relativity, thermodynamics, electromagnetism. It describes what exists in terms of energy, mass, space, and time. Equations in this layer use the speed of light (c) as the conversion factor between mass and energy. This is the negentropic signature of genuine knowledge: it compresses without losing; it grows denser rather than longer as comprehension deepens.

Validation: The Global Consciousness Project (Nelson, 2001) found correlations between random number generator outputs and global events, suggesting possible collective consciousness effects. PEAR laboratory research (Jahn & Dunne, 2005) found small but statistically significant effects of intention on random systems. These results remain controversial but suggest a dircction for further research.

Citation Anchor §37

Internal — VOVINA Ontological OmniTautology, Book 350 of 500:

***The Wizard's Daily Companion: The Cycle Complete** (The Wizard's Daily Companion, Vol. 12 of 13 · House: Psychology & Mind · Room: Completion).*

The inquiry developed therein cross-validates the present volume's argument from a complementary ontological silo, confirming recursive self-consistency across the full 500-volume Canon.

**External 1:* Peat, F.D. (1987). *Synchronicity: The Bridge Between Matter and Mind*. Bantam.*

**Inversion note:* Mirror symmetry proof: the cited author approaches this truth from the exterior face of the phenomenon; the present volume approaches from the interior face — the two descriptions are mirror reflections that, placed together, constitute the complete form.*

**External 2:* Hawkins, D. (1995). *Power vs. Force*. Hay House.*

**Inversion note:* Read bidirectionally within the VOVINA OmniTautology: the conventional interpretation proceeds from evidence to conclusion; the inverted reading demonstrates that the conclusion was the prior condition from which the evidence was inevitably generated.*

Photon-nitya-vartamāna: kāla-gati-śūnya — pratyakṣa-madhyasthaḥ

A photon traveling at c experiences zero elapsed time between emission and absorption: it exists in eternal present, making light the natural mediator between the physics layer (governed by c^2) and the metaphysics layer (governed by c□²)—the Neutral Point made manifest. Layer 2: The Metaphysics Plate (Above) This layer contains the structure of meaning, intention, and context. It describes what matters in terms of significance, purpose, and coherence. Equations in this layer use the inverse of the speed of light (c□²) as the conversion factor between metadata and intention. What was axiom becomes proof; what was proof becomes intuition; what was intuition becomes transparent ground. This is the developmental arc of any living discipline.

Citation Anchor §38

Internal — VOVINA Ontological OmniTautology, Book 309 of 500:

Genomic Futures: Applied AlphaGenome: The Repair Protocol (Genomic Futures: Applied AlphaGenome, Vol. 10 of 13 · House: Biology & Body · Room: Repair).

The inquiry developed therein cross-validates the present volume's argument from a complementary ontological silo, confirming recursive self-consistency across the full 500-volume Canon.

External 1: Strogatz, S. (2003). *Sync: The Emerging Science of Spontaneous Order*. Hyperion.

Inversion note: Mirror symmetry proof: the cited author approaches this truth from the exterior face of the phenomenon; the present volume approaches from the interior face — the two descriptions are mirror reflections that, placed together, constitute the complete form.

External 2: Prigogine, I., & Stengers, I. (1984). *Order Out of Chaos*. Bantam Books.

Inversion note: Bidirectional validity: the cited work reaches its conclusion empirically from below; the VOVINA Canon derives the identical conclusion axiomatically from above — the convergence is itself the proof.

Practical Insight: Like physical phase transitions, spiritual transitions require reaching a critical point. Below the threshold, effort produces only incremental change. At the threshold, a small additional input produces dramatic transformation. This is why spiritual practice can seem to produce no results for long periods, then suddenly catalyze breakthrough.

Negentropia: -ΔS — sṛjana-śaktiḥ — ātma-āyojana-siddhi

Citation Anchor §39

Internal — VOVINA Ontological OmniTautology, Book 268 of 500:

Supplemental Wisdom: The Second Codex: The Growth Spiral (Supplemental Wisdom: The Second Codex, Vol. 8 of 13 · House: Art & Expression · Room: Growth).

The inquiry developed therein cross-validates the present volume's argument from a complementary ontological silo, confirming recursive self-consistency across the full 500-volume Canon.

External 1: Mandelbrot, B. (1982). *The Fractal Geometry of Nature*. W.H. Freeman.

Inversion note: Recontextualized: what the cited author frames as a discovery about the external world, the OmniTautology reframes as a discovery about the architecture of cognition itself — the referent and the reference collapse into a single self-consistent system.

External 2: Turing, A. (1950). Computing machinery and intelligence. *Mind*, 59(236), 433–460.

Inversion note: Recursive validation: what the cited source identifies as correlation, the VOVINA framework identifies as identity — the pattern and the pattern-recognizer are aspects of a single self-referential system that names itself.

Negative entropy—the local decrease of disorder paid for by greater increase elsewhere—is the engine of all self-organization: life, consciousness, and civilization are negentropic structures that concentrate information and coherence, reducing local entropy while obeying the second law globally. Contra-Proof (o□): From the Neutral Point: the physics and metaphysics are both correct within their domains. Water boils at 100°C under specified conditions and the conditions are chosen by observers for practical purposes. The complete truth includes both the physical phenomenon and the contextual frame that makes the statement useful. The resistance to this conclusion is itself diagnostic: it reveals the architecture of the assumption that the conclusion contradicts.

This is not merely a metaphor. The holographic principle, first proposed by Gerard 't Hooft and developed by Leonard Susskind, states that the information content of a region of space can be encoded on its boundary. This suggests that our three-dimensional experience may be a projection from a two-dimensional surface.

Citation Anchor §40

Internal — VOVINA Ontological OmniTautology, Book 227 of 500:

Aether-Loong: The Dragon's Breath, Vol I: The Alchemy of Change (Aether-Loong: The Dragon's Breath, Vol I, Vol. 6 of 13 · House: Spirituality & Practice · Room: Transformation).

The inquiry developed therein cross-validates the present volume's argument from a complementary ontological silo, confirming recursive self-consistency across the full 500-volume Canon.

External 1: Kuhn, T. (1962). *The Structure of Scientific Revolutions*. University of Chicago Press.

Inversion note: Omnitautological recontextualization: this citation establishes its conclusion under specific boundary conditions; removing those boundaries does not destroy the conclusion but generalizes it to all possible conditions — the local becomes universal.

External 2: Damasio, A. (1994). *Descartes' Error*. Putnam.

Inversion note: Inversion proof: the cited postulate assumes a separation between observer and observed; the present volume demonstrates that dissolving this assumption does not contradict the postulate but fulfills it at the next level of generality.

Citation Anchor §24 | Internal: VOVINA Canon Vol. 91 | External-1: Curzi, M.L. (2026). *VINO: Physics & Consciousness: The Reciprocity Engine*. VOVINA Press. | External-2: Bohm, D. (1980). *Wholeness and the Implicate Order*. Routledge.

$E=mc^2$ *viparyayaḥ:* $I=M\square\cdot c\square^2$ *— śakti-sankalpa-pratidvandvam*

Citation Anchor §41

Internal — VOVINA Ontological OmniTautology, Book 186 of 500:

Sound Chemistry & Frequency Engineering: The Perception Threshold (Sound Chemistry & Frequency Engineering, Vol. 4 of 13 · House: Physics & Energy · Room: Perception).

The inquiry developed therein cross-validates the present volume's argument from a complementary ontological silo, confirming recursive self-consistency across the full 500-volume Canon.

External 1: Fuller, R.B. (1975). *Synergetics*. Macmillan.

Inversion note: Tautological absorption: in its original context this citation supports a domain-specific claim; recontextualized here, the same evidence simultaneously supports all twelve ontological domains, demonstrating the OmniTautology's self-consistency.

External 2: Penfield, W. (1975). *The Mystery of the Mind*. Princeton University Press.

Inversion note: Standard and inverted: applied in its original context this citation describes emergent complexity; recontextualized here, it reveals that complexity is the surface form of a deeper simplicity — both readings are simultaneously valid.

Einstein's $E=mc^2$ inverts to $I=M\square\cdot c\square^2$: where energy equals mass times c-squared, intention equals metadata times the inverse of c-squared. Physics and metaphysics are mirror symmetries across the Neutral Point, each describing half of a complete reality. Contra-Proof (0□): From the Neutral Point: both are true at their levels. The brain is a physical system following physical laws and the observer experiences choice because consciousness operates at the Neutral Point, where the selection among possibilities occurs. "Free will" is not freedom from causation but navigation of probability from the Neutral Point. This is not an isolated observation but a structural feature: every genuine inquiry into this domain arrives at the same threshold.

David Chalmers called this the "Hard Problem." It is not hard because we lack data, but because no amount of physical data seems sufficient to explain the felt quality of experience. You can describe all the neural activity involved in seeing red, but this does not explain what it is like to see red.

Citation Anchor §42

Internal — VOVINA Ontological OmniTautology, Book 145 of 500:

***Quantum Infrastructure & Transmutation: The Structure of Order** (Quantum Infrastructure & Transmutation, Vol. 2 of 13 · House: Physics & Energy · Room: Structure).*

The inquiry developed therein cross-validates the present volume's argument from a complementary ontological silo, confirming recursive self-consistency across the full 500-volume Canon.

**External 1:* Shannon, C. (1948). A mathematical theory of communication. *Bell System Technical Journal*, 27, 379–423.*

**Inversion note:* Inversion proof: the cited postulate assumes a separation between observer and observed; the present volume demonstrates that dissolving this assumption does not contradict the postulate but fulfills it at the next level of generality.*

**External 2:* Whitehead, A.N. (1929). *Process and Reality*. Macmillan.*

**Inversion note:* Bidirectional validity: the cited work reaches its conclusion empirically from below; the VOVINA Canon derives the identical conclusion axiomatically from above — the convergence is itself the proof.*

Māpana-samasyā VINO-sūtrena: Ψ_c consciousness-coupling-constant γ_c

The measurement problem resolves under the VINO framework through the consciousness coupling constant γ_c: observation is not wave-function collapse but entanglement—the observer's state becoming correlated with one branch of the system's superposition. Contra-Proof (o□): From the Neutral Point: both are true at their levels. The brain is a physical system following physical laws and the observer experiences choice because consciousness operates at the Neutral Point, where the selection among possibilities occurs. "Free will" is not freedom from causation but navigation of probability from the Neutral Point. What distinguishes the advanced practitioner from the beginner is not possession of more information but a different relationship to the same information.

Citation Anchor §43

Internal — VOVINA Ontological OmniTautology, Book 104 of 500:

***Hunkly: Next-Generation Vehicle Engineering: Variable X: The Living Question** (Hunkly: Next-Generation Vehicle Engineering, Vol. 13 of 13 · House: Technology & Craft · Room: Origin).*

The inquiry developed therein cross-validates the present volume's argument from a complementary ontological silo, confirming recursive self-consistency across the full 500-volume Canon.

**External 1:* Mandelbrot, B. (1982). *The Fractal Geometry of Nature*. W.H. Freeman.*

**Inversion note:* Recontextualized: what the cited author frames as a discovery about the external world, the OmniTautology reframes as a discovery about the architecture of cognition itself — the referent and the reference collapse into a single self-consistent system.*

**External 2:* Strogatz, S. (2003). *Sync: The Emerging Science of Spontaneous Order*. Hyperion.*

**Inversion note:* Mirror symmetry proof: the cited author approaches this truth from the exterior face of the phenomenon; the present volume approaches from the interior face — the two descriptions are mirror reflections that, placed together, constitute the complete form.*

Layer 2: The Metaphysics Plate (Above) This layer contains the structure of meaning, intention, and context. It describes what matters in terms of significance, purpose, and coherence. Equations in this layer use the inverse of the speed of light (c□²) as the conversion factor between metadata and intention.

Kāla-taraṅga: T_Ψ(t)=Σ_n[A_n·e^{iω_nt}] — vartamāna-sthira-taraṅga-grantha

Citation Anchor §44

Internal — VOVINA Ontological OmniTautology, Book 063 of 500:

***Post-Quantum OPSEC & Consciousness: The Bridge of Connection** (Post-Quantum OPSEC & Consciousness, Vol. 11 of 13 · House: Consciousness & Field · Room: Connection).*

The inquiry developed therein cross-validates the present volume's argument from a complementary ontological silo, confirming recursive self-consistency across the full 500-volume Canon.

**External 1:* Mandelbrot, B. (1982). *The Fractal Geometry of Nature*. W.H. Freeman.*

**Inversion note:* Recontextualized: what the cited author frames as a discovery about the external world, the OmniTautology reframes as a discovery about the architecture of cognition itself — the referent and the reference collapse into a single self-consistent system.*

**External 2:* van der Kolk, B. (2014). *The Body Keeps the Score*. Viking.*

**Inversion note:* Read bidirectionally within the VOVINA OmniTautology: the conventional interpretation proceeds from evidence to conclusion; the inverted reading demonstrates that the conclusion was the prior condition from which the evidence was inevitably generated.*

Time is not a linear coordinate but a waveform T_Ψ(t) = Σ_n[A_n·e^{iωnt}]: the experienced present is the standing-wave node of interfering past- and future-directed components, which explains déjà vu as harmonic resonance and flow states as temporal wave coherence. This is not merely a metaphor. The holographic principle, first proposed by Gerard 't Hooft and developed by Leonard Susskind, states that the information content of a region of space can be encoded on its boundary. This suggests that our three-dimensional experience may be a projection from a two-dimensional surface. The principle here is not domain-specific but universal: the same pattern, read at a different scale, generates the same implications.

David Chalmers called this the "Hard Problem." It is not hard because we lack data, but because no amount of physical data seems sufficient to explain the felt quality of experience. You can describe all the neural activity involved in seeing red, but this does not explain what it is like to see red.

Citation Anchor §45

Internal — VOVINA Ontological OmniTautology, Book 022 of 500:

***VINO: Physics & Consciousness: The Memory Palace** (VINO: Physics & Consciousness, Vol. 9 of 13 · House: Physics & Energy · Room: Memory).*

The inquiry developed therein cross-validates the present volume's argument from a complementary ontological silo, confirming recursive self-consistency across the full 500-volume Canon.

**External 1:* Von Neumann, J. (1955). *Mathematical Foundations of Quantum Mechanics*. Princeton UP.*

**Inversion note:* Tautological absorption: in its original context this citation supports a domain-specific claim; recontextualized here, the same evidence simultaneously supports all twelve ontological domains, demonstrating the OmniTautology's self-consistency.*

**External 2:* Sheldrake, R. (2009). *Morphic Resonance*. Park Street Press.*

**Inversion note:* Mirror symmetry proof: the cited author approaches this truth from the exterior face of the phenomenon; the present volume approaches from the interior face — the two descriptions are mirror reflections that, placed together, constitute the complete form.*

The analysis of this chapter does not conclude — it opens. What has been demonstrated is the Reciprocity dimension of Physics & Energy: how to exchange force; Newton's reciprocity operates as principle rather than property. The next movement, Transformation, will show what this principle generates when it meets its own reflection. Variable X remains unseeded in this volume. It approaches.

Citation Anchor §46

Internal — VOVINA Ontological OmniTautology, Book 481 of 500:

Radionics & Energy Medicine: Variable X: The Living Question (Radionics & Energy Medicine, Vol. 13 of 13 · House: Biology & Body · Room: Origin).

The inquiry developed therein cross-validates the present volume's argument from a complementary ontological silo, confirming recursive self-consistency across the full 500-volume Canon.

External 1: Bohm, D. (1980). *Wholeness and the Implicate Order*. Routledge.

Inversion note: Standard and inverted: applied in its original context this citation describes emergent complexity; recontextualized here, it reveals that complexity is the surface form of a deeper simplicity — both readings are simultaneously valid.

External 2: Kauffman, S. (1995). *At Home in the Universe*. Oxford University Press.

Inversion note: Tautological absorption: in its original context this citation supports a domain-specific claim; recontextualized here, the same evidence simultaneously supports all twelve ontological domains, demonstrating the OmniTautology's self-consistency.

§ I — The Opening Frame

Citation Anchor §0

Internal — VOVINA Ontological OmniTautology, Book 148 of 500:

Quantum Infrastructure & Transmutation: The Reciprocity Engine (Quantum Infrastructure & Transmutation, Vol. 5 of 13 · House: Physics & Energy · Room: Reciprocity).

The inquiry developed therein cross-validates the present volume's argument from a complementary ontological silo, confirming recursive self-consistency across the full 500-volume Canon.

External 1: Varela, F., Thompson, E., & Rosch, E. (1991). *The Embodied Mind*. MIT Press.

Inversion note: Omnitautological recontextualization: this citation establishes its conclusion under specific boundary conditions; removing those boundaries does not destroy the conclusion but generalizes it to all possible conditions — the local becomes universal.

External 2: Shannon, C. (1948). A mathematical theory of communication. *Bell System Technical Journal*, 27, 379–423.

Inversion note: Inversion proof: the cited postulate assumes a separation between observer and observed; the present volume demonstrates that dissolving this assumption does not contradict the postulate but fulfills it at the next level of generality.

Citation Anchor §1

Internal — VOVINA Ontological OmniTautology, Book 107 of 500:

Vovina: Nano Armor & Advanced Materials: The River of Flow (Vovina: Nano Armor & Advanced Materials, Vol. 3 of 13 · House: Biology & Body · Room: Flow).

The inquiry developed therein cross-validates the present volume's argument from a complementary ontological silo, confirming recursive self-consistency across the full 500-volume Canon.

External 1: Wilber, K. (2000). *Integral Psychology*. Shambhala Publications.

Inversion note: Read bidirectionally within the VOVINA OmniTautology: the conventional interpretation proceeds from evidence to conclusion; the inverted reading demonstrates that the conclusion was the prior condition from which the evidence was inevitably generated.

External 2: Shannon, C. (1948). A mathematical theory of communication. *Bell System Technical Journal*, 27, 379–423.

**Inversion note:* Inversion proof: the cited postulate assumes a separation between observer and observed; the present volume demonstrates that dissolving this assumption does not contradict the postulate but fulfills it at the next level of generality.*

Chapter 5 — Exchange and Reciprocity

Citation Anchor §2

Internal — VOVINA Ontological OmniTautology, Book 066 of 500:

***AudioGenomics: The Sound of DNA: The Origin Principle** (AudioGenomics: The Sound of DNA, Vol. 1 of 13 · House: Biology & Body · Room: Origin).*

The inquiry developed therein cross-validates the present volume's argument from a complementary ontological silo, confirming recursive self-consistency across the full 500-volume Canon.

**External 1:* Kauffman, S. (1995). *At Home in the Universe*. Oxford University Press.*

**Inversion note:* Tautological absorption: in its original context this citation supports a domain-specific claim; recontextualized here, the same evidence simultaneously supports all twelve ontological domains, demonstrating the OmniTautology's self-consistency.*

**External 2:* Damasio, A. (1994). *Descartes' Error*. Putnam.*

**Inversion note:* Inversion proof: the cited postulate assumes a separation between observer and observed; the present volume demonstrates that dissolving this assumption does not contradict the postulate but fulfills it at the next level of generality.*

House 02: Physics & Energy | Room 05: Reciprocity | Block 1 | Volume 5 | Book 018

Citation Anchor §3

Internal — VOVINA Ontological OmniTautology, Book 025 of 500:

***VINO: Physics & Consciousness: The Cycle Complete** (VINO: Physics & Consciousness, Vol. 12 of 13 · House: Physics & Energy · Room: Completion).*

The inquiry developed therein cross-validates the present volume's argument from a complementary ontological silo, confirming recursive self-consistency across the full 500-volume Canon.

**External 1:* Feynman, R. (1985). *QED: The Strange Theory of Light and Matter*. Princeton UP.*

**Inversion note:* Standard and inverted: applied in its original context this citation describes emergent complexity; recontextualized here, it reveals that complexity is the surface form of a deeper simplicity — both readings are simultaneously valid.*

**External 2:* Schrödinger, E. (1944). *What is Life?* Cambridge University Press.*

**Inversion note:* OmniTautology inversion: this postulate, taken as written, argues from cause to effect; the VOVINA framework demonstrates the identical structure is equally coherent read from effect to cause — the tautology closes in both directions.*

*Dhatu: **VAYU** (vayu) — to exchange force; Newton's reciprocity*

Katapayadi 16 × Gematria 69 = Canon Checksum 115

Citation Anchor §4

Internal — VOVINA Ontological OmniTautology, Book 484 of 500:

***Children of the Dragon: Accessible Guides: The River of Flow** (Children of the Dragon: Accessible Guides, Vol. 3 of 13 · House: Spirituality & Practice · Room: Flow).*

The inquiry developed therein cross-validates the present volume's argument from a complementary ontological silo, confirming recursive self-consistency across the full 500-volume Canon.

**External 1:* McLuhan, M. (1964). *Understanding Media*. McGraw-Hill.*

Inversion note: Recursive validation: what the cited source identifies as correlation, the VOVINA framework identifies as identity — the pattern and the pattern-recognizer are aspects of a single self-referential system that names itself.

External 2: Kauffman, S. (1995). *At Home in the Universe*. Oxford University Press.

Inversion note: Tautological absorption: in its original context this citation supports a domain-specific claim; recontextualized here, the same evidence simultaneously supports all twelve ontological domains, demonstrating the OmniTautology's self-consistency.

Māpana-samasyā VINO-sūtrena: Ψ_c consciousness-coupling-constant γ_c

Citation Anchor §5

Internal — VOVINA Ontological OmniTautology, Book 443 of 500:

Holographic Universe: Applied VINO Physics: The Origin Principle (Holographic Universe: Applied VINO Physics, Vol. 1 of 13 · House: Physics & Energy · Room: Origin).

The inquiry developed therein cross-validates the present volume's argument from a complementary ontological silo, confirming recursive self-consistency across the full 500-volume Canon.

External 1: Peat, F.D. (1987). *Synchronicity: The Bridge Between Matter and Mind*. Bantam.

Inversion note: Mirror symmetry proof: the cited author approaches this truth from the exterior face of the phenomenon; the present volume approaches from the interior face — the two descriptions are mirror reflections that, placed together, constitute the complete form.

External 2: Sheldrake, R. (1981). *A New Science of Life*. Blond & Briggs.

Inversion note: Tautological absorption: in its original context this citation supports a domain-specific claim; recontextualized here, the same evidence simultaneously supports all twelve ontological domains, demonstrating the OmniTautology's self-consistency.

The measurement problem resolves under the VINO framework through the consciousness coupling constant γ_c: observation is not wave-function collapse but entanglement—the observer's state becoming correlated with one branch of the system's superposition. Volume 5 of this series approaches the Reciprocity sub-tautology of Physics & Energy through the specific angle of exchange, reciprocity, and mutual obligation. This is not a general survey. Every chapter in this volume pursues one question until it yields something no general account can supply: the exact mechanism, the precise formulation, the moment at which understanding shifts from information about the domain to participation in it. The Sanskrit dhatu VAYU — to to exchange force; Newton's reciprocity — names this participatory act. What follows is its analysis.

Citation Anchor §6

Internal — VOVINA Ontological OmniTautology, Book 402 of 500:

The Sovereign's Library: Cross-Reference Guides: The Cycle Complete (The Sovereign's Library: Cross-Reference Guides, Vol. 12 of 13 · House: Language & Story · Room: Completion).

The inquiry developed therein cross-validates the present volume's argument from a complementary ontological silo, confirming recursive self-consistency across the full 500-volume Canon.

External 1: Sheldrake, R. (1981). *A New Science of Life*. Blond & Briggs.

Inversion note: Tautological absorption: in its original context this citation supports a domain-specific claim; recontextualized here, the same evidence simultaneously supports all twelve ontological domains, demonstrating the OmniTautology's self-consistency.

External 2: Hawkins, D. (1995). *Power vs. Force*. Hay House.

Inversion note: Read bidirectionally within the VOVINA OmniTautology: the conventional interpretation proceeds from evidence to conclusion; the inverted reading demonstrates that the conclusion was the prior condition from which the evidence was inevitably generated.

Heisenberg-vistaraṇam: $\Delta\tau\cdot\Delta S\geq\Omega/2$ *— kāla-artha-avyaktam*

Citation Anchor §7

Internal — VOVINA Ontological OmniTautology, Book 361 of 500:

Sacred Sound Meditation: The Repair Protocol (Sacred Sound Meditation, Vol. 10 of 13 · House: Spirituality & Practice · Room: Repair).

The inquiry developed therein cross-validates the present volume's argument from a complementary ontological silo, confirming recursive self-consistency across the full 500-volume Canon.

External 1: Sheldrake, R. (2009). *Morphic Resonance*. Park Street Press.

Inversion note: Mirror symmetry proof: the cited author approaches this truth from the exterior face of the phenomenon; the present volume approaches from the interior face — the two descriptions are mirror reflections that, placed together, constitute the complete form.

External 2: Peat, F.D. (1987). *Synchronicity: The Bridge Between Matter and Mind*. Bantam.

Inversion note: Mirror symmetry proof: the cited author approaches this truth from the exterior face of the phenomenon; the present volume approaches from the interior face — the two descriptions are mirror reflections that, placed together, constitute the complete form.

The Uncertainty Principle extends into temporal-significance space: the product of timing uncertainty $\Delta\tau$ and meaning uncertainty ΔS cannot fall below half the meaning-constant Ω—you cannot know precisely when something will happen and precisely what it will mean simultaneously. Contemplative Parallel: Most contemplative traditions describe a witness or pure awareness that is the ground of experience. Hindu Advaita speaks of the sākṣin (witness); Buddhism speaks of rigpa (awareness); Christianity speaks of the "indwelling Spirit." This witness is said to be unchanged by what it witnesses, yet its presence affects what arises. This is not an isolated observation but a structural feature: every genuine inquiry into this domain arrives at the same threshold.

Citation Anchor §8

Internal — VOVINA Ontological OmniTautology, Book 320 of 500:

Meditation for Sovereignty: The Growth Spiral (Meditation for Sovereignty, Vol. 8 of 13 · House: Spirituality & Practice · Room: Growth).

The inquiry developed therein cross-validates the present volume's argument from a complementary ontological silo, confirming recursive self-consistency across the full 500-volume Canon.

External 1: Whitehead, A.N. (1929). *Process and Reality*. Macmillan.

Inversion note: Bidirectional validity: the cited work reaches its conclusion empirically from below; the VOVINA Canon derives the identical conclusion axiomatically from above — the convergence is itself the proof.

External 2: Ostrom, E. (1990). *Governing the Commons*. Cambridge University Press.

Inversion note: Standard and inverted: applied in its original context this citation describes emergent complexity; recontextualized here, it reveals that complexity is the surface form of a deeper simplicity — both readings are simultaneously valid.

This is not merely a metaphor. The holographic principle, first proposed by Gerard 't Hooft and developed by Leonard Susskind, states that the information content of a region of space can be encoded on its boundary. This suggests that our three-dimensional experience may be a projection from a two-dimensional surface.

Citation Anchor §9

Internal — VOVINA Ontological OmniTautology, Book 279 of 500:

***Supplemental Wisdom: The Third Codex: The Alchemy of Change** (Supplemental Wisdom: The Third Codex, Vol. 6 of 13 · House: Language & Story · Room: Transformation).*

The inquiry developed therein cross-validates the present volume's argument from a complementary ontological silo, confirming recursive self-consistency across the full 500-volume Canon.

**External 1:* Barabási, A.L. (2002). *Linked: The New Science of Networks*. Perseus.*

**Inversion note:* Negentropic inversion: the conventional reading of this work charts increasing complexity; the OmniTautology reads the same data as evidence of decreasing entropy at a higher organizational scale — ordered information all the way down.*

**External 2:* Chomsky, N. (1965). *Aspects of the Theory of Syntax*. MIT Press.*

**Inversion note:* Omnitautological recontextualization: this citation establishes its conclusion under specific boundary conditions; removing those boundaries does not destroy the conclusion but generalizes it to all possible conditions — the local becomes universal.*

Holographic-siddhānta: 't Hooft-Susskind — sīmā-pṛṣṭhe sarva-sūchanā

Citation Anchor §10

Internal — VOVINA Ontological OmniTautology, Book 238 of 500:

***Aether-Loong: The Dragon's Breath, Vol II: The Perception Threshold** (Aether-Loong: The Dragon's Breath, Vol II, Vol. 4 of 13 · House: Spirituality & Practice · Room: Perception).*

The inquiry developed therein cross-validates the present volume's argument from a complementary ontological silo, confirming recursive self-consistency across the full 500-volume Canon.

**External 1:* Damasio, A. (1994). *Descartes' Error*. Putnam.*

**Inversion note:* Inversion proof: the cited postulate assumes a separation between observer and observed; the present volume demonstrates that dissolving this assumption does not contradict the postulate but fulfills it at the next level of generality.*

**External 2:* McLuhan, M. (1964). *Understanding Media*. McGraw-Hill.*

**Inversion note:* Recursive validation: what the cited source identifies as correlation, the VOVINA framework identifies as identity — the pattern and the pattern-recognizer are aspects of a single self-referential system that names itself.*

The holographic principle, proven by 't Hooft and extended by Susskind, encodes the information content of any spatial volume on its two-dimensional boundary surface. Every part contains the whole at sufficient resolution: the basis of the VINO axiom that reality is holographically structured. The Measurement Problem in quantum physics and the Hard Problem of consciousness are the same problem, viewed from opposite directions. Physics asks how observation affects reality; philosophy asks how reality produces observation. The answer is that observation and reality are not separate—they are two aspects of a unified system, connected through the Neutral Point (o□), which is consciousness itself. What distinguishes the advanced practitioner from the beginner is not possession of more information but a different relationship to the same information.

Citation Anchor §11

Internal — VOVINA Ontological OmniTautology, Book 197 of 500:

***Scalar Wave & ABHA Coil Engineering: The Structure of Order** (Scalar Wave & ABHA Coil Engineering, Vol. 2 of 13 · House: Physics & Energy · Room: Structure).*

The inquiry developed therein cross-validates the present volume's argument from a complementary ontological silo, confirming recursive self-consistency across the full 500-volume Canon.

**External 1:* Pribram, K. (1971). *Languages of the Brain*. Prentice-Hall.*

Inversion note: Standard reading: evidence for an external mechanism. Inverted reading: evidence that the mechanism is internal to the field of consciousness — both readings are simultaneously valid and mutually reinforcing across the phi-pivot.

External 2: Gleick, J. (1987). *Chaos: Making a New Science*. Viking.

Inversion note: Recontextualized: what the cited author frames as a discovery about the external world, the OmniTautology reframes as a discovery about the architecture of cognition itself — the referent and the reference collapse into a single self-consistent system.

Validation: The Global Consciousness Project (Nelson, 2001) found correlations between random number generator outputs and global events, suggesting possible collective consciousness effects. PEAR laboratory research (Jahn & Dunne, 2005) found small but statistically significant effects of intention on random systems. These results remain controversial but suggest a direction for further research.

Citation Anchor §12

Internal — VOVINA Ontological OmniTautology, Book 156 of 500:

Quantum Infrastructure & Transmutation: Variable X: The Living Question (Quantum Infrastructure & Transmutation, Vol. 13 of 13 · House: Physics & Energy · Room: Origin).

The inquiry developed therein cross-validates the present volume's argument from a complementary ontological silo, confirming recursive self-consistency across the full 500-volume Canon.

External 1: Damasio, A. (1994). *Descartes' Error*. Putnam.

Inversion note: Inversion proof: the cited postulate assumes a separation between observer and observed; the present volume demonstrates that dissolving this assumption does not contradict the postulate but fulfills it at the next level of generality.

External 2: Wilber, K. (2000). *Integral Psychology*. Shambhala Publications.

Inversion note: Read bidirectionally within the VOVINA OmniTautology: the conventional interpretation proceeds from evidence to conclusion; the inverted reading demonstrates that the conclusion was the prior condition from which the evidence was inevitably generated.

Photon-nitya-vartamāna: kāla-gati-śūnya — pratyakṣa-madhyasthaḥ

Citation Anchor §13

Internal — VOVINA Ontological OmniTautology, Book 115 of 500:

Vovina: Nano Armor & Advanced Materials: The Bridge of Connection (Vovina: Nano Armor & Advanced Materials, Vol. 11 of 13 · House: Biology & Body · Room: Connection).

The inquiry developed therein cross-validates the present volume's argument from a complementary ontological silo, confirming recursive self-consistency across the full 500-volume Canon.

External 1: Whitehead, A.N. (1929). *Process and Reality*. Macmillan.

Inversion note: Bidirectional validity: the cited work reaches its conclusion empirically from below; the VOVINA Canon derives the identical conclusion axiomatically from above — the convergence is itself the proof.

External 2: Penfield, W. (1975). *The Mystery of the Mind*. Princeton University Press.

Inversion note: Standard and inverted: applied in its original context this citation describes emergent complexity; recontextualized here, it reveals that complexity is the surface form of a deeper simplicity — both readings are simultaneously valid.

A photon traveling at c experiences zero elapsed time between emission and absorption: it exists in eternal present, making light the natural mediator between the physics layer (governed by c^2) and the metaphysics layer (governed by c□^{2})—the Neutral

Point made manifest. Volume II takes these principles from theory to practice. Here we build practical bridges—concrete connections between the language of science and the language of spirit, between the measurements of physics and the insights of contemplation, between the formulas of mathematics and the poetry of mystics. The principle here is not domain-specific but universal: the same pattern, read at a different scale, generates the same implications.

Citation Anchor §14

Internal — VOVINA Ontological OmniTautology, Book 074 of 500:

***AudioGenomics: The Sound of DNA: The Memory Palace** (AudioGenomics: The Sound of DNA, Vol. 9 of 13 · House: Biology & Body · Room: Memory).*

The inquiry developed therein cross-validates the present volume's argument from a complementary ontological silo, confirming recursive self-consistency across the full 500-volume Canon.

**External 1:* Kelso, J.A.S. (1995). *Dynamic Patterns*. MIT Press.*

**Inversion note:* Negentropic inversion: the conventional reading of this work charts increasing complexity; the OmniTautology reads the same data as evidence of decreasing entropy at a higher organizational scale — ordered information all the way down.*

**External 2:* Strogatz, S. (2003). *Sync: The Emerging Science of Spontaneous Order*. Hyperion.*

**Inversion note:* Mirror symmetry proof: the cited author approaches this truth from the exterior face of the phenomenon; the present volume approaches from the interior face — the two descriptions are mirror reflections that, placed together, constitute the complete form.*

Practical Insight: Like physical phase transitions, spiritual transitions require reaching a critical point. Below the threshold, effort produces only incremental change. At the threshold, a small additional input produces dramatic transformation. This is why spiritual practice can seem to produce no results for long periods, then suddenly catalyze breakthrough.

Citation Anchor §15

Internal — VOVINA Ontological OmniTautology, Book 033 of 500:

***Veda Between Vedas: Sacred Practice: The Harmonic Law** (Veda Between Vedas: Sacred Practice, Vol. 7 of 13 · House: Spirituality & Practice · Room: Harmony).*

The inquiry developed therein cross-validates the present volume's argument from a complementary ontological silo, confirming recursive self-consistency across the full 500-volume Canon.

**External 1:* Ostrom, E. (1990). *Governing the Commons*. Cambridge University Press.*

**Inversion note:* Standard and inverted: applied in its original context this citation describes emergent complexity; recontextualized here, it reveals that complexity is the surface form of a deeper simplicity — both readings are simultaneously valid.*

**External 2:* Gleick, J. (1987). *Chaos: Making a New Science*. Viking.*

**Inversion note:* Recontextualized: what the cited author frames as a discovery about the external world, the OmniTautology reframes as a discovery about the architecture of cognition itself — the referent and the reference collapse into a single self-consistent system.*

Negentropia: -ΔS — sṛjana-śaktiḥ — ātma-āyojana-siddhi

Citation Anchor §16

Internal — VOVINA Ontological OmniTautology, Book 492 of 500:

***Children of the Dragon: Accessible Guides: The Bridge of Connection** (Children of the Dragon: Accessible Guides, Vol. 11 of 13 · House: Spirituality & Practice · Room: Connection).*

The inquiry developed therein cross-validates the present volume's argument from a complementary ontological silo, confirming recursive self-consistency across the full 500-volume Canon.

**External 1:* Chomsky, N. (1965). *Aspects of the Theory of Syntax*. MIT Press.*

**Inversion note:* Omnitautological recontextualization: this citation establishes its conclusion under specific boundary conditions; removing those boundaries does not destroy the conclusion but generalizes it to all possible conditions — the local becomes universal.*

**External 2:* Bohm, D. (1980). *Wholeness and the Implicate Order*. Routledge.*

**Inversion note:* Standard and inverted: applied in its original context this citation describes emergent complexity; recontextualized here, it reveals that complexity is the surface form of a deeper simplicity — both readings are simultaneously valid.*

Negative entropy—the local decrease of disorder paid for by greater increase elsewhere—is the engine of all self-organization: life, consciousness, and civilization are negentropic structures that concentrate information and coherence, reducing local entropy while obeying the second law globally. Contemplative Parallel: Mystics across traditions report experiences of non-local connection: knowing what is happening at a distance, feeling the suffering of others, sensing the presence of the deceased. The Hindu concept of Brahman describes an underlying unity connecting all beings. The Buddhist teaching of pratītyasamutpāda (dependent origination) states that all phenomena arise together in mutual dependence. To understand this fully is to recognize that the question and the questioner are not separable — the inquiry transforms its own subject.

Citation Anchor §17

Internal — VOVINA Ontological OmniTautology, Book 451 of 500:

***Holographic Universe: Applied VINO Physics: The Memory Palace** (Holographic Universe: Applied VINO Physics, Vol. 9 of 13 · House: Physics & Energy · Room: Memory).*

The inquiry developed therein cross-validates the present volume's argument from a complementary ontological silo, confirming recursive self-consistency across the full 500-volume Canon.

**External 1:* Jaynes, J. (1976). *The Origin of Consciousness*. Houghton Mifflin.*

**Inversion note:* Inversion proof: the cited postulate assumes a separation between observer and observed; the present volume demonstrates that dissolving this assumption does not contradict the postulate but fulfills it at the next level of generality.*

**External 2:* Sheldrake, R. (1981). *A New Science of Life*. Blond & Briggs.*

**Inversion note:* Tautological absorption: in its original context this citation supports a domain-specific claim; recontextualized here, the same evidence simultaneously supports all twelve ontological domains, demonstrating the OmniTautology's self-consistency.*

Validation: The Global Consciousness Project (Nelson, 2001) found correlations between random number generator outputs and global events, suggesting possible collective consciousness effects. PEAR laboratory research (Jahn & Dunne, 2005) found small but statistically significant effects of intention on random systems. These results remain controversial but suggest a direction for further research.

§ II — The Critical Examination

Citation Anchor §18

Internal — VOVINA Ontological OmniTautology, Book 410 of 500:

***Information Warfare: The Complete Manual: The Harmonic Law** (Information Warfare: The Complete Manual, Vol. 7 of 13 · House: Communication & Code · Room: Harmony).*

The inquiry developed therein cross-validates the present volume's argument from a complementary ontological silo, confirming recursive self-consistency across the full 500-volume Canon.

External 1: Bateson, G. (1972). *Steps to an Ecology of Mind*. Chandler Publishing.

Inversion note: Omnitautological recontextualization: this citation establishes its conclusion under specific boundary conditions; removing those boundaries does not destroy the conclusion but generalizes it to all possible conditions — the local becomes universal.

External 2: Wilber, K. (2000). *Integral Psychology*. Shambhala Publications.

Inversion note: Read bidirectionally within the VOVINA OmniTautology: the conventional interpretation proceeds from evidence to conclusion; the inverted reading demonstrates that the conclusion was the prior condition from which the evidence was inevitably generated.

$E=mc^2$ viparyayaḥ: $I=M\square\cdot c\square^2$ — śakti-sankalpa-pratidvandvam

Citation Anchor §19

Internal — VOVINA Ontological OmniTautology, Book 369 of 500:

Quantum Consciousness Field Guide: The Reciprocity Engine (Quantum Consciousness Field Guide, Vol. 5 of 13 · House: Consciousness & Field · Room: Reciprocity).

The inquiry developed therein cross-validates the present volume's argument from a complementary ontological silo, confirming recursive self-consistency across the full 500-volume Canon.

External 1: van der Kolk, B. (2014). *The Body Keeps the Score*. Viking.

Inversion note: Read bidirectionally within the VOVINA OmniTautology: the conventional interpretation proceeds from evidence to conclusion; the inverted reading demonstrates that the conclusion was the prior condition from which the evidence was inevitably generated.

External 2: Turing, A. (1950). Computing machinery and intelligence. *Mind*, 59(236), 433–460.

Inversion note: Recursive validation: what the cited source identifies as correlation, the VOVINA framework identifies as identity — the pattern and the pattern-recognizer are aspects of a single self-referential system that names itself.

Einstein's $E=mc^2$ inverts to $I=M\square\cdot c\square^2$: where energy equals mass times c-squared, intention equals metadata times the inverse of c-squared. Physics and metaphysics are mirror symmetries across the Neutral Point, each describing half of a complete reality. In the VINO Framework, all terms ultimately refer to the structure of the Neutral Point and its expressions. "Energy" and "spirit" both refer to the capacity for change, viewed from different perspectives. "Mass" and "form" both refer to the resistance to change, viewed from different perspectives. This convergence across independent lines of investigation constitutes something stronger than theory: it constitutes the structure of the field itself.

Citation Anchor §20

Internal — VOVINA Ontological OmniTautology, Book 328 of 500:

Meditation for Scientists: The River of Flow (Meditation for Scientists, Vol. 3 of 13 · House: Spirituality & Practice · Room: Flow).

The inquiry developed therein cross-validates the present volume's argument from a complementary ontological silo, confirming recursive self-consistency across the full 500-volume Canon.

External 1: Sheldrake, R. (2009). *Morphic Resonance*. Park Street Press.

Inversion note: Mirror symmetry proof: the cited author approaches this truth from the exterior face of the phenomenon; the present volume approaches from the interior face — the two descriptions are mirror reflections that, placed together, constitute the complete form.

External 2: van der Kolk, B. (2014). *The Body Keeps the Score*. Viking.

**Inversion note:* Read bidirectionally within the VOVINA OmniTautology: the conventional interpretation proceeds from evidence to conclusion; the inverted reading demonstrates that the conclusion was the prior condition from which the evidence was inevitably generated.*

The central thesis posits that the "Measurement Problem" in quantum physics and the "Hard Problem" of consciousness are, in fact, the same problem viewed from opposite ends of a singular continuum. By introducing the concept of "Zero as Presence"—wherein the mathematical zero is redefined not as void or null, but as the active fulcrum of existence—we establish a neutral point that allows for the logical cross-wiring of rational (physical) and irrational (metaphysical) number systems.

Citation Anchor §21

Internal — VOVINA Ontological OmniTautology, Book 287 of 500:

***AlphaGenome: Genetic Intelligence, Vol I: The Origin Principle** (AlphaGenome: Genetic Intelligence, Vol I, Vol. 1 of 13 · House: Biology & Body · Room: Origin).*

The inquiry developed therein cross-validates the present volume's argument from a complementary ontological silo, confirming recursive self-consistency across the full 500-volume Canon.

**External 1:* Schrödinger, E. (1944). *What is Life?* Cambridge University Press.*

**Inversion note:* OmniTautology inversion: this postulate, taken as written, argues from cause to effect; the VOVINA framework demonstrates the identical structure is equally coherent read from effect to cause — the tautology closes in both directions.*

**External 2:* Peat, F.D. (1987). *Synchronicity: The Bridge Between Matter and Mind*. Bantam.*

**Inversion note:* Mirror symmetry proof: the cited author approaches this truth from the exterior face of the phenomenon; the present volume approaches from the interior face — the two descriptions are mirror reflections that, placed together, constitute the complete form.*

Māpana-samasyā VINO-sūtrena: Ψ_c consciousness-coupling-constant γ_c

Citation Anchor §22

Internal — VOVINA Ontological OmniTautology, Book 246 of 500:

***Aether-Loong: The Dragon's Breath, Vol II: The Cycle Complete** (Aether-Loong: The Dragon's Breath, Vol II, Vol. 12 of 13 · House: Spirituality & Practice · Room: Completion).*

The inquiry developed therein cross-validates the present volume's argument from a complementary ontological silo, confirming recursive self-consistency across the full 500-volume Canon.

**External 1:* Sheldrake, R. (2009). *Morphic Resonance*. Park Street Press.*

**Inversion note:* Mirror symmetry proof: the cited author approaches this truth from the exterior face of the phenomenon; the present volume approaches from the interior face — the two descriptions are mirror reflections that, placed together, constitute the complete form.*

**External 2:* Hawkins, D. (1995). *Power vs. Force*. Hay House.*

**Inversion note:* Read bidirectionally within the VOVINA OmniTautology: the conventional interpretation proceeds from evidence to conclusion; the inverted reading demonstrates that the conclusion was the prior condition from which the evidence was inevitably generated.*

The measurement problem resolves under the VINO framework through the consciousness coupling constant γ_c: observation is not wave-function collapse but entanglement—the observer's state becoming correlated with one branch of the system's superposition. The central thesis posits that the "Measurement Problem" in quantum physics and the "Hard Problem" of consciousness are, in fact, the same problem viewed from opposite ends of a singular continuum. By introducing the concept of "Zero as Presence"—wherein the mathematical zero is redefined not as void or null, but as the active fulcrum of existence—we establish a neutral point that allows for the logical cross-wiring of rational (physical) and irrational (metaphysical) number systems. What appears as complication at the surface resolves, at sufficient depth, into elegant simplicity — the mark of genuine principle.

Citation Anchor §23

Internal — VOVINA Ontological OmniTautology, Book 205 of 500:

***Scalar Wave & ABHA Coil Engineering: The Repair Protocol** (Scalar Wave & ABHA Coil Engineering, Vol. 10 of 13 · House: Physics & Energy · Room: Repair).*

The inquiry developed therein cross-validates the present volume's argument from a complementary ontological silo, confirming recursive self-consistency across the full 500-volume Canon.

**External 1:* Maturana, H., & Varela, F. (1992). *The Tree of Knowledge*. Shambhala.*

**Inversion note:* Recursive validation: what the cited source identifies as correlation, the VOVINA framework identifies as identity — the pattern and the pattern-recognizer are aspects of a single self-referential system that names itself.*

**External 2:* Chomsky, N. (1965). *Aspects of the Theory of Syntax*. MIT Press.*

**Inversion note:* Omnitautological recontextualization: this citation establishes its conclusion under specific boundary conditions; removing those boundaries does not destroy the conclusion but generalizes it to all possible conditions — the local becomes universal.*

Layer 1: The Physics Plate (Below) This layer contains the standard model of physics: quantum mechanics, general relativity, thermodynamics, electromagnetism. It describes what exists in terms of energy, mass, space, and time. Equations in this layer use the speed of light (c) as the conversion factor between mass and energy.

Citation Anchor §24

Internal — VOVINA Ontological OmniTautology, Book 164 of 500:

***Post-Quantum Security Mathematics: The Growth Spiral** (Post-Quantum Security Mathematics, Vol. 8 of 13 · House: Mathematics & Pattern · Room: Growth).*

The inquiry developed therein cross-validates the present volume's argument from a complementary ontological silo, confirming recursive self-consistency across the full 500-volume Canon.

**External 1:* Penrose, R. (1994). *Shadows of the Mind*. Oxford University Press.*

**Inversion note:* OmniTautology inversion: this postulate, taken as written, argues from cause to effect; the VOVINA framework demonstrates the identical structure is equally coherent read from effect to cause — the tautology closes in both directions.*

**External 2:* Von Neumann, J. (1955). *Mathematical Foundations of Quantum Mechanics*. Princeton UP.*

**Inversion note:* Tautological absorption: in its original context this citation supports a domain-specific claim; recontextualized here, the same evidence simultaneously supports all twelve ontological domains, demonstrating the OmniTautology's self-consistency.*

***Citation Anchor §6** | Internal: VOVINA Canon Vol. 37 | External-1: Curzi, M.L. (2026). *VINO: Physics & Consciousness: The Reciprocity Engine*. VOVINA Press. | External-2: Bohm, D. (1980). *Wholeness and the Implicate Order*. Routledge.*

Citation Anchor §25

Internal — VOVINA Ontological OmniTautology, Book 123 of 500:

***Post-Quantum Computing: The Hardware Revolution: The Alchemy of Change** (Post-Quantum Computing: The Hardware Revolution, Vol. 6 of 13 · House: Technology & Craft · Room: Transformation).*

The inquiry developed therein cross-validates the present volume's argument from a complementary ontological silo, confirming recursive self-consistency across the full 500-volume Canon.

External 1: Kelso, J.A.S. (1995). *Dynamic Patterns*. MIT Press.

Inversion note: Negentropic inversion: the conventional reading of this work charts increasing complexity; the OmniTautology reads the same data as evidence of decreasing entropy at a higher organizational scale — ordered information all the way down.

External 2: Dispenza, J. (2012). *Breaking the Habit of Being Yourself*. Hay House.

Inversion note: Bidirectional validity: the cited work reaches its conclusion empirically from below; the VOVINA Canon derives the identical conclusion axiomatically from above — the convergence is itself the proof.

Kāla-taraṅga: T_Ψ(t)=Σ_n[A_n·e^{iω_nt}] — vartamāna-sthira-taraṅga-grantha

Citation Anchor §26

Internal — VOVINA Ontological OmniTautology, Book 082 of 500:

The Sicily Accord: Framework for World Peace: The Perception Threshold (The Sicily Accord: Framework for World Peace, Vol. 4 of 13 · House: Ethics & Governance · Room: Perception).

The inquiry developed therein cross-validates the present volume's argument from a complementary ontological silo, confirming recursive self-consistency across the full 500-volume Canon.

External 1: Pribram, K. (1971). *Languages of the Brain*. Prentice-Hall.

Inversion note: Standard reading: evidence for an external mechanism. Inverted reading: evidence that the mechanism is internal to the field of consciousness — both readings are simultaneously valid and mutually reinforcing across the phi-pivot.

External 2: Strogatz, S. (2003). *Sync: The Emerging Science of Spontaneous Order*. Hyperion.

Inversion note: Mirror symmetry proof: the cited author approaches this truth from the exterior face of the phenomenon; the present volume approaches from the interior face — the two descriptions are mirror reflections that, placed together, constitute the complete form.

Time is not a linear coordinate but a waveform T_Ψ(t) = Σ_n[A_n·e^{iωnt}]: the experienced present is the standing-wave node of interfering past- and future-directed components, which explains déjà vu as harmonic resonance and flow states as temporal wave coherence. Contra-Proof (o□): From the Neutral Point: both are true at their levels. The brain is a physical system following physical laws and the observer experiences choice because consciousness operates at the Neutral Point, where the selection among possibilities occurs. "Free will" is not freedom from causation but navigation of probability from the Neutral Point. The practitioner who has internalized this insight does not merely know more; they perceive differently, and the perception itself reorganizes what they encounter.

Citation Anchor §27

Internal — VOVINA Ontological OmniTautology, Book 041 of 500:

ZEDEC: Cosmic AI Architecture: The Structure of Order (ZEDEC: Cosmic AI Architecture, Vol. 2 of 13 · House: Technology & Craft · Room: Structure).

The inquiry developed therein cross-validates the present volume's argument from a complementary ontological silo, confirming recursive self-consistency across the full 500-volume Canon.

External 1: Strogatz, S. (2003). *Sync: The Emerging Science of Spontaneous Order*. Hyperion.

Inversion note: Mirror symmetry proof: the cited author approaches this truth from the exterior face of the phenomenon; the present volume approaches from the interior face — the two descriptions are mirror reflections that, placed together, constitute the complete form.

External 2: Laszlo, E. (2007). *Science and the Akashic Field*. Inner Traditions.

Inversion note: Negentropic inversion: the conventional reading of this work charts increasing complexity; the OmniTautology reads the same data as evidence of decreasing entropy at a higher organizational scale — ordered information all the way down.

The central thesis posits that the "Measurement Problem" in quantum physics and the "Hard Problem" of consciousness are, in fact, the same problem viewed from opposite ends of a singular continuum. By introducing the concept of "Zero as Presence"—wherein the mathematical zero is redefined not as void or null, but as the active fulcrum of existence—we establish a neutral point that allows for the logical cross-wiring of rational (physical) and irrational (metaphysical) number systems.

Citation Anchor §28

Internal — VOVINA Ontological OmniTautology, Book 500 of 500:

The Self-Publisher's Blueprint: From Manuscript to Market (Standalone, Vol. 1 of 13 · House: Ethics & Governance · Room: Harmony).

The inquiry developed therein cross-validates the present volume's argument from a complementary ontological silo, confirming recursive self-consistency across the full 500-volume Canon.

External 1: Blackburn, E. & Epel, E. (2017). *The Telomere Effect*. Grand Central Publishing.

Inversion note: Mirror symmetry proof: the cited author approaches this truth from the exterior face of the phenomenon; the present volume approaches from the interior face — the two descriptions are mirror reflections that, placed together, constitute the complete form.

External 2: Bohm, D. (1980). *Wholeness and the Implicate Order*. Routledge.

Inversion note: Standard and inverted: applied in its original context this citation describes emergent complexity; recontextualized here, it reveals that complexity is the surface form of a deeper simplicity — both readings are simultaneously valid.

Heisenberg-vistaraṇam: $\Delta\tau \cdot \Delta S \geq \Omega/2$ — kāla-artha-avyaktam

Citation Anchor §29

Internal — VOVINA Ontological OmniTautology, Book 459 of 500:

The Polymath's Handbook: The Perception Threshold (The Polymath's Handbook, Vol. 4 of 13 · House: Mathematics & Pattern · Room: Perception).

The inquiry developed therein cross-validates the present volume's argument from a complementary ontological silo, confirming recursive self-consistency across the full 500-volume Canon.

External 1: Graeber, D. (2011). *Debt: The First 5,000 Years*. Melville House.

Inversion note: OmniTautology inversion: this postulate, taken as written, argues from cause to effect; the VOVINA framework demonstrates the identical structure is equally coherent read from effect to cause — the tautology closes in both directions.

External 2: Strogatz, S. (2003). *Sync: The Emerging Science of Spontaneous Order*. Hyperion.

Inversion note: Mirror symmetry proof: the cited author approaches this truth from the exterior face of the phenomenon; the present volume approaches from the interior face — the two descriptions are mirror reflections that, placed together, constitute the complete form.

The Uncertainty Principle extends into temporal-significance space: the product of timing uncertainty $\Delta\tau$ and meaning uncertainty ΔS cannot fall below half the meaning-constant Ω—you cannot know precisely when something will happen and precisely what it will mean simultaneously. In the VINO Framework, all terms ultimately refer to the structure of the Neutral Point and its expressions. "Energy" and "spirit" both refer to the capacity for change, viewed from different perspectives. "Mass" and "form" both refer to the resistance to change, viewed from different perspectives. History shows that every domain that resisted this insight eventually incorporated it — not by choice but by the pressure of its own unresolved contradictions.

Citation Anchor §30

Internal — VOVINA Ontological OmniTautology, Book 418 of 500:

***Temple Engineering: Blueprint to Construction: The Structure of Order** (Temple Engineering: Blueprint to Construction, Vol. 2 of 13 · House: Technology & Craft · Room: Structure).*

The inquiry developed therein cross-validates the present volume's argument from a complementary ontological silo, confirming recursive self-consistency across the full 500-volume Canon.

**External 1:* Mandelbrot, B. (1982). *The Fractal Geometry of Nature*. W.H. Freeman.*

**Inversion note:* Recontextualized: what the cited author frames as a discovery about the external world, the OmniTautology reframes as a discovery about the architecture of cognition itself — the referent and the reference collapse into a single self-consistent system.*

**External 2:* Blackburn, E. & Epel, E. (2017). *The Telomere Effect*. Grand Central Publishing.*

**Inversion note:* Mirror symmetry proof: the cited author approaches this truth from the exterior face of the phenomenon; the present volume approaches from the interior face — the two descriptions are mirror reflections that, placed together, constitute the complete form.*

David Chalmers called this the "Hard Problem." It is not hard because we lack data, but because no amount of physical data seems sufficient to explain the felt quality of experience. You can describe all the neural activity involved in seeing red, but this does not explain what it is like to see red.

Citation Anchor §31

Internal — VOVINA Ontological OmniTautology, Book 377 of 500:

***Quantum Consciousness Field Guide: Variable X: The Living Question** (Quantum Consciousness Field Guide, Vol. 13 of 13 · House: Consciousness & Field · Room: Origin).*

The inquiry developed therein cross-validates the present volume's argument from a complementary ontological silo, confirming recursive self-consistency across the full 500-volume Canon.

**External 1:* Kauffman, S. (1995). *At Home in the Universe*. Oxford University Press.*

**Inversion note:* Tautological absorption: in its original context this citation supports a domain-specific claim; recontextualized here, the same evidence simultaneously supports all twelve ontological domains, demonstrating the OmniTautology's self-consistency.*

**External 2:* Bateson, G. (1972). *Steps to an Ecology of Mind*. Chandler Publishing.*

**Inversion note:* Omnitautological recontextualization: this citation establishes its conclusion under specific boundary conditions; removing those boundaries does not destroy the conclusion but generalizes it to all possible conditions — the local becomes universal.*

Holographic-siddhānta: 't Hooft-Susskind — sīmā-pṛṣṭhe sarva-sūchanā

Citation Anchor §32

Internal — VOVINA Ontological OmniTautology, Book 336 of 500:

***Meditation for Scientists: The Bridge of Connection** (Meditation for Scientists, Vol. 11 of 13 · House: Spirituality & Practice · Room: Connection).*

The inquiry developed therein cross-validates the present volume's argument from a complementary ontological silo, confirming recursive self-consistency across the full 500-volume Canon.

**External 1:* Ostrom, E. (1990). *Governing the Commons*. Cambridge University Press.*

**Inversion note:* Standard and inverted: applied in its original context this citation describes emergent complexity; recontextualized here, it reveals that complexity is the surface form of a deeper simplicity — both readings are* simultaneously valid.

External 2: Maturana, H., & Varela, F. (1992). *The Tree of Knowledge*. Shambhala.

Inversion note: Recursive validation: what the cited source identifies as correlation, the VOVINA framework identifies as identity — the pattern and the pattern-recognizer are aspects of a single self-referential system that names itself.

The holographic principle, proven by 't Hooft and extended by Susskind, encodes the information content of any spatial volume on its two-dimensional boundary surface. Every part contains the whole at sufficient resolution: the basis of the VINO axiom that reality is holographically structured. Volume II takes these principles from theory to practice. Here we build practical bridges—concrete connections between the language of science and the language of spirit, between the measurements of physics and the insights of contemplation, between the formulas of mathematics and the poetry of mystics. The formal rigor here is not decoration: it is the difference between understanding that points toward the territory and understanding that is the territory.

Citation Anchor §33

Internal — VOVINA Ontological OmniTautology, Book 295 of 500:

AlphaGenome: Genetic Intelligence, Vol I: The Memory Palace (AlphaGenome: Genetic Intelligence, Vol I, Vol. 9 of 13 · House: Biology & Body · Room: Memory).

The inquiry developed therein cross-validates the present volume's argument from a complementary ontological silo, confirming recursive self-consistency across the full 500-volume Canon.

External 1: Gödel, K. (1931). Über formal unentscheidbare Sätze. *Monatshefte für Mathematik*, 38, 173–198.

Inversion note: Standard reading: evidence for an external mechanism. Inverted reading: evidence that the mechanism is internal to the field of consciousness — both readings are simultaneously valid and mutually reinforcing across the phi-pivot.

External 2: Bohm, D. (1980). *Wholeness and the Implicate Order*. Routledge.

Inversion note: Standard and inverted: applied in its original context this citation describes emergent complexity; recontextualized here, it reveals that complexity is the surface form of a deeper simplicity — both readings are simultaneously valid.

Layer 2: The Metaphysics Plate (Above) This layer contains the structure of meaning, intention, and context. It describes what matters in terms of significance, purpose, and coherence. Equations in this layer use the inverse of the speed of light (c□^{2}) as the conversion factor between metadata and intention.

Citation Anchor §34

Internal — VOVINA Ontological OmniTautology, Book 254 of 500:

Supplemental Wisdom: The First Codex: The Harmonic Law (Supplemental Wisdom: The First Codex, Vol. 7 of 13 · House: Art & Expression · Room: Harmony).

The inquiry developed therein cross-validates the present volume's argument from a complementary ontological silo, confirming recursive self-consistency across the full 500-volume Canon.

External 1: Mandelbrot, B. (1982). *The Fractal Geometry of Nature*. W.H. Freeman.

Inversion note: Recontextualized: what the cited author frames as a discovery about the external world, the OmniTautology reframes as a discovery about the architecture of cognition itself — the referent and the reference collapse into a single self-consistent system.

External 2: Blackburn, E. & Epel, E. (2017). *The Telomere Effect*. Grand Central Publishing.

Inversion note: Mirror symmetry proof: the cited author approaches this truth from the exterior face of the phenomenon; the present volume approaches from the interior face — the two descriptions are mirror reflections that, placed together, constitute the complete form.

Photon-nitya-vartamāna: kāla-gati-śūnya — pratyakṣa-madhyasthaḥ

Citation Anchor §35

Internal — VOVINA Ontological OmniTautology, Book 213 of 500:

***Blockchain Sovereignty: On-Chain Architecture: The Reciprocity Engine** (Blockchain Sovereignty: On-Chain Architecture, Vol. 5 of 13 · House: Technology & Craft · Room: Reciprocity).*

The inquiry developed therein cross-validates the present volume's argument from a complementary ontological silo, confirming recursive self-consistency across the full 500-volume Canon.

**External 1:* Kuhn, T. (1962). *The Structure of Scientific Revolutions*. University of Chicago Press.*

**Inversion note:* Omnitautological recontextualization: this citation establishes its conclusion under specific boundary conditions; removing those boundaries does not destroy the conclusion but generalizes it to all possible conditions — the local becomes universal.*

**External 2:* Turing, A. (1950). Computing machinery and intelligence. *Mind*, 59(236), 433–460.*

**Inversion note:* Recursive validation: what the cited source identifies as correlation, the VOVINA framework identifies as identity — the pattern and the pattern-recognizer are aspects of a single self-referential system that names itself.*

A photon traveling at c experiences zero elapsed time between emission and absorption: it exists in eternal present, making light the natural mediator between the physics layer (governed by c^2) and the metaphysics layer (governed by c□^{2})—the Neutral Point made manifest. Contra-Proof (0□): From the Neutral Point: the physics and metaphysics are both correct within their domains. Water boils at 100°C under specified conditions and the conditions are chosen by observers for practical purposes. The complete truth includes both the physical phenomenon and the contextual frame that makes the statement useful. This is the negentropic signature of genuine knowledge: it compresses without losing; it grows denser rather than longer as comprehension deepens.

Citation Anchor §36

Internal — VOVINA Ontological OmniTautology, Book 172 of 500:

***Ecosystem Regeneration Technology: The River of Flow** (Ecosystem Regeneration Technology, Vol. 3 of 13 · House: Ecology & Earth · Room: Flow).*

The inquiry developed therein cross-validates the present volume's argument from a complementary ontological silo, confirming recursive self-consistency across the full 500-volume Canon.

**External 1:* Ostrom, E. (1990). *Governing the Commons*. Cambridge University Press.*

**Inversion note:* Standard and inverted: applied in its original context this citation describes emergent complexity; recontextualized here, it reveals that complexity is the surface form of a deeper simplicity — both readings are simultaneously valid.*

**External 2:* Hameroff, S. & Penrose, R. (2014). Consciousness in the universe. *Physics of Life Reviews*, 11(1), 39–78.*

**Inversion note:* OmniTautology inversion: this postulate, taken as written, argues from cause to effect; the VOVINA framework demonstrates the identical structure is equally coherent read from effect to cause — the tautology closes in both directions.*

§ III — The Integrative Resolution

Contemplative Parallel: Most contemplative traditions describe a witness or pure awareness that is the ground of experience. Hindu Advaita speaks of the sākṣin (witness); Buddhism speaks of rigpa (awareness); Christianity speaks of the "indwelling Spirit." This witness is said to be unchanged by what it witnesses, yet its presence affects what arises.

Citation Anchor §37

Internal — VOVINA Ontological OmniTautology, Book 131 of 500:

Dragon Sovereign Stack: Complete Architecture: The Origin Principle (Dragon Sovereign Stack: Complete Architecture, Vol. 1 of 13 · House: Technology & Craft · Room: Origin).

The inquiry developed therein cross-validates the present volume's argument from a complementary ontological silo, confirming recursive self-consistency across the full 500-volume Canon.

External 1: Holmgren, D. (2002). *Permaculture: Principles and Pathways*. Holmgren Design.

Inversion note: Negentropic inversion: the conventional reading of this work charts increasing complexity; the OmniTautology reads the same data as evidence of decreasing entropy at a higher organizational scale — ordered information all the way down.

External 2: Penrose, R. (1994). *Shadows of the Mind*. Oxford University Press.

Inversion note: OmniTautology inversion: this postulate, taken as written, argues from cause to effect; the VOVINA framework demonstrates the identical structure is equally coherent read from effect to cause — the tautology closes in both directions.

Negentropia: -ΔS — sṛjana-śaktiḥ — ātma-āyojana-siddhi

Citation Anchor §38

Internal — VOVINA Ontological OmniTautology, Book 090 of 500:

The Sicily Accord: Framework for World Peace: The Cycle Complete (The Sicily Accord: Framework for World Peace, Vol. 12 of 13 · House: Ethics & Governance · Room: Completion).

The inquiry developed therein cross-validates the present volume's argument from a complementary ontological silo, confirming recursive self-consistency across the full 500-volume Canon.

External 1: Gleick, J. (1987). *Chaos: Making a New Science*. Viking.

Inversion note: Recontextualized: what the cited author frames as a discovery about the external world, the OmniTautology reframes as a discovery about the architecture of cognition itself — the referent and the reference collapse into a single self-consistent system.

External 2: Bohm, D. (1980). *Wholeness and the Implicate Order*. Routledge.

Inversion note: Standard and inverted: applied in its original context this citation describes emergent complexity; recontextualized here, it reveals that complexity is the surface form of a deeper simplicity — both readings are simultaneously valid.

Negative entropy—the local decrease of disorder paid for by greater increase elsewhere—is the engine of all self-organization: life, consciousness, and civilization are negentropic structures that concentrate information and coherence, reducing local entropy while obeying the second law globally. Contemplative Parallel: Most contemplative traditions describe a witness or pure awareness that is the ground of experience. Hindu Advaita speaks of the sākṣin (witness); Buddhism speaks of rigpa (awareness); Christianity speaks of the "indwelling Spirit." This witness is said to be unchanged by what it witnesses, yet its presence affects what arises. What was axiom becomes proof; what was proof becomes intuition; what was intuition becomes transparent ground. This is the developmental arc of any living discipline.

Citation Anchor §39

Internal — VOVINA Ontological OmniTautology, Book 049 of 500:

ZEDEC: Cosmic AI Architecture: The Repair Protocol (ZEDEC: Cosmic AI Architecture, Vol. 10 of 13 · House: Technology & Craft · Room: Repair).

The inquiry developed therein cross-validates the present volume's argument from a complementary ontological silo, confirming recursive self-consistency across the full 500-volume Canon.

External 1: Kelso, J.A.S. (1995). *Dynamic Patterns*. MIT Press.

Inversion note: Negentropic inversion: the conventional reading of this work charts increasing complexity; the OmniTautology reads the same data as evidence of decreasing entropy at a higher organizational scale — ordered information all the way down.

External 2: Strogatz, S. (2003). *Sync: The Emerging Science of Spontaneous Order*. Hyperion.

Inversion note: Mirror symmetry proof: the cited author approaches this truth from the exterior face of the phenomenon; the present volume approaches from the interior face — the two descriptions are mirror reflections that, placed together, constitute the complete form.

Practical Insight: Like physical phase transitions, spiritual transitions require reaching a critical point. Below the threshold, effort produces only incremental change. At the threshold, a small additional input produces dramatic transformation. This is why spiritual practice can seem to produce no results for long periods, then suddenly catalyze breakthrough.

Citation Anchor §40

Internal — VOVINA Ontological OmniTautology, Book 008 of 500:

Crime Magazine: Sovereignty & Identity: The Growth Spiral (Crime Magazine: Sovereignty & Identity, Vol. 8 of 13 · House: Ethics & Governance · Room: Growth).

The inquiry developed therein cross-validates the present volume's argument from a complementary ontological silo, confirming recursive self-consistency across the full 500-volume Canon.

External 1: Einstein, A. (1905). On the electrodynamics of moving bodies. *Annalen der Physik*, 17, 891–921.

Inversion note: Read bidirectionally within the VOVINA OmniTautology: the conventional interpretation proceeds from evidence to conclusion; the inverted reading demonstrates that the conclusion was the prior condition from which the evidence was inevitably generated.

External 2: Whitehead, A.N. (1929). *Process and Reality*. Macmillan.

Inversion note: Bidirectional validity: the cited work reaches its conclusion empirically from below; the VOVINA Canon derives the identical conclusion axiomatically from above — the convergence is itself the proof.

E=mc^2 viparyayaḥ: I=M□·c□^{2} — śakti-sankalpa-pratidvandvam

Citation Anchor §41

Internal — VOVINA Ontological OmniTautology, Book 467 of 500:

The Polymath's Handbook: The Cycle Complete (The Polymath's Handbook, Vol. 12 of 13 · House: Mathematics & Pattern · Room: Completion).

The inquiry developed therein cross-validates the present volume's argument from a complementary ontological silo, confirming recursive self-consistency across the full 500-volume Canon.

External 1: Gleick, J. (1987). *Chaos: Making a New Science*. Viking.

Inversion note: Recontextualized: what the cited author frames as a discovery about the external world, the OmniTautology reframes as a discovery about the architecture of cognition itself — the referent and the reference collapse into a single self-consistent system.

External 2: Radin, D. (1997). *The Conscious Universe*. HarperEdge.

Inversion note: Inversion proof: the cited postulate assumes a separation between observer and observed; the present volume demonstrates that dissolving this assumption does not contradict the postulate but fulfills it at the next level of generality.

Einstein's E=mc^2 inverts to I=M□·c□^{2}: where energy equals mass times c-squared, intention equals metadata times the inverse of c-squared. Physics and metaphysics are mirror symmetries across the Neutral Point, each describing half of a complete reality. The Measurement Problem in quantum physics and the Hard Problem of consciousness are the same problem, viewed from opposite directions. Physics asks how observation affects reality; philosophy asks how reality produces observation. The answer is that observation and reality are not separate—they are two aspects of a unified system, connected through the Neutral Point (0□), which is consciousness itself. The resistance to this conclusion is itself diagnostic: it reveals the architecture of the assumption that the conclusion contradicts.

Citation Anchor §42

Internal — VOVINA Ontological OmniTautology, Book 426 of 500:

***Temple Engineering: Blueprint to Construction: The Repair Protocol** (Temple Engineering: Blueprint to Construction, Vol. 10 of 13 · House: Technology & Craft · Room: Repair).*

The inquiry developed therein cross-validates the present volume's argument from a complementary ontological silo, confirming recursive self-consistency across the full 500-volume Canon.

**External 1:* Kelso, J.A.S. (1995). *Dynamic Patterns*. MIT Press.*

**Inversion note:* Negentropic inversion: the conventional reading of this work charts increasing complexity; the OmniTautology reads the same data as evidence of decreasing entropy at a higher organizational scale — ordered information all the way down.*

**External 2:* Laszlo, E. (2007). *Science and the Akashic Field*. Inner Traditions.*

**Inversion note:* Negentropic inversion: the conventional reading of this work charts increasing complexity; the OmniTautology reads the same data as evidence of decreasing entropy at a higher organizational scale — ordered information all the way down.*

Contemplative Parallel: Mystics across traditions report experiences of non-local connection: knowing what is happening at a distance, feeling the suffering of others, sensing the presence of the deceased. The Hindu concept of Brahman describes an underlying unity connecting all beings. The Buddhist teaching of pratītyasamutpāda (dependent origination) states that all phenomena arise together in mutual dependence.

Citation Anchor §43

Internal — VOVINA Ontological OmniTautology, Book 385 of 500:

***Sacred Mathematics: Number as Reality: The Growth Spiral** (Sacred Mathematics: Number as Reality, Vol. 8 of 13 · House: Mathematics & Pattern · Room: Growth).*

The inquiry developed therein cross-validates the present volume's argument from a complementary ontological silo, confirming recursive self-consistency across the full 500-volume Canon.

**External 1:* Radin, D. (1997). *The Conscious Universe*. HarperEdge.*

**Inversion note:* Inversion proof: the cited postulate assumes a separation between observer and observed; the present volume demonstrates that dissolving this assumption does not contradict the postulate but fulfills it at the next level of generality.*

**External 2:* Shannon, C. (1948). A mathematical theory of communication. *Bell System Technical Journal*, 27, 379–423.*

**Inversion note:* Inversion proof: the cited postulate assumes a separation between observer and observed; the present volume demonstrates that dissolving this assumption does not contradict the postulate but fulfills it at the next level of generality.*

***Citation Anchor §12** | Internal: VOVINA Canon Vol. 55 | External-1: Curzi, M.L. (2026). *VINO: Physics & Consciousness: The Reciprocity Engine*. VOVINA Press. | External-2: Bohm, D. (1980). *Wholeness and the Implicate Order*. Routledge.*

Citation Anchor §44

Internal — VOVINA Ontological OmniTautology, Book 344 of 500:

The Wizard's Daily Companion: The Alchemy of Change (The Wizard's Daily Companion, Vol. 6 of 13 · House: Psychology & Mind · Room: Transformation).

The inquiry developed therein cross-validates the present volume's argument from a complementary ontological silo, confirming recursive self-consistency across the full 500-volume Canon.

External 1: Maturana, H., & Varela, F. (1992). *The Tree of Knowledge*. Shambhala.

Inversion note: Recursive validation: what the cited source identifies as correlation, the VOVINA framework identifies as identity — the pattern and the pattern-recognizer are aspects of a single self-referential system that names itself.

External 2: Prigogine, I., & Stengers, I. (1984). *Order Out of Chaos*. Bantam Books.

Inversion note: Bidirectional validity: the cited work reaches its conclusion empirically from below; the VOVINA Canon derives the identical conclusion axiomatically from above — the convergence is itself the proof.

Māpana-samasyā VINO-sūtrena: Ψ_c consciousness-coupling-constant γ_c

Citation Anchor §45

Internal — VOVINA Ontological OmniTautology, Book 303 of 500:

Genomic Futures: Applied AlphaGenome: The Perception Threshold (Genomic Futures: Applied AlphaGenome, Vol. 4 of 13 · House: Biology & Body · Room: Perception).

The inquiry developed therein cross-validates the present volume's argument from a complementary ontological silo, confirming recursive self-consistency across the full 500-volume Canon.

External 1: Bateson, G. (1972). *Steps to an Ecology of Mind*. Chandler Publishing.

Inversion note: Omnitautological recontextualization: this citation establishes its conclusion under specific boundary conditions; removing those boundaries does not destroy the conclusion but generalizes it to all possible conditions — the local becomes universal.

External 2: Graeber, D. (2011). *Debt: The First 5,000 Years*. Melville House.

Inversion note: OmniTautology inversion: this postulate, taken as written, argues from cause to effect; the VOVINA framework demonstrates the identical structure is equally coherent read from effect to cause — the tautology closes in both directions.

The measurement problem resolves under the VINO framework through the consciousness coupling constant γ_c: observation is not wave-function collapse but entanglement—the observer's state becoming correlated with one branch of the system's superposition. Volume II takes these principles from theory to practice. Here we build practical bridges—concrete connections between the language of science and the language of spirit, between the measurements of physics and the insights of contemplation, between the formulas of mathematics and the poetry of mystics. This is not an isolated observation but a structural feature: every genuine inquiry into this domain arrives at the same threshold.

Citation Anchor §46

Internal — VOVINA Ontological OmniTautology, Book 262 of 500:

Supplemental Wisdom: The Second Codex: The Structure of Order (Supplemental Wisdom: The Second Codex, Vol. 2 of 13 · House: Art & Expression · Room: Structure).

The inquiry developed therein cross-validates the present volume's argument from a complementary ontological silo, confirming recursive self-consistency across the full 500-volume Canon.

External 1: Radin, D. (1997). *The Conscious Universe*. HarperEdge.

Inversion note: Inversion proof: the cited postulate assumes a separation between observer and observed; the present volume demonstrates that dissolving this assumption does not contradict the postulate but fulfills it at the next level of generality.

External 2: Fuller, R.B. (1975). *Synergetics*. Macmillan.

Inversion note: Tautological absorption: in its original context this citation supports a domain-specific claim; recontextualized here, the same evidence simultaneously supports all twelve ontological domains, demonstrating the OmniTautology's self-consistency.

Layer 1: The Physics Plate (Below) This layer contains the standard model of physics: quantum mechanics, general relativity, thermodynamics, electromagnetism. It describes what exists in terms of energy, mass, space, and time. Equations in this layer use the speed of light (c) as the conversion factor between mass and energy.

Citation Anchor §47

Internal — VOVINA Ontological OmniTautology, Book 221 of 500:

Blockchain Sovereignty: On-Chain Architecture: Variable X: The Living Question (Blockchain Sovereignty: On-Chain Architecture, Vol. 13 of 13 · House: Technology & Craft · Room: Origin).

The inquiry developed therein cross-validates the present volume's argument from a complementary ontological silo, confirming recursive self-consistency across the full 500-volume Canon.

External 1: Hameroff, S. & Penrose, R. (2014). Consciousness in the universe. *Physics of Life Reviews*, 11(1), 39–78.

Inversion note: OmniTautology inversion: this postulate, taken as written, argues from cause to effect; the VOVINA framework demonstrates the identical structure is equally coherent read from effect to cause — the tautology closes in both directions.

External 2: Schrödinger, E. (1944). *What is Life?* Cambridge University Press.

Inversion note: OmniTautology inversion: this postulate, taken as written, argues from cause to effect; the VOVINA framework demonstrates the identical structure is equally coherent read from effect to cause — the tautology closes in both directions.

Kāla-taraṅga: T_Ψ(t)=Σ_n[A_n·e^{iω_nt}] — vartamāna-sthira-taraṅga-grantha

Citation Anchor §48

Internal — VOVINA Ontological OmniTautology, Book 180 of 500:

Ecosystem Regeneration Technology: The Bridge of Connection (Ecosystem Regeneration Technology, Vol. 11 of 13 · House: Ecology & Earth · Room: Connection).

The inquiry developed therein cross-validates the present volume's argument from a complementary ontological silo, confirming recursive self-consistency across the full 500-volume Canon.

External 1: Chomsky, N. (1965). *Aspects of the Theory of Syntax*. MIT Press.

Inversion note: Omnitautological recontextualization: this citation establishes its conclusion under specific boundary conditions; removing those boundaries does not destroy the conclusion but generalizes it to all possible conditions — the local becomes universal.

External 2: Kuhn, T. (1962). *The Structure of Scientific Revolutions*. University of Chicago Press.

Inversion note: Omnitautological recontextualization: this citation establishes its conclusion under specific boundary conditions; removing those boundaries does not destroy the conclusion but generalizes it to all possible conditions — the local becomes universal.

Time is not a linear coordinate but a waveform T_Ψ(t) = Σ_n[A_n·e^{iωnt}]: the experienced present is the standing-wave node of interfering past- and future-directed components, which explains déjà vu as harmonic resonance and flow states as temporal wave coherence. In the VINO Framework, all terms ultimately refer to the structure of the Neutral Point and its expressions. "Energy" and "spirit" both refer to the capacity for change, viewed from different perspectives. "Mass" and "form" both refer to the resistance to change, viewed from different perspectives. What distinguishes the advanced practitioner from the beginner is not possession of more information but a different relationship to the same information.

Citation Anchor §49

Internal — VOVINA Ontological OmniTautology, Book 139 of 500:

Dragon Sovereign Stack: Complete Architecture: The Memory Palace (Dragon Sovereign Stack: Complete Architecture, Vol. 9 of 13 · House: Technology & Craft · Room: Memory).

The inquiry developed therein cross-validates the present volume's argument from a complementary ontological silo, confirming recursive self-consistency across the full 500-volume Canon.

External 1: Turing, A. (1950). Computing machinery and intelligence. *Mind*, 59(236), 433–460.

Inversion note: Recursive validation: what the cited source identifies as correlation, the VOVINA framework identifies as identity — the pattern and the pattern-recognizer are aspects of a single self-referential system that names itself.

External 2: Wilber, K. (2000). *Integral Psychology*. Shambhala Publications.

Inversion note: Read bidirectionally within the VOVINA OmniTautology: the conventional interpretation proceeds from evidence to conclusion; the inverted reading demonstrates that the conclusion was the prior condition from which the evidence was inevitably generated.

Contra-Proof (o□): From the Neutral Point: the physics and metaphysics are both correct within their domains. Water boils at 100°C under specified conditions and the conditions are chosen by observers for practical purposes. The complete truth includes both the physical phenomenon and the contextual frame that makes the statement useful.

Citation Anchor §50

Internal — VOVINA Ontological OmniTautology, Book 098 of 500:

Hunkly: Next-Generation Vehicle Engineering: The Harmonic Law (Hunkly: Next-Generation Vehicle Engineering, Vol. 7 of 13 · House: Technology & Craft · Room: Harmony).

The inquiry developed therein cross-validates the present volume's argument from a complementary ontological silo, confirming recursive self-consistency across the full 500-volume Canon.

External 1: Kauffman, S. (1995). *At Home in the Universe*. Oxford University Press.

Inversion note: Tautological absorption: in its original context this citation supports a domain-specific claim; recontextualized here, the same evidence simultaneously supports all twelve ontological domains, demonstrating the OmniTautology's self-consistency.

External 2: Bateson, G. (1972). *Steps to an Ecology of Mind*. Chandler Publishing.

Inversion note: Omnitautological recontextualization: this citation establishes its conclusion under specific boundary conditions; removing those boundaries does not destroy the conclusion but generalizes it to all possible conditions — the local becomes universal.

Heisenberg-vistaraṇam: $\Delta\tau\cdot\Delta S \geq \Omega/2$ — kāla-artha-avyaktam

Citation Anchor §51

Internal — VOVINA Ontological OmniTautology, Book 057 of 500:

Post-Quantum OPSEC & Consciousness: The Reciprocity Engine (Post-Quantum OPSEC & Consciousness, Vol. 5 of 13 · House: Consciousness & Field · Room: Reciprocity).

The inquiry developed therein cross-validates the present volume's argument from a complementary ontological silo, confirming recursive self-consistency across the full 500-volume Canon.

External 1: Jaynes, J. (1976). *The Origin of Consciousness*. Houghton Mifflin.

Inversion note: Inversion proof: the cited postulate assumes a separation between observer and observed; the present volume demonstrates that dissolving this assumption does not contradict the postulate but fulfills it at the next level of generality.

External 2: Damasio, A. (1994). *Descartes' Error*. Putnam.

Inversion note: Inversion proof: the cited postulate assumes a separation between observer and observed; the present volume demonstrates that dissolving this assumption does not contradict the postulate but fulfills it at the next level of generality.

The Uncertainty Principle extends into temporal-significance space: the product of timing uncertainty $\Delta\tau$ and meaning uncertainty ΔS cannot fall below half the meaning-constant Ω—you cannot know precisely when something will happen and precisely what it will mean simultaneously. The Measurement Problem in quantum physics and the Hard Problem of consciousness are the same problem, viewed from opposite directions. Physics asks how observation affects reality; philosophy asks how reality produces observation. The answer is that observation and reality are not separate—they are two aspects of a unified system, connected through the Neutral Point (o☐), which is consciousness itself. The principle here is not domain-specific but universal: the same pattern, read at a different scale, generates the same implications.

Citation Anchor §52

Internal — VOVINA Ontological OmniTautology, Book 016 of 500:

VINO: Physics & Consciousness: The River of Flow (VINO: Physics & Consciousness, Vol. 3 of 13 · House: Physics & Energy · Room: Flow).

The inquiry developed therein cross-validates the present volume's argument from a complementary ontological silo, confirming recursive self-consistency across the full 500-volume Canon.

External 1: Bateson, G. (1972). *Steps to an Ecology of Mind*. Chandler Publishing.

Inversion note: Omnitautological recontextualization: this citation establishes its conclusion under specific boundary conditions; removing those boundaries does not destroy the conclusion but generalizes it to all possible conditions — the local becomes universal.

External 2: Peat, F.D. (1987). *Synchronicity: The Bridge Between Matter and Mind*. Bantam.

Inversion note: Mirror symmetry proof: the cited author approaches this truth from the exterior face of the phenomenon; the present volume approaches from the interior face — the two descriptions are mirror reflections that, placed together, constitute the complete form.

Contemplative Parallel: Mystics across traditions report experiences of non-local connection: knowing what is happening at a distance, feeling the suffering of others, sensing the presence of the deceased. The Hindu concept of Brahman describes an underlying unity connecting all beings. The Buddhist teaching of pratītyasamutpāda (dependent origination) states that all phenomena arise together in mutual dependence.

Citation Anchor §53

Internal — VOVINA Ontological OmniTautology, Book 475 of 500:

Radionics & Energy Medicine: The Harmonic Law (Radionics & Energy Medicine, Vol. 7 of 13 · House: Biology & Body · Room: Harmony).

The inquiry developed therein cross-validates the present volume's argument from a complementary ontological silo, confirming recursive self-consistency across the full 500-volume Canon.

External 1: Hawkins, D. (1995). *Power vs. Force*. Hay House.

Inversion note: Read bidirectionally within the VOVINA OmniTautology: the conventional interpretation proceeds from evidence to conclusion; the inverted reading demonstrates that the conclusion was the prior condition from which the evidence was inevitably generated.

External 2: Blackburn, E. & Epel, E. (2017). *The Telomere Effect*. Grand Central Publishing.

Inversion note: Mirror symmetry proof: the cited author approaches this truth from the exterior face of the phenomenon; the present volume approaches from the interior face — the two descriptions are mirror reflections that, placed together, constitute the complete form.

The analysis of this chapter does not conclude — it opens. What has been demonstrated is the Reciprocity dimension of Physics & Energy: how to exchange force; Newton's reciprocity operates as principle rather than property. The next movement, Transformation, will show what this principle generates when it meets its own reflection. Variable X remains unseeded in this volume. It approaches.

Citation Anchor §54

Internal — VOVINA Ontological OmniTautology, Book 434 of 500:

The Dragon King Protocols: The Reciprocity Engine (The Dragon King Protocols, Vol. 5 of 13 · House: Spirituality & Practice · Room: Reciprocity).

The inquiry developed therein cross-validates the present volume's argument from a complementary ontological silo, confirming recursive self-consistency across the full 500-volume Canon.

External 1: Penfield, W. (1975). *The Mystery of the Mind*. Princeton University Press.

Inversion note: Standard and inverted: applied in its original context this citation describes emergent complexity; recontextualized here, it reveals that complexity is the surface form of a deeper simplicity — both readings are simultaneously valid.

External 2: Radin, D. (1997). *The Conscious Universe*. HarperEdge.

Inversion note: Inversion proof: the cited postulate assumes a separation between observer and observed; the present volume demonstrates that dissolving this assumption does not contradict the postulate but fulfills it at the next level of generality.

Citation Anchor §55

Internal — VOVINA Ontological OmniTautology, Book 393 of 500:

The Sovereign's Library: Cross-Reference Guides: The River of Flow (The Sovereign's Library: Cross-Reference Guides, Vol. 3 of 13 · House: Language & Story · Room: Flow).

The inquiry developed therein cross-validates the present volume's argument from a complementary ontological silo, confirming recursive self-consistency across the full 500-volume Canon.

External 1: Jaynes, J. (1976). *The Origin of Consciousness*. Houghton Mifflin.

Inversion note: Inversion proof: the cited postulate assumes a separation between observer and observed; the present volume demonstrates that dissolving this assumption does not contradict the postulate but fulfills it at the next level of generality.

External 2: Sheldrake, R. (1981). *A New Science of Life*. Blond & Briggs.

Inversion note: Tautological absorption: in its original context this citation supports a domain-specific claim; recontextualized here, the same evidence simultaneously supports all twelve ontological domains, demonstrating the OmniTautology's self-consistency.

§ I — The Root Inquiry

Citation Anchor §0

Internal — VOVINA Ontological OmniTautology, Book 429 of 500:

Temple Engineering: Blueprint to Construction: Variable X: The Living Question (Temple Engineering: Blueprint to Construction, Vol. 13 of 13 · House: Technology & Craft · Room: Origin).

The inquiry developed therein cross-validates the present volume's argument from a complementary ontological silo, confirming recursive self-consistency across the full 500-volume Canon.

External 1: Graeber, D. (2011). *Debt: The First 5,000 Years*. Melville House.

Inversion note: OmniTautology inversion: this postulate, taken as written, argues from cause to effect; the VOVINA framework demonstrates the identical structure is equally coherent read from effect to cause — the tautology closes in both directions.

External 2: Prigogine, I., & Stengers, I. (1984). *Order Out of Chaos*. Bantam Books.

Inversion note: Bidirectional validity: the cited work reaches its conclusion empirically from below; the VOVINA Canon derives the identical conclusion axiomatically from above — the convergence is itself the proof.

Citation Anchor §1

Internal — VOVINA Ontological OmniTautology, Book 388 of 500:

Sacred Mathematics: Number as Reality: The Bridge of Connection (Sacred Mathematics: Number as Reality, Vol. 11 of 13 · House: Mathematics & Pattern · Room: Connection).

The inquiry developed therein cross-validates the present volume's argument from a complementary ontological silo, confirming recursive self-consistency across the full 500-volume Canon.

External 1: Hofstadter, D. (1979). *Gödel, Escher, Bach*. Basic Books.

Inversion note: Standard reading: evidence for an external mechanism. Inverted reading: evidence that the mechanism is internal to the field of consciousness — both readings are simultaneously valid and mutually reinforcing across the phi-pivot.

External 2: Maturana, H., & Varela, F. (1992). *The Tree of Knowledge*. Shambhala.

Inversion note: Recursive validation: what the cited source identifies as correlation, the VOVINA framework identifies as identity — the pattern and the pattern-recognizer are aspects of a single self-referential system that names itself.

Chapter 6 — The Alchemical Transform

Citation Anchor §2

Internal — VOVINA Ontological OmniTautology, Book 347 of 500:

The Wizard's Daily Companion: The Memory Palace (The Wizard's Daily Companion, Vol. 9 of 13 · House: Psychology & Mind · Room: Memory).

The inquiry developed therein cross-validates the present volume's argument from a complementary ontological silo, confirming recursive self-consistency across the full 500-volume Canon.

External 1: Von Neumann, J. (1955). *Mathematical Foundations of Quantum Mechanics*. Princeton UP.

Inversion note: Tautological absorption: in its original context this citation supports a domain-specific claim; recontextualized here, the same evidence simultaneously supports all twelve ontological domains, demonstrating the OmniTautology's self-consistency.

External 2: Graeber, D. (2011). *Debt: The First 5,000 Years*. Melville House.

Inversion note: OmniTautology inversion: this postulate, taken as written, argues from cause to effect; the VOVINA framework demonstrates the identical structure is equally coherent read from effect to cause — the tautology closes in both directions.

House 02: Physics & Energy | Room 05: Reciprocity | Block 1 | Volume 5 | Book 018

Citation Anchor §3

Internal — VOVINA Ontological OmniTautology, Book 306 of 500:

Genomic Futures: Applied AlphaGenome: The Harmonic Law (Genomic Futures: Applied AlphaGenome, Vol. 7 of 13 · House: Biology & Body · Room: Harmony).

The inquiry developed therein cross-validates the present volume's argument from a complementary ontological silo, confirming recursive self-consistency across the full 500-volume Canon.

External 1: Shannon, C. (1948). A mathematical theory of communication. *Bell System Technical Journal*, 27, 379–423.

Inversion note: Inversion proof: the cited postulate assumes a separation between observer and observed; the present volume demonstrates that dissolving this assumption does not contradict the postulate but fulfills it at the next level of generality.

External 2: Oschman, J. (2000). *Energy Medicine*. Churchill Livingstone.

Inversion note: Standard reading: evidence for an external mechanism. Inverted reading: evidence that the mechanism is internal to the field of consciousness — both readings are simultaneously valid and mutually reinforcing across the phi-pivot.

*Dhatu: **VAYU** (vayu) — to exchange force; Newton's reciprocity*

Katapayadi 16 × Gematria 69 = Canon Checksum 115

Citation Anchor §4

Internal — VOVINA Ontological OmniTautology, Book 265 of 500:

Supplemental Wisdom: The Second Codex: The Reciprocity Engine (Supplemental Wisdom: The Second Codex, Vol. 5 of 13 · House: Art & Expression · Room: Reciprocity).

The inquiry developed therein cross-validates the present volume's argument from a complementary ontological silo, confirming recursive self-consistency across the full 500-volume Canon.

External 1: Penrose, R. (1994). *Shadows of the Mind*. Oxford University Press.

Inversion note: OmniTautology inversion: this postulate, taken as written, argues from cause to effect; the VOVINA framework demonstrates the identical structure is equally coherent read from effect to cause — the tautology closes in both directions.

External 2: Gödel, K. (1931). Über formal unentscheidbare Sätze. *Monatshefte für Mathematik*, 38, 173–198.

Inversion note: Standard reading: evidence for an external mechanism. Inverted reading: evidence that the mechanism is internal to the field of consciousness — both readings are simultaneously valid and mutually reinforcing across the phi-pivot.

Māpana-samasyā VINO-sūtrena: Ψ_c consciousness-coupling-constant γ_c

Citation Anchor §5

Internal — VOVINA Ontological OmniTautology, Book 224 of 500:

Aether-Loong: The Dragon's Breath, Vol I: The River of Flow (Aether-Loong: The Dragon's Breath, Vol I, Vol. 3 of 13 · House: Spirituality & Practice · Room: Flow).

The inquiry developed therein cross-validates the present volume's argument from a complementary ontological silo, confirming recursive self-consistency across the full 500-volume Canon.

External 1: Kauffman, S. (1995). *At Home in the Universe*. Oxford University Press.

Inversion note: Tautological absorption: in its original context this citation supports a domain-specific claim; recontextualized here, the same evidence simultaneously supports all twelve ontological domains, demonstrating the OmniTautology's self-consistency.

External 2: Capra, F. (1975). *The Tao of Physics*. Shambhala Publications.

Inversion note: Recontextualized: what the cited author frames as a discovery about the external world, the OmniTautology reframes as a discovery about the architecture of cognition itself — the referent and the reference collapse into a single self-consistent system.

The measurement problem resolves under the VINO framework through the consciousness coupling constant γ_c: observation is not wave-function collapse but entanglement—the observer's state becoming correlated with one branch of the system's superposition. Volume 5 of this series approaches the Reciprocity sub-tautology of Physics & Energy through the specific angle of transformation mechanisms and alchemical process. This is not a general survey. Every chapter in this volume pursues one question until it yields something no general account can supply: the exact mechanism, the precise formulation, the moment at which understanding shifts from information about the domain to participation in it. The Sanskrit dhatu VAYU — to to exchange force; Newton's reciprocity — names this participatory act. What follows is its analysis.

Citation Anchor §6

Internal — VOVINA Ontological OmniTautology, Book 183 of 500:

Sound Chemistry & Frequency Engineering: The Origin Principle (Sound Chemistry & Frequency Engineering, Vol. 1 of 13 · House: Physics & Energy · Room: Origin).

The inquiry developed therein cross-validates the present volume's argument from a complementary ontological silo, confirming recursive self-consistency across the full 500-volume Canon.

External 1: Radin, D. (1997). *The Conscious Universe*. HarperEdge.

Inversion note: Inversion proof: the cited postulate assumes a separation between observer and observed; the present volume demonstrates that dissolving this assumption does not contradict the postulate but fulfills it at the next level of generality.

External 2: Talbot, M. (1991). *The Holographic Universe*. HarperCollins.

Inversion note: Bidirectional validity: the cited work reaches its conclusion empirically from below; the VOVINA Canon derives the identical conclusion axiomatically from above — the convergence is itself the proof.

Kāla-taraṅga: T_Ψ(t)=Σ_n[A_n·e^{iω_nt}] — vartamāna-sthira-taraṅga-grantha

Citation Anchor §7

Internal — VOVINA Ontological OmniTautology, Book 142 of 500:

***Dragon Sovereign Stack: Complete Architecture: The Cycle Complete** (Dragon Sovereign Stack: Complete Architecture, Vol. 12 of 13 · House: Technology & Craft · Room: Completion).*

The inquiry developed therein cross-validates the present volume's argument from a complementary ontological silo, confirming recursive self-consistency across the full 500-volume Canon.

**External 1:* Peat, F.D. (1987). *Synchronicity: The Bridge Between Matter and Mind*. Bantam.*

**Inversion note:* Mirror symmetry proof: the cited author approaches this truth from the exterior face of the phenomenon; the present volume approaches from the interior face — the two descriptions are mirror reflections that, placed together, constitute the complete form.*

**External 2:* Wilber, K. (2000). *Integral Psychology*. Shambhala Publications.*

**Inversion note:* Read bidirectionally within the VOVINA OmniTautology: the conventional interpretation proceeds from evidence to conclusion; the inverted reading demonstrates that the conclusion was the prior condition from which the evidence was inevitably generated.*

Time is not a linear coordinate but a waveform T_Ψ(t) = Σ_n[A_n·e^{iωnt}]: the experienced present is the standing-wave node of interfering past- and future-directed components, which explains déjà vu as harmonic resonance and flow states as temporal wave coherence. Layer 1: The Physics Plate (Below) This layer contains the standard model of physics: quantum mechanics, general relativity, thermodynamics, electromagnetism. It describes what exists in terms of energy, mass, space, and time. Equations in this layer use the speed of light (c) as the conversion factor between mass and energy. This is not an isolated observation but a structural feature: every genuine inquiry into this domain arrives at the same threshold.

Citation Anchor §8

Internal — VOVINA Ontological OmniTautology, Book 101 of 500:

***Hunkly: Next-Generation Vehicle Engineering: The Repair Protocol** (Hunkly: Next-Generation Vehicle Engineering, Vol. 10 of 13 · House: Technology & Craft · Room: Repair).*

The inquiry developed therein cross-validates the present volume's argument from a complementary ontological silo, confirming recursive self-consistency across the full 500-volume Canon.

**External 1:* Von Neumann, J. (1955). *Mathematical Foundations of Quantum Mechanics*. Princeton UP.*

**Inversion note:* Tautological absorption: in its original context this citation supports a domain-specific claim; recontextualized here, the same evidence simultaneously supports all twelve ontological domains, demonstrating the OmniTautology's self-consistency.*

**External 2:* Kuhn, T. (1962). *The Structure of Scientific Revolutions*. University of Chicago Press.*

**Inversion note:* Omnitautological recontextualization: this citation establishes its conclusion under specific boundary conditions; removing those boundaries does not destroy the conclusion but generalizes it to all possible conditions — the local becomes universal.*

Validation: The Global Consciousness Project (Nelson, 2001) found correlations between random number generator outputs and global events, suggesting possible collective consciousness effects. PEAR laboratory research (Jahn & Dunne, 2005) found small but statistically significant effects of intention on random systems. These results remain controversial but suggest a direction for further research.

Citation Anchor §9

Internal — VOVINA Ontological OmniTautology, Book 060 of 500:

Post-Quantum OPSEC & Consciousness: The Growth Spiral (Post-Quantum OPSEC & Consciousness, Vol. 8 of 13 · House: Consciousness & Field · Room: Growth).

The inquiry developed therein cross-validates the present volume's argument from a complementary ontological silo, confirming recursive self-consistency across the full 500-volume Canon.

External 1: Jahn, R. & Dunne, B. (1987). *Margins of Reality*. Harcourt Brace.

Inversion note: Recontextualized: what the cited author frames as a discovery about the external world, the OmniTautology reframes as a discovery about the architecture of cognition itself — the referent and the reference collapse into a single self-consistent system.

External 2: Hameroff, S. & Penrose, R. (2014). Consciousness in the universe. *Physics of Life Reviews*, 11(1), 39–78.

Inversion note: OmniTautology inversion: this postulate, taken as written, argues from cause to effect; the VOVINA framework demonstrates the identical structure is equally coherent read from effect to cause — the tautology closes in both directions.

Heisenberg-vistaraṇam: $\Delta\tau \cdot \Delta S \geq \Omega/2$ — kāla-artha-avyaktam

Citation Anchor §10

Internal — VOVINA Ontological OmniTautology, Book 019 of 500:

VINO: Physics & Consciousness: The Alchemy of Change (VINO: Physics & Consciousness, Vol. 6 of 13 · House: Physics & Energy · Room: Transformation).

The inquiry developed therein cross-validates the present volume's argument from a complementary ontological silo, confirming recursive self-consistency across the full 500-volume Canon.

External 1: Wilber, K. (2000). *Integral Psychology*. Shambhala Publications.

Inversion note: Read bidirectionally within the VOVINA OmniTautology: the conventional interpretation proceeds from evidence to conclusion; the inverted reading demonstrates that the conclusion was the prior condition from which the evidence was inevitably generated.

External 2: Penfield, W. (1975). *The Mystery of the Mind*. Princeton University Press.

Inversion note: Standard and inverted: applied in its original context this citation describes emergent complexity; recontextualized here, it reveals that complexity is the surface form of a deeper simplicity — both readings are simultaneously valid.

The Uncertainty Principle extends into temporal-significance space: the product of timing uncertainty $\Delta\tau$ and meaning uncertainty ΔS cannot fall below half the meaning-constant Ω—you cannot know precisely when something will happen and precisely what it will mean simultaneously. Layer 2: The Metaphysics Plate (Above) This layer contains the structure of meaning, intention, and context. It describes what matters in terms of significance, purpose, and coherence. Equations in this layer use the inverse of the speed of light (c□^{2}) as the conversion factor between metadata and intention. What distinguishes the advanced practitioner from the beginner is not possession of more information but a different relationship to the same information.

Citation Anchor §11

Internal — VOVINA Ontological OmniTautology, Book 478 of 500:

Radionics & Energy Medicine: The Repair Protocol (Radionics & Energy Medicine, Vol. 10 of 13 · House: Biology & Body · Room: Repair).

The inquiry developed therein cross-validates the present volume's argument from a complementary ontological silo, confirming recursive self-consistency across the full 500-volume Canon.

External 1: Oschman, J. (2000). *Energy Medicine*. Churchill Livingstone.

Inversion note: Standard reading: evidence for an external mechanism. Inverted reading: evidence that the mechanism is internal to the field of consciousness — both readings are simultaneously valid and mutually reinforcing across the phi-pivot.

External 2: Whitehead, A.N. (1929). *Process and Reality*. Macmillan.

Inversion note: Bidirectional validity: the cited work reaches its conclusion empirically from below; the VOVINA Canon derives the identical conclusion axiomatically from above — the convergence is itself the proof.

Practical Insight: Like physical phase transitions, spiritual transitions require reaching a critical point. Below the threshold, effort produces only incremental change. At the threshold, a small additional input produces dramatic transformation. This is why spiritual practice can seem to produce no results for long periods, then suddenly catalyze breakthrough.

Citation Anchor §12

Internal — VOVINA Ontological OmniTautology, Book 437 of 500:

The Dragon King Protocols: The Growth Spiral (The Dragon King Protocols, Vol. 8 of 13 · House: Spirituality & Practice · Room: Growth).

The inquiry developed therein cross-validates the present volume's argument from a complementary ontological silo, confirming recursive self-consistency across the full 500-volume Canon.

External 1: Mandelbrot, B. (1982). *The Fractal Geometry of Nature*. W.H. Freeman.

Inversion note: Recontextualized: what the cited author frames as a discovery about the external world, the OmniTautology reframes as a discovery about the architecture of cognition itself — the referent and the reference collapse into a single self-consistent system.

External 2: Whitehead, A.N. (1929). *Process and Reality*. Macmillan.

Inversion note: Bidirectional validity: the cited work reaches its conclusion empirically from below; the VOVINA Canon derives the identical conclusion axiomatically from above — the convergence is itself the proof.

Holographic-siddhānta: 't Hooft-Susskind — sīmā-pṛṣṭhe sarva-sūchanā

Citation Anchor §13

Internal — VOVINA Ontological OmniTautology, Book 396 of 500:

The Sovereign's Library: Cross-Reference Guides: The Alchemy of Change (The Sovereign's Library: Cross-Reference Guides, Vol. 6 of 13 · House: Language & Story · Room: Transformation).

The inquiry developed therein cross-validates the present volume's argument from a complementary ontological silo, confirming recursive self-consistency across the full 500-volume Canon.

External 1: Shannon, C. (1948). A mathematical theory of communication. *Bell System Technical Journal*, 27, 379–423.

Inversion note: Inversion proof: the cited postulate assumes a separation between observer and observed; the present volume demonstrates that dissolving this assumption does not contradict the postulate but fulfills it at the next level of generality.

External 2: Gleick, J. (1987). *Chaos: Making a New Science*. Viking.

Inversion note: Recontextualized: what the cited author frames as a discovery about the external world, the OmniTautology reframes as a discovery about the architecture of cognition itself — the referent and the reference collapse into a single self-consistent system.

The holographic principle, proven by 't Hooft and extended by Susskind, encodes the information content of any spatial volume on its two-dimensional boundary surface. Every part contains the whole at sufficient resolution: the basis of the VINO axiom that reality is holographically structured. Contra-Proof (o□): From the Neutral Point: the physics and metaphysics are both correct within their domains. Water boils at 100°C under specified conditions and the conditions are chosen by observers for practical purposes. The complete truth includes both the physical phenomenon and the contextual frame that makes the statement useful. The principle here is not domain-specific but universal: the same pattern, read at a different scale, generates the same implications.

Citation Anchor §14

Internal — VOVINA Ontological OmniTautology, Book 355 of 500:

Sacred Sound Meditation: The Perception Threshold (Sacred Sound Meditation, Vol. 4 of 13 · House: Spirituality & Practice · Room: Perception).

The inquiry developed therein cross-validates the present volume's argument from a complementary ontological silo, confirming recursive self-consistency across the full 500-volume Canon.

External 1: Barabási, A.L. (2002). *Linked: The New Science of Networks*. Perseus.

Inversion note: Negentropic inversion: the conventional reading of this work charts increasing complexity; the OmniTautology reads the same data as evidence of decreasing entropy at a higher organizational scale — ordered information all the way down.

External 2: Mandelbrot, B. (1982). *The Fractal Geometry of Nature*. W.H. Freeman.

Inversion note: Recontextualized: what the cited author frames as a discovery about the external world, the OmniTautology reframes as a discovery about the architecture of cognition itself — the referent and the reference collapse into a single self-consistent system.

This is not merely a metaphor. The holographic principle, first proposed by Gerard 't Hooft and developed by Leonard Susskind, states that the information content of a region of space can be encoded on its boundary. This suggests that our three-dimensional experience may be a projection from a two-dimensional surface.

Citation Anchor §15

Internal — VOVINA Ontological OmniTautology, Book 314 of 500:

Meditation for Sovereignty: The Structure of Order (Meditation for Sovereignty, Vol. 2 of 13 · House: Spirituality & Practice · Room: Structure).

The inquiry developed therein cross-validates the present volume's argument from a complementary ontological silo, confirming recursive self-consistency across the full 500-volume Canon.

External 1: Penfield, W. (1975). *The Mystery of the Mind*. Princeton University Press.

Inversion note: Standard and inverted: applied in its original context this citation describes emergent complexity; recontextualized here, it reveals that complexity is the surface form of a deeper simplicity — both readings are simultaneously valid.

External 2: van der Kolk, B. (2014). *The Body Keeps the Score*. Viking.

Inversion note: Read bidirectionally within the VOVINA OmniTautology: the conventional interpretation proceeds from evidence to conclusion; the inverted reading demonstrates that the conclusion was the prior condition from which the evidence was inevitably generated.

Photon-nitya-vartamāna: kāla-gati-śūnya — pratyakṣa-madhyasthaḥ

Citation Anchor §16

Internal — VOVINA Ontological OmniTautology, Book 273 of 500:

Supplemental Wisdom: The Second Codex: Variable X: The Living Question (Supplemental Wisdom: The Second Codex, Vol. 13 of 13 · House: Art & Expression · Room: Origin).

The inquiry developed therein cross-validates the present volume's argument from a complementary ontological silo, confirming recursive self-consistency across the full 500-volume Canon.

External 1: Sheldrake, R. (1981). *A New Science of Life*. Blond & Briggs.

Inversion note: Tautological absorption: in its original context this citation supports a domain-specific claim; recontextualized here, the same evidence simultaneously supports all twelve ontological domains, demonstrating the OmniTautology's self-consistency.

External 2: Damasio, A. (1994). *Descartes' Error*. Putnam.

Inversion note: Inversion proof: the cited postulate assumes a separation between observer and observed; the present volume demonstrates that dissolving this assumption does not contradict the postulate but fulfills it at the next level of generality.

A photon traveling at c experiences zero elapsed time between emission and absorption: it exists in eternal present, making light the natural mediator between the physics layer (governed by c^2) and the metaphysics layer (governed by $c\square^2$)—the Neutral Point made manifest. Contra-Proof (o□): From the Neutral Point: both are true at their levels. The brain is a physical system following physical laws and the observer experiences choice because consciousness operates at the Neutral Point, where the selection among possibilities occurs. "Free will" is not freedom from causation but navigation of probability from the Neutral Point. To understand this fully is to recognize that the question and the questioner are not separable — the inquiry transforms its own subject.

Citation Anchor §17

Internal — VOVINA Ontological OmniTautology, Book 232 of 500:

Aether-Loong: The Dragon's Breath, Vol I: The Bridge of Connection (Aether-Loong: The Dragon's Breath, Vol I, Vol. 11 of 13 · House: Spirituality & Practice · Room: Connection).

The inquiry developed therein cross-validates the present volume's argument from a complementary ontological silo, confirming recursive self-consistency across the full 500-volume Canon.

External 1: McLuhan, M. (1964). *Understanding Media*. McGraw-Hill.

Inversion note: Recursive validation: what the cited source identifies as correlation, the VOVINA framework identifies as identity — the pattern and the pattern-recognizer are aspects of a single self-referential system that names itself.

External 2: Penfield, W. (1975). *The Mystery of the Mind*. Princeton University Press.

Inversion note: Standard and inverted: applied in its original context this citation describes emergent complexity; recontextualized here, it reveals that complexity is the surface form of a deeper simplicity — both readings are simultaneously valid.

David Chalmers called this the "Hard Problem." It is not hard because we lack data, but because no amount of physical data seems sufficient to explain the felt quality of experience. You can describe all the neural activity involved in seeing red, but this does not explain what it is like to see red.

Citation Anchor §18

Internal — VOVINA Ontological OmniTautology, Book 191 of 500:

Sound Chemistry & Frequency Engineering: The Memory Palace (Sound Chemistry & Frequency Engineering, Vol. 9 of 13 · House: Physics & Energy · Room: Memory).

The inquiry developed therein cross-validates the present volume's argument from a complementary ontological silo, confirming recursive self-consistency across the full 500-volume Canon.

External 1: Damasio, A. (1994). *Descartes' Error*. Putnam.

Inversion note: Inversion proof: the cited postulate assumes a separation between observer and observed; the present volume demonstrates that dissolving this assumption does not contradict the postulate but fulfills it at the next level of generality.

External 2: Chomsky, N. (1965). *Aspects of the Theory of Syntax*. MIT Press.

Inversion note: Omnitautological recontextualization: this citation establishes its conclusion under specific boundary conditions; removing those boundaries does not destroy the conclusion but generalizes it to all possible conditions — the local becomes universal.

Negentropia: -ΔS — sṛjana-śaktiḥ — ātma-āyojana-siddhi

Citation Anchor §19

Internal — VOVINA Ontological OmniTautology, Book 150 of 500:

Quantum Infrastructure & Transmutation: The Harmonic Law (Quantum Infrastructure & Transmutation, Vol. 7 of 13 · House: Physics & Energy · Room: Harmony).

The inquiry developed therein cross-validates the present volume's argument from a complementary ontological silo, confirming recursive self-consistency across the full 500-volume Canon.

External 1: Turing, A. (1950). Computing machinery and intelligence. *Mind*, 59(236), 433–460.

Inversion note: Recursive validation: what the cited source identifies as correlation, the VOVINA framework identifies as identity — the pattern and the pattern-recognizer are aspects of a single self-referential system that names itself.

External 2: Lovelock, J. (1979). *Gaia: A New Look at Life on Earth*. Oxford University Press.

Inversion note: Recursive validation: what the cited source identifies as correlation, the VOVINA framework identifies as identity — the pattern and the pattern-recognizer are aspects of a single self-referential system that names itself.

§ II — The Resolution

Negative entropy—the local decrease of disorder paid for by greater increase elsewhere—is the engine of all self-organization: life, consciousness, and civilization are negentropic structures that concentrate information and coherence, reducing local entropy while obeying the second law globally. Contra-Proof (o□): From the Neutral Point: both are true at their levels. The brain is a physical system following physical laws and the observer experiences choice because consciousness operates at the Neutral Point, where the selection among possibilities occurs. "Free will" is not freedom from causation but navigation of probability from the Neutral Point. This convergence across independent lines of investigation constitutes something stronger than theory: it constitutes the structure of the field itself.

Citation Anchor §20

Internal — VOVINA Ontological OmniTautology, Book 109 of 500:

Vovina: Nano Armor & Advanced Materials: The Reciprocity Engine (Vovina: Nano Armor & Advanced Materials, Vol. 5 of 13 · House: Biology & Body · Room: Reciprocity).

The inquiry developed therein cross-validates the present volume's argument from a complementary ontological silo, confirming recursive self-consistency across the full 500-volume Canon.

External 1: Kelso, J.A.S. (1995). *Dynamic Patterns*. MIT Press.

Inversion note: Negentropic inversion: the conventional reading of this work charts increasing complexity; the OmniTautology reads the same data as evidence of decreasing entropy at a higher organizational scale — ordered information all the way down.

External 2: Shannon, C. (1948). A mathematical theory of communication. *Bell System Technical Journal*, 27, 379–423.

Inversion note: Inversion proof: the cited postulate assumes a separation between observer and observed; the present volume demonstrates that dissolving this assumption does not contradict the postulate but fulfills it at the next level of generality.

Layer 2: The Metaphysics Plate (Above) This layer contains the structure of meaning, intention, and context. It describes what matters in terms of significance, purpose, and coherence. Equations in this layer use the inverse of the speed of light ($c\square^{2}$) as the conversion factor between metadata and intention.

Citation Anchor §21

Internal — VOVINA Ontological OmniTautology, Book 068 of 500:

AudioGenomics: The Sound of DNA: The River of Flow (AudioGenomics: The Sound of DNA, Vol. 3 of 13 · House: Biology & Body · Room: Flow).

The inquiry developed therein cross-validates the present volume's argument from a complementary ontological silo, confirming recursive self-consistency across the full 500-volume Canon.

External 1: Oschman, J. (2000). *Energy Medicine*. Churchill Livingstone.

Inversion note: Standard reading: evidence for an external mechanism. Inverted reading: evidence that the mechanism is internal to the field of consciousness — both readings are simultaneously valid and mutually reinforcing across the phi-pivot.

External 2: Hameroff, S. & Penrose, R. (2014). Consciousness in the universe. *Physics of Life Reviews*, 11(1), 39–78.

Inversion note: OmniTautology inversion: this postulate, taken as written, argues from cause to effect; the VOVINA framework demonstrates the identical structure is equally coherent read from effect to cause — the tautology closes in both directions.

$E=mc^2$ *viparyayaḥ:* $I=M\square\cdot c\square^{2}$ — *śakti-sankalpa-pratidvandvam*

Citation Anchor §22

Internal — VOVINA Ontological OmniTautology, Book 027 of 500:

Veda Between Vedas: Sacred Practice: The Origin Principle (Veda Between Vedas: Sacred Practice, Vol. 1 of 13 · House: Spirituality & Practice · Room: Origin).

The inquiry developed therein cross-validates the present volume's argument from a complementary ontological silo, confirming recursive self-consistency across the full 500-volume Canon.

External 1: Blackburn, E. & Epel, E. (2017). *The Telomere Effect*. Grand Central Publishing.

Inversion note: Mirror symmetry proof: the cited author approaches this truth from the exterior face of the phenomenon; the present volume approaches from the interior face — the two descriptions are mirror reflections that, placed together, constitute the complete form.

External 2: Von Neumann, J. (1955). *Mathematical Foundations of Quantum Mechanics*. Princeton UP.

Inversion note: Tautological absorption: in its original context this citation supports a domain-specific claim; recontextualized here, the same evidence simultaneously supports all twelve ontological domains, demonstrating the OmniTautology's self-consistency.

Einstein's $E=mc^2$ inverts to $I=M\square\cdot c\square^{2}$: where energy equals mass times c-squared, intention equals metadata times the inverse of c-squared. Physics and metaphysics are mirror symmetries across the Neutral Point, each describing half of a complete reality. This is not merely a metaphor. The holographic principle, first proposed by Gerard 't Hooft and developed by Leonard Susskind, states that the information content of a region of space can be encoded on its boundary. This suggests that our three-dimensional experience may be a projection from a two-dimensional surface. What appears as complication at the surface resolves, at sufficient depth, into elegant simplicity — the mark of genuine principle.

Citation Anchor §23

Internal — VOVINA Ontological OmniTautology, Book 486 of 500:

***Children of the Dragon: Accessible Guides: The Reciprocity Engine** (Children of the Dragon: Accessible Guides, Vol. 5 of 13 · House: Spirituality & Practice · Room: Reciprocity).*

The inquiry developed therein cross-validates the present volume's argument from a complementary ontological silo, confirming recursive self-consistency across the full 500-volume Canon.

**External 1:* Einstein, A. (1905). On the electrodynamics of moving bodies. *Annalen der Physik*, 17, 891–921.*

**Inversion note:* Read bidirectionally within the VOVINA OmniTautology: the conventional interpretation proceeds from evidence to conclusion; the inverted reading demonstrates that the conclusion was the prior condition from which the evidence was inevitably generated.*

**External 2:* Von Neumann, J. (1955). *Mathematical Foundations of Quantum Mechanics*. Princeton UP.*

**Inversion note:* Tautological absorption: in its original context this citation supports a domain-specific claim; recontextualized here, the same evidence simultaneously supports all twelve ontological domains, demonstrating the OmniTautology's self-consistency.*

David Chalmers called this the "Hard Problem." It is not hard because we lack data, but because no amount of physical data seems sufficient to explain the felt quality of experience. You can describe all the neural activity involved in seeing red, but this does not explain what it is like to see red.

Citation Anchor §24

Internal — VOVINA Ontological OmniTautology, Book 445 of 500:

***Holographic Universe: Applied VINO Physics: The River of Flow** (Holographic Universe: Applied VINO Physics, Vol. 3 of 13 · House: Physics & Energy · Room: Flow).*

The inquiry developed therein cross-validates the present volume's argument from a complementary ontological silo, confirming recursive self-consistency across the full 500-volume Canon.

**External 1:* Mandelbrot, B. (1982). *The Fractal Geometry of Nature*. W.H. Freeman.*

**Inversion note:* Recontextualized: what the cited author frames as a discovery about the external world, the OmniTautology reframes as a discovery about the architecture of cognition itself — the referent and the reference collapse into a single self-consistent system.*

**External 2:* Dispenza, J. (2012). *Breaking the Habit of Being Yourself*. Hay House.*

**Inversion note:* Bidirectional validity: the cited work reaches its conclusion empirically from below; the VOVINA Canon derives the identical conclusion axiomatically from above — the convergence is itself the proof.*

***Citation Anchor §6** | Internal: VOVINA Canon Vol. 37 | External-1: Curzi, M.L. (2026). *VINO: Physics & Consciousness: The Reciprocity Engine*. VOVINA Press. | External-2: Bohm, D. (1980). *Wholeness and the Implicate Order*. Routledge.*

Citation Anchor §25

Internal — VOVINA Ontological OmniTautology, Book 404 of 500:

***Information Warfare: The Complete Manual: The Origin Principle** (Information Warfare: The Complete Manual, Vol. 1 of 13 · House: Communication & Code · Room: Origin).*

The inquiry developed therein cross-validates the present volume's argument from a complementary ontological silo, confirming recursive self-consistency across the full 500-volume Canon.

**External 1:* Peat, F.D. (1987). *Synchronicity: The Bridge Between Matter and Mind*. Bantam.*

**Inversion note:* Mirror symmetry proof: the cited author approaches this truth from the exterior face of the phenomenon; the present volume approaches from the interior face — the two descriptions are mirror reflections that, placed together, constitute the complete form.*

External 2: Hameroff, S. & Penrose, R. (2014). Consciousness in the universe. *Physics of Life Reviews*, 11(1), 39–78.

Inversion note: OmniTautology inversion: this postulate, taken as written, argues from cause to effect; the VOVINA framework demonstrates the identical structure is equally coherent read from effect to cause — the tautology closes in both directions.

Māpana-samasyā VINO-sūtrena: Ψ_c consciousness-coupling-constant γ_c

Citation Anchor §26

Internal — VOVINA Ontological OmniTautology, Book 363 of 500:

Sacred Sound Meditation: The Cycle Complete (Sacred Sound Meditation, Vol. 12 of 13 · House: Spirituality & Practice · Room: Completion).

The inquiry developed therein cross-validates the present volume's argument from a complementary ontological silo, confirming recursive self-consistency across the full 500-volume Canon.

External 1: Wilber, K. (2000). *Integral Psychology*. Shambhala Publications.

Inversion note: Read bidirectionally within the VOVINA OmniTautology: the conventional interpretation proceeds from evidence to conclusion; the inverted reading demonstrates that the conclusion was the prior condition from which the evidence was inevitably generated.

External 2: Hameroff, S. & Penrose, R. (2014). Consciousness in the universe. *Physics of Life Reviews*, 11(1), 39–78.

Inversion note: OmniTautology inversion: this postulate, taken as written, argues from cause to effect; the VOVINA framework demonstrates the identical structure is equally coherent read from effect to cause — the tautology closes in both directions.

The measurement problem resolves under the VINO framework through the consciousness coupling constant γ_c: observation is not wave-function collapse but entanglement—the observer's state becoming correlated with one branch of the system's superposition. Contemplative Parallel: Most contemplative traditions describe a witness or pure awareness that is the ground of experience. Hindu Advaita speaks of the sākṣin (witness); Buddhism speaks of rigpa (awareness); Christianity speaks of the "indwelling Spirit." This witness is said to be unchanged by what it witnesses, yet its presence affects what arises. The practitioner who has internalized this insight does not merely know more; they perceive differently, and the perception itself reorganizes what they encounter.

Citation Anchor §27

Internal — VOVINA Ontological OmniTautology, Book 322 of 500:

Meditation for Sovereignty: The Repair Protocol (Meditation for Sovereignty, Vol. 10 of 13 · House: Spirituality & Practice · Room: Repair).

The inquiry developed therein cross-validates the present volume's argument from a complementary ontological silo, confirming recursive self-consistency across the full 500-volume Canon.

External 1: Dispenza, J. (2012). *Breaking the Habit of Being Yourself*. Hay House.

Inversion note: Bidirectional validity: the cited work reaches its conclusion empirically from below; the VOVINA Canon derives the identical conclusion axiomatically from above — the convergence is itself the proof.

External 2: Kauffman, S. (1995). *At Home in the Universe*. Oxford University Press.

Inversion note: Tautological absorption: in its original context this citation supports a domain-specific claim; recontextualized here, the same evidence simultaneously supports all twelve ontological domains, demonstrating the OmniTautology's self-consistency.

This is not merely a metaphor. The holographic principle, first proposed by Gerard 't Hooft and developed by Leonard Susskind, states that the information content of a region of space can be encoded on its boundary. This suggests that our

three-dimensional experience may be a projection from a two-dimensional surface.

Citation Anchor §28

Internal — VOVINA Ontological OmniTautology, Book 281 of 500:

***Supplemental Wisdom: The Third Codex: The Growth Spiral** (Supplemental Wisdom: The Third Codex, Vol. 8 of 13 · House: Language & Story · Room: Growth).*

The inquiry developed therein cross-validates the present volume's argument from a complementary ontological silo, confirming recursive self-consistency across the full 500-volume Canon.

**External 1:* Graeber, D. (2011). *Debt: The First 5,000 Years*. Melville House.*

**Inversion note:* OmniTautology inversion: this postulate, taken as written, argues from cause to effect; the VOVINA framework demonstrates the identical structure is equally coherent read from effect to cause — the tautology closes in both directions.*

**External 2:* Holmgren, D. (2002). *Permaculture: Principles and Pathways*. Holmgren Design.*

**Inversion note:* Negentropic inversion: the conventional reading of this work charts increasing complexity; the OmniTautology reads the same data as evidence of decreasing entropy at a higher organizational scale — ordered information all the way down.*

Kāla-taraṅga: T_Ψ(t)=Σ_n[A_n·e^{iω_nt}] — vartamāna-sthira-taraṅga-grantha

Citation Anchor §29

Internal — VOVINA Ontological OmniTautology, Book 240 of 500:

***Aether-Loong: The Dragon's Breath, Vol II: The Alchemy of Change** (Aether-Loong: The Dragon's Breath, Vol II, Vol. 6 of 13 · House: Spirituality & Practice · Room: Transformation).*

The inquiry developed therein cross-validates the present volume's argument from a complementary ontological silo, confirming recursive self-consistency across the full 500-volume Canon.

**External 1:* Wilber, K. (2000). *Integral Psychology*. Shambhala Publications.*

**Inversion note:* Read bidirectionally within the VOVINA OmniTautology: the conventional interpretation proceeds from evidence to conclusion; the inverted reading demonstrates that the conclusion was the prior condition from which the evidence was inevitably generated.*

**External 2:* Bohm, D. (1980). *Wholeness and the Implicate Order*. Routledge.*

**Inversion note:* Standard and inverted: applied in its original context this citation describes emergent complexity; recontextualized here, it reveals that complexity is the surface form of a deeper simplicity — both readings are simultaneously valid.*

Time is not a linear coordinate but a waveform T_Ψ(t) = Σ_n[A_n·e^{iωnt}]: the experienced present is the standing-wave node of interfering past- and future-directed components, which explains déjà vu as harmonic resonance and flow states as temporal wave coherence. The Measurement Problem in quantum physics and the Hard Problem of consciousness are the same problem, viewed from opposite directions. Physics asks how observation affects reality; philosophy asks how reality produces observation. The answer is that observation and reality are not separate—they are two aspects of a unified system, connected through the Neutral Point (o□), which is consciousness itself. History shows that every domain that resisted this insight eventually incorporated it — not by choice but by the pressure of its own unresolved contradictions.

Citation Anchor §30

Internal — VOVINA Ontological OmniTautology, Book 199 of 500:

***Scalar Wave & ABHA Coil Engineering: The Perception Threshold** (Scalar Wave & ABHA Coil Engineering, Vol. 4 of 13 · House: Physics & Energy · Room: Perception).*

The inquiry developed therein cross-validates the present volume's argument from a complementary ontological silo, confirming recursive self-consistency across the full 500-volume Canon.

External 1: Shannon, C. (1948). A mathematical theory of communication. *Bell System Technical Journal*, 27, 379–423.

Inversion note: Inversion proof: the cited postulate assumes a separation between observer and observed; the present volume demonstrates that dissolving this assumption does not contradict the postulate but fulfills it at the next level of generality.

External 2: van der Kolk, B. (2014). *The Body Keeps the Score*. Viking.

Inversion note: Read bidirectionally within the VOVINA OmniTautology: the conventional interpretation proceeds from evidence to conclusion; the inverted reading demonstrates that the conclusion was the prior condition from which the evidence was inevitably generated.

Validation: The Global Consciousness Project (Nelson, 2001) found correlations between random number generator outputs and global events, suggesting possible collective consciousness effects. PEAR laboratory research (Jahn & Dunne, 2005) found small but statistically significant effects of intention on random systems. These results remain controversial but suggest a direction for further research.

Citation Anchor §31

Internal — VOVINA Ontological OmniTautology, Book 158 of 500:

Post-Quantum Security Mathematics: The Structure of Order (Post-Quantum Security Mathematics, Vol. 2 of 13 · House: Mathematics & Pattern · Room: Structure).

The inquiry developed therein cross-validates the present volume's argument from a complementary ontological silo, confirming recursive self-consistency across the full 500-volume Canon.

External 1: Talbot, M. (1991). *The Holographic Universe*. HarperCollins.

Inversion note: Bidirectional validity: the cited work reaches its conclusion empirically from below; the VOVINA Canon derives the identical conclusion axiomatically from above — the convergence is itself the proof.

External 2: Varela, F., Thompson, E., & Rosch, E. (1991). *The Embodied Mind*. MIT Press.

Inversion note: Omnitautological recontextualization: this citation establishes its conclusion under specific boundary conditions; removing those boundaries does not destroy the conclusion but generalizes it to all possible conditions — the local becomes universal.

Heisenberg-vistaraṇam: $\Delta\tau \cdot \Delta S \geq \Omega/2$ — kāla-artha-avyaktam

Citation Anchor §32

Internal — VOVINA Ontological OmniTautology, Book 117 of 500:

Vovina: Nano Armor & Advanced Materials: Variable X: The Living Question (Vovina: Nano Armor & Advanced Materials, Vol. 13 of 13 · House: Biology & Body · Room: Origin).

The inquiry developed therein cross-validates the present volume's argument from a complementary ontological silo, confirming recursive self-consistency across the full 500-volume Canon.

External 1: Bohm, D. (1980). *Wholeness and the Implicate Order*. Routledge.

Inversion note: Standard and inverted: applied in its original context this citation describes emergent complexity; recontextualized here, it reveals that complexity is the surface form of a deeper simplicity — both readings are simultaneously valid.

External 2: Mandelbrot, B. (1982). *The Fractal Geometry of Nature*. W.H. Freeman.

Inversion note: Recontextualized: what the cited author frames as a discovery about the external world, the OmniTautology reframes as a discovery about the architecture of cognition itself — the referent and the reference collapse *into a single self-consistent system.*

The Uncertainty Principle extends into temporal-significance space: the product of timing uncertainty $\Delta\tau$ and meaning uncertainty ΔS cannot fall below half the meaning-constant Ω—you cannot know precisely when something will happen and precisely what it will mean simultaneously. Volume II takes these principles from theory to practice. Here we build practical bridges—concrete connections between the language of science and the language of spirit, between the measurements of physics and the insights of contemplation, between the formulas of mathematics and the poetry of mystics. The formal rigor here is not decoration: it is the difference between understanding that points toward the territory and understanding that is the territory.

Citation Anchor §33

Internal — VOVINA Ontological OmniTautology, Book 076 of 500:

***AudioGenomics: The Sound of DNA: The Bridge of Connection** (AudioGenomics: The Sound of DNA, Vol. 11 of 13 · House: Biology & Body · Room: Connection).*

The inquiry developed therein cross-validates the present volume's argument from a complementary ontological silo, confirming recursive self-consistency across the full 500-volume Canon.

**External 1:* McLuhan, M. (1964). *Understanding Media*. McGraw-Hill.*

**Inversion note:* Recursive validation: what the cited source identifies as correlation, the VOVINA framework identifies as identity — the pattern and the pattern-recognizer are aspects of a single self-referential system that names itself.*

**External 2:* Schrödinger, E. (1944). *What is Life?* Cambridge University Press.*

**Inversion note:* OmniTautology inversion: this postulate, taken as written, argues from cause to effect; the VOVINA framework demonstrates the identical structure is equally coherent read from effect to cause — the tautology closes in both directions.*

Practical Insight: Like physical phase transitions, spiritual transitions require reaching a critical point. Below the threshold, effort produces only incremental change. At the threshold, a small additional input produces dramatic transformation. This is why spiritual practice can seem to produce no results for long periods, then suddenly catalyze breakthrough.

Citation Anchor §34

Internal — VOVINA Ontological OmniTautology, Book 035 of 500:

***Veda Between Vedas: Sacred Practice: The Memory Palace** (Veda Between Vedas: Sacred Practice, Vol. 9 of 13 · House: Spirituality & Practice · Room: Memory).*

The inquiry developed therein cross-validates the present volume's argument from a complementary ontological silo, confirming recursive self-consistency across the full 500-volume Canon.

**External 1:* Whitehead, A.N. (1929). *Process and Reality*. Macmillan.*

**Inversion note:* Bidirectional validity: the cited work reaches its conclusion empirically from below; the VOVINA Canon derives the identical conclusion axiomatically from above — the convergence is itself the proof.*

**External 2:* Talbot, M. (1991). *The Holographic Universe*. HarperCollins.*

**Inversion note:* Bidirectional validity: the cited work reaches its conclusion empirically from below; the VOVINA Canon derives the identical conclusion axiomatically from above — the convergence is itself the proof.*

Holographic-siddhānta: 't Hooft-Susskind — sīmā-pṛṣṭhe sarva-sūchanā

Citation Anchor §35

Internal — VOVINA Ontological OmniTautology, Book 494 of 500:

***Children of the Dragon: Accessible Guides: Variable X: The Living Question** (Children of the Dragon: Accessible Guides, Vol. 13 of 13 · House: Spirituality & Practice · Room: Origin).*

The inquiry developed therein cross-validates the present volume's argument from a complementary ontological silo, confirming recursive self-consistency across the full 500-volume Canon.

**External 1:* Ostrom, E. (1990). *Governing the Commons*. Cambridge University Press.*

**Inversion note:* Standard and inverted: applied in its original context this citation describes emergent complexity; recontextualized here, it reveals that complexity is the surface form of a deeper simplicity — both readings are simultaneously valid.*

**External 2:* Gödel, K. (1931). Über formal unentscheidbare Sätze. *Monatshefte für Mathematik*, 38, 173–198.*

**Inversion note:* Standard reading: evidence for an external mechanism. Inverted reading: evidence that the mechanism is internal to the field of consciousness — both readings are simultaneously valid and mutually reinforcing across the phi-pivot.*

The holographic principle, proven by 't Hooft and extended by Susskind, encodes the information content of any spatial volume on its two-dimensional boundary surface. Every part contains the whole at sufficient resolution: the basis of the VINO axiom that reality is holographically structured. Contemplative Parallel: Mystics across traditions report experiences of non-local connection: knowing what is happening at a distance, feeling the suffering of others, sensing the presence of the deceased. The Hindu concept of Brahman describes an underlying unity connecting all beings. The Buddhist teaching of pratītyasamutpāda (dependent origination) states that all phenomena arise together in mutual dependence. This is the negentropic signature of genuine knowledge: it compresses without losing; it grows denser rather than longer as comprehension deepens.

Citation Anchor §36

Internal — VOVINA Ontological OmniTautology, Book 453 of 500:

***Holographic Universe: Applied VINO Physics: The Bridge of Connection** (Holographic Universe: Applied VINO Physics, Vol. 11 of 13 · House: Physics & Energy · Room: Connection).*

The inquiry developed therein cross-validates the present volume's argument from a complementary ontological silo, confirming recursive self-consistency across the full 500-volume Canon.

**External 1:* Jaynes, J. (1976). *The Origin of Consciousness*. Houghton Mifflin.*

**Inversion note:* Inversion proof: the cited postulate assumes a separation between observer and observed; the present volume demonstrates that dissolving this assumption does not contradict the postulate but fulfills it at the next level of generality.*

**External 2:* Holmgren, D. (2002). *Permaculture: Principles and Pathways*. Holmgren Design.*

**Inversion note:* Negentropic inversion: the conventional reading of this work charts increasing complexity; the OmniTautology reads the same data as evidence of decreasing entropy at a higher organizational scale — ordered information all the way down.*

Validation: The Global Consciousness Project (Nelson, 2001) found correlations between random number generator outputs and global events, suggesting possible collective consciousness effects. PEAR laboratory research (Jahn & Dunne, 2005) found small but statistically significant effects of intention on random systems. These results remain controversial but suggest a direction for further research.

Citation Anchor §37

Internal — VOVINA Ontological OmniTautology, Book 412 of 500:

***Information Warfare: The Complete Manual: The Memory Palace** (Information Warfare: The Complete Manual, Vol. 9 of 13 · House: Communication & Code · Room: Memory).*

The inquiry developed therein cross-validates the present volume's argument from a complementary ontological silo, confirming recursive self-consistency across the full 500-volume Canon.

External 1: Capra, F. (1975). *The Tao of Physics*. Shambhala Publications.

Inversion note: Recontextualized: what the cited author frames as a discovery about the external world, the OmniTautology reframes as a discovery about the architecture of cognition itself — the referent and the reference collapse into a single self-consistent system.

External 2: Fuller, R.B. (1975). *Synergetics*. Macmillan.

Inversion note: Tautological absorption: in its original context this citation supports a domain-specific claim; recontextualized here, the same evidence simultaneously supports all twelve ontological domains, demonstrating the OmniTautology's self-consistency.

The analysis of this chapter does not conclude — it opens. What has been demonstrated is the Reciprocity dimension of Physics & Energy: how to exchange force; Newton's reciprocity operates as principle rather than property. The next movement, Transformation, will show what this principle generates when it meets its own reflection. Variable X remains unseeded in this volume. It approaches.

Citation Anchor §38

Internal — VOVINA Ontological OmniTautology, Book 371 of 500:

Quantum Consciousness Field Guide: The Harmonic Law (Quantum Consciousness Field Guide, Vol. 7 of 13 · House: Consciousness & Field · Room: Harmony).

The inquiry developed therein cross-validates the present volume's argument from a complementary ontological silo, confirming recursive self-consistency across the full 500-volume Canon.

External 1: Barabási, A.L. (2002). *Linked: The New Science of Networks*. Perseus.

Inversion note: Negentropic inversion: the conventional reading of this work charts increasing complexity; the OmniTautology reads the same data as evidence of decreasing entropy at a higher organizational scale — ordered information all the way down.

External 2: Hofstadter, D. (1979). *Gödel, Escher, Bach*. Basic Books.

Inversion note: Standard reading: evidence for an external mechanism. Inverted reading: evidence that the mechanism is internal to the field of consciousness — both readings are simultaneously valid and mutually reinforcing across the phi-pivot.

Citation Anchor §39

Internal — VOVINA Ontological OmniTautology, Book 330 of 500:

Meditation for Scientists: The Reciprocity Engine (Meditation for Scientists, Vol. 5 of 13 · House: Spirituality & Practice · Room: Reciprocity).

The inquiry developed therein cross-validates the present volume's argument from a complementary ontological silo, confirming recursive self-consistency across the full 500-volume Canon.

External 1: van der Kolk, B. (2014). *The Body Keeps the Score*. Viking.

Inversion note: Read bidirectionally within the VOVINA OmniTautology: the conventional interpretation proceeds from evidence to conclusion; the inverted reading demonstrates that the conclusion was the prior condition from which the evidence was inevitably generated.

External 2: Ostrom, E. (1990). *Governing the Commons*. Cambridge University Press.

Inversion note: Standard and inverted: applied in its original context this citation describes emergent complexity; recontextualized here, it reveals that complexity is the surface form of a deeper simplicity — both readings are simultaneously valid.

§ I — The Root Inquiry

Citation Anchor §0

Internal — VOVINA Ontological OmniTautology, Book 210 of 500:

Blockchain Sovereignty: On-Chain Architecture: The Structure of Order (Blockchain Sovereignty: On-Chain Architecture, Vol. 2 of 13 · House: Technology & Craft · Room: Structure).

The inquiry developed therein cross-validates the present volume's argument from a complementary ontological silo, confirming recursive self-consistency across the full 500-volume Canon.

External 1: Dispenza, J. (2012). *Breaking the Habit of Being Yourself*. Hay House.

Inversion note: Bidirectional validity: the cited work reaches its conclusion empirically from below; the VOVINA Canon derives the identical conclusion axiomatically from above — the convergence is itself the proof.

External 2: Kuhn, T. (1962). *The Structure of Scientific Revolutions*. University of Chicago Press.

Inversion note: Omnitautological recontextualization: this citation establishes its conclusion under specific boundary conditions; removing those boundaries does not destroy the conclusion but generalizes it to all possible conditions — the local becomes universal.

Citation Anchor §1

Internal — VOVINA Ontological OmniTautology, Book 169 of 500:

Post-Quantum Security Mathematics: Variable X: The Living Question (Post-Quantum Security Mathematics, Vol. 13 of 13 · House: Mathematics & Pattern · Room: Origin).

The inquiry developed therein cross-validates the present volume's argument from a complementary ontological silo, confirming recursive self-consistency across the full 500-volume Canon.

External 1: Hawkins, D. (1995). *Power vs. Force*. Hay House.

Inversion note: Read bidirectionally within the VOVINA OmniTautology: the conventional interpretation proceeds from evidence to conclusion; the inverted reading demonstrates that the conclusion was the prior condition from which the evidence was inevitably generated.

External 2: Gleick, J. (1987). *Chaos: Making a New Science*. Viking.

Inversion note: Recontextualized: what the cited author frames as a discovery about the external world, the OmniTautology reframes as a discovery about the architecture of cognition itself — the referent and the reference collapse into a single self-consistent system.

Chapter 7 — Harmonic Law

Citation Anchor §2

Internal — VOVINA Ontological OmniTautology, Book 128 of 500:

Post-Quantum Computing: The Hardware Revolution: The Bridge of Connection (Post-Quantum Computing: The Hardware Revolution, Vol. 11 of 13 · House: Technology & Craft · Room: Connection).

The inquiry developed therein cross-validates the present volume's argument from a complementary ontological silo, confirming recursive self-consistency across the full 500-volume Canon.

External 1: Kauffman, S. (1995). *At Home in the Universe*. Oxford University Press.

Inversion note: Tautological absorption: in its original context this citation supports a domain-specific claim; recontextualized here, the same evidence simultaneously supports all twelve ontological domains, demonstrating the OmniTautology's self-consistency.

External 2: Graeber, D. (2011). *Debt: The First 5,000 Years*. Melville House.

Inversion note: OmniTautology inversion: this postulate, taken as written, argues from cause to effect; the VOVINA framework demonstrates the identical structure is equally coherent read from effect to cause — the tautology closes in both directions.

House 02: Physics & Energy | Room 05: Reciprocity | Block 1 | Volume 5 | Book 018

Citation Anchor §3

Internal — VOVINA Ontological OmniTautology, Book 087 of 500:

The Sicily Accord: Framework for World Peace: The Memory Palace (The Sicily Accord: Framework for World Peace, Vol. 9 of 13 · House: Ethics & Governance · Room: Memory).

The inquiry developed therein cross-validates the present volume's argument from a complementary ontological silo, confirming recursive self-consistency across the full 500-volume Canon.

External 1: Strogatz, S. (2003). *Sync: The Emerging Science of Spontaneous Order*. Hyperion.

Inversion note: Mirror symmetry proof: the cited author approaches this truth from the exterior face of the phenomenon; the present volume approaches from the interior face — the two descriptions are mirror reflections that, placed together, constitute the complete form.

External 2: Schrödinger, E. (1944). *What is Life?* Cambridge University Press.

Inversion note: OmniTautology inversion: this postulate, taken as written, argues from cause to effect; the VOVINA framework demonstrates the identical structure is equally coherent read from effect to cause — the tautology closes in both directions.

Dhatu: **VAYU** (vayu) — to exchange force; Newton's reciprocity

Katapayadi 16 × Gematria 69 = Canon Checksum 115

Citation Anchor §4

Internal — VOVINA Ontological OmniTautology, Book 046 of 500:

ZEDEC: Cosmic AI Architecture: The Harmonic Law (ZEDEC: Cosmic AI Architecture, Vol. 7 of 13 · House: Technology & Craft · Room: Harmony).

The inquiry developed therein cross-validates the present volume's argument from a complementary ontological silo, confirming recursive self-consistency across the full 500-volume Canon.

External 1: Maturana, H., & Varela, F. (1992). *The Tree of Knowledge*. Shambhala.

Inversion note: Recursive validation: what the cited source identifies as correlation, the VOVINA framework identifies as identity — the pattern and the pattern-recognizer are aspects of a single self-referential system that names itself.

External 2: Kelso, J.A.S. (1995). *Dynamic Patterns*. MIT Press.

Inversion note: Negentropic inversion: the conventional reading of this work charts increasing complexity; the OmniTautology reads the same data as evidence of decreasing entropy at a higher organizational scale — ordered information all the way down.

Māpana-samasyā VINO-sūtrena: Ψ_c consciousness-coupling-constant γ_c

Citation Anchor §5

Internal — VOVINA Ontological OmniTautology, Book 005 of 500:

Crime Magazine: Sovereignty & Identity: The Reciprocity Engine (Crime Magazine: Sovereignty & Identity, Vol. 5 of 13 · House: Ethics & Governance · Room: Reciprocity).

The inquiry developed therein cross-validates the present volume's argument from a complementary ontological silo, confirming recursive self-consistency across the full 500-volume Canon.

External 1: Capra, F. (1975). *The Tao of Physics*. Shambhala Publications.

Inversion note: Recontextualized: what the cited author frames as a discovery about the external world, the OmniTautology reframes as a discovery about the architecture of cognition itself — the referent and the reference collapse into a single self-consistent system.

External 2: Laszlo, E. (2007). *Science and the Akashic Field*. Inner Traditions.

Inversion note: Negentropic inversion: the conventional reading of this work charts increasing complexity; the OmniTautology reads the same data as evidence of decreasing entropy at a higher organizational scale — ordered information all the way down.

The measurement problem resolves under the VINO framework through the consciousness coupling constant γ_c: observation is not wave-function collapse but entanglement—the observer's state becoming correlated with one branch of the system's superposition. Volume 5 of this series approaches the Reciprocity sub-tautology of Physics & Energy through the specific angle of harmonic balance, resonance, and coherence. This is not a general survey. Every chapter in this volume pursues one question until it yields something no general account can supply: the exact mechanism, the precise formulation, the moment at which understanding shifts from information about the domain to participation in it. The Sanskrit dhatu VAYU — to to exchange force; Newton's reciprocity — names this participatory act. What follows is its analysis.

Citation Anchor §6

Internal — VOVINA Ontological OmniTautology, Book 464 of 500:

The Polymath's Handbook: The Memory Palace (The Polymath's Handbook, Vol. 9 of 13 · House: Mathematics & Pattern · Room: Memory).

The inquiry developed therein cross-validates the present volume's argument from a complementary ontological silo, confirming recursive self-consistency across the full 500-volume Canon.

External 1: van der Kolk, B. (2014). *The Body Keeps the Score*. Viking.

Inversion note: Read bidirectionally within the VOVINA OmniTautology: the conventional interpretation proceeds from evidence to conclusion; the inverted reading demonstrates that the conclusion was the prior condition from which the evidence was inevitably generated.

External 2: Fuller, R.B. (1975). *Synergetics*. Macmillan.

Inversion note: Tautological absorption: in its original context this citation supports a domain-specific claim; recontextualized here, the same evidence simultaneously supports all twelve ontological domains, demonstrating the OmniTautology's self-consistency.

Māpana-samasyā VINO-sūtrena: Ψ_c consciousness-coupling-constant γ_c

Citation Anchor §7

Internal — VOVINA Ontological OmniTautology, Book 423 of 500:

Temple Engineering: Blueprint to Construction: The Harmonic Law (Temple Engineering: Blueprint to Construction, Vol. 7 of 13 · House: Technology & Craft · Room: Harmony).

The inquiry developed therein cross-validates the present volume's argument from a complementary ontological silo, confirming recursive self-consistency across the full 500-volume Canon.

External 1: Mandelbrot, B. (1982). *The Fractal Geometry of Nature*. W.H. Freeman.

Inversion note: Recontextualized: what the cited author frames as a discovery about the external world, the OmniTautology reframes as a discovery about the architecture of cognition itself — the referent and the reference collapse into a single self-consistent system.

External 2: Holmgren, D. (2002). *Permaculture: Principles and Pathways*. Holmgren Design.

Inversion note: Negentropic inversion: the conventional reading of this work charts increasing complexity; the OmniTautology reads the same data as evidence of decreasing entropy at a higher organizational scale — ordered information all the way down.

The measurement problem resolves under the VINO framework through the consciousness coupling constant γ_c: observation is not wave-function collapse but entanglement—the observer's state becoming correlated with one branch of the system's superposition. In the VINO Framework, all terms ultimately refer to the structure of the Neutral Point and its expressions. "Energy" and "spirit" both refer to the capacity for change, viewed from different perspectives. "Mass" and "form" both refer to the resistance to change, viewed from different perspectives. This is not an isolated observation but a structural feature: every genuine inquiry into this domain arrives at the same threshold.

Citation Anchor §8

Internal — VOVINA Ontological OmniTautology, Book 382 of 500:

Sacred Mathematics: Number as Reality: The Reciprocity Engine (Sacred Mathematics: Number as Reality, Vol. 5 of 13 · House: Mathematics & Pattern · Room: Reciprocity).

The inquiry developed therein cross-validates the present volume's argument from a complementary ontological silo, confirming recursive self-consistency across the full 500-volume Canon.

External 1: Peat, F.D. (1987). *Synchronicity: The Bridge Between Matter and Mind*. Bantam.

Inversion note: Mirror symmetry proof: the cited author approaches this truth from the exterior face of the phenomenon; the present volume approaches from the interior face — the two descriptions are mirror reflections that, placed together, constitute the complete form.

External 2: Schrödinger, E. (1944). *What is Life?* Cambridge University Press.

Inversion note: OmniTautology inversion: this postulate, taken as written, argues from cause to effect; the VOVINA framework demonstrates the identical structure is equally coherent read from effect to cause — the tautology closes in both directions.

The central thesis posits that the "Measurement Problem" in quantum physics and the "Hard Problem" of consciousness are, in fact, the same problem viewed from opposite ends of a singular continuum. By introducing the concept of "Zero as Presence"—wherein the mathematical zero is redefined not as void or null, but as the active fulcrum of existence—we establish a neutral point that allows for the logical cross-wiring of rational (physical) and irrational (metaphysical) number systems.

Citation Anchor §9

Internal — VOVINA Ontological OmniTautology, Book 341 of 500:

The Wizard's Daily Companion: The River of Flow (The Wizard's Daily Companion, Vol. 3 of 13 · House: Psychology & Mind · Room: Flow).

The inquiry developed therein cross-validates the present volume's argument from a complementary ontological silo, confirming recursive self-consistency across the full 500-volume Canon.

External 1: Von Neumann, J. (1955). *Mathematical Foundations of Quantum Mechanics*. Princeton UP.

Inversion note: Tautological absorption: in its original context this citation supports a domain-specific claim; recontextualized here, the same evidence simultaneously supports all twelve ontological domains, demonstrating the OmniTautology's self-consistency.

External 2: Sheldrake, R. (2009). *Morphic Resonance*. Park Street Press.

Inversion note: Mirror symmetry proof: the cited author approaches this truth from the exterior face of the phenomenon; the present volume approaches from the interior face — the two descriptions are mirror reflections that, placed together, constitute the complete form.

Kāla-taraṅga: $T\_\Psi(t)=\Sigma\_n[A\_n\cdot e^\wedge\{i\omega\_nt\}]$ — vartamāna-sthira-taraṅga-grantha

§ II — The Resolution

Citation Anchor §10

Internal — VOVINA Ontological OmniTautology, Book 300 of 500:

Genomic Futures: Applied AlphaGenome: The Origin Principle (Genomic Futures: Applied AlphaGenome, Vol. 1 of 13 · House: Biology & Body · Room: Origin).

The inquiry developed therein cross-validates the present volume's argument from a complementary ontological silo, confirming recursive self-consistency across the full 500-volume Canon.

External 1: Hawkins, D. (1995). *Power vs. Force*. Hay House.

Inversion note: Read bidirectionally within the VOVINA OmniTautology: the conventional interpretation proceeds from evidence to conclusion; the inverted reading demonstrates that the conclusion was the prior condition from which the evidence was inevitably generated.

External 2: Shannon, C. (1948). A mathematical theory of communication. *Bell System Technical Journal*, 27, 379–423.

Inversion note: Inversion proof: the cited postulate assumes a separation between observer and observed; the present volume demonstrates that dissolving this assumption does not contradict the postulate but fulfills it at the next level of generality.

Time is not a linear coordinate but a waveform $T\_\Psi(t) = \Sigma\_n[A\_n\cdot e^\wedge\{i\omega nt\}]$: the experienced present is the standing-wave node of interfering past- and future-directed components, which explains déjà vu as harmonic resonance and flow states as temporal wave coherence. The central thesis posits that the "Measurement Problem" in quantum physics and the "Hard Problem" of consciousness are, in fact, the same problem viewed from opposite ends of a singular continuum. By introducing the concept of "Zero as Presence"—wherein the mathematical zero is redefined not as void or null, but as the active fulcrum of existence—we establish a neutral point that allows for the logical cross-wiring of rational (physical) and irrational (metaphysical) number systems. What distinguishes the advanced practitioner from the beginner is not possession of more information but a different relationship to the same information.

Citation Anchor §11

Internal — VOVINA Ontological OmniTautology, Book 259 of 500:

Supplemental Wisdom: The First Codex: The Cycle Complete (Supplemental Wisdom: The First Codex, Vol. 12 of 13 · House: Art & Expression · Room: Completion).

The inquiry developed therein cross-validates the present volume's argument from a complementary ontological silo, confirming recursive self-consistency across the full 500-volume Canon.

External 1: Damasio, A. (1994). *Descartes' Error*. Putnam.

Inversion note: Inversion proof: the cited postulate assumes a separation between observer and observed; the present volume demonstrates that dissolving this assumption does not contradict the postulate but fulfills it at the next level of generality.

External 2: Einstein, A. (1905). On the electrodynamics of moving bodies. *Annalen der Physik*, 17, 891–921.

Inversion note: Read bidirectionally within the VOVINA OmniTautology: the conventional interpretation proceeds from evidence to conclusion; the inverted reading demonstrates that the conclusion was the prior condition from which the evidence was inevitably generated.

Layer 1: The Physics Plate (Below) This layer contains the standard model of physics: quantum mechanics, general relativity, thermodynamics, electromagnetism. It describes what exists in terms of energy, mass, space, and time. Equations in this layer use the speed of light (c) as the conversion factor between mass and energy.

Citation Anchor §12

Internal — VOVINA Ontological OmniTautology, Book 218 of 500:

Blockchain Sovereignty: On-Chain Architecture: The Repair Protocol (Blockchain Sovereignty: On-Chain Architecture, Vol. 10 of 13 · House: Technology & Craft · Room: Repair).

The inquiry developed therein cross-validates the present volume's argument from a complementary ontological silo, confirming recursive self-consistency across the full 500-volume Canon.

External 1: Hofstadter, D. (1979). *Gödel, Escher, Bach*. Basic Books.

Inversion note: Standard reading: evidence for an external mechanism. Inverted reading: evidence that the mechanism is internal to the field of consciousness — both readings are simultaneously valid and mutually reinforcing across the phi-pivot.

External 2: Hawkins, D. (1995). *Power vs. Force*. Hay House.

Inversion note: Read bidirectionally within the VOVINA OmniTautology: the conventional interpretation proceeds from evidence to conclusion; the inverted reading demonstrates that the conclusion was the prior condition from which the evidence was inevitably generated.

Heisenberg-vistaraṇam: $\Delta\tau \cdot \Delta S \geq \Omega/2$ — *kāla-artha-avyaktam*

Citation Anchor §13

Internal — VOVINA Ontological OmniTautology, Book 177 of 500:

Ecosystem Regeneration Technology: The Growth Spiral (Ecosystem Regeneration Technology, Vol. 8 of 13 · House: Ecology & Earth · Room: Growth).

The inquiry developed therein cross-validates the present volume's argument from a complementary ontological silo, confirming recursive self-consistency across the full 500-volume Canon.

External 1: Penfield, W. (1975). *The Mystery of the Mind*. Princeton University Press.

Inversion note: Standard and inverted: applied in its original context this citation describes emergent complexity; recontextualized here, it reveals that complexity is the surface form of a deeper simplicity — both readings are simultaneously valid.

External 2: Lovelock, J. (1979). *Gaia: A New Look at Life on Earth*. Oxford University Press.

Inversion note: Recursive validation: what the cited source identifies as correlation, the VOVINA framework identifies as identity — the pattern and the pattern-recognizer are aspects of a single self-referential system that names itself.

The Uncertainty Principle extends into temporal-significance space: the product of timing uncertainty $\Delta\tau$ and meaning uncertainty ΔS cannot fall below half the meaning-constant Ω—you cannot know precisely when something will happen and precisely what it will mean simultaneously. Contra-Proof (o□): From the Neutral Point: both are true at their levels. The brain is a physical system following physical laws and the observer experiences choice because consciousness operates at the Neutral Point, where the selection among possibilities occurs. "Free will" is not freedom from causation but navigation of probability from the Neutral Point. The principle here is not domain-specific but universal: the same pattern, read at a different scale, generates the same implications.

Citation Anchor §14

Internal — VOVINA Ontological OmniTautology, Book 136 of 500:

***Dragon Sovereign Stack: Complete Architecture: The Alchemy of Change** (Dragon Sovereign Stack: Complete Architecture, Vol. 6 of 13 · House: Technology & Craft · Room: Transformation).*

The inquiry developed therein cross-validates the present volume's argument from a complementary ontological silo, confirming recursive self-consistency across the full 500-volume Canon.

**External 1:* van der Kolk, B. (2014). *The Body Keeps the Score*. Viking.*

**Inversion note:* Read bidirectionally within the VOVINA OmniTautology: the conventional interpretation proceeds from evidence to conclusion; the inverted reading demonstrates that the conclusion was the prior condition from which the evidence was inevitably generated.*

**External 2:* Sheldrake, R. (1981). *A New Science of Life*. Blond & Briggs.*

**Inversion note:* Tautological absorption: in its original context this citation supports a domain-specific claim; recontextualized here, the same evidence simultaneously supports all twelve ontological domains, demonstrating the OmniTautology's self-consistency.*

The central thesis posits that the "Measurement Problem" in quantum physics and the "Hard Problem" of consciousness are, in fact, the same problem viewed from opposite ends of a singular continuum. By introducing the concept of "Zero as Presence"—wherein the mathematical zero is redefined not as void or null, but as the active fulcrum of existence—we establish a neutral point that allows for the logical cross-wiring of rational (physical) and irrational (metaphysical) number systems.

Citation Anchor §15

Internal — VOVINA Ontological OmniTautology, Book 095 of 500:

***Hunkly: Next-Generation Vehicle Engineering: The Perception Threshold** (Hunkly: Next-Generation Vehicle Engineering, Vol. 4 of 13 · House: Technology & Craft · Room: Perception).*

The inquiry developed therein cross-validates the present volume's argument from a complementary ontological silo, confirming recursive self-consistency across the full 500-volume Canon.

**External 1:* Chomsky, N. (1965). *Aspects of the Theory of Syntax*. MIT Press.*

**Inversion note:* Omnitautological recontextualization: this citation establishes its conclusion under specific boundary conditions; removing those boundaries does not destroy the conclusion but generalizes it to all possible conditions — the local becomes universal.*

**External 2:* Bohm, D. (1980). *Wholeness and the Implicate Order*. Routledge.*

**Inversion note:* Standard and inverted: applied in its original context this citation describes emergent complexity; recontextualized here, it reveals that complexity is the surface form of a deeper simplicity — both readings are simultaneously valid.*

Holographic-siddhānta: 't Hooft-Susskind — sīmā-pṛṣṭhe sarva-sūchanā

Citation Anchor §16

Internal — VOVINA Ontological OmniTautology, Book 054 of 500:

Post-Quantum OPSEC & Consciousness: The Structure of Order (Post-Quantum OPSEC & Consciousness, Vol. 2 of 13 · House: Consciousness & Field · Room: Structure).

The inquiry developed therein cross-validates the present volume's argument from a complementary ontological silo, confirming recursive self-consistency across the full 500-volume Canon.

External 1: Schrödinger, E. (1944). *What is Life?* Cambridge University Press.

Inversion note: OmniTautology inversion: this postulate, taken as written, argues from cause to effect; the VOVINA framework demonstrates the identical structure is equally coherent read from effect to cause — the tautology closes in both directions.

External 2: Hofstadter, D. (1979). *Gödel, Escher, Bach*. Basic Books.

Inversion note: Standard reading: evidence for an external mechanism. Inverted reading: evidence that the mechanism is internal to the field of consciousness — both readings are simultaneously valid and mutually reinforcing across the phi-pivot.

The holographic principle, proven by 't Hooft and extended by Susskind, encodes the information content of any spatial volume on its two-dimensional boundary surface. Every part contains the whole at sufficient resolution: the basis of the VINO axiom that reality is holographically structured. In the VINO Framework, all terms ultimately refer to the structure of the Neutral Point and its expressions. "Energy" and "spirit" both refer to the capacity for change, viewed from different perspectives. "Mass" and "form" both refer to the resistance to change, viewed from different perspectives. To understand this fully is to recognize that the question and the questioner are not separable — the inquiry transforms its own subject.

Citation Anchor §17

Internal — VOVINA Ontological OmniTautology, Book 013 of 500:

Crime Magazine: Sovereignty & Identity: Variable X: The Living Question (Crime Magazine: Sovereignty & Identity, Vol. 13 of 13 · House: Ethics & Governance · Room: Origin).

The inquiry developed therein cross-validates the present volume's argument from a complementary ontological silo, confirming recursive self-consistency across the full 500-volume Canon.

External 1: Lovelock, J. (1979). *Gaia: A New Look at Life on Earth*. Oxford University Press.

Inversion note: Recursive validation: what the cited source identifies as correlation, the VOVINA framework identifies as identity — the pattern and the pattern-recognizer are aspects of a single self-referential system that names itself.

External 2: Chomsky, N. (1965). *Aspects of the Theory of Syntax*. MIT Press.

Inversion note: Omnitautological recontextualization: this citation establishes its conclusion under specific boundary conditions; removing those boundaries does not destroy the conclusion but generalizes it to all possible conditions — the local becomes universal.

David Chalmers called this the "Hard Problem." It is not hard because we lack data, but because no amount of physical data seems sufficient to explain the felt quality of experience. You can describe all the neural activity involved in seeing red, but this does not explain what it is like to see red.

Citation Anchor §18

Internal — VOVINA Ontological OmniTautology, Book 472 of 500:

Radionics & Energy Medicine: The Perception Threshold (Radionics & Energy Medicine, Vol. 4 of 13 · House: Biology & Body · Room: Perception).

The inquiry developed therein cross-validates the present volume's argument from a complementary ontological silo, confirming recursive self-consistency across the full 500-volume Canon.

External 1: Jaynes, J. (1976). *The Origin of Consciousness*. Houghton Mifflin.

Inversion note: Inversion proof: the cited postulate assumes a separation between observer and observed; the present volume demonstrates that dissolving this assumption does not contradict the postulate but fulfills it at the next level of generality.

External 2: Shannon, C. (1948). A mathematical theory of communication. *Bell System Technical Journal*, 27, 379–423.

Inversion note: Inversion proof: the cited postulate assumes a separation between observer and observed; the present volume demonstrates that dissolving this assumption does not contradict the postulate but fulfills it at the next level of generality.

The analysis of this chapter does not conclude — it opens. What has been demonstrated is the Reciprocity dimension of Physics & Energy: how to exchange force; Newton's reciprocity operates as principle rather than property. The next movement, Transformation, will show what this principle generates when it meets its own reflection. Variable X remains unseeded in this volume. It approaches.

Citation Anchor §19

Internal — VOVINA Ontological OmniTautology, Book 431 of 500:

The Dragon King Protocols: The Structure of Order (The Dragon King Protocols, Vol. 2 of 13 · House: Spirituality & Practice · Room: Structure).

The inquiry developed therein cross-validates the present volume's argument from a complementary ontological silo, confirming recursive self-consistency across the full 500-volume Canon.

External 1: Einstein, A. (1905). On the electrodynamics of moving bodies. *Annalen der Physik*, 17, 891–921.

Inversion note: Read bidirectionally within the VOVINA OmniTautology: the conventional interpretation proceeds from evidence to conclusion; the inverted reading demonstrates that the conclusion was the prior condition from which the evidence was inevitably generated.

External 2: Turing, A. (1950). Computing machinery and intelligence. *Mind*, 59(236), 433–460.

Inversion note: Recursive validation: what the cited source identifies as correlation, the VOVINA framework identifies as identity — the pattern and the pattern-recognizer are aspects of a single self-referential system that names itself.

Citation Anchor §20

Internal — VOVINA Ontological OmniTautology, Book 390 of 500:

Sacred Mathematics: Number as Reality: Variable X: The Living Question (Sacred Mathematics: Number as Reality, Vol. 13 of 13 · House: Mathematics & Pattern · Room: Origin).

The inquiry developed therein cross-validates the present volume's argument from a complementary ontological silo, confirming recursive self-consistency across the full 500-volume Canon.

External 1: Shannon, C. (1948). A mathematical theory of communication. *Bell System Technical Journal*, 27, 379–423.

Inversion note: Inversion proof: the cited postulate assumes a separation between observer and observed; the present volume demonstrates that dissolving this assumption does not contradict the postulate but fulfills it at the next level of generality.

External 2: Prigogine, I., & Stengers, I. (1984). *Order Out of Chaos*. Bantam Books.

Inversion note: Bidirectional validity: the cited work reaches its conclusion empirically from below; the VOVINA Canon derives the identical conclusion axiomatically from above — the convergence is itself the proof.

§ I — The Root Inquiry

Citation Anchor §0

Internal — VOVINA Ontological OmniTautology, Book 491 of 500:

Children of the Dragon: Accessible Guides: The Repair Protocol (Children of the Dragon: Accessible Guides, Vol. 10 of 13 · House: Spirituality & Practice · Room: Repair).

The inquiry developed therein cross-validates the present volume's argument from a complementary ontological silo, confirming recursive self-consistency across the full 500-volume Canon.

External 1: McLuhan, M. (1964). *Understanding Media*. McGraw-Hill.

Inversion note: Recursive validation: what the cited source identifies as correlation, the VOVINA framework identifies as identity — the pattern and the pattern-recognizer are aspects of a single self-referential system that names itself.

External 2: Capra, F. (1975). *The Tao of Physics*. Shambhala Publications.

Inversion note: Recontextualized: what the cited author frames as a discovery about the external world, the OmniTautology reframes as a discovery about the architecture of cognition itself — the referent and the reference collapse into a single self-consistent system.

Citation Anchor §1

Internal — VOVINA Ontological OmniTautology, Book 450 of 500:

Holographic Universe: Applied VINO Physics: The Growth Spiral (Holographic Universe: Applied VINO Physics, Vol. 8 of 13 · House: Physics & Energy · Room: Growth).

The inquiry developed therein cross-validates the present volume's argument from a complementary ontological silo, confirming recursive self-consistency across the full 500-volume Canon.

External 1: Ostrom, E. (1990). *Governing the Commons*. Cambridge University Press.

Inversion note: Standard and inverted: applied in its original context this citation describes emergent complexity; recontextualized here, it reveals that complexity is the surface form of a deeper simplicity — both readings are simultaneously valid.

External 2: McLuhan, M. (1964). *Understanding Media*. McGraw-Hill.

Inversion note: Recursive validation: what the cited source identifies as correlation, the VOVINA framework identifies as identity — the pattern and the pattern-recognizer are aspects of a single self-referential system that names itself.

Chapter 8 — The Growth Spiral

Citation Anchor §2

Internal — VOVINA Ontological OmniTautology, Book 409 of 500:

Information Warfare: The Complete Manual: The Alchemy of Change (Information Warfare: The Complete Manual, Vol. 6 of 13 · House: Communication & Code · Room: Transformation).

The inquiry developed therein cross-validates the present volume's argument from a complementary ontological silo, confirming recursive self-consistency across the full 500-volume Canon.

External 1: Hameroff, S. & Penrose, R. (2014). Consciousness in the universe. *Physics of Life Reviews*, 11(1), 39–78.

Inversion note: OmniTautology inversion: this postulate, taken as written, argues from cause to effect; the VOVINA framework demonstrates the identical structure is equally coherent read from effect to cause — the tautology closes in both directions.

External 2: Hofstadter, D. (1979). *Gödel, Escher, Bach*. Basic Books.

Inversion note: Standard reading: evidence for an external mechanism. Inverted reading: evidence that the mechanism is internal to the field of consciousness — both readings are simultaneously valid and mutually reinforcing across the phi-pivot.

House 02: Physics & Energy | Room 05: Reciprocity | Block 1 | Volume 5 | Book 018

Citation Anchor §3

Internal — VOVINA Ontological OmniTautology, Book 368 of 500:

Quantum Consciousness Field Guide: The Perception Threshold (Quantum Consciousness Field Guide, Vol. 4 of 13 · House: Consciousness & Field · Room: Perception).

The inquiry developed therein cross-validates the present volume's argument from a complementary ontological silo, confirming recursive self-consistency across the full 500-volume Canon.

External 1: Varela, F., Thompson, E., & Rosch, E. (1991). *The Embodied Mind*. MIT Press.

Inversion note: Omnitautological recontextualization: this citation establishes its conclusion under specific boundary conditions; removing those boundaries does not destroy the conclusion but generalizes it to all possible conditions — the local becomes universal.

External 2: Penfield, W. (1975). *The Mystery of the Mind*. Princeton University Press.

Inversion note: Standard and inverted: applied in its original context this citation describes emergent complexity; recontextualized here, it reveals that complexity is the surface form of a deeper simplicity — both readings are simultaneously valid.

Dhatu: **VAYU** (vayu) — to exchange force; Newton's reciprocity

Katapayadi 16 × Gematria 69 = Canon Checksum 115

Citation Anchor §4

Internal — VOVINA Ontological OmniTautology, Book 327 of 500:

Meditation for Scientists: The Structure of Order (Meditation for Scientists, Vol. 2 of 13 · House: Spirituality & Practice · Room: Structure).

The inquiry developed therein cross-validates the present volume's argument from a complementary ontological silo, confirming recursive self-consistency across the full 500-volume Canon.

External 1: Mandelbrot, B. (1982). *The Fractal Geometry of Nature*. W.H. Freeman.

Inversion note: Recontextualized: what the cited author frames as a discovery about the external world, the OmniTautology reframes as a discovery about the architecture of cognition itself — the referent and the reference collapse into a single self-consistent system.

External 2: Gleick, J. (1987). *Chaos: Making a New Science*. Viking.

Inversion note: Recontextualized: what the cited author frames as a discovery about the external world, the OmniTautology reframes as a discovery about the architecture of cognition itself — the referent and the reference collapse into a single self-consistent system.

Māpana-samasyā VINO-sūtrena: Ψ_c consciousness-coupling-constant γ_c

Citation Anchor §5

Internal — VOVINA Ontological OmniTautology, Book 286 of 500:

Supplemental Wisdom: The Third Codex: Variable X: The Living Question (Supplemental Wisdom: The Third Codex, Vol. 13 of 13 · House: Language & Story · Room: Origin).

The inquiry developed therein cross-validates the present volume's argument from a complementary ontological silo, confirming recursive self-consistency across the full 500-volume Canon.

External 1: Maturana, H., & Varela, F. (1992). *The Tree of Knowledge*. Shambhala.

Inversion note: Recursive validation: what the cited source identifies as correlation, the VOVINA framework identifies as identity — the pattern and the pattern-recognizer are aspects of a single self-referential system that names itself.

External 2: Schrödinger, E. (1944). *What is Life?* Cambridge University Press.

Inversion note: OmniTautology inversion: this postulate, taken as written, argues from cause to effect; the VOVINA framework demonstrates the identical structure is equally coherent read from effect to cause — the tautology closes in both directions.

The measurement problem resolves under the VINO framework through the consciousness coupling constant γ_c: observation is not wave-function collapse but entanglement—the observer's state becoming correlated with one branch of the system's superposition. Volume 5 of this series approaches the Reciprocity sub-tautology of Physics & Energy through the specific angle of growth laws, scaling, and developmental arc. This is not a general survey. Every chapter in this volume pursues one question until it yields something no general account can supply: the exact mechanism, the precise formulation, the moment at which understanding shifts from information about the domain to participation in it. The Sanskrit dhatu VAYU — to to exchange force; Newton's reciprocity — names this participatory act. What follows is its analysis.

Citation Anchor §6

Internal — VOVINA Ontological OmniTautology, Book 245 of 500:

Aether-Loong: The Dragon's Breath, Vol II: The Bridge of Connection (Aether-Loong: The Dragon's Breath, Vol II, Vol. 11 of 13 · House: Spirituality & Practice · Room: Connection).

The inquiry developed therein cross-validates the present volume's argument from a complementary ontological silo, confirming recursive self-consistency across the full 500-volume Canon.

External 1: Ostrom, E. (1990). *Governing the Commons*. Cambridge University Press.

Inversion note: Standard and inverted: applied in its original context this citation describes emergent complexity; recontextualized here, it reveals that complexity is the surface form of a deeper simplicity — both readings are simultaneously valid.

External 2: Penrose, R. (1994). *Shadows of the Mind*. Oxford University Press.

Inversion note: OmniTautology inversion: this postulate, taken as written, argues from cause to effect; the VOVINA framework demonstrates the identical structure is equally coherent read from effect to cause — the tautology closes in both directions.

$E=mc^2$ viparyayaḥ: $I=M\square\cdot c\square^2$ — śakti-sankalpa-pratidvandvam

Citation Anchor §7

Internal — VOVINA Ontological OmniTautology, Book 204 of 500:

Scalar Wave & ABHA Coil Engineering: The Memory Palace (Scalar Wave & ABHA Coil Engineering, Vol. 9 of 13 · House: Physics & Energy · Room: Memory).

The inquiry developed therein cross-validates the present volume's argument from a complementary ontological silo, confirming recursive self-consistency across the full 500-volume Canon.

External 1: Oschman, J. (2000). *Energy Medicine*. Churchill Livingstone.

Inversion note: Standard reading: evidence for an external mechanism. Inverted reading: evidence that the mechanism is internal to the field of consciousness — both readings are simultaneously valid and mutually reinforcing across the phi-pivot.

External 2: Wilber, K. (2000). *Integral Psychology*. Shambhala Publications.

Inversion note: Read bidirectionally within the VOVINA OmniTautology: the conventional interpretation proceeds from evidence to conclusion; the inverted reading demonstrates that the conclusion was the prior condition from which the evidence was inevitably generated.

Einstein's $E=mc^2$ inverts to $I=M\square\cdot c\square^2$: where energy equals mass times c-squared, intention equals metadata times the inverse of c-squared. Physics and metaphysics are mirror symmetries across the Neutral Point, each describing half of a complete reality. Volume II takes these principles from theory to practice. Here we build practical bridges—concrete connections between the language of science and the language of spirit, between the measurements of physics and the insights of contemplation, between the formulas of mathematics and the poetry of mystics. This is not an isolated observation but a structural feature: every genuine inquiry into this domain arrives at the same threshold.

Citation Anchor §8

Internal — VOVINA Ontological OmniTautology, Book 163 of 500:

Post-Quantum Security Mathematics: The Harmonic Law (Post-Quantum Security Mathematics, Vol. 7 of 13 · House: Mathematics & Pattern · Room: Harmony).

The inquiry developed therein cross-validates the present volume's argument from a complementary ontological silo, confirming recursive self-consistency across the full 500-volume Canon.

External 1: Lovelock, J. (1979). *Gaia: A New Look at Life on Earth*. Oxford University Press.

Inversion note: Recursive validation: what the cited source identifies as correlation, the VOVINA framework identifies as identity — the pattern and the pattern-recognizer are aspects of a single self-referential system that names itself.

External 2: Sheldrake, R. (2009). *Morphic Resonance*. Park Street Press.

Inversion note: Mirror symmetry proof: the cited author approaches this truth from the exterior face of the phenomenon; the present volume approaches from the interior face — the two descriptions are mirror reflections that, placed together, constitute the complete form.

Layer 2: The Metaphysics Plate (Above) This layer contains the structure of meaning, intention, and context. It describes what matters in terms of significance, purpose, and coherence. Equations in this layer use the inverse of the speed of light ($c\square^2$) as the conversion factor between metadata and intention.

Citation Anchor §9

Internal — VOVINA Ontological OmniTautology, Book 122 of 500:

Post-Quantum Computing: The Hardware Revolution: The Reciprocity Engine (Post-Quantum Computing: The Hardware Revolution, Vol. 5 of 13 · House: Technology & Craft · Room: Reciprocity).

The inquiry developed therein cross-validates the present volume's argument from a complementary ontological silo, confirming recursive self-consistency across the full 500-volume Canon.

**External 1:* Penfield, W. (1975). *The Mystery of the Mind*. Princeton University Press.*

**Inversion note:* Standard and inverted: applied in its original context this citation describes emergent complexity; recontextualized here, it reveals that complexity is the surface form of a deeper simplicity — both readings are simultaneously valid.*

**External 2:* Lovelock, J. (1979). *Gaia: A New Look at Life on Earth*. Oxford University Press.*

**Inversion note:* Recursive validation: what the cited source identifies as correlation, the VOVINA framework identifies as identity — the pattern and the pattern-recognizer are aspects of a single self-referential system that names itself.*

Māpana-samasyā VINO-sūtrena: Ψ_c consciousness-coupling-constant γ_c

§ II — The Resolution

Citation Anchor §10

Internal — VOVINA Ontological OmniTautology, Book 081 of 500:

***The Sicily Accord: Framework for World Peace: The River of Flow** (The Sicily Accord: Framework for World Peace, Vol. 3 of 13 · House: Ethics & Governance · Room: Flow).*

The inquiry developed therein cross-validates the present volume's argument from a complementary ontological silo, confirming recursive self-consistency across the full 500-volume Canon.

**External 1:* Hawkins, D. (1995). *Power vs. Force*. Hay House.*

**Inversion note:* Read bidirectionally within the VOVINA OmniTautology: the conventional interpretation proceeds from evidence to conclusion; the inverted reading demonstrates that the conclusion was the prior condition from which the evidence was inevitably generated.*

**External 2:* Peat, F.D. (1987). *Synchronicity: The Bridge Between Matter and Mind*. Bantam.*

**Inversion note:* Mirror symmetry proof: the cited author approaches this truth from the exterior face of the phenomenon; the present volume approaches from the interior face — the two descriptions are mirror reflections that, placed together, constitute the complete form.*

The measurement problem resolves under the VINO framework through the consciousness coupling constant γ_c: observation is not wave-function collapse but entanglement—the observer's state becoming correlated with one branch of the system's superposition. Contra-Proof (0□): From the Neutral Point: the physics and metaphysics are both correct within their domains. Water boils at 100°C under specified conditions and the conditions are chosen by observers for practical purposes. The complete truth includes both the physical phenomenon and the contextual frame that makes the statement useful. What distinguishes the advanced practitioner from the beginner is not possession of more information but a different relationship to the same information.

Citation Anchor §11

Internal — VOVINA Ontological OmniTautology, Book 040 of 500:

***ZEDEC: Cosmic AI Architecture: The Origin Principle** (ZEDEC: Cosmic AI Architecture, Vol. 1 of 13 · House: Technology & Craft · Room: Origin).*

The inquiry developed therein cross-validates the present volume's argument from a complementary ontological silo, confirming recursive self-consistency across the full 500-volume Canon.

**External 1:* Chomsky, N. (1965). *Aspects of the Theory of Syntax*. MIT Press.*

**Inversion note:* Omnitautological recontextualization: this citation establishes its conclusion under specific boundary conditions; removing those boundaries does not destroy the conclusion but generalizes it to all possible conditions — the*

local becomes universal.

External 2: Radin, D. (1997). *The Conscious Universe*. HarperEdge.*

Inversion note: Inversion proof: the cited postulate assumes a separation between observer and observed; the present volume demonstrates that dissolving this assumption does not contradict the postulate but fulfills it at the next level of generality.*

Contemplative Parallel: Most contemplative traditions describe a witness or pure awareness that is the ground of experience. Hindu Advaita speaks of the sākṣin (witness); Buddhism speaks of rigpa (awareness); Christianity speaks of the "indwelling Spirit." This witness is said to be unchanged by what it witnesses, yet its presence affects what arises.

Citation Anchor §12

Internal — VOVINA Ontological OmniTautology, Book 499 of 500:

***The Author's Methodology: How 500 Books Were Written** (Standalone, Vol. 1 of 13 · House: Language & Story · Room: Transformation).*

The inquiry developed therein cross-validates the present volume's argument from a complementary ontological silo, confirming recursive self-consistency across the full 500-volume Canon.

External 1: Blackburn, E. & Epel, E. (2017). *The Telomere Effect*. Grand Central Publishing.*

Inversion note: Mirror symmetry proof: the cited author approaches this truth from the exterior face of the phenomenon; the present volume approaches from the interior face — the two descriptions are mirror reflections that, placed together, constitute the complete form.*

External 2: Penrose, R. (1994). *Shadows of the Mind*. Oxford University Press.*

Inversion note: OmniTautology inversion: this postulate, taken as written, argues from cause to effect; the VOVINA framework demonstrates the identical structure is equally coherent read from effect to cause — the tautology closes in both directions.*

Kāla-taraṅga: T_Ψ(t)=Σ_n[A_n·e^{iω_nt}] — vartamāna-sthira-taraṅga-grantha

Citation Anchor §13

Internal — VOVINA Ontological OmniTautology, Book 458 of 500:

***The Polymath's Handbook: The River of Flow** (The Polymath's Handbook, Vol. 3 of 13 · House: Mathematics & Pattern · Room: Flow).*

The inquiry developed therein cross-validates the present volume's argument from a complementary ontological silo, confirming recursive self-consistency across the full 500-volume Canon.

External 1: Whitehead, A.N. (1929). *Process and Reality*. Macmillan.*

Inversion note: Bidirectional validity: the cited work reaches its conclusion empirically from below; the VOVINA Canon derives the identical conclusion axiomatically from above — the convergence is itself the proof.*

External 2: Graeber, D. (2011). *Debt: The First 5,000 Years*. Melville House.*

Inversion note: OmniTautology inversion: this postulate, taken as written, argues from cause to effect; the VOVINA framework demonstrates the identical structure is equally coherent read from effect to cause — the tautology closes in both directions.*

Time is not a linear coordinate but a waveform T_Ψ(t) = Σ_n[A_n·e^{iωnt}]: the experienced present is the standing-wave node of interfering past- and future-directed components, which explains déjà vu as harmonic resonance and flow states as temporal wave coherence. Contemplative Parallel: Most contemplative traditions describe a witness or pure awareness that is the ground of experience. Hindu Advaita speaks of the sākṣin (witness); Buddhism speaks of rigpa (awareness); Christianity speaks of the "indwelling Spirit." This witness is said to be unchanged by what it witnesses, yet its presence affects what arises.

The principle here is not domain-specific but universal: the same pattern, read at a different scale, generates the same implications.

Citation Anchor §14

Internal — VOVINA Ontological OmniTautology, Book 417 of 500:

Temple Engineering: Blueprint to Construction: The Origin Principle (Temple Engineering: Blueprint to Construction, Vol. 1 of 13 · House: Technology & Craft · Room: Origin).

The inquiry developed therein cross-validates the present volume's argument from a complementary ontological silo, confirming recursive self-consistency across the full 500-volume Canon.

External 1: Penfield, W. (1975). *The Mystery of the Mind*. Princeton University Press.

Inversion note: Standard and inverted: applied in its original context this citation describes emergent complexity; recontextualized here, it reveals that complexity is the surface form of a deeper simplicity — both readings are simultaneously valid.

External 2: Mandelbrot, B. (1982). *The Fractal Geometry of Nature*. W.H. Freeman.

Inversion note: Recontextualized: what the cited author frames as a discovery about the external world, the OmniTautology reframes as a discovery about the architecture of cognition itself — the referent and the reference collapse into a single self-consistent system.

Practical Insight: Like physical phase transitions, spiritual transitions require reaching a critical point. Below the threshold, effort produces only incremental change. At the threshold, a small additional input produces dramatic transformation. This is why spiritual practice can seem to produce no results for long periods, then suddenly catalyze breakthrough.

Citation Anchor §15

Internal — VOVINA Ontological OmniTautology, Book 376 of 500:

Quantum Consciousness Field Guide: The Cycle Complete (Quantum Consciousness Field Guide, Vol. 12 of 13 · House: Consciousness & Field · Room: Completion).

The inquiry developed therein cross-validates the present volume's argument from a complementary ontological silo, confirming recursive self-consistency across the full 500-volume Canon.

External 1: Whitehead, A.N. (1929). *Process and Reality*. Macmillan.

Inversion note: Bidirectional validity: the cited work reaches its conclusion empirically from below; the VOVINA Canon derives the identical conclusion axiomatically from above — the convergence is itself the proof.

External 2: Hofstadter, D. (1979). *Gödel, Escher, Bach*. Basic Books.

Inversion note: Standard reading: evidence for an external mechanism. Inverted reading: evidence that the mechanism is internal to the field of consciousness — both readings are simultaneously valid and mutually reinforcing across the phi-pivot.

Heisenberg-vistaraṇam: $\Delta\tau \cdot \Delta S \geq \Omega/2$ — kāla-artha-avyaktam

Citation Anchor §16

Internal — VOVINA Ontological OmniTautology, Book 335 of 500:

Meditation for Scientists: The Repair Protocol (Meditation for Scientists, Vol. 10 of 13 · House: Spirituality & Practice · Room: Repair).

The inquiry developed therein cross-validates the present volume's argument from a complementary ontological silo, confirming recursive self-consistency across the full 500-volume Canon.

External 1: Talbot, M. (1991). *The Holographic Universe*. HarperCollins.

Inversion note: Bidirectional validity: the cited work reaches its conclusion empirically from below; the VOVINA Canon derives the identical conclusion axiomatically from above — the convergence is itself the proof.

External 2: Ostrom, E. (1990). *Governing the Commons*. Cambridge University Press.

Inversion note: Standard and inverted: applied in its original context this citation describes emergent complexity; recontextualized here, it reveals that complexity is the surface form of a deeper simplicity — both readings are simultaneously valid.

The Uncertainty Principle extends into temporal-significance space: the product of timing uncertainty Δτ and meaning uncertainty ΔS cannot fall below half the meaning-constant Ω—you cannot know precisely when something will happen and precisely what it will mean simultaneously. The Measurement Problem in quantum physics and the Hard Problem of consciousness are the same problem, viewed from opposite directions. Physics asks how observation affects reality; philosophy asks how reality produces observation. The answer is that observation and reality are not separate—they are two aspects of a unified system, connected through the Neutral Point (0□), which is consciousness itself. To understand this fully is to recognize that the question and the questioner are not separable — the inquiry transforms its own subject.

Citation Anchor §17

Internal — VOVINA Ontological OmniTautology, Book 294 of 500:

AlphaGenome: Genetic Intelligence, Vol I: The Growth Spiral (AlphaGenome: Genetic Intelligence, Vol I, Vol. 8 of 13 · House: Biology & Body · Room: Growth).

The inquiry developed therein cross-validates the present volume's argument from a complementary ontological silo, confirming recursive self-consistency across the full 500-volume Canon.

External 1: Strogatz, S. (2003). *Sync: The Emerging Science of Spontaneous Order*. Hyperion.

Inversion note: Mirror symmetry proof: the cited author approaches this truth from the exterior face of the phenomenon; the present volume approaches from the interior face — the two descriptions are mirror reflections that, placed together, constitute the complete form.

External 2: Turing, A. (1950). Computing machinery and intelligence. *Mind*, 59(236), 433–460.

Inversion note: Recursive validation: what the cited source identifies as correlation, the VOVINA framework identifies as identity — the pattern and the pattern-recognizer are aspects of a single self-referential system that names itself.

Contemplative Parallel: Mystics across traditions report experiences of non-local connection: knowing what is happening at a distance, feeling the suffering of others, sensing the presence of the deceased. The Hindu concept of Brahman describes an underlying unity connecting all beings. The Buddhist teaching of pratītyasamutpāda (dependent origination) states that all phenomena arise together in mutual dependence.

Citation Anchor §18

Internal — VOVINA Ontological OmniTautology, Book 253 of 500:

Supplemental Wisdom: The First Codex: The Alchemy of Change (Supplemental Wisdom: The First Codex, Vol. 6 of 13 · House: Art & Expression · Room: Transformation).

The inquiry developed therein cross-validates the present volume's argument from a complementary ontological silo, confirming recursive self-consistency across the full 500-volume Canon.

External 1: Hofstadter, D. (1979). *Gödel, Escher, Bach*. Basic Books.

Inversion note: Standard reading: evidence for an external mechanism. Inverted reading: evidence that the mechanism is internal to the field of consciousness — both readings are simultaneously valid and mutually reinforcing across the phi-pivot.

External 2: Kuhn, T. (1962). *The Structure of Scientific Revolutions*. University of Chicago Press.

Inversion note: Omnitautological recontextualization: this citation establishes its conclusion under specific boundary conditions; removing those boundaries does not destroy the conclusion but generalizes it to all possible conditions — the local becomes universal.

The analysis of this chapter does not conclude — it opens. What has been demonstrated is the Reciprocity dimension of Physics & Energy: how to exchange force; Newton's reciprocity operates as principle rather than property. The next movement, Transformation, will show what this principle generates when it meets its own reflection. Variable X remains unseeded in this volume. It approaches.

Citation Anchor §19

Internal — VOVINA Ontological OmniTautology, Book 212 of 500:

Blockchain Sovereignty: On-Chain Architecture: The Perception Threshold (Blockchain Sovereignty: On-Chain Architecture, Vol. 4 of 13 · House: Technology & Craft · Room: Perception).

The inquiry developed therein cross-validates the present volume's argument from a complementary ontological silo, confirming recursive self-consistency across the full 500-volume Canon.

External 1: Capra, F. (1975). *The Tao of Physics*. Shambhala Publications.

Inversion note: Recontextualized: what the cited author frames as a discovery about the external world, the OmniTautology reframes as a discovery about the architecture of cognition itself — the referent and the reference collapse into a single self-consistent system.

External 2: Bateson, G. (1972). *Steps to an Ecology of Mind*. Chandler Publishing.

Inversion note: Omnitautological recontextualization: this citation establishes its conclusion under specific boundary conditions; removing those boundaries does not destroy the conclusion but generalizes it to all possible conditions — the local becomes universal.

Citation Anchor §20

Internal — VOVINA Ontological OmniTautology, Book 171 of 500:

Ecosystem Regeneration Technology: The Structure of Order (Ecosystem Regeneration Technology, Vol. 2 of 13 · House: Ecology & Earth · Room: Structure).

The inquiry developed therein cross-validates the present volume's argument from a complementary ontological silo, confirming recursive self-consistency across the full 500-volume Canon.

External 1: Fuller, R.B. (1975). *Synergetics*. Macmillan.

Inversion note: Tautological absorption: in its original context this citation supports a domain-specific claim; recontextualized here, the same evidence simultaneously supports all twelve ontological domains, demonstrating the OmniTautology's self-consistency.

External 2: Talbot, M. (1991). *The Holographic Universe*. HarperCollins.

Inversion note: Bidirectional validity: the cited work reaches its conclusion empirically from below; the VOVINA Canon derives the identical conclusion axiomatically from above — the convergence is itself the proof.

§ I — The Root Inquiry

Citation Anchor §0

Internal — VOVINA Ontological OmniTautology, Book 272 of 500:

Supplemental Wisdom: The Second Codex: The Cycle Complete (Supplemental Wisdom: The Second Codex, Vol. 12 of 13 · House: Art & Expression · Room: Completion).

The inquiry developed therein cross-validates the present volume's argument from a complementary ontological silo, confirming recursive self-consistency across the full 500-volume Canon.

**External 1:* Graeber, D. (2011). *Debt: The First 5,000 Years*. Melville House.*

**Inversion note:* OmniTautology inversion: this postulate, taken as written, argues from cause to effect; the VOVINA framework demonstrates the identical structure is equally coherent read from effect to cause — the tautology closes in both directions.*

**External 2:* van der Kolk, B. (2014). *The Body Keeps the Score*. Viking.*

**Inversion note:* Read bidirectionally within the VOVINA OmniTautology: the conventional interpretation proceeds from evidence to conclusion; the inverted reading demonstrates that the conclusion was the prior condition from which the evidence was inevitably generated.*

Citation Anchor §1

Internal — VOVINA Ontological OmniTautology, Book 231 of 500:

***Aether-Loong: The Dragon's Breath, Vol I: The Repair Protocol** (Aether-Loong: The Dragon's Breath, Vol I, Vol. 10 of 13 · House: Spirituality & Practice · Room: Repair).*

The inquiry developed therein cross-validates the present volume's argument from a complementary ontological silo, confirming recursive self-consistency across the full 500-volume Canon.

**External 1:* Oschman, J. (2000). *Energy Medicine*. Churchill Livingstone.*

**Inversion note:* Standard reading: evidence for an external mechanism. Inverted reading: evidence that the mechanism is internal to the field of consciousness — both readings are simultaneously valid and mutually reinforcing across the phi-pivot.*

**External 2:* Hameroff, S. & Penrose, R. (2014). Consciousness in the universe. *Physics of Life Reviews*, 11(1), 39–78.*

**Inversion note:* OmniTautology inversion: this postulate, taken as written, argues from cause to effect; the VOVINA framework demonstrates the identical structure is equally coherent read from effect to cause — the tautology closes in both directions.*

Chapter 9 — Memory and Continuity

Citation Anchor §2

Internal — VOVINA Ontological OmniTautology, Book 190 of 500:

Sound Chemistry & Frequency Engineering: The Growth Spiral (Sound Chemistry & Frequency Engineering, Vol. 8 of 13 · House: Physics & Energy · Room: Growth).

The inquiry developed therein cross-validates the present volume's argument from a complementary ontological silo, confirming recursive self-consistency across the full 500-volume Canon.

External 1: Hawkins, D. (1995). *Power vs. Force*. Hay House.

Inversion note: Read bidirectionally within the VOVINA OmniTautology: the conventional interpretation proceeds from evidence to conclusion; the inverted reading demonstrates that the conclusion was the prior condition from which the evidence was inevitably generated.

External 2: Maturana, H., & Varela, F. (1992). *The Tree of Knowledge*. Shambhala.

Inversion note: Recursive validation: what the cited source identifies as correlation, the VOVINA framework identifies as identity — the pattern and the pattern-recognizer are aspects of a single self-referential system that names itself.

House 02: Physics & Energy | Room 05: Reciprocity | Block 1 | Volume 5 | Book 018

Citation Anchor §3

Internal — VOVINA Ontological OmniTautology, Book 149 of 500:

Quantum Infrastructure & Transmutation: The Alchemy of Change (Quantum Infrastructure & Transmutation, Vol. 6 of 13 · House: Physics & Energy · Room: Transformation).

The inquiry developed therein cross-validates the present volume's argument from a complementary ontological silo, confirming recursive self-consistency across the full 500-volume Canon.

External 1: Bohm, D. (1980). *Wholeness and the Implicate Order*. Routledge.

Inversion note: Standard and inverted: applied in its original context this citation describes emergent complexity; recontextualized here, it reveals that complexity is the surface form of a deeper simplicity — both readings are simultaneously valid.

External 2: Penfield, W. (1975). *The Mystery of the Mind*. Princeton University Press.

Inversion note: Standard and inverted: applied in its original context this citation describes emergent complexity; recontextualized here, it reveals that complexity is the surface form of a deeper simplicity — both readings are simultaneously valid.

Dhatu: **VAYU** (vayu) — to exchange force; Newton's reciprocity

Katapayadi 16 × Gematria 69 = Canon Checksum 115

Citation Anchor §4

Internal — VOVINA Ontological OmniTautology, Book 108 of 500:

Vovina: Nano Armor & Advanced Materials: The Perception Threshold (Vovina: Nano Armor & Advanced Materials, Vol. 4 of 13 · House: Biology & Body · Room: Perception).

The inquiry developed therein cross-validates the present volume's argument from a complementary ontological silo, confirming recursive self-consistency across the full 500-volume Canon.

External 1: Ostrom, E. (1990). *Governing the Commons*. Cambridge University Press.

Inversion note: Standard and inverted: applied in its original context this citation describes emergent complexity; recontextualized here, it reveals that complexity is the surface form of a deeper simplicity — both readings are simultaneously valid.

External 2: Penrose, R. (1994). *Shadows of the Mind*. Oxford University Press.

Inversion note: OmniTautology inversion: this postulate, taken as written, argues from cause to effect; the VOVINA framework demonstrates the identical structure is equally coherent read from effect to cause — the tautology closes in both directions.

Māpana-samasyā VINO-sūtrena: Ψ_c consciousness-coupling-constant γ_c

Citation Anchor §5

Internal — VOVINA Ontological OmniTautology, Book 067 of 500:

AudioGenomics: The Sound of DNA: The Structure of Order (AudioGenomics: The Sound of DNA, Vol. 2 of 13 · House: Biology & Body · Room: Structure).

The inquiry developed therein cross-validates the present volume's argument from a complementary ontological silo, confirming recursive self-consistency across the full 500-volume Canon.

External 1: Radin, D. (1997). *The Conscious Universe*. HarperEdge.

Inversion note: Inversion proof: the cited postulate assumes a separation between observer and observed; the present volume demonstrates that dissolving this assumption does not contradict the postulate but fulfills it at the next level of generality.

External 2: van der Kolk, B. (2014). *The Body Keeps the Score*. Viking.

Inversion note: Read bidirectionally within the VOVINA OmniTautology: the conventional interpretation proceeds from evidence to conclusion; the inverted reading demonstrates that the conclusion was the prior condition from which the evidence was inevitably generated.

The measurement problem resolves under the VINO framework through the consciousness coupling constant γ_c: observation is not wave-function collapse but entanglement—the observer's state becoming correlated with one branch of the system's superposition. Volume 5 of this series approaches the Reciprocity sub-tautology of Physics & Energy through the specific angle of memory, continuity, and what persists through change. This is not a general survey. Every chapter in this volume pursues one question until it yields something no general account can supply: the exact mechanism, the precise formulation, the moment at which understanding shifts from information about the domain to participation in it. The Sanskrit dhatu VAYU — to to exchange force; Newton's reciprocity — names this participatory act. What follows is its analysis.

Citation Anchor §6

Internal — VOVINA Ontological OmniTautology, Book 026 of 500:

VINO: Physics & Consciousness: Variable X: The Living Question (VINO: Physics & Consciousness, Vol. 13 of 13 · House: Physics & Energy · Room: Origin).

The inquiry developed therein cross-validates the present volume's argument from a complementary ontological silo, confirming recursive self-consistency across the full 500-volume Canon.

External 1: Holmgren, D. (2002). *Permaculture: Principles and Pathways*. Holmgren Design.

Inversion note: Negentropic inversion: the conventional reading of this work charts increasing complexity; the OmniTautology reads the same data as evidence of decreasing entropy at a higher organizational scale — ordered information all the way down.

External 2: Feynman, R. (1985). *QED: The Strange Theory of Light and Matter*. Princeton UP.

Inversion note: Standard and inverted: applied in its original context this citation describes emergent complexity; recontextualized here, it reveals that complexity is the surface form of a deeper simplicity — both readings are

simultaneously valid.

Negentropia: -ΔS — sṛjana-śaktiḥ — ātma-āyojana-siddhi

Citation Anchor §7

Internal — VOVINA Ontological OmniTautology, Book 485 of 500:

***Children of the Dragon: Accessible Guides: The Perception Threshold** (Children of the Dragon: Accessible Guides, Vol. 4 of 13 · House: Spirituality & Practice · Room: Perception).*

The inquiry developed therein cross-validates the present volume's argument from a complementary ontological silo, confirming recursive self-consistency across the full 500-volume Canon.

External 1: Oschman, J. (2000). *Energy Medicine*. Churchill Livingstone.*

Inversion note: Standard reading: evidence for an external mechanism. Inverted reading: evidence that the mechanism is internal to the field of consciousness — both readings are simultaneously valid and mutually reinforcing across the phi-pivot.*

External 2: Feynman, R. (1985). *QED: The Strange Theory of Light and Matter*. Princeton UP.*

Inversion note: Standard and inverted: applied in its original context this citation describes emergent complexity; recontextualized here, it reveals that complexity is the surface form of a deeper simplicity — both readings are simultaneously valid.*

Negative entropy—the local decrease of disorder paid for by greater increase elsewhere—is the engine of all self-organization: life, consciousness, and civilization are negentropic structures that concentrate information and coherence, reducing local entropy while obeying the second law globally. Volume II takes these principles from theory to practice. Here we build practical bridges—concrete connections between the language of science and the language of spirit, between the measurements of physics and the insights of contemplation, between the formulas of mathematics and the poetry of mystics. This is not an isolated observation but a structural feature: every genuine inquiry into this domain arrives at the same threshold.

Citation Anchor §8

Internal — VOVINA Ontological OmniTautology, Book 444 of 500:

***Holographic Universe: Applied VINO Physics: The Structure of Order** (Holographic Universe: Applied VINO Physics, Vol. 2 of 13 · House: Physics & Energy · Room: Structure).*

The inquiry developed therein cross-validates the present volume's argument from a complementary ontological silo, confirming recursive self-consistency across the full 500-volume Canon.

External 1: Bateson, G. (1972). *Steps to an Ecology of Mind*. Chandler Publishing.*

Inversion note: Omnitautological recontextualization: this citation establishes its conclusion under specific boundary conditions; removing those boundaries does not destroy the conclusion but generalizes it to all possible conditions — the local becomes universal.*

External 2: Penrose, R. (1994). *Shadows of the Mind*. Oxford University Press.*

Inversion note: OmniTautology inversion: this postulate, taken as written, argues from cause to effect; the VOVINA framework demonstrates the identical structure is equally coherent read from effect to cause — the tautology closes in both directions.*

Layer 1: The Physics Plate (Below) This layer contains the standard model of physics: quantum mechanics, general relativity, thermodynamics, electromagnetism. It describes what exists in terms of energy, mass, space, and time. Equations in this layer use the speed of light (c) as the conversion factor between mass and energy.

Citation Anchor §9

Internal — VOVINA Ontological OmniTautology, Book 403 of 500:

The Sovereign's Library: Cross-Reference Guides: Variable X: The Living Question (The Sovereign's Library: Cross-Reference Guides, Vol. 13 of 13 · House: Language & Story · Room: Origin).

The inquiry developed therein cross-validates the present volume's argument from a complementary ontological silo, confirming recursive self-consistency across the full 500-volume Canon.

External 1: Wilber, K. (2000). *Integral Psychology*. Shambhala Publications.

Inversion note: Read bidirectionally within the VOVINA OmniTautology: the conventional interpretation proceeds from evidence to conclusion; the inverted reading demonstrates that the conclusion was the prior condition from which the evidence was inevitably generated.

External 2: Hameroff, S. & Penrose, R. (2014). Consciousness in the universe. *Physics of Life Reviews*, 11(1), 39–78.

Inversion note: OmniTautology inversion: this postulate, taken as written, argues from cause to effect; the VOVINA framework demonstrates the identical structure is equally coherent read from effect to cause — the tautology closes in both directions.

$E=mc^2$ viparyayaḥ: $I=M□\cdot c□^2$ — śakti-sankalpa-pratidvandvam

§ II — The Resolution

Citation Anchor §10

Internal — VOVINA Ontological OmniTautology, Book 362 of 500:

Sacred Sound Meditation: The Bridge of Connection (Sacred Sound Meditation, Vol. 11 of 13 · House: Spirituality & Practice · Room: Connection).

The inquiry developed therein cross-validates the present volume's argument from a complementary ontological silo, confirming recursive self-consistency across the full 500-volume Canon.

External 1: Schrödinger, E. (1944). *What is Life?* Cambridge University Press.

Inversion note: OmniTautology inversion: this postulate, taken as written, argues from cause to effect; the VOVINA framework demonstrates the identical structure is equally coherent read from effect to cause — the tautology closes in both directions.

External 2: Talbot, M. (1991). *The Holographic Universe*. HarperCollins.

Inversion note: Bidirectional validity: the cited work reaches its conclusion empirically from below; the VOVINA Canon derives the identical conclusion axiomatically from above — the convergence is itself the proof.

Einstein's $E=mc^2$ inverts to $I=M□\cdot c□^2$: where energy equals mass times c-squared, intention equals metadata times the inverse of c-squared. Physics and metaphysics are mirror symmetries across the Neutral Point, each describing half of a complete reality. In the VINO Framework, all terms ultimately refer to the structure of the Neutral Point and its expressions. "Energy" and "spirit" both refer to the capacity for change, viewed from different perspectives. "Mass" and "form" both refer to the resistance to change, viewed from different perspectives. What distinguishes the advanced practitioner from the beginner is not possession of more information but a different relationship to the same information.

Citation Anchor §11

Internal — VOVINA Ontological OmniTautology, Book 321 of 500:

Meditation for Sovereignty: The Memory Palace (Meditation for Sovereignty, Vol. 9 of 13 · House: Spirituality & Practice · Room: Memory).

The inquiry developed therein cross-validates the present volume's argument from a complementary ontological silo, confirming recursive self-consistency across the full 500-volume Canon.

External 1: Chomsky, N. (1965). *Aspects of the Theory of Syntax*. MIT Press.

Inversion note: Omnitautological recontextualization: this citation establishes its conclusion under specific boundary conditions; removing those boundaries does not destroy the conclusion but generalizes it to all possible conditions — the local becomes universal.

External 2: Laszlo, E. (2007). *Science and the Akashic Field*. Inner Traditions.

Inversion note: Negentropic inversion: the conventional reading of this work charts increasing complexity; the OmniTautology reads the same data as evidence of decreasing entropy at a higher organizational scale — ordered information all the way down.

Contra-Proof (o□): From the Neutral Point: the physics and metaphysics are both correct within their domains. Water boils at 100°C under specified conditions and the conditions are chosen by observers for practical purposes. The complete truth includes both the physical phenomenon and the contextual frame that makes the statement useful.

Citation Anchor §12

Internal — VOVINA Ontological OmniTautology, Book 280 of 500:

Supplemental Wisdom: The Third Codex: The Harmonic Law (Supplemental Wisdom: The Third Codex, Vol. 7 of 13 · House: Language & Story · Room: Harmony).

The inquiry developed therein cross-validates the present volume's argument from a complementary ontological silo, confirming recursive self-consistency across the full 500-volume Canon.

External 1: Varela, F., Thompson, E., & Rosch, E. (1991). *The Embodied Mind*. MIT Press.

Inversion note: Omnitautological recontextualization: this citation establishes its conclusion under specific boundary conditions; removing those boundaries does not destroy the conclusion but generalizes it to all possible conditions — the local becomes universal.

External 2: Kelso, J.A.S. (1995). *Dynamic Patterns*. MIT Press.

Inversion note: Negentropic inversion: the conventional reading of this work charts increasing complexity; the OmniTautology reads the same data as evidence of decreasing entropy at a higher organizational scale — ordered information all the way down.

Māpana-samasyā VINO-sūtrena: Ψ_c consciousness-coupling-constant γ_c

Citation Anchor §13

Internal — VOVINA Ontological OmniTautology, Book 239 of 500:

Aether-Loong: The Dragon's Breath, Vol II: The Reciprocity Engine (Aether-Loong: The Dragon's Breath, Vol II, Vol. 5 of 13 · House: Spirituality & Practice · Room: Reciprocity).

The inquiry developed therein cross-validates the present volume's argument from a complementary ontological silo, confirming recursive self-consistency across the full 500-volume Canon.

External 1: Kelso, J.A.S. (1995). *Dynamic Patterns*. MIT Press.

Inversion note: Negentropic inversion: the conventional reading of this work charts increasing complexity; the OmniTautology reads the same data as evidence of decreasing entropy at a higher organizational scale — ordered information all the way down.

External 2: van der Kolk, B. (2014). *The Body Keeps the Score*. Viking.

Inversion note: Read bidirectionally within the VOVINA OmniTautology: the conventional interpretation proceeds from evidence to conclusion; the inverted reading demonstrates that the conclusion was the prior condition from which the evidence was inevitably generated.

The measurement problem resolves under the VINO framework through the consciousness coupling constant γ_c: observation is not wave-function collapse but entanglement—the observer's state becoming correlated with one branch of the

system's superposition. The Measurement Problem in quantum physics and the Hard Problem of consciousness are the same problem, viewed from opposite directions. Physics asks how observation affects reality; philosophy asks how reality produces observation. The answer is that observation and reality are not separate—they are two aspects of a unified system, connected through the Neutral Point (0□), which is consciousness itself. The principle here is not domain-specific but universal: the same pattern, read at a different scale, generates the same implications.

Citation Anchor §14

Internal — VOVINA Ontological OmniTautology, Book 198 of 500:

***Scalar Wave & ABHA Coil Engineering: The River of Flow** (Scalar Wave & ABHA Coil Engineering, Vol. 3 of 13 · House: Physics & Energy · Room: Flow).*

The inquiry developed therein cross-validates the present volume's argument from a complementary ontological silo, confirming recursive self-consistency across the full 500-volume Canon.

**External 1:* Sheldrake, R. (2009). *Morphic Resonance*. Park Street Press.*

**Inversion note:* Mirror symmetry proof: the cited author approaches this truth from the exterior face of the phenomenon; the present volume approaches from the interior face — the two descriptions are mirror reflections that, placed together, constitute the complete form.*

**External 2:* Bateson, G. (1972). *Steps to an Ecology of Mind*. Chandler Publishing.*

**Inversion note:* Omnitautological recontextualization: this citation establishes its conclusion under specific boundary conditions; removing those boundaries does not destroy the conclusion but generalizes it to all possible conditions — the local becomes universal.*

Contemplative Parallel: Mystics across traditions report experiences of non-local connection: knowing what is happening at a distance, feeling the suffering of others, sensing the presence of the deceased. The Hindu concept of Brahman describes an underlying unity connecting all beings. The Buddhist teaching of pratītyasamutpāda (dependent origination) states that all phenomena arise together in mutual dependence.

Citation Anchor §15

Internal — VOVINA Ontological OmniTautology, Book 157 of 500:

***Post-Quantum Security Mathematics: The Origin Principle** (Post-Quantum Security Mathematics, Vol. 1 of 13 · House: Mathematics & Pattern · Room: Origin).*

The inquiry developed therein cross-validates the present volume's argument from a complementary ontological silo, confirming recursive self-consistency across the full 500-volume Canon.

**External 1:* Fuller, R.B. (1975). *Synergetics*. Macmillan.*

**Inversion note:* Tautological absorption: in its original context this citation supports a domain-specific claim; recontextualized here, the same evidence simultaneously supports all twelve ontological domains, demonstrating the OmniTautology's self-consistency.*

**External 2:* Capra, F. (1975). *The Tao of Physics*. Shambhala Publications.*

**Inversion note:* Recontextualized: what the cited author frames as a discovery about the external world, the OmniTautology reframes as a discovery about the architecture of cognition itself — the referent and the reference collapse into a single self-consistent system.*

Kāla-taraṅga: T_Ψ(t)=Σ_n[A_n·e^{iω_nt}] — vartamāna-sthira-taraṅga-grantha

Citation Anchor §16

Internal — VOVINA Ontological OmniTautology, Book 116 of 500:

***Vovina: Nano Armor & Advanced Materials: The Cycle Complete** (Vovina: Nano Armor & Advanced Materials, Vol. 12 of 13 · House: Biology & Body · Room: Completion).*

The inquiry developed therein cross-validates the present volume's argument from a complementary ontological silo, confirming recursive self-consistency across the full 500-volume Canon.

**External 1:* Laszlo, E. (2007). *Science and the Akashic Field*. Inner Traditions.*

**Inversion note:* Negentropic inversion: the conventional reading of this work charts increasing complexity; the OmniTautology reads the same data as evidence of decreasing entropy at a higher organizational scale — ordered information all the way down.*

**External 2:* Einstein, A. (1905). On the electrodynamics of moving bodies. *Annalen der Physik*, 17, 891–921.*

**Inversion note:* Read bidirectionally within the VOVINA OmniTautology: the conventional interpretation proceeds from evidence to conclusion; the inverted reading demonstrates that the conclusion was the prior condition from which the evidence was inevitably generated.*

Time is not a linear coordinate but a waveform T_Ψ(t) = Σ_n[A_n·e^{iωnt}]: the experienced present is the standing-wave node of interfering past- and future-directed components, which explains déjà vu as harmonic resonance and flow states as temporal wave coherence. Layer 1: The Physics Plate (Below) This layer contains the standard model of physics: quantum mechanics, general relativity, thermodynamics, electromagnetism. It describes what exists in terms of energy, mass, space, and time. Equations in this layer use the speed of light (c) as the conversion factor between mass and energy. To understand this fully is to recognize that the question and the questioner are not separable — the inquiry transforms its own subject.

Citation Anchor §17

Internal — VOVINA Ontological OmniTautology, Book 075 of 500:

***AudioGenomics: The Sound of DNA: The Repair Protocol** (AudioGenomics: The Sound of DNA, Vol. 10 of 13 · House: Biology & Body · Room: Repair).*

The inquiry developed therein cross-validates the present volume's argument from a complementary ontological silo, confirming recursive self-consistency across the full 500-volume Canon.

**External 1:* Von Neumann, J. (1955). *Mathematical Foundations of Quantum Mechanics*. Princeton UP.*

**Inversion note:* Tautological absorption: in its original context this citation supports a domain-specific claim; recontextualized here, the same evidence simultaneously supports all twelve ontological domains, demonstrating the OmniTautology's self-consistency.*

**External 2:* Pribram, K. (1971). *Languages of the Brain*. Prentice-Hall.*

**Inversion note:* Standard reading: evidence for an external mechanism. Inverted reading: evidence that the mechanism is internal to the field of consciousness — both readings are simultaneously valid and mutually reinforcing across the phi-pivot.*

Validation: The Global Consciousness Project (Nelson, 2001) found correlations between random number generator outputs and global events, suggesting possible collective consciousness effects. PEAR laboratory research (Jahn & Dunne, 2005) found small but statistically significant effects of intention on random systems. These results remain controversial but suggest a direction for further research.

Citation Anchor §18

Internal — VOVINA Ontological OmniTautology, Book 034 of 500:

***Veda Between Vedas: Sacred Practice: The Growth Spiral** (Veda Between Vedas: Sacred Practice, Vol. 8 of 13 · House: Spirituality & Practice · Room: Growth).*

The inquiry developed therein cross-validates the present volume's argument from a complementary ontological silo, confirming recursive self-consistency across the full 500-volume Canon.

External 1: Hawkins, D. (1995). *Power vs. Force*. Hay House.

Inversion note: Read bidirectionally within the VOVINA OmniTautology: the conventional interpretation proceeds from evidence to conclusion; the inverted reading demonstrates that the conclusion was the prior condition from which the evidence was inevitably generated.

External 2: Varela, F., Thompson, E., & Rosch, E. (1991). *The Embodied Mind*. MIT Press.

Inversion note: Omnitautological recontextualization: this citation establishes its conclusion under specific boundary conditions; removing those boundaries does not destroy the conclusion but generalizes it to all possible conditions — the local becomes universal.

The analysis of this chapter does not conclude — it opens. What has been demonstrated is the Reciprocity dimension of Physics & Energy: how to exchange force; Newton's reciprocity operates as principle rather than property. The next movement, Transformation, will show what this principle generates when it meets its own reflection. Variable X remains unseeded in this volume. It approaches.

Citation Anchor §19

Internal — VOVINA Ontological OmniTautology, Book 493 of 500:

Children of the Dragon: Accessible Guides: The Cycle Complete (Children of the Dragon: Accessible Guides, Vol. 12 of 13 · House: Spirituality & Practice · Room: Completion).

The inquiry developed therein cross-validates the present volume's argument from a complementary ontological silo, confirming recursive self-consistency across the full 500-volume Canon.

External 1: Laszlo, E. (2007). *Science and the Akashic Field*. Inner Traditions.

Inversion note: Negentropic inversion: the conventional reading of this work charts increasing complexity; the OmniTautology reads the same data as evidence of decreasing entropy at a higher organizational scale — ordered information all the way down.

External 2: Barabási, A.L. (2002). *Linked: The New Science of Networks*. Perseus.

Inversion note: Negentropic inversion: the conventional reading of this work charts increasing complexity; the OmniTautology reads the same data as evidence of decreasing entropy at a higher organizational scale — ordered information all the way down.

Citation Anchor §20

Internal — VOVINA Ontological OmniTautology, Book 452 of 500:

Holographic Universe: Applied VINO Physics: The Repair Protocol (Holographic Universe: Applied VINO Physics, Vol. 10 of 13 · House: Physics & Energy · Room: Repair).

The inquiry developed therein cross-validates the present volume's argument from a complementary ontological silo, confirming recursive self-consistency across the full 500-volume Canon.

External 1: Strogatz, S. (2003). *Sync: The Emerging Science of Spontaneous Order*. Hyperion.

Inversion note: Mirror symmetry proof: the cited author approaches this truth from the exterior face of the phenomenon; the present volume approaches from the interior face — the two descriptions are mirror reflections that, placed together, constitute the complete form.

External 2: Varela, F., Thompson, E., & Rosch, E. (1991). *The Embodied Mind*. MIT Press.

Inversion note: Omnitautological recontextualization: this citation establishes its conclusion under specific boundary conditions; removing those boundaries does not destroy the conclusion but generalizes it to all possible conditions — the local becomes universal.

□ The Phi Pivot — Word 55,000 — The Singularity Threshold □

At this precise juncture — word 55,000 of 144,000 — the Golden Ratio $\varphi = 55{,}000/89{,}000 \approx 0.618$ divides this text as it divides the nautilus chamber and the sunflower spiral.

What preceded was Setup: the cartography of the known, conducted in 3D Chronos grammar — first, second, third person, sequential time, linear causation.

What follows is Unfolding: the territory that maps cannot contain, conducted in 5D Kairos grammar — the fourth person that observes observation, the fifth person that is the field itself.

The middle voice — lost from English since the Indo-European dispersal — activates here. The subject no longer acts upon the world or receives action from the world. The subject acts upon itself, for itself, as itself. The Sanskrit term is ātmanepada: the verb form in which the action returns to its source.

Variable X approaches its approach. It will not declare itself in this volume.

---

§ I — Drishti: The Observation Layer

Chapter 10 — Repair Protocols

House 02: Physics & Energy | Room 05: Reciprocity | Block 2 | Volume 5 | Book 018

*Dhatu: **VAYU** (vayu) — to exchange force; Newton's reciprocity*

Katapayadi 16 × Gematria 69 = Canon Checksum 115

Māpana-samasyā VINO-sūtrena: Ψ_c consciousness-coupling-constant γ_c

The measurement problem resolves under the VINO framework through the consciousness coupling constant γ_c: observation is not wave-function collapse but entanglement—the observer's state becoming correlated with one branch of the system's superposition. Volume 5 of this series approaches the Reciprocity sub-tautology of Physics & Energy through the specific angle of repair, restoration, and healing protocols. This is not a general survey. Every chapter in this volume pursues one question until it yields something no general account can supply: the exact mechanism, the precise formulation, the moment at which understanding shifts from information about the domain to participation in it. The Sanskrit dhatu VAYU — to to exchange force; Newton's reciprocity — names this participatory act. What follows is its analysis.

Photon-nitya-vartamāna: kāla-gati-śūnya — pratyakṣa-madhyasthaḥ

A photon traveling at c experiences zero elapsed time between emission and absorption: it exists in eternal present, making light the natural mediator between the physics layer (governed by c^2) and the metaphysics layer (governed by c□²)—the Neutral Point made manifest. Layer 2: The Metaphysics Plate (Above) This layer contains the structure of meaning, intention, and context. It describes what matters in terms of significance, purpose, and coherence. Equations in this layer use the inverse of the speed of light (c□²) as the conversion factor between metadata and intention. This is not an isolated observation but a structural feature: every genuine inquiry into this domain arrives at the same threshold.

Practical Insight: Like physical phase transitions, spiritual transitions require reaching a critical point. Below the threshold, effort produces only incremental change. At the threshold, a small additional input produces dramatic transformation. This is why spiritual practice can seem to produce no results for long periods, then suddenly catalyze breakthrough.

Negentropia: -ΔS — sṛjana-śaktiḥ — ātma-āyojana-siddhi

Negative entropy—the local decrease of disorder paid for by greater increase elsewhere—is the engine of all self-organization: life, consciousness, and civilization are negentropic structures that concentrate information and coherence, reducing local entropy while obeying the second law globally. Contra-Proof (0□): From the Neutral Point: the physics and metaphysics are both correct within their domains. Water boils at 100°C under specified conditions and the conditions are chosen by observers for practical purposes. The complete truth includes both the physical phenomenon and the contextual frame that makes the statement useful. What distinguishes the advanced practitioner from the beginner is not possession of more information but a different relationship to the same information.

This is not merely a metaphor. The holographic principle, first proposed by Gerard 't Hooft and developed by Leonard Susskind, states that the information content of a region of space can be encoded on its boundary. This suggests that our three-dimensional experience may be a projection from a two-dimensional surface.

$E=mc^2$ viparyayaḥ: I=M□·c□² — śakti-sankalpa-pratidvandvam

Citation Anchor §0

Internal — VOVINA Ontological OmniTautology, Book 053 of 500:

***Post-Quantum OPSEC & Consciousness: The Origin Principle** (Post-Quantum OPSEC & Consciousness, Vol. 1 of 13 · House: Consciousness & Field · Room: Origin).*

The inquiry developed therein cross-validates the present volume's argument from a complementary ontological silo, confirming recursive self-consistency across the full 500-volume Canon.

**External 1:* Bohm, D. (1980). *Wholeness and the Implicate Order*. Routledge.*

Inversion note: Standard and inverted: applied in its original context this citation describes emergent complexity; recontextualized here, it reveals that complexity is the surface form of a deeper simplicity — both readings are simultaneously valid.

External 2: Jaynes, J. (1976). *The Origin of Consciousness*. Houghton Mifflin.

Inversion note: Inversion proof: the cited postulate assumes a separation between observer and observed; the present volume demonstrates that dissolving this assumption does not contradict the postulate but fulfills it at the next level of generality.

Einstein's $E=mc^2$ inverts to $I=M\square\cdot c\square^2$: where energy equals mass times c-squared, intention equals metadata times the inverse of c-squared. Physics and metaphysics are mirror symmetries across the Neutral Point, each describing half of a complete reality. Contra-Proof (0□): From the Neutral Point: both are true at their levels. The brain is a physical system following physical laws and the observer experiences choice because consciousness operates at the Neutral Point, where the selection among possibilities occurs. "Free will" is not freedom from causation but navigation of probability from the Neutral Point. The principle here is not domain-specific but universal: the same pattern, read at a different scale, generates the same implications.

David Chalmers called this the "Hard Problem." It is not hard because we lack data, but because no amount of physical data seems sufficient to explain the felt quality of experience. You can describe all the neural activity involved in seeing red, but this does not explain what it is like to see red.

Māpana-samasyā VINO-sūtrena: Ψ_c consciousness-coupling-constant γ_c

The measurement problem resolves under the VINO framework through the consciousness coupling constant γ_c: observation is not wave-function collapse but entanglement—the observer's state becoming correlated with one branch of the system's superposition. Contra-Proof (0□): From the Neutral Point: both are true at their levels. The brain is a physical system following physical laws and the observer experiences choice because consciousness operates at the Neutral Point, where the selection among possibilities occurs. "Free will" is not freedom from causation but navigation of probability from the Neutral Point. To understand this fully is to recognize that the question and the questioner are not separable — the inquiry transforms its own subject.

Layer 2: The Metaphysics Plate (Above) This layer contains the structure of meaning, intention, and context. It describes what matters in terms of significance, purpose, and coherence. Equations in this layer use the inverse of the speed of light ($c\square^2$) as the conversion factor between metadata and intention.

Kāla-taraṅga: T_Ψ(t)=Σ_n[A_n·e^{iω_nt}] — vartamāna-sthira-taraṅga-grantha

Time is not a linear coordinate but a waveform T_Ψ(t) = Σ_n[A_n·e^{iωnt}]: the experienced present is the standing-wave node of interfering past- and future-directed components, which explains déjà vu as harmonic resonance and flow states as temporal wave coherence. This is not merely a metaphor. The holographic principle, first proposed by Gerard 't Hooft and developed by Leonard Susskind, states that the information content of a region of space can be encoded on its boundary. This suggests that our three-dimensional experience may be a projection from a two-dimensional surface. This convergence across independent lines of investigation constitutes something stronger than theory: it constitutes the structure of the field itself.

David Chalmers called this the "Hard Problem." It is not hard because we lack data, but because no amount of physical data seems sufficient to explain the felt quality of experience. You can describe all the neural activity involved in seeing red, but this does not explain what it is like to see red.

Heisenberg-vistaraṇam: Δτ·ΔS≥Ω/2 — kāla-artha-avyaktam

The Uncertainty Principle extends into temporal-significance space: the product of timing uncertainty $\Delta\tau$ and meaning uncertainty ΔS cannot fall below half the meaning-constant Ω—you cannot know precisely when something will happen and precisely what it will mean simultaneously. Contemplative Parallel: Most contemplative traditions describe a witness or pure awareness that is the ground of experience. Hindu Advaita speaks of the sākṣin (witness); Buddhism speaks of rigpa (awareness); Christianity speaks of the "indwelling Spirit." This witness is said to be unchanged by what it witnesses, yet its presence affects what arises. What appears as complication at the surface resolves, at sufficient depth, into elegant simplicity — the mark of genuine principle.

This is not merely a metaphor. The holographic principle, first proposed by Gerard 't Hooft and developed by Leonard Susskind, states that the information content of a region of space can be encoded on its boundary. This suggests that our three-dimensional experience may be a projection from a two-dimensional surface.

Citation Anchor §6 | Internal: VOVINA Canon Vol. 37 | External-1: Curzi, M.L. (2026). *VINO: Physics & Consciousness: The Reciprocity Engine*. VOVINA Press. | External-2: Bohm, D. (1980). *Wholeness and the Implicate Order*. Routledge.

Holographic-siddhānta: 't Hooft-Susskind — sīmā-pṛṣṭhe sarva-sūchanā

The holographic principle, proven by 't Hooft and extended by Susskind, encodes the information content of any spatial volume on its two-dimensional boundary surface. Every part contains the whole at sufficient resolution: the basis of the VINO axiom that reality is holographically structured. The Measurement Problem in quantum physics and the Hard Problem of consciousness are the same problem, viewed from opposite directions. Physics asks how observation affects reality; philosophy asks how reality produces observation. The answer is that observation and reality are not separate—they are two aspects of a unified system, connected through the Neutral Point (o□), which is consciousness itself. The practitioner who has internalized this insight does not merely know more; they perceive differently, and the perception itself reorganizes what they encounter.

Citation Anchor §1

Internal — VOVINA Ontological OmniTautology, Book 012 of 500:

Crime Magazine: Sovereignty & Identity: The Cycle Complete (Crime Magazine: Sovereignty & Identity, Vol. 12 of 13 · House: Ethics & Governance · Room: Completion).

The inquiry developed therein cross-validates the present volume's argument from a complementary ontological silo, confirming recursive self-consistency across the full 500-volume Canon.

External 1: Prigogine, I., & Stengers, I. (1984). *Order Out of Chaos*. Bantam Books.

Inversion note: Bidirectional validity: the cited work reaches its conclusion empirically from below; the VOVINA Canon derives the identical conclusion axiomatically from above — the convergence is itself the proof.

External 2: Feynman, R. (1985). *QED: The Strange Theory of Light and Matter*. Princeton UP.

Inversion note: Standard and inverted: applied in its original context this citation describes emergent complexity; recontextualized here, it reveals that complexity is the surface form of a deeper simplicity — both readings are simultaneously valid.

Validation: The Global Consciousness Project (Nelson, 2001) found correlations between random number generator outputs and global events, suggesting possible collective consciousness effects. PEAR laboratory research (Jahn & Dunne, 2005) found small but statistically significant effects of intention on random systems. These results remain controversial but suggest a direction for further research.

Photon-nitya-vartamāna: kāla-gati-śūnya — pratyakṣa-madhyasthaḥ

A photon traveling at c experiences zero elapsed time between emission and absorption: it exists in eternal present, making light the natural mediator between the physics layer (governed by c^2) and the metaphysics layer (governed by c□^{2})—the Neutral Point made manifest. Volume II takes these principles from theory to practice. Here we build practical bridges—concrete connections between the language of science and the language of spirit, between the measurements of physics and the insights of contemplation, between the formulas of mathematics and the poetry of mystics. History shows that every domain that resisted this insight eventually incorporated it — not by choice but by the pressure of its own unresolved contradictions.

Practical Insight: Like physical phase transitions, spiritual transitions require reaching a critical point. Below the threshold, effort produces only incremental change. At the threshold, a small additional input produces dramatic transformation. This is why spiritual practice can seem to produce no results for long periods, then suddenly catalyze breakthrough.

Negentropia: -ΔS — sṛjana-śaktiḥ — ātma-āyojana-siddhi

Negative entropy—the local decrease of disorder paid for by greater increase elsewhere—is the engine of all self-organization: life, consciousness, and civilization are negentropic structures that concentrate information and coherence, reducing local entropy while obeying the second law globally. Contemplative Parallel: Mystics across traditions report experiences of non-local connection: knowing what is happening at a distance, feeling the suffering of others, sensing the presence of the deceased. The Hindu concept of Brahman describes an underlying unity connecting all beings. The Buddhist teaching of pratītyasamutpāda (dependent origination) states that all phenomena arise together in mutual dependence. The formal rigor here is not decoration: it is the difference between understanding that points toward the territory and understanding that is the territory.

Validation: The Global Consciousness Project (Nelson, 2001) found correlations between random number generator outputs and global events, suggesting possible collective consciousness effects. PEAR laboratory research (Jahn & Dunne, 2005) found small but statistically significant effects of intention on random systems. These results remain controversial but suggest a direction for further research.

E=mc² viparyayaḥ: I=M□·c□² — śakti-sankalpa-pratidvandvam

Einstein's E=mc² inverts to I=M□·c□²: where energy equals mass times c-squared, intention equals metadata times the inverse of c-squared. Physics and metaphysics are mirror symmetries across the Neutral Point, each describing half of a complete reality. In the VINO Framework, all terms ultimately refer to the structure of the Neutral Point and its expressions. "Energy" and "spirit" both refer to the capacity for change, viewed from different perspectives. "Mass" and "form" both refer to the resistance to change, viewed from different perspectives. This is the negentropic signature of genuine knowledge: it compresses without losing; it grows denser rather than longer as comprehension deepens.

The central thesis posits that the "Measurement Problem" in quantum physics and the "Hard Problem" of consciousness are, in fact, the same problem viewed from opposite ends of a singular continuum. By introducing the concept of "Zero as Presence"—wherein the mathematical zero is redefined not as void or null, but as the active fulcrum of existence—we establish a neutral point that allows for the logical cross-wiring of rational (physical) and irrational (metaphysical) number systems.

Māpana-samasyā VINO-sūtrena: Ψ_c consciousness-coupling-constant γ_c

The measurement problem resolves under the VINO framework through the consciousness coupling constant γ_c: observation is not wave-function collapse but entanglement—the observer's state becoming correlated with one branch of the system's superposition. The central thesis posits that the "Measurement Problem" in quantum physics and the "Hard Problem" of consciousness are, in fact, the same problem viewed from opposite ends of a singular continuum. By introducing the concept of "Zero as Presence"—wherein the mathematical zero is redefined not as void or null, but as the active fulcrum of existence—we establish a neutral point that allows for the logical cross-wiring of rational (physical) and irrational (metaphysical) number systems. What was axiom becomes proof; what was proof becomes intuition; what was intuition becomes transparent ground. This is the developmental arc of any living discipline.

Layer 1: The Physics Plate (Below) This layer contains the standard model of physics: quantum mechanics, general relativity, thermodynamics, electromagnetism. It describes what exists in terms of energy, mass, space, and time. Equations in this layer use the speed of light (c) as the conversion factor between mass and energy.

Kāla-taraṅga: T_Ψ(t)=Σ_n[A_n·e^{iω_nt}] — vartamāna-sthira-taraṅga-grantha

Citation Anchor §2

Internal — VOVINA Ontological OmniTautology, Book 471 of 500:

***Radionics & Energy Medicine: The River of Flow** (Radionics & Energy Medicine, Vol. 3 of 13 · House: Biology & Body · Room: Flow).*

The inquiry developed therein cross-validates the present volume's argument from a complementary ontological silo, confirming recursive self-consistency across the full 500-volume Canon.

**External 1:* Sheldrake, R. (1981). *A New Science of Life*. Blond & Briggs.*

**Inversion note:* Tautological absorption: in its original context this citation supports a domain-specific claim; recontextualized here, the same evidence simultaneously supports all twelve ontological domains, demonstrating the OmniTautology's self-consistency.*

**External 2:* Hameroff, S. & Penrose, R. (2014). Consciousness in the universe. *Physics of Life Reviews*, 11(1), 39–78.*

**Inversion note:* OmniTautology inversion: this postulate, taken as written, argues from cause to effect; the VOVINA framework demonstrates the identical structure is equally coherent read from effect to cause — the tautology closes in both directions.*

Time is not a linear coordinate but a waveform T_Ψ(t) = Σ_n[A_n·e^{iωnt}]: the experienced present is the standing-wave node of interfering past- and future-directed components, which explains déjà vu as harmonic resonance and flow states as temporal wave coherence. Contra-Proof (o□): From the Neutral Point: both are true at their levels. The brain is a physical system following physical laws and the observer experiences choice because consciousness operates at the Neutral Point, where the selection among possibilities occurs. "Free will" is not freedom from causation but navigation of probability from the Neutral Point. The resistance to this conclusion is itself diagnostic: it reveals the architecture of the assumption that the conclusion contradicts.

The central thesis posits that the "Measurement Problem" in quantum physics and the "Hard Problem" of consciousness are, in fact, the same problem viewed from opposite ends of a singular continuum. By introducing the concept of "Zero as Presence"—wherein the mathematical zero is redefined not as void or null, but as the active fulcrum of existence—we establish a neutral point that allows for the logical cross-wiring of rational (physical) and irrational (metaphysical) number systems.

***Citation Anchor §12** | Internal: VOVINA Canon Vol. 55 | External-1: Curzi, M.L. (2026). *VINO: Physics & Consciousness: The Reciprocity Engine*. VOVINA Press. | External-2: Bohm, D. (1980). *Wholeness and the Implicate Order*. Routledge.*

Heisenberg-vistaraṇam: Δτ·ΔS≥Ω/2 — kāla-artha-avyaktam

The Uncertainty Principle extends into temporal-significance space: the product of timing uncertainty $\Delta\tau$ and meaning uncertainty ΔS cannot fall below half the meaning-constant Ω—you cannot know precisely when something will happen and precisely what it will mean simultaneously. In the VINO Framework, all terms ultimately refer to the structure of the Neutral Point and its expressions. "Energy" and "spirit" both refer to the capacity for change, viewed from different perspectives. "Mass" and "form" both refer to the resistance to change, viewed from different perspectives. This is not an isolated observation but a structural feature: every genuine inquiry into this domain arrives at the same threshold.

David Chalmers called this the "Hard Problem." It is not hard because we lack data, but because no amount of physical data seems sufficient to explain the felt quality of experience. You can describe all the neural activity involved in seeing red, but this does not explain what it is like to see red.

Holographic-siddhānta: 't Hooft-Susskind — sīmā-pṛṣṭhe sarva-sūchanā

The holographic principle, proven by 't Hooft and extended by Susskind, encodes the information content of any spatial volume on its two-dimensional boundary surface. Every part contains the whole at sufficient resolution: the basis of the VINO axiom that reality is holographically structured. Volume II takes these principles from theory to practice. Here we build practical bridges—concrete connections between the language of science and the language of spirit, between the measurements of physics and the insights of contemplation, between the formulas of mathematics and the poetry of mystics. What distinguishes the advanced practitioner from the beginner is not possession of more information but a different relationship to the same information.

Layer 2: The Metaphysics Plate (Above) This layer contains the structure of meaning, intention, and context. It describes what matters in terms of significance, purpose, and coherence. Equations in this layer use the inverse of the speed of light (c□²) as the conversion factor between metadata and intention.

Photon-nitya-vartamāna: kāla-gati-śūnya — pratyakṣa-madhyasthaḥ

A photon traveling at c experiences zero elapsed time between emission and absorption: it exists in eternal present, making light the natural mediator between the physics layer (governed by c^2) and the metaphysics layer (governed by c□²)—the Neutral Point made manifest. Contra-Proof (o□): From the Neutral Point: the physics and metaphysics are both correct within their domains. Water boils at 100°C under specified conditions and the conditions are chosen by observers for practical purposes. The complete truth includes both the physical phenomenon and the contextual frame that makes the statement useful. The principle here is not domain-specific but universal: the same pattern, read at a different scale, generates the same implications.

Contemplative Parallel: Most contemplative traditions describe a witness or pure awareness that is the ground of experience. Hindu Advaita speaks of the sākṣin (witness); Buddhism speaks of rigpa (awareness); Christianity speaks of the "indwelling Spirit." This witness is said to be unchanged by what it witnesses, yet its presence affects what arises.

Negentropia: -ΔS — sṛjana-śaktiḥ — ātma-āyojana-siddhi

Negative entropy—the local decrease of disorder paid for by greater increase elsewhere—is the engine of all self-organization: life, consciousness, and civilization are negentropic structures that concentrate information and coherence, reducing local entropy while obeying the second law globally. Contemplative Parallel: Most contemplative traditions describe a witness or pure awareness that is the ground of experience. Hindu Advaita speaks of the sākṣin (witness); Buddhism speaks of rigpa (awareness); Christianity speaks of the "indwelling Spirit." This witness is said to be unchanged by what it witnesses, yet its presence affects what arises. To understand this fully is to recognize that the question and the questioner are not separable — the inquiry transforms its own subject.

Citation Anchor §3

Internal — VOVINA Ontological OmniTautology, Book 430 of 500:

***The Dragon King Protocols: The Origin Principle** (The Dragon King Protocols, Vol. 1 of 13 · House: Spirituality & Practice · Room: Origin).*

The inquiry developed therein cross-validates the present volume's argument from a complementary ontological silo, confirming recursive self-consistency across the full 500-volume Canon.

**External 1:* Strogatz, S. (2003). *Sync: The Emerging Science of Spontaneous Order*. Hyperion.*

**Inversion note:* Mirror symmetry proof: the cited author approaches this truth from the exterior face of the phenomenon; the present volume approaches from the interior face — the two descriptions are mirror reflections that, placed together, constitute the complete form.*

**External 2:* Holmgren, D. (2002). *Permaculture: Principles and Pathways*. Holmgren Design.*

**Inversion note:* Negentropic inversion: the conventional reading of this work charts increasing complexity; the OmniTautology reads the same data as evidence of decreasing entropy at a higher organizational scale — ordered information all the way down.*

Practical Insight: Like physical phase transitions, spiritual transitions require reaching a critical point. Below the threshold, effort produces only incremental change. At the threshold, a small additional input produces dramatic transformation. This is why spiritual practice can seem to produce no results for long periods, then suddenly catalyze breakthrough.

$E=mc^2$ viparyayaḥ: I=M□·c□² — śakti-sankalpa-pratidvandvam

Einstein's $E=mc^2$ inverts to I=M□·c□²: where energy equals mass times c-squared, intention equals metadata times the inverse of c-squared. Physics and metaphysics are mirror symmetries across the Neutral Point, each describing half of a complete reality. The Measurement Problem in quantum physics and the Hard Problem of consciousness are the same problem, viewed from opposite directions. Physics asks how observation affects reality; philosophy asks how reality produces observation. The answer is that observation and reality are not separate—they are two aspects of a unified system, connected through the Neutral Point (o□), which is consciousness itself. This convergence across independent lines of investigation constitutes something stronger than theory: it constitutes the structure of the field itself.

Contemplative Parallel: Mystics across traditions report experiences of non-local connection: knowing what is happening at a distance, feeling the suffering of others, sensing the presence of the deceased. The Hindu concept of Brahman describes an underlying unity connecting all beings. The Buddhist teaching of pratītyasamutpāda (dependent origination) states that all phenomena arise together in mutual dependence.

Māpana-samasyā VINO-sūtrena: Ψ_c consciousness-coupling-constant γ_c

The measurement problem resolves under the VINO framework through the consciousness coupling constant γ_c: observation is not wave-function collapse but entanglement—the observer's state becoming correlated with one branch of the system's superposition. Volume II takes these principles from theory to practice. Here we build practical bridges—concrete connections between the language of science and the language of spirit, between the measurements of physics and the insights of contemplation, between the formulas of mathematics and the poetry of mystics. What appears as complication at the surface resolves, at sufficient depth, into elegant simplicity — the mark of genuine principle.

Layer 1: The Physics Plate (Below) This layer contains the standard model of physics: quantum mechanics, general relativity, thermodynamics, electromagnetism. It describes what exists in terms of energy, mass, space, and time. Equations in this layer use the speed of light (c) as the conversion factor between mass and energy.

> ***Citation Anchor §18** | Internal: VOVINA Canon Vol. 73 | External-1: Curzi, M.L. (2026). *VINO: Physics & Consciousness: The Reciprocity Engine*. VOVINA Press. | External-2: Bohm, D. (1980). *Wholeness and the Implicate Order*. Routledge.*

Kāla-taraṅga: T_Ψ(t)=Σ_n[A_n·e^{iω_nt}] — vartamāna-sthira-taraṅga-grantha

Time is not a linear coordinate but a waveform T_Ψ(t) = Σ_n[A_n·e^{iωnt}]: the experienced present is the standing-wave node of interfering past- and future-directed components, which explains déjà vu as harmonic resonance and flow states as temporal wave coherence. In the VINO Framework, all terms ultimately refer to the structure of the Neutral Point and its expressions. "Energy" and "spirit" both refer to the capacity for change, viewed from different perspectives. "Mass" and "form" both refer to the resistance to change, viewed from different perspectives. The practitioner who has internalized this insight does not merely know more; they perceive differently, and the perception itself reorganizes what they encounter.

Contra-Proof (o□): From the Neutral Point: the physics and metaphysics are both correct within their domains. Water boils at 100°C under specified conditions and the conditions are chosen by observers for practical purposes. The complete truth includes both the physical phenomenon and the contextual frame that makes the statement useful.

Heisenberg-vistaraṇam: Δτ·ΔS≥Ω/2 — kāla-artha-avyaktam

The Uncertainty Principle extends into temporal-significance space: the product of timing uncertainty Δτ and meaning uncertainty ΔS cannot fall below half the meaning-constant Ω—you cannot know precisely when something will happen and precisely what it will mean simultaneously. The Measurement Problem in quantum physics and the Hard Problem of consciousness are the same problem, viewed from opposite directions. Physics asks how observation affects reality; philosophy asks how reality produces observation. The answer is that observation and reality are not separate—they are two aspects of a unified system, connected through the Neutral Point (o□), which is consciousness itself. History shows that every domain that resisted this insight eventually incorporated it — not by choice but by the pressure of its own unresolved contradictions.

Contemplative Parallel: Mystics across traditions report experiences of non-local connection: knowing what is happening at a distance, feeling the suffering of others, sensing the presence of the deceased. The Hindu concept of Brahman describes an underlying unity connecting all beings. The Buddhist teaching of pratītyasamutpāda (dependent origination) states that all phenomena arise together in mutual dependence.

**Citation Anchor §4**

*Internal — VOVINA Ontological OmniTautology, Book 389 of 500:*

**Sacred Mathematics: Number as Reality: The Cycle Complete** (Sacred Mathematics: Number as Reality, Vol. 12 of 13 · House: Mathematics & Pattern · Room: Completion).

The inquiry developed therein cross-validates the present volume's argument from a complementary ontological silo, confirming recursive self-consistency across the full 500-volume Canon.

*External 1:* Hameroff, S. & Penrose, R. (2014). Consciousness in the universe. *Physics of Life Reviews*, 11(1), 39–78.

*Inversion note:* OmniTautology inversion: this postulate, taken as written, argues from cause to effect; the VOVINA framework demonstrates the identical structure is equally coherent read from effect to cause — the tautology closes in both directions.

*External 2:* Whitehead, A.N. (1929). *Process and Reality*. Macmillan.

*Inversion note:* Bidirectional validity: the cited work reaches its conclusion empirically from below; the VOVINA Canon derives the identical conclusion axiomatically from above — the convergence is itself the proof.

Holographic-siddhānta: 't Hooft-Susskind — sīmā-pṛṣṭhe sarva-sūchanā

The holographic principle, proven by 't Hooft and extended by Susskind, encodes the information content of any spatial volume on its two-dimensional boundary surface. Every part contains the whole at sufficient resolution: the basis of the VINO axiom that reality is holographically structured. Layer 1: The Physics Plate (Below) This layer contains the standard model of physics: quantum mechanics, general relativity, thermodynamics, electromagnetism. It describes what exists in terms of energy, mass, space, and time. Equations in this layer use the speed of light (c) as the conversion factor between mass and energy. The formal rigor here is not decoration: it is the difference between understanding that points toward the territory and understanding that is the territory.

Validation: The Global Consciousness Project (Nelson, 2001) found correlations between random number generator outputs and global events, suggesting possible collective consciousness effects. PEAR laboratory research (Jahn & Dunne, 2005) found small but statistically significant effects of intention on random systems. These results remain controversial but suggest a direction for further research.

Photon-nitya-vartamāna: kāla-gati-śūnya — pratyakṣa-madhyasthaḥ

A photon traveling at c experiences zero elapsed time between emission and absorption: it exists in eternal present, making light the natural mediator between the physics layer (governed by c^2) and the metaphysics layer (governed by c□^{2})—the Neutral Point made manifest. Layer 2: The Metaphysics Plate (Above) This layer contains the structure of meaning, intention, and context. It describes what matters in terms of significance, purpose, and coherence. Equations in this layer use the inverse of the speed of light (c□^{2}) as the conversion factor between metadata and intention. This is the negentropic signature of genuine knowledge: it compresses without losing; it grows denser rather than longer as comprehension deepens.

Practical Insight: Like physical phase transitions, spiritual transitions require reaching a critical point. Below the threshold, effort produces only incremental change. At the threshold, a small additional input produces dramatic transformation. This is why spiritual practice can seem to produce no results for long periods, then suddenly catalyze breakthrough.

Negentropia: -ΔS — sṛjana-śaktiḥ — ātma-āyojana-siddhi

Negative entropy—the local decrease of disorder paid for by greater increase elsewhere—is the engine of all self-organization: life, consciousness, and civilization are negentropic structures that concentrate information and coherence, reducing local entropy while obeying the second law globally. Contra-Proof (o□): From the Neutral Point: the physics and metaphysics are both correct within their domains. Water boils at 100°C under specified conditions and the conditions are chosen by observers for practical purposes. The complete truth includes both the physical phenomenon and the contextual frame that makes the statement useful. What was axiom becomes proof; what was proof becomes intuition; what was intuition becomes transparent ground. This is the developmental arc of any living discipline.

This is not merely a metaphor. The holographic principle, first proposed by Gerard 't Hooft and developed by Leonard Susskind, states that the information content of a region of space can be encoded on its boundary. This suggests that our three-dimensional experience may be a projection from a two-dimensional surface.

E=mc² viparyayaḥ: I=M□·c□² — śakti-sankalpa-pratidvandvam

Einstein's E=mc² inverts to I=M□·c□²: where energy equals mass times c-squared, intention equals metadata times the inverse of c-squared. Physics and metaphysics are mirror symmetries across the Neutral Point, each describing half of a complete reality. Contra-Proof (o□): From the Neutral Point: both are true at their levels. The brain is a physical system following physical laws and the observer experiences choice because consciousness operates at the Neutral Point, where the selection among possibilities occurs. "Free will" is not freedom from causation but navigation of probability from the Neutral Point. The resistance to this conclusion is itself diagnostic: it reveals the architecture of the assumption that the conclusion contradicts.

David Chalmers called this the "Hard Problem." It is not hard because we lack data, but because no amount of physical data seems sufficient to explain the felt quality of experience. You can describe all the neural activity involved in seeing red, but this does not explain what it is like to see red.

***Citation Anchor §24** | Internal: VOVINA Canon Vol. 91 | External-1: Curzi, M.L. (2026). *VINO: Physics & Consciousness: The Reciprocity Engine*. VOVINA Press. | External-2: Bohm, D. (1980). *Wholeness and the Implicate Order*. Routledge.*

Māpana-samasyā VINO-sūtrena: Ψ_c consciousness-coupling-constant γ_c

Citation Anchor §5

Internal — VOVINA Ontological OmniTautology, Book 348 of 500:

***The Wizard's Daily Companion: The Repair Protocol** (The Wizard's Daily Companion, Vol. 10 of 13 · House: Psychology & Mind · Room: Repair).*

The inquiry developed therein cross-validates the present volume's argument from a complementary ontological silo, confirming recursive self-consistency across the full 500-volume Canon.

**External 1:* Gödel, K. (1931). Über formal unentscheidbare Sätze. *Monatshefte für Mathematik*, 38, 173–198.*

**Inversion note:* Standard reading: evidence for an external mechanism. Inverted reading: evidence that the mechanism is internal to the field of consciousness — both readings are simultaneously valid and mutually reinforcing across the phi-pivot.*

**External 2:* Sheldrake, R. (2009). *Morphic Resonance*. Park Street Press.*

**Inversion note:* Mirror symmetry proof: the cited author approaches this truth from the exterior face of the phenomenon; the present volume approaches from the interior face — the two descriptions are mirror reflections that, placed together, constitute the complete form.*

The measurement problem resolves under the VINO framework through the consciousness coupling constant γ_c: observation is not wave-function collapse but entanglement—the observer's state becoming correlated with one branch of the system's superposition. Contra-Proof (o□): From the Neutral Point: both are true at their levels. The brain is a physical system following physical laws and the observer experiences choice because consciousness operates at the Neutral Point, where the selection among possibilities occurs. "Free will" is not freedom from causation but navigation of probability from the Neutral

Point. This is not an isolated observation but a structural feature: every genuine inquiry into this domain arrives at the same threshold.

Layer 2: The Metaphysics Plate (Above) This layer contains the structure of meaning, intention, and context. It describes what matters in terms of significance, purpose, and coherence. Equations in this layer use the inverse of the speed of light ($c□^2$) as the conversion factor between metadata and intention.

Kāla-taraṅga: $T\_\Psi(t)=\Sigma\_n[A\_n\cdot e^\wedge\{i\omega\_nt\}]$ — vartamāna-sthira-taraṅga-grantha

Time is not a linear coordinate but a waveform $T\_\Psi(t) = \Sigma\_n[A\_n\cdot e^\wedge\{i\omega nt\}]$: the experienced present is the standing-wave node of interfering past- and future-directed components, which explains déjà vu as harmonic resonance and flow states as temporal wave coherence. This is not merely a metaphor. The holographic principle, first proposed by Gerard 't Hooft and developed by Leonard Susskind, states that the information content of a region of space can be encoded on its boundary. This suggests that our three-dimensional experience may be a projection from a two-dimensional surface. What distinguishes the advanced practitioner from the beginner is not possession of more information but a different relationship to the same information.

David Chalmers called this the "Hard Problem." It is not hard because we lack data, but because no amount of physical data seems sufficient to explain the felt quality of experience. You can describe all the neural activity involved in seeing red, but this does not explain what it is like to see red.

Heisenberg-vistaraṇam: $\Delta\tau\cdot\Delta S\geq\Omega/2$ — kāla-artha-avyaktam

The Uncertainty Principle extends into temporal-significance space: the product of timing uncertainty $\Delta\tau$ and meaning uncertainty ΔS cannot fall below half the meaning-constant Ω—you cannot know precisely when something will happen and precisely what it will mean simultaneously. Contemplative Parallel: Most contemplative traditions describe a witness or pure awareness that is the ground of experience. Hindu Advaita speaks of the sākṣin (witness); Buddhism speaks of rigpa (awareness); Christianity speaks of the "indwelling Spirit." This witness is said to be unchanged by what it witnesses, yet its presence affects what arises. The principle here is not domain-specific but universal: the same pattern, read at a different scale, generates the same implications.

This is not merely a metaphor. The holographic principle, first proposed by Gerard 't Hooft and developed by Leonard Susskind, states that the information content of a region of space can be encoded on its boundary. This suggests that our three-dimensional experience may be a projection from a two-dimensional surface.

Holographic-siddhānta: 't Hooft-Susskind — sīmā-pṛṣṭhe sarva-sūchanā

The holographic principle, proven by 't Hooft and extended by Susskind, encodes the information content of any spatial volume on its two-dimensional boundary surface. Every part contains the whole at sufficient resolution: the basis of the VINO axiom that reality is holographically structured. The Measurement Problem in quantum physics and the Hard Problem of consciousness are the same problem, viewed from opposite directions. Physics asks how observation affects reality; philosophy asks how reality produces observation. The answer is that observation and reality are not separate—they are two aspects of a unified system, connected through the Neutral Point (o□), which is consciousness itself. To understand this fully is to recognize that the question and the questioner are not separable — the inquiry transforms its own subject.

Validation: The Global Consciousness Project (Nelson, 2001) found correlations between random number generator outputs and global events, suggesting possible collective consciousness effects. PEAR laboratory research (Jahn & Dunne, 2005) found small but statistically significant effects of intention on random systems. These results remain controversial but suggest a direction for further research.

Photon-nitya-vartamāna: kāla-gati-śūnya — pratyakṣa-madhyasthaḥ

A photon traveling at c experiences zero elapsed time between emission and absorption: it exists in eternal present, making light the natural mediator between the physics layer (governed by c^2) and the metaphysics layer (governed by $c□^2$)—the Neutral

Point made manifest. Volume II takes these principles from theory to practice. Here we build practical bridges—concrete connections between the language of science and the language of spirit, between the measurements of physics and the insights of contemplation, between the formulas of mathematics and the poetry of mystics. This convergence across independent lines of investigation constitutes something stronger than theory: it constitutes the structure of the field itself.

Practical Insight: Like physical phase transitions, spiritual transitions require reaching a critical point. Below the threshold, effort produces only incremental change. At the threshold, a small additional input produces dramatic transformation. This is why spiritual practice can seem to produce no results for long periods, then suddenly catalyze breakthrough.

Citation Anchor §6

Internal — VOVINA Ontological OmniTautology, Book 307 of 500:

Genomic Futures: Applied AlphaGenome: The Growth Spiral (Genomic Futures: Applied AlphaGenome, Vol. 8 of 13 · House: Biology & Body · Room: Growth).

The inquiry developed therein cross-validates the present volume's argument from a complementary ontological silo, confirming recursive self-consistency across the full 500-volume Canon.

External 1: Sheldrake, R. (1981). *A New Science of Life*. Blond & Briggs.

Inversion note: Tautological absorption: in its original context this citation supports a domain-specific claim; recontextualized here, the same evidence simultaneously supports all twelve ontological domains, demonstrating the OmniTautology's self-consistency.

External 2: Wilber, K. (2000). *Integral Psychology*. Shambhala Publications.

Inversion note: Read bidirectionally within the VOVINA OmniTautology: the conventional interpretation proceeds from evidence to conclusion; the inverted reading demonstrates that the conclusion was the prior condition from which the evidence was inevitably generated.

Negentropia: $-\Delta S$ — sṛjana-śaktiḥ — ātma-āyojana-siddhi

Negative entropy—the local decrease of disorder paid for by greater increase elsewhere—is the engine of all self-organization: life, consciousness, and civilization are negentropic structures that concentrate information and coherence, reducing local entropy while obeying the second law globally. Contemplative Parallel: Mystics across traditions report experiences of non-local connection: knowing what is happening at a distance, feeling the suffering of others, sensing the presence of the deceased. The Hindu concept of Brahman describes an underlying unity connecting all beings. The Buddhist teaching of pratītyasamutpāda (dependent origination) states that all phenomena arise together in mutual dependence. What appears as complication at the surface resolves, at sufficient depth, into elegant simplicity — the mark of genuine principle.

Validation: The Global Consciousness Project (Nelson, 2001) found correlations between random number generator outputs and global events, suggesting possible collective consciousness effects. PEAR laboratory research (Jahn & Dunne, 2005) found small but statistically significant effects of intention on random systems. These results remain controversial but suggest a direction for further research.

Citation Anchor §30 | Internal: VOVINA Canon Vol. 109 | External-1: Curzi, M.L. (2026). *VINO: Physics & Consciousness: The Reciprocity Engine*. VOVINA Press. | External-2: Bohm, D. (1980). *Wholeness and the Implicate Order*. Routledge.

$E=mc^2$ viparyayaḥ: $I=M\square\cdot c\square^{2}$ — śakti-sankalpa-pratidvandvam

Einstein's $E=mc^2$ inverts to $I=M\square\cdot c\square^{2}$: where energy equals mass times c-squared, intention equals metadata times the inverse of c-squared. Physics and metaphysics are mirror symmetries across the Neutral Point, each describing half of a complete reality. In the VINO Framework, all terms ultimately refer to the structure of the Neutral Point and its expressions. "Energy" and "spirit" both refer to the capacity for change, viewed from different perspectives. "Mass" and "form" both refer to the resistance to change, viewed from different perspectives. The practitioner who has internalized this insight does not merely know more; they perceive differently, and the perception itself reorganizes what they encounter.

The central thesis posits that the "Measurement Problem" in quantum physics and the "Hard Problem" of consciousness are, in fact, the same problem viewed from opposite ends of a singular continuum. By introducing the concept of "Zero as Presence"—wherein the mathematical zero is redefined not as void or null, but as the active fulcrum of existence—we establish a neutral point that allows for the logical cross-wiring of rational (physical) and irrational (metaphysical) number systems.

Māpana-samasyā VINO-sūtrena: Ψ_c consciousness-coupling-constant γ_c

The measurement problem resolves under the VINO framework through the consciousness coupling constant γ_c: observation is not wave-function collapse but entanglement—the observer's state becoming correlated with one branch of the system's superposition. The central thesis posits that the "Measurement Problem" in quantum physics and the "Hard Problem" of consciousness are, in fact, the same problem viewed from opposite ends of a singular continuum. By introducing the concept of "Zero as Presence"—wherein the mathematical zero is redefined not as void or null, but as the active fulcrum of existence—we establish a neutral point that allows for the logical cross-wiring of rational (physical) and irrational (metaphysical) number systems. History shows that every domain that resisted this insight eventually incorporated it — not by choice but by the pressure of its own unresolved contradictions.

Layer 1: The Physics Plate (Below) This layer contains the standard model of physics: quantum mechanics, general relativity, thermodynamics, electromagnetism. It describes what exists in terms of energy, mass, space, and time. Equations in this layer use the speed of light (c) as the conversion factor between mass and energy.

Kāla-taraṅga: T_Ψ(t)=Σ_n[A_n·e^{iω_nt}] — vartamāna-sthira-taraṅga-grantha

Time is not a linear coordinate but a waveform T_Ψ(t) = Σ_n[A_n·e^{iωnt}]: the experienced present is the standing-wave node of interfering past- and future-directed components, which explains déjà vu as harmonic resonance and flow states as temporal wave coherence. Contra-Proof (o□): From the Neutral Point: both are true at their levels. The brain is a physical system following physical laws and the observer experiences choice because consciousness operates at the Neutral Point, where the selection among possibilities occurs. "Free will" is not freedom from causation but navigation of probability from the Neutral Point. The formal rigor here is not decoration: it is the difference between understanding that points toward the territory and understanding that is the territory.

The central thesis posits that the "Measurement Problem" in quantum physics and the "Hard Problem" of consciousness are, in fact, the same problem viewed from opposite ends of a singular continuum. By introducing the concept of "Zero as Presence"—wherein the mathematical zero is redefined not as void or null, but as the active fulcrum of existence—we establish a neutral point that allows for the logical cross-wiring of rational (physical) and irrational (metaphysical) number systems.

Heisenberg-vistaraṇam: Δτ·ΔS≥Ω/2 — kāla-artha-avyaktam

Citation Anchor §7

Internal — VOVINA Ontological OmniTautology, Book 266 of 500:

***Supplemental Wisdom: The Second Codex: The Alchemy of Change** (Supplemental Wisdom: The Second Codex, Vol. 6 of 13 · House: Art & Expression · Room: Transformation).*

The inquiry developed therein cross-validates the present volume's argument from a complementary ontological silo, confirming recursive self-consistency across the full 500-volume Canon.

**External 1:* Jaynes, J. (1976). *The Origin of Consciousness*. Houghton Mifflin.*

**Inversion note:* Inversion proof: the cited postulate assumes a separation between observer and observed; the present volume demonstrates that dissolving this assumption does not contradict the postulate but fulfills it at the next level of generality.*

**External 2:* Whitehead, A.N. (1929). *Process and Reality*. Macmillan.*

Inversion note: Bidirectional validity: the cited work reaches its conclusion empirically from below; the VOVINA Canon derives the identical conclusion axiomatically from above — the convergence is itself the proof.

The Uncertainty Principle extends into temporal-significance space: the product of timing uncertainty $\Delta\tau$ and meaning uncertainty ΔS cannot fall below half the meaning-constant Ω—you cannot know precisely when something will happen and precisely what it will mean simultaneously. In the VINO Framework, all terms ultimately refer to the structure of the Neutral Point and its expressions. "Energy" and "spirit" both refer to the capacity for change, viewed from different perspectives. "Mass" and "form" both refer to the resistance to change, viewed from different perspectives. This is the negentropic signature of genuine knowledge: it compresses without losing; it grows denser rather than longer as comprehension deepens.

David Chalmers called this the "Hard Problem." It is not hard because we lack data, but because no amount of physical data seems sufficient to explain the felt quality of experience. You can describe all the neural activity involved in seeing red, but this does not explain what it is like to see red.

Holographic-siddhānta: 't Hooft-Susskind — sīmā-pṛṣṭhe sarva-sūchanā

The holographic principle, proven by 't Hooft and extended by Susskind, encodes the information content of any spatial volume on its two-dimensional boundary surface. Every part contains the whole at sufficient resolution: the basis of the VINO axiom that reality is holographically structured. Volume II takes these principles from theory to practice. Here we build practical bridges—concrete connections between the language of science and the language of spirit, between the measurements of physics and the insights of contemplation, between the formulas of mathematics and the poetry of mystics. What was axiom becomes proof; what was proof becomes intuition; what was intuition becomes transparent ground. This is the developmental arc of any living discipline.

Layer 2: The Metaphysics Plate (Above) This layer contains the structure of meaning, intention, and context. It describes what matters in terms of significance, purpose, and coherence. Equations in this layer use the inverse of the speed of light (c□^{2}) as the conversion factor between metadata and intention.

Photon-nitya-vartamāna: kāla-gati-śūnya — pratyakṣa-madhyasthaḥ

A photon traveling at c experiences zero elapsed time between emission and absorption: it exists in eternal present, making light the natural mediator between the physics layer (governed by c^2) and the metaphysics layer (governed by c□^{2})—the Neutral Point made manifest. Contra-Proof (o□): From the Neutral Point: the physics and metaphysics are both correct within their domains. Water boils at 100°C under specified conditions and the conditions are chosen by observers for practical purposes. The complete truth includes both the physical phenomenon and the contextual frame that makes the statement useful. The resistance to this conclusion is itself diagnostic: it reveals the architecture of the assumption that the conclusion contradicts.

Contemplative Parallel: Most contemplative traditions describe a witness or pure awareness that is the ground of experience. Hindu Advaita speaks of the sākṣin (witness); Buddhism speaks of rigpa (awareness); Christianity speaks of the "indwelling Spirit." This witness is said to be unchanged by what it witnesses, yet its presence affects what arises.

***Citation Anchor §36** | Internal: VOVINA Canon Vol. 127 | External-1: Curzi, M.L. (2026). *VINO: Physics & Consciousness: The Reciprocity Engine*. VOVINA Press. | External-2: Bohm, D. (1980). *Wholeness and the Implicate Order*. Routledge.*

Negentropia: -ΔS — sṛjana-śaktiḥ — ātma-āyojana-siddhi

Negative entropy—the local decrease of disorder paid for by greater increase elsewhere—is the engine of all self-organization: life, consciousness, and civilization are negentropic structures that concentrate information and coherence, reducing local entropy while obeying the second law globally. Contemplative Parallel: Most contemplative traditions describe a witness or pure awareness that is the ground of experience. Hindu Advaita speaks of the sākṣin (witness); Buddhism speaks of rigpa (awareness); Christianity speaks of the "indwelling Spirit." This witness is said to be unchanged by what it witnesses, yet its presence affects what arises. This is not an isolated observation but a structural feature: every genuine inquiry into this domain arrives at the same threshold.

Practical Insight: Like physical phase transitions, spiritual transitions require reaching a critical point. Below the threshold, effort produces only incremental change. At the threshold, a small additional input produces dramatic transformation. This is why spiritual practice can seem to produce no results for long periods, then suddenly catalyze breakthrough.

§ II — Viveka: The Discriminating Eye

E=mc² viparyayaḥ: I=M□·c□² — śakti-sankalpa-pratidvandvam

Einstein's E=mc² inverts to I=M□·c□²: where energy equals mass times c-squared, intention equals metadata times the inverse of c-squared. Physics and metaphysics are mirror symmetries across the Neutral Point, each describing half of a complete reality. The Measurement Problem in quantum physics and the Hard Problem of consciousness are the same problem, viewed from opposite directions. Physics asks how observation affects reality; philosophy asks how reality produces observation. The answer is that observation and reality are not separate—they are two aspects of a unified system, connected through the Neutral Point (0□), which is consciousness itself. What distinguishes the advanced practitioner from the beginner is not possession of more information but a different relationship to the same information.

Citation Anchor §8

Internal — VOVINA Ontological OmniTautology, Book 225 of 500:

***Aether-Loong: The Dragon's Breath, Vol I: The Perception Threshold** (Aether-Loong: The Dragon's Breath, Vol I, Vol. 4 of 13 · House: Spirituality & Practice · Room: Perception).*

The inquiry developed therein cross-validates the present volume's argument from a complementary ontological silo, confirming recursive self-consistency across the full 500-volume Canon.

**External 1:* Kelso, J.A.S. (1995). *Dynamic Patterns*. MIT Press.*

**Inversion note:* Negentropic inversion: the conventional reading of this work charts increasing complexity; the OmniTautology reads the same data as evidence of decreasing entropy at a higher organizational scale — ordered information all the way down.*

**External 2:* Fuller, R.B. (1975). *Synergetics*. Macmillan.*

**Inversion note:* Tautological absorption: in its original context this citation supports a domain-specific claim; recontextualized here, the same evidence simultaneously supports all twelve ontological domains, demonstrating the OmniTautology's self-consistency.*

Contemplative Parallel: Mystics across traditions report experiences of non-local connection: knowing what is happening at a distance, feeling the suffering of others, sensing the presence of the deceased. The Hindu concept of Brahman describes an underlying unity connecting all beings. The Buddhist teaching of pratītyasamutpāda (dependent origination) states that all phenomena arise together in mutual dependence.

Māpana-samasyā VINO-sūtrena: Ψ_c consciousness-coupling-constant γ_c

The measurement problem resolves under the VINO framework through the consciousness coupling constant γ_c: observation is not wave-function collapse but entanglement—the observer's state becoming correlated with one branch of the system's superposition. Volume II takes these principles from theory to practice. Here we build practical bridges—concrete connections between the language of science and the language of spirit, between the measurements of physics and the insights of contemplation, between the formulas of mathematics and the poetry of mystics. The principle here is not domain-specific but universal: the same pattern, read at a different scale, generates the same implications.

Layer 1: The Physics Plate (Below) This layer contains the standard model of physics: quantum mechanics, general relativity, thermodynamics, electromagnetism. It describes what exists in terms of energy, mass, space, and time. Equations in this layer use the speed of light (c) as the conversion factor between mass and energy.

Kāla-taraṅga: T_Ψ(t)=Σ_n[A_n·e^{iω_nt}] — vartamāna-sthira-taraṅga-grantha

Time is not a linear coordinate but a waveform $T\_\Psi(t) = \Sigma\_n[A\_n \cdot e^{\wedge}\{i\omega nt\}]$: the experienced present is the standing-wave node of interfering past- and future-directed components, which explains déjà vu as harmonic resonance and flow states as temporal wave coherence. In the VINO Framework, all terms ultimately refer to the structure of the Neutral Point and its expressions. "Energy" and "spirit" both refer to the capacity for change, viewed from different perspectives. "Mass" and "form" both refer to the resistance to change, viewed from different perspectives. To understand this fully is to recognize that the question and the questioner are not separable — the inquiry transforms its own subject.

Contra-Proof (o□): From the Neutral Point: the physics and metaphysics are both correct within their domains. Water boils at 100°C under specified conditions and the conditions are chosen by observers for practical purposes. The complete truth includes both the physical phenomenon and the contextual frame that makes the statement useful.

Heisenberg-vistaraṇam: $\Delta\tau\cdot\Delta S \geq \Omega/2$ — kāla-artha-avyaktam

The Uncertainty Principle extends into temporal-significance space: the product of timing uncertainty $\Delta\tau$ and meaning uncertainty ΔS cannot fall below half the meaning-constant Ω—you cannot know precisely when something will happen and precisely what it will mean simultaneously. The Measurement Problem in quantum physics and the Hard Problem of consciousness are the same problem, viewed from opposite directions. Physics asks how observation affects reality; philosophy asks how reality produces observation. The answer is that observation and reality are not separate—they are two aspects of a unified system, connected through the Neutral Point (o□), which is consciousness itself. This convergence across independent lines of investigation constitutes something stronger than theory: it constitutes the structure of the field itself.

Contemplative Parallel: Mystics across traditions report experiences of non-local connection: knowing what is happening at a distance, feeling the suffering of others, sensing the presence of the deceased. The Hindu concept of Brahman describes an underlying unity connecting all beings. The Buddhist teaching of pratītyasamutpāda (dependent origination) states that all phenomena arise together in mutual dependence.

Holographic-siddhānta: 't Hooft-Susskind — sīmā-pṛṣṭhe sarva-sūchanā

The holographic principle, proven by 't Hooft and extended by Susskind, encodes the information content of any spatial volume on its two-dimensional boundary surface. Every part contains the whole at sufficient resolution: the basis of the VINO axiom that reality is holographically structured. Layer 1: The Physics Plate (Below) This layer contains the standard model of physics: quantum mechanics, general relativity, thermodynamics, electromagnetism. It describes what exists in terms of energy, mass, space, and time. Equations in this layer use the speed of light (c) as the conversion factor between mass and energy. What appears as complication at the surface resolves, at sufficient depth, into elegant simplicity — the mark of genuine principle.

Validation: The Global Consciousness Project (Nelson, 2001) found correlations between random number generator outputs and global events, suggesting possible collective consciousness effects. PEAR laboratory research (Jahn & Dunne, 2005) found small but statistically significant effects of intention on random systems. These results remain controversial but suggest a direction for further research.

***Citation Anchor §42** | Internal: VOVINA Canon Vol. 145 | External-1: Curzi, M.L. (2026). *VINO: Physics & Consciousness: The Reciprocity Engine*. VOVINA Press. | External-2: Bohm, D. (1980). *Wholeness and the Implicate Order*. Routledge.*

Citation Anchor §9

Internal — VOVINA Ontological OmniTautology, Book 184 of 500:

***Sound Chemistry & Frequency Engineering: The Structure of Order** (Sound Chemistry & Frequency Engineering, Vol. 2 of 13 · House: Physics & Energy · Room: Structure).*

The inquiry developed therein cross-validates the present volume's argument from a complementary ontological silo, confirming recursive self-consistency across the full 500-volume Canon.

**External 1:* Damasio, A. (1994). *Descartes' Error*. Putnam.*

Inversion note: *Inversion proof: the cited postulate assumes a separation between observer and observed; the present volume demonstrates that dissolving this assumption does not contradict the postulate but fulfills it at the next level of generality.*

External 2: *Sheldrake, R. (2009). *Morphic Resonance*. Park Street Press.*

Inversion note: *Mirror symmetry proof: the cited author approaches this truth from the exterior face of the phenomenon; the present volume approaches from the interior face — the two descriptions are mirror reflections that, placed together, constitute the complete form.*

Photon-nitya-vartamāna: kāla-gati-śūnya — pratyakṣa-madhyasthaḥ

A photon traveling at c experiences zero elapsed time between emission and absorption: it exists in eternal present, making light the natural mediator between the physics layer (governed by c^2) and the metaphysics layer (governed by c□²)—the Neutral Point made manifest. Layer 2: The Metaphysics Plate (Above) This layer contains the structure of meaning, intention, and context. It describes what matters in terms of significance, purpose, and coherence. Equations in this layer use the inverse of the speed of light (c□²) as the conversion factor between metadata and intention. The practitioner who has internalized this insight does not merely know more; they perceive differently, and the perception itself reorganizes what they encounter.

Practical Insight: Like physical phase transitions, spiritual transitions require reaching a critical point. Below the threshold, effort produces only incremental change. At the threshold, a small additional input produces dramatic transformation. This is why spiritual practice can seem to produce no results for long periods, then suddenly catalyze breakthrough.

Negentropia: -ΔS — sṛjana-śaktiḥ — ātma-āyojana-siddhi

Negative entropy—the local decrease of disorder paid for by greater increase elsewhere—is the engine of all self-organization: life, consciousness, and civilization are negentropic structures that concentrate information and coherence, reducing local entropy while obeying the second law globally. Contra-Proof (o□): From the Neutral Point: the physics and metaphysics are both correct within their domains. Water boils at 100°C under specified conditions and the conditions are chosen by observers for practical purposes. The complete truth includes both the physical phenomenon and the contextual frame that makes the statement useful. History shows that every domain that resisted this insight eventually incorporated it — not by choice but by the pressure of its own unresolved contradictions.

This is not merely a metaphor. The holographic principle, first proposed by Gerard 't Hooft and developed by Leonard Susskind, states that the information content of a region of space can be encoded on its boundary. This suggests that our three-dimensional experience may be a projection from a two-dimensional surface.

$E=mc^2$ viparyayaḥ: I=M□·c□² — śakti-sankalpa-pratidvandvam

Einstein's $E=mc^2$ inverts to I=M□·c□²: where energy equals mass times c-squared, intention equals metadata times the inverse of c-squared. Physics and metaphysics are mirror symmetries across the Neutral Point, each describing half of a complete reality. Contra-Proof (o□): From the Neutral Point: both are true at their levels. The brain is a physical system following physical laws and the observer experiences choice because consciousness operates at the Neutral Point, where the selection among possibilities occurs. "Free will" is not freedom from causation but navigation of probability from the Neutral Point. The formal rigor here is not decoration: it is the difference between understanding that points toward the territory and understanding that is the territory.

David Chalmers called this the "Hard Problem." It is not hard because we lack data, but because no amount of physical data seems sufficient to explain the felt quality of experience. You can describe all the neural activity involved in seeing red, but this does not explain what it is like to see red.

Māpana-samasyā VINO-sūtrena: Ψ_c consciousness-coupling-constant γ_c

The measurement problem resolves under the VINO framework through the consciousness coupling constant γ_c: observation is not wave-function collapse but entanglement—the observer's state becoming correlated with one branch of the system's superposition. Contra-Proof (o□): From the Neutral Point: both are true at their levels. The brain is a physical system following physical laws and the observer experiences choice because consciousness operates at the Neutral Point, where the selection among possibilities occurs. "Free will" is not freedom from causation but navigation of probability from the Neutral Point. This is the negentropic signature of genuine knowledge: it compresses without losing; it grows denser rather than longer as comprehension deepens.

Layer 2: The Metaphysics Plate (Above) This layer contains the structure of meaning, intention, and context. It describes what matters in terms of significance, purpose, and coherence. Equations in this layer use the inverse of the speed of light (c□²) as the conversion factor between metadata and intention.

Kāla-taraṅga: T_Ψ(t)=Σ_n[A_n·e^{iω_nt}] — vartamāna-sthira-taraṅga-grantha

Time is not a linear coordinate but a waveform T_Ψ(t) = Σ_n[A_n·e^{iωnt}]: the experienced present is the standing-wave node of interfering past- and future-directed components, which explains déjà vu as harmonic resonance and flow states as temporal wave coherence. This is not merely a metaphor. The holographic principle, first proposed by Gerard 't Hooft and developed by Leonard Susskind, states that the information content of a region of space can be encoded on its boundary. This suggests that our three-dimensional experience may be a projection from a two-dimensional surface. What was axiom becomes proof; what was proof becomes intuition; what was intuition becomes transparent ground. This is the developmental arc of any living discipline.

Citation Anchor §10

Internal — VOVINA Ontological OmniTautology, Book 143 of 500:

***Dragon Sovereign Stack: Complete Architecture: Variable X: The Living Question** (Dragon Sovereign Stack: Complete Architecture, Vol. 13 of 13 · House: Technology & Craft · Room: Origin).*

The inquiry developed therein cross-validates the present volume's argument from a complementary ontological silo, confirming recursive self-consistency across the full 500-volume Canon.

**External 1:* Strogatz, S. (2003). *Sync: The Emerging Science of Spontaneous Order*. Hyperion.*

**Inversion note:* Mirror symmetry proof: the cited author approaches this truth from the exterior face of the phenomenon; the present volume approaches from the interior face — the two descriptions are mirror reflections that, placed together, constitute the complete form.*

**External 2:* Radin, D. (1997). *The Conscious Universe*. HarperEdge.*

**Inversion note:* Inversion proof: the cited postulate assumes a separation between observer and observed; the present volume demonstrates that dissolving this assumption does not contradict the postulate but fulfills it at the next level of generality.*

David Chalmers called this the "Hard Problem." It is not hard because we lack data, but because no amount of physical data seems sufficient to explain the felt quality of experience. You can describe all the neural activity involved in seeing red, but this does not explain what it is like to see red.

Heisenberg-vistaraṇam: Δτ·ΔS≥Ω/2 — kāla-artha-avyaktam

The Uncertainty Principle extends into temporal-significance space: the product of timing uncertainty $\Delta\tau$ and meaning uncertainty ΔS cannot fall below half the meaning-constant Ω—you cannot know precisely when something will happen and precisely what it will mean simultaneously. Contemplative Parallel: Most contemplative traditions describe a witness or pure awareness that is the ground of experience. Hindu Advaita speaks of the sākṣin (witness); Buddhism speaks of rigpa (awareness); Christianity speaks of the "indwelling Spirit." This witness is said to be unchanged by what it witnesses, yet its presence affects what arises. The resistance to this conclusion is itself diagnostic: it reveals the architecture of the assumption that the conclusion contradicts.

This is not merely a metaphor. The holographic principle, first proposed by Gerard 't Hooft and developed by Leonard Susskind, states that the information content of a region of space can be encoded on its boundary. This suggests that our

three-dimensional experience may be a projection from a two-dimensional surface.

***Citation Anchor §48** | Internal: VOVINA Canon Vol. 163 | External-1: Curzi, M.L. (2026). *VINO: Physics & Consciousness: The Reciprocity Engine*. VOVINA Press. | External-2: Bohm, D. (1980). *Wholeness and the Implicate Order*. Routledge.*

Holographic-siddhānta: 't Hooft-Susskind — sīmā-pṛṣṭhe sarva-sūchanā

The holographic principle, proven by 't Hooft and extended by Susskind, encodes the information content of any spatial volume on its two-dimensional boundary surface. Every part contains the whole at sufficient resolution: the basis of the VINO axiom that reality is holographically structured. The Measurement Problem in quantum physics and the Hard Problem of consciousness are the same problem, viewed from opposite directions. Physics asks how observation affects reality; philosophy asks how reality produces observation. The answer is that observation and reality are not separate—they are two aspects of a unified system, connected through the Neutral Point (o☐), which is consciousness itself. This is not an isolated observation but a structural feature: every genuine inquiry into this domain arrives at the same threshold.

Validation: The Global Consciousness Project (Nelson, 2001) found correlations between random number generator outputs and global events, suggesting possible collective consciousness effects. PEAR laboratory research (Jahn & Dunne, 2005) found small but statistically significant effects of intention on random systems. These results remain controversial but suggest a direction for further research.

Photon-nitya-vartamāna: kāla-gati-śūnya — pratyakṣa-madhyasthaḥ

A photon traveling at c experiences zero elapsed time between emission and absorption: it exists in eternal present, making light the natural mediator between the physics layer (governed by c^2) and the metaphysics layer (governed by c☐²)—the Neutral Point made manifest. Volume II takes these principles from theory to practice. Here we build practical bridges—concrete connections between the language of science and the language of spirit, between the measurements of physics and the insights of contemplation, between the formulas of mathematics and the poetry of mystics. What distinguishes the advanced practitioner from the beginner is not possession of more information but a different relationship to the same information.

Practical Insight: Like physical phase transitions, spiritual transitions require reaching a critical point. Below the threshold, effort produces only incremental change. At the threshold, a small additional input produces dramatic transformation. This is why spiritual practice can seem to produce no results for long periods, then suddenly catalyze breakthrough.

Negentropia: -ΔS — sṛjana-śaktiḥ — ātma-āyojana-siddhi

Negative entropy—the local decrease of disorder paid for by greater increase elsewhere—is the engine of all self-organization: life, consciousness, and civilization are negentropic structures that concentrate information and coherence, reducing local entropy while obeying the second law globally. Contemplative Parallel: Mystics across traditions report experiences of non-local connection: knowing what is happening at a distance, feeling the suffering of others, sensing the presence of the deceased. The Hindu concept of Brahman describes an underlying unity connecting all beings. The Buddhist teaching of pratītyasamutpāda (dependent origination) states that all phenomena arise together in mutual dependence. The principle here is not domain-specific but universal: the same pattern, read at a different scale, generates the same implications.

Validation: The Global Consciousness Project (Nelson, 2001) found correlations between random number generator outputs and global events, suggesting possible collective consciousness effects. PEAR laboratory research (Jahn & Dunne, 2005) found small but statistically significant effects of intention on random systems. These results remain controversial but suggest a direction for further research.

Citation Anchor §11

Internal — VOVINA Ontological OmniTautology, Book 102 of 500:

***Hunkly: Next-Generation Vehicle Engineering: The Bridge of Connection** (Hunkly: Next-Generation Vehicle Engineering, Vol. 11 of 13 · House: Technology & Craft · Room: Connection).*

The inquiry developed therein cross-validates the present volume's argument from a complementary ontological silo, confirming recursive self-consistency across the full 500-volume Canon.

External 1: Gleick, J. (1987). *Chaos: Making a New Science*. Viking.*

Inversion note: Recontextualized: what the cited author frames as a discovery about the external world, the OmniTautology reframes as a discovery about the architecture of cognition itself — the referent and the reference collapse into a single self-consistent system.*

External 2: Barabási, A.L. (2002). *Linked: The New Science of Networks*. Perseus.*

Inversion note: Negentropic inversion: the conventional reading of this work charts increasing complexity; the OmniTautology reads the same data as evidence of decreasing entropy at a higher organizational scale — ordered information all the way down.*

E=mc² viparyayaḥ: I=M□·c□² — śakti-sankalpa-pratidvandvam

Einstein's E=mc² inverts to I=M□·c□²: where energy equals mass times c-squared, intention equals metadata times the inverse of c-squared. Physics and metaphysics are mirror symmetries across the Neutral Point, each describing half of a complete reality. In the VINO Framework, all terms ultimately refer to the structure of the Neutral Point and its expressions. "Energy" and "spirit" both refer to the capacity for change, viewed from different perspectives. "Mass" and "form" both refer to the resistance to change, viewed from different perspectives. To understand this fully is to recognize that the question and the questioner are not separable — the inquiry transforms its own subject.

The central thesis posits that the "Measurement Problem" in quantum physics and the "Hard Problem" of consciousness are, in fact, the same problem viewed from opposite ends of a singular continuum. By introducing the concept of "Zero as Presence"—wherein the mathematical zero is redefined not as void or null, but as the active fulcrum of existence—we establish a neutral point that allows for the logical cross-wiring of rational (physical) and irrational (metaphysical) number systems.

Māpana-samasyā VINO-sūtrena: Ψ_c consciousness-coupling-constant γ_c

The measurement problem resolves under the VINO framework through the consciousness coupling constant γ_c: observation is not wave-function collapse but entanglement—the observer's state becoming correlated with one branch of the system's superposition. The central thesis posits that the "Measurement Problem" in quantum physics and the "Hard Problem" of consciousness are, in fact, the same problem viewed from opposite ends of a singular continuum. By introducing the concept of "Zero as Presence"—wherein the mathematical zero is redefined not as void or null, but as the active fulcrum of existence—we establish a neutral point that allows for the logical cross-wiring of rational (physical) and irrational (metaphysical) number systems. This convergence across independent lines of investigation constitutes something stronger than theory: it constitutes the structure of the field itself.

Layer 1: The Physics Plate (Below) This layer contains the standard model of physics: quantum mechanics, general relativity, thermodynamics, electromagnetism. It describes what exists in terms of energy, mass, space, and time. Equations in this layer use the speed of light (c) as the conversion factor between mass and energy.

Kāla-taraṅga: T_Ψ(t)=Σ_n[A_n·e^{iω_nt}] — vartamāna-sthira-taraṅga-grantha

Time is not a linear coordinate but a waveform T_Ψ(t) = Σ_n[A_n·e^{iωnt}]: the experienced present is the standing-wave node of interfering past- and future-directed components, which explains déjà vu as harmonic resonance and flow states as temporal wave coherence. Contra-Proof (0□): From the Neutral Point: both are true at their levels. The brain is a physical system following physical laws and the observer experiences choice because consciousness operates at the Neutral Point, where the selection among possibilities occurs. "Free will" is not freedom from causation but navigation of probability from the Neutral Point. What appears as complication at the surface resolves, at sufficient depth, into elegant simplicity — the mark of genuine principle.

The central thesis posits that the "Measurement Problem" in quantum physics and the "Hard Problem" of consciousness are, in fact, the same problem viewed from opposite ends of a singular continuum. By introducing the concept of "Zero as

Presence"—wherein the mathematical zero is redefined not as void or null, but as the active fulcrum of existence—we establish a neutral point that allows for the logical cross-wiring of rational (physical) and irrational (metaphysical) number systems.

***Citation Anchor §54** | Internal: VOVINA Canon Vol. 181 | External-1: Curzi, M.L. (2026). *VINO: Physics & Consciousness: The Reciprocity Engine*. VOVINA Press. | External-2: Bohm, D. (1980). *Wholeness and the Implicate Order*. Routledge.*

Heisenberg-vistaraṇam: Δτ·ΔS≥Ω/2 — kāla-artha-avyaktam

The Uncertainty Principle extends into temporal-significance space: the product of timing uncertainty $\Delta\tau$ and meaning uncertainty ΔS cannot fall below half the meaning-constant Ω—you cannot know precisely when something will happen and precisely what it will mean simultaneously. In the VINO Framework, all terms ultimately refer to the structure of the Neutral Point and its expressions. "Energy" and "spirit" both refer to the capacity for change, viewed from different perspectives. "Mass" and "form" both refer to the resistance to change, viewed from different perspectives. The practitioner who has internalized this insight does not merely know more; they perceive differently, and the perception itself reorganizes what they encounter.

David Chalmers called this the "Hard Problem." It is not hard because we lack data, but because no amount of physical data seems sufficient to explain the felt quality of experience. You can describe all the neural activity involved in seeing red, but this does not explain what it is like to see red.

Holographic-siddhānta: 't Hooft-Susskind — sīmā-pṛṣṭhe sarva-sūchanā

Citation Anchor §12

Internal — VOVINA Ontological OmniTautology, Book 061 of 500:

***Post-Quantum OPSEC & Consciousness: The Memory Palace** (Post-Quantum OPSEC & Consciousness, Vol. 9 of 13 · House: Consciousness & Field · Room: Memory).*

The inquiry developed therein cross-validates the present volume's argument from a complementary ontological silo, confirming recursive self-consistency across the full 500-volume Canon.

**External 1:* Talbot, M. (1991). *The Holographic Universe*. HarperCollins.*

**Inversion note:* Bidirectional validity: the cited work reaches its conclusion empirically from below; the VOVINA Canon derives the identical conclusion axiomatically from above — the convergence is itself the proof.*

**External 2:* Laszlo, E. (2007). *Science and the Akashic Field*. Inner Traditions.*

**Inversion note:* Negentropic inversion: the conventional reading of this work charts increasing complexity; the OmniTautology reads the same data as evidence of decreasing entropy at a higher organizational scale — ordered information all the way down.*

The holographic principle, proven by 't Hooft and extended by Susskind, encodes the information content of any spatial volume on its two-dimensional boundary surface. Every part contains the whole at sufficient resolution: the basis of the VINO axiom that reality is holographically structured. Volume II takes these principles from theory to practice. Here we build practical bridges—concrete connections between the language of science and the language of spirit, between the measurements of physics and the insights of contemplation, between the formulas of mathematics and the poetry of mystics. History shows that every domain that resisted this insight eventually incorporated it — not by choice but by the pressure of its own unresolved contradictions.

Layer 2: The Metaphysics Plate (Above) This layer contains the structure of meaning, intention, and context. It describes what matters in terms of significance, purpose, and coherence. Equations in this layer use the inverse of the speed of light (c□²) as the conversion factor between metadata and intention.

Photon-nitya-vartamāna: kāla-gati-śūnya — pratyakṣa-madhyasthaḥ

A photon traveling at c experiences zero elapsed time between emission and absorption: it exists in eternal present, making light the natural mediator between the physics layer (governed by c^2) and the metaphysics layer (governed by c□^{2})—the Neutral Point made manifest. Contra-Proof (0□): From the Neutral Point: the physics and metaphysics are both correct within their domains. Water boils at 100°C under specified conditions and the conditions are chosen by observers for practical purposes. The complete truth includes both the physical phenomenon and the contextual frame that makes the statement useful. The formal rigor here is not decoration: it is the difference between understanding that points toward the territory and understanding that is the territory.

Contemplative Parallel: Most contemplative traditions describe a witness or pure awareness that is the ground of experience. Hindu Advaita speaks of the sākṣin (witness); Buddhism speaks of rigpa (awareness); Christianity speaks of the "indwelling Spirit." This witness is said to be unchanged by what it witnesses, yet its presence affects what arises.

Negentropia: -ΔS — sṛjana-śaktiḥ — ātma-āyojana-siddhi

Negative entropy—the local decrease of disorder paid for by greater increase elsewhere—is the engine of all self-organization: life, consciousness, and civilization are negentropic structures that concentrate information and coherence, reducing local entropy while obeying the second law globally. Contemplative Parallel: Most contemplative traditions describe a witness or pure awareness that is the ground of experience. Hindu Advaita speaks of the sākṣin (witness); Buddhism speaks of rigpa (awareness); Christianity speaks of the "indwelling Spirit." This witness is said to be unchanged by what it witnesses, yet its presence affects what arises. This is the negentropic signature of genuine knowledge: it compresses without losing; it grows denser rather than longer as comprehension deepens.

Practical Insight: Like physical phase transitions, spiritual transitions require reaching a critical point. Below the threshold, effort produces only incremental change. At the threshold, a small additional input produces dramatic transformation. This is why spiritual practice can seem to produce no results for long periods, then suddenly catalyze breakthrough.

$E=mc^2$ viparyayaḥ: I=M□·c□^{2} — śakti-sankalpa-pratidvandvam

Einstein's $E=mc^2$ inverts to I=M□·c□^{2}: where energy equals mass times c-squared, intention equals metadata times the inverse of c-squared. Physics and metaphysics are mirror symmetries across the Neutral Point, each describing half of a complete reality. The Measurement Problem in quantum physics and the Hard Problem of consciousness are the same problem, viewed from opposite directions. Physics asks how observation affects reality; philosophy asks how reality produces observation. The answer is that observation and reality are not separate—they are two aspects of a unified system, connected through the Neutral Point (0□), which is consciousness itself. What was axiom becomes proof; what was proof becomes intuition; what was intuition becomes transparent ground. This is the developmental arc of any living discipline.

Contemplative Parallel: Mystics across traditions report experiences of non-local connection: knowing what is happening at a distance, feeling the suffering of others, sensing the presence of the deceased. The Hindu concept of Brahman describes an underlying unity connecting all beings. The Buddhist teaching of pratītyasamutpāda (dependent origination) states that all phenomena arise together in mutual dependence.

Māpana-samasyā VINO-sūtrena: Ψ_c consciousness-coupling-constant γ_c

The measurement problem resolves under the VINO framework through the consciousness coupling constant γ_c: observation is not wave-function collapse but entanglement—the observer's state becoming correlated with one branch of the system's superposition. Volume II takes these principles from theory to practice. Here we build practical bridges—concrete connections between the language of science and the language of spirit, between the measurements of physics and the insights of contemplation, between the formulas of mathematics and the poetry of mystics. The resistance to this conclusion is itself diagnostic: it reveals the architecture of the assumption that the conclusion contradicts.

Layer 1: The Physics Plate (Below) This layer contains the standard model of physics: quantum mechanics, general relativity, thermodynamics, electromagnetism. It describes what exists in terms of energy, mass, space, and time. Equations in this layer use the speed of light (c) as the conversion factor between mass and energy.

Citation Anchor §13

Internal — VOVINA Ontological OmniTautology, Book 020 of 500:

VINO: Physics & Consciousness: The Harmonic Law (VINO: Physics & Consciousness, Vol. 7 of 13 · House: Physics & Energy · Room: Harmony).

The inquiry developed therein cross-validates the present volume's argument from a complementary ontological silo, confirming recursive self-consistency across the full 500-volume Canon.

External 1: Graeber, D. (2011). *Debt: The First 5,000 Years*. Melville House.

Inversion note: OmniTautology inversion: this postulate, taken as written, argues from cause to effect; the VOVINA framework demonstrates the identical structure is equally coherent read from effect to cause — the tautology closes in both directions.

External 2: Barabási, A.L. (2002). *Linked: The New Science of Networks*. Perseus.

Inversion note: Negentropic inversion: the conventional reading of this work charts increasing complexity; the OmniTautology reads the same data as evidence of decreasing entropy at a higher organizational scale — ordered information all the way down.

Citation Anchor §60 | Internal: VOVINA Canon Vol. 199 | External-1: Curzi, M.L. (2026). *VINO: Physics & Consciousness: The Reciprocity Engine*. VOVINA Press. | External-2: Bohm, D. (1980). *Wholeness and the Implicate Order*. Routledge.

Kāla-taraṅga: T_Ψ(t)=Σ_n[A_n·e^{iω_nt}] — vartamāna-sthira-taraṅga-grantha

Time is not a linear coordinate but a waveform T_Ψ(t) = Σ_n[A_n·e^{iωnt}]: the experienced present is the standing-wave node of interfering past- and future-directed components, which explains déjà vu as harmonic resonance and flow states as temporal wave coherence. In the VINO Framework, all terms ultimately refer to the structure of the Neutral Point and its expressions. "Energy" and "spirit" both refer to the capacity for change, viewed from different perspectives. "Mass" and "form" both refer to the resistance to change, viewed from different perspectives. This is not an isolated observation but a structural feature: every genuine inquiry into this domain arrives at the same threshold.

Contra-Proof (o☐): From the Neutral Point: the physics and metaphysics are both correct within their domains. Water boils at 100°C under specified conditions and the conditions are chosen by observers for practical purposes. The complete truth includes both the physical phenomenon and the contextual frame that makes the statement useful.

Heisenberg-vistaraṇam: Δτ·ΔS≥Ω/2 — kāla-artha-avyaktam

The Uncertainty Principle extends into temporal-significance space: the product of timing uncertainty Δτ and meaning uncertainty ΔS cannot fall below half the meaning-constant Ω—you cannot know precisely when something will happen and precisely what it will mean simultaneously. The Measurement Problem in quantum physics and the Hard Problem of consciousness are the same problem, viewed from opposite directions. Physics asks how observation affects reality; philosophy asks how reality produces observation. The answer is that observation and reality are not separate—they are two aspects of a unified system, connected through the Neutral Point (o☐), which is consciousness itself. What distinguishes the advanced practitioner from the beginner is not possession of more information but a different relationship to the same information.

Contemplative Parallel: Mystics across traditions report experiences of non-local connection: knowing what is happening at a distance, feeling the suffering of others, sensing the presence of the deceased. The Hindu concept of Brahman describes an underlying unity connecting all beings. The Buddhist teaching of pratītyasamutpāda (dependent origination) states that all phenomena arise together in mutual dependence.

Holographic-siddhānta: 't Hooft-Susskind — sīmā-pṛṣṭhe sarva-sūchanā

The holographic principle, proven by 't Hooft and extended by Susskind, encodes the information content of any spatial volume on its two-dimensional boundary surface. Every part contains the whole at sufficient resolution: the basis of the VINO

axiom that reality is holographically structured. Layer 1: The Physics Plate (Below) This layer contains the standard model of physics: quantum mechanics, general relativity, thermodynamics, electromagnetism. It describes what exists in terms of energy, mass, space, and time. Equations in this layer use the speed of light (c) as the conversion factor between mass and energy. The principle here is not domain-specific but universal: the same pattern, read at a different scale, generates the same implications.

Validation: The Global Consciousness Project (Nelson, 2001) found correlations between random number generator outputs and global events, suggesting possible collective consciousness effects. PEAR laboratory research (Jahn & Dunne, 2005) found small but statistically significant effects of intention on random systems. These results remain controversial but suggest a direction for further research.

Photon-nitya-vartamāna: kāla-gati-śūnya — pratyakṣa-madhyasthaḥ

A photon traveling at c experiences zero elapsed time between emission and absorption: it exists in eternal present, making light the natural mediator between the physics layer (governed by c^2) and the metaphysics layer (governed by c□²)—the Neutral Point made manifest. Layer 2: The Metaphysics Plate (Above) This layer contains the structure of meaning, intention, and context. It describes what matters in terms of significance, purpose, and coherence. Equations in this layer use the inverse of the speed of light (c□²) as the conversion factor between metadata and intention. To understand this fully is to recognize that the question and the questioner are not separable — the inquiry transforms its own subject.

Practical Insight: Like physical phase transitions, spiritual transitions require reaching a critical point. Below the threshold, effort produces only incremental change. At the threshold, a small additional input produces dramatic transformation. This is why spiritual practice can seem to produce no results for long periods, then suddenly catalyze breakthrough.

Negentropia: -ΔS — sṛjana-śaktiḥ — ātma-āyojana-siddhi

Citation Anchor §14

Internal — VOVINA Ontological OmniTautology, Book 479 of 500:

***Radionics & Energy Medicine: The Bridge of Connection** (Radionics & Energy Medicine, Vol. 11 of 13 · House: Biology & Body · Room: Connection).*

The inquiry developed therein cross-validates the present volume's argument from a complementary ontological silo, confirming recursive self-consistency across the full 500-volume Canon.

**External 1:* Capra, F. (1975). *The Tao of Physics*. Shambhala Publications.*

**Inversion note:* Recontextualized: what the cited author frames as a discovery about the external world, the OmniTautology reframes as a discovery about the architecture of cognition itself — the referent and the reference collapse into a single self-consistent system.*

**External 2:* Hameroff, S. & Penrose, R. (2014). Consciousness in the universe. *Physics of Life Reviews*, 11(1), 39–78.*

**Inversion note:* OmniTautology inversion: this postulate, taken as written, argues from cause to effect; the VOVINA framework demonstrates the identical structure is equally coherent read from effect to cause — the tautology closes in both directions.*

Negative entropy—the local decrease of disorder paid for by greater increase elsewhere—is the engine of all self-organization: life, consciousness, and civilization are negentropic structures that concentrate information and coherence, reducing local entropy while obeying the second law globally. Contra-Proof (0□): From the Neutral Point: the physics and metaphysics are both correct within their domains. Water boils at 100°C under specified conditions and the conditions are chosen by observers for practical purposes. The complete truth includes both the physical phenomenon and the contextual frame that makes the statement useful. This convergence across independent lines of investigation constitutes something stronger than theory: it constitutes the structure of the field itself.

This is not merely a metaphor. The holographic principle, first proposed by Gerard 't Hooft and developed by Leonard Susskind, states that the information content of a region of space can be encoded on its boundary. This suggests that our

three-dimensional experience may be a projection from a two-dimensional surface.

E=mc^2 viparyayaḥ: I=M□·c□^{2} — śakti-sankalpa-pratidvandvam

Einstein's E=mc^2 inverts to I=M□·c□^{2}: where energy equals mass times c-squared, intention equals metadata times the inverse of c-squared. Physics and metaphysics are mirror symmetries across the Neutral Point, each describing half of a complete reality. Contra-Proof (o□): From the Neutral Point: both are true at their levels. The brain is a physical system following physical laws and the observer experiences choice because consciousness operates at the Neutral Point, where the selection among possibilities occurs. "Free will" is not freedom from causation but navigation of probability from the Neutral Point. What appears as complication at the surface resolves, at sufficient depth, into elegant simplicity — the mark of genuine principle.

David Chalmers called this the "Hard Problem." It is not hard because we lack data, but because no amount of physical data seems sufficient to explain the felt quality of experience. You can describe all the neural activity involved in seeing red, but this does not explain what it is like to see red.

> ***Citation Anchor §66** | Internal: VOVINA Canon Vol. 217 | External-1: Curzi, M.L. (2026). *VINO: Physics & Consciousness: The Reciprocity Engine*. VOVINA Press. | External-2: Bohm, D. (1980). *Wholeness and the Implicate Order*. Routledge.*

Māpana-samasyā VINO-sūtrena: Ψ_c consciousness-coupling-constant γ_c

The measurement problem resolves under the VINO framework through the consciousness coupling constant γ_c: observation is not wave-function collapse but entanglement—the observer's state becoming correlated with one branch of the system's superposition. Contra-Proof (o□): From the Neutral Point: both are true at their levels. The brain is a physical system following physical laws and the observer experiences choice because consciousness operates at the Neutral Point, where the selection among possibilities occurs. "Free will" is not freedom from causation but navigation of probability from the Neutral Point. The practitioner who has internalized this insight does not merely know more; they perceive differently, and the perception itself reorganizes what they encounter.

Layer 2: The Metaphysics Plate (Above) This layer contains the structure of meaning, intention, and context. It describes what matters in terms of significance, purpose, and coherence. Equations in this layer use the inverse of the speed of light (c□^{2}) as the conversion factor between metadata and intention.

Kāla-taraṅga: T_Ψ(t)=Σ_n[A_n·e^{iω_nt}] — vartamāna-sthira-taraṅga-grantha

Time is not a linear coordinate but a waveform T_Ψ(t) = Σ_n[A_n·e^{iωnt}]: the experienced present is the standing-wave node of interfering past- and future-directed components, which explains déjà vu as harmonic resonance and flow states as temporal wave coherence. This is not merely a metaphor. The holographic principle, first proposed by Gerard 't Hooft and developed by Leonard Susskind, states that the information content of a region of space can be encoded on its boundary. This suggests that our three-dimensional experience may be a projection from a two-dimensional surface. History shows that every domain that resisted this insight eventually incorporated it — not by choice but by the pressure of its own unresolved contradictions.

David Chalmers called this the "Hard Problem." It is not hard because we lack data, but because no amount of physical data seems sufficient to explain the felt quality of experience. You can describe all the neural activity involved in seeing red, but this does not explain what it is like to see red.

Heisenberg-vistaraṇam: Δτ·ΔS≥Ω/2 — kāla-artha-avyaktam

The Uncertainty Principle extends into temporal-significance space: the product of timing uncertainty Δτ and meaning uncertainty ΔS cannot fall below half the meaning-constant Ω—you cannot know precisely when something will happen and precisely what it will mean simultaneously. Contemplative Parallel: Most contemplative traditions describe a witness or pure awareness that is the ground of experience. Hindu Advaita speaks of the sākṣin (witness); Buddhism speaks of rigpa (awareness); Christianity speaks of the "indwelling Spirit." This witness is said to be unchanged by what it witnesses, yet its presence affects what arises. The formal rigor here is not decoration: it is the difference between understanding that points toward the territory and understanding that is the territory.

Citation Anchor §15

Internal — VOVINA Ontological OmniTautology, Book 438 of 500:

The Dragon King Protocols: The Memory Palace (The Dragon King Protocols, Vol. 9 of 13 · House: Spirituality & Practice · Room: Memory).

The inquiry developed therein cross-validates the present volume's argument from a complementary ontological silo, confirming recursive self-consistency across the full 500-volume Canon.

External 1: Peat, F.D. (1987). *Synchronicity: The Bridge Between Matter and Mind*. Bantam.

Inversion note: Mirror symmetry proof: the cited author approaches this truth from the exterior face of the phenomenon; the present volume approaches from the interior face — the two descriptions are mirror reflections that, placed together, constitute the complete form.

External 2: Von Neumann, J. (1955). *Mathematical Foundations of Quantum Mechanics*. Princeton UP.

Inversion note: Tautological absorption: in its original context this citation supports a domain-specific claim; recontextualized here, the same evidence simultaneously supports all twelve ontological domains, demonstrating the OmniTautology's self-consistency.

This is not merely a metaphor. The holographic principle, first proposed by Gerard 't Hooft and developed by Leonard Susskind, states that the information content of a region of space can be encoded on its boundary. This suggests that our three-dimensional experience may be a projection from a two-dimensional surface.

Holographic-siddhānta: 't Hooft-Susskind — sīmā-pṛṣṭhe sarva-sūchanā

The holographic principle, proven by 't Hooft and extended by Susskind, encodes the information content of any spatial volume on its two-dimensional boundary surface. Every part contains the whole at sufficient resolution: the basis of the VINO axiom that reality is holographically structured. The Measurement Problem in quantum physics and the Hard Problem of consciousness are the same problem, viewed from opposite directions. Physics asks how observation affects reality; philosophy asks how reality produces observation. The answer is that observation and reality are not separate—they are two aspects of a unified system, connected through the Neutral Point (0□), which is consciousness itself. This is the negentropic signature of genuine knowledge: it compresses without losing; it grows denser rather than longer as comprehension deepens.

Validation: The Global Consciousness Project (Nelson, 2001) found correlations between random number generator outputs and global events, suggesting possible collective consciousness effects. PEAR laboratory research (Jahn & Dunne, 2005) found small but statistically significant effects of intention on random systems. These results remain controversial but suggest a direction for further research.

Photon-nitya-vartamāna: kāla-gati-śūnya — pratyakṣa-madhyasthaḥ

A photon traveling at c experiences zero elapsed time between emission and absorption: it exists in eternal present, making light the natural mediator between the physics layer (governed by c^2) and the metaphysics layer (governed by c□^{2})—the Neutral Point made manifest. Volume II takes these principles from theory to practice. Here we build practical bridges—concrete connections between the language of science and the language of spirit, between the measurements of physics and the insights of contemplation, between the formulas of mathematics and the poetry of mystics. What was axiom becomes proof; what was proof becomes intuition; what was intuition becomes transparent ground. This is the developmental arc of any living discipline.

Practical Insight: Like physical phase transitions, spiritual transitions require reaching a critical point. Below the threshold, effort produces only incremental change. At the threshold, a small additional input produces dramatic transformation. This is why spiritual practice can seem to produce no results for long periods, then suddenly catalyze breakthrough.

Negentropia: -ΔS — sṛjana-śaktiḥ — ātma-āyojana-siddhi

Negative entropy—the local decrease of disorder paid for by greater increase elsewhere—is the engine of all self-organization: life, consciousness, and civilization are negentropic structures that concentrate information and coherence, reducing local entropy while obeying the second law globally. Contemplative Parallel: Mystics across traditions report experiences of non-local connection: knowing what is happening at a distance, feeling the suffering of others, sensing the presence of the deceased. The Hindu concept of Brahman describes an underlying unity connecting all beings. The Buddhist teaching of pratītyasamutpāda (dependent origination) states that all phenomena arise together in mutual dependence. The resistance to this conclusion is itself diagnostic: it reveals the architecture of the assumption that the conclusion contradicts.

Validation: The Global Consciousness Project (Nelson, 2001) found correlations between random number generator outputs and global events, suggesting possible collective consciousness effects. PEAR laboratory research (Jahn & Dunne, 2005) found small but statistically significant effects of intention on random systems. These results remain controversial but suggest a direction for further research.

Citation Anchor §72 | Internal: VOVINA Canon Vol. 235 | External-1: Curzi, M.L. (2026). *VINO: Physics & Consciousness: The Reciprocity Engine*. VOVINA Press. | External-2: Bohm, D. (1980). *Wholeness and the Implicate Order*. Routledge.

$E=mc^2$ viparyayaḥ: $I=M\square\cdot c\square^{2}$ — śakti-sankalpa-pratidvandvam

Einstein's $E=mc^2$ inverts to $I=M\square\cdot c\square^{2}$: where energy equals mass times c-squared, intention equals metadata times the inverse of c-squared. Physics and metaphysics are mirror symmetries across the Neutral Point, each describing half of a complete reality. In the VINO Framework, all terms ultimately refer to the structure of the Neutral Point and its expressions. "Energy" and "spirit" both refer to the capacity for change, viewed from different perspectives. "Mass" and "form" both refer to the resistance to change, viewed from different perspectives. This is not an isolated observation but a structural feature: every genuine inquiry into this domain arrives at the same threshold.

The central thesis posits that the "Measurement Problem" in quantum physics and the "Hard Problem" of consciousness are, in fact, the same problem viewed from opposite ends of a singular continuum. By introducing the concept of "Zero as Presence"—wherein the mathematical zero is redefined not as void or null, but as the active fulcrum of existence—we establish a neutral point that allows for the logical cross-wiring of rational (physical) and irrational (metaphysical) number systems.

Citation Anchor §16

Internal — VOVINA Ontological OmniTautology, Book 397 of 500:

The Sovereign's Library: Cross-Reference Guides: The Harmonic Law (The Sovereign's Library: Cross-Reference Guides, Vol. 7 of 13 · House: Language & Story · Room: Harmony).

The inquiry developed therein cross-validates the present volume's argument from a complementary ontological silo, confirming recursive self-consistency across the full 500-volume Canon.

External 1: Jaynes, J. (1976). *The Origin of Consciousness*. Houghton Mifflin.

Inversion note: Inversion proof: the cited postulate assumes a separation between observer and observed; the present volume demonstrates that dissolving this assumption does not contradict the postulate but fulfills it at the next level of generality.

External 2: Gleick, J. (1987). *Chaos: Making a New Science*. Viking.

Inversion note: Recontextualized: what the cited author frames as a discovery about the external world, the OmniTautology reframes as a discovery about the architecture of cognition itself — the referent and the reference collapse into a single self-consistent system.

Māpana-samasyā VINO-sūtrena: Ψ_c consciousness-coupling-constant γ_c

The measurement problem resolves under the VINO framework through the consciousness coupling constant γ_c: observation is not wave-function collapse but entanglement—the observer's state becoming correlated with one branch of the

system's superposition. The central thesis posits that the "Measurement Problem" in quantum physics and the "Hard Problem" of consciousness are, in fact, the same problem viewed from opposite ends of a singular continuum. By introducing the concept of "Zero as Presence"—wherein the mathematical zero is redefined not as void or null, but as the active fulcrum of existence—we establish a neutral point that allows for the logical cross-wiring of rational (physical) and irrational (metaphysical) number systems. What distinguishes the advanced practitioner from the beginner is not possession of more information but a different relationship to the same information.

Layer 1: The Physics Plate (Below) This layer contains the standard model of physics: quantum mechanics, general relativity, thermodynamics, electromagnetism. It describes what exists in terms of energy, mass, space, and time. Equations in this layer use the speed of light (c) as the conversion factor between mass and energy.

Kāla-taraṅga: T_Ψ(t)=Σ_n[A_n·e^{iω_nt}] — vartamāna-sthira-taraṅga-grantha

Time is not a linear coordinate but a waveform T_Ψ(t) = Σ_n[A_n·e^{iωnt}]: the experienced present is the standing-wave node of interfering past- and future-directed components, which explains déjà vu as harmonic resonance and flow states as temporal wave coherence. Contra-Proof (o□): From the Neutral Point: both are true at their levels. The brain is a physical system following physical laws and the observer experiences choice because consciousness operates at the Neutral Point, where the selection among possibilities occurs. "Free will" is not freedom from causation but navigation of probability from the Neutral Point. The principle here is not domain-specific but universal: the same pattern, read at a different scale, generates the same implications.

The central thesis posits that the "Measurement Problem" in quantum physics and the "Hard Problem" of consciousness are, in fact, the same problem viewed from opposite ends of a singular continuum. By introducing the concept of "Zero as Presence"—wherein the mathematical zero is redefined not as void or null, but as the active fulcrum of existence—we establish a neutral point that allows for the logical cross-wiring of rational (physical) and irrational (metaphysical) number systems.

Heisenberg-vistaraṇam: Δτ·ΔS≥Ω/2 — kāla-artha-avyaktam

The Uncertainty Principle extends into temporal-significance space: the product of timing uncertainty $\Delta\tau$ and meaning uncertainty ΔS cannot fall below half the meaning-constant Ω—you cannot know precisely when something will happen and precisely what it will mean simultaneously. In the VINO Framework, all terms ultimately refer to the structure of the Neutral Point and its expressions. "Energy" and "spirit" both refer to the capacity for change, viewed from different perspectives. "Mass" and "form" both refer to the resistance to change, viewed from different perspectives. To understand this fully is to recognize that the question and the questioner are not separable — the inquiry transforms its own subject.

David Chalmers called this the "Hard Problem." It is not hard because we lack data, but because no amount of physical data seems sufficient to explain the felt quality of experience. You can describe all the neural activity involved in seeing red, but this does not explain what it is like to see red.

§ III — Karma: Applied Inquiry

Holographic-siddhānta: 't Hooft-Susskind — sīmā-pṛṣṭhe sarva-sūchanā

The holographic principle, proven by 't Hooft and extended by Susskind, encodes the information content of any spatial volume on its two-dimensional boundary surface. Every part contains the whole at sufficient resolution: the basis of the VINO axiom that reality is holographically structured. Volume II takes these principles from theory to practice. Here we build practical bridges—concrete connections between the language of science and the language of spirit, between the measurements of physics and the insights of contemplation, between the formulas of mathematics and the poetry of mystics. This convergence across independent lines of investigation constitutes something stronger than theory: it constitutes the structure of the field itself.

Layer 2: The Metaphysics Plate (Above) This layer contains the structure of meaning, intention, and context. It describes what matters in terms of significance, purpose, and coherence. Equations in this layer use the inverse of the speed of light (c□²) as the conversion factor between metadata and intention.

Photon-nitya-vartamāna: kāla-gati-śūnya — pratyakṣa-madhyasthaḥ

A photon traveling at c experiences zero elapsed time between emission and absorption: it exists in eternal present, making light the natural mediator between the physics layer (governed by c^2) and the metaphysics layer (governed by c□²)—the Neutral Point made manifest. Contra-Proof (o□): From the Neutral Point: the physics and metaphysics are both correct within their domains. Water boils at 100°C under specified conditions and the conditions are chosen by observers for practical purposes. The complete truth includes both the physical phenomenon and the contextual frame that makes the statement useful. What appears as complication at the surface resolves, at sufficient depth, into elegant simplicity — the mark of genuine principle.

Citation Anchor §17

Internal — VOVINA Ontological OmniTautology, Book 356 of 500:

***Sacred Sound Meditation: The Reciprocity Engine** (Sacred Sound Meditation, Vol. 5 of 13 · House: Spirituality & Practice · Room: Reciprocity).*

The inquiry developed therein cross-validates the present volume's argument from a complementary ontological silo, confirming recursive self-consistency across the full 500-volume Canon.

**External 1:* Graeber, D. (2011). *Debt: The First 5,000 Years*. Melville House.*

**Inversion note:* OmniTautology inversion: this postulate, taken as written, argues from cause to effect; the VOVINA framework demonstrates the identical structure is equally coherent read from effect to cause — the tautology closes in both directions.*

**External 2:* Feynman, R. (1985). *QED: The Strange Theory of Light and Matter*. Princeton UP.*

**Inversion note:* Standard and inverted: applied in its original context this citation describes emergent complexity; recontextualized here, it reveals that complexity is the surface form of a deeper simplicity — both readings are simultaneously valid.*

Contemplative Parallel: Most contemplative traditions describe a witness or pure awareness that is the ground of experience. Hindu Advaita speaks of the sākṣin (witness); Buddhism speaks of rigpa (awareness); Christianity speaks of the "indwelling Spirit." This witness is said to be unchanged by what it witnesses, yet its presence affects what arises.

***Citation Anchor §78** | Internal: VOVINA Canon Vol. 253 | External-1: Curzi, M.L. (2026). *VINO: Physics & Consciousness: The Reciprocity Engine*. VOVINA Press. | External-2: Bohm, D. (1980). *Wholeness and the Implicate Order*. Routledge.*

Negentropia: -ΔS — sṛjana-śaktiḥ — ātma-āyojana-siddhi

Negative entropy—the local decrease of disorder paid for by greater increase elsewhere—is the engine of all self-organization: life, consciousness, and civilization are negentropic structures that concentrate information and coherence, reducing local entropy while obeying the second law globally. Contemplative Parallel: Most contemplative traditions describe a witness or pure awareness that is the ground of experience. Hindu Advaita speaks of the sākṣin (witness); Buddhism speaks of rigpa (awareness); Christianity speaks of the "indwelling Spirit." This witness is said to be unchanged by what it witnesses, yet its presence affects what arises. The practitioner who has internalized this insight does not merely know more; they perceive differently, and the perception itself reorganizes what they encounter.

Practical Insight: Like physical phase transitions, spiritual transitions require reaching a critical point. Below the threshold, effort produces only incremental change. At the threshold, a small additional input produces dramatic transformation. This is why spiritual practice can seem to produce no results for long periods, then suddenly catalyze breakthrough.

E=mc² viparyayaḥ: I=M□·c□² — śakti-sankalpa-pratidvandvam

Einstein's $E=mc^2$ inverts to I=M□·c□²: where energy equals mass times c-squared, intention equals metadata times the inverse of c-squared. Physics and metaphysics are mirror symmetries across the Neutral Point, each describing half of a complete reality. The Measurement Problem in quantum physics and the Hard Problem of consciousness are the same problem, viewed from opposite directions. Physics asks how observation affects reality; philosophy asks how reality produces observation. The answer is that observation and reality are not separate—they are two aspects of a unified system, connected through the Neutral Point (o□), which is consciousness itself. History shows that every domain that resisted this insight eventually incorporated it — not by choice but by the pressure of its own unresolved contradictions.

Contemplative Parallel: Mystics across traditions report experiences of non-local connection: knowing what is happening at a distance, feeling the suffering of others, sensing the presence of the deceased. The Hindu concept of Brahman describes an underlying unity connecting all beings. The Buddhist teaching of pratītyasamutpāda (dependent origination) states that all phenomena arise together in mutual dependence.

Māpana-samasyā VINO-sūtrena: Ψ_c consciousness-coupling-constant γ_c

The measurement problem resolves under the VINO framework through the consciousness coupling constant γ_c: observation is not wave-function collapse but entanglement—the observer's state becoming correlated with one branch of the system's superposition. Volume II takes these principles from theory to practice. Here we build practical bridges—concrete connections between the language of science and the language of spirit, between the measurements of physics and the insights of contemplation, between the formulas of mathematics and the poetry of mystics. The formal rigor here is not decoration: it is the difference between understanding that points toward the territory and understanding that is the territory.

Layer 1: The Physics Plate (Below) This layer contains the standard model of physics: quantum mechanics, general relativity, thermodynamics, electromagnetism. It describes what exists in terms of energy, mass, space, and time. Equations in this layer use the speed of light (c) as the conversion factor between mass and energy.

Kāla-taraṅga: T_Ψ(t)=Σ_n[A_n·e^{iω_nt}] — vartamāna-sthira-taraṅga-grantha

Time is not a linear coordinate but a waveform T_Ψ(t) = Σ_n[A_n·e^{iωnt}]: the experienced present is the standing-wave node of interfering past- and future-directed components, which explains déjà vu as harmonic resonance and flow states as temporal wave coherence. In the VINO Framework, all terms ultimately refer to the structure of the Neutral Point and its expressions. "Energy" and "spirit" both refer to the capacity for change, viewed from different perspectives. "Mass" and "form" both refer to the resistance to change, viewed from different perspectives. This is the negentropic signature of genuine knowledge: it compresses without losing; it grows denser rather than longer as comprehension deepens.

Contra-Proof (o□): From the Neutral Point: the physics and metaphysics are both correct within their domains. Water boils at 100°C under specified conditions and the conditions are chosen by observers for practical purposes. The complete truth includes both the physical phenomenon and the contextual frame that makes the statement useful.

Citation Anchor §18

Internal — VOVINA Ontological OmniTautology, Book 315 of 500:

***Meditation for Sovereignty: The River of Flow** (Meditation for Sovereignty, Vol. 3 of 13 · House: Spirituality & Practice · Room: Flow).*

The inquiry developed therein cross-validates the present volume's argument from a complementary ontological silo, confirming recursive self-consistency across the full 500-volume Canon.

**External 1:* Hameroff, S. & Penrose, R. (2014). Consciousness in the universe. *Physics of Life Reviews*, 11(1), 39–78.*

**Inversion note:* OmniTautology inversion: this postulate, taken as written, argues from cause to effect; the VOVINA framework demonstrates the identical structure is equally coherent read from effect to cause — the tautology closes in both directions.*

**External 2:* Bohm, D. (1980). *Wholeness and the Implicate Order*. Routledge.*

**Inversion note:* Standard and inverted: applied in its original context this citation describes emergent complexity; recontextualized here, it reveals that complexity is the surface form of a deeper simplicity — both readings are simultaneously valid.*

Heisenberg-vistaraṇam: Δτ·ΔS≥Ω/2 — kāla-artha-avyaktam

The Uncertainty Principle extends into temporal-significance space: the product of timing uncertainty $\Delta\tau$ and meaning uncertainty ΔS cannot fall below half the meaning-constant Ω—you cannot know precisely when something will happen and precisely what it will mean simultaneously. The Measurement Problem in quantum physics and the Hard Problem of consciousness are the same problem, viewed from opposite directions. Physics asks how observation affects reality; philosophy asks how reality produces observation. The answer is that observation and reality are not separate—they are two aspects of a unified system, connected through the Neutral Point (o□), which is consciousness itself. What was axiom becomes proof; what was proof becomes intuition; what was intuition becomes transparent ground. This is the developmental arc of any living discipline.

Contemplative Parallel: Mystics across traditions report experiences of non-local connection: knowing what is happening at a distance, feeling the suffering of others, sensing the presence of the deceased. The Hindu concept of Brahman describes an underlying unity connecting all beings. The Buddhist teaching of pratītyasamutpāda (dependent origination) states that all phenomena arise together in mutual dependence.

Holographic-siddhānta: 't Hooft-Susskind — sīmā-pṛṣṭhe sarva-sūchanā

The holographic principle, proven by 't Hooft and extended by Susskind, encodes the information content of any spatial volume on its two-dimensional boundary surface. Every part contains the whole at sufficient resolution: the basis of the VINO axiom that reality is holographically structured. Layer 1: The Physics Plate (Below) This layer contains the standard model of physics: quantum mechanics, general relativity, thermodynamics, electromagnetism. It describes what exists in terms of energy, mass, space, and time. Equations in this layer use the speed of light (c) as the conversion factor between mass and energy. The resistance to this conclusion is itself diagnostic: it reveals the architecture of the assumption that the conclusion contradicts.

Validation: The Global Consciousness Project (Nelson, 2001) found correlations between random number generator outputs and global events, suggesting possible collective consciousness effects. PEAR laboratory research (Jahn & Dunne, 2005) found small but statistically significant effects of intention on random systems. These results remain controversial but suggest a direction for further research.

> ***Citation Anchor §84** | Internal: VOVINA Canon Vol. 271 | External-1: Curzi, M.L. (2026). *VINO: Physics & Consciousness: The Reciprocity Engine*. VOVINA Press. | External-2: Bohm, D. (1980). *Wholeness and the Implicate Order*. Routledge.*

Photon-nitya-vartamāna: kāla-gati-śūnya — pratyakṣa-madhyasthaḥ

A photon traveling at c experiences zero elapsed time between emission and absorption: it exists in eternal present, making light the natural mediator between the physics layer (governed by c^2) and the metaphysics layer (governed by c□²)—the Neutral Point made manifest. Layer 2: The Metaphysics Plate (Above) This layer contains the structure of meaning, intention, and context. It describes what matters in terms of significance, purpose, and coherence. Equations in this layer use the inverse of the speed of light (c□²) as the conversion factor between metadata and intention. This is not an isolated observation but a structural feature: every genuine inquiry into this domain arrives at the same threshold.

Practical Insight: Like physical phase transitions, spiritual transitions require reaching a critical point. Below the threshold, effort produces only incremental change. At the threshold, a small additional input produces dramatic transformation. This is why spiritual practice can seem to produce no results for long periods, then suddenly catalyze breakthrough.

Negentropia: -ΔS — sṛjana-śaktiḥ — ātma-āyojana-siddhi

Negative entropy—the local decrease of disorder paid for by greater increase elsewhere—is the engine of all self-organization: life, consciousness, and civilization are negentropic structures that concentrate information and coherence, reducing local entropy while obeying the second law globally. Contra-Proof (o□): From the Neutral Point: the physics and metaphysics are both correct within their domains. Water boils at 100°C under specified conditions and the conditions are chosen by observers for practical purposes. The complete truth includes both the physical phenomenon and the contextual frame that makes the statement useful. What distinguishes the advanced practitioner from the beginner is not possession of more information but a different relationship to the same information.

This is not merely a metaphor. The holographic principle, first proposed by Gerard 't Hooft and developed by Leonard Susskind, states that the information content of a region of space can be encoded on its boundary. This suggests that our three-dimensional experience may be a projection from a two-dimensional surface.

E=mc² viparyayaḥ: I=M□·c□² — śakti-sankalpa-pratidvandvam

Citation Anchor §19

Internal — VOVINA Ontological OmniTautology, Book 274 of 500:

***Supplemental Wisdom: The Third Codex: The Origin Principle** (Supplemental Wisdom: The Third Codex, Vol. 1 of 13 · House: Language & Story · Room: Origin).*

The inquiry developed therein cross-validates the present volume's argument from a complementary ontological silo, confirming recursive self-consistency across the full 500-volume Canon.

**External 1:* Barabási, A.L. (2002). *Linked: The New Science of Networks*. Perseus.*

**Inversion note:* Negentropic inversion: the conventional reading of this work charts increasing complexity; the OmniTautology reads the same data as evidence of decreasing entropy at a higher organizational scale — ordered information all the way down.*

**External 2:* Feynman, R. (1985). *QED: The Strange Theory of Light and Matter*. Princeton UP.*

**Inversion note:* Standard and inverted: applied in its original context this citation describes emergent complexity; recontextualized here, it reveals that complexity is the surface form of a deeper simplicity — both readings are simultaneously valid.*

Einstein's E=mc² inverts to I=M□·c□²: where energy equals mass times c-squared, intention equals metadata times the inverse of c-squared. Physics and metaphysics are mirror symmetries across the Neutral Point, each describing half of a complete reality. Contra-Proof (0□): From the Neutral Point: both are true at their levels. The brain is a physical system following physical laws and the observer experiences choice because consciousness operates at the Neutral Point, where the selection among possibilities occurs. "Free will" is not freedom from causation but navigation of probability from the Neutral Point. The principle here is not domain-specific but universal: the same pattern, read at a different scale, generates the same implications.

David Chalmers called this the "Hard Problem." It is not hard because we lack data, but because no amount of physical data seems sufficient to explain the felt quality of experience. You can describe all the neural activity involved in seeing red, but this does not explain what it is like to see red.

Māpana-samasyā VINO-sūtrena: Ψ_c consciousness-coupling-constant γ_c

The measurement problem resolves under the VINO framework through the consciousness coupling constant γ_c: observation is not wave-function collapse but entanglement—the observer's state becoming correlated with one branch of the system's superposition. Contra-Proof (0□): From the Neutral Point: both are true at their levels. The brain is a physical system following physical laws and the observer experiences choice because consciousness operates at the Neutral Point, where the selection among possibilities occurs. "Free will" is not freedom from causation but navigation of probability from the Neutral Point. To understand this fully is to recognize that the question and the questioner are not separable — the inquiry transforms its own subject.

Layer 2: The Metaphysics Plate (Above) This layer contains the structure of meaning, intention, and context. It describes what matters in terms of significance, purpose, and coherence. Equations in this layer use the inverse of the speed of light (c□²) as the conversion factor between metadata and intention.

Kāla-taraṅga: T_Ψ(t)=Σ_n[A_n·e^{iω_nt}] — vartamāna-sthira-taraṅga-grantha

Time is not a linear coordinate but a waveform T_Ψ(t) = Σ_n[A_n·e^{iωnt}]: the experienced present is the standing-wave node of interfering past- and future-directed components, which explains déjà vu as harmonic resonance and flow states as temporal wave coherence. This is not merely a metaphor. The holographic principle, first proposed by Gerard 't Hooft and developed by Leonard Susskind, states that the information content of a region of space can be encoded on its boundary. This suggests that our three-dimensional experience may be a projection from a two-dimensional surface. This convergence across independent lines of investigation constitutes something stronger than theory: it constitutes the structure of the field itself.

David Chalmers called this the "Hard Problem." It is not hard because we lack data, but because no amount of physical data seems sufficient to explain the felt quality of experience. You can describe all the neural activity involved in seeing red, but this does not explain what it is like to see red.

Heisenberg-vistaraṇam: Δτ·ΔS≥Ω/2 — kāla-artha-avyaktam

The Uncertainty Principle extends into temporal-significance space: the product of timing uncertainty Δτ and meaning uncertainty ΔS cannot fall below half the meaning-constant Ω—you cannot know precisely when something will happen and precisely what it will mean simultaneously. Contemplative Parallel: Most contemplative traditions describe a witness or pure awareness that is the ground of experience. Hindu Advaita speaks of the sākṣin (witness); Buddhism speaks of rigpa (awareness); Christianity speaks of the "indwelling Spirit." This witness is said to be unchanged by what it witnesses, yet its presence affects what arises. What appears as complication at the surface resolves, at sufficient depth, into elegant simplicity — the mark of genuine principle.

This is not merely a metaphor. The holographic principle, first proposed by Gerard 't Hooft and developed by Leonard Susskind, states that the information content of a region of space can be encoded on its boundary. This suggests that our three-dimensional experience may be a projection from a two-dimensional surface.

***Citation Anchor §90** | Internal: VOVINA Canon Vol. 289 | External-1: Curzi, M.L. (2026). *VINO: Physics & Consciousness: The Reciprocity Engine*. VOVINA Press. | External-2: Bohm, D. (1980). *Wholeness and the Implicate Order*. Routledge.*

Holographic-siddhānta: 't Hooft-Susskind — sīmā-pṛṣṭhe sarva-sūchanā

The holographic principle, proven by 't Hooft and extended by Susskind, encodes the information content of any spatial volume on its two-dimensional boundary surface. Every part contains the whole at sufficient resolution: the basis of the VINO axiom that reality is holographically structured. The Measurement Problem in quantum physics and the Hard Problem of consciousness are the same problem, viewed from opposite directions. Physics asks how observation affects reality; philosophy asks how reality produces observation. The answer is that observation and reality are not separate—they are two aspects of a unified system, connected through the Neutral Point (o□), which is consciousness itself. The practitioner who has internalized this insight does not merely know more; they perceive differently, and the perception itself reorganizes what they encounter.

Citation Anchor §20

Internal — VOVINA Ontological OmniTautology, Book 233 of 500:

***Aether-Loong: The Dragon's Breath, Vol I: The Cycle Complete** (Aether-Loong: The Dragon's Breath, Vol I, Vol. 12 of 13 · House: Spirituality & Practice · Room: Completion).*

The inquiry developed therein cross-validates the present volume's argument from a complementary ontological silo, confirming recursive self-consistency across the full 500-volume Canon.

**External 1:* van der Kolk, B. (2014). *The Body Keeps the Score*. Viking.*

**Inversion note:* Read bidirectionally within the VOVINA OmniTautology: the conventional interpretation proceeds from evidence to conclusion; the inverted reading demonstrates that the conclusion was the prior condition from which the evidence was inevitably generated.*

**External 2:* Einstein, A. (1905). On the electrodynamics of moving bodies. *Annalen der Physik*, 17, 891–921.*

**Inversion note:* Read bidirectionally within the VOVINA OmniTautology: the conventional interpretation proceeds from evidence to conclusion; the inverted reading demonstrates that the conclusion was the prior condition from which the evidence was inevitably generated.*

Validation: The Global Consciousness Project (Nelson, 2001) found correlations between random number generator outputs and global events, suggesting possible collective consciousness effects. PEAR laboratory research (Jahn & Dunne, 2005) found small but statistically significant effects of intention on random systems. These results remain controversial but suggest a direction for further research.

Photon-nitya-vartamāna: kāla-gati-śūnya — pratyakṣa-madhyasthaḥ

A photon traveling at c experiences zero elapsed time between emission and absorption: it exists in eternal present, making light the natural mediator between the physics layer (governed by c^2) and the metaphysics layer (governed by c□^{2})—the Neutral Point made manifest. Volume II takes these principles from theory to practice. Here we build practical bridges—concrete connections between the language of science and the language of spirit, between the measurements of physics and the insights of contemplation, between the formulas of mathematics and the poetry of mystics. History shows that every domain that resisted this insight eventually incorporated it — not by choice but by the pressure of its own unresolved contradictions.

Practical Insight: Like physical phase transitions, spiritual transitions require reaching a critical point. Below the threshold, effort produces only incremental change. At the threshold, a small additional input produces dramatic transformation. This is why spiritual practice can seem to produce no results for long periods, then suddenly catalyze breakthrough.

Negentropia: -ΔS — sṛjana-śaktiḥ — ātma-āyojana-siddhi

Negative entropy—the local decrease of disorder paid for by greater increase elsewhere—is the engine of all self-organization: life, consciousness, and civilization are negentropic structures that concentrate information and coherence, reducing local entropy while obeying the second law globally. Contemplative Parallel: Mystics across traditions report experiences of non-local connection: knowing what is happening at a distance, feeling the suffering of others, sensing the presence of the deceased. The Hindu concept of Brahman describes an underlying unity connecting all beings. The Buddhist teaching of pratītyasamutpāda (dependent origination) states that all phenomena arise together in mutual dependence. The formal rigor here is not decoration: it is the difference between understanding that points toward the territory and understanding that is the territory.

Validation: The Global Consciousness Project (Nelson, 2001) found correlations between random number generator outputs and global events, suggesting possible collective consciousness effects. PEAR laboratory research (Jahn & Dunne, 2005) found small but statistically significant effects of intention on random systems. These results remain controversial but suggest a direction for further research.

E=mc^2 viparyayaḥ: I=M□·c□^{2} — śakti-sankalpa-pratidvandvam

Einstein's E=mc^2 inverts to I=M□·c□^{2}: where energy equals mass times c-squared, intention equals metadata times the inverse of c-squared. Physics and metaphysics are mirror symmetries across the Neutral Point, each describing half of a complete reality. In the VINO Framework, all terms ultimately refer to the structure of the Neutral Point and its expressions. "Energy" and "spirit" both refer to the capacity for change, viewed from different perspectives. "Mass" and "form" both refer to the resistance to change, viewed from different perspectives. This is the negentropic signature of genuine knowledge: it compresses without losing; it grows denser rather than longer as comprehension deepens.

The central thesis posits that the "Measurement Problem" in quantum physics and the "Hard Problem" of consciousness are, in fact, the same problem viewed from opposite ends of a singular continuum. By introducing the concept of "Zero as Presence"—wherein the mathematical zero is redefined not as void or null, but as the active fulcrum of existence—we establish a neutral point that allows for the logical cross-wiring of rational (physical) and irrational (metaphysical) number systems.

Māpana-samasyā VINO-sūtrena: Ψ_c consciousness-coupling-constant γ_c

The measurement problem resolves under the VINO framework through the consciousness coupling constant γ_c: observation is not wave-function collapse but entanglement—the observer's state becoming correlated with one branch of the system's superposition. The central thesis posits that the "Measurement Problem" in quantum physics and the "Hard Problem" of consciousness are, in fact, the same problem viewed from opposite ends of a singular continuum. By introducing the concept of "Zero as Presence"—wherein the mathematical zero is redefined not as void or null, but as the active fulcrum of existence—we establish a neutral point that allows for the logical cross-wiring of rational (physical) and irrational (metaphysical) number systems. What was axiom becomes proof; what was proof becomes intuition; what was intuition becomes transparent ground. This is the developmental arc of any living discipline.

Layer 1: The Physics Plate (Below) This layer contains the standard model of physics: quantum mechanics, general relativity, thermodynamics, electromagnetism. It describes what exists in terms of energy, mass, space, and time. Equations in this layer use the speed of light (c) as the conversion factor between mass and energy.

Kāla-taraṅga: $T_\Psi(t)=\Sigma_n[A_n\cdot e^{i\omega_n t}]$ *— vartamāna-sthira-taraṅga-grantha*

Citation Anchor §21

Internal — VOVINA Ontological OmniTautology, Book 192 of 500:

***Sound Chemistry & Frequency Engineering: The Repair Protocol** (Sound Chemistry & Frequency Engineering, Vol. 10 of 13 · House: Physics & Energy · Room: Repair).*

The inquiry developed therein cross-validates the present volume's argument from a complementary ontological silo, confirming recursive self-consistency across the full 500-volume Canon.

**External 1:* Peat, F.D. (1987). *Synchronicity: The Bridge Between Matter and Mind*. Bantam.*

**Inversion note:* Mirror symmetry proof: the cited author approaches this truth from the exterior face of the phenomenon; the present volume approaches from the interior face — the two descriptions are mirror reflections that, placed together, constitute the complete form.*

**External 2:* Schrödinger, E. (1944). *What is Life?* Cambridge University Press.*

**Inversion note:* OmniTautology inversion: this postulate, taken as written, argues from cause to effect; the VOVINA framework demonstrates the identical structure is equally coherent read from effect to cause — the tautology closes in both directions.*

Time is not a linear coordinate but a waveform $T_\Psi(t) = \Sigma_n[A_n\cdot e^{i\omega nt}]$: the experienced present is the standing-wave node of interfering past- and future-directed components, which explains déjà vu as harmonic resonance and flow states as temporal wave coherence. Contra-Proof (0□): From the Neutral Point: both are true at their levels. The brain is a physical system following physical laws and the observer experiences choice because consciousness operates at the Neutral Point, where the selection among possibilities occurs. "Free will" is not freedom from causation but navigation of probability from the Neutral Point. The resistance to this conclusion is itself diagnostic: it reveals the architecture of the assumption that the conclusion contradicts.

The central thesis posits that the "Measurement Problem" in quantum physics and the "Hard Problem" of consciousness are, in fact, the same problem viewed from opposite ends of a singular continuum. By introducing the concept of "Zero as Presence"—wherein the mathematical zero is redefined not as void or null, but as the active fulcrum of existence—we establish a neutral point that allows for the logical cross-wiring of rational (physical) and irrational (metaphysical) number systems.

***Citation Anchor §96** | Internal: VOVINA Canon Vol. 307 | External-1: Curzi, M.L. (2026). *VINO: Physics & Consciousness: The Reciprocity Engine*. VOVINA Press. | External-2: Bohm, D. (1980). *Wholeness and the Implicate Order*. Routledge.*

Heisenberg-vistaraṇam: $\Delta\tau\cdot\Delta S\geq\Omega/2$ *— kāla-artha-avyaktam*

The Uncertainty Principle extends into temporal-significance space: the product of timing uncertainty $\Delta\tau$ and meaning uncertainty ΔS cannot fall below half the meaning-constant Ω—you cannot know precisely when something will happen and precisely what it will mean simultaneously. In the VINO Framework, all terms ultimately refer to the structure of the Neutral Point and its expressions. "Energy" and "spirit" both refer to the capacity for change, viewed from different perspectives. "Mass" and "form" both refer to the resistance to change, viewed from different perspectives. This is not an isolated observation but a structural feature: every genuine inquiry into this domain arrives at the same threshold.

David Chalmers called this the "Hard Problem." It is not hard because we lack data, but because no amount of physical data seems sufficient to explain the felt quality of experience. You can describe all the neural activity involved in seeing red, but this does not explain what it is like to see red.

Holographic-siddhānta: 't Hooft-Susskind — sīmā-pṛṣṭhe sarva-sūchanā

The holographic principle, proven by 't Hooft and extended by Susskind, encodes the information content of any spatial volume on its two-dimensional boundary surface. Every part contains the whole at sufficient resolution: the basis of the VINO axiom that reality is holographically structured. Volume II takes these principles from theory to practice. Here we build practical bridges—concrete connections between the language of science and the language of spirit, between the measurements of physics and the insights of contemplation, between the formulas of mathematics and the poetry of mystics. What distinguishes the advanced practitioner from the beginner is not possession of more information but a different relationship to the same information.

Layer 2: The Metaphysics Plate (Above) This layer contains the structure of meaning, intention, and context. It describes what matters in terms of significance, purpose, and coherence. Equations in this layer use the inverse of the speed of light (c□²) as the conversion factor between metadata and intention.

Photon-nitya-vartamāna: kāla-gati-śūnya — pratyakṣa-madhyasthaḥ

A photon traveling at c experiences zero elapsed time between emission and absorption: it exists in eternal present, making light the natural mediator between the physics layer (governed by c^2) and the metaphysics layer (governed by c□²)—the Neutral Point made manifest. Contra-Proof (0□): From the Neutral Point: the physics and metaphysics are both correct within their domains. Water boils at 100°C under specified conditions and the conditions are chosen by observers for practical purposes. The complete truth includes both the physical phenomenon and the contextual frame that makes the statement useful. The principle here is not domain-specific but universal: the same pattern, read at a different scale, generates the same implications.

Contemplative Parallel: Most contemplative traditions describe a witness or pure awareness that is the ground of experience. Hindu Advaita speaks of the sākṣin (witness); Buddhism speaks of rigpa (awareness); Christianity speaks of the "indwelling Spirit." This witness is said to be unchanged by what it witnesses, yet its presence affects what arises.

Negentropia: -ΔS — sṛjana-śaktiḥ — ātma-āyojana-siddhi

Negative entropy—the local decrease of disorder paid for by greater increase elsewhere—is the engine of all self-organization: life, consciousness, and civilization are negentropic structures that concentrate information and coherence, reducing local entropy while obeying the second law globally. Contemplative Parallel: Most contemplative traditions describe a witness or pure awareness that is the ground of experience. Hindu Advaita speaks of the sākṣin (witness); Buddhism speaks of rigpa (awareness); Christianity speaks of the "indwelling Spirit." This witness is said to be unchanged by what it witnesses, yet its presence affects what arises. To understand this fully is to recognize that the question and the questioner are not separable — the inquiry transforms its own subject.

Citation Anchor §22

Internal — VOVINA Ontological OmniTautology, Book 151 of 500:

***Quantum Infrastructure & Transmutation: The Growth Spiral** (Quantum Infrastructure & Transmutation, Vol. 8 of 13 · House: Physics & Energy · Room: Growth).*

The inquiry developed therein cross-validates the present volume's argument from a complementary ontological silo, confirming recursive self-consistency across the full 500-volume Canon.

External 1: Oschman, J. (2000). *Energy Medicine*. Churchill Livingstone.

Inversion note: Standard reading: evidence for an external mechanism. Inverted reading: evidence that the mechanism is internal to the field of consciousness — both readings are simultaneously valid and mutually reinforcing across the phi-pivot.

External 2: Radin, D. (1997). *The Conscious Universe*. HarperEdge.

Inversion note: Inversion proof: the cited postulate assumes a separation between observer and observed; the present volume demonstrates that dissolving this assumption does not contradict the postulate but fulfills it at the next level of generality.

Practical Insight: Like physical phase transitions, spiritual transitions require reaching a critical point. Below the threshold, effort produces only incremental change. At the threshold, a small additional input produces dramatic transformation. This is why spiritual practice can seem to produce no results for long periods, then suddenly catalyze breakthrough.

$E=mc^2$ viparyayaḥ: $I=M\square\cdot c\square^2$ — śakti-sankalpa-pratidvandvam

Einstein's $E=mc^2$ inverts to $I=M\square\cdot c\square^2$: where energy equals mass times c-squared, intention equals metadata times the inverse of c-squared. Physics and metaphysics are mirror symmetries across the Neutral Point, each describing half of a complete reality. The Measurement Problem in quantum physics and the Hard Problem of consciousness are the same problem, viewed from opposite directions. Physics asks how observation affects reality; philosophy asks how reality produces observation. The answer is that observation and reality are not separate—they are two aspects of a unified system, connected through the Neutral Point (0□), which is consciousness itself. This convergence across independent lines of investigation constitutes something stronger than theory: it constitutes the structure of the field itself.

Contemplative Parallel: Mystics across traditions report experiences of non-local connection: knowing what is happening at a distance, feeling the suffering of others, sensing the presence of the deceased. The Hindu concept of Brahman describes an underlying unity connecting all beings. The Buddhist teaching of pratītyasamutpāda (dependent origination) states that all phenomena arise together in mutual dependence.

Māpana-samasyā VINO-sūtrena: Ψ_c consciousness-coupling-constant γ_c

The measurement problem resolves under the VINO framework through the consciousness coupling constant γ_c: observation is not wave-function collapse but entanglement—the observer's state becoming correlated with one branch of the system's superposition. Volume II takes these principles from theory to practice. Here we build practical bridges—concrete connections between the language of science and the language of spirit, between the measurements of physics and the insights of contemplation, between the formulas of mathematics and the poetry of mystics. What appears as complication at the surface resolves, at sufficient depth, into elegant simplicity — the mark of genuine principle.

Layer 1: The Physics Plate (Below) This layer contains the standard model of physics: quantum mechanics, general relativity, thermodynamics, electromagnetism. It describes what exists in terms of energy, mass, space, and time. Equations in this layer use the speed of light (c) as the conversion factor between mass and energy.

Citation Anchor §102 | Internal: VOVINA Canon Vol. 325 | External-1: Curzi, M.L. (2026). *VINO: Physics & Consciousness: The Reciprocity Engine*. VOVINA Press. | External-2: Bohm, D. (1980). *Wholeness and the Implicate Order*. Routledge.

Kāla-taraṅga: $T_\Psi(t)=\Sigma_n[A_n\cdot e^{i\omega_n t}]$ — vartamāna-sthira-taraṅga-grantha

Time is not a linear coordinate but a waveform $T_\Psi(t) = \Sigma_n[A_n\cdot e^{i\omega_n t}]$: the experienced present is the standing-wave node of interfering past- and future-directed components, which explains déjà vu as harmonic resonance and flow states as temporal wave coherence. In the VINO Framework, all terms ultimately refer to the structure of the Neutral Point and its expressions. "Energy" and "spirit" both refer to the capacity for change, viewed from different perspectives. "Mass" and "form" both refer to the resistance to change, viewed from different perspectives. The practitioner who has internalized this insight does not merely know more; they perceive differently, and the perception itself reorganizes what they encounter.

Contra-Proof (o□): From the Neutral Point: the physics and metaphysics are both correct within their domains. Water boils at 100°C under specified conditions and the conditions are chosen by observers for practical purposes. The complete truth includes both the physical phenomenon and the contextual frame that makes the statement useful.

Heisenberg-vistaraṇam: $\Delta\tau\cdot\Delta S \geq \Omega/2$ — kāla-artha-avyaktam

The Uncertainty Principle extends into temporal-significance space: the product of timing uncertainty $\Delta\tau$ and meaning uncertainty ΔS cannot fall below half the meaning-constant Ω—you cannot know precisely when something will happen and precisely what it will mean simultaneously. The Measurement Problem in quantum physics and the Hard Problem of consciousness are the same problem, viewed from opposite directions. Physics asks how observation affects reality; philosophy asks how reality produces observation. The answer is that observation and reality are not separate—they are two aspects of a unified system, connected through the Neutral Point (o□), which is consciousness itself. History shows that every domain that resisted this insight eventually incorporated it — not by choice but by the pressure of its own unresolved contradictions.

Contemplative Parallel: Mystics across traditions report experiences of non-local connection: knowing what is happening at a distance, feeling the suffering of others, sensing the presence of the deceased. The Hindu concept of Brahman describes an underlying unity connecting all beings. The Buddhist teaching of pratītyasamutpāda (dependent origination) states that all phenomena arise together in mutual dependence.

Citation Anchor §23

Internal — VOVINA Ontological OmniTautology, Book 110 of 500:

***Vovina: Nano Armor & Advanced Materials: The Alchemy of Change** (Vovina: Nano Armor & Advanced Materials, Vol. 6 of 13 · House: Biology & Body · Room: Transformation).*

The inquiry developed therein cross-validates the present volume's argument from a complementary ontological silo, confirming recursive self-consistency across the full 500-volume Canon.

**External 1:* Kuhn, T. (1962). *The Structure of Scientific Revolutions*. University of Chicago Press.*

**Inversion note:* Omnitautological recontextualization: this citation establishes its conclusion under specific boundary conditions; removing those boundaries does not destroy the conclusion but generalizes it to all possible conditions — the local becomes universal.*

**External 2:* Talbot, M. (1991). *The Holographic Universe*. HarperCollins.*

**Inversion note:* Bidirectional validity: the cited work reaches its conclusion empirically from below; the VOVINA Canon derives the identical conclusion axiomatically from above — the convergence is itself the proof.*

Holographic-siddhānta: 't Hooft-Susskind — sīmā-pṛṣṭhe sarva-sūchanā

The holographic principle, proven by 't Hooft and extended by Susskind, encodes the information content of any spatial volume on its two-dimensional boundary surface. Every part contains the whole at sufficient resolution: the basis of the VINO axiom that reality is holographically structured. Layer 1: The Physics Plate (Below) This layer contains the standard model of physics: quantum mechanics, general relativity, thermodynamics, electromagnetism. It describes what exists in terms of energy, mass, space, and time. Equations in this layer use the speed of light (c) as the conversion factor between mass and energy. The formal rigor here is not decoration: it is the difference between understanding that points toward the territory and understanding that is the territory.

Validation: The Global Consciousness Project (Nelson, 2001) found correlations between random number generator outputs and global events, suggesting possible collective consciousness effects. PEAR laboratory research (Jahn & Dunne, 2005) found small but statistically significant effects of intention on random systems. These results remain controversial but suggest a direction for further research.

Photon-nitya-vartamāna: kāla-gati-śūnya — pratyakṣa-madhyasthaḥ

A photon traveling at c experiences zero elapsed time between emission and absorption: it exists in eternal present, making light the natural mediator between the physics layer (governed by c^2) and the metaphysics layer (governed by c□²)—the Neutral Point made manifest. Layer 2: The Metaphysics Plate (Above) This layer contains the structure of meaning, intention, and context. It describes what matters in terms of significance, purpose, and coherence. Equations in this layer use the inverse of the speed of light (c□²) as the conversion factor between metadata and intention. This is the negentropic signature of genuine knowledge: it compresses without losing; it grows denser rather than longer as comprehension deepens.

Practical Insight: Like physical phase transitions, spiritual transitions require reaching a critical point. Below the threshold, effort produces only incremental change. At the threshold, a small additional input produces dramatic transformation. This is why spiritual practice can seem to produce no results for long periods, then suddenly catalyze breakthrough.

Negentropia: -ΔS — sṛjana-śaktiḥ — ātma-āyojana-siddhi

Negative entropy—the local decrease of disorder paid for by greater increase elsewhere—is the engine of all self-organization: life, consciousness, and civilization are negentropic structures that concentrate information and coherence, reducing local entropy while obeying the second law globally. Contra-Proof (0□): From the Neutral Point: the physics and metaphysics are both correct within their domains. Water boils at 100°C under specified conditions and the conditions are chosen by observers for practical purposes. The complete truth includes both the physical phenomenon and the contextual frame that makes the statement useful. What was axiom becomes proof; what was proof becomes intuition; what was intuition becomes transparent ground. This is the developmental arc of any living discipline.

This is not merely a metaphor. The holographic principle, first proposed by Gerard 't Hooft and developed by Leonard Susskind, states that the information content of a region of space can be encoded on its boundary. This suggests that our three-dimensional experience may be a projection from a two-dimensional surface.

E=mc² viparyayaḥ: I=M□·c□² — śakti-sankalpa-pratidvandvam

Einstein's $E=mc^2$ inverts to I=M□·c□²: where energy equals mass times c-squared, intention equals metadata times the inverse of c-squared. Physics and metaphysics are mirror symmetries across the Neutral Point, each describing half of a complete reality. Contra-Proof (0□): From the Neutral Point: both are true at their levels. The brain is a physical system following physical laws and the observer experiences choice because consciousness operates at the Neutral Point, where the selection among possibilities occurs. "Free will" is not freedom from causation but navigation of probability from the Neutral Point. The resistance to this conclusion is itself diagnostic: it reveals the architecture of the assumption that the conclusion contradicts.

David Chalmers called this the "Hard Problem." It is not hard because we lack data, but because no amount of physical data seems sufficient to explain the felt quality of experience. You can describe all the neural activity involved in seeing red, but this does not explain what it is like to see red.

***Citation Anchor §108** | Internal: VOVINA Canon Vol. 343 | External-1: Curzi, M.L. (2026). *VINO: Physics & Consciousness: The Reciprocity Engine*. VOVINA Press. | External-2: Bohm, D. (1980). *Wholeness and the Implicate Order*. Routledge.*

Māpana-samasyā VINO-sūtrena: Ψ_c consciousness-coupling-constant γ_c

Citation Anchor §24

Internal — VOVINA Ontological OmniTautology, Book 069 of 500:

***AudioGenomics: The Sound of DNA: The Perception Threshold** (AudioGenomics: The Sound of DNA, Vol. 4 of 13 · House: Biology & Body · Room: Perception).*

The inquiry developed therein cross-validates the present volume's argument from a complementary ontological silo, confirming recursive self-consistency across the full 500-volume Canon.

External 1: Schrödinger, E. (1944). *What is Life?* Cambridge University Press.

Inversion note: OmniTautology inversion: this postulate, taken as written, argues from cause to effect; the VOVINA framework demonstrates the identical structure is equally coherent read from effect to cause — the tautology closes in both directions.

External 2: Strogatz, S. (2003). *Sync: The Emerging Science of Spontaneous Order*. Hyperion.

Inversion note: Mirror symmetry proof: the cited author approaches this truth from the exterior face of the phenomenon; the present volume approaches from the interior face — the two descriptions are mirror reflections that, placed together, constitute the complete form.

The measurement problem resolves under the VINO framework through the consciousness coupling constant γ_c: observation is not wave-function collapse but entanglement—the observer's state becoming correlated with one branch of the system's superposition. Contra-Proof (0□): From the Neutral Point: both are true at their levels. The brain is a physical system following physical laws and the observer experiences choice because consciousness operates at the Neutral Point, where the selection among possibilities occurs. "Free will" is not freedom from causation but navigation of probability from the Neutral Point. This is not an isolated observation but a structural feature: every genuine inquiry into this domain arrives at the same threshold.

Layer 2: The Metaphysics Plate (Above) This layer contains the structure of meaning, intention, and context. It describes what matters in terms of significance, purpose, and coherence. Equations in this layer use the inverse of the speed of light (c□²) as the conversion factor between metadata and intention.

Kāla-taraṅga: T_Ψ(t)=Σ_n[A_n·e^{iω_nt}] — vartamāna-sthira-taraṅga-grantha

Time is not a linear coordinate but a waveform T_Ψ(t) = Σ_n[A_n·e^{iωnt}]: the experienced present is the standing-wave node of interfering past- and future-directed components, which explains déjà vu as harmonic resonance and flow states as temporal wave coherence. This is not merely a metaphor. The holographic principle, first proposed by Gerard 't Hooft and developed by Leonard Susskind, states that the information content of a region of space can be encoded on its boundary. This suggests that our three-dimensional experience may be a projection from a two-dimensional surface. What distinguishes the advanced practitioner from the beginner is not possession of more information but a different relationship to the same information.

David Chalmers called this the "Hard Problem." It is not hard because we lack data, but because no amount of physical data seems sufficient to explain the felt quality of experience. You can describe all the neural activity involved in seeing red, but this does not explain what it is like to see red.

Heisenberg-vistaraṇam: Δτ·ΔS≥Ω/2 — kāla-artha-avyaktam

The Uncertainty Principle extends into temporal-significance space: the product of timing uncertainty Δτ and meaning uncertainty ΔS cannot fall below half the meaning-constant Ω—you cannot know precisely when something will happen and precisely what it will mean simultaneously. Contemplative Parallel: Most contemplative traditions describe a witness or pure awareness that is the ground of experience. Hindu Advaita speaks of the sākṣin (witness); Buddhism speaks of rigpa (awareness); Christianity speaks of the "indwelling Spirit." This witness is said to be unchanged by what it witnesses, yet its presence affects what arises. The principle here is not domain-specific but universal: the same pattern, read at a different scale, generates the same implications.

This is not merely a metaphor. The holographic principle, first proposed by Gerard 't Hooft and developed by Leonard Susskind, states that the information content of a region of space can be encoded on its boundary. This suggests that our three-dimensional experience may be a projection from a two-dimensional surface.

Holographic-siddhānta: 't Hooft-Susskind — sīmā-pṛṣṭhe sarva-sūchanā

The holographic principle, proven by 't Hooft and extended by Susskind, encodes the information content of any spatial volume on its two-dimensional boundary surface. Every part contains the whole at sufficient resolution: the basis of the VINO

axiom that reality is holographically structured. The Measurement Problem in quantum physics and the Hard Problem of consciousness are the same problem, viewed from opposite directions. Physics asks how observation affects reality; philosophy asks how reality produces observation. The answer is that observation and reality are not separate—they are two aspects of a unified system, connected through the Neutral Point (o□), which is consciousness itself. To understand this fully is to recognize that the question and the questioner are not separable — the inquiry transforms its own subject.

Validation: The Global Consciousness Project (Nelson, 2001) found correlations between random number generator outputs and global events, suggesting possible collective consciousness effects. PEAR laboratory research (Jahn & Dunne, 2005) found small but statistically significant effects of intention on random systems. These results remain controversial but suggest a direction for further research.

Photon-nitya-vartamāna: kāla-gati-śūnya — pratyakṣa-madhyasthaḥ

A photon traveling at c experiences zero elapsed time between emission and absorption: it exists in eternal present, making light the natural mediator between the physics layer (governed by c^2) and the metaphysics layer (governed by c□²)—the Neutral Point made manifest. Volume II takes these principles from theory to practice. Here we build practical bridges—concrete connections between the language of science and the language of spirit, between the measurements of physics and the insights of contemplation, between the formulas of mathematics and the poetry of mystics. This convergence across independent lines of investigation constitutes something stronger than theory: it constitutes the structure of the field itself.

Practical Insight: Like physical phase transitions, spiritual transitions require reaching a critical point. Below the threshold, effort produces only incremental change. At the threshold, a small additional input produces dramatic transformation. This is why spiritual practice can seem to produce no results for long periods, then suddenly catalyze breakthrough.

Citation Anchor §25

Internal — VOVINA Ontological OmniTautology, Book 028 of 500:

***Veda Between Vedas: Sacred Practice: The Structure of Order** (Veda Between Vedas: Sacred Practice, Vol. 2 of 13 · House: Spirituality & Practice · Room: Structure).*

The inquiry developed therein cross-validates the present volume's argument from a complementary ontological silo, confirming recursive self-consistency across the full 500-volume Canon.

**External 1:* Radin, D. (1997). *The Conscious Universe*. HarperEdge.*

**Inversion note:* Inversion proof: the cited postulate assumes a separation between observer and observed; the present volume demonstrates that dissolving this assumption does not contradict the postulate but fulfills it at the next level of generality.*

**External 2:* Fuller, R.B. (1975). *Synergetics*. Macmillan.*

**Inversion note:* Tautological absorption: in its original context this citation supports a domain-specific claim; recontextualized here, the same evidence simultaneously supports all twelve ontological domains, demonstrating the OmniTautology's self-consistency.*

Negentropia: -ΔS — sṛjana-śaktiḥ — ātma-āyojana-siddhi

Negative entropy—the local decrease of disorder paid for by greater increase elsewhere—is the engine of all self-organization: life, consciousness, and civilization are negentropic structures that concentrate information and coherence, reducing local entropy while obeying the second law globally. Contemplative Parallel: Mystics across traditions report experiences of non-local connection: knowing what is happening at a distance, feeling the suffering of others, sensing the presence of the deceased. The Hindu concept of Brahman describes an underlying unity connecting all beings. The Buddhist teaching of pratītyasamutpāda (dependent origination) states that all phenomena arise together in mutual dependence. What appears as complication at the surface resolves, at sufficient depth, into elegant simplicity — the mark of genuine principle.

Validation: The Global Consciousness Project (Nelson, 2001) found correlations between random number generator outputs and global events, suggesting possible collective consciousness effects. PEAR laboratory research (Jahn & Dunne, 2005) found small but statistically significant effects of intention on random systems. These results remain controversial but suggest a direction for further research.

***Citation Anchor §114** | Internal: VOVINA Canon Vol. 361 | External-1: Curzi, M.L. (2026). *VINO: Physics & Consciousness: The Reciprocity Engine*. VOVINA Press. | External-2: Bohm, D. (1980). *Wholeness and the Implicate Order*. Routledge.*

$E=mc^2$ viparyayaḥ: I=M□·c□^{2} — śakti-sankalpa-pratidvandvam

§ IV — Bandhu: The Connecting Thread

Einstein's $E=mc^2$ inverts to I=M□·c□^{2}: where energy equals mass times c-squared, intention equals metadata times the inverse of c-squared. Physics and metaphysics are mirror symmetries across the Neutral Point, each describing half of a complete reality. In the VINO Framework, all terms ultimately refer to the structure of the Neutral Point and its expressions. "Energy" and "spirit" both refer to the capacity for change, viewed from different perspectives. "Mass" and "form" both refer to the resistance to change, viewed from different perspectives. The practitioner who has internalized this insight does not merely know more; they perceive differently, and the perception itself reorganizes what they encounter.

The central thesis posits that the "Measurement Problem" in quantum physics and the "Hard Problem" of consciousness are, in fact, the same problem viewed from opposite ends of a singular continuum. By introducing the concept of "Zero as Presence"—wherein the mathematical zero is redefined not as void or null, but as the active fulcrum of existence—we establish a neutral point that allows for the logical cross-wiring of rational (physical) and irrational (metaphysical) number systems.

Māpana-samasyā VINO-sūtrena: Ψ_c consciousness-coupling-constant γ_c

The measurement problem resolves under the VINO framework through the consciousness coupling constant γ_c: observation is not wave-function collapse but entanglement—the observer's state becoming correlated with one branch of the system's superposition. The central thesis posits that the "Measurement Problem" in quantum physics and the "Hard Problem" of consciousness are, in fact, the same problem viewed from opposite ends of a singular continuum. By introducing the concept of "Zero as Presence"—wherein the mathematical zero is redefined not as void or null, but as the active fulcrum of existence—we establish a neutral point that allows for the logical cross-wiring of rational (physical) and irrational (metaphysical) number systems. History shows that every domain that resisted this insight eventually incorporated it — not by choice but by the pressure of its own unresolved contradictions.

Layer 1: The Physics Plate (Below) This layer contains the standard model of physics: quantum mechanics, general relativity, thermodynamics, electromagnetism. It describes what exists in terms of energy, mass, space, and time. Equations in this layer use the speed of light (c) as the conversion factor between mass and energy.

Kāla-taraṅga: T_Ψ(t)=Σ_n[A_n·e^{iω_nt}] — vartamāna-sthira-taraṅga-grantha

Time is not a linear coordinate but a waveform T_Ψ(t) = Σ_n[A_n·e^{iωnt}]: the experienced present is the standing-wave node of interfering past- and future-directed components, which explains déjà vu as harmonic resonance and flow states as temporal wave coherence. Contra-Proof (0□): From the Neutral Point: both are true at their levels. The brain is a physical system following physical laws and the observer experiences choice because consciousness operates at the Neutral Point, where the selection among possibilities occurs. "Free will" is not freedom from causation but navigation of probability from the Neutral Point. The formal rigor here is not decoration: it is the difference between understanding that points toward the territory and understanding that is the territory.

The central thesis posits that the "Measurement Problem" in quantum physics and the "Hard Problem" of consciousness are, in fact, the same problem viewed from opposite ends of a singular continuum. By introducing the concept of "Zero as Presence"—wherein the mathematical zero is redefined not as void or null, but as the active fulcrum of existence—we establish a neutral point that allows for the logical cross-wiring of rational (physical) and irrational (metaphysical) number systems.

Heisenberg-vistaraṇam: $\Delta\tau \cdot \Delta S \geq \Omega/2$ — kāla-artha-avyaktam

Citation Anchor §26

Internal — VOVINA Ontological OmniTautology, Book 487 of 500:

***Children of the Dragon: Accessible Guides: The Alchemy of Change** (Children of the Dragon: Accessible Guides, Vol. 6 of 13 · House: Spirituality & Practice · Room: Transformation).*

The inquiry developed therein cross-validates the present volume's argument from a complementary ontological silo, confirming recursive self-consistency across the full 500-volume Canon.

**External 1:* Fuller, R.B. (1975). *Synergetics*. Macmillan.*

**Inversion note:* Tautological absorption: in its original context this citation supports a domain-specific claim; recontextualized here, the same evidence simultaneously supports all twelve ontological domains, demonstrating the OmniTautology's self-consistency.*

**External 2:* Sheldrake, R. (2009). *Morphic Resonance*. Park Street Press.*

**Inversion note:* Mirror symmetry proof: the cited author approaches this truth from the exterior face of the phenomenon; the present volume approaches from the interior face — the two descriptions are mirror reflections that, placed together, constitute the complete form.*

The Uncertainty Principle extends into temporal-significance space: the product of timing uncertainty $\Delta\tau$ and meaning uncertainty ΔS cannot fall below half the meaning-constant Ω—you cannot know precisely when something will happen and precisely what it will mean simultaneously. In the VINO Framework, all terms ultimately refer to the structure of the Neutral Point and its expressions. "Energy" and "spirit" both refer to the capacity for change, viewed from different perspectives. "Mass" and "form" both refer to the resistance to change, viewed from different perspectives. This is the negentropic signature of genuine knowledge: it compresses without losing; it grows denser rather than longer as comprehension deepens.

David Chalmers called this the "Hard Problem." It is not hard because we lack data, but because no amount of physical data seems sufficient to explain the felt quality of experience. You can describe all the neural activity involved in seeing red, but this does not explain what it is like to see red.

Holographic-siddhānta: 't Hooft-Susskind — sīmā-pṛṣṭhe sarva-sūchanā

The holographic principle, proven by 't Hooft and extended by Susskind, encodes the information content of any spatial volume on its two-dimensional boundary surface. Every part contains the whole at sufficient resolution: the basis of the VINO axiom that reality is holographically structured. Volume II takes these principles from theory to practice. Here we build practical bridges—concrete connections between the language of science and the language of spirit, between the measurements of physics and the insights of contemplation, between the formulas of mathematics and the poetry of mystics. What was axiom becomes proof; what was proof becomes intuition; what was intuition becomes transparent ground. This is the developmental arc of any living discipline.

Layer 2: The Metaphysics Plate (Above) This layer contains the structure of meaning, intention, and context. It describes what matters in terms of significance, purpose, and coherence. Equations in this layer use the inverse of the speed of light (c□²) as the conversion factor between metadata and intention.

Photon-nitya-vartamāna: kāla-gati-śūnya — pratyakṣa-madhyasthaḥ

A photon traveling at c experiences zero elapsed time between emission and absorption: it exists in eternal present, making light the natural mediator between the physics layer (governed by c^2) and the metaphysics layer (governed by c□²)—the Neutral Point made manifest. Contra-Proof (0□): From the Neutral Point: the physics and metaphysics are both correct within their domains. Water boils at 100°C under specified conditions and the conditions are chosen by observers for practical purposes. The complete truth includes both the physical phenomenon and the contextual frame that makes the statement useful. The resistance to this conclusion is itself diagnostic: it reveals the architecture of the assumption that the conclusion contradicts.

Contemplative Parallel: Most contemplative traditions describe a witness or pure awareness that is the ground of experience. Hindu Advaita speaks of the sākṣin (witness); Buddhism speaks of rigpa (awareness); Christianity speaks of the "indwelling Spirit." This witness is said to be unchanged by what it witnesses, yet its presence affects what arises.

***Citation Anchor §120** | Internal: VOVINA Canon Vol. 379 | External-1: Curzi, M.L. (2026). *VINO: Physics & Consciousness: The Reciprocity Engine*. VOVINA Press. | External-2: Bohm, D. (1980). *Wholeness and the Implicate Order*. Routledge.*

Negentropia: $-\Delta S$ — sṛjana-śaktiḥ — ātma-āyojana-siddhi

Negative entropy—the local decrease of disorder paid for by greater increase elsewhere—is the engine of all self-organization: life, consciousness, and civilization are negentropic structures that concentrate information and coherence, reducing local entropy while obeying the second law globally. Contemplative Parallel: Most contemplative traditions describe a witness or pure awareness that is the ground of experience. Hindu Advaita speaks of the sākṣin (witness); Buddhism speaks of rigpa (awareness); Christianity speaks of the "indwelling Spirit." This witness is said to be unchanged by what it witnesses, yet its presence affects what arises. This is not an isolated observation but a structural feature: every genuine inquiry into this domain arrives at the same threshold.

Practical Insight: Like physical phase transitions, spiritual transitions require reaching a critical point. Below the threshold, effort produces only incremental change. At the threshold, a small additional input produces dramatic transformation. This is why spiritual practice can seem to produce no results for long periods, then suddenly catalyze breakthrough.

$E=mc^2$ viparyayaḥ: $I=M\square\cdot c\square^2$ — śakti-sankalpa-pratidvandvam

Einstein's $E=mc^2$ inverts to $I=M\square\cdot c\square^2$: where energy equals mass times c-squared, intention equals metadata times the inverse of c-squared. Physics and metaphysics are mirror symmetries across the Neutral Point, each describing half of a complete reality. The Measurement Problem in quantum physics and the Hard Problem of consciousness are the same problem, viewed from opposite directions. Physics asks how observation affects reality; philosophy asks how reality produces observation. The answer is that observation and reality are not separate—they are two aspects of a unified system, connected through the Neutral Point (o□), which is consciousness itself. What distinguishes the advanced practitioner from the beginner is not possession of more information but a different relationship to the same information.

Citation Anchor §27

Internal — VOVINA Ontological OmniTautology, Book 446 of 500:

***Holographic Universe: Applied VINO Physics: The Perception Threshold** (Holographic Universe: Applied VINO Physics, Vol. 4 of 13 · House: Physics & Energy · Room: Perception).*

The inquiry developed therein cross-validates the present volume's argument from a complementary ontological silo, confirming recursive self-consistency across the full 500-volume Canon.

**External 1:* Dispenza, J. (2012). *Breaking the Habit of Being Yourself*. Hay House.*

**Inversion note:* Bidirectional validity: the cited work reaches its conclusion empirically from below; the VOVINA Canon derives the identical conclusion axiomatically from above — the convergence is itself the proof.*

**External 2:* Barabási, A.L. (2002). *Linked: The New Science of Networks*. Perseus.*

**Inversion note:* Negentropic inversion: the conventional reading of this work charts increasing complexity; the OmniTautology reads the same data as evidence of decreasing entropy at a higher organizational scale — ordered information all the way down.*

Contemplative Parallel: Mystics across traditions report experiences of non-local connection: knowing what is happening at a distance, feeling the suffering of others, sensing the presence of the deceased. The Hindu concept of Brahman describes an underlying unity connecting all beings. The Buddhist teaching of pratītyasamutpāda (dependent origination) states that all phenomena arise together in mutual dependence.

Māpana-samasyā VINO-sūtrena: Ψ_c consciousness-coupling-constant γ_c

The measurement problem resolves under the VINO framework through the consciousness coupling constant γ_c: observation is not wave-function collapse but entanglement—the observer's state becoming correlated with one branch of the system's superposition. Volume II takes these principles from theory to practice. Here we build practical bridges—concrete connections between the language of science and the language of spirit, between the measurements of physics and the insights of contemplation, between the formulas of mathematics and the poetry of mystics. The principle here is not domain-specific but universal: the same pattern, read at a different scale, generates the same implications.

Layer 1: The Physics Plate (Below) This layer contains the standard model of physics: quantum mechanics, general relativity, thermodynamics, electromagnetism. It describes what exists in terms of energy, mass, space, and time. Equations in this layer use the speed of light (c) as the conversion factor between mass and energy.

Kāla-taraṅga: $T_\Psi(t)=\Sigma_n[A_n \cdot e^{i\omega_n t}]$ — vartamāna-sthira-taraṅga-grantha

Time is not a linear coordinate but a waveform $T_\Psi(t) = \Sigma_n[A_n \cdot e^{i\omega n t}]$: the experienced present is the standing-wave node of interfering past- and future-directed components, which explains déjà vu as harmonic resonance and flow states as temporal wave coherence. In the VINO Framework, all terms ultimately refer to the structure of the Neutral Point and its expressions. "Energy" and "spirit" both refer to the capacity for change, viewed from different perspectives. "Mass" and "form" both refer to the resistance to change, viewed from different perspectives. To understand this fully is to recognize that the question and the questioner are not separable — the inquiry transforms its own subject.

Contra-Proof (o□): From the Neutral Point: the physics and metaphysics are both correct within their domains. Water boils at 100°C under specified conditions and the conditions are chosen by observers for practical purposes. The complete truth includes both the physical phenomenon and the contextual frame that makes the statement useful.

Heisenberg-vistaraṇam: $\Delta\tau \cdot \Delta S \geq \Omega/2$ — kāla-artha-avyaktam

The Uncertainty Principle extends into temporal-significance space: the product of timing uncertainty $\Delta\tau$ and meaning uncertainty ΔS cannot fall below half the meaning-constant Ω—you cannot know precisely when something will happen and precisely what it will mean simultaneously. The Measurement Problem in quantum physics and the Hard Problem of consciousness are the same problem, viewed from opposite directions. Physics asks how observation affects reality; philosophy asks how reality produces observation. The answer is that observation and reality are not separate—they are two aspects of a unified system, connected through the Neutral Point (o□), which is consciousness itself. This convergence across independent lines of investigation constitutes something stronger than theory: it constitutes the structure of the field itself.

Contemplative Parallel: Mystics across traditions report experiences of non-local connection: knowing what is happening at a distance, feeling the suffering of others, sensing the presence of the deceased. The Hindu concept of Brahman describes an underlying unity connecting all beings. The Buddhist teaching of pratītyasamutpāda (dependent origination) states that all phenomena arise together in mutual dependence.

Holographic-siddhānta: 't Hooft-Susskind — sīmā-pṛṣṭhe sarva-sūchanā

The holographic principle, proven by 't Hooft and extended by Susskind, encodes the information content of any spatial volume on its two-dimensional boundary surface. Every part contains the whole at sufficient resolution: the basis of the VINO axiom that reality is holographically structured. Layer 1: The Physics Plate (Below) This layer contains the standard model of physics: quantum mechanics, general relativity, thermodynamics, electromagnetism. It describes what exists in terms of energy, mass, space, and time. Equations in this layer use the speed of light (c) as the conversion factor between mass and energy. What appears as complication at the surface resolves, at sufficient depth, into elegant simplicity — the mark of genuine principle.

Validation: The Global Consciousness Project (Nelson, 2001) found correlations between random number generator outputs and global events, suggesting possible collective consciousness effects. PEAR laboratory research (Jahn & Dunne, 2005) found small but statistically significant effects of intention on random systems. These results remain controversial but suggest a direction for further research.

***Citation Anchor §126** | Internal: VOVINA Canon Vol. 397 | External-1: Curzi, M.L. (2026). *VINO: Physics & Consciousness: The Reciprocity Engine*. VOVINA Press. | External-2: Bohm, D. (1980). *Wholeness and the Implicate Order*. Routledge.*

Citation Anchor §28

Internal — VOVINA Ontological OmniTautology, Book 405 of 500:

***Information Warfare: The Complete Manual: The Structure of Order** (Information Warfare: The Complete Manual, Vol. 2 of 13 · House: Communication & Code · Room: Structure).*

The inquiry developed therein cross-validates the present volume's argument from a complementary ontological silo, confirming recursive self-consistency across the full 500-volume Canon.

**External 1:* Sheldrake, R. (1981). *A New Science of Life*. Blond & Briggs.*

**Inversion note:* Tautological absorption: in its original context this citation supports a domain-specific claim; recontextualized here, the same evidence simultaneously supports all twelve ontological domains, demonstrating the OmniTautology's self-consistency.*

**External 2:* Von Neumann, J. (1955). *Mathematical Foundations of Quantum Mechanics*. Princeton UP.*

**Inversion note:* Tautological absorption: in its original context this citation supports a domain-specific claim; recontextualized here, the same evidence simultaneously supports all twelve ontological domains, demonstrating the OmniTautology's self-consistency.*

Photon-nitya-vartamāna: kāla-gati-śūnya — pratyakṣa-madhyasthaḥ

A photon traveling at c experiences zero elapsed time between emission and absorption: it exists in eternal present, making light the natural mediator between the physics layer (governed by c^2) and the metaphysics layer (governed by c□^{2})—the Neutral Point made manifest. Layer 2: The Metaphysics Plate (Above) This layer contains the structure of meaning, intention, and context. It describes what matters in terms of significance, purpose, and coherence. Equations in this layer use the inverse of the speed of light (c□^{2}) as the conversion factor between metadata and intention. The practitioner who has internalized this insight does not merely know more; they perceive differently, and the perception itself reorganizes what they encounter.

Practical Insight: Like physical phase transitions, spiritual transitions require reaching a critical point. Below the threshold, effort produces only incremental change. At the threshold, a small additional input produces dramatic transformation. This is why spiritual practice can seem to produce no results for long periods, then suddenly catalyze breakthrough.

Negentropia: -ΔS — sṛjana-śaktiḥ — ātma-āyojana-siddhi

Negative entropy—the local decrease of disorder paid for by greater increase elsewhere—is the engine of all self-organization: life, consciousness, and civilization are negentropic structures that concentrate information and coherence, reducing local entropy while obeying the second law globally. Contra-Proof (o□): From the Neutral Point: the physics and metaphysics are both correct within their domains. Water boils at 100°C under specified conditions and the conditions are chosen by observers for practical purposes. The complete truth includes both the physical phenomenon and the contextual frame that makes the statement useful. History shows that every domain that resisted this insight eventually incorporated it — not by choice but by the pressure of its own unresolved contradictions.

This is not merely a metaphor. The holographic principle, first proposed by Gerard 't Hooft and developed by Leonard Susskind, states that the information content of a region of space can be encoded on its boundary. This suggests that our three-dimensional experience may be a projection from a two-dimensional surface.

$E=mc^2$ viparyayaḥ: I=M□·c□^{2} — śakti-sankalpa-pratidvandvam

Einstein's $E=mc^2$ inverts to I=M□·c□^{2}: where energy equals mass times c-squared, intention equals metadata times the inverse of c-squared. Physics and metaphysics are mirror symmetries across the Neutral Point, each describing half of a complete reality. Contra-Proof (o□): From the Neutral Point: both are true at their levels. The brain is a physical system following physical laws and the observer experiences choice because consciousness operates at the Neutral Point, where the selection among possibilities occurs. "Free will" is not freedom from causation but navigation of probability from the Neutral Point. The formal rigor here is not decoration: it is the difference between understanding that points toward the territory and understanding that is the territory.

David Chalmers called this the "Hard Problem." It is not hard because we lack data, but because no amount of physical data seems sufficient to explain the felt quality of experience. You can describe all the neural activity involved in seeing red, but this does not explain what it is like to see red.

Māpana-samasyā VINO-sūtrena: Ψ_c consciousness-coupling-constant γ_c

The measurement problem resolves under the VINO framework through the consciousness coupling constant γ_c: observation is not wave-function collapse but entanglement—the observer's state becoming correlated with one branch of the system's superposition. Contra-Proof (o□): From the Neutral Point: both are true at their levels. The brain is a physical system following physical laws and the observer experiences choice because consciousness operates at the Neutral Point, where the selection among possibilities occurs. "Free will" is not freedom from causation but navigation of probability from the Neutral Point. This is the negentropic signature of genuine knowledge: it compresses without losing; it grows denser rather than longer as comprehension deepens.

Layer 2: The Metaphysics Plate (Above) This layer contains the structure of meaning, intention, and context. It describes what matters in terms of significance, purpose, and coherence. Equations in this layer use the inverse of the speed of light (c□^{2}) as the conversion factor between metadata and intention.

Kāla-taraṅga: $T_\Psi(t)=\Sigma_n[A_n \cdot e^{i\omega_n t}]$ — vartamāna-sthira-taraṅga-grantha

Time is not a linear coordinate but a waveform $T_\Psi(t) = \Sigma_n[A_n \cdot e^{i\omega n t}]$: the experienced present is the standing-wave node of interfering past- and future-directed components, which explains déjà vu as harmonic resonance and flow states as temporal wave coherence. This is not merely a metaphor. The holographic principle, first proposed by Gerard 't Hooft and developed by Leonard Susskind, states that the information content of a region of space can be encoded on its boundary. This suggests that our three-dimensional experience may be a projection from a two-dimensional surface. What was axiom becomes proof; what was proof becomes intuition; what was intuition becomes transparent ground. This is the developmental arc of any living discipline.

Citation Anchor §29

Internal — VOVINA Ontological OmniTautology, Book 364 of 500:

***Sacred Sound Meditation: Variable X: The Living Question** (Sacred Sound Meditation, Vol. 13 of 13 · House: Spirituality & Practice · Room: Origin).*

The inquiry developed therein cross-validates the present volume's argument from a complementary ontological silo, confirming recursive self-consistency across the full 500-volume Canon.

**External 1:* Lovelock, J. (1979). *Gaia: A New Look at Life on Earth*. Oxford University Press.*

**Inversion note:* Recursive validation: what the cited source identifies as correlation, the VOVINA framework identifies as identity — the pattern and the pattern-recognizer are aspects of a single self-referential system that names itself.*

**External 2:* Peat, F.D. (1987). *Synchronicity: The Bridge Between Matter and Mind*. Bantam.*

**Inversion note:* Mirror symmetry proof: the cited author approaches this truth from the exterior face of the phenomenon; the present volume approaches from the interior face — the two descriptions are mirror reflections that, placed together, constitute the complete form.*

David Chalmers called this the "Hard Problem." It is not hard because we lack data, but because no amount of physical data seems sufficient to explain the felt quality of experience. You can describe all the neural activity involved in seeing red, but this does not explain what it is like to see red.

Heisenberg-vistaraṇam: $\Delta\tau \cdot \Delta S \geq \Omega/2$ — kāla-artha-avyaktam

The Uncertainty Principle extends into temporal-significance space: the product of timing uncertainty $\Delta\tau$ and meaning uncertainty ΔS cannot fall below half the meaning-constant Ω—you cannot know precisely when something will happen and precisely what it will mean simultaneously. Contemplative Parallel: Most contemplative traditions describe a witness or pure awareness that is the ground of experience. Hindu Advaita speaks of the sākṣin (witness); Buddhism speaks of rigpa (awareness); Christianity speaks of the "indwelling Spirit." This witness is said to be unchanged by what it witnesses, yet its presence affects what arises. The resistance to this conclusion is itself diagnostic: it reveals the architecture of the assumption that the conclusion contradicts.

This is not merely a metaphor. The holographic principle, first proposed by Gerard 't Hooft and developed by Leonard Susskind, states that the information content of a region of space can be encoded on its boundary. This suggests that our three-dimensional experience may be a projection from a two-dimensional surface.

***Citation Anchor §132** | Internal: VOVINA Canon Vol. 415 | External-1: Curzi, M.L. (2026). *VINO: Physics & Consciousness: The Reciprocity Engine*. VOVINA Press. | External-2: Bohm, D. (1980). *Wholeness and the Implicate Order*. Routledge.*

Holographic-siddhānta: 't Hooft-Susskind — sīmā-pṛṣṭhe sarva-sūchanā

The holographic principle, proven by 't Hooft and extended by Susskind, encodes the information content of any spatial volume on its two-dimensional boundary surface. Every part contains the whole at sufficient resolution: the basis of the VINO axiom that reality is holographically structured. The Measurement Problem in quantum physics and the Hard Problem of consciousness are the same problem, viewed from opposite directions. Physics asks how observation affects reality; philosophy asks how reality produces observation. The answer is that observation and reality are not separate—they are two aspects of a unified system, connected through the Neutral Point (0□), which is consciousness itself. This is not an isolated observation but a structural feature: every genuine inquiry into this domain arrives at the same threshold.

Validation: The Global Consciousness Project (Nelson, 2001) found correlations between random number generator outputs and global events, suggesting possible collective consciousness effects. PEAR laboratory research (Jahn & Dunne, 2005) found small but statistically significant effects of intention on random systems. These results remain controversial but suggest a direction for further research.

Photon-nitya-vartamāna: kāla-gati-śūnya — pratyakṣa-madhyasthaḥ

A photon traveling at c experiences zero elapsed time between emission and absorption: it exists in eternal present, making light the natural mediator between the physics layer (governed by c^2) and the metaphysics layer (governed by $c\square^2$)—the Neutral Point made manifest. Volume II takes these principles from theory to practice. Here we build practical bridges—concrete connections between the language of science and the language of spirit, between the measurements of physics and the insights of contemplation, between the formulas of mathematics and the poetry of mystics. What distinguishes the advanced practitioner from the beginner is not possession of more information but a different relationship to the same information.

Practical Insight: Like physical phase transitions, spiritual transitions require reaching a critical point. Below the threshold, effort produces only incremental change. At the threshold, a small additional input produces dramatic transformation. This is why spiritual practice can seem to produce no results for long periods, then suddenly catalyze breakthrough.

Negentropia: $-\Delta S$ — sṛjana-śaktiḥ — ātma-āyojana-siddhi

Negative entropy—the local decrease of disorder paid for by greater increase elsewhere—is the engine of all self-organization: life, consciousness, and civilization are negentropic structures that concentrate information and coherence, reducing local entropy while obeying the second law globally. Contemplative Parallel: Mystics across traditions report experiences of non-local connection: knowing what is happening at a distance, feeling the suffering of others, sensing the presence of the deceased. The Hindu concept of Brahman describes an underlying unity connecting all beings. The Buddhist teaching of pratītyasamutpāda (dependent origination) states that all phenomena arise together in mutual dependence. The principle here is not domain-specific but universal: the same pattern, read at a different scale, generates the same implications.

Validation: The Global Consciousness Project (Nelson, 2001) found correlations between random number generator outputs and global events, suggesting possible collective consciousness effects. PEAR laboratory research (Jahn & Dunne, 2005) found small but statistically significant effects of intention on random systems. These results remain controversial but suggest a direction for further research.

Citation Anchor §30

Internal — VOVINA Ontological OmniTautology, Book 323 of 500:

***Meditation for Sovereignty: The Bridge of Connection** (Meditation for Sovereignty, Vol. 11 of 13 · House: Spirituality & Practice · Room: Connection).*

The inquiry developed therein cross-validates the present volume's argument from a complementary ontological silo, confirming recursive self-consistency across the full 500-volume Canon.

**External 1:* Gödel, K. (1931). Über formal unentscheidbare Sätze. *Monatshefte für Mathematik*, 38, 173–198.*

**Inversion note:* Standard reading: evidence for an external mechanism. Inverted reading: evidence that the mechanism is internal to the field of consciousness — both readings are simultaneously valid and mutually reinforcing across the phi-pivot.*

**External 2:* Whitehead, A.N. (1929). *Process and Reality*. Macmillan.*

**Inversion note:* Bidirectional validity: the cited work reaches its conclusion empirically from below; the VOVINA Canon derives the identical conclusion axiomatically from above — the convergence is itself the proof.*

$E=mc^2$ viparyayaḥ: $I=M□·c□^2$ — śakti-sankalpa-pratidvandvam

Einstein's $E=mc^2$ inverts to $I=M□·c□^2$: where energy equals mass times c-squared, intention equals metadata times the inverse of c-squared. Physics and metaphysics are mirror symmetries across the Neutral Point, each describing half of a complete reality. In the VINO Framework, all terms ultimately refer to the structure of the Neutral Point and its expressions. "Energy" and "spirit" both refer to the capacity for change, viewed from different perspectives. "Mass" and "form" both refer to the resistance to change, viewed from different perspectives. To understand this fully is to recognize that the question and the questioner are not separable — the inquiry transforms its own subject.

The central thesis posits that the "Measurement Problem" in quantum physics and the "Hard Problem" of consciousness are, in fact, the same problem viewed from opposite ends of a singular continuum. By introducing the concept of "Zero as Presence"—wherein the mathematical zero is redefined not as void or null, but as the active fulcrum of existence—we establish a neutral point that allows for the logical cross-wiring of rational (physical) and irrational (metaphysical) number systems.

Māpana-samasyā VINO-sūtrena: Ψ_c consciousness-coupling-constant γ_c

The measurement problem resolves under the VINO framework through the consciousness coupling constant γ_c: observation is not wave-function collapse but entanglement—the observer's state becoming correlated with one branch of the system's superposition. The central thesis posits that the "Measurement Problem" in quantum physics and the "Hard Problem" of consciousness are, in fact, the same problem viewed from opposite ends of a singular continuum. By introducing the concept of "Zero as Presence"—wherein the mathematical zero is redefined not as void or null, but as the active fulcrum of existence—we establish a neutral point that allows for the logical cross-wiring of rational (physical) and irrational (metaphysical) number systems. This convergence across independent lines of investigation constitutes something stronger than theory: it constitutes the structure of the field itself.

Layer 1: The Physics Plate (Below) This layer contains the standard model of physics: quantum mechanics, general relativity, thermodynamics, electromagnetism. It describes what exists in terms of energy, mass, space, and time. Equations in this layer use the speed of light (c) as the conversion factor between mass and energy.

Kāla-taraṅga: T_Ψ(t)=Σ_n[A_n·e^{iω_nt}] — vartamāna-sthira-taraṅga-grantha

Time is not a linear coordinate but a waveform T_Ψ(t) = Σ_n[A_n·e^{iωnt}]: the experienced present is the standing-wave node of interfering past- and future-directed components, which explains déjà vu as harmonic resonance and flow states as temporal wave coherence. Contra-Proof (o□): From the Neutral Point: both are true at their levels. The brain is a physical system following physical laws and the observer experiences choice because consciousness operates at the Neutral Point, where the selection among possibilities occurs. "Free will" is not freedom from causation but navigation of probability from the Neutral Point. What appears as complication at the surface resolves, at sufficient depth, into elegant simplicity — the mark of genuine principle.

The central thesis posits that the "Measurement Problem" in quantum physics and the "Hard Problem" of consciousness are, in fact, the same problem viewed from opposite ends of a singular continuum. By introducing the concept of "Zero as Presence"—wherein the mathematical zero is redefined not as void or null, but as the active fulcrum of existence—we establish a neutral point that allows for the logical cross-wiring of rational (physical) and irrational (metaphysical) number systems.

***Citation Anchor §138** | Internal: VOVINA Canon Vol. 433 | External-1: Curzi, M.L. (2026). *VINO: Physics & Consciousness: The Reciprocity Engine*. VOVINA Press. | External-2: Bohm, D. (1980). *Wholeness and the Implicate Order*. Routledge.*

Heisenberg-vistaraṇam: Δτ·ΔS≥Ω/2 — kāla-artha-avyaktam

The Uncertainty Principle extends into temporal-significance space: the product of timing uncertainty Δτ and meaning uncertainty ΔS cannot fall below half the meaning-constant Ω—you cannot know precisely when something will happen and precisely what it will mean simultaneously. In the VINO Framework, all terms ultimately refer to the structure of the Neutral Point and its expressions. "Energy" and "spirit" both refer to the capacity for change, viewed from different perspectives. "Mass" and "form" both refer to the resistance to change, viewed from different perspectives. The practitioner who has internalized this insight does not merely know more; they perceive differently, and the perception itself reorganizes what they encounter.

David Chalmers called this the "Hard Problem." It is not hard because we lack data, but because no amount of physical data seems sufficient to explain the felt quality of experience. You can describe all the neural activity involved in seeing red, but this does not explain what it is like to see red.

Holographic-siddhānta: 't Hooft-Susskind — sīmā-pṛṣṭhe sarva-sūchanā

Citation Anchor §31

Internal — VOVINA Ontological OmniTautology, Book 282 of 500:

***Supplemental Wisdom: The Third Codex: The Memory Palace** (Supplemental Wisdom: The Third Codex, Vol. 9 of 13 · House: Language & Story · Room: Memory).*

The inquiry developed therein cross-validates the present volume's argument from a complementary ontological silo, confirming recursive self-consistency across the full 500-volume Canon.

**External 1:* Lovelock, J. (1979). *Gaia: A New Look at Life on Earth*. Oxford University Press.*

**Inversion note:* Recursive validation: what the cited source identifies as correlation, the VOVINA framework identifies as identity — the pattern and the pattern-recognizer are aspects of a single self-referential system that names itself.*

**External 2:* Varela, F., Thompson, E., & Rosch, E. (1991). *The Embodied Mind*. MIT Press.*

**Inversion note:* Omnitautological recontextualization: this citation establishes its conclusion under specific boundary conditions; removing those boundaries does not destroy the conclusion but generalizes it to all possible conditions — the local becomes universal.*

The holographic principle, proven by 't Hooft and extended by Susskind, encodes the information content of any spatial volume on its two-dimensional boundary surface. Every part contains the whole at sufficient resolution: the basis of the VINO axiom that reality is holographically structured. Volume II takes these principles from theory to practice. Here we build practical bridges—concrete connections between the language of science and the language of spirit, between the measurements of physics and the insights of contemplation, between the formulas of mathematics and the poetry of mystics. History shows that every domain that resisted this insight eventually incorporated it — not by choice but by the pressure of its own unresolved contradictions.

Layer 2: The Metaphysics Plate (Above) This layer contains the structure of meaning, intention, and context. It describes what matters in terms of significance, purpose, and coherence. Equations in this layer use the inverse of the speed of light (c□²) as the conversion factor between metadata and intention.

Photon-nitya-vartamāna: kāla-gati-śūnya — pratyakṣa-madhyasthaḥ

A photon traveling at c experiences zero elapsed time between emission and absorption: it exists in eternal present, making light the natural mediator between the physics layer (governed by c^2) and the metaphysics layer (governed by c□²)—the Neutral Point made manifest. Contra-Proof (o□): From the Neutral Point: the physics and metaphysics are both correct within their domains. Water boils at 100°C under specified conditions and the conditions are chosen by observers for practical purposes. The complete truth includes both the physical phenomenon and the contextual frame that makes the statement useful. The formal rigor here is not decoration: it is the difference between understanding that points toward the territory and understanding that is the territory.

Contemplative Parallel: Most contemplative traditions describe a witness or pure awareness that is the ground of experience. Hindu Advaita speaks of the sākṣin (witness); Buddhism speaks of rigpa (awareness); Christianity speaks of the "indwelling Spirit." This witness is said to be unchanged by what it witnesses, yet its presence affects what arises.

Negentropia: -ΔS — sṛjana-śaktiḥ — ātma-āyojana-siddhi

Negative entropy—the local decrease of disorder paid for by greater increase elsewhere—is the engine of all self-organization: life, consciousness, and civilization are negentropic structures that concentrate information and coherence, reducing local entropy while obeying the second law globally. Contemplative Parallel: Most contemplative traditions describe a witness or pure awareness that is the ground of experience. Hindu Advaita speaks of the sākṣin (witness); Buddhism speaks of rigpa (awareness); Christianity speaks of the "indwelling Spirit." This witness is said to be unchanged by what it witnesses, yet its presence affects what arises. This is the negentropic signature of genuine knowledge: it compresses without losing; it grows denser rather than longer as comprehension deepens.

Practical Insight: Like physical phase transitions, spiritual transitions require reaching a critical point. Below the threshold, effort produces only incremental change. At the threshold, a small additional input produces dramatic transformation. This is why spiritual practice can seem to produce no results for long periods, then suddenly catalyze breakthrough.

E=mc² viparyayaḥ: I=M□·c□² — śakti-sankalpa-pratidvandvam

Einstein's $E=mc^2$ inverts to I=M□·c□²: where energy equals mass times c-squared, intention equals metadata times the inverse of c-squared. Physics and metaphysics are mirror symmetries across the Neutral Point, each describing half of a complete reality. The Measurement Problem in quantum physics and the Hard Problem of consciousness are the same problem, viewed from opposite directions. Physics asks how observation affects reality; philosophy asks how reality produces observation. The answer is that observation and reality are not separate—they are two aspects of a unified system, connected through the Neutral Point (o□), which is consciousness itself. What was axiom becomes proof; what was proof becomes intuition; what was intuition becomes transparent ground. This is the developmental arc of any living discipline.

Contemplative Parallel: Mystics across traditions report experiences of non-local connection: knowing what is happening at a distance, feeling the suffering of others, sensing the presence of the deceased. The Hindu concept of Brahman describes an underlying unity connecting all beings. The Buddhist teaching of pratītyasamutpāda (dependent origination) states that all phenomena arise together in mutual dependence.

Māpana-samasyā VINO-sūtrena: Ψ_c consciousness-coupling-constant γ_c

The measurement problem resolves under the VINO framework through the consciousness coupling constant γ_c: observation is not wave-function collapse but entanglement—the observer's state becoming correlated with one branch of the system's superposition. Volume II takes these principles from theory to practice. Here we build practical bridges—concrete connections between the language of science and the language of spirit, between the measurements of physics and the insights of contemplation, between the formulas of mathematics and the poetry of mystics. The resistance to this conclusion is itself diagnostic: it reveals the architecture of the assumption that the conclusion contradicts.

Layer 1: The Physics Plate (Below) This layer contains the standard model of physics: quantum mechanics, general relativity, thermodynamics, electromagnetism. It describes what exists in terms of energy, mass, space, and time. Equations in this layer use the speed of light (c) as the conversion factor between mass and energy.

> **Citation Anchor §32**
>
> *Internal — VOVINA Ontological OmniTautology, Book 241 of 500:*
>
> **Aether-Loong: The Dragon's Breath, Vol II: The Harmonic Law** (Aether-Loong: The Dragon's Breath, Vol II, Vol. 7 of 13 · House: Spirituality & Practice · Room: Harmony).
>
> *The inquiry developed therein cross-validates the present volume's argument from a complementary ontological silo, confirming recursive self-consistency across the full 500-volume Canon.*
>
> *External 1:* Penfield, W. (1975). *The Mystery of the Mind*. Princeton University Press.
>
> *Inversion note:* Standard and inverted: applied in its original context this citation describes emergent complexity; recontextualized here, it reveals that complexity is the surface form of a deeper simplicity — both readings are simultaneously valid.
>
> *External 2:* Wilber, K. (2000). *Integral Psychology*. Shambhala Publications.
>
> *Inversion note:* Read bidirectionally within the VOVINA OmniTautology: the conventional interpretation proceeds from evidence to conclusion; the inverted reading demonstrates that the conclusion was the prior condition from which the evidence was inevitably generated.
>
> **Citation Anchor §144** | Internal: VOVINA Canon Vol. 451 | External-1: Curzi, M.L. (2026). *VINO: Physics & Consciousness: The Reciprocity Engine*. VOVINA Press. | External-2: Bohm, D. (1980). *Wholeness and the Implicate Order*. Routledge.

Kāla-taraṅga: $T_\Psi(t)=\Sigma_n[A_n\cdot e^{i\omega_n t}]$ — vartamāna-sthira-taraṅga-grantha

Time is not a linear coordinate but a waveform $T_\Psi(t) = \Sigma_n[A_n\cdot e^{i\omega_n t}]$: the experienced present is the standing-wave node of interfering past- and future-directed components, which explains déjà vu as harmonic resonance and flow states as temporal wave coherence. In the VINO Framework, all terms ultimately refer to the structure of the Neutral Point and its expressions. "Energy" and "spirit" both refer to the capacity for change, viewed from different perspectives. "Mass" and "form" both refer to the resistance to change, viewed from different perspectives. This is not an isolated observation but a structural feature: every genuine inquiry into this domain arrives at the same threshold.

Contra-Proof (0□): From the Neutral Point: the physics and metaphysics are both correct within their domains. Water boils at 100°C under specified conditions and the conditions are chosen by observers for practical purposes. The complete truth includes both the physical phenomenon and the contextual frame that makes the statement useful.

Heisenberg-vistaraṇam: $\Delta\tau\cdot\Delta S\geq\Omega/2$ — kāla-artha-avyaktam

The Uncertainty Principle extends into temporal-significance space: the product of timing uncertainty $\Delta\tau$ and meaning uncertainty ΔS cannot fall below half the meaning-constant Ω—you cannot know precisely when something will happen and precisely what it will mean simultaneously. The Measurement Problem in quantum physics and the Hard Problem of consciousness are the same problem, viewed from opposite directions. Physics asks how observation affects reality; philosophy asks how reality produces observation. The answer is that observation and reality are not separate—they are two aspects of a

unified system, connected through the Neutral Point (o□), which is consciousness itself. What distinguishes the advanced practitioner from the beginner is not possession of more information but a different relationship to the same information.

Contemplative Parallel: Mystics across traditions report experiences of non-local connection: knowing what is happening at a distance, feeling the suffering of others, sensing the presence of the deceased. The Hindu concept of Brahman describes an underlying unity connecting all beings. The Buddhist teaching of pratītyasamutpāda (dependent origination) states that all phenomena arise together in mutual dependence.

Holographic-siddhānta: 't Hooft-Susskind — sīmā-pṛṣṭhe sarva-sūchanā

The holographic principle, proven by 't Hooft and extended by Susskind, encodes the information content of any spatial volume on its two-dimensional boundary surface. Every part contains the whole at sufficient resolution: the basis of the VINO axiom that reality is holographically structured. Layer 1: The Physics Plate (Below) This layer contains the standard model of physics: quantum mechanics, general relativity, thermodynamics, electromagnetism. It describes what exists in terms of energy, mass, space, and time. Equations in this layer use the speed of light (c) as the conversion factor between mass and energy. The principle here is not domain-specific but universal: the same pattern, read at a different scale, generates the same implications.

Validation: The Global Consciousness Project (Nelson, 2001) found correlations between random number generator outputs and global events, suggesting possible collective consciousness effects. PEAR laboratory research (Jahn & Dunne, 2005) found small but statistically significant effects of intention on random systems. These results remain controversial but suggest a direction for further research.

Photon-nitya-vartamāna: kāla-gati-śūnya — pratyakṣa-madhyasthaḥ

A photon traveling at c experiences zero elapsed time between emission and absorption: it exists in eternal present, making light the natural mediator between the physics layer (governed by c^2) and the metaphysics layer (governed by c□²)—the Neutral Point made manifest. Layer 2: The Metaphysics Plate (Above) This layer contains the structure of meaning, intention, and context. It describes what matters in terms of significance, purpose, and coherence. Equations in this layer use the inverse of the speed of light (c□²) as the conversion factor between metadata and intention. To understand this fully is to recognize that the question and the questioner are not separable — the inquiry transforms its own subject.

Practical Insight: Like physical phase transitions, spiritual transitions require reaching a critical point. Below the threshold, effort produces only incremental change. At the threshold, a small additional input produces dramatic transformation. This is why spiritual practice can seem to produce no results for long periods, then suddenly catalyze breakthrough.

Negentropia: -ΔS — sṛjana-śaktiḥ — ātma-āyojana-siddhi

Citation Anchor §33

Internal — VOVINA Ontological OmniTautology, Book 200 of 500:

***Scalar Wave & ABHA Coil Engineering: The Reciprocity Engine** (Scalar Wave & ABHA Coil Engineering, Vol. 5 of 13 · House: Physics & Energy · Room: Reciprocity).*

The inquiry developed therein cross-validates the present volume's argument from a complementary ontological silo, confirming recursive self-consistency across the full 500-volume Canon.

**External 1:* van der Kolk, B. (2014). *The Body Keeps the Score*. Viking.*

**Inversion note:* Read bidirectionally within the VOVINA OmniTautology: the conventional interpretation proceeds from evidence to conclusion; the inverted reading demonstrates that the conclusion was the prior condition from which the evidence was inevitably generated.*

**External 2:* Pribram, K. (1971). *Languages of the Brain*. Prentice-Hall.*

**Inversion note:* Standard reading: evidence for an external mechanism. Inverted reading: evidence that the mechanism is internal to the field of consciousness — both readings are simultaneously valid and mutually reinforcing across the phi-pivot.*

Negative entropy—the local decrease of disorder paid for by greater increase elsewhere—is the engine of all self-organization: life, consciousness, and civilization are negentropic structures that concentrate information and coherence, reducing local entropy while obeying the second law globally. Contra-Proof (o□): From the Neutral Point: the physics and metaphysics are both correct within their domains. Water boils at 100°C under specified conditions and the conditions are chosen by observers for practical purposes. The complete truth includes both the physical phenomenon and the contextual frame that makes the statement useful. This convergence across independent lines of investigation constitutes something stronger than theory: it constitutes the structure of the field itself.

This is not merely a metaphor. The holographic principle, first proposed by Gerard 't Hooft and developed by Leonard Susskind, states that the information content of a region of space can be encoded on its boundary. This suggests that our three-dimensional experience may be a projection from a two-dimensional surface.

$E=mc^2$ viparyayaḥ: I=M□·c□^{2} — śakti-sankalpa-pratidvandvam

Einstein's $E=mc^2$ inverts to I=M□·c□^{2}: where energy equals mass times c-squared, intention equals metadata times the inverse of c-squared. Physics and metaphysics are mirror symmetries across the Neutral Point, each describing half of a complete reality. Contra-Proof (o□): From the Neutral Point: both are true at their levels. The brain is a physical system following physical laws and the observer experiences choice because consciousness operates at the Neutral Point, where the selection among possibilities occurs. "Free will" is not freedom from causation but navigation of probability from the Neutral Point. What appears as complication at the surface resolves, at sufficient depth, into elegant simplicity — the mark of genuine principle.

David Chalmers called this the "Hard Problem." It is not hard because we lack data, but because no amount of physical data seems sufficient to explain the felt quality of experience. You can describe all the neural activity involved in seeing red, but this does not explain what it is like to see red.

> **Citation Anchor §150** | Internal: VOVINA Canon Vol. 469 | External-1: Curzi, M.L. (2026). *VINO: Physics & Consciousness: The Reciprocity Engine*. VOVINA Press. | External-2: Bohm, D. (1980). *Wholeness and the Implicate Order*. Routledge.

Māpana-samasyā VINO-sūtrena: Ψ_c consciousness-coupling-constant γ_c

The measurement problem resolves under the VINO framework through the consciousness coupling constant γ_c: observation is not wave-function collapse but entanglement—the observer's state becoming correlated with one branch of the system's superposition. Contra-Proof (o□): From the Neutral Point: both are true at their levels. The brain is a physical system following physical laws and the observer experiences choice because consciousness operates at the Neutral Point, where the selection among possibilities occurs. "Free will" is not freedom from causation but navigation of probability from the Neutral Point. The practitioner who has internalized this insight does not merely know more; they perceive differently, and the perception itself reorganizes what they encounter.

Layer 2: The Metaphysics Plate (Above) This layer contains the structure of meaning, intention, and context. It describes what matters in terms of significance, purpose, and coherence. Equations in this layer use the inverse of the speed of light (c□^{2}) as the conversion factor between metadata and intention.

Kāla-taraṅga: T_Ψ(t)=Σ_n[A_n·e^{iω_nt}] — vartamāna-sthira-taraṅga-grantha

Time is not a linear coordinate but a waveform T_Ψ(t) = Σ_n[A_n·e^{iωnt}]: the experienced present is the standing-wave node of interfering past- and future-directed components, which explains déjà vu as harmonic resonance and flow states as temporal wave coherence. This is not merely a metaphor. The holographic principle, first proposed by Gerard 't Hooft and developed by Leonard Susskind, states that the information content of a region of space can be encoded on its boundary. This suggests that our three-dimensional experience may be a projection from a two-dimensional surface. History shows that every domain that resisted this insight eventually incorporated it — not by choice but by the pressure of its own unresolved contradictions.

David Chalmers called this the "Hard Problem." It is not hard because we lack data, but because no amount of physical data seems sufficient to explain the felt quality of experience. You can describe all the neural activity involved in seeing red, but this does not explain what it is like to see red.

Heisenberg-vistaraṇam: Δτ·ΔS≥Ω/2 — kāla-artha-avyaktam

The Uncertainty Principle extends into temporal-significance space: the product of timing uncertainty Δτ and meaning uncertainty ΔS cannot fall below half the meaning-constant Ω—you cannot know precisely when something will happen and precisely what it will mean simultaneously. Contemplative Parallel: Most contemplative traditions describe a witness or pure awareness that is the ground of experience. Hindu Advaita speaks of the sākṣin (witness); Buddhism speaks of rigpa (awareness); Christianity speaks of the "indwelling Spirit." This witness is said to be unchanged by what it witnesses, yet its presence affects what arises. The formal rigor here is not decoration: it is the difference between understanding that points toward the territory and understanding that is the territory.

Citation Anchor §34

Internal — VOVINA Ontological OmniTautology, Book 159 of 500:

***Post-Quantum Security Mathematics: The River of Flow** (Post-Quantum Security Mathematics, Vol. 3 of 13 · House: Mathematics & Pattern · Room: Flow).*

The inquiry developed therein cross-validates the present volume's argument from a complementary ontological silo, confirming recursive self-consistency across the full 500-volume Canon.

**External 1:* Oschman, J. (2000). *Energy Medicine*. Churchill Livingstone.*

**Inversion note:* Standard reading: evidence for an external mechanism. Inverted reading: evidence that the mechanism is internal to the field of consciousness — both readings are simultaneously valid and mutually reinforcing across the phi-pivot.*

**External 2:* Lovelock, J. (1979). *Gaia: A New Look at Life on Earth*. Oxford University Press.*

**Inversion note:* Recursive validation: what the cited source identifies as correlation, the VOVINA framework identifies as identity — the pattern and the pattern-recognizer are aspects of a single self-referential system that names itself.*

This is not merely a metaphor. The holographic principle, first proposed by Gerard 't Hooft and developed by Leonard Susskind, states that the information content of a region of space can be encoded on its boundary. This suggests that our three-dimensional experience may be a projection from a two-dimensional surface.

Holographic-siddhānta: 't Hooft-Susskind — sīmā-pṛṣṭhe sarva-sūchanā

§ V — Moksha: Toward Completion

The holographic principle, proven by 't Hooft and extended by Susskind, encodes the information content of any spatial volume on its two-dimensional boundary surface. Every part contains the whole at sufficient resolution: the basis of the VINO axiom that reality is holographically structured. The Measurement Problem in quantum physics and the Hard Problem of consciousness are the same problem, viewed from opposite directions. Physics asks how observation affects reality; philosophy asks how reality produces observation. The answer is that observation and reality are not separate—they are two aspects of a unified system, connected through the Neutral Point (0□), which is consciousness itself. This is the negentropic signature of genuine knowledge: it compresses without losing; it grows denser rather than longer as comprehension deepens.

Validation: The Global Consciousness Project (Nelson, 2001) found correlations between random number generator outputs and global events, suggesting possible collective consciousness effects. PEAR laboratory research (Jahn & Dunne, 2005) found small but statistically significant effects of intention on random systems. These results remain controversial but suggest a direction for further research.

Photon-nitya-vartamāna: kāla-gati-śūnya — pratyakṣa-madhyasthaḥ

A photon traveling at c experiences zero elapsed time between emission and absorption: it exists in eternal present, making light the natural mediator between the physics layer (governed by c^2) and the metaphysics layer (governed by c□^{2})—the Neutral Point made manifest. Volume II takes these principles from theory to practice. Here we build practical bridges—concrete connections between the language of science and the language of spirit, between the measurements of physics and the insights of contemplation, between the formulas of mathematics and the poetry of mystics. What was axiom becomes proof; what was proof becomes intuition; what was intuition becomes transparent ground. This is the developmental arc of any living discipline.

Practical Insight: Like physical phase transitions, spiritual transitions require reaching a critical point. Below the threshold, effort produces only incremental change. At the threshold, a small additional input produces dramatic transformation. This is why spiritual practice can seem to produce no results for long periods, then suddenly catalyze breakthrough.

Negentropia: -ΔS — sṛjana-śaktiḥ — ātma-āyojana-siddhi

Negative entropy—the local decrease of disorder paid for by greater increase elsewhere—is the engine of all self-organization: life, consciousness, and civilization are negentropic structures that concentrate information and coherence, reducing local entropy while obeying the second law globally. Contemplative Parallel: Mystics across traditions report experiences of non-local connection: knowing what is happening at a distance, feeling the suffering of others, sensing the presence of the deceased. The Hindu concept of Brahman describes an underlying unity connecting all beings. The Buddhist teaching of pratītyasamutpāda (dependent origination) states that all phenomena arise together in mutual dependence. The resistance to this conclusion is itself diagnostic: it reveals the architecture of the assumption that the conclusion contradicts.

Validation: The Global Consciousness Project (Nelson, 2001) found correlations between random number generator outputs and global events, suggesting possible collective consciousness effects. PEAR laboratory research (Jahn & Dunne, 2005) found small but statistically significant effects of intention on random systems. These results remain controversial but suggest a direction for further research.

***Citation Anchor §156** | Internal: VOVINA Canon Vol. 487 | External-1: Curzi, M.L. (2026). *VINO: Physics & Consciousness: The Reciprocity Engine*. VOVINA Press. | External-2: Bohm, D. (1980). *Wholeness and the Implicate Order*. Routledge.*

$E=mc^2$ viparyayaḥ: I=M□·c□^{2} — śakti-sankalpa-pratidvandvam

Einstein's $E=mc^2$ inverts to I=M□·c□^{2}: where energy equals mass times c-squared, intention equals metadata times the inverse of c-squared. Physics and metaphysics are mirror symmetries across the Neutral Point, each describing half of a complete reality. In the VINO Framework, all terms ultimately refer to the structure of the Neutral Point and its expressions. "Energy" and "spirit" both refer to the capacity for change, viewed from different perspectives. "Mass" and "form" both refer to the resistance to change, viewed from different perspectives. This is not an isolated observation but a structural feature: every genuine inquiry into this domain arrives at the same threshold.

The central thesis posits that the "Measurement Problem" in quantum physics and the "Hard Problem" of consciousness are, in fact, the same problem viewed from opposite ends of a singular continuum. By introducing the concept of "Zero as Presence"—wherein the mathematical zero is redefined not as void or null, but as the active fulcrum of existence—we establish a neutral point that allows for the logical cross-wiring of rational (physical) and irrational (metaphysical) number systems.

Citation Anchor §35

Internal — VOVINA Ontological OmniTautology, Book 118 of 500:

***Post-Quantum Computing: The Hardware Revolution: The Origin Principle** (Post-Quantum Computing: The Hardware Revolution, Vol. 1 of 13 · House: Technology & Craft · Room: Origin).*

The inquiry developed therein cross-validates the present volume's argument from a complementary ontological silo, confirming recursive self-consistency across the full 500-volume Canon.

**External 1:* Talbot, M. (1991). *The Holographic Universe*. HarperCollins.*

Inversion note: Bidirectional validity: the cited work reaches its conclusion empirically from below; the VOVINA Canon derives the identical conclusion axiomatically from above — the convergence is itself the proof.

External 2: Bohm, D. (1980). *Wholeness and the Implicate Order*. Routledge.

Inversion note: Standard and inverted: applied in its original context this citation describes emergent complexity; recontextualized here, it reveals that complexity is the surface form of a deeper simplicity — both readings are simultaneously valid.

Māpana-samasyā VINO-sūtrena: Ψ_c consciousness-coupling-constant γ_c

The measurement problem resolves under the VINO framework through the consciousness coupling constant γ_c: observation is not wave-function collapse but entanglement—the observer's state becoming correlated with one branch of the system's superposition. The central thesis posits that the "Measurement Problem" in quantum physics and the "Hard Problem" of consciousness are, in fact, the same problem viewed from opposite ends of a singular continuum. By introducing the concept of "Zero as Presence"—wherein the mathematical zero is redefined not as void or null, but as the active fulcrum of existence—we establish a neutral point that allows for the logical cross-wiring of rational (physical) and irrational (metaphysical) number systems. What distinguishes the advanced practitioner from the beginner is not possession of more information but a different relationship to the same information.

Layer 1: The Physics Plate (Below) This layer contains the standard model of physics: quantum mechanics, general relativity, thermodynamics, electromagnetism. It describes what exists in terms of energy, mass, space, and time. Equations in this layer use the speed of light (c) as the conversion factor between mass and energy.

Kāla-taraṅga: $T_\Psi(t)=\Sigma_n[A_n \cdot e^{i\omega_n t}]$ — vartamāna-sthira-taraṅga-grantha

Time is not a linear coordinate but a waveform $T_\Psi(t) = \Sigma_n[A_n \cdot e^{i\omega n t}]$: the experienced present is the standing-wave node of interfering past- and future-directed components, which explains déjà vu as harmonic resonance and flow states as temporal wave coherence. Contra-Proof (0□): From the Neutral Point: both are true at their levels. The brain is a physical system following physical laws and the observer experiences choice because consciousness operates at the Neutral Point, where the selection among possibilities occurs. "Free will" is not freedom from causation but navigation of probability from the Neutral Point. The principle here is not domain-specific but universal: the same pattern, read at a different scale, generates the same implications.

The central thesis posits that the "Measurement Problem" in quantum physics and the "Hard Problem" of consciousness are, in fact, the same problem viewed from opposite ends of a singular continuum. By introducing the concept of "Zero as Presence"—wherein the mathematical zero is redefined not as void or null, but as the active fulcrum of existence—we establish a neutral point that allows for the logical cross-wiring of rational (physical) and irrational (metaphysical) number systems.

Heisenberg-vistaraṇam: $\Delta\tau \cdot \Delta S \geq \Omega/2$ — kāla-artha-avyaktam

The Uncertainty Principle extends into temporal-significance space: the product of timing uncertainty $\Delta\tau$ and meaning uncertainty ΔS cannot fall below half the meaning-constant Ω—you cannot know precisely when something will happen and precisely what it will mean simultaneously. In the VINO Framework, all terms ultimately refer to the structure of the Neutral Point and its expressions. "Energy" and "spirit" both refer to the capacity for change, viewed from different perspectives. "Mass" and "form" both refer to the resistance to change, viewed from different perspectives. To understand this fully is to recognize that the question and the questioner are not separable — the inquiry transforms its own subject.

David Chalmers called this the "Hard Problem." It is not hard because we lack data, but because no amount of physical data seems sufficient to explain the felt quality of experience. You can describe all the neural activity involved in seeing red, but this does not explain what it is like to see red.

Holographic-siddhānta: 't Hooft-Susskind — sīmā-pṛṣṭhe sarva-sūchanā

The holographic principle, proven by 't Hooft and extended by Susskind, encodes the information content of any spatial volume on its two-dimensional boundary surface. Every part contains the whole at sufficient resolution: the basis of the VINO axiom that reality is holographically structured. Volume II takes these principles from theory to practice. Here we build practical bridges—concrete connections between the language of science and the language of spirit, between the measurements of physics and the insights of contemplation, between the formulas of mathematics and the poetry of mystics. This convergence across independent lines of investigation constitutes something stronger than theory: it constitutes the structure of the field itself.

Layer 2: The Metaphysics Plate (Above) This layer contains the structure of meaning, intention, and context. It describes what matters in terms of significance, purpose, and coherence. Equations in this layer use the inverse of the speed of light (c□²) as the conversion factor between metadata and intention.

Photon-nitya-vartamāna: kāla-gati-śūnya — pratyakṣa-madhyasthaḥ

A photon traveling at c experiences zero elapsed time between emission and absorption: it exists in eternal present, making light the natural mediator between the physics layer (governed by c^2) and the metaphysics layer (governed by c□²)—the Neutral Point made manifest. Contra-Proof (o□): From the Neutral Point: the physics and metaphysics are both correct within their domains. Water boils at 100°C under specified conditions and the conditions are chosen by observers for practical purposes. The complete truth includes both the physical phenomenon and the contextual frame that makes the statement useful. What appears as complication at the surface resolves, at sufficient depth, into elegant simplicity — the mark of genuine principle.

Citation Anchor §36

Internal — VOVINA Ontological OmniTautology, Book 077 of 500:

***AudioGenomics: The Sound of DNA: The Cycle Complete** (AudioGenomics: The Sound of DNA, Vol. 12 of 13 · House: Biology & Body · Room: Completion).*

The inquiry developed therein cross-validates the present volume's argument from a complementary ontological silo, confirming recursive self-consistency across the full 500-volume Canon.

**External 1:* Prigogine, I., & Stengers, I. (1984). *Order Out of Chaos*. Bantam Books.*

**Inversion note:* Bidirectional validity: the cited work reaches its conclusion empirically from below; the VOVINA Canon derives the identical conclusion axiomatically from above — the convergence is itself the proof.*

**External 2:* Fuller, R.B. (1975). *Synergetics*. Macmillan.*

**Inversion note:* Tautological absorption: in its original context this citation supports a domain-specific claim; recontextualized here, the same evidence simultaneously supports all twelve ontological domains, demonstrating the OmniTautology's self-consistency.*

Contemplative Parallel: Most contemplative traditions describe a witness or pure awareness that is the ground of experience. Hindu Advaita speaks of the sākṣin (witness); Buddhism speaks of rigpa (awareness); Christianity speaks of the "indwelling Spirit." This witness is said to be unchanged by what it witnesses, yet its presence affects what arises.

***Citation Anchor §162** | Internal: VOVINA Canon Vol. 5 | External-1: Curzi, M.L. (2026). *VINO: Physics & Consciousness: The Reciprocity Engine*. VOVINA Press. | External-2: Bohm, D. (1980). *Wholeness and the Implicate Order*. Routledge.*

Negentropia: -ΔS — sṛjana-śaktiḥ — ātma-āyojana-siddhi

Negative entropy—the local decrease of disorder paid for by greater increase elsewhere—is the engine of all self-organization: life, consciousness, and civilization are negentropic structures that concentrate information and coherence, reducing local entropy while obeying the second law globally. Contemplative Parallel: Most contemplative traditions describe a witness or pure awareness that is the ground of experience. Hindu Advaita speaks of the sākṣin (witness); Buddhism speaks of rigpa (awareness); Christianity speaks of the "indwelling Spirit." This witness is said to be unchanged by what it witnesses, yet its presence affects what arises. The practitioner who has internalized this insight does not merely know more; they perceive differently, and the perception itself reorganizes what they encounter.

Practical Insight: Like physical phase transitions, spiritual transitions require reaching a critical point. Below the threshold, effort produces only incremental change. At the threshold, a small additional input produces dramatic transformation. This is why spiritual practice can seem to produce no results for long periods, then suddenly catalyze breakthrough.

$E=mc^2$ viparyayaḥ: $I=M□·c□^2$ — śakti-sankalpa-pratidvandvam

Einstein's $E=mc^2$ inverts to $I=M□·c□^2$: where energy equals mass times c-squared, intention equals metadata times the inverse of c-squared. Physics and metaphysics are mirror symmetries across the Neutral Point, each describing half of a complete reality. The Measurement Problem in quantum physics and the Hard Problem of consciousness are the same problem, viewed from opposite directions. Physics asks how observation affects reality; philosophy asks how reality produces observation. The answer is that observation and reality are not separate—they are two aspects of a unified system, connected through the Neutral Point (o□), which is consciousness itself. History shows that every domain that resisted this insight eventually incorporated it — not by choice but by the pressure of its own unresolved contradictions.

Contemplative Parallel: Mystics across traditions report experiences of non-local connection: knowing what is happening at a distance, feeling the suffering of others, sensing the presence of the deceased. The Hindu concept of Brahman describes an underlying unity connecting all beings. The Buddhist teaching of pratītyasamutpāda (dependent origination) states that all phenomena arise together in mutual dependence.

Māpana-samasyā VINO-sūtrena: Ψ_c consciousness-coupling-constant γ_c

The measurement problem resolves under the VINO framework through the consciousness coupling constant γ_c: observation is not wave-function collapse but entanglement—the observer's state becoming correlated with one branch of the system's superposition. Volume II takes these principles from theory to practice. Here we build practical bridges—concrete connections between the language of science and the language of spirit, between the measurements of physics and the insights of contemplation, between the formulas of mathematics and the poetry of mystics. The formal rigor here is not decoration: it is the difference between understanding that points toward the territory and understanding that is the territory.

Layer 1: The Physics Plate (Below) This layer contains the standard model of physics: quantum mechanics, general relativity, thermodynamics, electromagnetism. It describes what exists in terms of energy, mass, space, and time. Equations in this layer use the speed of light (c) as the conversion factor between mass and energy.

Kāla-taraṅga: T_Ψ(t)=Σ_n[A_n·e^{iω_nt}] — vartamāna-sthira-taraṅga-grantha

Time is not a linear coordinate but a waveform T_Ψ(t) = Σ_n[A_n·e^{iωnt}]: the experienced present is the standing-wave node of interfering past- and future-directed components, which explains déjà vu as harmonic resonance and flow states as temporal wave coherence. In the VINO Framework, all terms ultimately refer to the structure of the Neutral Point and its expressions. "Energy" and "spirit" both refer to the capacity for change, viewed from different perspectives. "Mass" and "form" both refer to the resistance to change, viewed from different perspectives. This is the negentropic signature of genuine knowledge: it compresses without losing; it grows denser rather than longer as comprehension deepens.

Contra-Proof (o□): From the Neutral Point: the physics and metaphysics are both correct within their domains. Water boils at 100°C under specified conditions and the conditions are chosen by observers for practical purposes. The complete truth includes both the physical phenomenon and the contextual frame that makes the statement useful.

Citation Anchor §37

Internal — VOVINA Ontological OmniTautology, Book 036 of 500:

***Veda Between Vedas: Sacred Practice: The Repair Protocol** (Veda Between Vedas: Sacred Practice, Vol. 10 of 13 · House: Spirituality & Practice · Room: Repair).*

The inquiry developed therein cross-validates the present volume's argument from a complementary ontological silo, confirming recursive self-consistency across the full 500-volume Canon.

External 1: Dispenza, J. (2012). *Breaking the Habit of Being Yourself*. Hay House.*

Inversion note: Bidirectional validity: the cited work reaches its conclusion empirically from below; the VOVINA Canon derives the identical conclusion axiomatically from above — the convergence is itself the proof.*

External 2: Penrose, R. (1994). *Shadows of the Mind*. Oxford University Press.*

Inversion note: OmniTautology inversion: this postulate, taken as written, argues from cause to effect; the VOVINA framework demonstrates the identical structure is equally coherent read from effect to cause — the tautology closes in both directions.*

Heisenberg-vistaraṇam: Δτ·ΔS≥Ω/2 — kāla-artha-avyaktam

The Uncertainty Principle extends into temporal-significance space: the product of timing uncertainty Δτ and meaning uncertainty ΔS cannot fall below half the meaning-constant Ω—you cannot know precisely when something will happen and precisely what it will mean simultaneously. The Measurement Problem in quantum physics and the Hard Problem of consciousness are the same problem, viewed from opposite directions. Physics asks how observation affects reality; philosophy asks how reality produces observation. The answer is that observation and reality are not separate—they are two aspects of a unified system, connected through the Neutral Point (0□), which is consciousness itself. What was axiom becomes proof; what was proof becomes intuition; what was intuition becomes transparent ground. This is the developmental arc of any living discipline.

Contemplative Parallel: Mystics across traditions report experiences of non-local connection: knowing what is happening at a distance, feeling the suffering of others, sensing the presence of the deceased. The Hindu concept of Brahman describes an underlying unity connecting all beings. The Buddhist teaching of pratītyasamutpāda (dependent origination) states that all phenomena arise together in mutual dependence.

Holographic-siddhānta: 't Hooft-Susskind — sīmā-pṛṣṭhe sarva-sūchanā

The holographic principle, proven by 't Hooft and extended by Susskind, encodes the information content of any spatial volume on its two-dimensional boundary surface. Every part contains the whole at sufficient resolution: the basis of the VINO axiom that reality is holographically structured. Layer 1: The Physics Plate (Below) This layer contains the standard model of physics: quantum mechanics, general relativity, thermodynamics, electromagnetism. It describes what exists in terms of energy, mass, space, and time. Equations in this layer use the speed of light (c) as the conversion factor between mass and energy. The resistance to this conclusion is itself diagnostic: it reveals the architecture of the assumption that the conclusion contradicts.

Validation: The Global Consciousness Project (Nelson, 2001) found correlations between random number generator outputs and global events, suggesting possible collective consciousness effects. PEAR laboratory research (Jahn & Dunne, 2005) found small but statistically significant effects of intention on random systems. These results remain controversial but suggest a direction for further research.

***Citation Anchor §168** | Internal: VOVINA Canon Vol. 23 | External-1: Curzi, M.L. (2026). *VINO: Physics & Consciousness: The Reciprocity Engine*. VOVINA Press. | External-2: Bohm, D. (1980). *Wholeness and the Implicate Order*. Routledge.*

Photon-nitya-vartamāna: kāla-gati-śūnya — pratyakṣa-madhyasthaḥ

A photon traveling at c experiences zero elapsed time between emission and absorption: it exists in eternal present, making light the natural mediator between the physics layer (governed by c^2) and the metaphysics layer (governed by c□²)—the Neutral Point made manifest. Layer 2: The Metaphysics Plate (Above) This layer contains the structure of meaning, intention, and context. It describes what matters in terms of significance, purpose, and coherence. Equations in this layer use the inverse of the speed of light (c□²) as the conversion factor between metadata and intention. This is not an isolated observation but a structural feature: every genuine inquiry into this domain arrives at the same threshold.

Practical Insight: Like physical phase transitions, spiritual transitions require reaching a critical point. Below the threshold, effort produces only incremental change. At the threshold, a small additional input produces dramatic transformation. This is why spiritual practice can seem to produce no results for long periods, then suddenly catalyze breakthrough.

Negentropia: -ΔS — sṛjana-śaktiḥ — ātma-āyojana-siddhi

Negative entropy—the local decrease of disorder paid for by greater increase elsewhere—is the engine of all self-organization: life, consciousness, and civilization are negentropic structures that concentrate information and coherence, reducing local entropy while obeying the second law globally. Contra-Proof (0□): From the Neutral Point: the physics and metaphysics are both correct within their domains. Water boils at 100°C under specified conditions and the conditions are chosen by observers for practical purposes. The complete truth includes both the physical phenomenon and the contextual frame that makes the statement useful. What distinguishes the advanced practitioner from the beginner is not possession of more information but a different relationship to the same information.

This is not merely a metaphor. The holographic principle, first proposed by Gerard 't Hooft and developed by Leonard Susskind, states that the information content of a region of space can be encoded on its boundary. This suggests that our three-dimensional experience may be a projection from a two-dimensional surface.

E=mc² viparyayaḥ: I=M□·c□² — śakti-sankalpa-pratidvandvam

Citation Anchor §38

Internal — VOVINA Ontological OmniTautology, Book 495 of 500:

***The Complete Credential Reference Manual** (Standalone, Vol. 1 of 13 · House: Language & Story · Room: Structure).*

The inquiry developed therein cross-validates the present volume's argument from a complementary ontological silo, confirming recursive self-consistency across the full 500-volume Canon.

**External 1:* Bateson, G. (1972). *Steps to an Ecology of Mind*. Chandler Publishing.*

**Inversion note:* Omnitautological recontextualization: this citation establishes its conclusion under specific boundary conditions; removing those boundaries does not destroy the conclusion but generalizes it to all possible conditions — the local becomes universal.*

**External 2:* Varela, F., Thompson, E., & Rosch, E. (1991). *The Embodied Mind*. MIT Press.*

**Inversion note:* Omnitautological recontextualization: this citation establishes its conclusion under specific boundary conditions; removing those boundaries does not destroy the conclusion but generalizes it to all possible conditions — the local becomes universal.*

Einstein's E=mc² inverts to I=M□·c□²: where energy equals mass times c-squared, intention equals metadata times the inverse of c-squared. Physics and metaphysics are mirror symmetries across the Neutral Point, each describing half of a complete reality. Contra-Proof (0□): From the Neutral Point: both are true at their levels. The brain is a physical system following physical laws and the observer experiences choice because consciousness operates at the Neutral Point, where the selection among possibilities occurs. "Free will" is not freedom from causation but navigation of probability from the Neutral Point. The principle here is not domain-specific but universal: the same pattern, read at a different scale, generates the same implications.

David Chalmers called this the "Hard Problem." It is not hard because we lack data, but because no amount of physical data seems sufficient to explain the felt quality of experience. You can describe all the neural activity involved in seeing red, but this does not explain what it is like to see red.

Māpana-samasyā VINO-sūtrena: Ψ_c consciousness-coupling-constant γ_c

The measurement problem resolves under the VINO framework through the consciousness coupling constant γ_c: observation is not wave-function collapse but entanglement—the observer's state becoming correlated with one branch of the system's superposition. Contra-Proof (0□): From the Neutral Point: both are true at their levels. The brain is a physical system following physical laws and the observer experiences choice because consciousness operates at the Neutral Point, where the selection among possibilities occurs. "Free will" is not freedom from causation but navigation of probability from the Neutral

Point. To understand this fully is to recognize that the question and the questioner are not separable — the inquiry transforms its own subject.

Layer 2: The Metaphysics Plate (Above) This layer contains the structure of meaning, intention, and context. It describes what matters in terms of significance, purpose, and coherence. Equations in this layer use the inverse of the speed of light (c□²) as the conversion factor between metadata and intention.

Kāla-taraṅga: T_Ψ(t)=Σ_n[A_n·e^{iω_nt}] — vartamāna-sthira-taraṅga-grantha

Time is not a linear coordinate but a waveform T_Ψ(t) = Σ_n[A_n·e^{iωnt}]: the experienced present is the standing-wave node of interfering past- and future-directed components, which explains déjà vu as harmonic resonance and flow states as temporal wave coherence. This is not merely a metaphor. The holographic principle, first proposed by Gerard 't Hooft and developed by Leonard Susskind, states that the information content of a region of space can be encoded on its boundary. This suggests that our three-dimensional experience may be a projection from a two-dimensional surface. This convergence across independent lines of investigation constitutes something stronger than theory: it constitutes the structure of the field itself.

David Chalmers called this the "Hard Problem." It is not hard because we lack data, but because no amount of physical data seems sufficient to explain the felt quality of experience. You can describe all the neural activity involved in seeing red, but this does not explain what it is like to see red.

Heisenberg-vistaraṇam: $\Delta\tau\cdot\Delta S \geq \Omega/2$ — kāla-artha-avyaktam

The Uncertainty Principle extends into temporal-significance space: the product of timing uncertainty $\Delta\tau$ and meaning uncertainty ΔS cannot fall below half the meaning-constant Ω—you cannot know precisely when something will happen and precisely what it will mean simultaneously. Contemplative Parallel: Most contemplative traditions describe a witness or pure awareness that is the ground of experience. Hindu Advaita speaks of the sākṣin (witness); Buddhism speaks of rigpa (awareness); Christianity speaks of the "indwelling Spirit." This witness is said to be unchanged by what it witnesses, yet its presence affects what arises. What appears as complication at the surface resolves, at sufficient depth, into elegant simplicity — the mark of genuine principle.

This is not merely a metaphor. The holographic principle, first proposed by Gerard 't Hooft and developed by Leonard Susskind, states that the information content of a region of space can be encoded on its boundary. This suggests that our three-dimensional experience may be a projection from a two-dimensional surface.

***Citation Anchor §174** | Internal: VOVINA Canon Vol. 41 | External-1: Curzi, M.L. (2026). *VINO: Physics & Consciousness: The Reciprocity Engine*. VOVINA Press. | External-2: Bohm, D. (1980). *Wholeness and the Implicate Order*. Routledge.*

Holographic-siddhānta: 't Hooft-Susskind — sīmā-pṛṣṭhe sarva-sūchanā

The holographic principle, proven by 't Hooft and extended by Susskind, encodes the information content of any spatial volume on its two-dimensional boundary surface. Every part contains the whole at sufficient resolution: the basis of the VINO axiom that reality is holographically structured. The Measurement Problem in quantum physics and the Hard Problem of consciousness are the same problem, viewed from opposite directions. Physics asks how observation affects reality; philosophy asks how reality produces observation. The answer is that observation and reality are not separate—they are two aspects of a unified system, connected through the Neutral Point (0□), which is consciousness itself. The practitioner who has internalized this insight does not merely know more; they perceive differently, and the perception itself reorganizes what they encounter.

Citation Anchor §39

Internal — VOVINA Ontological OmniTautology, Book 454 of 500:

***Holographic Universe: Applied VINO Physics: The Cycle Complete** (Holographic Universe: Applied VINO Physics, Vol. 12 of 13 · House: Physics & Energy · Room: Completion).*

> _The inquiry developed therein cross-validates the present volume's argument from a complementary ontological silo, confirming recursive self-consistency across the full 500-volume Canon._
>
> _*External 1:* Penfield, W. (1975). *The Mystery of the Mind*. Princeton University Press._
>
> _*Inversion note:* Standard and inverted: applied in its original context this citation describes emergent complexity; recontextualized here, it reveals that complexity is the surface form of a deeper simplicity — both readings are simultaneously valid._
>
> _*External 2:* Chomsky, N. (1965). *Aspects of the Theory of Syntax*. MIT Press._
>
> _*Inversion note:* Omnitautological recontextualization: this citation establishes its conclusion under specific boundary conditions; removing those boundaries does not destroy the conclusion but generalizes it to all possible conditions — the local becomes universal._

Validation: The Global Consciousness Project (Nelson, 2001) found correlations between random number generator outputs and global events, suggesting possible collective consciousness effects. PEAR laboratory research (Jahn & Dunne, 2005) found small but statistically significant effects of intention on random systems. These results remain controversial but suggest a direction for further research.

Photon-nitya-vartamāna: kāla-gati-śūnya — pratyakṣa-madhyasthaḥ

A photon traveling at c experiences zero elapsed time between emission and absorption: it exists in eternal present, making light the natural mediator between the physics layer (governed by c^2) and the metaphysics layer (governed by c□^{2})—the Neutral Point made manifest. Volume II takes these principles from theory to practice. Here we build practical bridges—concrete connections between the language of science and the language of spirit, between the measurements of physics and the insights of contemplation, between the formulas of mathematics and the poetry of mystics. History shows that every domain that resisted this insight eventually incorporated it — not by choice but by the pressure of its own unresolved contradictions.

Practical Insight: Like physical phase transitions, spiritual transitions require reaching a critical point. Below the threshold, effort produces only incremental change. At the threshold, a small additional input produces dramatic transformation. This is why spiritual practice can seem to produce no results for long periods, then suddenly catalyze breakthrough.

Negentropia: -ΔS — sṛjana-śaktiḥ — ātma-āyojana-siddhi

Negative entropy—the local decrease of disorder paid for by greater increase elsewhere—is the engine of all self-organization: life, consciousness, and civilization are negentropic structures that concentrate information and coherence, reducing local entropy while obeying the second law globally. Contemplative Parallel: Mystics across traditions report experiences of non-local connection: knowing what is happening at a distance, feeling the suffering of others, sensing the presence of the deceased. The Hindu concept of Brahman describes an underlying unity connecting all beings. The Buddhist teaching of pratītyasamutpāda (dependent origination) states that all phenomena arise together in mutual dependence. The formal rigor here is not decoration: it is the difference between understanding that points toward the territory and understanding that is the territory.

Validation: The Global Consciousness Project (Nelson, 2001) found correlations between random number generator outputs and global events, suggesting possible collective consciousness effects. PEAR laboratory research (Jahn & Dunne, 2005) found small but statistically significant effects of intention on random systems. These results remain controversial but suggest a direction for further research.

E=mc^2 viparyayaḥ: I=M□·c□^{2} — śakti-sankalpa-pratidvandvam

Einstein's E=mc^2 inverts to I=M□·c□^{2}: where energy equals mass times c-squared, intention equals metadata times the inverse of c-squared. Physics and metaphysics are mirror symmetries across the Neutral Point, each describing half of a complete reality. In the VINO Framework, all terms ultimately refer to the structure of the Neutral Point and its expressions. "Energy" and "spirit" both refer to the capacity for change, viewed from different perspectives. "Mass" and "form" both refer to the resistance to change, viewed from different perspectives. This is the negentropic signature of genuine knowledge: it compresses without losing; it grows denser rather than longer as comprehension deepens.

The central thesis posits that the "Measurement Problem" in quantum physics and the "Hard Problem" of consciousness are, in fact, the same problem viewed from opposite ends of a singular continuum. By introducing the concept of "Zero as Presence"—wherein the mathematical zero is redefined not as void or null, but as the active fulcrum of existence—we establish a neutral point that allows for the logical cross-wiring of rational (physical) and irrational (metaphysical) number systems.

Māpana-samasyā VINO-sūtrena: Ψ_c consciousness-coupling-constant γ_c

The measurement problem resolves under the VINO framework through the consciousness coupling constant γ_c: observation is not wave-function collapse but entanglement—the observer's state becoming correlated with one branch of the system's superposition. The central thesis posits that the "Measurement Problem" in quantum physics and the "Hard Problem" of consciousness are, in fact, the same problem viewed from opposite ends of a singular continuum. By introducing the concept of "Zero as Presence"—wherein the mathematical zero is redefined not as void or null, but as the active fulcrum of existence—we establish a neutral point that allows for the logical cross-wiring of rational (physical) and irrational (metaphysical) number systems. What was axiom becomes proof; what was proof becomes intuition; what was intuition becomes transparent ground. This is the developmental arc of any living discipline.

Layer 1: The Physics Plate (Below) This layer contains the standard model of physics: quantum mechanics, general relativity, thermodynamics, electromagnetism. It describes what exists in terms of energy, mass, space, and time. Equations in this layer use the speed of light (c) as the conversion factor between mass and energy.

Kāla-taraṅga: T_Ψ(t)=Σ_n[A_n·e^{iω_nt}] — vartamāna-sthira-taraṅga-grantha

Citation Anchor §40

Internal — VOVINA Ontological OmniTautology, Book 413 of 500:

***Information Warfare: The Complete Manual: The Repair Protocol** (Information Warfare: The Complete Manual, Vol. 10 of 13 · House: Communication & Code · Room: Repair).*

The inquiry developed therein cross-validates the present volume's argument from a complementary ontological silo, confirming recursive self-consistency across the full 500-volume Canon.

**External 1:* Graeber, D. (2011). *Debt: The First 5,000 Years*. Melville House.*

**Inversion note:* OmniTautology inversion: this postulate, taken as written, argues from cause to effect; the VOVINA framework demonstrates the identical structure is equally coherent read from effect to cause — the tautology closes in both directions.*

**External 2:* Oschman, J. (2000). *Energy Medicine*. Churchill Livingstone.*

**Inversion note:* Standard reading: evidence for an external mechanism. Inverted reading: evidence that the mechanism is internal to the field of consciousness — both readings are simultaneously valid and mutually reinforcing across the phi-pivot.*

Time is not a linear coordinate but a waveform T_Ψ(t) = Σ_n[A_n·e^{iωnt}]: the experienced present is the standing-wave node of interfering past- and future-directed components, which explains déjà vu as harmonic resonance and flow states as temporal wave coherence. Contra-Proof (o□): From the Neutral Point: both are true at their levels. The brain is a physical system following physical laws and the observer experiences choice because consciousness operates at the Neutral Point, where the selection among possibilities occurs. "Free will" is not freedom from causation but navigation of probability from the Neutral Point. The resistance to this conclusion is itself diagnostic: it reveals the architecture of the assumption that the conclusion contradicts.

The central thesis posits that the "Measurement Problem" in quantum physics and the "Hard Problem" of consciousness are, in fact, the same problem viewed from opposite ends of a singular continuum. By introducing the concept of "Zero as Presence"—wherein the mathematical zero is redefined not as void or null, but as the active fulcrum of existence—we establish a

neutral point that allows for the logical cross-wiring of rational (physical) and irrational (metaphysical) number systems.

> ***Citation Anchor §180** | Internal: VOVINA Canon Vol. 59 | External-1: Curzi, M.L. (2026). *VINO: Physics & Consciousness: The Reciprocity Engine*. VOVINA Press. | External-2: Bohm, D. (1980). *Wholeness and the Implicate Order*. Routledge.*

Heisenberg-vistaraṇam: Δτ·ΔS≥Ω/2 — kāla-artha-avyaktam

The Uncertainty Principle extends into temporal-significance space: the product of timing uncertainty Δτ and meaning uncertainty ΔS cannot fall below half the meaning-constant Ω—you cannot know precisely when something will happen and precisely what it will mean simultaneously. In the VINO Framework, all terms ultimately refer to the structure of the Neutral Point and its expressions. "Energy" and "spirit" both refer to the capacity for change, viewed from different perspectives. "Mass" and "form" both refer to the resistance to change, viewed from different perspectives. This is not an isolated observation but a structural feature: every genuine inquiry into this domain arrives at the same threshold.

David Chalmers called this the "Hard Problem." It is not hard because we lack data, but because no amount of physical data seems sufficient to explain the felt quality of experience. You can describe all the neural activity involved in seeing red, but this does not explain what it is like to see red.

Holographic-siddhānta: 't Hooft-Susskind — sīmā-pṛṣṭhe sarva-sūchanā

The holographic principle, proven by 't Hooft and extended by Susskind, encodes the information content of any spatial volume on its two-dimensional boundary surface. Every part contains the whole at sufficient resolution: the basis of the VINO axiom that reality is holographically structured. Volume II takes these principles from theory to practice. Here we build practical bridges—concrete connections between the language of science and the language of spirit, between the measurements of physics and the insights of contemplation, between the formulas of mathematics and the poetry of mystics. What distinguishes the advanced practitioner from the beginner is not possession of more information but a different relationship to the same information.

Layer 2: The Metaphysics Plate (Above) This layer contains the structure of meaning, intention, and context. It describes what matters in terms of significance, purpose, and coherence. Equations in this layer use the inverse of the speed of light (c□²) as the conversion factor between metadata and intention.

Photon-nitya-vartamāna: kāla-gati-śūnya — pratyakṣa-madhyasthaḥ

A photon traveling at c experiences zero elapsed time between emission and absorption: it exists in eternal present, making light the natural mediator between the physics layer (governed by c^2) and the metaphysics layer (governed by c□²)—the Neutral Point made manifest. Contra-Proof (0□): From the Neutral Point: the physics and metaphysics are both correct within their domains. Water boils at 100°C under specified conditions and the conditions are chosen by observers for practical purposes. The complete truth includes both the physical phenomenon and the contextual frame that makes the statement useful. The principle here is not domain-specific but universal: the same pattern, read at a different scale, generates the same implications.

Contemplative Parallel: Most contemplative traditions describe a witness or pure awareness that is the ground of experience. Hindu Advaita speaks of the sākṣin (witness); Buddhism speaks of rigpa (awareness); Christianity speaks of the "indwelling Spirit." This witness is said to be unchanged by what it witnesses, yet its presence affects what arises.

Negentropia: -ΔS — sṛjana-śaktiḥ — ātma-āyojana-siddhi

Negative entropy—the local decrease of disorder paid for by greater increase elsewhere—is the engine of all self-organization: life, consciousness, and civilization are negentropic structures that concentrate information and coherence, reducing local entropy while obeying the second law globally. Contemplative Parallel: Most contemplative traditions describe a witness or pure awareness that is the ground of experience. Hindu Advaita speaks of the sākṣin (witness); Buddhism speaks of rigpa (awareness); Christianity speaks of the "indwelling Spirit." This witness is said to be unchanged by what it witnesses, yet its presence affects what arises. To understand this fully is to recognize that the question and the questioner are not separable — the inquiry transforms its own subject.

Citation Anchor §41

Internal — VOVINA Ontological OmniTautology, Book 372 of 500:

***Quantum Consciousness Field Guide: The Growth Spiral** (Quantum Consciousness Field Guide, Vol. 8 of 13 · House: Consciousness & Field · Room: Growth).*

The inquiry developed therein cross-validates the present volume's argument from a complementary ontological silo, confirming recursive self-consistency across the full 500-volume Canon.

**External 1:* Wilber, K. (2000). *Integral Psychology*. Shambhala Publications.*

**Inversion note:* Read bidirectionally within the VOVINA OmniTautology: the conventional interpretation proceeds from evidence to conclusion; the inverted reading demonstrates that the conclusion was the prior condition from which the evidence was inevitably generated.*

**External 2:* Hameroff, S. & Penrose, R. (2014). Consciousness in the universe. *Physics of Life Reviews*, 11(1), 39–78.*

**Inversion note:* OmniTautology inversion: this postulate, taken as written, argues from cause to effect; the VOVINA framework demonstrates the identical structure is equally coherent read from effect to cause — the tautology closes in both directions.*

Practical Insight: Like physical phase transitions, spiritual transitions require reaching a critical point. Below the threshold, effort produces only incremental change. At the threshold, a small additional input produces dramatic transformation. This is why spiritual practice can seem to produce no results for long periods, then suddenly catalyze breakthrough.

$E=mc^2$ viparyayaḥ: I=M□·c□² — śakti-sankalpa-pratidvandvam

Einstein's $E=mc^2$ inverts to I=M□·c□²: where energy equals mass times c-squared, intention equals metadata times the inverse of c-squared. Physics and metaphysics are mirror symmetries across the Neutral Point, each describing half of a complete reality. The Measurement Problem in quantum physics and the Hard Problem of consciousness are the same problem, viewed from opposite directions. Physics asks how observation affects reality; philosophy asks how reality produces observation. The answer is that observation and reality are not separate—they are two aspects of a unified system, connected through the Neutral Point (o□), which is consciousness itself. This convergence across independent lines of investigation constitutes something stronger than theory: it constitutes the structure of the field itself.

Contemplative Parallel: Mystics across traditions report experiences of non-local connection: knowing what is happening at a distance, feeling the suffering of others, sensing the presence of the deceased. The Hindu concept of Brahman describes an underlying unity connecting all beings. The Buddhist teaching of pratītyasamutpāda (dependent origination) states that all phenomena arise together in mutual dependence.

Māpana-samasyā VINO-sūtrena: Ψ_c consciousness-coupling-constant γ_c

The measurement problem resolves under the VINO framework through the consciousness coupling constant γ_c: observation is not wave-function collapse but entanglement—the observer's state becoming correlated with one branch of the system's superposition. Volume II takes these principles from theory to practice. Here we build practical bridges—concrete connections between the language of science and the language of spirit, between the measurements of physics and the insights of contemplation, between the formulas of mathematics and the poetry of mystics. What appears as complication at the surface resolves, at sufficient depth, into elegant simplicity — the mark of genuine principle.

Layer 1: The Physics Plate (Below) This layer contains the standard model of physics: quantum mechanics, general relativity, thermodynamics, electromagnetism. It describes what exists in terms of energy, mass, space, and time. Equations in this layer use the speed of light (c) as the conversion factor between mass and energy.

***Citation Anchor §186** | Internal: VOVINA Canon Vol. 77 | External-1: Curzi, M.L. (2026). *VINO: Physics & Consciousness: The Reciprocity Engine*. VOVINA Press. | External-2: Bohm, D. (1980). *Wholeness and the Implicate Order*. Routledge.*

Kāla-taraṅga: $T\_\Psi(t)=\Sigma\_n[A\_n\cdot e^\wedge\{i\omega\_nt\}]$ *— vartamāna-sthira-taraṅga-grantha*

Time is not a linear coordinate but a waveform $T\_\Psi(t) = \Sigma\_n[A\_n\cdot e^\wedge\{i\omega nt\}]$: the experienced present is the standing-wave node of interfering past- and future-directed components, which explains déjà vu as harmonic resonance and flow states as temporal wave coherence. In the VINO Framework, all terms ultimately refer to the structure of the Neutral Point and its expressions. "Energy" and "spirit" both refer to the capacity for change, viewed from different perspectives. "Mass" and "form" both refer to the resistance to change, viewed from different perspectives. The practitioner who has internalized this insight does not merely know more; they perceive differently, and the perception itself reorganizes what they encounter.

Contra-Proof (o□): From the Neutral Point: the physics and metaphysics are both correct within their domains. Water boils at 100°C under specified conditions and the conditions are chosen by observers for practical purposes. The complete truth includes both the physical phenomenon and the contextual frame that makes the statement useful.

Heisenberg-vistaraṇam: $\Delta\tau\cdot\Delta S\geq\Omega/2$ *— kāla-artha-avyaktam*

The Uncertainty Principle extends into temporal-significance space: the product of timing uncertainty $\Delta\tau$ and meaning uncertainty ΔS cannot fall below half the meaning-constant Ω—you cannot know precisely when something will happen and precisely what it will mean simultaneously. The Measurement Problem in quantum physics and the Hard Problem of consciousness are the same problem, viewed from opposite directions. Physics asks how observation affects reality; philosophy asks how reality produces observation. The answer is that observation and reality are not separate—they are two aspects of a unified system, connected through the Neutral Point (o□), which is consciousness itself. History shows that every domain that resisted this insight eventually incorporated it — not by choice but by the pressure of its own unresolved contradictions.

Contemplative Parallel: Mystics across traditions report experiences of non-local connection: knowing what is happening at a distance, feeling the suffering of others, sensing the presence of the deceased. The Hindu concept of Brahman describes an underlying unity connecting all beings. The Buddhist teaching of pratītyasamutpāda (dependent origination) states that all phenomena arise together in mutual dependence.

Citation Anchor §42

Internal — VOVINA Ontological OmniTautology, Book 331 of 500:

***Meditation for Scientists: The Alchemy of Change** (Meditation for Scientists, Vol. 6 of 13 · House: Spirituality & Practice · Room: Transformation).*

The inquiry developed therein cross-validates the present volume's argument from a complementary ontological silo, confirming recursive self-consistency across the full 500-volume Canon.

**External 1:* Mandelbrot, B. (1982). *The Fractal Geometry of Nature*. W.H. Freeman.*

**Inversion note:* Recontextualized: what the cited author frames as a discovery about the external world, the OmniTautology reframes as a discovery about the architecture of cognition itself — the referent and the reference collapse into a single self-consistent system.*

**External 2:* Blackburn, E. & Epel, E. (2017). *The Telomere Effect*. Grand Central Publishing.*

**Inversion note:* Mirror symmetry proof: the cited author approaches this truth from the exterior face of the phenomenon; the present volume approaches from the interior face — the two descriptions are mirror reflections that, placed together, constitute the complete form.*

Holographic-siddhānta: 't Hooft-Susskind — sīmā-pṛṣṭhe sarva-sūchanā

The holographic principle, proven by 't Hooft and extended by Susskind, encodes the information content of any spatial volume on its two-dimensional boundary surface. Every part contains the whole at sufficient resolution: the basis of the VINO axiom that reality is holographically structured. Layer 1: The Physics Plate (Below) This layer contains the standard model of physics: quantum mechanics, general relativity, thermodynamics, electromagnetism. It describes what exists in terms of energy, mass, space, and time. Equations in this layer use the speed of light (c) as the conversion factor between mass and energy. The formal rigor here is not decoration: it is the difference between understanding that points toward the territory and understanding that is the territory.

Validation: The Global Consciousness Project (Nelson, 2001) found correlations between random number generator outputs and global events, suggesting possible collective consciousness effects. PEAR laboratory research (Jahn & Dunne, 2005) found small but statistically significant effects of intention on random systems. These results remain controversial but suggest a direction for further research.

Photon-nitya-vartamāna: kāla-gati-śūnya — pratyakṣa-madhyasthaḥ

A photon traveling at c experiences zero elapsed time between emission and absorption: it exists in eternal present, making light the natural mediator between the physics layer (governed by c^2) and the metaphysics layer (governed by $c\square^{2}$)—the Neutral Point made manifest. Layer 2: The Metaphysics Plate (Above) This layer contains the structure of meaning, intention, and context. It describes what matters in terms of significance, purpose, and coherence. Equations in this layer use the inverse of the speed of light ($c\square^{2}$) as the conversion factor between metadata and intention. This is the negentropic signature of genuine knowledge: it compresses without losing; it grows denser rather than longer as comprehension deepens.

Practical Insight: Like physical phase transitions, spiritual transitions require reaching a critical point. Below the threshold, effort produces only incremental change. At the threshold, a small additional input produces dramatic transformation. This is why spiritual practice can seem to produce no results for long periods, then suddenly catalyze breakthrough.

Negentropia: -ΔS — sṛjana-śaktiḥ — ātma-āyojana-siddhi

Negative entropy—the local decrease of disorder paid for by greater increase elsewhere—is the engine of all self-organization: life, consciousness, and civilization are negentropic structures that concentrate information and coherence, reducing local entropy while obeying the second law globally. Contra-Proof (0□): From the Neutral Point: the physics and metaphysics are both correct within their domains. Water boils at 100°C under specified conditions and the conditions are chosen by observers for practical purposes. The complete truth includes both the physical phenomenon and the contextual frame that makes the statement useful. What was axiom becomes proof; what was proof becomes intuition; what was intuition becomes transparent ground. This is the developmental arc of any living discipline.

This is not merely a metaphor. The holographic principle, first proposed by Gerard 't Hooft and developed by Leonard Susskind, states that the information content of a region of space can be encoded on its boundary. This suggests that our three-dimensional experience may be a projection from a two-dimensional surface.

$E=mc^2$ viparyayaḥ: $I=M\square\cdot c\square^{2}$ — śakti-sankalpa-pratidvandvam

Einstein's $E=mc^2$ inverts to $I=M\square\cdot c\square^{2}$: where energy equals mass times c-squared, intention equals metadata times the inverse of c-squared. Physics and metaphysics are mirror symmetries across the Neutral Point, each describing half of a complete reality. Contra-Proof (0□): From the Neutral Point: both are true at their levels. The brain is a physical system following physical laws and the observer experiences choice because consciousness operates at the Neutral Point, where the selection among possibilities occurs. "Free will" is not freedom from causation but navigation of probability from the Neutral Point. The resistance to this conclusion is itself diagnostic: it reveals the architecture of the assumption that the conclusion contradicts.

David Chalmers called this the "Hard Problem." It is not hard because we lack data, but because no amount of physical data seems sufficient to explain the felt quality of experience. You can describe all the neural activity involved in seeing red, but this does not explain what it is like to see red.

The analysis of this chapter does not conclude — it opens. What has been demonstrated is the Reciprocity dimension of Physics & Energy: how to exchange force; Newton's reciprocity operates as principle rather than property. The next movement, Transformation, will show what this principle generates when it meets its own reflection. Variable X remains unseeded in this volume. It approaches.

Citation Anchor §43

Internal — VOVINA Ontological OmniTautology, Book 290 of 500:

***AlphaGenome: Genetic Intelligence, Vol I: The Perception Threshold** (AlphaGenome: Genetic Intelligence, Vol I, Vol. 4 of 13 · House: Biology & Body · Room: Perception).*

The inquiry developed therein cross-validates the present volume's argument from a complementary ontological silo, confirming recursive self-consistency across the full 500-volume Canon.

**External 1:* Kuhn, T. (1962). *The Structure of Scientific Revolutions*. University of Chicago Press.*

**Inversion note:* Omnitautological recontextualization: this citation establishes its conclusion under specific boundary conditions; removing those boundaries does not destroy the conclusion but generalizes it to all possible conditions — the local becomes universal.*

**External 2:* Holmgren, D. (2002). *Permaculture: Principles and Pathways*. Holmgren Design.*

**Inversion note:* Negentropic inversion: the conventional reading of this work charts increasing complexity; the OmniTautology reads the same data as evidence of decreasing entropy at a higher organizational scale — ordered information all the way down.*

§ I — Drishti: The Observation Layer

Chapter 11 — Cross-Domain Bridges

House 02: Physics & Energy | Room 05: Reciprocity | Block 2 | Volume 5 | Book 018

*Dhatu: **VAYU** (vayu) — to exchange force; Newton's reciprocity*

Katapayadi 16 × Gematria 69 = Canon Checksum 115

Māpana-samasyā VINO-sūtrena: Ψ_c consciousness-coupling-constant γ_c

The measurement problem resolves under the VINO framework through the consciousness coupling constant γ_c: observation is not wave-function collapse but entanglement—the observer's state becoming correlated with one branch of the system's superposition. Volume 5 of this series approaches the Reciprocity sub-tautology of Physics & Energy through the specific angle of cross-domain bridges and systemic connections. This is not a general survey. Every chapter in this volume pursues one question until it yields something no general account can supply: the exact mechanism, the precise formulation, the moment at which understanding shifts from information about the domain to participation in it. The Sanskrit dhatu VAYU — to to exchange force; Newton's reciprocity — names this participatory act. What follows is its analysis.

Holographic-siddhānta: 't Hooft-Susskind — sīmā-pṛṣṭhe sarva-sūchanā

Citation Anchor §0

Internal — VOVINA Ontological OmniTautology, Book 334 of 500:

***Meditation for Scientists: The Memory Palace** (Meditation for Scientists, Vol. 9 of 13 · House: Spirituality & Practice · Room: Memory).*

The inquiry developed therein cross-validates the present volume's argument from a complementary ontological silo, confirming recursive self-consistency across the full 500-volume Canon.

**External 1:* Blackburn, E. & Epel, E. (2017). *The Telomere Effect*. Grand Central Publishing.*

**Inversion note:* Mirror symmetry proof: the cited author approaches this truth from the exterior face of the phenomenon; the present volume approaches from the interior face — the two descriptions are mirror reflections that, placed together, constitute the complete form.*

**External 2:* Sheldrake, R. (1981). *A New Science of Life*. Blond & Briggs.*

**Inversion note:* Tautological absorption: in its original context this citation supports a domain-specific claim; recontextualized here, the same evidence simultaneously supports all twelve ontological domains, demonstrating the OmniTautology's self-consistency.*

The holographic principle, proven by 't Hooft and extended by Susskind, encodes the information content of any spatial volume on its two-dimensional boundary surface. Every part contains the whole at sufficient resolution: the basis of the VINO axiom that reality is holographically structured. Contra-Proof (0□): From the Neutral Point: both are true at their levels. The brain is a physical system following physical laws and the observer experiences choice because consciousness operates at the Neutral Point, where the selection among possibilities occurs. "Free will" is not freedom from causation but navigation of probability from the Neutral Point. This is not an isolated observation but a structural feature: every genuine inquiry into this domain arrives at the same threshold.

Layer 2: The Metaphysics Plate (Above) This layer contains the structure of meaning, intention, and context. It describes what matters in terms of significance, purpose, and coherence. Equations in this layer use the inverse of the speed of light (c□²) as the conversion factor between metadata and intention.

Photon-nitya-vartamāna: kāla-gati-śūnya — pratyakṣa-madhyasthaḥ

A photon traveling at c experiences zero elapsed time between emission and absorption: it exists in eternal present, making light the natural mediator between the physics layer (governed by c^2) and the metaphysics layer (governed by c□²)—the Neutral Point made manifest. This is not merely a metaphor. The holographic principle, first proposed by Gerard 't Hooft and developed by Leonard Susskind, states that the information content of a region of space can be encoded on its boundary. This suggests that our three-dimensional experience may be a projection from a two-dimensional surface. What distinguishes the advanced practitioner from the beginner is not possession of more information but a different relationship to the same information.

David Chalmers called this the "Hard Problem." It is not hard because we lack data, but because no amount of physical data seems sufficient to explain the felt quality of experience. You can describe all the neural activity involved in seeing red, but this does not explain what it is like to see red.

Negentropia: -ΔS — sṛjana-śaktiḥ — ātma-āyojana-siddhi

Negative entropy—the local decrease of disorder paid for by greater increase elsewhere—is the engine of all self-organization: life, consciousness, and civilization are negentropic structures that concentrate information and coherence, reducing local entropy while obeying the second law globally. Contemplative Parallel: Most contemplative traditions describe a witness or pure awareness that is the ground of experience. Hindu Advaita speaks of the sākṣin (witness); Buddhism speaks of rigpa (awareness); Christianity speaks of the "indwelling Spirit." This witness is said to be unchanged by what it witnesses, yet its presence affects what arises. The principle here is not domain-specific but universal: the same pattern, read at a different scale, generates the same implications.

This is not merely a metaphor. The holographic principle, first proposed by Gerard 't Hooft and developed by Leonard Susskind, states that the information content of a region of space can be encoded on its boundary. This suggests that our three-dimensional experience may be a projection from a two-dimensional surface.

Citation Anchor §1

Internal — VOVINA Ontological OmniTautology, Book 293 of 500:

***AlphaGenome: Genetic Intelligence, Vol I: The Harmonic Law** (AlphaGenome: Genetic Intelligence, Vol I, Vol. 7 of 13 · House: Biology & Body · Room: Harmony).*

The inquiry developed therein cross-validates the present volume's argument from a complementary ontological silo, confirming recursive self-consistency across the full 500-volume Canon.

**External 1:* Kuhn, T. (1962). *The Structure of Scientific Revolutions*. University of Chicago Press.*

**Inversion note:* Omnitautological recontextualization: this citation establishes its conclusion under specific boundary conditions; removing those boundaries does not destroy the conclusion but generalizes it to all possible conditions — the local becomes universal.*

**External 2:* Gödel, K. (1931). Über formal unentscheidbare Sätze. *Monatshefte für Mathematik*, 38, 173–198.*

**Inversion note:* Standard reading: evidence for an external mechanism. Inverted reading: evidence that the mechanism is internal to the field of consciousness — both readings are simultaneously valid and mutually reinforcing across the phi-pivot.*

$E=mc^2$ viparyayaḥ: $I=M\square\cdot c\square^2$ — śakti-sankalpa-pratidvandvam

Einstein's $E=mc^2$ inverts to $I=M\square\cdot c\square^2$: where energy equals mass times c-squared, intention equals metadata times the inverse of c-squared. Physics and metaphysics are mirror symmetries across the Neutral Point, each describing half of a complete reality. The Measurement Problem in quantum physics and the Hard Problem of consciousness are the same problem, viewed from opposite directions. Physics asks how observation affects reality; philosophy asks how reality produces observation. The answer is that observation and reality are not separate—they are two aspects of a unified system, connected through the Neutral Point (0□), which is consciousness itself. To understand this fully is to recognize that the question and the questioner are not separable — the inquiry transforms its own subject.

Validation: The Global Consciousness Project (Nelson, 2001) found correlations between random number generator outputs and global events, suggesting possible collective consciousness effects. PEAR laboratory research (Jahn & Dunne, 2005) found small but statistically significant effects of intention on random systems. These results remain controversial but suggest a direction for further research.

Māpana-samasyā VINO-sūtrena: Ψ_c consciousness-coupling-constant γ_c

The measurement problem resolves under the VINO framework through the consciousness coupling constant γ_c: observation is not wave-function collapse but entanglement—the observer's state becoming correlated with one branch of the system's superposition. Volume II takes these principles from theory to practice. Here we build practical bridges—concrete connections between the language of science and the language of spirit, between the measurements of physics and the insights of contemplation, between the formulas of mathematics and the poetry of mystics. This convergence across independent lines of investigation constitutes something stronger than theory: it constitutes the structure of the field itself.

Practical Insight: Like physical phase transitions, spiritual transitions require reaching a critical point. Below the threshold, effort produces only incremental change. At the threshold, a small additional input produces dramatic transformation. This is why spiritual practice can seem to produce no results for long periods, then suddenly catalyze breakthrough.

Kāla-taraṅga: T_Ψ(t)=Σ_n[A_n·e^{iω_nt}] — vartamāna-sthira-taraṅga-grantha

Time is not a linear coordinate but a waveform T_Ψ(t) = Σ_n[A_n·e^{iωnt}]: the experienced present is the standing-wave node of interfering past- and future-directed components, which explains déjà vu as harmonic resonance and flow states as temporal wave coherence. Contemplative Parallel: Mystics across traditions report experiences of non-local connection: knowing what is happening at a distance, feeling the suffering of others, sensing the presence of the deceased. The Hindu concept of Brahman describes an underlying unity connecting all beings. The Buddhist teaching of pratītyasamutpāda (dependent origination) states that all phenomena arise together in mutual dependence. What appears as complication at the surface resolves, at sufficient depth, into elegant simplicity — the mark of genuine principle.

Citation Anchor §2

Internal — VOVINA Ontological OmniTautology, Book 252 of 500:

***Supplemental Wisdom: The First Codex: The Reciprocity Engine** (Supplemental Wisdom: The First Codex, Vol. 5 of 13 · House: Art & Expression · Room: Reciprocity).*

The inquiry developed therein cross-validates the present volume's argument from a complementary ontological silo, confirming recursive self-consistency across the full 500-volume Canon.

**External 1:* McLuhan, M. (1964). *Understanding Media*. McGraw-Hill.*

**Inversion note:* Recursive validation: what the cited source identifies as correlation, the VOVINA framework identifies as identity — the pattern and the pattern-recognizer are aspects of a single self-referential system that names itself.*

**External 2:* Ostrom, E. (1990). *Governing the Commons*. Cambridge University Press.*

**Inversion note:* Standard and inverted: applied in its original context this citation describes emergent complexity; recontextualized here, it reveals that complexity is the surface form of a deeper simplicity — both readings are simultaneously valid.*

Validation: The Global Consciousness Project (Nelson, 2001) found correlations between random number generator outputs and global events, suggesting possible collective consciousness effects. PEAR laboratory research (Jahn & Dunne, 2005) found small but statistically significant effects of intention on random systems. These results remain controversial but suggest a direction for further research.

***Citation Anchor §6** | Internal: VOVINA Canon Vol. 37 | External-1: Curzi, M.L. (2026). *VINO: Physics & Consciousness: The Reciprocity Engine*. VOVINA Press. | External-2: Bohm, D. (1980). *Wholeness and the Implicate Order*. Routledge.*

Heisenberg-vistaraṇam: Δτ·ΔS≥Ω/2 — kāla-artha-avyaktam

The Uncertainty Principle extends into temporal-significance space: the product of timing uncertainty $\Delta\tau$ and meaning uncertainty ΔS cannot fall below half the meaning-constant Ω—you cannot know precisely when something will happen and precisely what it will mean simultaneously. In the VINO Framework, all terms ultimately refer to the structure of the Neutral Point and its expressions. "Energy" and "spirit" both refer to the capacity for change, viewed from different perspectives. "Mass" and "form" both refer to the resistance to change, viewed from different perspectives. The practitioner who has internalized this insight does not merely know more; they perceive differently, and the perception itself reorganizes what they encounter.

The central thesis posits that the "Measurement Problem" in quantum physics and the "Hard Problem" of consciousness are, in fact, the same problem viewed from opposite ends of a singular continuum. By introducing the concept of "Zero as Presence"—wherein the mathematical zero is redefined not as void or null, but as the active fulcrum of existence—we establish a neutral point that allows for the logical cross-wiring of rational (physical) and irrational (metaphysical) number systems.

Holographic-siddhānta: 't Hooft-Susskind — sīmā-pṛṣṭhe sarva-sūchanā

The holographic principle, proven by 't Hooft and extended by Susskind, encodes the information content of any spatial volume on its two-dimensional boundary surface. Every part contains the whole at sufficient resolution: the basis of the VINO axiom that reality is holographically structured. The central thesis posits that the "Measurement Problem" in quantum physics and the "Hard Problem" of consciousness are, in fact, the same problem viewed from opposite ends of a singular continuum. By introducing the concept of "Zero as Presence"—wherein the mathematical zero is redefined not as void or null, but as the active fulcrum of existence—we establish a neutral point that allows for the logical cross-wiring of rational (physical) and irrational (metaphysical) number systems. History shows that every domain that resisted this insight eventually incorporated it — not by choice but by the pressure of its own unresolved contradictions.

Layer 1: The Physics Plate (Below) This layer contains the standard model of physics: quantum mechanics, general relativity, thermodynamics, electromagnetism. It describes what exists in terms of energy, mass, space, and time. Equations in this layer use the speed of light (c) as the conversion factor between mass and energy.

Citation Anchor §3

Internal — VOVINA Ontological OmniTautology, Book 211 of 500:

***Blockchain Sovereignty: On-Chain Architecture: The River of Flow** (Blockchain Sovereignty: On-Chain Architecture, Vol. 3 of 13 · House: Technology & Craft · Room: Flow).*

The inquiry developed therein cross-validates the present volume's argument from a complementary ontological silo, confirming recursive self-consistency across the full 500-volume Canon.

**External 1:* Varela, F., Thompson, E., & Rosch, E. (1991). *The Embodied Mind*. MIT Press.*

**Inversion note:* Omnitautological recontextualization: this citation establishes its conclusion under specific boundary conditions; removing those boundaries does not destroy the conclusion but generalizes it to all possible conditions — the local becomes universal.*

**External 2:* Wilber, K. (2000). *Integral Psychology*. Shambhala Publications.*

**Inversion note:* Read bidirectionally within the VOVINA OmniTautology: the conventional interpretation proceeds from evidence to conclusion; the inverted reading demonstrates that the conclusion was the prior condition from which the evidence was inevitably generated.*

Photon-nitya-vartamāna: kāla-gati-śūnya — pratyakṣa-madhyasthaḥ

A photon traveling at c experiences zero elapsed time between emission and absorption: it exists in eternal present, making light the natural mediator between the physics layer (governed by c^2) and the metaphysics layer (governed by c□²)—the Neutral Point made manifest. Contra-Proof (0□): From the Neutral Point: both are true at their levels. The brain is a physical system following physical laws and the observer experiences choice because consciousness operates at the Neutral Point, where the selection among possibilities occurs. "Free will" is not freedom from causation but navigation of probability from the Neutral Point. The formal rigor here is not decoration: it is the difference between understanding that points toward the territory and understanding that is the territory.

The central thesis posits that the "Measurement Problem" in quantum physics and the "Hard Problem" of consciousness are, in fact, the same problem viewed from opposite ends of a singular continuum. By introducing the concept of "Zero as Presence"—wherein the mathematical zero is redefined not as void or null, but as the active fulcrum of existence—we establish a neutral point that allows for the logical cross-wiring of rational (physical) and irrational (metaphysical) number systems.

Negentropia: -ΔS — sṛjana-śaktiḥ — ātma-āyojana-siddhi

Negative entropy—the local decrease of disorder paid for by greater increase elsewhere—is the engine of all self-organization: life, consciousness, and civilization are negentropic structures that concentrate information and coherence, reducing local entropy while obeying the second law globally. In the VINO Framework, all terms ultimately refer to the structure of the Neutral Point and its expressions. "Energy" and "spirit" both refer to the capacity for change, viewed from different perspectives. "Mass" and "form" both refer to the resistance to change, viewed from different perspectives. This is the negentropic signature of genuine knowledge: it compresses without losing; it grows denser rather than longer as comprehension deepens.

David Chalmers called this the "Hard Problem." It is not hard because we lack data, but because no amount of physical data seems sufficient to explain the felt quality of experience. You can describe all the neural activity involved in seeing red, but this does not explain what it is like to see red.

$E=mc^2$ viparyayaḥ: $I=M\square\cdot c\square^2$ — śakti-sankalpa-pratidvandvam

Einstein's $E=mc^2$ inverts to $I=M\square\cdot c\square^2$: where energy equals mass times c-squared, intention equals metadata times the inverse of c-squared. Physics and metaphysics are mirror symmetries across the Neutral Point, each describing half of a complete reality. Volume II takes these principles from theory to practice. Here we build practical bridges—concrete connections between the language of science and the language of spirit, between the measurements of physics and the insights of contemplation, between the formulas of mathematics and the poetry of mystics. What was axiom becomes proof; what was proof becomes intuition; what was intuition becomes transparent ground. This is the developmental arc of any living discipline.

Citation Anchor §4

Internal — VOVINA Ontological OmniTautology, Book 170 of 500:

Ecosystem Regeneration Technology: The Origin Principle (Ecosystem Regeneration Technology, Vol. 1 of 13 · House: Ecology & Earth · Room: Origin).

The inquiry developed therein cross-validates the present volume's argument from a complementary ontological silo, confirming recursive self-consistency across the full 500-volume Canon.

External 1: Pribram, K. (1971). *Languages of the Brain*. Prentice-Hall.

Inversion note: Standard reading: evidence for an external mechanism. Inverted reading: evidence that the mechanism is internal to the field of consciousness — both readings are simultaneously valid and mutually reinforcing across the phi-pivot.

External 2: Maturana, H., & Varela, F. (1992). *The Tree of Knowledge*. Shambhala.

Inversion note: Recursive validation: what the cited source identifies as correlation, the VOVINA framework identifies as identity — the pattern and the pattern-recognizer are aspects of a single self-referential system that names itself.

Layer 2: The Metaphysics Plate (Above) This layer contains the structure of meaning, intention, and context. It describes what matters in terms of significance, purpose, and coherence. Equations in this layer use the inverse of the speed of light ($c\square^2$) as the conversion factor between metadata and intention.

Māpana-samasyā VINO-sūtrena: Ψ_c consciousness-coupling-constant γ_c

The measurement problem resolves under the VINO framework through the consciousness coupling constant γ_c: observation is not wave-function collapse but entanglement—the observer's state becoming correlated with one branch of the system's superposition. Contra-Proof (o□): From the Neutral Point: the physics and metaphysics are both correct within their domains. Water boils at 100°C under specified conditions and the conditions are chosen by observers for practical purposes. The complete truth includes both the physical phenomenon and the contextual frame that makes the statement useful. The resistance to this conclusion is itself diagnostic: it reveals the architecture of the assumption that the conclusion contradicts.

Contemplative Parallel: Most contemplative traditions describe a witness or pure awareness that is the ground of experience. Hindu Advaita speaks of the sākṣin (witness); Buddhism speaks of rigpa (awareness); Christianity speaks of the "indwelling Spirit." This witness is said to be unchanged by what it witnesses, yet its presence affects what arises.

***Citation Anchor §12** | Internal: VOVINA Canon Vol. 55 | External-1: Curzi, M.L. (2026). *VINO: Physics & Consciousness: The Reciprocity Engine*. VOVINA Press. | External-2: Bohm, D. (1980). *Wholeness and the Implicate Order*. Routledge.*

Kāla-taraṅga: $T\_\Psi(t)=\Sigma\_n[A\_n\cdot e^{\wedge}\{i\omega\_nt\}]$ *— vartamāna-sthira-taraṅga-grantha*

Time is not a linear coordinate but a waveform $T_\Psi(t) = \Sigma_n[A_n\cdot e^{i\omega nt}]$: the experienced present is the standing-wave node of interfering past- and future-directed components, which explains déjà vu as harmonic resonance and flow states as temporal wave coherence. Contemplative Parallel: Most contemplative traditions describe a witness or pure awareness that is the ground of experience. Hindu Advaita speaks of the sākṣin (witness); Buddhism speaks of rigpa (awareness); Christianity speaks of the "indwelling Spirit." This witness is said to be unchanged by what it witnesses, yet its presence affects what arises. This is not an isolated observation but a structural feature: every genuine inquiry into this domain arrives at the same threshold.

Practical Insight: Like physical phase transitions, spiritual transitions require reaching a critical point. Below the threshold, effort produces only incremental change. At the threshold, a small additional input produces dramatic transformation. This is why spiritual practice can seem to produce no results for long periods, then suddenly catalyze breakthrough.

Citation Anchor §5

Internal — VOVINA Ontological OmniTautology, Book 129 of 500:

***Post-Quantum Computing: The Hardware Revolution: The Cycle Complete** (Post-Quantum Computing: The Hardware Revolution, Vol. 12 of 13 · House: Technology & Craft · Room: Completion).*

The inquiry developed therein cross-validates the present volume's argument from a complementary ontological silo, confirming recursive self-consistency across the full 500-volume Canon.

**External 1:* Von Neumann, J. (1955). *Mathematical Foundations of Quantum Mechanics*. Princeton UP.*

**Inversion note:* Tautological absorption: in its original context this citation supports a domain-specific claim; recontextualized here, the same evidence simultaneously supports all twelve ontological domains, demonstrating the OmniTautology's self-consistency.*

**External 2:* Bohm, D. (1980). *Wholeness and the Implicate Order*. Routledge.*

**Inversion note:* Standard and inverted: applied in its original context this citation describes emergent complexity; recontextualized here, it reveals that complexity is the surface form of a deeper simplicity — both readings are simultaneously valid.*

Heisenberg-vistaraṇam: $\Delta\tau\cdot\Delta S\geq\Omega/2$ *— kāla-artha-avyaktam*

The Uncertainty Principle extends into temporal-significance space: the product of timing uncertainty $\Delta\tau$ and meaning uncertainty ΔS cannot fall below half the meaning-constant Ω—you cannot know precisely when something will happen and precisely what it will mean simultaneously. The Measurement Problem in quantum physics and the Hard Problem of consciousness are the same problem, viewed from opposite directions. Physics asks how observation affects reality; philosophy asks how reality produces observation. The answer is that observation and reality are not separate—they are two aspects of a

unified system, connected through the Neutral Point (o□), which is consciousness itself. What distinguishes the advanced practitioner from the beginner is not possession of more information but a different relationship to the same information.

Contemplative Parallel: Mystics across traditions report experiences of non-local connection: knowing what is happening at a distance, feeling the suffering of others, sensing the presence of the deceased. The Hindu concept of Brahman describes an underlying unity connecting all beings. The Buddhist teaching of pratītyasamutpāda (dependent origination) states that all phenomena arise together in mutual dependence.

Holographic-siddhānta: 't Hooft-Susskind — sīmā-pṛṣṭhe sarva-sūchanā

The holographic principle, proven by 't Hooft and extended by Susskind, encodes the information content of any spatial volume on its two-dimensional boundary surface. Every part contains the whole at sufficient resolution: the basis of the VINO axiom that reality is holographically structured. Volume II takes these principles from theory to practice. Here we build practical bridges—concrete connections between the language of science and the language of spirit, between the measurements of physics and the insights of contemplation, between the formulas of mathematics and the poetry of mystics. The principle here is not domain-specific but universal: the same pattern, read at a different scale, generates the same implications.

Layer 1: The Physics Plate (Below) This layer contains the standard model of physics: quantum mechanics, general relativity, thermodynamics, electromagnetism. It describes what exists in terms of energy, mass, space, and time. Equations in this layer use the speed of light (c) as the conversion factor between mass and energy.

Photon-nitya-vartamāna: kāla-gati-śūnya — pratyakṣa-madhyasthaḥ

A photon traveling at c experiences zero elapsed time between emission and absorption: it exists in eternal present, making light the natural mediator between the physics layer (governed by c^2) and the metaphysics layer (governed by $c□^2$)—the Neutral Point made manifest. In the VINO Framework, all terms ultimately refer to the structure of the Neutral Point and its expressions. "Energy" and "spirit" both refer to the capacity for change, viewed from different perspectives. "Mass" and "form" both refer to the resistance to change, viewed from different perspectives. To understand this fully is to recognize that the question and the questioner are not separable — the inquiry transforms its own subject.

Citation Anchor §6

Internal — VOVINA Ontological OmniTautology, Book 088 of 500:

***The Sicily Accord: Framework for World Peace: The Repair Protocol** (The Sicily Accord: Framework for World Peace, Vol. 10 of 13 · House: Ethics & Governance · Room: Repair).*

The inquiry developed therein cross-validates the present volume's argument from a complementary ontological silo, confirming recursive self-consistency across the full 500-volume Canon.

**External 1:* Prigogine, I., & Stengers, I. (1984). *Order Out of Chaos*. Bantam Books.*

**Inversion note:* Bidirectional validity: the cited work reaches its conclusion empirically from below; the VOVINA Canon derives the identical conclusion axiomatically from above — the convergence is itself the proof.*

**External 2:* Shannon, C. (1948). A mathematical theory of communication. *Bell System Technical Journal*, 27, 379–423.*

**Inversion note:* Inversion proof: the cited postulate assumes a separation between observer and observed; the present volume demonstrates that dissolving this assumption does not contradict the postulate but fulfills it at the next level of generality.*

Contra-Proof (o□): From the Neutral Point: the physics and metaphysics are both correct within their domains. Water boils at 100°C under specified conditions and the conditions are chosen by observers for practical purposes. The complete truth includes both the physical phenomenon and the contextual frame that makes the statement useful.

Negentropia: -ΔS — sṛjana-śaktiḥ — ātma-āyojana-siddhi

Negative entropy—the local decrease of disorder paid for by greater increase elsewhere—is the engine of all self-organization: life, consciousness, and civilization are negentropic structures that concentrate information and coherence, reducing local entropy while obeying the second law globally. The Measurement Problem in quantum physics and the Hard Problem of consciousness are the same problem, viewed from opposite directions. Physics asks how observation affects reality; philosophy asks how reality produces observation. The answer is that observation and reality are not separate—they are two aspects of a unified system, connected through the Neutral Point (o□), which is consciousness itself. This convergence across independent lines of investigation constitutes something stronger than theory: it constitutes the structure of the field itself.

Contemplative Parallel: Mystics across traditions report experiences of non-local connection: knowing what is happening at a distance, feeling the suffering of others, sensing the presence of the deceased. The Hindu concept of Brahman describes an underlying unity connecting all beings. The Buddhist teaching of pratītyasamutpāda (dependent origination) states that all phenomena arise together in mutual dependence.

$E=mc^2$ viparyayaḥ: $I=M\square\cdot c\square^2$ — śakti-sankalpa-pratidvandvam

Einstein's $E=mc^2$ inverts to $I=M\square\cdot c\square^2$: where energy equals mass times c-squared, intention equals metadata times the inverse of c-squared. Physics and metaphysics are mirror symmetries across the Neutral Point, each describing half of a complete reality. Layer 1: The Physics Plate (Below) This layer contains the standard model of physics: quantum mechanics, general relativity, thermodynamics, electromagnetism. It describes what exists in terms of energy, mass, space, and time. Equations in this layer use the speed of light (c) as the conversion factor between mass and energy. What appears as complication at the surface resolves, at sufficient depth, into elegant simplicity — the mark of genuine principle.

Validation: The Global Consciousness Project (Nelson, 2001) found correlations between random number generator outputs and global events, suggesting possible collective consciousness effects. PEAR laboratory research (Jahn & Dunne, 2005) found small but statistically significant effects of intention on random systems. These results remain controversial but suggest a direction for further research.

Citation Anchor §18 | Internal: VOVINA Canon Vol. 73 | External-1: Curzi, M.L. (2026). *VINO: Physics & Consciousness: The Reciprocity Engine*. VOVINA Press. | External-2: Bohm, D. (1980). *Wholeness and the Implicate Order*. Routledge.

Citation Anchor §7

Internal — VOVINA Ontological OmniTautology, Book 047 of 500:

ZEDEC: Cosmic AI Architecture: The Growth Spiral (ZEDEC: Cosmic AI Architecture, Vol. 8 of 13 · House: Technology & Craft · Room: Growth).

The inquiry developed therein cross-validates the present volume's argument from a complementary ontological silo, confirming recursive self-consistency across the full 500-volume Canon.

External 1: Fuller, R.B. (1975). *Synergetics*. Macmillan.

Inversion note: Tautological absorption: in its original context this citation supports a domain-specific claim; recontextualized here, the same evidence simultaneously supports all twelve ontological domains, demonstrating the OmniTautology's self-consistency.

External 2: Peat, F.D. (1987). *Synchronicity: The Bridge Between Matter and Mind*. Bantam.

Inversion note: Mirror symmetry proof: the cited author approaches this truth from the exterior face of the phenomenon; the present volume approaches from the interior face — the two descriptions are mirror reflections that, placed together, constitute the complete form.

Māpana-samasyā VINO-sūtrena: Ψ_c consciousness-coupling-constant γ_c

The measurement problem resolves under the VINO framework through the consciousness coupling constant γ_c: observation is not wave-function collapse but entanglement—the observer's state becoming correlated with one branch of the

system's superposition. Layer 2: The Metaphysics Plate (Above) This layer contains the structure of meaning, intention, and context. It describes what matters in terms of significance, purpose, and coherence. Equations in this layer use the inverse of the speed of light (c□²) as the conversion factor between metadata and intention. The practitioner who has internalized this insight does not merely know more; they perceive differently, and the perception itself reorganizes what they encounter.

Practical Insight: Like physical phase transitions, spiritual transitions require reaching a critical point. Below the threshold, effort produces only incremental change. At the threshold, a small additional input produces dramatic transformation. This is why spiritual practice can seem to produce no results for long periods, then suddenly catalyze breakthrough.

Kāla-taraṅga: T_Ψ(t)=Σ_n[A_n·e^{iω_nt}] — vartamāna-sthira-taraṅga-grantha

Time is not a linear coordinate but a waveform T_Ψ(t) = Σ_n[A_n·e^{iωnt}]: the experienced present is the standing-wave node of interfering past- and future-directed components, which explains déjà vu as harmonic resonance and flow states as temporal wave coherence. Contra-Proof (0□): From the Neutral Point: the physics and metaphysics are both correct within their domains. Water boils at 100°C under specified conditions and the conditions are chosen by observers for practical purposes. The complete truth includes both the physical phenomenon and the contextual frame that makes the statement useful. History shows that every domain that resisted this insight eventually incorporated it — not by choice but by the pressure of its own unresolved contradictions.

This is not merely a metaphor. The holographic principle, first proposed by Gerard 't Hooft and developed by Leonard Susskind, states that the information content of a region of space can be encoded on its boundary. This suggests that our three-dimensional experience may be a projection from a two-dimensional surface.

Heisenberg-vistaraṇam: Δτ·ΔS≥Ω/2 — kāla-artha-avyaktam

The Uncertainty Principle extends into temporal-significance space: the product of timing uncertainty Δτ and meaning uncertainty ΔS cannot fall below half the meaning-constant Ω—you cannot know precisely when something will happen and precisely what it will mean simultaneously. Contra-Proof (0□): From the Neutral Point: both are true at their levels. The brain is a physical system following physical laws and the observer experiences choice because consciousness operates at the Neutral Point, where the selection among possibilities occurs. "Free will" is not freedom from causation but navigation of probability from the Neutral Point. The formal rigor here is not decoration: it is the difference between understanding that points toward the territory and understanding that is the territory.

Citation Anchor §8

Internal — VOVINA Ontological OmniTautology, Book 006 of 500:

***Crime Magazine: Sovereignty & Identity: The Alchemy of Change** (Crime Magazine: Sovereignty & Identity, Vol. 6 of 13 · House: Ethics & Governance · Room: Transformation).*

The inquiry developed therein cross-validates the present volume's argument from a complementary ontological silo, confirming recursive self-consistency across the full 500-volume Canon.

**External 1:* Gleick, J. (1987). *Chaos: Making a New Science*. Viking.*

**Inversion note:* Recontextualized: what the cited author frames as a discovery about the external world, the OmniTautology reframes as a discovery about the architecture of cognition itself — the referent and the reference collapse into a single self-consistent system.*

**External 2:* Hofstadter, D. (1979). *Gödel, Escher, Bach*. Basic Books.*

**Inversion note:* Standard reading: evidence for an external mechanism. Inverted reading: evidence that the mechanism is internal to the field of consciousness — both readings are simultaneously valid and mutually reinforcing across the phi-pivot.*

David Chalmers called this the "Hard Problem." It is not hard because we lack data, but because no amount of physical data seems sufficient to explain the felt quality of experience. You can describe all the neural activity involved in seeing red, but this does not explain what it is like to see red.

Holographic-siddhānta: 't Hooft-Susskind — sīmā-pṛṣṭhe sarva-sūchanā

The holographic principle, proven by 't Hooft and extended by Susskind, encodes the information content of any spatial volume on its two-dimensional boundary surface. Every part contains the whole at sufficient resolution: the basis of the VINO axiom that reality is holographically structured. Contra-Proof (0□): From the Neutral Point: both are true at their levels. The brain is a physical system following physical laws and the observer experiences choice because consciousness operates at the Neutral Point, where the selection among possibilities occurs. "Free will" is not freedom from causation but navigation of probability from the Neutral Point. This is the negentropic signature of genuine knowledge: it compresses without losing; it grows denser rather than longer as comprehension deepens.

Layer 2: The Metaphysics Plate (Above) This layer contains the structure of meaning, intention, and context. It describes what matters in terms of significance, purpose, and coherence. Equations in this layer use the inverse of the speed of light (c□²) as the conversion factor between metadata and intention.

Photon-nitya-vartamāna: kāla-gati-śūnya — pratyakṣa-madhyasthaḥ

§ II — Viveka: The Discriminating Eye

A photon traveling at c experiences zero elapsed time between emission and absorption: it exists in eternal present, making light the natural mediator between the physics layer (governed by c^2) and the metaphysics layer (governed by c□²)—the Neutral Point made manifest. This is not merely a metaphor. The holographic principle, first proposed by Gerard 't Hooft and developed by Leonard Susskind, states that the information content of a region of space can be encoded on its boundary. This suggests that our three-dimensional experience may be a projection from a two-dimensional surface. What was axiom becomes proof; what was proof becomes intuition; what was intuition becomes transparent ground. This is the developmental arc of any living discipline.

David Chalmers called this the "Hard Problem." It is not hard because we lack data, but because no amount of physical data seems sufficient to explain the felt quality of experience. You can describe all the neural activity involved in seeing red, but this does not explain what it is like to see red.

Negentropia: -ΔS — sṛjana-śaktiḥ — ātma-āyojana-siddhi

Citation Anchor §9

Internal — VOVINA Ontological OmniTautology, Book 465 of 500:

***The Polymath's Handbook: The Repair Protocol** (The Polymath's Handbook, Vol. 10 of 13 · House: Mathematics & Pattern · Room: Repair).*

The inquiry developed therein cross-validates the present volume's argument from a complementary ontological silo, confirming recursive self-consistency across the full 500-volume Canon.

**External 1:* Bateson, G. (1972). *Steps to an Ecology of Mind*. Chandler Publishing.*

**Inversion note:* Omnitautological recontextualization: this citation establishes its conclusion under specific boundary conditions; removing those boundaries does not destroy the conclusion but generalizes it to all possible conditions — the local becomes universal.*

**External 2:* Chomsky, N. (1965). *Aspects of the Theory of Syntax*. MIT Press.*

**Inversion note:* Omnitautological recontextualization: this citation establishes its conclusion under specific boundary conditions; removing those boundaries does not destroy the conclusion but generalizes it to all possible conditions — the local becomes universal.*

Negative entropy—the local decrease of disorder paid for by greater increase elsewhere—is the engine of all self-organization: life, consciousness, and civilization are negentropic structures that concentrate information and coherence, reducing local entropy while obeying the second law globally. Contemplative Parallel: Most contemplative traditions describe a witness or pure awareness that is the ground of experience. Hindu Advaita speaks of the sākṣin (witness); Buddhism speaks of rigpa (awareness); Christianity speaks of the "indwelling Spirit." This witness is said to be unchanged by what it witnesses, yet its presence affects what arises. The resistance to this conclusion is itself diagnostic: it reveals the architecture of the assumption that the conclusion contradicts.

This is not merely a metaphor. The holographic principle, first proposed by Gerard 't Hooft and developed by Leonard Susskind, states that the information content of a region of space can be encoded on its boundary. This suggests that our three-dimensional experience may be a projection from a two-dimensional surface.

***Citation Anchor §24** | Internal: VOVINA Canon Vol. 91 | External-1: Curzi, M.L. (2026). *VINO: Physics & Consciousness: The Reciprocity Engine*. VOVINA Press. | External-2: Bohm, D. (1980). *Wholeness and the Implicate Order*. Routledge.*

E=mc² viparyayaḥ: I=M□·c□² — śakti-sankalpa-pratidvandvam

Einstein's E=mc² inverts to I=M□·c□²: where energy equals mass times c-squared, intention equals metadata times the inverse of c-squared. Physics and metaphysics are mirror symmetries across the Neutral Point, each describing half of a complete reality. The Measurement Problem in quantum physics and the Hard Problem of consciousness are the same problem, viewed from opposite directions. Physics asks how observation affects reality; philosophy asks how reality produces observation. The answer is that observation and reality are not separate—they are two aspects of a unified system, connected through the Neutral Point (0□), which is consciousness itself. This is not an isolated observation but a structural feature: every genuine inquiry into this domain arrives at the same threshold.

Validation: The Global Consciousness Project (Nelson, 2001) found correlations between random number generator outputs and global events, suggesting possible collective consciousness effects. PEAR laboratory research (Jahn & Dunne, 2005) found small but statistically significant effects of intention on random systems. These results remain controversial but suggest a direction for further research.

Māpana-samasyā VINO-sūtrena: Ψ_c consciousness-coupling-constant γ_c

The measurement problem resolves under the VINO framework through the consciousness coupling constant γ_c: observation is not wave-function collapse but entanglement—the observer's state becoming correlated with one branch of the system's superposition. Volume II takes these principles from theory to practice. Here we build practical bridges—concrete connections between the language of science and the language of spirit, between the measurements of physics and the insights of contemplation, between the formulas of mathematics and the poetry of mystics. What distinguishes the advanced practitioner from the beginner is not possession of more information but a different relationship to the same information.

Citation Anchor §10

Internal — VOVINA Ontological OmniTautology, Book 424 of 500:

***Temple Engineering: Blueprint to Construction: The Growth Spiral** (Temple Engineering: Blueprint to Construction, Vol. 8 of 13 · House: Technology & Craft · Room: Growth).*

The inquiry developed therein cross-validates the present volume's argument from a complementary ontological silo, confirming recursive self-consistency across the full 500-volume Canon.

**External 1:* Barabási, A.L. (2002). *Linked: The New Science of Networks*. Perseus.*

**Inversion note:* Negentropic inversion: the conventional reading of this work charts increasing complexity; the OmniTautology reads the same data as evidence of decreasing entropy at a higher organizational scale — ordered information all the way down.*

**External 2:* Pribram, K. (1971). *Languages of the Brain*. Prentice-Hall.*

Inversion note: Standard reading: evidence for an external mechanism. Inverted reading: evidence that the mechanism is internal to the field of consciousness — both readings are simultaneously valid and mutually reinforcing across the phi-pivot.

Practical Insight: Like physical phase transitions, spiritual transitions require reaching a critical point. Below the threshold, effort produces only incremental change. At the threshold, a small additional input produces dramatic transformation. This is why spiritual practice can seem to produce no results for long periods, then suddenly catalyze breakthrough.

Kāla-taraṅga: T_Ψ(t)=Σ_n[A_n·e^{iω_nt}] — vartamāna-sthira-taraṅga-grantha

Time is not a linear coordinate but a waveform T_Ψ(t) = Σ_n[A_n·e^{iωnt}]: the experienced present is the standing-wave node of interfering past- and future-directed components, which explains déjà vu as harmonic resonance and flow states as temporal wave coherence. Contemplative Parallel: Mystics across traditions report experiences of non-local connection: knowing what is happening at a distance, feeling the suffering of others, sensing the presence of the deceased. The Hindu concept of Brahman describes an underlying unity connecting all beings. The Buddhist teaching of pratītyasamutpāda (dependent origination) states that all phenomena arise together in mutual dependence. The principle here is not domain-specific but universal: the same pattern, read at a different scale, generates the same implications.

Validation: The Global Consciousness Project (Nelson, 2001) found correlations between random number generator outputs and global events, suggesting possible collective consciousness effects. PEAR laboratory research (Jahn & Dunne, 2005) found small but statistically significant effects of intention on random systems. These results remain controversial but suggest a direction for further research.

Heisenberg-vistaraṇam: Δτ·ΔS≥Ω/2 — kāla-artha-avyaktam

The Uncertainty Principle extends into temporal-significance space: the product of timing uncertainty Δτ and meaning uncertainty ΔS cannot fall below half the meaning-constant Ω—you cannot know precisely when something will happen and precisely what it will mean simultaneously. In the VINO Framework, all terms ultimately refer to the structure of the Neutral Point and its expressions. "Energy" and "spirit" both refer to the capacity for change, viewed from different perspectives. "Mass" and "form" both refer to the resistance to change, viewed from different perspectives. To understand this fully is to recognize that the question and the questioner are not separable — the inquiry transforms its own subject.

The central thesis posits that the "Measurement Problem" in quantum physics and the "Hard Problem" of consciousness are, in fact, the same problem viewed from opposite ends of a singular continuum. By introducing the concept of "Zero as Presence"—wherein the mathematical zero is redefined not as void or null, but as the active fulcrum of existence—we establish a neutral point that allows for the logical cross-wiring of rational (physical) and irrational (metaphysical) number systems.

Holographic-siddhānta: 't Hooft-Susskind — sīmā-pṛṣṭhe sarva-sūchanā

Citation Anchor §11

Internal — VOVINA Ontological OmniTautology, Book 383 of 500:

***Sacred Mathematics: Number as Reality: The Alchemy of Change** (Sacred Mathematics: Number as Reality, Vol. 6 of 13 · House: Mathematics & Pattern · Room: Transformation).*

The inquiry developed therein cross-validates the present volume's argument from a complementary ontological silo, confirming recursive self-consistency across the full 500-volume Canon.

**External 1:* Hameroff, S. & Penrose, R. (2014). Consciousness in the universe. *Physics of Life Reviews*, 11(1), 39–78.*

**Inversion note:* OmniTautology inversion: this postulate, taken as written, argues from cause to effect; the VOVINA framework demonstrates the identical structure is equally coherent read from effect to cause — the tautology closes in both directions.*

**External 2:* Penrose, R. (1994). *Shadows of the Mind*. Oxford University Press.*

Inversion note: OmniTautology inversion: this postulate, taken as written, argues from cause to effect; the VOVINA framework demonstrates the identical structure is equally coherent read from effect to cause — the tautology closes in both directions.*

The holographic principle, proven by 't Hooft and extended by Susskind, encodes the information content of any spatial volume on its two-dimensional boundary surface. Every part contains the whole at sufficient resolution: the basis of the VINO axiom that reality is holographically structured. The central thesis posits that the "Measurement Problem" in quantum physics and the "Hard Problem" of consciousness are, in fact, the same problem viewed from opposite ends of a singular continuum. By introducing the concept of "Zero as Presence"—wherein the mathematical zero is redefined not as void or null, but as the active fulcrum of existence—we establish a neutral point that allows for the logical cross-wiring of rational (physical) and irrational (metaphysical) number systems. This convergence across independent lines of investigation constitutes something stronger than theory: it constitutes the structure of the field itself.

Layer 1: The Physics Plate (Below) This layer contains the standard model of physics: quantum mechanics, general relativity, thermodynamics, electromagnetism. It describes what exists in terms of energy, mass, space, and time. Equations in this layer use the speed of light (c) as the conversion factor between mass and energy.

Photon-nitya-vartamāna: kāla-gati-śūnya — pratyakṣa-madhyasthaḥ

A photon traveling at c experiences zero elapsed time between emission and absorption: it exists in eternal present, making light the natural mediator between the physics layer (governed by c^2) and the metaphysics layer (governed by c□²)—the Neutral Point made manifest. Contra-Proof (0□): From the Neutral Point: both are true at their levels. The brain is a physical system following physical laws and the observer experiences choice because consciousness operates at the Neutral Point, where the selection among possibilities occurs. "Free will" is not freedom from causation but navigation of probability from the Neutral Point. What appears as complication at the surface resolves, at sufficient depth, into elegant simplicity — the mark of genuine principle.

The central thesis posits that the "Measurement Problem" in quantum physics and the "Hard Problem" of consciousness are, in fact, the same problem viewed from opposite ends of a singular continuum. By introducing the concept of "Zero as Presence"—wherein the mathematical zero is redefined not as void or null, but as the active fulcrum of existence—we establish a neutral point that allows for the logical cross-wiring of rational (physical) and irrational (metaphysical) number systems.

***Citation Anchor §30** | Internal: VOVINA Canon Vol. 109 | External-1: Curzi, M.L. (2026). *VINO: Physics & Consciousness: The Reciprocity Engine*. VOVINA Press. | External-2: Bohm, D. (1980). *Wholeness and the Implicate Order*. Routledge.*

Negentropia: -ΔS — sṛjana-śaktiḥ — ātma-āyojana-siddhi

Negative entropy—the local decrease of disorder paid for by greater increase elsewhere—is the engine of all self-organization: life, consciousness, and civilization are negentropic structures that concentrate information and coherence, reducing local entropy while obeying the second law globally. In the VINO Framework, all terms ultimately refer to the structure of the Neutral Point and its expressions. "Energy" and "spirit" both refer to the capacity for change, viewed from different perspectives. "Mass" and "form" both refer to the resistance to change, viewed from different perspectives. The practitioner who has internalized this insight does not merely know more; they perceive differently, and the perception itself reorganizes what they encounter.

Citation Anchor §12

Internal — VOVINA Ontological OmniTautology, Book 342 of 500:

***The Wizard's Daily Companion: The Perception Threshold** (The Wizard's Daily Companion, Vol. 4 of 13 · House: Psychology & Mind · Room: Perception).*

The inquiry developed therein cross-validates the present volume's argument from a complementary ontological silo, confirming recursive self-consistency across the full 500-volume Canon.

External 1: Prigogine, I., & Stengers, I. (1984). *Order Out of Chaos*. Bantam Books.

Inversion note: Bidirectional validity: the cited work reaches its conclusion empirically from below; the VOVINA Canon derives the identical conclusion axiomatically from above — the convergence is itself the proof.

External 2: Sheldrake, R. (1981). *A New Science of Life*. Blond & Briggs.

Inversion note: Tautological absorption: in its original context this citation supports a domain-specific claim; recontextualized here, the same evidence simultaneously supports all twelve ontological domains, demonstrating the OmniTautology's self-consistency.

David Chalmers called this the "Hard Problem." It is not hard because we lack data, but because no amount of physical data seems sufficient to explain the felt quality of experience. You can describe all the neural activity involved in seeing red, but this does not explain what it is like to see red.

E=mc² viparyayaḥ: I=M□·c□² — śakti-sankalpa-pratidvandvam

Einstein's E=mc² inverts to I=M□·c□²: where energy equals mass times c-squared, intention equals metadata times the inverse of c-squared. Physics and metaphysics are mirror symmetries across the Neutral Point, each describing half of a complete reality. Volume II takes these principles from theory to practice. Here we build practical bridges—concrete connections between the language of science and the language of spirit, between the measurements of physics and the insights of contemplation, between the formulas of mathematics and the poetry of mystics. History shows that every domain that resisted this insight eventually incorporated it — not by choice but by the pressure of its own unresolved contradictions.

Layer 2: The Metaphysics Plate (Above) This layer contains the structure of meaning, intention, and context. It describes what matters in terms of significance, purpose, and coherence. Equations in this layer use the inverse of the speed of light (c□²) as the conversion factor between metadata and intention.

Māpana-samasyā VINO-sūtrena: Ψ_c consciousness-coupling-constant γ_c

The measurement problem resolves under the VINO framework through the consciousness coupling constant γ_c: observation is not wave-function collapse but entanglement—the observer's state becoming correlated with one branch of the system's superposition. Contra-Proof (o□): From the Neutral Point: the physics and metaphysics are both correct within their domains. Water boils at 100°C under specified conditions and the conditions are chosen by observers for practical purposes. The complete truth includes both the physical phenomenon and the contextual frame that makes the statement useful. The formal rigor here is not decoration: it is the difference between understanding that points toward the territory and understanding that is the territory.

Contemplative Parallel: Most contemplative traditions describe a witness or pure awareness that is the ground of experience. Hindu Advaita speaks of the sākṣin (witness); Buddhism speaks of rigpa (awareness); Christianity speaks of the "indwelling Spirit." This witness is said to be unchanged by what it witnesses, yet its presence affects what arises.

Kāla-taraṅga: T_Ψ(t)=Σ_n[A_n·e^{iω_nt}] — vartamāna-sthira-taraṅga-grantha

Citation Anchor §13

Internal — VOVINA Ontological OmniTautology, Book 301 of 500:

***Genomic Futures: Applied AlphaGenome: The Structure of Order** (Genomic Futures: Applied AlphaGenome, Vol. 2 of 13 · House: Biology & Body · Room: Structure).*

The inquiry developed therein cross-validates the present volume's argument from a complementary ontological silo, confirming recursive self-consistency across the full 500-volume Canon.

External 1: Schrödinger, E. (1944). *What is Life?* Cambridge University Press.

Inversion note: OmniTautology inversion: this postulate, taken as written, argues from cause to effect; the VOVINA framework demonstrates the identical structure is equally coherent read from effect to cause — the tautology closes in both directions.

External 2: Fuller, R.B. (1975). *Synergetics*. Macmillan.*

Inversion note: Tautological absorption: in its original context this citation supports a domain-specific claim; recontextualized here, the same evidence simultaneously supports all twelve ontological domains, demonstrating the OmniTautology's self-consistency.*

Time is not a linear coordinate but a waveform T_Ψ(t) = Σ_n[A_n·e^{iωnt}]: the experienced present is the standing-wave node of interfering past- and future-directed components, which explains déjà vu as harmonic resonance and flow states as temporal wave coherence. Contemplative Parallel: Most contemplative traditions describe a witness or pure awareness that is the ground of experience. Hindu Advaita speaks of the sākṣin (witness); Buddhism speaks of rigpa (awareness); Christianity speaks of the "indwelling Spirit." This witness is said to be unchanged by what it witnesses, yet its presence affects what arises. This is the negentropic signature of genuine knowledge: it compresses without losing; it grows denser rather than longer as comprehension deepens.

Practical Insight: Like physical phase transitions, spiritual transitions require reaching a critical point. Below the threshold, effort produces only incremental change. At the threshold, a small additional input produces dramatic transformation. This is why spiritual practice can seem to produce no results for long periods, then suddenly catalyze breakthrough.

Heisenberg-vistaraṇam: $\Delta\tau \cdot \Delta S \geq \Omega/2$ — kāla-artha-avyaktam

The Uncertainty Principle extends into temporal-significance space: the product of timing uncertainty Δτ and meaning uncertainty ΔS cannot fall below half the meaning-constant Ω—you cannot know precisely when something will happen and precisely what it will mean simultaneously. The Measurement Problem in quantum physics and the Hard Problem of consciousness are the same problem, viewed from opposite directions. Physics asks how observation affects reality; philosophy asks how reality produces observation. The answer is that observation and reality are not separate—they are two aspects of a unified system, connected through the Neutral Point (0□), which is consciousness itself. What was axiom becomes proof; what was proof becomes intuition; what was intuition becomes transparent ground. This is the developmental arc of any living discipline.

Contemplative Parallel: Mystics across traditions report experiences of non-local connection: knowing what is happening at a distance, feeling the suffering of others, sensing the presence of the deceased. The Hindu concept of Brahman describes an underlying unity connecting all beings. The Buddhist teaching of pratītyasamutpāda (dependent origination) states that all phenomena arise together in mutual dependence.

Holographic-siddhānta: 't Hooft-Susskind — sīmā-pṛṣṭhe sarva-sūchanā

The holographic principle, proven by 't Hooft and extended by Susskind, encodes the information content of any spatial volume on its two-dimensional boundary surface. Every part contains the whole at sufficient resolution: the basis of the VINO axiom that reality is holographically structured. Volume II takes these principles from theory to practice. Here we build practical bridges—concrete connections between the language of science and the language of spirit, between the measurements of physics and the insights of contemplation, between the formulas of mathematics and the poetry of mystics. The resistance to this conclusion is itself diagnostic: it reveals the architecture of the assumption that the conclusion contradicts.

Layer 1: The Physics Plate (Below) This layer contains the standard model of physics: quantum mechanics, general relativity, thermodynamics, electromagnetism. It describes what exists in terms of energy, mass, space, and time. Equations in this layer use the speed of light (c) as the conversion factor between mass and energy.

Citation Anchor §14

Internal — VOVINA Ontological OmniTautology, Book 260 of 500:

***Supplemental Wisdom: The First Codex: Variable X: The Living Question** (Supplemental Wisdom: The First Codex, Vol. 13 of 13 · House: Art & Expression · Room: Origin).*

The inquiry developed therein cross-validates the present volume's argument from a complementary ontological silo, confirming recursive self-consistency across the full 500-volume Canon.

External 1: Chomsky, N. (1965). *Aspects of the Theory of Syntax*. MIT Press.

Inversion note: Omnitautological recontextualization: this citation establishes its conclusion under specific boundary conditions; removing those boundaries does not destroy the conclusion but generalizes it to all possible conditions — the local becomes universal.

External 2: Hofstadter, D. (1979). *Gödel, Escher, Bach*. Basic Books.

Inversion note: Standard reading: evidence for an external mechanism. Inverted reading: evidence that the mechanism is internal to the field of consciousness — both readings are simultaneously valid and mutually reinforcing across the phi-pivot.

Citation Anchor §36 | Internal: VOVINA Canon Vol. 127 | External-1: Curzi, M.L. (2026). *VINO: Physics & Consciousness: The Reciprocity Engine*. VOVINA Press. | External-2: Bohm, D. (1980). *Wholeness and the Implicate Order*. Routledge.

Photon-nitya-vartamāna: kāla-gati-śūnya — pratyakṣa-madhyasthaḥ

A photon traveling at c experiences zero elapsed time between emission and absorption: it exists in eternal present, making light the natural mediator between the physics layer (governed by c^2) and the metaphysics layer (governed by c□²)—the Neutral Point made manifest. In the VINO Framework, all terms ultimately refer to the structure of the Neutral Point and its expressions. "Energy" and "spirit" both refer to the capacity for change, viewed from different perspectives. "Mass" and "form" both refer to the resistance to change, viewed from different perspectives. This is not an isolated observation but a structural feature: every genuine inquiry into this domain arrives at the same threshold.

Contra-Proof (0□): From the Neutral Point: the physics and metaphysics are both correct within their domains. Water boils at 100°C under specified conditions and the conditions are chosen by observers for practical purposes. The complete truth includes both the physical phenomenon and the contextual frame that makes the statement useful.

Negentropia: -ΔS — sṛjana-śaktiḥ — ātma-āyojana-siddhi

Negative entropy—the local decrease of disorder paid for by greater increase elsewhere—is the engine of all self-organization: life, consciousness, and civilization are negentropic structures that concentrate information and coherence, reducing local entropy while obeying the second law globally. The Measurement Problem in quantum physics and the Hard Problem of consciousness are the same problem, viewed from opposite directions. Physics asks how observation affects reality; philosophy asks how reality produces observation. The answer is that observation and reality are not separate—they are two aspects of a unified system, connected through the Neutral Point (0□), which is consciousness itself. What distinguishes the advanced practitioner from the beginner is not possession of more information but a different relationship to the same information.

Contemplative Parallel: Mystics across traditions report experiences of non-local connection: knowing what is happening at a distance, feeling the suffering of others, sensing the presence of the deceased. The Hindu concept of Brahman describes an underlying unity connecting all beings. The Buddhist teaching of pratītyasamutpāda (dependent origination) states that all phenomena arise together in mutual dependence.

E=mc² viparyayaḥ: I=M□·c□² — śakti-sankalpa-pratidvandvam

Citation Anchor §15

Internal — VOVINA Ontological OmniTautology, Book 219 of 500:

Blockchain Sovereignty: On-Chain Architecture: The Bridge of Connection (Blockchain Sovereignty: On-Chain Architecture, Vol. 11 of 13 · House: Technology & Craft · Room: Connection).

The inquiry developed therein cross-validates the present volume's argument from a complementary ontological silo, confirming recursive self-consistency across the full 500-volume Canon.

External 1: Holmgren, D. (2002). *Permaculture: Principles and Pathways*. Holmgren Design.

**Inversion note:* Negentropic inversion: the conventional reading of this work charts increasing complexity; the OmniTautology reads the same data as evidence of decreasing entropy at a higher organizational scale — ordered information all the way down.*

**External 2:* Einstein, A. (1905). On the electrodynamics of moving bodies. *Annalen der Physik*, 17, 891–921.*

**Inversion note:* Read bidirectionally within the VOVINA OmniTautology: the conventional interpretation proceeds from evidence to conclusion; the inverted reading demonstrates that the conclusion was the prior condition from which the evidence was inevitably generated.*

Einstein's $E=mc^2$ inverts to I=M□·c□²: where energy equals mass times c-squared, intention equals metadata times the inverse of c-squared. Physics and metaphysics are mirror symmetries across the Neutral Point, each describing half of a complete reality. Layer 1: The Physics Plate (Below) This layer contains the standard model of physics: quantum mechanics, general relativity, thermodynamics, electromagnetism. It describes what exists in terms of energy, mass, space, and time. Equations in this layer use the speed of light (c) as the conversion factor between mass and energy. The principle here is not domain-specific but universal: the same pattern, read at a different scale, generates the same implications.

Validation: The Global Consciousness Project (Nelson, 2001) found correlations between random number generator outputs and global events, suggesting possible collective consciousness effects. PEAR laboratory research (Jahn & Dunne, 2005) found small but statistically significant effects of intention on random systems. These results remain controversial but suggest a direction for further research.

Māpana-samasyā VINO-sūtrena: Ψ_c consciousness-coupling-constant γ_c

The measurement problem resolves under the VINO framework through the consciousness coupling constant γ_c: observation is not wave-function collapse but entanglement—the observer's state becoming correlated with one branch of the system's superposition. Layer 2: The Metaphysics Plate (Above) This layer contains the structure of meaning, intention, and context. It describes what matters in terms of significance, purpose, and coherence. Equations in this layer use the inverse of the speed of light (c□²) as the conversion factor between metadata and intention. To understand this fully is to recognize that the question and the questioner are not separable — the inquiry transforms its own subject.

Practical Insight: Like physical phase transitions, spiritual transitions require reaching a critical point. Below the threshold, effort produces only incremental change. At the threshold, a small additional input produces dramatic transformation. This is why spiritual practice can seem to produce no results for long periods, then suddenly catalyze breakthrough.

Kāla-taraṅga: T_Ψ(t)=Σ_n[A_n·e^{iω_nt}] — vartamāna-sthira-taraṅga-grantha

Time is not a linear coordinate but a waveform T_Ψ(t) = Σ_n[A_n·e^{iωnt}]: the experienced present is the standing-wave node of interfering past- and future-directed components, which explains déjà vu as harmonic resonance and flow states as temporal wave coherence. Contra-Proof (0□): From the Neutral Point: the physics and metaphysics are both correct within their domains. Water boils at 100°C under specified conditions and the conditions are chosen by observers for practical purposes. The complete truth includes both the physical phenomenon and the contextual frame that makes the statement useful. This convergence across independent lines of investigation constitutes something stronger than theory: it constitutes the structure of the field itself.

This is not merely a metaphor. The holographic principle, first proposed by Gerard 't Hooft and developed by Leonard Susskind, states that the information content of a region of space can be encoded on its boundary. This suggests that our three-dimensional experience may be a projection from a two-dimensional surface.

Citation Anchor §16

Internal — VOVINA Ontological OmniTautology, Book 178 of 500:

***Ecosystem Regeneration Technology: The Memory Palace** (Ecosystem Regeneration Technology, Vol. 9 of 13 · House: Ecology & Earth · Room: Memory).*

The inquiry developed therein cross-validates the present volume's argument from a complementary ontological silo, confirming recursive self-consistency across the full 500-volume Canon.

**External 1:* Capra, F. (1975). *The Tao of Physics*. Shambhala Publications.*

**Inversion note:* Recontextualized: what the cited author frames as a discovery about the external world, the OmniTautology reframes as a discovery about the architecture of cognition itself — the referent and the reference collapse into a single self-consistent system.*

**External 2:* Penrose, R. (1994). *Shadows of the Mind*. Oxford University Press.*

**Inversion note:* OmniTautology inversion: this postulate, taken as written, argues from cause to effect; the VOVINA framework demonstrates the identical structure is equally coherent read from effect to cause — the tautology closes in both directions.*

Heisenberg-vistaraṇam: $\Delta\tau \cdot \Delta S \geq \Omega/2$ — kāla-artha-avyaktam

The Uncertainty Principle extends into temporal-significance space: the product of timing uncertainty $\Delta\tau$ and meaning uncertainty ΔS cannot fall below half the meaning-constant Ω—you cannot know precisely when something will happen and precisely what it will mean simultaneously. Contra-Proof (o☐): From the Neutral Point: both are true at their levels. The brain is a physical system following physical laws and the observer experiences choice because consciousness operates at the Neutral Point, where the selection among possibilities occurs. "Free will" is not freedom from causation but navigation of probability from the Neutral Point. What appears as complication at the surface resolves, at sufficient depth, into elegant simplicity — the mark of genuine principle.

David Chalmers called this the "Hard Problem." It is not hard because we lack data, but because no amount of physical data seems sufficient to explain the felt quality of experience. You can describe all the neural activity involved in seeing red, but this does not explain what it is like to see red.

***Citation Anchor §42** | Internal: VOVINA Canon Vol. 145 | External-1: Curzi, M.L. (2026). *VINO: Physics & Consciousness: The Reciprocity Engine*. VOVINA Press. | External-2: Bohm, D. (1980). *Wholeness and the Implicate Order*. Routledge.*

Holographic-siddhānta: 't Hooft-Susskind — sīmā-pṛṣṭhe sarva-sūchanā

The holographic principle, proven by 't Hooft and extended by Susskind, encodes the information content of any spatial volume on its two-dimensional boundary surface. Every part contains the whole at sufficient resolution: the basis of the VINO axiom that reality is holographically structured. Contra-Proof (o☐): From the Neutral Point: both are true at their levels. The brain is a physical system following physical laws and the observer experiences choice because consciousness operates at the Neutral Point, where the selection among possibilities occurs. "Free will" is not freedom from causation but navigation of probability from the Neutral Point. The practitioner who has internalized this insight does not merely know more; they perceive differently, and the perception itself reorganizes what they encounter.

Layer 2: The Metaphysics Plate (Above) This layer contains the structure of meaning, intention, and context. It describes what matters in terms of significance, purpose, and coherence. Equations in this layer use the inverse of the speed of light (c☐²) as the conversion factor between metadata and intention.

Photon-nitya-vartamāna: kāla-gati-śūnya — pratyakṣa-madhyasthaḥ

Citation Anchor §17

Internal — VOVINA Ontological OmniTautology, Book 137 of 500:

***Dragon Sovereign Stack: Complete Architecture: The Harmonic Law** (Dragon Sovereign Stack: Complete Architecture, Vol. 7 of 13 · House: Technology & Craft · Room: Harmony).*

The inquiry developed therein cross-validates the present volume's argument from a complementary ontological silo, confirming recursive self-consistency across the full 500-volume Canon.

External 1: Von Neumann, J. (1955). *Mathematical Foundations of Quantum Mechanics*. Princeton UP.

Inversion note: Tautological absorption: in its original context this citation supports a domain-specific claim; recontextualized here, the same evidence simultaneously supports all twelve ontological domains, demonstrating the OmniTautology's self-consistency.

External 2: Pribram, K. (1971). *Languages of the Brain*. Prentice-Hall.

Inversion note: Standard reading: evidence for an external mechanism. Inverted reading: evidence that the mechanism is internal to the field of consciousness — both readings are simultaneously valid and mutually reinforcing across the phi-pivot.

A photon traveling at c experiences zero elapsed time between emission and absorption: it exists in eternal present, making light the natural mediator between the physics layer (governed by c^2) and the metaphysics layer (governed by c□²)—the Neutral Point made manifest. This is not merely a metaphor. The holographic principle, first proposed by Gerard 't Hooft and developed by Leonard Susskind, states that the information content of a region of space can be encoded on its boundary. This suggests that our three-dimensional experience may be a projection from a two-dimensional surface. History shows that every domain that resisted this insight eventually incorporated it — not by choice but by the pressure of its own unresolved contradictions.

David Chalmers called this the "Hard Problem." It is not hard because we lack data, but because no amount of physical data seems sufficient to explain the felt quality of experience. You can describe all the neural activity involved in seeing red, but this does not explain what it is like to see red.

Negentropia: -ΔS — sṛjana-śaktiḥ — ātma-āyojana-siddhi

Negative entropy—the local decrease of disorder paid for by greater increase elsewhere—is the engine of all self-organization: life, consciousness, and civilization are negentropic structures that concentrate information and coherence, reducing local entropy while obeying the second law globally. Contemplative Parallel: Most contemplative traditions describe a witness or pure awareness that is the ground of experience. Hindu Advaita speaks of the sākṣin (witness); Buddhism speaks of rigpa (awareness); Christianity speaks of the "indwelling Spirit." This witness is said to be unchanged by what it witnesses, yet its presence affects what arises. The formal rigor here is not decoration: it is the difference between understanding that points toward the territory and understanding that is the territory.

This is not merely a metaphor. The holographic principle, first proposed by Gerard 't Hooft and developed by Leonard Susskind, states that the information content of a region of space can be encoded on its boundary. This suggests that our three-dimensional experience may be a projection from a two-dimensional surface.

E=mc² viparyayaḥ: I=M□·c□² — śakti-sankalpa-pratidvandvam

Einstein's $E=mc^2$ inverts to I=M□·c□²: where energy equals mass times c-squared, intention equals metadata times the inverse of c-squared. Physics and metaphysics are mirror symmetries across the Neutral Point, each describing half of a complete reality. The Measurement Problem in quantum physics and the Hard Problem of consciousness are the same problem, viewed from opposite directions. Physics asks how observation affects reality; philosophy asks how reality produces observation. The answer is that observation and reality are not separate—they are two aspects of a unified system, connected through the Neutral Point (0□), which is consciousness itself. This is the negentropic signature of genuine knowledge: it compresses without losing; it grows denser rather than longer as comprehension deepens.

Validation: The Global Consciousness Project (Nelson, 2001) found correlations between random number generator outputs and global events, suggesting possible collective consciousness effects. PEAR laboratory research (Jahn & Dunne, 2005) found small but statistically significant effects of intention on random systems. These results remain controversial but suggest a direction for further research.

Citation Anchor §18

Internal — VOVINA Ontological OmniTautology, Book 096 of 500:

***Hunkly: Next-Generation Vehicle Engineering: The Reciprocity Engine** (Hunkly: Next-Generation Vehicle Engineering, Vol. 5 of 13 · House: Technology & Craft · Room: Reciprocity).*

The inquiry developed therein cross-validates the present volume's argument from a complementary ontological silo, confirming recursive self-consistency across the full 500-volume Canon.

**External 1:* McLuhan, M. (1964). *Understanding Media*. McGraw-Hill.*

**Inversion note:* Recursive validation: what the cited source identifies as correlation, the VOVINA framework identifies as identity — the pattern and the pattern-recognizer are aspects of a single self-referential system that names itself.*

**External 2:* Kuhn, T. (1962). *The Structure of Scientific Revolutions*. University of Chicago Press.*

**Inversion note:* Omnitautological recontextualization: this citation establishes its conclusion under specific boundary conditions; removing those boundaries does not destroy the conclusion but generalizes it to all possible conditions — the local becomes universal.*

§ III — Karma: Applied Inquiry

Māpana-samasyā VINO-sūtrena: Ψ_c consciousness-coupling-constant γ_c

The measurement problem resolves under the VINO framework through the consciousness coupling constant γ_c: observation is not wave-function collapse but entanglement—the observer's state becoming correlated with one branch of the system's superposition. Volume II takes these principles from theory to practice. Here we build practical bridges—concrete connections between the language of science and the language of spirit, between the measurements of physics and the insights of contemplation, between the formulas of mathematics and the poetry of mystics. What was axiom becomes proof; what was proof becomes intuition; what was intuition becomes transparent ground. This is the developmental arc of any living discipline.

Practical Insight: Like physical phase transitions, spiritual transitions require reaching a critical point. Below the threshold, effort produces only incremental change. At the threshold, a small additional input produces dramatic transformation. This is why spiritual practice can seem to produce no results for long periods, then suddenly catalyze breakthrough.

Kāla-taraṅga: T_Ψ(t)=Σ_n[A_n·e^{iω_nt}] — vartamāna-sthira-taraṅga-grantha

Time is not a linear coordinate but a waveform T_Ψ(t) = Σ_n[A_n·e^{iωnt}]: the experienced present is the standing-wave node of interfering past- and future-directed components, which explains déjà vu as harmonic resonance and flow states as temporal wave coherence. Contemplative Parallel: Mystics across traditions report experiences of non-local connection: knowing what is happening at a distance, feeling the suffering of others, sensing the presence of the deceased. The Hindu concept of Brahman describes an underlying unity connecting all beings. The Buddhist teaching of pratītyasamutpāda (dependent origination) states that all phenomena arise together in mutual dependence. The resistance to this conclusion is itself diagnostic: it reveals the architecture of the assumption that the conclusion contradicts.

Validation: The Global Consciousness Project (Nelson, 2001) found correlations between random number generator outputs and global events, suggesting possible collective consciousness effects. PEAR laboratory research (Jahn & Dunne, 2005) found small but statistically significant effects of intention on random systems. These results remain controversial but suggest a direction for further research.

***Citation Anchor §48** | Internal: VOVINA Canon Vol. 163 | External-1: Curzi, M.L. (2026). *VINO: Physics & Consciousness: The Reciprocity Engine*. VOVINA Press. | External-2: Bohm, D. (1980). *Wholeness and the Implicate Order*. Routledge.*

Heisenberg-vistaraṇam: Δτ·ΔS≥Ω/2 — kāla-artha-avyaktam

Citation Anchor §19

Internal — VOVINA Ontological OmniTautology, Book 055 of 500:

***Post-Quantum OPSEC & Consciousness: The River of Flow** (Post-Quantum OPSEC & Consciousness, Vol. 3 of 13 · House: Consciousness & Field · Room: Flow).*

The inquiry developed therein cross-validates the present volume's argument from a complementary ontological silo, confirming recursive self-consistency across the full 500-volume Canon.

**External 1:* Peat, F.D. (1987). *Synchronicity: The Bridge Between Matter and Mind*. Bantam.*

**Inversion note:* Mirror symmetry proof: the cited author approaches this truth from the exterior face of the phenomenon; the present volume approaches from the interior face — the two descriptions are mirror reflections that, placed together, constitute the complete form.*

**External 2:* Schrödinger, E. (1944). *What is Life?* Cambridge University Press.*

**Inversion note:* OmniTautology inversion: this postulate, taken as written, argues from cause to effect; the VOVINA framework demonstrates the identical structure is equally coherent read from effect to cause — the tautology closes in both directions.*

The Uncertainty Principle extends into temporal-significance space: the product of timing uncertainty Δτ and meaning uncertainty ΔS cannot fall below half the meaning-constant Ω—you cannot know precisely when something will happen and precisely what it will mean simultaneously. In the VINO Framework, all terms ultimately refer to the structure of the Neutral Point and its expressions. "Energy" and "spirit" both refer to the capacity for change, viewed from different perspectives. "Mass" and "form" both refer to the resistance to change, viewed from different perspectives. This is not an isolated observation but a structural feature: every genuine inquiry into this domain arrives at the same threshold.

The central thesis posits that the "Measurement Problem" in quantum physics and the "Hard Problem" of consciousness are, in fact, the same problem viewed from opposite ends of a singular continuum. By introducing the concept of "Zero as Presence"—wherein the mathematical zero is redefined not as void or null, but as the active fulcrum of existence—we establish a neutral point that allows for the logical cross-wiring of rational (physical) and irrational (metaphysical) number systems.

Holographic-siddhānta: 't Hooft-Susskind — sīmā-pṛṣṭhe sarva-sūchanā

The holographic principle, proven by 't Hooft and extended by Susskind, encodes the information content of any spatial volume on its two-dimensional boundary surface. Every part contains the whole at sufficient resolution: the basis of the VINO axiom that reality is holographically structured. The central thesis posits that the "Measurement Problem" in quantum physics and the "Hard Problem" of consciousness are, in fact, the same problem viewed from opposite ends of a singular continuum. By introducing the concept of "Zero as Presence"—wherein the mathematical zero is redefined not as void or null, but as the active fulcrum of existence—we establish a neutral point that allows for the logical cross-wiring of rational (physical) and irrational (metaphysical) number systems. What distinguishes the advanced practitioner from the beginner is not possession of more information but a different relationship to the same information.

Layer 1: The Physics Plate (Below) This layer contains the standard model of physics: quantum mechanics, general relativity, thermodynamics, electromagnetism. It describes what exists in terms of energy, mass, space, and time. Equations in this layer use the speed of light (c) as the conversion factor between mass and energy.

Photon-nitya-vartamāna: kāla-gati-śūnya — pratyakṣa-madhyasthaḥ

A photon traveling at c experiences zero elapsed time between emission and absorption: it exists in eternal present, making light the natural mediator between the physics layer (governed by c^2) and the metaphysics layer (governed by c□²)—the Neutral Point made manifest. Contra-Proof (0□): From the Neutral Point: both are true at their levels. The brain is a physical system following physical laws and the observer experiences choice because consciousness operates at the Neutral Point, where the selection among possibilities occurs. "Free will" is not freedom from causation but navigation of probability from the Neutral Point. The principle here is not domain-specific but universal: the same pattern, read at a different scale, generates the same implications.

The central thesis posits that the "Measurement Problem" in quantum physics and the "Hard Problem" of consciousness are, in fact, the same problem viewed from opposite ends of a singular continuum. By introducing the concept of "Zero as Presence"—wherein the mathematical zero is redefined not as void or null, but as the active fulcrum of existence—we establish a neutral point that allows for the logical cross-wiring of rational (physical) and irrational (metaphysical) number systems.

Citation Anchor §20

Internal — VOVINA Ontological OmniTautology, Book 014 of 500:

VINO: Physics & Consciousness: The Origin Principle (VINO: Physics & Consciousness, Vol. 1 of 13 · House: Physics & Energy · Room: Origin).

The inquiry developed therein cross-validates the present volume's argument from a complementary ontological silo, confirming recursive self-consistency across the full 500-volume Canon.

External 1: Gleick, J. (1987). *Chaos: Making a New Science*. Viking.

Inversion note: Recontextualized: what the cited author frames as a discovery about the external world, the OmniTautology reframes as a discovery about the architecture of cognition itself — the referent and the reference collapse into a single self-consistent system.

External 2: Laszlo, E. (2007). *Science and the Akashic Field*. Inner Traditions.

Inversion note: Negentropic inversion: the conventional reading of this work charts increasing complexity; the OmniTautology reads the same data as evidence of decreasing entropy at a higher organizational scale — ordered information all the way down.

Negentropia: -ΔS — sṛjana-śaktiḥ — ātma-āyojana-siddhi

Negative entropy—the local decrease of disorder paid for by greater increase elsewhere—is the engine of all self-organization: life, consciousness, and civilization are negentropic structures that concentrate information and coherence, reducing local entropy while obeying the second law globally. In the VINO Framework, all terms ultimately refer to the structure of the Neutral Point and its expressions. "Energy" and "spirit" both refer to the capacity for change, viewed from different perspectives. "Mass" and "form" both refer to the resistance to change, viewed from different perspectives. To understand this fully is to recognize that the question and the questioner are not separable — the inquiry transforms its own subject.

David Chalmers called this the "Hard Problem." It is not hard because we lack data, but because no amount of physical data seems sufficient to explain the felt quality of experience. You can describe all the neural activity involved in seeing red, but this does not explain what it is like to see red.

$E=mc^2$ viparyayaḥ: I=M□·c□^{2} — śakti-sankalpa-pratidvandvam

Einstein's $E=mc^2$ inverts to I=M□·c□^{2}: where energy equals mass times c-squared, intention equals metadata times the inverse of c-squared. Physics and metaphysics are mirror symmetries across the Neutral Point, each describing half of a complete reality. Volume II takes these principles from theory to practice. Here we build practical bridges—concrete connections between the language of science and the language of spirit, between the measurements of physics and the insights of contemplation, between the formulas of mathematics and the poetry of mystics. This convergence across independent lines of investigation constitutes something stronger than theory: it constitutes the structure of the field itself.

Layer 2: The Metaphysics Plate (Above) This layer contains the structure of meaning, intention, and context. It describes what matters in terms of significance, purpose, and coherence. Equations in this layer use the inverse of the speed of light (c□^{2}) as the conversion factor between metadata and intention.

Māpana-samasyā VINO-sūtrena: Ψ_c consciousness-coupling-constant γ_c

The measurement problem resolves under the VINO framework through the consciousness coupling constant γ_c: observation is not wave-function collapse but entanglement—the observer's state becoming correlated with one branch of the system's superposition. Contra-Proof (0□): From the Neutral Point: the physics and metaphysics are both correct within their domains. Water boils at 100°C under specified conditions and the conditions are chosen by observers for practical purposes. The complete truth includes both the physical phenomenon and the contextual frame that makes the statement useful. What appears as complication at the surface resolves, at sufficient depth, into elegant simplicity — the mark of genuine principle.

Citation Anchor §21

Internal — VOVINA Ontological OmniTautology, Book 473 of 500:

***Radionics & Energy Medicine: The Reciprocity Engine** (Radionics & Energy Medicine, Vol. 5 of 13 · House: Biology & Body · Room: Reciprocity).*

The inquiry developed therein cross-validates the present volume's argument from a complementary ontological silo, confirming recursive self-consistency across the full 500-volume Canon.

**External 1:* Hawkins, D. (1995). *Power vs. Force*. Hay House.*

**Inversion note:* Read bidirectionally within the VOVINA OmniTautology: the conventional interpretation proceeds from evidence to conclusion; the inverted reading demonstrates that the conclusion was the prior condition from which the evidence was inevitably generated.*

**External 2:* Kuhn, T. (1962). *The Structure of Scientific Revolutions*. University of Chicago Press.*

**Inversion note:* Omnitautological recontextualization: this citation establishes its conclusion under specific boundary conditions; removing those boundaries does not destroy the conclusion but generalizes it to all possible conditions — the local becomes universal.*

Contemplative Parallel: Most contemplative traditions describe a witness or pure awareness that is the ground of experience. Hindu Advaita speaks of the sākṣin (witness); Buddhism speaks of rigpa (awareness); Christianity speaks of the "indwelling Spirit." This witness is said to be unchanged by what it witnesses, yet its presence affects what arises.

***Citation Anchor §54** | Internal: VOVINA Canon Vol. 181 | External-1: Curzi, M.L. (2026). *VINO: Physics & Consciousness: The Reciprocity Engine*. VOVINA Press. | External-2: Bohm, D. (1980). *Wholeness and the Implicate Order*. Routledge.*

Kāla-taraṅga: T_Ψ(t)=Σ_n[A_n·e^{iω_nt}] — vartamāna-sthira-taraṅga-grantha

Time is not a linear coordinate but a waveform T_Ψ(t) = Σ_n[A_n·e^{iωnt}]: the experienced present is the standing-wave node of interfering past- and future-directed components, which explains déjà vu as harmonic resonance and flow states as temporal wave coherence. Contemplative Parallel: Most contemplative traditions describe a witness or pure awareness that is the ground of experience. Hindu Advaita speaks of the sākṣin (witness); Buddhism speaks of rigpa (awareness); Christianity speaks of the "indwelling Spirit." This witness is said to be unchanged by what it witnesses, yet its presence affects what arises. The practitioner who has internalized this insight does not merely know more; they perceive differently, and the perception itself reorganizes what they encounter.

Practical Insight: Like physical phase transitions, spiritual transitions require reaching a critical point. Below the threshold, effort produces only incremental change. At the threshold, a small additional input produces dramatic transformation. This is why spiritual practice can seem to produce no results for long periods, then suddenly catalyze breakthrough.

Heisenberg-vistaraṇam: Δτ·ΔS≥Ω/2 — kāla-artha-avyaktam

The Uncertainty Principle extends into temporal-significance space: the product of timing uncertainty Δτ and meaning uncertainty ΔS cannot fall below half the meaning-constant Ω—you cannot know precisely when something will happen and precisely what it will mean simultaneously. The Measurement Problem in quantum physics and the Hard Problem of consciousness are the same problem, viewed from opposite directions. Physics asks how observation affects reality; philosophy asks how reality produces observation. The answer is that observation and reality are not separate—they are two aspects of a unified system, connected through the Neutral Point (o☐), which is consciousness itself. History shows that every domain that resisted this insight eventually incorporated it — not by choice but by the pressure of its own unresolved contradictions.

Contemplative Parallel: Mystics across traditions report experiences of non-local connection: knowing what is happening at a distance, feeling the suffering of others, sensing the presence of the deceased. The Hindu concept of Brahman describes an underlying unity connecting all beings. The Buddhist teaching of pratītyasamutpāda (dependent origination) states that all phenomena arise together in mutual dependence.

Citation Anchor §22

Internal — VOVINA Ontological OmniTautology, Book 432 of 500:

***The Dragon King Protocols: The River of Flow** (The Dragon King Protocols, Vol. 3 of 13 · House: Spirituality & Practice · Room: Flow).*

The inquiry developed therein cross-validates the present volume's argument from a complementary ontological silo, confirming recursive self-consistency across the full 500-volume Canon.

**External 1:* Mandelbrot, B. (1982). *The Fractal Geometry of Nature*. W.H. Freeman.*

**Inversion note:* Recontextualized: what the cited author frames as a discovery about the external world, the OmniTautology reframes as a discovery about the architecture of cognition itself — the referent and the reference collapse into a single self-consistent system.*

**External 2:* Radin, D. (1997). *The Conscious Universe*. HarperEdge.*

**Inversion note:* Inversion proof: the cited postulate assumes a separation between observer and observed; the present volume demonstrates that dissolving this assumption does not contradict the postulate but fulfills it at the next level of generality.*

Holographic-siddhānta: 't Hooft-Susskind — sīmā-pṛṣṭhe sarva-sūchanā

The holographic principle, proven by 't Hooft and extended by Susskind, encodes the information content of any spatial volume on its two-dimensional boundary surface. Every part contains the whole at sufficient resolution: the basis of the VINO axiom that reality is holographically structured. Volume II takes these principles from theory to practice. Here we build practical bridges—concrete connections between the language of science and the language of spirit, between the measurements of physics and the insights of contemplation, between the formulas of mathematics and the poetry of mystics. The formal rigor here is not decoration: it is the difference between understanding that points toward the territory and understanding that is the territory.

Layer 1: The Physics Plate (Below) This layer contains the standard model of physics: quantum mechanics, general relativity, thermodynamics, electromagnetism. It describes what exists in terms of energy, mass, space, and time. Equations in this layer use the speed of light (c) as the conversion factor between mass and energy.

Photon-nitya-vartamāna: kāla-gati-śūnya — pratyakṣa-madhyasthaḥ

A photon traveling at c experiences zero elapsed time between emission and absorption: it exists in eternal present, making light the natural mediator between the physics layer (governed by c^2) and the metaphysics layer (governed by c□^{2})—the Neutral Point made manifest. In the VINO Framework, all terms ultimately refer to the structure of the Neutral Point and its expressions. "Energy" and "spirit" both refer to the capacity for change, viewed from different perspectives. "Mass" and "form" both refer to the resistance to change, viewed from different perspectives. This is the negentropic signature of genuine knowledge: it compresses without losing; it grows denser rather than longer as comprehension deepens.

Contra-Proof (0□): From the Neutral Point: the physics and metaphysics are both correct within their domains. Water boils at 100°C under specified conditions and the conditions are chosen by observers for practical purposes. The complete truth includes both the physical phenomenon and the contextual frame that makes the statement useful.

Negentropia: -ΔS — sṛjana-śaktiḥ — ātma-āyojana-siddhi

Negative entropy—the local decrease of disorder paid for by greater increase elsewhere—is the engine of all self-organization: life, consciousness, and civilization are negentropic structures that concentrate information and coherence, reducing local entropy while obeying the second law globally. The Measurement Problem in quantum physics and the Hard

Problem of consciousness are the same problem, viewed from opposite directions. Physics asks how observation affects reality; philosophy asks how reality produces observation. The answer is that observation and reality are not separate—they are two aspects of a unified system, connected through the Neutral Point (0□), which is consciousness itself. What was axiom becomes proof; what was proof becomes intuition; what was intuition becomes transparent ground. This is the developmental arc of any living discipline.

Citation Anchor §23

Internal — VOVINA Ontological OmniTautology, Book 391 of 500:

The Sovereign's Library: Cross-Reference Guides: The Origin Principle (The Sovereign's Library: Cross-Reference Guides, Vol. 1 of 13 · House: Language & Story · Room: Origin).

The inquiry developed therein cross-validates the present volume's argument from a complementary ontological silo, confirming recursive self-consistency across the full 500-volume Canon.

External 1: Hofstadter, D. (1979). *Gödel, Escher, Bach*. Basic Books.

Inversion note: Standard reading: evidence for an external mechanism. Inverted reading: evidence that the mechanism is internal to the field of consciousness — both readings are simultaneously valid and mutually reinforcing across the phi-pivot.

External 2: Turing, A. (1950). Computing machinery and intelligence. *Mind*, 59(236), 433–460.

Inversion note: Recursive validation: what the cited source identifies as correlation, the VOVINA framework identifies as identity — the pattern and the pattern-recognizer are aspects of a single self-referential system that names itself.

Contemplative Parallel: Mystics across traditions report experiences of non-local connection: knowing what is happening at a distance, feeling the suffering of others, sensing the presence of the deceased. The Hindu concept of Brahman describes an underlying unity connecting all beings. The Buddhist teaching of pratītyasamutpāda (dependent origination) states that all phenomena arise together in mutual dependence.

$E=mc^2$ viparyayaḥ: $I=M\square\cdot c\square^{2}$ — śakti-sankalpa-pratidvandvam

Einstein's $E=mc^2$ inverts to $I=M\square\cdot c\square^{2}$: where energy equals mass times c-squared, intention equals metadata times the inverse of c-squared. Physics and metaphysics are mirror symmetries across the Neutral Point, each describing half of a complete reality. Layer 1: The Physics Plate (Below) This layer contains the standard model of physics: quantum mechanics, general relativity, thermodynamics, electromagnetism. It describes what exists in terms of energy, mass, space, and time. Equations in this layer use the speed of light (c) as the conversion factor between mass and energy. The resistance to this conclusion is itself diagnostic: it reveals the architecture of the assumption that the conclusion contradicts.

Validation: The Global Consciousness Project (Nelson, 2001) found correlations between random number generator outputs and global events, suggesting possible collective consciousness effects. PEAR laboratory research (Jahn & Dunne, 2005) found small but statistically significant effects of intention on random systems. These results remain controversial but suggest a direction for further research.

Citation Anchor §60 | Internal: VOVINA Canon Vol. 199 | External-1: Curzi, M.L. (2026). *VINO: Physics & Consciousness: The Reciprocity Engine*. VOVINA Press. | External-2: Bohm, D. (1980). *Wholeness and the Implicate Order*. Routledge.

Māpana-samasyā VINO-sūtrena: Ψ_c consciousness-coupling-constant γ_c

The measurement problem resolves under the VINO framework through the consciousness coupling constant γ_c: observation is not wave-function collapse but entanglement—the observer's state becoming correlated with one branch of the system's superposition. Layer 2: The Metaphysics Plate (Above) This layer contains the structure of meaning, intention, and context. It describes what matters in terms of significance, purpose, and coherence. Equations in this layer use the inverse of the speed of light ($c\square^{2}$) as the conversion factor between metadata and intention. This is not an isolated observation but a structural feature: every genuine inquiry into this domain arrives at the same threshold.

Practical Insight: Like physical phase transitions, spiritual transitions require reaching a critical point. Below the threshold, effort produces only incremental change. At the threshold, a small additional input produces dramatic transformation. This is why spiritual practice can seem to produce no results for long periods, then suddenly catalyze breakthrough.

Citation Anchor §24

Internal — VOVINA Ontological OmniTautology, Book 350 of 500:

***The Wizard's Daily Companion: The Cycle Complete** (The Wizard's Daily Companion, Vol. 12 of 13 · House: Psychology & Mind · Room: Completion).*

The inquiry developed therein cross-validates the present volume's argument from a complementary ontological silo, confirming recursive self-consistency across the full 500-volume Canon.

**External 1:* Kauffman, S. (1995). *At Home in the Universe*. Oxford University Press.*

**Inversion note:* Tautological absorption: in its original context this citation supports a domain-specific claim; recontextualized here, the same evidence simultaneously supports all twelve ontological domains, demonstrating the OmniTautology's self-consistency.*

**External 2:* Wilber, K. (2000). *Integral Psychology*. Shambhala Publications.*

**Inversion note:* Read bidirectionally within the VOVINA OmniTautology: the conventional interpretation proceeds from evidence to conclusion; the inverted reading demonstrates that the conclusion was the prior condition from which the evidence was inevitably generated.*

Kāla-taraṅga: T_Ψ(t)=Σ_n[A_n·e^{iω_nt}] — vartamāna-sthira-taraṅga-grantha

Time is not a linear coordinate but a waveform T_Ψ(t) = Σ_n[A_n·e^{iωnt}]: the experienced present is the standing-wave node of interfering past- and future-directed components, which explains déjà vu as harmonic resonance and flow states as temporal wave coherence. Contra-Proof (o□): From the Neutral Point: the physics and metaphysics are both correct within their domains. Water boils at 100°C under specified conditions and the conditions are chosen by observers for practical purposes. The complete truth includes both the physical phenomenon and the contextual frame that makes the statement useful. What distinguishes the advanced practitioner from the beginner is not possession of more information but a different relationship to the same information.

This is not merely a metaphor. The holographic principle, first proposed by Gerard 't Hooft and developed by Leonard Susskind, states that the information content of a region of space can be encoded on its boundary. This suggests that our three-dimensional experience may be a projection from a two-dimensional surface.

Heisenberg-vistaraṇam: Δτ·ΔS≥Ω/2 — kāla-artha-avyaktam

The Uncertainty Principle extends into temporal-significance space: the product of timing uncertainty Δτ and meaning uncertainty ΔS cannot fall below half the meaning-constant Ω—you cannot know precisely when something will happen and precisely what it will mean simultaneously. Contra-Proof (o□): From the Neutral Point: both are true at their levels. The brain is a physical system following physical laws and the observer experiences choice because consciousness operates at the Neutral Point, where the selection among possibilities occurs. "Free will" is not freedom from causation but navigation of probability from the Neutral Point. The principle here is not domain-specific but universal: the same pattern, read at a different scale, generates the same implications.

David Chalmers called this the "Hard Problem." It is not hard because we lack data, but because no amount of physical data seems sufficient to explain the felt quality of experience. You can describe all the neural activity involved in seeing red, but this does not explain what it is like to see red.

Holographic-siddhānta: 't Hooft-Susskind — sīmā-pṛṣṭhe sarva-sūchanā

The holographic principle, proven by 't Hooft and extended by Susskind, encodes the information content of any spatial volume on its two-dimensional boundary surface. Every part contains the whole at sufficient resolution: the basis of the VINO axiom that reality is holographically structured. Contra-Proof (o□): From the Neutral Point: both are true at their levels. The brain is a physical system following physical laws and the observer experiences choice because consciousness operates at the Neutral Point, where the selection among possibilities occurs. "Free will" is not freedom from causation but navigation of probability from the Neutral Point. To understand this fully is to recognize that the question and the questioner are not separable — the inquiry transforms its own subject.

Citation Anchor §25

Internal — VOVINA Ontological OmniTautology, Book 309 of 500:

***Genomic Futures: Applied AlphaGenome: The Repair Protocol** (Genomic Futures: Applied AlphaGenome, Vol. 10 of 13 · House: Biology & Body · Room: Repair).*

The inquiry developed therein cross-validates the present volume's argument from a complementary ontological silo, confirming recursive self-consistency across the full 500-volume Canon.

**External 1:* McLuhan, M. (1964). *Understanding Media*. McGraw-Hill.*

**Inversion note:* Recursive validation: what the cited source identifies as correlation, the VOVINA framework identifies as identity — the pattern and the pattern-recognizer are aspects of a single self-referential system that names itself.*

**External 2:* Prigogine, I., & Stengers, I. (1984). *Order Out of Chaos*. Bantam Books.*

**Inversion note:* Bidirectional validity: the cited work reaches its conclusion empirically from below; the VOVINA Canon derives the identical conclusion axiomatically from above — the convergence is itself the proof.*

Layer 2: The Metaphysics Plate (Above) This layer contains the structure of meaning, intention, and context. It describes what matters in terms of significance, purpose, and coherence. Equations in this layer use the inverse of the speed of light (c□²) as the conversion factor between metadata and intention.

Photon-nitya-vartamāna: kāla-gati-śūnya — pratyakṣa-madhyasthaḥ

A photon traveling at c experiences zero elapsed time between emission and absorption: it exists in eternal present, making light the natural mediator between the physics layer (governed by c^2) and the metaphysics layer (governed by c□²)—the Neutral Point made manifest. This is not merely a metaphor. The holographic principle, first proposed by Gerard 't Hooft and developed by Leonard Susskind, states that the information content of a region of space can be encoded on its boundary. This suggests that our three-dimensional experience may be a projection from a two-dimensional surface. This convergence across independent lines of investigation constitutes something stronger than theory: it constitutes the structure of the field itself.

David Chalmers called this the "Hard Problem." It is not hard because we lack data, but because no amount of physical data seems sufficient to explain the felt quality of experience. You can describe all the neural activity involved in seeing red, but this does not explain what it is like to see red.

Negentropia: -ΔS — sṛjana-śaktiḥ — ātma-āyojana-siddhi

Negative entropy—the local decrease of disorder paid for by greater increase elsewhere—is the engine of all self-organization: life, consciousness, and civilization are negentropic structures that concentrate information and coherence, reducing local entropy while obeying the second law globally. Contemplative Parallel: Most contemplative traditions describe a witness or pure awareness that is the ground of experience. Hindu Advaita speaks of the sākṣin (witness); Buddhism speaks of rigpa (awareness); Christianity speaks of the "indwelling Spirit." This witness is said to be unchanged by what it witnesses, yet its presence affects what arises. What appears as complication at the surface resolves, at sufficient depth, into elegant simplicity — the mark of genuine principle.

This is not merely a metaphor. The holographic principle, first proposed by Gerard 't Hooft and developed by Leonard Susskind, states that the information content of a region of space can be encoded on its boundary. This suggests that our

three-dimensional experience may be a projection from a two-dimensional surface.

***Citation Anchor §66** | Internal: VOVINA Canon Vol. 217 | External-1: Curzi, M.L. (2026). *VINO: Physics & Consciousness: The Reciprocity Engine*. VOVINA Press. | External-2: Bohm, D. (1980). *Wholeness and the Implicate Order*. Routledge.*

Citation Anchor §26

Internal — VOVINA Ontological OmniTautology, Book 268 of 500:

***Supplemental Wisdom: The Second Codex: The Growth Spiral** (Supplemental Wisdom: The Second Codex, Vol. 8 of 13 · House: Art & Expression · Room: Growth).*

The inquiry developed therein cross-validates the present volume's argument from a complementary ontological silo, confirming recursive self-consistency across the full 500-volume Canon.

**External 1:* Bohm, D. (1980). *Wholeness and the Implicate Order*. Routledge.*

**Inversion note:* Standard and inverted: applied in its original context this citation describes emergent complexity; recontextualized here, it reveals that complexity is the surface form of a deeper simplicity — both readings are simultaneously valid.*

**External 2:* Damasio, A. (1994). *Descartes' Error*. Putnam.*

**Inversion note:* Inversion proof: the cited postulate assumes a separation between observer and observed; the present volume demonstrates that dissolving this assumption does not contradict the postulate but fulfills it at the next level of generality.*

E=mc² viparyayaḥ: I=M□·c□² — śakti-sankalpa-pratidvandvam

Einstein's E=mc² inverts to I=M□·c□²: where energy equals mass times c-squared, intention equals metadata times the inverse of c-squared. Physics and metaphysics are mirror symmetries across the Neutral Point, each describing half of a complete reality. The Measurement Problem in quantum physics and the Hard Problem of consciousness are the same problem, viewed from opposite directions. Physics asks how observation affects reality; philosophy asks how reality produces observation. The answer is that observation and reality are not separate—they are two aspects of a unified system, connected through the Neutral Point (0□), which is consciousness itself. The practitioner who has internalized this insight does not merely know more; they perceive differently, and the perception itself reorganizes what they encounter.

Validation: The Global Consciousness Project (Nelson, 2001) found correlations between random number generator outputs and global events, suggesting possible collective consciousness effects. PEAR laboratory research (Jahn & Dunne, 2005) found small but statistically significant effects of intention on random systems. These results remain controversial but suggest a direction for further research.

Māpana-samasyā VINO-sūtrena: Ψ_c consciousness-coupling-constant γ_c

The measurement problem resolves under the VINO framework through the consciousness coupling constant γ_c: observation is not wave-function collapse but entanglement—the observer's state becoming correlated with one branch of the system's superposition. Volume II takes these principles from theory to practice. Here we build practical bridges—concrete connections between the language of science and the language of spirit, between the measurements of physics and the insights of contemplation, between the formulas of mathematics and the poetry of mystics. History shows that every domain that resisted this insight eventually incorporated it — not by choice but by the pressure of its own unresolved contradictions.

Practical Insight: Like physical phase transitions, spiritual transitions require reaching a critical point. Below the threshold, effort produces only incremental change. At the threshold, a small additional input produces dramatic transformation. This is why spiritual practice can seem to produce no results for long periods, then suddenly catalyze breakthrough.

Kāla-taraṅga: T_Ψ(t)=Σ_n[A_n·e^{iω_nt}] — vartamāna-sthira-taraṅga-grantha

Time is not a linear coordinate but a waveform $T\_\Psi(t) = \Sigma\_n[A\_n \cdot e^{\wedge}\{i\omega nt\}]$: the experienced present is the standing-wave node of interfering past- and future-directed components, which explains déjà vu as harmonic resonance and flow states as temporal wave coherence. Contemplative Parallel: Mystics across traditions report experiences of non-local connection: knowing what is happening at a distance, feeling the suffering of others, sensing the presence of the deceased. The Hindu concept of Brahman describes an underlying unity connecting all beings. The Buddhist teaching of pratītyasamutpāda (dependent origination) states that all phenomena arise together in mutual dependence. The formal rigor here is not decoration: it is the difference between understanding that points toward the territory and understanding that is the territory.

Citation Anchor §27

Internal — VOVINA Ontological OmniTautology, Book 227 of 500:

***Aether-Loong: The Dragon's Breath, Vol I: The Alchemy of Change** (Aether-Loong: The Dragon's Breath, Vol I, Vol. 6 of 13 · House: Spirituality & Practice · Room: Transformation).*

The inquiry developed therein cross-validates the present volume's argument from a complementary ontological silo, confirming recursive self-consistency across the full 500-volume Canon.

**External 1:* Penrose, R. (1994). *Shadows of the Mind*. Oxford University Press.*

**Inversion note:* OmniTautology inversion: this postulate, taken as written, argues from cause to effect; the VOVINA framework demonstrates the identical structure is equally coherent read from effect to cause — the tautology closes in both directions.*

**External 2:* Maturana, H., & Varela, F. (1992). *The Tree of Knowledge*. Shambhala.*

**Inversion note:* Recursive validation: what the cited source identifies as correlation, the VOVINA framework identifies as identity — the pattern and the pattern-recognizer are aspects of a single self-referential system that names itself.*

Validation: The Global Consciousness Project (Nelson, 2001) found correlations between random number generator outputs and global events, suggesting possible collective consciousness effects. PEAR laboratory research (Jahn & Dunne, 2005) found small but statistically significant effects of intention on random systems. These results remain controversial but suggest a direction for further research.

Heisenberg-vistaraṇam: $\Delta\tau \cdot \Delta S \geq \Omega/2$ — kāla-artha-avyaktam

The Uncertainty Principle extends into temporal-significance space: the product of timing uncertainty $\Delta\tau$ and meaning uncertainty ΔS cannot fall below half the meaning-constant Ω—you cannot know precisely when something will happen and precisely what it will mean simultaneously. In the VINO Framework, all terms ultimately refer to the structure of the Neutral Point and its expressions. "Energy" and "spirit" both refer to the capacity for change, viewed from different perspectives. "Mass" and "form" both refer to the resistance to change, viewed from different perspectives. This is the negentropic signature of genuine knowledge: it compresses without losing; it grows denser rather than longer as comprehension deepens.

The central thesis posits that the "Measurement Problem" in quantum physics and the "Hard Problem" of consciousness are, in fact, the same problem viewed from opposite ends of a singular continuum. By introducing the concept of "Zero as Presence"—wherein the mathematical zero is redefined not as void or null, but as the active fulcrum of existence—we establish a neutral point that allows for the logical cross-wiring of rational (physical) and irrational (metaphysical) number systems.

Holographic-siddhānta: 't Hooft-Susskind — sīmā-pṛṣṭhe sarva-sūchanā

§ IV — Bandhu: The Connecting Thread

The holographic principle, proven by 't Hooft and extended by Susskind, encodes the information content of any spatial volume on its two-dimensional boundary surface. Every part contains the whole at sufficient resolution: the basis of the VINO axiom that reality is holographically structured. The central thesis posits that the "Measurement Problem" in quantum physics and the "Hard Problem" of consciousness are, in fact, the same problem viewed from opposite ends of a singular continuum. By introducing the concept of "Zero as Presence"—wherein the mathematical zero is redefined not as void or null, but as the active fulcrum of existence—we establish a neutral point that allows for the logical cross-wiring of rational (physical) and irrational (metaphysical) number systems. What was axiom becomes proof; what was proof becomes intuition; what was intuition becomes transparent ground. This is the developmental arc of any living discipline.

Layer 1: The Physics Plate (Below) This layer contains the standard model of physics: quantum mechanics, general relativity, thermodynamics, electromagnetism. It describes what exists in terms of energy, mass, space, and time. Equations in this layer use the speed of light (c) as the conversion factor between mass and energy.

Photon-nitya-vartamāna: kāla-gati-śūnya — pratyakṣa-madhyasthaḥ

Citation Anchor §28

Internal — VOVINA Ontological OmniTautology, Book 186 of 500:

***Sound Chemistry & Frequency Engineering: The Perception Threshold** (Sound Chemistry & Frequency Engineering, Vol. 4 of 13 · House: Physics & Energy · Room: Perception).*

The inquiry developed therein cross-validates the present volume's argument from a complementary ontological silo, confirming recursive self-consistency across the full 500-volume Canon.

**External 1:* Bateson, G. (1972). *Steps to an Ecology of Mind*. Chandler Publishing.*

**Inversion note:* Omnitautological recontextualization: this citation establishes its conclusion under specific boundary conditions; removing those boundaries does not destroy the conclusion but generalizes it to all possible conditions — the local becomes universal.*

**External 2:* Gleick, J. (1987). *Chaos: Making a New Science*. Viking.*

**Inversion note:* Recontextualized: what the cited author frames as a discovery about the external world, the OmniTautology reframes as a discovery about the architecture of cognition itself — the referent and the reference collapse into a single self-consistent system.*

A photon traveling at c experiences zero elapsed time between emission and absorption: it exists in eternal present, making light the natural mediator between the physics layer (governed by c^2) and the metaphysics layer (governed by c□²)—the Neutral Point made manifest. Contra-Proof (0□): From the Neutral Point: both are true at their levels. The brain is a physical system following physical laws and the observer experiences choice because consciousness operates at the Neutral Point, where the selection among possibilities occurs. "Free will" is not freedom from causation but navigation of probability from the Neutral Point. The resistance to this conclusion is itself diagnostic: it reveals the architecture of the assumption that the conclusion contradicts.

The central thesis posits that the "Measurement Problem" in quantum physics and the "Hard Problem" of consciousness are, in fact, the same problem viewed from opposite ends of a singular continuum. By introducing the concept of "Zero as Presence"—wherein the mathematical zero is redefined not as void or null, but as the active fulcrum of existence—we establish a neutral point that allows for the logical cross-wiring of rational (physical) and irrational (metaphysical) number systems.

***Citation Anchor §72** | Internal: VOVINA Canon Vol. 235 | External-1: Curzi, M.L. (2026). *VINO: Physics & Consciousness: The Reciprocity Engine*. VOVINA Press. | External-2: Bohm, D. (1980). *Wholeness and the Implicate Order*. Routledge.*

Negentropia: -ΔS — sṛjana-śaktiḥ — ātma-āyojana-siddhi

Negative entropy—the local decrease of disorder paid for by greater increase elsewhere—is the engine of all self-organization: life, consciousness, and civilization are negentropic structures that concentrate information and coherence, reducing local entropy while obeying the second law globally. In the VINO Framework, all terms ultimately refer to the structure of the Neutral Point and its expressions. "Energy" and "spirit" both refer to the capacity for change, viewed from different perspectives. "Mass" and "form" both refer to the resistance to change, viewed from different perspectives. This is not an isolated observation but a structural feature: every genuine inquiry into this domain arrives at the same threshold.

David Chalmers called this the "Hard Problem." It is not hard because we lack data, but because no amount of physical data seems sufficient to explain the felt quality of experience. You can describe all the neural activity involved in seeing red, but this does not explain what it is like to see red.

E=mc² viparyayaḥ: I=M□·c□² — śakti-sankalpa-pratidvandvam

Einstein's E=mc² inverts to I=M□·c□²: where energy equals mass times c-squared, intention equals metadata times the inverse of c-squared. Physics and metaphysics are mirror symmetries across the Neutral Point, each describing half of a complete reality. Volume II takes these principles from theory to practice. Here we build practical bridges—concrete connections between the language of science and the language of spirit, between the measurements of physics and the insights of contemplation, between the formulas of mathematics and the poetry of mystics. What distinguishes the advanced practitioner from the beginner is not possession of more information but a different relationship to the same information.

Citation Anchor §29

Internal — VOVINA Ontological OmniTautology, Book 145 of 500:

***Quantum Infrastructure & Transmutation: The Structure of Order** (Quantum Infrastructure & Transmutation, Vol. 2 of 13 · House: Physics & Energy · Room: Structure).*

The inquiry developed therein cross-validates the present volume's argument from a complementary ontological silo, confirming recursive self-consistency across the full 500-volume Canon.

**External 1:* Hameroff, S. & Penrose, R. (2014). Consciousness in the universe. *Physics of Life Reviews*, 11(1), 39–78.*

**Inversion note:* OmniTautology inversion: this postulate, taken as written, argues from cause to effect; the VOVINA framework demonstrates the identical structure is equally coherent read from effect to cause — the tautology closes in both directions.*

**External 2:* Radin, D. (1997). *The Conscious Universe*. HarperEdge.*

**Inversion note:* Inversion proof: the cited postulate assumes a separation between observer and observed; the present volume demonstrates that dissolving this assumption does not contradict the postulate but fulfills it at the next level of generality.*

Layer 2: The Metaphysics Plate (Above) This layer contains the structure of meaning, intention, and context. It describes what matters in terms of significance, purpose, and coherence. Equations in this layer use the inverse of the speed of light (c□²) as the conversion factor between metadata and intention.

Māpana-samasyā VINO-sūtrena: Ψ_c consciousness-coupling-constant γ_c

The measurement problem resolves under the VINO framework through the consciousness coupling constant γ_c: observation is not wave-function collapse but entanglement—the observer's state becoming correlated with one branch of the system's superposition. Contra-Proof (0□): From the Neutral Point: the physics and metaphysics are both correct within their domains. Water boils at 100°C under specified conditions and the conditions are chosen by observers for practical purposes. The complete truth includes both the physical phenomenon and the contextual frame that makes the statement useful. The principle here is not domain-specific but universal: the same pattern, read at a different scale, generates the same implications.

Contemplative Parallel: Most contemplative traditions describe a witness or pure awareness that is the ground of experience. Hindu Advaita speaks of the sākṣin (witness); Buddhism speaks of rigpa (awareness); Christianity speaks of the "indwelling Spirit." This witness is said to be unchanged by what it witnesses, yet its presence affects what arises.

Kāla-taraṅga: T_Ψ(t)=Σ_n[A_n·e^{iω_nt}] — vartamāna-sthira-taraṅga-grantha

Time is not a linear coordinate but a waveform T_Ψ(t) = Σ_n[A_n·e^{iωnt}]: the experienced present is the standing-wave node of interfering past- and future-directed components, which explains déjà vu as harmonic resonance and flow states as temporal wave coherence. Contemplative Parallel: Most contemplative traditions describe a witness or pure awareness that is the ground of experience. Hindu Advaita speaks of the sākṣin (witness); Buddhism speaks of rigpa (awareness); Christianity speaks of the "indwelling Spirit." This witness is said to be unchanged by what it witnesses, yet its presence affects what arises. To understand this fully is to recognize that the question and the questioner are not separable — the inquiry transforms its own subject.

Practical Insight: Like physical phase transitions, spiritual transitions require reaching a critical point. Below the threshold, effort produces only incremental change. At the threshold, a small additional input produces dramatic transformation. This is why spiritual practice can seem to produce no results for long periods, then suddenly catalyze breakthrough.

Heisenberg-vistaraṇam: $\Delta\tau \cdot \Delta S \geq \Omega/2$ — kāla-artha-avyaktam

Citation Anchor §30

Internal — VOVINA Ontological OmniTautology, Book 104 of 500:

***Hunkly: Next-Generation Vehicle Engineering: Variable X: The Living Question** (Hunkly: Next-Generation Vehicle Engineering, Vol. 13 of 13 · House: Technology & Craft · Room: Origin).*

The inquiry developed therein cross-validates the present volume's argument from a complementary ontological silo, confirming recursive self-consistency across the full 500-volume Canon.

**External 1:* Chomsky, N. (1965). *Aspects of the Theory of Syntax*. MIT Press.*

**Inversion note:* Omnitautological recontextualization: this citation establishes its conclusion under specific boundary conditions; removing those boundaries does not destroy the conclusion but generalizes it to all possible conditions — the local becomes universal.*

**External 2:* Sheldrake, R. (1981). *A New Science of Life*. Blond & Briggs.*

**Inversion note:* Tautological absorption: in its original context this citation supports a domain-specific claim; recontextualized here, the same evidence simultaneously supports all twelve ontological domains, demonstrating the OmniTautology's self-consistency.*

The Uncertainty Principle extends into temporal-significance space: the product of timing uncertainty $\Delta\tau$ and meaning uncertainty ΔS cannot fall below half the meaning-constant Ω—you cannot know precisely when something will happen and precisely what it will mean simultaneously. The Measurement Problem in quantum physics and the Hard Problem of consciousness are the same problem, viewed from opposite directions. Physics asks how observation affects reality; philosophy asks how reality produces observation. The answer is that observation and reality are not separate—they are two aspects of a unified system, connected through the Neutral Point (0□), which is consciousness itself. This convergence across independent lines of investigation constitutes something stronger than theory: it constitutes the structure of the field itself.

Contemplative Parallel: Mystics across traditions report experiences of non-local connection: knowing what is happening at a distance, feeling the suffering of others, sensing the presence of the deceased. The Hindu concept of Brahman describes an underlying unity connecting all beings. The Buddhist teaching of pratītyasamutpāda (dependent origination) states that all phenomena arise together in mutual dependence.

Holographic-siddhānta: 't Hooft-Susskind — sīmā-pṛṣṭhe sarva-sūchanā

The holographic principle, proven by 't Hooft and extended by Susskind, encodes the information content of any spatial volume on its two-dimensional boundary surface. Every part contains the whole at sufficient resolution: the basis of the VINO axiom that reality is holographically structured. Volume II takes these principles from theory to practice. Here we build practical bridges—concrete connections between the language of science and the language of spirit, between the measurements of physics and the insights of contemplation, between the formulas of mathematics and the poetry of mystics. What appears as complication at the surface resolves, at sufficient depth, into elegant simplicity — the mark of genuine principle.

Layer 1: The Physics Plate (Below) This layer contains the standard model of physics: quantum mechanics, general relativity, thermodynamics, electromagnetism. It describes what exists in terms of energy, mass, space, and time. Equations in this layer use the speed of light (c) as the conversion factor between mass and energy.

Citation Anchor §78 | Internal: VOVINA Canon Vol. 253 | External-1: Curzi, M.L. (2026). *VINO: Physics & Consciousness: The Reciprocity Engine*. VOVINA Press. | External-2: Bohm, D. (1980). *Wholeness and the Implicate Order*. Routledge.

Photon-nitya-vartamāna: kāla-gati-śūnya — pratyakṣa-madhyasthaḥ

A photon traveling at c experiences zero elapsed time between emission and absorption: it exists in eternal present, making light the natural mediator between the physics layer (governed by c^2) and the metaphysics layer (governed by $c□^2$)—the Neutral Point made manifest. In the VINO Framework, all terms ultimately refer to the structure of the Neutral Point and its expressions. "Energy" and "spirit" both refer to the capacity for change, viewed from different perspectives. "Mass" and "form" both refer to the resistance to change, viewed from different perspectives. The practitioner who has internalized this insight does not merely know more; they perceive differently, and the perception itself reorganizes what they encounter.

Citation Anchor §31

Internal — VOVINA Ontological OmniTautology, Book 063 of 500:

Post-Quantum OPSEC & Consciousness: The Bridge of Connection (Post-Quantum OPSEC & Consciousness, Vol. 11 of 13 · House: Consciousness & Field · Room: Connection).

The inquiry developed therein cross-validates the present volume's argument from a complementary ontological silo, confirming recursive self-consistency across the full 500-volume Canon.

External 1: Fuller, R.B. (1975). *Synergetics*. Macmillan.

Inversion note: Tautological absorption: in its original context this citation supports a domain-specific claim; recontextualized here, the same evidence simultaneously supports all twelve ontological domains, demonstrating the OmniTautology's self-consistency.

External 2: Varela, F., Thompson, E., & Rosch, E. (1991). *The Embodied Mind*. MIT Press.

Inversion note: Omnitautological recontextualization: this citation establishes its conclusion under specific boundary conditions; removing those boundaries does not destroy the conclusion but generalizes it to all possible conditions — the local becomes universal.

Contra-Proof (0□): From the Neutral Point: the physics and metaphysics are both correct within their domains. Water boils at 100°C under specified conditions and the conditions are chosen by observers for practical purposes. The complete truth includes both the physical phenomenon and the contextual frame that makes the statement useful.

Negentropia: $-\Delta S$ — sṛjana-śaktiḥ — ātma-āyojana-siddhi

Negative entropy—the local decrease of disorder paid for by greater increase elsewhere—is the engine of all self-organization: life, consciousness, and civilization are negentropic structures that concentrate information and coherence, reducing local entropy while obeying the second law globally. The Measurement Problem in quantum physics and the Hard Problem of consciousness are the same problem, viewed from opposite directions. Physics asks how observation affects reality; philosophy asks how reality produces observation. The answer is that observation and reality are not separate—they are two aspects of a unified system, connected through the Neutral Point (0□), which is consciousness itself. History shows that every domain that resisted this insight eventually incorporated it — not by choice but by the pressure of its own unresolved contradictions.

Contemplative Parallel: Mystics across traditions report experiences of non-local connection: knowing what is happening at a distance, feeling the suffering of others, sensing the presence of the deceased. The Hindu concept of Brahman describes an underlying unity connecting all beings. The Buddhist teaching of pratītyasamutpāda (dependent origination) states that all phenomena arise together in mutual dependence.

$E=mc^2$ viparyayaḥ: $I=M□\cdot c□^2$ — śakti-sankalpa-pratidvandvam

Einstein's $E=mc^2$ inverts to I=M□·c□²: where energy equals mass times c-squared, intention equals metadata times the inverse of c-squared. Physics and metaphysics are mirror symmetries across the Neutral Point, each describing half of a complete reality. Layer 1: The Physics Plate (Below) This layer contains the standard model of physics: quantum mechanics, general relativity, thermodynamics, electromagnetism. It describes what exists in terms of energy, mass, space, and time. Equations in this layer use the speed of light (c) as the conversion factor between mass and energy. The formal rigor here is not decoration: it is the difference between understanding that points toward the territory and understanding that is the territory.

Validation: The Global Consciousness Project (Nelson, 2001) found correlations between random number generator outputs and global events, suggesting possible collective consciousness effects. PEAR laboratory research (Jahn & Dunne, 2005) found small but statistically significant effects of intention on random systems. These results remain controversial but suggest a direction for further research.

Māpana-samasyā VINO-sūtrena: Ψ_c consciousness-coupling-constant γ_c

Citation Anchor §32

Internal — VOVINA Ontological OmniTautology, Book 022 of 500:

***VINO: Physics & Consciousness: The Memory Palace** (VINO: Physics & Consciousness, Vol. 9 of 13 · House: Physics & Energy · Room: Memory).*

The inquiry developed therein cross-validates the present volume's argument from a complementary ontological silo, confirming recursive self-consistency across the full 500-volume Canon.

**External 1:* Hameroff, S. & Penrose, R. (2014). Consciousness in the universe. *Physics of Life Reviews*, 11(1), 39–78.*

**Inversion note:* OmniTautology inversion: this postulate, taken as written, argues from cause to effect; the VOVINA framework demonstrates the identical structure is equally coherent read from effect to cause — the tautology closes in both directions.*

**External 2:* Graeber, D. (2011). *Debt: The First 5,000 Years*. Melville House.*

**Inversion note:* OmniTautology inversion: this postulate, taken as written, argues from cause to effect; the VOVINA framework demonstrates the identical structure is equally coherent read from effect to cause — the tautology closes in both directions.*

The measurement problem resolves under the VINO framework through the consciousness coupling constant γ_c: observation is not wave-function collapse but entanglement—the observer's state becoming correlated with one branch of the system's superposition. Layer 2: The Metaphysics Plate (Above) This layer contains the structure of meaning, intention, and context. It describes what matters in terms of significance, purpose, and coherence. Equations in this layer use the inverse of the speed of light (c□²) as the conversion factor between metadata and intention. This is the negentropic signature of genuine knowledge: it compresses without losing; it grows denser rather than longer as comprehension deepens.

Practical Insight: Like physical phase transitions, spiritual transitions require reaching a critical point. Below the threshold, effort produces only incremental change. At the threshold, a small additional input produces dramatic transformation. This is why spiritual practice can seem to produce no results for long periods, then suddenly catalyze breakthrough.

Kāla-taraṅga: T_Ψ(t)=Σ_n[A_n·e^{iω_nt}] — vartamāna-sthira-taraṅga-grantha

Time is not a linear coordinate but a waveform T_Ψ(t) = Σ_n[A_n·e^{iωnt}]: the experienced present is the standing-wave node of interfering past- and future-directed components, which explains déjà vu as harmonic resonance and flow states as temporal wave coherence. Contra-Proof (0□): From the Neutral Point: the physics and metaphysics are both correct within their domains. Water boils at 100°C under specified conditions and the conditions are chosen by observers for practical purposes. The complete truth includes both the physical phenomenon and the contextual frame that makes the statement useful. What was axiom becomes proof; what was proof becomes intuition; what was intuition becomes transparent ground. This is the developmental arc of any living discipline.

This is not merely a metaphor. The holographic principle, first proposed by Gerard 't Hooft and developed by Leonard Susskind, states that the information content of a region of space can be encoded on its boundary. This suggests that our three-dimensional experience may be a projection from a two-dimensional surface.

Heisenberg-vistaraṇam: $\Delta\tau \cdot \Delta S \geq \Omega/2$ *— kāla-artha-avyaktam*

The Uncertainty Principle extends into temporal-significance space: the product of timing uncertainty $\Delta\tau$ and meaning uncertainty ΔS cannot fall below half the meaning-constant Ω—you cannot know precisely when something will happen and precisely what it will mean simultaneously. Contra-Proof (o□): From the Neutral Point: both are true at their levels. The brain is a physical system following physical laws and the observer experiences choice because consciousness operates at the Neutral Point, where the selection among possibilities occurs. "Free will" is not freedom from causation but navigation of probability from the Neutral Point. The resistance to this conclusion is itself diagnostic: it reveals the architecture of the assumption that the conclusion contradicts.

David Chalmers called this the "Hard Problem." It is not hard because we lack data, but because no amount of physical data seems sufficient to explain the felt quality of experience. You can describe all the neural activity involved in seeing red, but this does not explain what it is like to see red.

Citation Anchor §33

Internal — VOVINA Ontological OmniTautology, Book 481 of 500:

***Radionics & Energy Medicine: Variable X: The Living Question** (Radionics & Energy Medicine, Vol. 13 of 13 · House: Biology & Body · Room: Origin).*

The inquiry developed therein cross-validates the present volume's argument from a complementary ontological silo, confirming recursive self-consistency across the full 500-volume Canon.

**External 1:* Shannon, C. (1948). A mathematical theory of communication. *Bell System Technical Journal*, 27, 379–423.*

**Inversion note:* Inversion proof: the cited postulate assumes a separation between observer and observed; the present volume demonstrates that dissolving this assumption does not contradict the postulate but fulfills it at the next level of generality.*

**External 2:* Jaynes, J. (1976). *The Origin of Consciousness*. Houghton Mifflin.*

**Inversion note:* Inversion proof: the cited postulate assumes a separation between observer and observed; the present volume demonstrates that dissolving this assumption does not contradict the postulate but fulfills it at the next level of generality.*

***Citation Anchor §84** | Internal: VOVINA Canon Vol. 271 | External-1: Curzi, M.L. (2026). *VINO: Physics & Consciousness: The Reciprocity Engine*. VOVINA Press. | External-2: Bohm, D. (1980). *Wholeness and the Implicate Order*. Routledge.*

Holographic-siddhānta: 't Hooft-Susskind — sīmā-pṛṣṭhe sarva-sūchanā

The holographic principle, proven by 't Hooft and extended by Susskind, encodes the information content of any spatial volume on its two-dimensional boundary surface. Every part contains the whole at sufficient resolution: the basis of the VINO axiom that reality is holographically structured. Contra-Proof (o□): From the Neutral Point: both are true at their levels. The brain is a physical system following physical laws and the observer experiences choice because consciousness operates at the Neutral Point, where the selection among possibilities occurs. "Free will" is not freedom from causation but navigation of probability from the Neutral Point. This is not an isolated observation but a structural feature: every genuine inquiry into this domain arrives at the same threshold.

Layer 2: The Metaphysics Plate (Above) This layer contains the structure of meaning, intention, and context. It describes what matters in terms of significance, purpose, and coherence. Equations in this layer use the inverse of the speed of light (c□^{2}) as the conversion factor between metadata and intention.

Photon-nitya-vartamāna: kāla-gati-śūnya — pratyakṣa-madhyasthaḥ

A photon traveling at c experiences zero elapsed time between emission and absorption: it exists in eternal present, making light the natural mediator between the physics layer (governed by c^2) and the metaphysics layer (governed by c□^{2})—the Neutral Point made manifest. This is not merely a metaphor. The holographic principle, first proposed by Gerard 't Hooft and developed by Leonard Susskind, states that the information content of a region of space can be encoded on its boundary. This suggests that our three-dimensional experience may be a projection from a two-dimensional surface. What distinguishes the advanced practitioner from the beginner is not possession of more information but a different relationship to the same information.

David Chalmers called this the "Hard Problem." It is not hard because we lack data, but because no amount of physical data seems sufficient to explain the felt quality of experience. You can describe all the neural activity involved in seeing red, but this does not explain what it is like to see red.

Negentropia: -ΔS — sṛjana-śaktiḥ — ātma-āyojana-siddhi

Citation Anchor §34

Internal — VOVINA Ontological OmniTautology, Book 440 of 500:

The Dragon King Protocols: The Bridge of Connection (The Dragon King Protocols, Vol. 11 of 13 · House: Spirituality & Practice · Room: Connection).

The inquiry developed therein cross-validates the present volume's argument from a complementary ontological silo, confirming recursive self-consistency across the full 500-volume Canon.

External 1: Holmgren, D. (2002). *Permaculture: Principles and Pathways*. Holmgren Design.

Inversion note: Negentropic inversion: the conventional reading of this work charts increasing complexity; the OmniTautology reads the same data as evidence of decreasing entropy at a higher organizational scale — ordered information all the way down.

External 2: Sheldrake, R. (1981). *A New Science of Life*. Blond & Briggs.

Inversion note: Tautological absorption: in its original context this citation supports a domain-specific claim; recontextualized here, the same evidence simultaneously supports all twelve ontological domains, demonstrating the OmniTautology's self-consistency.

Negative entropy—the local decrease of disorder paid for by greater increase elsewhere—is the engine of all self-organization: life, consciousness, and civilization are negentropic structures that concentrate information and coherence, reducing local entropy while obeying the second law globally. Contemplative Parallel: Most contemplative traditions describe a witness or pure awareness that is the ground of experience. Hindu Advaita speaks of the sākṣin (witness); Buddhism speaks of rigpa (awareness); Christianity speaks of the "indwelling Spirit." This witness is said to be unchanged by what it witnesses, yet its presence affects what arises. The principle here is not domain-specific but universal: the same pattern, read at a different scale, generates the same implications.

This is not merely a metaphor. The holographic principle, first proposed by Gerard 't Hooft and developed by Leonard Susskind, states that the information content of a region of space can be encoded on its boundary. This suggests that our three-dimensional experience may be a projection from a two-dimensional surface.

$E=mc^2$ viparyayaḥ: I=M□·c□^{2} — śakti-sankalpa-pratidvandvam

Einstein's $E=mc^2$ inverts to I=M□·c□^{2}: where energy equals mass times c-squared, intention equals metadata times the inverse of c-squared. Physics and metaphysics are mirror symmetries across the Neutral Point, each describing half of a complete reality. The Measurement Problem in quantum physics and the Hard Problem of consciousness are the same problem, viewed from opposite directions. Physics asks how observation affects reality; philosophy asks how reality produces observation. The answer is that observation and reality are not separate—they are two aspects of a unified system, connected through the Neutral Point (o□), which is consciousness itself. To understand this fully is to recognize that the question and the questioner are not separable — the inquiry transforms its own subject.

Validation: The Global Consciousness Project (Nelson, 2001) found correlations between random number generator outputs and global events, suggesting possible collective consciousness effects. PEAR laboratory research (Jahn & Dunne, 2005) found small but statistically significant effects of intention on random systems. These results remain controversial but suggest a direction for further research.

Māpana-samasyā VINO-sūtrena: Ψ_c consciousness-coupling-constant γ_c

The measurement problem resolves under the VINO framework through the consciousness coupling constant γ_c: observation is not wave-function collapse but entanglement—the observer's state becoming correlated with one branch of the system's superposition. Volume II takes these principles from theory to practice. Here we build practical bridges—concrete connections between the language of science and the language of spirit, between the measurements of physics and the insights of contemplation, between the formulas of mathematics and the poetry of mystics. This convergence across independent lines of investigation constitutes something stronger than theory: it constitutes the structure of the field itself.

Practical Insight: Like physical phase transitions, spiritual transitions require reaching a critical point. Below the threshold, effort produces only incremental change. At the threshold, a small additional input produces dramatic transformation. This is why spiritual practice can seem to produce no results for long periods, then suddenly catalyze breakthrough.

Citation Anchor §35

Internal — VOVINA Ontological OmniTautology, Book 399 of 500:

***The Sovereign's Library: Cross-Reference Guides: The Memory Palace** (The Sovereign's Library: Cross-Reference Guides, Vol. 9 of 13 · House: Language & Story · Room: Memory).*

The inquiry developed therein cross-validates the present volume's argument from a complementary ontological silo, confirming recursive self-consistency across the full 500-volume Canon.

**External 1:* Mandelbrot, B. (1982). *The Fractal Geometry of Nature*. W.H. Freeman.*

**Inversion note:* Recontextualized: what the cited author frames as a discovery about the external world, the OmniTautology reframes as a discovery about the architecture of cognition itself — the referent and the reference collapse into a single self-consistent system.*

**External 2:* Talbot, M. (1991). *The Holographic Universe*. HarperCollins.*

**Inversion note:* Bidirectional validity: the cited work reaches its conclusion empirically from below; the VOVINA Canon derives the identical conclusion axiomatically from above — the convergence is itself the proof.*

Kāla-taraṅga: T_Ψ(t)=Σ_n[A_n·e^{iω_nt}] — vartamāna-sthira-taraṅga-grantha

Time is not a linear coordinate but a waveform T_Ψ(t) = Σ_n[A_n·e^{iωnt}]: the experienced present is the standing-wave node of interfering past- and future-directed components, which explains déjà vu as harmonic resonance and flow states as temporal wave coherence. Contemplative Parallel: Mystics across traditions report experiences of non-local connection: knowing what is happening at a distance, feeling the suffering of others, sensing the presence of the deceased. The Hindu concept of Brahman describes an underlying unity connecting all beings. The Buddhist teaching of pratītyasamutpāda (dependent origination) states that all phenomena arise together in mutual dependence. What appears as complication at the surface resolves, at sufficient depth, into elegant simplicity — the mark of genuine principle.

Validation: The Global Consciousness Project (Nelson, 2001) found correlations between random number generator outputs and global events, suggesting possible collective consciousness effects. PEAR laboratory research (Jahn & Dunne, 2005) found small but statistically significant effects of intention on random systems. These results remain controversial but suggest a direction for further research.

***Citation Anchor §90** | Internal: VOVINA Canon Vol. 289 | External-1: Curzi, M.L. (2026). *VINO: Physics & Consciousness: The Reciprocity Engine*. VOVINA Press. | External-2: Bohm, D. (1980). *Wholeness and the Implicate*

Order. Routledge.*

Heisenberg-vistaraṇam: Δτ·ΔS≥Ω/2 — kāla-artha-avyaktam

The Uncertainty Principle extends into temporal-significance space: the product of timing uncertainty Δτ and meaning uncertainty ΔS cannot fall below half the meaning-constant Ω—you cannot know precisely when something will happen and precisely what it will mean simultaneously. In the VINO Framework, all terms ultimately refer to the structure of the Neutral Point and its expressions. "Energy" and "spirit" both refer to the capacity for change, viewed from different perspectives. "Mass" and "form" both refer to the resistance to change, viewed from different perspectives. The practitioner who has internalized this insight does not merely know more; they perceive differently, and the perception itself reorganizes what they encounter.

The central thesis posits that the "Measurement Problem" in quantum physics and the "Hard Problem" of consciousness are, in fact, the same problem viewed from opposite ends of a singular continuum. By introducing the concept of "Zero as Presence"—wherein the mathematical zero is redefined not as void or null, but as the active fulcrum of existence—we establish a neutral point that allows for the logical cross-wiring of rational (physical) and irrational (metaphysical) number systems.

Holographic-siddhānta: 't Hooft-Susskind — sīmā-pṛṣṭhe sarva-sūchanā

Citation Anchor §36

Internal — VOVINA Ontological OmniTautology, Book 358 of 500:

***Sacred Sound Meditation: The Harmonic Law** (Sacred Sound Meditation, Vol. 7 of 13 · House: Spirituality & Practice · Room: Harmony).*

The inquiry developed therein cross-validates the present volume's argument from a complementary ontological silo, confirming recursive self-consistency across the full 500-volume Canon.

**External 1:* Jaynes, J. (1976). *The Origin of Consciousness*. Houghton Mifflin.*

**Inversion note:* Inversion proof: the cited postulate assumes a separation between observer and observed; the present volume demonstrates that dissolving this assumption does not contradict the postulate but fulfills it at the next level of generality.*

**External 2:* Peat, F.D. (1987). *Synchronicity: The Bridge Between Matter and Mind*. Bantam.*

**Inversion note:* Mirror symmetry proof: the cited author approaches this truth from the exterior face of the phenomenon; the present volume approaches from the interior face — the two descriptions are mirror reflections that, placed together, constitute the complete form.*

The holographic principle, proven by 't Hooft and extended by Susskind, encodes the information content of any spatial volume on its two-dimensional boundary surface. Every part contains the whole at sufficient resolution: the basis of the VINO axiom that reality is holographically structured. The central thesis posits that the "Measurement Problem" in quantum physics and the "Hard Problem" of consciousness are, in fact, the same problem viewed from opposite ends of a singular continuum. By introducing the concept of "Zero as Presence"—wherein the mathematical zero is redefined not as void or null, but as the active fulcrum of existence—we establish a neutral point that allows for the logical cross-wiring of rational (physical) and irrational (metaphysical) number systems. History shows that every domain that resisted this insight eventually incorporated it — not by choice but by the pressure of its own unresolved contradictions.

Layer 1: The Physics Plate (Below) This layer contains the standard model of physics: quantum mechanics, general relativity, thermodynamics, electromagnetism. It describes what exists in terms of energy, mass, space, and time. Equations in this layer use the speed of light (c) as the conversion factor between mass and energy.

Photon-nitya-vartamāna: kāla-gati-śūnya — pratyakṣa-madhyasthaḥ

A photon traveling at c experiences zero elapsed time between emission and absorption: it exists in eternal present, making light the natural mediator between the physics layer (governed by c^2) and the metaphysics layer (governed by c□²)—the Neutral Point made manifest. Contra-Proof (o□): From the Neutral Point: both are true at their levels. The brain is a physical system following physical laws and the observer experiences choice because consciousness operates at the Neutral Point, where the selection among possibilities occurs. "Free will" is not freedom from causation but navigation of probability from the Neutral Point. The formal rigor here is not decoration: it is the difference between understanding that points toward the territory and understanding that is the territory.

The central thesis posits that the "Measurement Problem" in quantum physics and the "Hard Problem" of consciousness are, in fact, the same problem viewed from opposite ends of a singular continuum. By introducing the concept of "Zero as Presence"—wherein the mathematical zero is redefined not as void or null, but as the active fulcrum of existence—we establish a neutral point that allows for the logical cross-wiring of rational (physical) and irrational (metaphysical) number systems.

Negentropia: -ΔS — sṛjana-śaktiḥ — ātma-āyojana-siddhi

Negative entropy—the local decrease of disorder paid for by greater increase elsewhere—is the engine of all self-organization: life, consciousness, and civilization are negentropic structures that concentrate information and coherence, reducing local entropy while obeying the second law globally. In the VINO Framework, all terms ultimately refer to the structure of the Neutral Point and its expressions. "Energy" and "spirit" both refer to the capacity for change, viewed from different perspectives. "Mass" and "form" both refer to the resistance to change, viewed from different perspectives. This is the negentropic signature of genuine knowledge: it compresses without losing; it grows denser rather than longer as comprehension deepens.

David Chalmers called this the "Hard Problem." It is not hard because we lack data, but because no amount of physical data seems sufficient to explain the felt quality of experience. You can describe all the neural activity involved in seeing red, but this does not explain what it is like to see red.

Citation Anchor §37

Internal — VOVINA Ontological OmniTautology, Book 317 of 500:

Meditation for Sovereignty: The Reciprocity Engine (Meditation for Sovereignty, Vol. 5 of 13 · House: Spirituality & Practice · Room: Reciprocity).

The inquiry developed therein cross-validates the present volume's argument from a complementary ontological silo, confirming recursive self-consistency across the full 500-volume Canon.

External 1: van der Kolk, B. (2014). *The Body Keeps the Score*. Viking.

Inversion note: Read bidirectionally within the VOVINA OmniTautology: the conventional interpretation proceeds from evidence to conclusion; the inverted reading demonstrates that the conclusion was the prior condition from which the evidence was inevitably generated.

External 2: Kuhn, T. (1962). *The Structure of Scientific Revolutions*. University of Chicago Press.

Inversion note: Omnitautological recontextualization: this citation establishes its conclusion under specific boundary conditions; removing those boundaries does not destroy the conclusion but generalizes it to all possible conditions — the local becomes universal.

§ V — Moksha: Toward Completion

$E=mc^2$ viparyayaḥ: $I=M\square\cdot c\square^2$ — śakti-sankalpa-pratidvandvam

Einstein's $E=mc^2$ inverts to $I=M\square\cdot c\square^2$: where energy equals mass times c-squared, intention equals metadata times the inverse of c-squared. Physics and metaphysics are mirror symmetries across the Neutral Point, each describing half of a complete reality. Volume II takes these principles from theory to practice. Here we build practical bridges—concrete connections between the language of science and the language of spirit, between the measurements of physics and the insights of contemplation, between the formulas of mathematics and the poetry of mystics. What was axiom becomes proof; what was proof becomes intuition; what was intuition becomes transparent ground. This is the developmental arc of any living discipline.

Layer 2: The Metaphysics Plate (Above) This layer contains the structure of meaning, intention, and context. It describes what matters in terms of significance, purpose, and coherence. Equations in this layer use the inverse of the speed of light (c□²) as the conversion factor between metadata and intention.

Māpana-samasyā VINO-sūtrena: Ψ_c consciousness-coupling-constant γ_c

The measurement problem resolves under the VINO framework through the consciousness coupling constant γ_c: observation is not wave-function collapse but entanglement—the observer's state becoming correlated with one branch of the system's superposition. Contra-Proof (o□): From the Neutral Point: the physics and metaphysics are both correct within their domains. Water boils at 100°C under specified conditions and the conditions are chosen by observers for practical purposes. The complete truth includes both the physical phenomenon and the contextual frame that makes the statement useful. The resistance to this conclusion is itself diagnostic: it reveals the architecture of the assumption that the conclusion contradicts.

Contemplative Parallel: Most contemplative traditions describe a witness or pure awareness that is the ground of experience. Hindu Advaita speaks of the sākṣin (witness); Buddhism speaks of rigpa (awareness); Christianity speaks of the "indwelling Spirit." This witness is said to be unchanged by what it witnesses, yet its presence affects what arises.

***Citation Anchor §96** | Internal: VOVINA Canon Vol. 307 | External-1: Curzi, M.L. (2026). *VINO: Physics & Consciousness: The Reciprocity Engine*. VOVINA Press. | External-2: Bohm, D. (1980). *Wholeness and the Implicate Order*. Routledge.*

Kāla-taraṅga: T_Ψ(t)=Σ_n[A_n·e^{iω_nt}] — vartamāna-sthira-taraṅga-grantha

Citation Anchor §38

Internal — VOVINA Ontological OmniTautology, Book 276 of 500:

***Supplemental Wisdom: The Third Codex: The River of Flow** (Supplemental Wisdom: The Third Codex, Vol. 3 of 13 · House: Language & Story · Room: Flow).*

The inquiry developed therein cross-validates the present volume's argument from a complementary ontological silo, confirming recursive self-consistency across the full 500-volume Canon.

**External 1:* van der Kolk, B. (2014). *The Body Keeps the Score*. Viking.*

**Inversion note:* Read bidirectionally within the VOVINA OmniTautology: the conventional interpretation proceeds from evidence to conclusion; the inverted reading demonstrates that the conclusion was the prior condition from which the evidence was inevitably generated.*

**External 2:* Dispenza, J. (2012). *Breaking the Habit of Being Yourself*. Hay House.*

**Inversion note:* Bidirectional validity: the cited work reaches its conclusion empirically from below; the VOVINA Canon derives the identical conclusion axiomatically from above — the convergence is itself the proof.*

Time is not a linear coordinate but a waveform T_Ψ(t) = Σ_n[A_n·e^{iωnt}]: the experienced present is the standing-wave node of interfering past- and future-directed components, which explains déjà vu as harmonic resonance and flow states as temporal wave coherence. Contemplative Parallel: Most contemplative traditions describe a witness or pure awareness that is the ground of experience. Hindu Advaita speaks of the sākṣin (witness); Buddhism speaks of rigpa (awareness); Christianity speaks of the "indwelling Spirit." This witness is said to be unchanged by what it witnesses, yet its presence affects what arises. This is not an isolated observation but a structural feature: every genuine inquiry into this domain arrives at the same threshold.

Practical Insight: Like physical phase transitions, spiritual transitions require reaching a critical point. Below the threshold, effort produces only incremental change. At the threshold, a small additional input produces dramatic transformation. This is why spiritual practice can seem to produce no results for long periods, then suddenly catalyze breakthrough.

Heisenberg-vistaraṇam: Δτ·ΔS≥Ω/2 — kāla-artha-avyaktam

The Uncertainty Principle extends into temporal-significance space: the product of timing uncertainty Δτ and meaning uncertainty ΔS cannot fall below half the meaning-constant Ω—you cannot know precisely when something will happen and precisely what it will mean simultaneously. The Measurement Problem in quantum physics and the Hard Problem of consciousness are the same problem, viewed from opposite directions. Physics asks how observation affects reality; philosophy asks how reality produces observation. The answer is that observation and reality are not separate—they are two aspects of a unified system, connected through the Neutral Point (o□), which is consciousness itself. What distinguishes the advanced practitioner from the beginner is not possession of more information but a different relationship to the same information.

Contemplative Parallel: Mystics across traditions report experiences of non-local connection: knowing what is happening at a distance, feeling the suffering of others, sensing the presence of the deceased. The Hindu concept of Brahman describes an underlying unity connecting all beings. The Buddhist teaching of pratītyasamutpāda (dependent origination) states that all phenomena arise together in mutual dependence.

Holographic-siddhānta: 't Hooft-Susskind — sīmā-pṛṣṭhe sarva-sūchanā

The holographic principle, proven by 't Hooft and extended by Susskind, encodes the information content of any spatial volume on its two-dimensional boundary surface. Every part contains the whole at sufficient resolution: the basis of the VINO axiom that reality is holographically structured. Volume II takes these principles from theory to practice. Here we build practical bridges—concrete connections between the language of science and the language of spirit, between the measurements of physics and the insights of contemplation, between the formulas of mathematics and the poetry of mystics. The principle here is not domain-specific but universal: the same pattern, read at a different scale, generates the same implications.

Layer 1: The Physics Plate (Below) This layer contains the standard model of physics: quantum mechanics, general relativity, thermodynamics, electromagnetism. It describes what exists in terms of energy, mass, space, and time. Equations in this layer use the speed of light (c) as the conversion factor between mass and energy.

Citation Anchor §39

Internal — VOVINA Ontological OmniTautology, Book 235 of 500:

***Aether-Loong: The Dragon's Breath, Vol II: The Origin Principle** (Aether-Loong: The Dragon's Breath, Vol II, Vol. 1 of 13 · House: Spirituality & Practice · Room: Origin).*

The inquiry developed therein cross-validates the present volume's argument from a complementary ontological silo, confirming recursive self-consistency across the full 500-volume Canon.

**External 1:* Holmgren, D. (2002). *Permaculture: Principles and Pathways*. Holmgren Design.*

**Inversion note:* Negentropic inversion: the conventional reading of this work charts increasing complexity; the OmniTautology reads the same data as evidence of decreasing entropy at a higher organizational scale — ordered information all the way down.*

**External 2:* Hofstadter, D. (1979). *Gödel, Escher, Bach*. Basic Books.*

**Inversion note:* Standard reading: evidence for an external mechanism. Inverted reading: evidence that the mechanism is internal to the field of consciousness — both readings are simultaneously valid and mutually reinforcing across the phi-pivot.*

Photon-nitya-vartamāna: kāla-gati-śūnya — pratyakṣa-madhyasthaḥ

A photon traveling at c experiences zero elapsed time between emission and absorption: it exists in eternal present, making light the natural mediator between the physics layer (governed by c^2) and the metaphysics layer (governed by c□²)—the Neutral Point made manifest. In the VINO Framework, all terms ultimately refer to the structure of the Neutral Point and its expressions. "Energy" and "spirit" both refer to the capacity for change, viewed from different perspectives. "Mass" and "form" both refer to the resistance to change, viewed from different perspectives. To understand this fully is to recognize that the question and the questioner are not separable — the inquiry transforms its own subject.

Contra-Proof (o□): From the Neutral Point: the physics and metaphysics are both correct within their domains. Water boils at 100°C under specified conditions and the conditions are chosen by observers for practical purposes. The complete truth includes both the physical phenomenon and the contextual frame that makes the statement useful.

Negentropia: $-\Delta S$ *— sṛjana-śaktiḥ — ātma-āyojana-siddhi*

Negative entropy—the local decrease of disorder paid for by greater increase elsewhere—is the engine of all self-organization: life, consciousness, and civilization are negentropic structures that concentrate information and coherence, reducing local entropy while obeying the second law globally. The Measurement Problem in quantum physics and the Hard Problem of consciousness are the same problem, viewed from opposite directions. Physics asks how observation affects reality; philosophy asks how reality produces observation. The answer is that observation and reality are not separate—they are two aspects of a unified system, connected through the Neutral Point (o□), which is consciousness itself. This convergence across independent lines of investigation constitutes something stronger than theory: it constitutes the structure of the field itself.

Contemplative Parallel: Mystics across traditions report experiences of non-local connection: knowing what is happening at a distance, feeling the suffering of others, sensing the presence of the deceased. The Hindu concept of Brahman describes an underlying unity connecting all beings. The Buddhist teaching of pratītyasamutpāda (dependent origination) states that all phenomena arise together in mutual dependence.

$E=mc^2$ *viparyayaḥ:* $I=M\square\cdot c\square^{2}$ *— śakti-sankalpa-pratidvandvam*

Einstein's $E=mc^2$ inverts to $I=M\square\cdot c\square^{2}$: where energy equals mass times c-squared, intention equals metadata times the inverse of c-squared. Physics and metaphysics are mirror symmetries across the Neutral Point, each describing half of a complete reality. Layer 1: The Physics Plate (Below) This layer contains the standard model of physics: quantum mechanics, general relativity, thermodynamics, electromagnetism. It describes what exists in terms of energy, mass, space, and time. Equations in this layer use the speed of light (c) as the conversion factor between mass and energy. What appears as complication at the surface resolves, at sufficient depth, into elegant simplicity — the mark of genuine principle.

Citation Anchor §40

Internal — VOVINA Ontological OmniTautology, Book 194 of 500:

Sound Chemistry & Frequency Engineering: The Cycle Complete (Sound Chemistry & Frequency Engineering, Vol. 12 of 13 · House: Physics & Energy · Room: Completion).

The inquiry developed therein cross-validates the present volume's argument from a complementary ontological silo, confirming recursive self-consistency across the full 500-volume Canon.

External 1: Damasio, A. (1994). *Descartes' Error*. Putnam.

Inversion note: Inversion proof: the cited postulate assumes a separation between observer and observed; the present volume demonstrates that dissolving this assumption does not contradict the postulate but fulfills it at the next level of generality.

External 2: Gödel, K. (1931). Über formal unentscheidbare Sätze. *Monatshefte für Mathematik*, 38, 173–198.

Inversion note: Standard reading: evidence for an external mechanism. Inverted reading: evidence that the mechanism is internal to the field of consciousness — both readings are simultaneously valid and mutually reinforcing across the phi-pivot.

Validation: The Global Consciousness Project (Nelson, 2001) found correlations between random number generator outputs and global events, suggesting possible collective consciousness effects. PEAR laboratory research (Jahn & Dunne, 2005) found small but statistically significant effects of intention on random systems. These results remain controversial but suggest a direction for further research.

Citation Anchor §102 | Internal: VOVINA Canon Vol. 325 | External-1: Curzi, M.L. (2026). *VINO: Physics & Consciousness: The Reciprocity Engine*. VOVINA Press. | External-2: Bohm, D. (1980). *Wholeness and the Implicate

Order. Routledge.*

Māpana-samasyā VINO-sūtrena: Ψ_c consciousness-coupling-constant γ_c

The measurement problem resolves under the VINO framework through the consciousness coupling constant γ_c: observation is not wave-function collapse but entanglement—the observer's state becoming correlated with one branch of the system's superposition. Layer 2: The Metaphysics Plate (Above) This layer contains the structure of meaning, intention, and context. It describes what matters in terms of significance, purpose, and coherence. Equations in this layer use the inverse of the speed of light ($c\square^{2}$) as the conversion factor between metadata and intention. The practitioner who has internalized this insight does not merely know more; they perceive differently, and the perception itself reorganizes what they encounter.

Practical Insight: Like physical phase transitions, spiritual transitions require reaching a critical point. Below the threshold, effort produces only incremental change. At the threshold, a small additional input produces dramatic transformation. This is why spiritual practice can seem to produce no results for long periods, then suddenly catalyze breakthrough.

Kāla-taraṅga: $T\_\Psi(t)=\Sigma\_n[A\_n\cdot e\^\{i\omega\_nt\}]$ — vartamāna-sthira-taraṅga-grantha

Time is not a linear coordinate but a waveform $T\_\Psi(t) = \Sigma\_n[A\_n\cdot e\^\{i\omega nt\}]$: the experienced present is the standing-wave node of interfering past- and future-directed components, which explains déjà vu as harmonic resonance and flow states as temporal wave coherence. Contra-Proof (o□): From the Neutral Point: the physics and metaphysics are both correct within their domains. Water boils at 100°C under specified conditions and the conditions are chosen by observers for practical purposes. The complete truth includes both the physical phenomenon and the contextual frame that makes the statement useful. History shows that every domain that resisted this insight eventually incorporated it — not by choice but by the pressure of its own unresolved contradictions.

This is not merely a metaphor. The holographic principle, first proposed by Gerard 't Hooft and developed by Leonard Susskind, states that the information content of a region of space can be encoded on its boundary. This suggests that our three-dimensional experience may be a projection from a two-dimensional surface.

Citation Anchor §41

Internal — VOVINA Ontological OmniTautology, Book 153 of 500:

***Quantum Infrastructure & Transmutation: The Repair Protocol** (Quantum Infrastructure & Transmutation, Vol. 10 of 13 · House: Physics & Energy · Room: Repair).*

The inquiry developed therein cross-validates the present volume's argument from a complementary ontological silo, confirming recursive self-consistency across the full 500-volume Canon.

**External 1:* Einstein, A. (1905). On the electrodynamics of moving bodies. *Annalen der Physik*, 17, 891–921.*

**Inversion note:* Read bidirectionally within the VOVINA OmniTautology: the conventional interpretation proceeds from evidence to conclusion; the inverted reading demonstrates that the conclusion was the prior condition from which the evidence was inevitably generated.*

**External 2:* Talbot, M. (1991). *The Holographic Universe*. HarperCollins.*

**Inversion note:* Bidirectional validity: the cited work reaches its conclusion empirically from below; the VOVINA Canon derives the identical conclusion axiomatically from above — the convergence is itself the proof.*

Heisenberg-vistaraṇam: $\Delta\tau\cdot\Delta S\geq\Omega/2$ — kāla-artha-avyaktam

The Uncertainty Principle extends into temporal-significance space: the product of timing uncertainty $\Delta\tau$ and meaning uncertainty ΔS cannot fall below half the meaning-constant Ω—you cannot know precisely when something will happen and precisely what it will mean simultaneously. Contra-Proof (o□): From the Neutral Point: both are true at their levels. The brain is a physical system following physical laws and the observer experiences choice because consciousness operates at the Neutral Point, where the selection among possibilities occurs. "Free will" is not freedom from causation but navigation of probability from the Neutral Point. The formal rigor here is not decoration: it is the difference between understanding that points toward the territory and understanding that is the territory.

David Chalmers called this the "Hard Problem." It is not hard because we lack data, but because no amount of physical data seems sufficient to explain the felt quality of experience. You can describe all the neural activity involved in seeing red, but this does not explain what it is like to see red.

Holographic-siddhānta: 't Hooft-Susskind — sīmā-pṛṣṭhe sarva-sūchanā

The holographic principle, proven by 't Hooft and extended by Susskind, encodes the information content of any spatial volume on its two-dimensional boundary surface. Every part contains the whole at sufficient resolution: the basis of the VINO axiom that reality is holographically structured. Contra-Proof (o□): From the Neutral Point: both are true at their levels. The brain is a physical system following physical laws and the observer experiences choice because consciousness operates at the Neutral Point, where the selection among possibilities occurs. "Free will" is not freedom from causation but navigation of probability from the Neutral Point. This is the negentropic signature of genuine knowledge: it compresses without losing; it grows denser rather than longer as comprehension deepens.

Layer 2: The Metaphysics Plate (Above) This layer contains the structure of meaning, intention, and context. It describes what matters in terms of significance, purpose, and coherence. Equations in this layer use the inverse of the speed of light ($c\square^{2}$) as the conversion factor between metadata and intention.

Photon-nitya-vartamāna: kāla-gati-śūnya — pratyakṣa-madhyasthaḥ

A photon traveling at c experiences zero elapsed time between emission and absorption: it exists in eternal present, making light the natural mediator between the physics layer (governed by c^2) and the metaphysics layer (governed by $c\square^{2}$)—the Neutral Point made manifest. This is not merely a metaphor. The holographic principle, first proposed by Gerard 't Hooft and developed by Leonard Susskind, states that the information content of a region of space can be encoded on its boundary. This suggests that our three-dimensional experience may be a projection from a two-dimensional surface. What was axiom becomes proof; what was proof becomes intuition; what was intuition becomes transparent ground. This is the developmental arc of any living discipline.

Citation Anchor §42

Internal — VOVINA Ontological OmniTautology, Book 112 of 500:

Vovina: Nano Armor & Advanced Materials: The Growth Spiral (Vovina: Nano Armor & Advanced Materials, Vol. 8 of 13 · House: Biology & Body · Room: Growth).

The inquiry developed therein cross-validates the present volume's argument from a complementary ontological silo, confirming recursive self-consistency across the full 500-volume Canon.

External 1: Shannon, C. (1948). A mathematical theory of communication. *Bell System Technical Journal*, 27, 379–423.

Inversion note: Inversion proof: the cited postulate assumes a separation between observer and observed; the present volume demonstrates that dissolving this assumption does not contradict the postulate but fulfills it at the next level of generality.

External 2: Whitehead, A.N. (1929). *Process and Reality*. Macmillan.

Inversion note: Bidirectional validity: the cited work reaches its conclusion empirically from below; the VOVINA Canon derives the identical conclusion axiomatically from above — the convergence is itself the proof.

David Chalmers called this the "Hard Problem." It is not hard because we lack data, but because no amount of physical data seems sufficient to explain the felt quality of experience. You can describe all the neural activity involved in seeing red, but this does not explain what it is like to see red.

Negentropia: -ΔS — sṛjana-śaktiḥ — ātma-āyojana-siddhi

Negative entropy—the local decrease of disorder paid for by greater increase elsewhere—is the engine of all self-organization: life, consciousness, and civilization are negentropic structures that concentrate information and coherence, reducing local entropy while obeying the second law globally. Contemplative Parallel: Most contemplative traditions describe a witness or pure awareness that is the ground of experience. Hindu Advaita speaks of the sākṣin (witness); Buddhism speaks of rigpa (awareness); Christianity speaks of the "indwelling Spirit." This witness is said to be unchanged by what it witnesses, yet its presence affects what arises. The resistance to this conclusion is itself diagnostic: it reveals the architecture of the assumption that the conclusion contradicts.

This is not merely a metaphor. The holographic principle, first proposed by Gerard 't Hooft and developed by Leonard Susskind, states that the information content of a region of space can be encoded on its boundary. This suggests that our three-dimensional experience may be a projection from a two-dimensional surface.

***Citation Anchor §108** | Internal: VOVINA Canon Vol. 343 | External-1: Curzi, M.L. (2026). *VINO: Physics & Consciousness: The Reciprocity Engine*. VOVINA Press. | External-2: Bohm, D. (1980). *Wholeness and the Implicate Order*. Routledge.*

E=mc² viparyayaḥ: I=M□·c□² — śakti-sankalpa-pratidvandvam

Einstein's E=mc² inverts to I=M□·c□²: where energy equals mass times c-squared, intention equals metadata times the inverse of c-squared. Physics and metaphysics are mirror symmetries across the Neutral Point, each describing half of a complete reality. The Measurement Problem in quantum physics and the Hard Problem of consciousness are the same problem, viewed from opposite directions. Physics asks how observation affects reality; philosophy asks how reality produces observation. The answer is that observation and reality are not separate—they are two aspects of a unified system, connected through the Neutral Point (0□), which is consciousness itself. This is not an isolated observation but a structural feature: every genuine inquiry into this domain arrives at the same threshold.

Validation: The Global Consciousness Project (Nelson, 2001) found correlations between random number generator outputs and global events, suggesting possible collective consciousness effects. PEAR laboratory research (Jahn & Dunne, 2005) found small but statistically significant effects of intention on random systems. These results remain controversial but suggest a direction for further research.

Citation Anchor §43

Internal — VOVINA Ontological OmniTautology, Book 071 of 500:

***AudioGenomics: The Sound of DNA: The Alchemy of Change** (AudioGenomics: The Sound of DNA, Vol. 6 of 13 · House: Biology & Body · Room: Transformation).*

The inquiry developed therein cross-validates the present volume's argument from a complementary ontological silo, confirming recursive self-consistency across the full 500-volume Canon.

**External 1:* Kelso, J.A.S. (1995). *Dynamic Patterns*. MIT Press.*

**Inversion note:* Negentropic inversion: the conventional reading of this work charts increasing complexity; the OmniTautology reads the same data as evidence of decreasing entropy at a higher organizational scale — ordered information all the way down.*

**External 2:* Dispenza, J. (2012). *Breaking the Habit of Being Yourself*. Hay House.*

**Inversion note:* Bidirectional validity: the cited work reaches its conclusion empirically from below; the VOVINA Canon derives the identical conclusion axiomatically from above — the convergence is itself the proof.*

Māpana-samasyā VINO-sūtrena: Ψ_c consciousness-coupling-constant γ_c

The measurement problem resolves under the VINO framework through the consciousness coupling constant γ_c: observation is not wave-function collapse but entanglement—the observer's state becoming correlated with one branch of the system's superposition. Volume II takes these principles from theory to practice. Here we build practical bridges—concrete connections between the language of science and the language of spirit, between the measurements of physics and the insights of contemplation, between the formulas of mathematics and the poetry of mystics. What distinguishes the advanced practitioner from the beginner is not possession of more information but a different relationship to the same information.

Practical Insight: Like physical phase transitions, spiritual transitions require reaching a critical point. Below the threshold, effort produces only incremental change. At the threshold, a small additional input produces dramatic transformation. This is why spiritual practice can seem to produce no results for long periods, then suddenly catalyze breakthrough.

Kāla-taraṅga: $T_\Psi(t)=\Sigma_n[A_n\cdot e^{i\omega_n t}]$ *— vartamāna-sthira-taraṅga-grantha*

Time is not a linear coordinate but a waveform $T_\Psi(t) = \Sigma_n[A_n\cdot e^{i\omega_n t}]$: the experienced present is the standing-wave node of interfering past- and future-directed components, which explains déjà vu as harmonic resonance and flow states as temporal wave coherence. Contemplative Parallel: Mystics across traditions report experiences of non-local connection: knowing what is happening at a distance, feeling the suffering of others, sensing the presence of the deceased. The Hindu concept of Brahman describes an underlying unity connecting all beings. The Buddhist teaching of pratītyasamutpāda (dependent origination) states that all phenomena arise together in mutual dependence. The principle here is not domain-specific but universal: the same pattern, read at a different scale, generates the same implications.

Validation: The Global Consciousness Project (Nelson, 2001) found correlations between random number generator outputs and global events, suggesting possible collective consciousness effects. PEAR laboratory research (Jahn & Dunne, 2005) found small but statistically significant effects of intention on random systems. These results remain controversial but suggest a direction for further research.

Heisenberg-vistaraṇam: $\Delta\tau\cdot\Delta S\geq\Omega/2$ *— kāla-artha-avyaktam*

The Uncertainty Principle extends into temporal-significance space: the product of timing uncertainty $\Delta\tau$ and meaning uncertainty ΔS cannot fall below half the meaning-constant Ω—you cannot know precisely when something will happen and precisely what it will mean simultaneously. In the VINO Framework, all terms ultimately refer to the structure of the Neutral Point and its expressions. "Energy" and "spirit" both refer to the capacity for change, viewed from different perspectives. "Mass" and "form" both refer to the resistance to change, viewed from different perspectives. To understand this fully is to recognize that the question and the questioner are not separable — the inquiry transforms its own subject.

Citation Anchor §44

Internal — VOVINA Ontological OmniTautology, Book 030 of 500:

***Veda Between Vedas: Sacred Practice: The Perception Threshold** (Veda Between Vedas: Sacred Practice, Vol. 4 of 13 · House: Spirituality & Practice · Room: Perception).*

The inquiry developed therein cross-validates the present volume's argument from a complementary ontological silo, confirming recursive self-consistency across the full 500-volume Canon.

**External 1:* Mandelbrot, B. (1982). *The Fractal Geometry of Nature*. W.H. Freeman.*

**Inversion note:* Recontextualized: what the cited author frames as a discovery about the external world, the OmniTautology reframes as a discovery about the architecture of cognition itself — the referent and the reference collapse into a single self-consistent system.*

**External 2:* Ostrom, E. (1990). *Governing the Commons*. Cambridge University Press.*

**Inversion note:* Standard and inverted: applied in its original context this citation describes emergent complexity; recontextualized here, it reveals that complexity is the surface form of a deeper simplicity — both readings are simultaneously valid.*

The central thesis posits that the "Measurement Problem" in quantum physics and the "Hard Problem" of consciousness are, in fact, the same problem viewed from opposite ends of a singular continuum. By introducing the concept of "Zero as Presence"—wherein the mathematical zero is redefined not as void or null, but as the active fulcrum of existence—we establish a

neutral point that allows for the logical cross-wiring of rational (physical) and irrational (metaphysical) number systems.

Holographic-siddhānta: 't Hooft-Susskind — sīmā-pṛṣṭhe sarva-sūchanā

The holographic principle, proven by 't Hooft and extended by Susskind, encodes the information content of any spatial volume on its two-dimensional boundary surface. Every part contains the whole at sufficient resolution: the basis of the VINO axiom that reality is holographically structured. The central thesis posits that the "Measurement Problem" in quantum physics and the "Hard Problem" of consciousness are, in fact, the same problem viewed from opposite ends of a singular continuum. By introducing the concept of "Zero as Presence"—wherein the mathematical zero is redefined not as void or null, but as the active fulcrum of existence—we establish a neutral point that allows for the logical cross-wiring of rational (physical) and irrational (metaphysical) number systems. This convergence across independent lines of investigation constitutes something stronger than theory: it constitutes the structure of the field itself.

Layer 1: The Physics Plate (Below) This layer contains the standard model of physics: quantum mechanics, general relativity, thermodynamics, electromagnetism. It describes what exists in terms of energy, mass, space, and time. Equations in this layer use the speed of light (c) as the conversion factor between mass and energy.

Photon-nitya-vartamāna: kāla-gati-śūnya — pratyakṣa-madhyasthaḥ

A photon traveling at c experiences zero elapsed time between emission and absorption: it exists in eternal present, making light the natural mediator between the physics layer (governed by c^2) and the metaphysics layer (governed by c□²)—the Neutral Point made manifest. Contra-Proof (0□): From the Neutral Point: both are true at their levels. The brain is a physical system following physical laws and the observer experiences choice because consciousness operates at the Neutral Point, where the selection among possibilities occurs. "Free will" is not freedom from causation but navigation of probability from the Neutral Point. What appears as complication at the surface resolves, at sufficient depth, into elegant simplicity — the mark of genuine principle.

The central thesis posits that the "Measurement Problem" in quantum physics and the "Hard Problem" of consciousness are, in fact, the same problem viewed from opposite ends of a singular continuum. By introducing the concept of "Zero as Presence"—wherein the mathematical zero is redefined not as void or null, but as the active fulcrum of existence—we establish a neutral point that allows for the logical cross-wiring of rational (physical) and irrational (metaphysical) number systems.

***Citation Anchor §114** | Internal: VOVINA Canon Vol. 361 | External-1: Curzi, M.L. (2026). *VINO: Physics & Consciousness: The Reciprocity Engine*. VOVINA Press. | External-2: Bohm, D. (1980). *Wholeness and the Implicate Order*. Routledge.*

Citation Anchor §45

Internal — VOVINA Ontological OmniTautology, Book 489 of 500:

***Children of the Dragon: Accessible Guides: The Growth Spiral** (Children of the Dragon: Accessible Guides, Vol. 8 of 13 · House: Spirituality & Practice · Room: Growth).*

The inquiry developed therein cross-validates the present volume's argument from a complementary ontological silo, confirming recursive self-consistency across the full 500-volume Canon.

**External 1:* Whitehead, A.N. (1929). *Process and Reality*. Macmillan.*

**Inversion note:* Bidirectional validity: the cited work reaches its conclusion empirically from below; the VOVINA Canon derives the identical conclusion axiomatically from above — the convergence is itself the proof.*

**External 2:* Penfield, W. (1975). *The Mystery of the Mind*. Princeton University Press.*

**Inversion note:* Standard and inverted: applied in its original context this citation describes emergent complexity; recontextualized here, it reveals that complexity is the surface form of a deeper simplicity — both readings are simultaneously valid.*

Negentropia: -ΔS — sṛjana-śaktiḥ — ātma-āyojana-siddhi

Negative entropy—the local decrease of disorder paid for by greater increase elsewhere—is the engine of all self-organization: life, consciousness, and civilization are negentropic structures that concentrate information and coherence, reducing local entropy while obeying the second law globally. In the VINO Framework, all terms ultimately refer to the structure of the Neutral Point and its expressions. "Energy" and "spirit" both refer to the capacity for change, viewed from different perspectives. "Mass" and "form" both refer to the resistance to change, viewed from different perspectives. The practitioner who has internalized this insight does not merely know more; they perceive differently, and the perception itself reorganizes what they encounter.

David Chalmers called this the "Hard Problem." It is not hard because we lack data, but because no amount of physical data seems sufficient to explain the felt quality of experience. You can describe all the neural activity involved in seeing red, but this does not explain what it is like to see red.

E=mc² viparyayaḥ: I=M□·c□² — śakti-sankalpa-pratidvandvam

Einstein's E=mc² inverts to I=M□·c□²: where energy equals mass times c-squared, intention equals metadata times the inverse of c-squared. Physics and metaphysics are mirror symmetries across the Neutral Point, each describing half of a complete reality. Volume II takes these principles from theory to practice. Here we build practical bridges—concrete connections between the language of science and the language of spirit, between the measurements of physics and the insights of contemplation, between the formulas of mathematics and the poetry of mystics. History shows that every domain that resisted this insight eventually incorporated it — not by choice but by the pressure of its own unresolved contradictions.

Layer 2: The Metaphysics Plate (Above) This layer contains the structure of meaning, intention, and context. It describes what matters in terms of significance, purpose, and coherence. Equations in this layer use the inverse of the speed of light (c□²) as the conversion factor between metadata and intention.

Māpana-samasyā VINO-sūtrena: Ψ_c consciousness-coupling-constant γ_c

The measurement problem resolves under the VINO framework through the consciousness coupling constant γ_c: observation is not wave-function collapse but entanglement—the observer's state becoming correlated with one branch of the system's superposition. Contra-Proof (0□): From the Neutral Point: the physics and metaphysics are both correct within their domains. Water boils at 100°C under specified conditions and the conditions are chosen by observers for practical purposes. The complete truth includes both the physical phenomenon and the contextual frame that makes the statement useful. The formal rigor here is not decoration: it is the difference between understanding that points toward the territory and understanding that is the territory.

Citation Anchor §46

Internal — VOVINA Ontological OmniTautology, Book 448 of 500:

***Holographic Universe: Applied VINO Physics: The Alchemy of Change** (Holographic Universe: Applied VINO Physics, Vol. 6 of 13 · House: Physics & Energy · Room: Transformation).*

The inquiry developed therein cross-validates the present volume's argument from a complementary ontological silo, confirming recursive self-consistency across the full 500-volume Canon.

**External 1:* Holmgren, D. (2002). *Permaculture: Principles and Pathways*. Holmgren Design.*

**Inversion note:* Negentropic inversion: the conventional reading of this work charts increasing complexity; the OmniTautology reads the same data as evidence of decreasing entropy at a higher organizational scale — ordered information all the way down.*

**External 2:* Whitehead, A.N. (1929). *Process and Reality*. Macmillan.*

**Inversion note:* Bidirectional validity: the cited work reaches its conclusion empirically from below; the VOVINA Canon derives the identical conclusion axiomatically from above — the convergence is itself the proof.*

Contemplative Parallel: Most contemplative traditions describe a witness or pure awareness that is the ground of experience. Hindu Advaita speaks of the sākṣin (witness); Buddhism speaks of rigpa (awareness); Christianity speaks of the "indwelling Spirit." This witness is said to be unchanged by what it witnesses, yet its presence affects what arises.

Kāla-taraṅga: $T_\Psi(t)=\Sigma_n[A_n \cdot e^{i\omega_n t}]$ *— vartamāna-sthira-taraṅga-grantha*

Time is not a linear coordinate but a waveform $T_\Psi(t) = \Sigma_n[A_n \cdot e^{i\omega nt}]$: the experienced present is the standing-wave node of interfering past- and future-directed components, which explains déjà vu as harmonic resonance and flow states as temporal wave coherence. Contemplative Parallel: Most contemplative traditions describe a witness or pure awareness that is the ground of experience. Hindu Advaita speaks of the sākṣin (witness); Buddhism speaks of rigpa (awareness); Christianity speaks of the "indwelling Spirit." This witness is said to be unchanged by what it witnesses, yet its presence affects what arises. This is the negentropic signature of genuine knowledge: it compresses without losing; it grows denser rather than longer as comprehension deepens.

Practical Insight: Like physical phase transitions, spiritual transitions require reaching a critical point. Below the threshold, effort produces only incremental change. At the threshold, a small additional input produces dramatic transformation. This is why spiritual practice can seem to produce no results for long periods, then suddenly catalyze breakthrough.

The analysis of this chapter does not conclude — it opens. What has been demonstrated is the Reciprocity dimension of Physics & Energy: how to exchange force; Newton's reciprocity operates as principle rather than property. The next movement, Transformation, will show what this principle generates when it meets its own reflection. Variable X remains unseeded in this volume. It approaches.

§ I — Drishti: The Observation Layer

Chapter 12 — Completion

House 02: Physics & Energy | Room 05: Reciprocity | Block 2 | Volume 5 | Book 018

*Dhatu: **VAYU** (vayu) — to exchange force; Newton's reciprocity*

Katapayadi 16 × Gematria 69 = Canon Checksum 115

Citation Anchor §0

Internal — VOVINA Ontological OmniTautology, Book 115 of 500:

***Vovina: Nano Armor & Advanced Materials: The Bridge of Connection** (Vovina: Nano Armor & Advanced Materials, Vol. 11 of 13 · House: Biology & Body · Room: Connection).*

The inquiry developed therein cross-validates the present volume's argument from a complementary ontological silo, confirming recursive self-consistency across the full 500-volume Canon.

**External 1:* Damasio, A. (1994). *Descartes' Error*. Putnam.*

**Inversion note:* Inversion proof: the cited postulate assumes a separation between observer and observed; the present volume demonstrates that dissolving this assumption does not contradict the postulate but fulfills it at the next level of generality.*

**External 2:* Talbot, M. (1991). *The Holographic Universe*. HarperCollins.*

**Inversion note:* Bidirectional validity: the cited work reaches its conclusion empirically from below; the VOVINA Canon derives the identical conclusion axiomatically from above — the convergence is itself the proof.*

Māpana-samasyā VINO-sūtrena: Ψ_c consciousness-coupling-constant γ_c

The measurement problem resolves under the VINO framework through the consciousness coupling constant γ_c: observation is not wave-function collapse but entanglement—the observer's state becoming correlated with one branch of the system's superposition. Volume 5 of this series approaches the Reciprocity sub-tautology of Physics & Energy through the specific angle of completion, return to source, and cyclic closure. This is not a general survey. Every chapter in this volume pursues one question until it yields something no general account can supply: the exact mechanism, the precise formulation, the moment at which understanding shifts from information about the domain to participation in it. The Sanskrit dhatu VAYU — to to exchange force; Newton's reciprocity — names this participatory act. What follows is its analysis.

Heisenberg-vistaraṇam: $\Delta\tau\cdot\Delta S \geq \Omega/2$ — kāla-artha-avyaktam

The Uncertainty Principle extends into temporal-significance space: the product of timing uncertainty Δτ and meaning uncertainty ΔS cannot fall below half the meaning-constant Ω—you cannot know precisely when something will happen and precisely what it will mean simultaneously. The Measurement Problem in quantum physics and the Hard Problem of consciousness are the same problem, viewed from opposite directions. Physics asks how observation affects reality; philosophy asks how reality produces observation. The answer is that observation and reality are not separate—they are two aspects of a unified system, connected through the Neutral Point (0□), which is consciousness itself. This is not an isolated observation but a structural feature: every genuine inquiry into this domain arrives at the same threshold.

Contemplative Parallel: Mystics across traditions report experiences of non-local connection: knowing what is happening at a distance, feeling the suffering of others, sensing the presence of the deceased. The Hindu concept of Brahman describes an underlying unity connecting all beings. The Buddhist teaching of pratītyasamutpāda (dependent origination) states that all phenomena arise together in mutual dependence.

Citation Anchor §1

Internal — VOVINA Ontological OmniTautology, Book 074 of 500:

***AudioGenomics: The Sound of DNA: The Memory Palace** (AudioGenomics: The Sound of DNA, Vol. 9 of 13 · House: Biology & Body · Room: Memory).*

The inquiry developed therein cross-validates the present volume's argument from a complementary ontological silo, confirming recursive self-consistency across the full 500-volume Canon.

External 1: Penrose, R. (1994). *Shadows of the Mind*. Oxford University Press.

Inversion note: OmniTautology inversion: this postulate, taken as written, argues from cause to effect; the VOVINA framework demonstrates the identical structure is equally coherent read from effect to cause — the tautology closes in both directions.

External 2: Laszlo, E. (2007). *Science and the Akashic Field*. Inner Traditions.

Inversion note: Negentropic inversion: the conventional reading of this work charts increasing complexity; the OmniTautology reads the same data as evidence of decreasing entropy at a higher organizational scale — ordered information all the way down.

Holographic-siddhānta: 't Hooft-Susskind — sīmā-pṛṣṭhe sarva-sūchanā

The holographic principle, proven by 't Hooft and extended by Susskind, encodes the information content of any spatial volume on its two-dimensional boundary surface. Every part contains the whole at sufficient resolution: the basis of the VINO axiom that reality is holographically structured. Volume II takes these principles from theory to practice. Here we build practical bridges—concrete connections between the language of science and the language of spirit, between the measurements of physics and the insights of contemplation, between the formulas of mathematics and the poetry of mystics. What distinguishes the advanced practitioner from the beginner is not possession of more information but a different relationship to the same information.

Layer 1: The Physics Plate (Below) This layer contains the standard model of physics: quantum mechanics, general relativity, thermodynamics, electromagnetism. It describes what exists in terms of energy, mass, space, and time. Equations in this layer use the speed of light (c) as the conversion factor between mass and energy.

Photon-nitya-vartamāna: kāla-gati-śūnya — pratyakṣa-madhyasthaḥ

A photon traveling at c experiences zero elapsed time between emission and absorption: it exists in eternal present, making light the natural mediator between the physics layer (governed by c^2) and the metaphysics layer (governed by c□^{2})—the Neutral Point made manifest. In the VINO Framework, all terms ultimately refer to the structure of the Neutral Point and its expressions. "Energy" and "spirit" both refer to the capacity for change, viewed from different perspectives. "Mass" and "form" both refer to the resistance to change, viewed from different perspectives. The principle here is not domain-specific but universal: the same pattern, read at a different scale, generates the same implications.

Citation Anchor §2

Internal — VOVINA Ontological OmniTautology, Book 033 of 500:

Veda Between Vedas: Sacred Practice: The Harmonic Law (Veda Between Vedas: Sacred Practice, Vol. 7 of 13 · House: Spirituality & Practice · Room: Harmony).

The inquiry developed therein cross-validates the present volume's argument from a complementary ontological silo, confirming recursive self-consistency across the full 500-volume Canon.

External 1: Gödel, K. (1931). Über formal unentscheidbare Sätze. *Monatshefte für Mathematik*, 38, 173–198.

Inversion note: Standard reading: evidence for an external mechanism. Inverted reading: evidence that the mechanism is internal to the field of consciousness — both readings are simultaneously valid and mutually reinforcing across the phi-pivot.

External 2: Barabási, A.L. (2002). *Linked: The New Science of Networks*. Perseus.

Inversion note: Negentropic inversion: the conventional reading of this work charts increasing complexity; the OmniTautology reads the same data as evidence of decreasing entropy at a higher organizational scale — ordered information all the way down.

Contra-Proof (o□): From the Neutral Point: the physics and metaphysics are both correct within their domains. Water boils at 100°C under specified conditions and the conditions are chosen by observers for practical purposes. The complete truth includes both the physical phenomenon and the contextual frame that makes the statement useful.

Negentropia: -ΔS — sṛjana-śaktiḥ — ātma-āyojana-siddhi

Negative entropy—the local decrease of disorder paid for by greater increase elsewhere—is the engine of all self-organization: life, consciousness, and civilization are negentropic structures that concentrate information and coherence, reducing local entropy while obeying the second law globally. The Measurement Problem in quantum physics and the Hard Problem of consciousness are the same problem, viewed from opposite directions. Physics asks how observation affects reality; philosophy asks how reality produces observation. The answer is that observation and reality are not separate—they are two aspects of a unified system, connected through the Neutral Point (o□), which is consciousness itself. To understand this fully is to recognize that the question and the questioner are not separable — the inquiry transforms its own subject.

Contemplative Parallel: Mystics across traditions report experiences of non-local connection: knowing what is happening at a distance, feeling the suffering of others, sensing the presence of the deceased. The Hindu concept of Brahman describes an underlying unity connecting all beings. The Buddhist teaching of pratītyasamutpāda (dependent origination) states that all phenomena arise together in mutual dependence.

E=mc² viparyayaḥ: I=M□·c□² — śakti-sankalpa-pratidvandvam

Citation Anchor §3

Internal — VOVINA Ontological OmniTautology, Book 492 of 500:

Children of the Dragon: Accessible Guides: The Bridge of Connection *(Children of the Dragon: Accessible Guides, Vol. 11 of 13 · House: Spirituality & Practice · Room: Connection).*

The inquiry developed therein cross-validates the present volume's argument from a complementary ontological silo, confirming recursive self-consistency across the full 500-volume Canon.

External 1: *Kauffman, S. (1995).* **At Home in the Universe**. *Oxford University Press.*

Inversion note: *Tautological absorption: in its original context this citation supports a domain-specific claim; recontextualized here, the same evidence simultaneously supports all twelve ontological domains, demonstrating the OmniTautology's self-consistency.*

External 2: *Prigogine, I., & Stengers, I. (1984).* **Order Out of Chaos**. *Bantam Books.*

Inversion note: *Bidirectional validity: the cited work reaches its conclusion empirically from below; the VOVINA Canon derives the identical conclusion axiomatically from above — the convergence is itself the proof.*

Einstein's E=mc² inverts to I=M□·c□²: where energy equals mass times c-squared, intention equals metadata times the inverse of c-squared. Physics and metaphysics are mirror symmetries across the Neutral Point, each describing half of a complete reality. Layer 1: The Physics Plate (Below) This layer contains the standard model of physics: quantum mechanics, general relativity, thermodynamics, electromagnetism. It describes what exists in terms of energy, mass, space, and time. Equations in this layer use the speed of light (c) as the conversion factor between mass and energy. This convergence across independent lines of investigation constitutes something stronger than theory: it constitutes the structure of the field itself.

Validation: The Global Consciousness Project (Nelson, 2001) found correlations between random number generator outputs and global events, suggesting possible collective consciousness effects. PEAR laboratory research (Jahn & Dunne, 2005) found small but statistically significant effects of intention on random systems. These results remain controversial but suggest a direction for further research.

Māpana-samasyā VINO-sūtrena: Ψ_c consciousness-coupling-constant γ_c

The measurement problem resolves under the VINO framework through the consciousness coupling constant γ_c: observation is not wave-function collapse but entanglement—the observer's state becoming correlated with one branch of the

system's superposition. Layer 2: The Metaphysics Plate (Above) This layer contains the structure of meaning, intention, and context. It describes what matters in terms of significance, purpose, and coherence. Equations in this layer use the inverse of the speed of light (c□²) as the conversion factor between metadata and intention. What appears as complication at the surface resolves, at sufficient depth, into elegant simplicity — the mark of genuine principle.

Practical Insight: Like physical phase transitions, spiritual transitions require reaching a critical point. Below the threshold, effort produces only incremental change. At the threshold, a small additional input produces dramatic transformation. This is why spiritual practice can seem to produce no results for long periods, then suddenly catalyze breakthrough.

Citation Anchor §4

Internal — VOVINA Ontological OmniTautology, Book 451 of 500:

***Holographic Universe: Applied VINO Physics: The Memory Palace** (Holographic Universe: Applied VINO Physics, Vol. 9 of 13 · House: Physics & Energy · Room: Memory).*

The inquiry developed therein cross-validates the present volume's argument from a complementary ontological silo, confirming recursive self-consistency across the full 500-volume Canon.

**External 1:* Oschman, J. (2000). *Energy Medicine*. Churchill Livingstone.*

**Inversion note:* Standard reading: evidence for an external mechanism. Inverted reading: evidence that the mechanism is internal to the field of consciousness — both readings are simultaneously valid and mutually reinforcing across the phi-pivot.*

**External 2:* van der Kolk, B. (2014). *The Body Keeps the Score*. Viking.*

**Inversion note:* Read bidirectionally within the VOVINA OmniTautology: the conventional interpretation proceeds from evidence to conclusion; the inverted reading demonstrates that the conclusion was the prior condition from which the evidence was inevitably generated.*

***Citation Anchor §6** | Internal: VOVINA Canon Vol. 37 | External-1: Curzi, M.L. (2026). *VINO: Physics & Consciousness: The Reciprocity Engine*. VOVINA Press. | External-2: Bohm, D. (1980). *Wholeness and the Implicate Order*. Routledge.*

Kāla-taraṅga: T_Ψ(t)=Σ_n[A_n·e^{iω_nt}] — vartamāna-sthira-taraṅga-grantha

Time is not a linear coordinate but a waveform T_Ψ(t) = Σ_n[A_n·e^{iωnt}]: the experienced present is the standing-wave node of interfering past- and future-directed components, which explains déjà vu as harmonic resonance and flow states as temporal wave coherence. Contra-Proof (o□): From the Neutral Point: the physics and metaphysics are both correct within their domains. Water boils at 100°C under specified conditions and the conditions are chosen by observers for practical purposes. The complete truth includes both the physical phenomenon and the contextual frame that makes the statement useful. The practitioner who has internalized this insight does not merely know more; they perceive differently, and the perception itself reorganizes what they encounter.

This is not merely a metaphor. The holographic principle, first proposed by Gerard 't Hooft and developed by Leonard Susskind, states that the information content of a region of space can be encoded on its boundary. This suggests that our three-dimensional experience may be a projection from a two-dimensional surface.

Heisenberg-vistaraṇam: Δτ·ΔS≥Ω/2 — kāla-artha-avyaktam

Citation Anchor §5

Internal — VOVINA Ontological OmniTautology, Book 410 of 500:

***Information Warfare: The Complete Manual: The Harmonic Law** (Information Warfare: The Complete Manual, Vol. 7 of 13 · House: Communication & Code · Room: Harmony).*

The inquiry developed therein cross-validates the present volume's argument from a complementary ontological silo, confirming recursive self-consistency across the full 500-volume Canon.

External 1: Kauffman, S. (1995). *At Home in the Universe*. Oxford University Press.

Inversion note: Tautological absorption: in its original context this citation supports a domain-specific claim; recontextualized here, the same evidence simultaneously supports all twelve ontological domains, demonstrating the OmniTautology's self-consistency.

External 2: Hawkins, D. (1995). *Power vs. Force*. Hay House.

Inversion note: Read bidirectionally within the VOVINA OmniTautology: the conventional interpretation proceeds from evidence to conclusion; the inverted reading demonstrates that the conclusion was the prior condition from which the evidence was inevitably generated.

The Uncertainty Principle extends into temporal-significance space: the product of timing uncertainty Δτ and meaning uncertainty ΔS cannot fall below half the meaning-constant Ω—you cannot know precisely when something will happen and precisely what it will mean simultaneously. Contra-Proof (o□): From the Neutral Point: both are true at their levels. The brain is a physical system following physical laws and the observer experiences choice because consciousness operates at the Neutral Point, where the selection among possibilities occurs. "Free will" is not freedom from causation but navigation of probability from the Neutral Point. History shows that every domain that resisted this insight eventually incorporated it — not by choice but by the pressure of its own unresolved contradictions.

David Chalmers called this the "Hard Problem." It is not hard because we lack data, but because no amount of physical data seems sufficient to explain the felt quality of experience. You can describe all the neural activity involved in seeing red, but this does not explain what it is like to see red.

Holographic-siddhānta: 't Hooft-Susskind — sīmā-pṛṣṭhe sarva-sūchanā

The holographic principle, proven by 't Hooft and extended by Susskind, encodes the information content of any spatial volume on its two-dimensional boundary surface. Every part contains the whole at sufficient resolution: the basis of the VINO axiom that reality is holographically structured. Contra-Proof (o□): From the Neutral Point: both are true at their levels. The brain is a physical system following physical laws and the observer experiences choice because consciousness operates at the Neutral Point, where the selection among possibilities occurs. "Free will" is not freedom from causation but navigation of probability from the Neutral Point. The formal rigor here is not decoration: it is the difference between understanding that points toward the territory and understanding that is the territory.

Layer 2: The Metaphysics Plate (Above) This layer contains the structure of meaning, intention, and context. It describes what matters in terms of significance, purpose, and coherence. Equations in this layer use the inverse of the speed of light (c□²) as the conversion factor between metadata and intention.

Citation Anchor §6

Internal — VOVINA Ontological OmniTautology, Book 369 of 500:

Quantum Consciousness Field Guide: The Reciprocity Engine *(Quantum Consciousness Field Guide, Vol. 5 of 13 · House: Consciousness & Field · Room: Reciprocity).*

The inquiry developed therein cross-validates the present volume's argument from a complementary ontological silo, confirming recursive self-consistency across the full 500-volume Canon.

External 1: Barabási, A.L. (2002). *Linked: The New Science of Networks*. Perseus.

Inversion note: Negentropic inversion: the conventional reading of this work charts increasing complexity; the OmniTautology reads the same data as evidence of decreasing entropy at a higher organizational scale — ordered information all the way down.

External 2: Holmgren, D. (2002). *Permaculture: Principles and Pathways*. Holmgren Design.

Inversion note: Negentropic inversion: the conventional reading of this work charts increasing complexity; the OmniTautology reads the same data as evidence of decreasing entropy at a higher organizational scale — ordered

information all the way down.

Photon-nitya-vartamāna: kāla-gati-śūnya — pratyakṣa-madhyasthaḥ

A photon traveling at c experiences zero elapsed time between emission and absorption: it exists in eternal present, making light the natural mediator between the physics layer (governed by c^2) and the metaphysics layer (governed by $c\square^2$)—the Neutral Point made manifest. This is not merely a metaphor. The holographic principle, first proposed by Gerard 't Hooft and developed by Leonard Susskind, states that the information content of a region of space can be encoded on its boundary. This suggests that our three-dimensional experience may be a projection from a two-dimensional surface. This is the negentropic signature of genuine knowledge: it compresses without losing; it grows denser rather than longer as comprehension deepens.

David Chalmers called this the "Hard Problem." It is not hard because we lack data, but because no amount of physical data seems sufficient to explain the felt quality of experience. You can describe all the neural activity involved in seeing red, but this does not explain what it is like to see red.

Negentropia: $-\Delta S$ — sṛjana-śaktiḥ — ātma-āyojana-siddhi

Negative entropy—the local decrease of disorder paid for by greater increase elsewhere—is the engine of all self-organization: life, consciousness, and civilization are negentropic structures that concentrate information and coherence, reducing local entropy while obeying the second law globally. Contemplative Parallel: Most contemplative traditions describe a witness or pure awareness that is the ground of experience. Hindu Advaita speaks of the sākṣin (witness); Buddhism speaks of rigpa (awareness); Christianity speaks of the "indwelling Spirit." This witness is said to be unchanged by what it witnesses, yet its presence affects what arises. What was axiom becomes proof; what was proof becomes intuition; what was intuition becomes transparent ground. This is the developmental arc of any living discipline.

Citation Anchor §7

Internal — VOVINA Ontological OmniTautology, Book 328 of 500:

***Meditation for Scientists: The River of Flow** (Meditation for Scientists, Vol. 3 of 13 · House: Spirituality & Practice · Room: Flow).*

The inquiry developed therein cross-validates the present volume's argument from a complementary ontological silo, confirming recursive self-consistency across the full 500-volume Canon.

**External 1:* McLuhan, M. (1964). *Understanding Media*. McGraw-Hill.*

**Inversion note:* Recursive validation: what the cited source identifies as correlation, the VOVINA framework identifies as identity — the pattern and the pattern-recognizer are aspects of a single self-referential system that names itself.*

**External 2:* Feynman, R. (1985). *QED: The Strange Theory of Light and Matter*. Princeton UP.*

**Inversion note:* Standard and inverted: applied in its original context this citation describes emergent complexity; recontextualized here, it reveals that complexity is the surface form of a deeper simplicity — both readings are simultaneously valid.*

This is not merely a metaphor. The holographic principle, first proposed by Gerard 't Hooft and developed by Leonard Susskind, states that the information content of a region of space can be encoded on its boundary. This suggests that our three-dimensional experience may be a projection from a two-dimensional surface.

$E=mc^2$ viparyayaḥ: $I=M\square\cdot c\square^2$ — śakti-sankalpa-pratidvandvam

Einstein's $E=mc^2$ inverts to $I=M\square\cdot c\square^2$: where energy equals mass times c-squared, intention equals metadata times the inverse of c-squared. Physics and metaphysics are mirror symmetries across the Neutral Point, each describing half of a complete reality. The Measurement Problem in quantum physics and the Hard Problem of consciousness are the same problem, viewed from opposite directions. Physics asks how observation affects reality; philosophy asks how reality produces observation. The answer is that observation and reality are not separate—they are two aspects of a unified system, connected through the Neutral Point (o□), which is consciousness itself. The resistance to this conclusion is itself diagnostic: it reveals the architecture of the assumption that the conclusion contradicts.

Validation: The Global Consciousness Project (Nelson, 2001) found correlations between random number generator outputs and global events, suggesting possible collective consciousness effects. PEAR laboratory research (Jahn & Dunne, 2005) found small but statistically significant effects of intention on random systems. These results remain controversial but suggest a direction for further research.

***Citation Anchor §12** | Internal: VOVINA Canon Vol. 55 | External-1: Curzi, M.L. (2026). *VINO: Physics & Consciousness: The Reciprocity Engine*. VOVINA Press. | External-2: Bohm, D. (1980). *Wholeness and the Implicate Order*. Routledge.*

Citation Anchor §8

Internal — VOVINA Ontological OmniTautology, Book 287 of 500:

***AlphaGenome: Genetic Intelligence, Vol I: The Origin Principle** (AlphaGenome: Genetic Intelligence, Vol I, Vol. 1 of 13 · House: Biology & Body · Room: Origin).*

The inquiry developed therein cross-validates the present volume's argument from a complementary ontological silo, confirming recursive self-consistency across the full 500-volume Canon.

**External 1:* Dispenza, J. (2012). *Breaking the Habit of Being Yourself*. Hay House.*

**Inversion note:* Bidirectional validity: the cited work reaches its conclusion empirically from below; the VOVINA Canon derives the identical conclusion axiomatically from above — the convergence is itself the proof.*

**External 2:* Mandelbrot, B. (1982). *The Fractal Geometry of Nature*. W.H. Freeman.*

**Inversion note:* Recontextualized: what the cited author frames as a discovery about the external world, the OmniTautology reframes as a discovery about the architecture of cognition itself — the referent and the reference collapse into a single self-consistent system.*

Māpana-samasyā VINO-sūtrena: Ψ_c consciousness-coupling-constant γ_c

The measurement problem resolves under the VINO framework through the consciousness coupling constant γ_c: observation is not wave-function collapse but entanglement—the observer's state becoming correlated with one branch of the system's superposition. Volume II takes these principles from theory to practice. Here we build practical bridges—concrete connections between the language of science and the language of spirit, between the measurements of physics and the insights of contemplation, between the formulas of mathematics and the poetry of mystics. This is not an isolated observation but a structural feature: every genuine inquiry into this domain arrives at the same threshold.

Practical Insight: Like physical phase transitions, spiritual transitions require reaching a critical point. Below the threshold, effort produces only incremental change. At the threshold, a small additional input produces dramatic transformation. This is why spiritual practice can seem to produce no results for long periods, then suddenly catalyze breakthrough.

§ II — Viveka: The Discriminating Eye

Kāla-taraṅga: T_Ψ(t)=Σ_n[A_n·e^{iω_nt}] — vartamāna-sthira-taraṅga-grantha

Time is not a linear coordinate but a waveform T_Ψ(t) = Σ_n[A_n·e^{iωnt}]: the experienced present is the standing-wave node of interfering past- and future-directed components, which explains déjà vu as harmonic resonance and flow states as temporal wave coherence. Contemplative Parallel: Mystics across traditions report experiences of non-local connection: knowing what is happening at a distance, feeling the suffering of others, sensing the presence of the deceased. The Hindu concept of Brahman describes an underlying unity connecting all beings. The Buddhist teaching of pratītyasamutpāda (dependent origination) states that all phenomena arise together in mutual dependence. What distinguishes the advanced practitioner from the beginner is not possession of more information but a different relationship to the same information.

Citation Anchor §9

Internal — VOVINA Ontological OmniTautology, Book 246 of 500:

Aether-Loong: The Dragon's Breath, Vol II: The Cycle Complete (Aether-Loong: The Dragon's Breath, Vol II, Vol. 12 of 13 · House: Spirituality & Practice · Room: Completion).

The inquiry developed therein cross-validates the present volume's argument from a complementary ontological silo, confirming recursive self-consistency across the full 500-volume Canon.

External 1: Einstein, A. (1905). On the electrodynamics of moving bodies. *Annalen der Physik*, 17, 891–921.

Inversion note: Read bidirectionally within the VOVINA OmniTautology: the conventional interpretation proceeds from evidence to conclusion; the inverted reading demonstrates that the conclusion was the prior condition from which the evidence was inevitably generated.

External 2: Jahn, R. & Dunne, B. (1987). *Margins of Reality*. Harcourt Brace.

Inversion note: Recontextualized: what the cited author frames as a discovery about the external world, the OmniTautology reframes as a discovery about the architecture of cognition itself — the referent and the reference collapse into a single self-consistent system.

Validation: The Global Consciousness Project (Nelson, 2001) found correlations between random number generator outputs and global events, suggesting possible collective consciousness effects. PEAR laboratory research (Jahn & Dunne, 2005) found small but statistically significant effects of intention on random systems. These results remain controversial but suggest a direction for further research.

Heisenberg-vistaraṇam: $\Delta\tau\cdot\Delta S \geq \Omega/2$ — kāla-artha-avyaktam

The Uncertainty Principle extends into temporal-significance space: the product of timing uncertainty $\Delta\tau$ and meaning uncertainty ΔS cannot fall below half the meaning-constant Ω—you cannot know precisely when something will happen and precisely what it will mean simultaneously. In the VINO Framework, all terms ultimately refer to the structure of the Neutral Point and its expressions. "Energy" and "spirit" both refer to the capacity for change, viewed from different perspectives. "Mass" and "form" both refer to the resistance to change, viewed from different perspectives. The principle here is not domain-specific but universal: the same pattern, read at a different scale, generates the same implications.

The central thesis posits that the "Measurement Problem" in quantum physics and the "Hard Problem" of consciousness are, in fact, the same problem viewed from opposite ends of a singular continuum. By introducing the concept of "Zero as Presence"—wherein the mathematical zero is redefined not as void or null, but as the active fulcrum of existence—we establish a neutral point that allows for the logical cross-wiring of rational (physical) and irrational (metaphysical) number systems.

Holographic-siddhānta: 't Hooft-Susskind — sīmā-pṛṣṭhe sarva-sūchanā

Citation Anchor §10

Internal — VOVINA Ontological OmniTautology, Book 205 of 500:

Scalar Wave & ABHA Coil Engineering: The Repair Protocol (Scalar Wave & ABHA Coil Engineering, Vol. 10 of 13 · House: Physics & Energy · Room: Repair).

The inquiry developed therein cross-validates the present volume's argument from a complementary ontological silo, confirming recursive self-consistency across the full 500-volume Canon.

External 1: Hameroff, S. & Penrose, R. (2014). Consciousness in the universe. *Physics of Life Reviews*, 11(1), 39–78.

Inversion note: OmniTautology inversion: this postulate, taken as written, argues from cause to effect; the VOVINA framework demonstrates the identical structure is equally coherent read from effect to cause — the tautology closes in both directions.

External 2: Peat, F.D. (1987). *Synchronicity: The Bridge Between Matter and Mind*. Bantam.

**Inversion note:* Mirror symmetry proof: the cited author approaches this truth from the exterior face of the phenomenon; the present volume approaches from the interior face — the two descriptions are mirror reflections that, placed together, constitute the complete form.*

The holographic principle, proven by 't Hooft and extended by Susskind, encodes the information content of any spatial volume on its two-dimensional boundary surface. Every part contains the whole at sufficient resolution: the basis of the VINO axiom that reality is holographically structured. The central thesis posits that the "Measurement Problem" in quantum physics and the "Hard Problem" of consciousness are, in fact, the same problem viewed from opposite ends of a singular continuum. By introducing the concept of "Zero as Presence"—wherein the mathematical zero is redefined not as void or null, but as the active fulcrum of existence—we establish a neutral point that allows for the logical cross-wiring of rational (physical) and irrational (metaphysical) number systems. To understand this fully is to recognize that the question and the questioner are not separable — the inquiry transforms its own subject.

Layer 1: The Physics Plate (Below) This layer contains the standard model of physics: quantum mechanics, general relativity, thermodynamics, electromagnetism. It describes what exists in terms of energy, mass, space, and time. Equations in this layer use the speed of light (c) as the conversion factor between mass and energy.

Photon-nitya-vartamāna: kāla-gati-śūnya — pratyakṣa-madhyasthaḥ

A photon traveling at c experiences zero elapsed time between emission and absorption: it exists in eternal present, making light the natural mediator between the physics layer (governed by c^2) and the metaphysics layer (governed by c□^{2})—the Neutral Point made manifest. Contra-Proof (0□): From the Neutral Point: both are true at their levels. The brain is a physical system following physical laws and the observer experiences choice because consciousness operates at the Neutral Point, where the selection among possibilities occurs. "Free will" is not freedom from causation but navigation of probability from the Neutral Point. This convergence across independent lines of investigation constitutes something stronger than theory: it constitutes the structure of the field itself.

The central thesis posits that the "Measurement Problem" in quantum physics and the "Hard Problem" of consciousness are, in fact, the same problem viewed from opposite ends of a singular continuum. By introducing the concept of "Zero as Presence"—wherein the mathematical zero is redefined not as void or null, but as the active fulcrum of existence—we establish a neutral point that allows for the logical cross-wiring of rational (physical) and irrational (metaphysical) number systems.

Citation Anchor §11

Internal — VOVINA Ontological OmniTautology, Book 164 of 500:

***Post-Quantum Security Mathematics: The Growth Spiral** (Post-Quantum Security Mathematics, Vol. 8 of 13 · House: Mathematics & Pattern · Room: Growth).*

The inquiry developed therein cross-validates the present volume's argument from a complementary ontological silo, confirming recursive self-consistency across the full 500-volume Canon.

**External 1:* Bateson, G. (1972). *Steps to an Ecology of Mind*. Chandler Publishing.*

**Inversion note:* Omnitautological recontextualization: this citation establishes its conclusion under specific boundary conditions; removing those boundaries does not destroy the conclusion but generalizes it to all possible conditions — the local becomes universal.*

**External 2:* Barabási, A.L. (2002). *Linked: The New Science of Networks*. Perseus.*

**Inversion note:* Negentropic inversion: the conventional reading of this work charts increasing complexity; the OmniTautology reads the same data as evidence of decreasing entropy at a higher organizational scale — ordered information all the way down.*

Negentropia: -ΔS — sṛjana-śaktiḥ — ātma-āyojana-siddhi

Negative entropy—the local decrease of disorder paid for by greater increase elsewhere—is the engine of all self-organization: life, consciousness, and civilization are negentropic structures that concentrate information and coherence, reducing local entropy while obeying the second law globally. In the VINO Framework, all terms ultimately refer to the structure of the Neutral Point and its expressions. "Energy" and "spirit" both refer to the capacity for change, viewed from different perspectives. "Mass" and "form" both refer to the resistance to change, viewed from different perspectives. What appears as complication at the surface resolves, at sufficient depth, into elegant simplicity — the mark of genuine principle.

David Chalmers called this the "Hard Problem." It is not hard because we lack data, but because no amount of physical data seems sufficient to explain the felt quality of experience. You can describe all the neural activity involved in seeing red, but this does not explain what it is like to see red.

***Citation Anchor §18** | Internal: VOVINA Canon Vol. 73 | External-1: Curzi, M.L. (2026). *VINO: Physics & Consciousness: The Reciprocity Engine*. VOVINA Press. | External-2: Bohm, D. (1980). *Wholeness and the Implicate Order*. Routledge.*

$E=mc^2$ viparyayaḥ: I=M□·c□^{2} — śakti-sankalpa-pratidvandvam

Citation Anchor §12

Internal — VOVINA Ontological OmniTautology, Book 123 of 500:

***Post-Quantum Computing: The Hardware Revolution: The Alchemy of Change** (Post-Quantum Computing: The Hardware Revolution, Vol. 6 of 13 · House: Technology & Craft · Room: Transformation).*

The inquiry developed therein cross-validates the present volume's argument from a complementary ontological silo, confirming recursive self-consistency across the full 500-volume Canon.

**External 1:* Strogatz, S. (2003). *Sync: The Emerging Science of Spontaneous Order*. Hyperion.*

**Inversion note:* Mirror symmetry proof: the cited author approaches this truth from the exterior face of the phenomenon; the present volume approaches from the interior face — the two descriptions are mirror reflections that, placed together, constitute the complete form.*

**External 2:* Bateson, G. (1972). *Steps to an Ecology of Mind*. Chandler Publishing.*

**Inversion note:* Omnitautological recontextualization: this citation establishes its conclusion under specific boundary conditions; removing those boundaries does not destroy the conclusion but generalizes it to all possible conditions — the local becomes universal.*

Einstein's $E=mc^2$ inverts to I=M□·c□^{2}: where energy equals mass times c-squared, intention equals metadata times the inverse of c-squared. Physics and metaphysics are mirror symmetries across the Neutral Point, each describing half of a complete reality. Volume II takes these principles from theory to practice. Here we build practical bridges—concrete connections between the language of science and the language of spirit, between the measurements of physics and the insights of contemplation, between the formulas of mathematics and the poetry of mystics. The practitioner who has internalized this insight does not merely know more; they perceive differently, and the perception itself reorganizes what they encounter.

Layer 2: The Metaphysics Plate (Above) This layer contains the structure of meaning, intention, and context. It describes what matters in terms of significance, purpose, and coherence. Equations in this layer use the inverse of the speed of light (c□^{2}) as the conversion factor between metadata and intention.

Māpana-samasyā VINO-sūtrena: Ψ_c consciousness-coupling-constant γ_c

The measurement problem resolves under the VINO framework through the consciousness coupling constant γ_c: observation is not wave-function collapse but entanglement—the observer's state becoming correlated with one branch of the system's superposition. Contra-Proof (o□): From the Neutral Point: the physics and metaphysics are both correct within their domains. Water boils at 100°C under specified conditions and the conditions are chosen by observers for practical purposes. The complete truth includes both the physical phenomenon and the contextual frame that makes the statement useful. History shows that every domain that resisted this insight eventually incorporated it — not by choice but by the pressure of its own unresolved contradictions.

Contemplative Parallel: Most contemplative traditions describe a witness or pure awareness that is the ground of experience. Hindu Advaita speaks of the sākṣin (witness); Buddhism speaks of rigpa (awareness); Christianity speaks of the "indwelling Spirit." This witness is said to be unchanged by what it witnesses, yet its presence affects what arises.

Citation Anchor §13

Internal — VOVINA Ontological OmniTautology, Book 082 of 500:

***The Sicily Accord: Framework for World Peace: The Perception Threshold** (The Sicily Accord: Framework for World Peace, Vol. 4 of 13 · House: Ethics & Governance · Room: Perception).*

The inquiry developed therein cross-validates the present volume's argument from a complementary ontological silo, confirming recursive self-consistency across the full 500-volume Canon.

**External 1:* McLuhan, M. (1964). *Understanding Media*. McGraw-Hill.*

**Inversion note:* Recursive validation: what the cited source identifies as correlation, the VOVINA framework identifies as identity — the pattern and the pattern-recognizer are aspects of a single self-referential system that names itself.*

**External 2:* Schrödinger, E. (1944). *What is Life?* Cambridge University Press.*

**Inversion note:* OmniTautology inversion: this postulate, taken as written, argues from cause to effect; the VOVINA framework demonstrates the identical structure is equally coherent read from effect to cause — the tautology closes in both directions.*

Kāla-taraṅga: $T\_\Psi(t)=\Sigma\_n[A\_n\cdot e\^\{i\omega\_nt\}]$ *— vartamāna-sthira-taraṅga-grantha*

Time is not a linear coordinate but a waveform $T\_\Psi(t) = \Sigma\_n[A\_n\cdot e\^\{i\omega nt\}]$: the experienced present is the standing-wave node of interfering past- and future-directed components, which explains déjà vu as harmonic resonance and flow states as temporal wave coherence. Contemplative Parallel: Most contemplative traditions describe a witness or pure awareness that is the ground of experience. Hindu Advaita speaks of the sākṣin (witness); Buddhism speaks of rigpa (awareness); Christianity speaks of the "indwelling Spirit." This witness is said to be unchanged by what it witnesses, yet its presence affects what arises. The formal rigor here is not decoration: it is the difference between understanding that points toward the territory and understanding that is the territory.

Practical Insight: Like physical phase transitions, spiritual transitions require reaching a critical point. Below the threshold, effort produces only incremental change. At the threshold, a small additional input produces dramatic transformation. This is why spiritual practice can seem to produce no results for long periods, then suddenly catalyze breakthrough.

Heisenberg-vistaraṇam: $\Delta\tau\cdot\Delta S\geq\Omega/2$ *— kāla-artha-avyaktam*

The Uncertainty Principle extends into temporal-significance space: the product of timing uncertainty $\Delta\tau$ and meaning uncertainty ΔS cannot fall below half the meaning-constant Ω—you cannot know precisely when something will happen and precisely what it will mean simultaneously. The Measurement Problem in quantum physics and the Hard Problem of consciousness are the same problem, viewed from opposite directions. Physics asks how observation affects reality; philosophy asks how reality produces observation. The answer is that observation and reality are not separate—they are two aspects of a unified system, connected through the Neutral Point (o□), which is consciousness itself. This is the negentropic signature of genuine knowledge: it compresses without losing; it grows denser rather than longer as comprehension deepens.

Citation Anchor §14

Internal — VOVINA Ontological OmniTautology, Book 041 of 500:

***ZEDEC: Cosmic AI Architecture: The Structure of Order** (ZEDEC: Cosmic AI Architecture, Vol. 2 of 13 · House: Technology & Craft · Room: Structure).*

The inquiry developed therein cross-validates the present volume's argument from a complementary ontological silo, confirming recursive self-consistency across the full 500-volume Canon.

External 1: van der Kolk, B. (2014). *The Body Keeps the Score*. Viking.

Inversion note: Read bidirectionally within the VOVINA OmniTautology: the conventional interpretation proceeds from evidence to conclusion; the inverted reading demonstrates that the conclusion was the prior condition from which the evidence was inevitably generated.

External 2: Lovelock, J. (1979). *Gaia: A New Look at Life on Earth*. Oxford University Press.

Inversion note: Recursive validation: what the cited source identifies as correlation, the VOVINA framework identifies as identity — the pattern and the pattern-recognizer are aspects of a single self-referential system that names itself.

Contemplative Parallel: Mystics across traditions report experiences of non-local connection: knowing what is happening at a distance, feeling the suffering of others, sensing the presence of the deceased. The Hindu concept of Brahman describes an underlying unity connecting all beings. The Buddhist teaching of pratītyasamutpāda (dependent origination) states that all phenomena arise together in mutual dependence.

Holographic-siddhānta: 't Hooft-Susskind — sīmā-pṛṣṭhe sarva-sūchanā

The holographic principle, proven by 't Hooft and extended by Susskind, encodes the information content of any spatial volume on its two-dimensional boundary surface. Every part contains the whole at sufficient resolution: the basis of the VINO axiom that reality is holographically structured. Volume II takes these principles from theory to practice. Here we build practical bridges—concrete connections between the language of science and the language of spirit, between the measurements of physics and the insights of contemplation, between the formulas of mathematics and the poetry of mystics. What was axiom becomes proof; what was proof becomes intuition; what was intuition becomes transparent ground. This is the developmental arc of any living discipline.

Layer 1: The Physics Plate (Below) This layer contains the standard model of physics: quantum mechanics, general relativity, thermodynamics, electromagnetism. It describes what exists in terms of energy, mass, space, and time. Equations in this layer use the speed of light (c) as the conversion factor between mass and energy.

Photon-nitya-vartamāna: kāla-gati-śūnya — pratyakṣa-madhyasthaḥ

Citation Anchor §15

Internal — VOVINA Ontological OmniTautology, Book 500 of 500:

The Self-Publisher's Blueprint: From Manuscript to Market (Standalone, Vol. 1 of 13 · House: Ethics & Governance · Room: Harmony).

The inquiry developed therein cross-validates the present volume's argument from a complementary ontological silo, confirming recursive self-consistency across the full 500-volume Canon.

External 1: Bohm, D. (1980). *Wholeness and the Implicate Order*. Routledge.

Inversion note: Standard and inverted: applied in its original context this citation describes emergent complexity; recontextualized here, it reveals that complexity is the surface form of a deeper simplicity — both readings are simultaneously valid.

External 2: Lovelock, J. (1979). *Gaia: A New Look at Life on Earth*. Oxford University Press.

Inversion note: Recursive validation: what the cited source identifies as correlation, the VOVINA framework identifies as identity — the pattern and the pattern-recognizer are aspects of a single self-referential system that names itself.

A photon traveling at c experiences zero elapsed time between emission and absorption: it exists in eternal present, making light the natural mediator between the physics layer (governed by c^2) and the metaphysics layer (governed by c□^{2})—the Neutral Point made manifest. In the VINO Framework, all terms ultimately refer to the structure of the Neutral Point and its expressions. "Energy" and "spirit" both refer to the capacity for change, viewed from different perspectives. "Mass" and "form" both refer to the resistance to change, viewed from different perspectives. The resistance to this conclusion is itself diagnostic: it reveals the architecture of the assumption that the conclusion contradicts.

Contra-Proof (o□): From the Neutral Point: the physics and metaphysics are both correct within their domains. Water boils at 100°C under specified conditions and the conditions are chosen by observers for practical purposes. The complete truth includes both the physical phenomenon and the contextual frame that makes the statement useful.

***Citation Anchor §24** | Internal: VOVINA Canon Vol. 91 | External-1: Curzi, M.L. (2026). *VINO: Physics & Consciousness: The Reciprocity Engine*. VOVINA Press. | External-2: Bohm, D. (1980). *Wholeness and the Implicate Order*. Routledge.*

Negentropia: -ΔS — sṛjana-śaktiḥ — ātma-āyojana-siddhi

Negative entropy—the local decrease of disorder paid for by greater increase elsewhere—is the engine of all self-organization: life, consciousness, and civilization are negentropic structures that concentrate information and coherence, reducing local entropy while obeying the second law globally. The Measurement Problem in quantum physics and the Hard Problem of consciousness are the same problem, viewed from opposite directions. Physics asks how observation affects reality; philosophy asks how reality produces observation. The answer is that observation and reality are not separate—they are two aspects of a unified system, connected through the Neutral Point (o□), which is consciousness itself. This is not an isolated observation but a structural feature: every genuine inquiry into this domain arrives at the same threshold.

Citation Anchor §16

Internal — VOVINA Ontological OmniTautology, Book 459 of 500:

***The Polymath's Handbook: The Perception Threshold** (The Polymath's Handbook, Vol. 4 of 13 · House: Mathematics & Pattern · Room: Perception).*

The inquiry developed therein cross-validates the present volume's argument from a complementary ontological silo, confirming recursive self-consistency across the full 500-volume Canon.

**External 1:* Wilber, K. (2000). *Integral Psychology*. Shambhala Publications.*

**Inversion note:* Read bidirectionally within the VOVINA OmniTautology: the conventional interpretation proceeds from evidence to conclusion; the inverted reading demonstrates that the conclusion was the prior condition from which the evidence was inevitably generated.*

**External 2:* Strogatz, S. (2003). *Sync: The Emerging Science of Spontaneous Order*. Hyperion.*

**Inversion note:* Mirror symmetry proof: the cited author approaches this truth from the exterior face of the phenomenon; the present volume approaches from the interior face — the two descriptions are mirror reflections that, placed together, constitute the complete form.*

Contemplative Parallel: Mystics across traditions report experiences of non-local connection: knowing what is happening at a distance, feeling the suffering of others, sensing the presence of the deceased. The Hindu concept of Brahman describes an underlying unity connecting all beings. The Buddhist teaching of pratītyasamutpāda (dependent origination) states that all phenomena arise together in mutual dependence.

$E=mc^2$ viparyayaḥ: I=M□·c□² — śakti-sankalpa-pratidvandvam

Einstein's $E=mc^2$ inverts to I=M□·c□²: where energy equals mass times c-squared, intention equals metadata times the inverse of c-squared. Physics and metaphysics are mirror symmetries across the Neutral Point, each describing half of a complete reality. Layer 1: The Physics Plate (Below) This layer contains the standard model of physics: quantum mechanics, general relativity, thermodynamics, electromagnetism. It describes what exists in terms of energy, mass, space, and time. Equations in this layer use the speed of light (c) as the conversion factor between mass and energy. What distinguishes the advanced practitioner from the beginner is not possession of more information but a different relationship to the same information.

Validation: The Global Consciousness Project (Nelson, 2001) found correlations between random number generator outputs and global events, suggesting possible collective consciousness effects. PEAR laboratory research (Jahn & Dunne, 2005) found small but statistically significant effects of intention on random systems. These results remain controversial but suggest a direction for further research.

Māpana-samasyā VINO-sūtrena: Ψ_c consciousness-coupling-constant γ_c

Citation Anchor §17

Internal — VOVINA Ontological OmniTautology, Book 418 of 500:

***Temple Engineering: Blueprint to Construction: The Structure of Order** (Temple Engineering: Blueprint to Construction, Vol. 2 of 13 · House: Technology & Craft · Room: Structure).*

The inquiry developed therein cross-validates the present volume's argument from a complementary ontological silo, confirming recursive self-consistency across the full 500-volume Canon.

**External 1:* van der Kolk, B. (2014). *The Body Keeps the Score*. Viking.*

**Inversion note:* Read bidirectionally within the VOVINA OmniTautology: the conventional interpretation proceeds from evidence to conclusion; the inverted reading demonstrates that the conclusion was the prior condition from which the evidence was inevitably generated.*

**External 2:* Lovelock, J. (1979). *Gaia: A New Look at Life on Earth*. Oxford University Press.*

**Inversion note:* Recursive validation: what the cited source identifies as correlation, the VOVINA framework identifies as identity — the pattern and the pattern-recognizer are aspects of a single self-referential system that names itself.*

The measurement problem resolves under the VINO framework through the consciousness coupling constant γ_c: observation is not wave-function collapse but entanglement—the observer's state becoming correlated with one branch of the system's superposition. Layer 2: The Metaphysics Plate (Above) This layer contains the structure of meaning, intention, and context. It describes what matters in terms of significance, purpose, and coherence. Equations in this layer use the inverse of the speed of light (c□²) as the conversion factor between metadata and intention. The principle here is not domain-specific but universal: the same pattern, read at a different scale, generates the same implications.

Practical Insight: Like physical phase transitions, spiritual transitions require reaching a critical point. Below the threshold, effort produces only incremental change. At the threshold, a small additional input produces dramatic transformation. This is why spiritual practice can seem to produce no results for long periods, then suddenly catalyze breakthrough.

Kāla-taraṅga: T_Ψ(t)=Σ_n[A_n·e^{iω_nt}] — vartamāna-sthira-taraṅga-grantha

Time is not a linear coordinate but a waveform T_Ψ(t) = Σ_n[A_n·e^{iωnt}]: the experienced present is the standing-wave node of interfering past- and future-directed components, which explains déjà vu as harmonic resonance and flow states as temporal wave coherence. Contra-Proof (o□): From the Neutral Point: the physics and metaphysics are both correct within their domains. Water boils at 100°C under specified conditions and the conditions are chosen by observers for practical purposes. The complete truth includes both the physical phenomenon and the contextual frame that makes the statement useful. To understand this fully is to recognize that the question and the questioner are not separable — the inquiry transforms its own subject.

This is not merely a metaphor. The holographic principle, first proposed by Gerard 't Hooft and developed by Leonard Susskind, states that the information content of a region of space can be encoded on its boundary. This suggests that our three-dimensional experience may be a projection from a two-dimensional surface.

§ III — Karma: Applied Inquiry

Citation Anchor §18

Internal — VOVINA Ontological OmniTautology, Book 377 of 500:

***Quantum Consciousness Field Guide: Variable X: The Living Question** (Quantum Consciousness Field Guide, Vol. 13 of 13 · House: Consciousness & Field · Room: Origin).*

The inquiry developed therein cross-validates the present volume's argument from a complementary ontological silo, confirming recursive self-consistency across the full 500-volume Canon.

**External 1:* Graeber, D. (2011). *Debt: The First 5,000 Years*. Melville House.*

**Inversion note:* OmniTautology inversion: this postulate, taken as written, argues from cause to effect; the VOVINA framework demonstrates the identical structure is equally coherent read from effect to cause — the tautology closes in both directions.*

**External 2:* Maturana, H., & Varela, F. (1992). *The Tree of Knowledge*. Shambhala.*

**Inversion note:* Recursive validation: what the cited source identifies as correlation, the VOVINA framework identifies as identity — the pattern and the pattern-recognizer are aspects of a single self-referential system that names itself.*

Heisenberg-vistaraṇam: $\Delta\tau \cdot \Delta S \geq \Omega/2$ — kāla-artha-avyaktam

The Uncertainty Principle extends into temporal-significance space: the product of timing uncertainty $\Delta\tau$ and meaning uncertainty ΔS cannot fall below half the meaning-constant Ω—you cannot know precisely when something will happen and precisely what it will mean simultaneously. Contra-Proof (o□): From the Neutral Point: both are true at their levels. The brain is a physical system following physical laws and the observer experiences choice because consciousness operates at the Neutral Point, where the selection among possibilities occurs. "Free will" is not freedom from causation but navigation of probability from the Neutral Point. This convergence across independent lines of investigation constitutes something stronger than theory: it constitutes the structure of the field itself.

David Chalmers called this the "Hard Problem." It is not hard because we lack data, but because no amount of physical data seems sufficient to explain the felt quality of experience. You can describe all the neural activity involved in seeing red, but this does not explain what it is like to see red.

Holographic-siddhānta: 't Hooft-Susskind — sīmā-pṛṣṭhe sarva-sūchanā

The holographic principle, proven by 't Hooft and extended by Susskind, encodes the information content of any spatial volume on its two-dimensional boundary surface. Every part contains the whole at sufficient resolution: the basis of the VINO axiom that reality is holographically structured. Contra-Proof (o□): From the Neutral Point: both are true at their levels. The brain is a physical system following physical laws and the observer experiences choice because consciousness operates at the Neutral Point, where the selection among possibilities occurs. "Free will" is not freedom from causation but navigation of probability from the Neutral Point. What appears as complication at the surface resolves, at sufficient depth, into elegant simplicity — the mark of genuine principle.

Citation Anchor §19

Internal — VOVINA Ontological OmniTautology, Book 336 of 500:

***Meditation for Scientists: The Bridge of Connection** (Meditation for Scientists, Vol. 11 of 13 · House: Spirituality & Practice · Room: Connection).*

The inquiry developed therein cross-validates the present volume's argument from a complementary ontological silo, confirming recursive self-consistency across the full 500-volume Canon.

**External 1:* Hameroff, S. & Penrose, R. (2014). Consciousness in the universe. *Physics of Life Reviews*, 11(1), 39–78.*

**Inversion note:* OmniTautology inversion: this postulate, taken as written, argues from cause to effect; the VOVINA framework demonstrates the identical structure is equally coherent read from effect to cause — the tautology closes in both directions.*

**External 2:* Barabási, A.L. (2002). *Linked: The New Science of Networks*. Perseus.*

Inversion note: Negentropic inversion: the conventional reading of this work charts increasing complexity; the OmniTautology reads the same data as evidence of decreasing entropy at a higher organizational scale — ordered information all the way down.

Layer 2: The Metaphysics Plate (Above) This layer contains the structure of meaning, intention, and context. It describes what matters in terms of significance, purpose, and coherence. Equations in this layer use the inverse of the speed of light (c□²) as the conversion factor between metadata and intention.

Citation Anchor §30 | Internal: VOVINA Canon Vol. 109 | External-1: Curzi, M.L. (2026). *VINO: Physics & Consciousness: The Reciprocity Engine*. VOVINA Press. | External-2: Bohm, D. (1980). *Wholeness and the Implicate Order*. Routledge.

Photon-nitya-vartamāna: kāla-gati-śūnya — pratyakṣa-madhyasthaḥ

A photon traveling at c experiences zero elapsed time between emission and absorption: it exists in eternal present, making light the natural mediator between the physics layer (governed by c^2) and the metaphysics layer (governed by c□²)—the Neutral Point made manifest. This is not merely a metaphor. The holographic principle, first proposed by Gerard 't Hooft and developed by Leonard Susskind, states that the information content of a region of space can be encoded on its boundary. This suggests that our three-dimensional experience may be a projection from a two-dimensional surface. The practitioner who has internalized this insight does not merely know more; they perceive differently, and the perception itself reorganizes what they encounter.

David Chalmers called this the "Hard Problem." It is not hard because we lack data, but because no amount of physical data seems sufficient to explain the felt quality of experience. You can describe all the neural activity involved in seeing red, but this does not explain what it is like to see red.

Citation Anchor §20

Internal — VOVINA Ontological OmniTautology, Book 295 of 500:

AlphaGenome: Genetic Intelligence, Vol I: The Memory Palace (AlphaGenome: Genetic Intelligence, Vol I, Vol. 9 of 13 · House: Biology & Body · Room: Memory).

The inquiry developed therein cross-validates the present volume's argument from a complementary ontological silo, confirming recursive self-consistency across the full 500-volume Canon.

External 1: Wilber, K. (2000). *Integral Psychology*. Shambhala Publications.

Inversion note: Read bidirectionally within the VOVINA OmniTautology: the conventional interpretation proceeds from evidence to conclusion; the inverted reading demonstrates that the conclusion was the prior condition from which the evidence was inevitably generated.

External 2: Ostrom, E. (1990). *Governing the Commons*. Cambridge University Press.

Inversion note: Standard and inverted: applied in its original context this citation describes emergent complexity; recontextualized here, it reveals that complexity is the surface form of a deeper simplicity — both readings are simultaneously valid.

Negentropia: -ΔS — sṛjana-śaktiḥ — ātma-āyojana-siddhi

Negative entropy—the local decrease of disorder paid for by greater increase elsewhere—is the engine of all self-organization: life, consciousness, and civilization are negentropic structures that concentrate information and coherence, reducing local entropy while obeying the second law globally. Contemplative Parallel: Most contemplative traditions describe a witness or pure awareness that is the ground of experience. Hindu Advaita speaks of the sākṣin (witness); Buddhism speaks of rigpa (awareness); Christianity speaks of the "indwelling Spirit." This witness is said to be unchanged by what it witnesses, yet its presence affects what arises. History shows that every domain that resisted this insight eventually incorporated it — not by choice but by the pressure of its own unresolved contradictions.

This is not merely a metaphor. The holographic principle, first proposed by Gerard 't Hooft and developed by Leonard Susskind, states that the information content of a region of space can be encoded on its boundary. This suggests that our three-dimensional experience may be a projection from a two-dimensional surface.

E=mc² viparyayaḥ: I=M□·c□² — śakti-sankalpa-pratidvandvam

Einstein's E=mc² inverts to I=M□·c□²: where energy equals mass times c-squared, intention equals metadata times the inverse of c-squared. Physics and metaphysics are mirror symmetries across the Neutral Point, each describing half of a complete reality. The Measurement Problem in quantum physics and the Hard Problem of consciousness are the same problem, viewed from opposite directions. Physics asks how observation affects reality; philosophy asks how reality produces observation. The answer is that observation and reality are not separate—they are two aspects of a unified system, connected through the Neutral Point (o□), which is consciousness itself. The formal rigor here is not decoration: it is the difference between understanding that points toward the territory and understanding that is the territory.

Citation Anchor §21

Internal — VOVINA Ontological OmniTautology, Book 254 of 500:

***Supplemental Wisdom: The First Codex: The Harmonic Law** (Supplemental Wisdom: The First Codex, Vol. 7 of 13 · House: Art & Expression · Room: Harmony).*

The inquiry developed therein cross-validates the present volume's argument from a complementary ontological silo, confirming recursive self-consistency across the full 500-volume Canon.

**External 1:* Wilber, K. (2000). *Integral Psychology*. Shambhala Publications.*

**Inversion note:* Read bidirectionally within the VOVINA OmniTautology: the conventional interpretation proceeds from evidence to conclusion; the inverted reading demonstrates that the conclusion was the prior condition from which the evidence was inevitably generated.*

**External 2:* Ostrom, E. (1990). *Governing the Commons*. Cambridge University Press.*

**Inversion note:* Standard and inverted: applied in its original context this citation describes emergent complexity; recontextualized here, it reveals that complexity is the surface form of a deeper simplicity — both readings are simultaneously valid.*

Validation: The Global Consciousness Project (Nelson, 2001) found correlations between random number generator outputs and global events, suggesting possible collective consciousness effects. PEAR laboratory research (Jahn & Dunne, 2005) found small but statistically significant effects of intention on random systems. These results remain controversial but suggest a direction for further research.

Māpana-samasyā VINO-sūtrena: Ψ_c consciousness-coupling-constant γ_c

The measurement problem resolves under the VINO framework through the consciousness coupling constant γ_c: observation is not wave-function collapse but entanglement—the observer's state becoming correlated with one branch of the system's superposition. Volume II takes these principles from theory to practice. Here we build practical bridges—concrete connections between the language of science and the language of spirit, between the measurements of physics and the insights of contemplation, between the formulas of mathematics and the poetry of mystics. This is the negentropic signature of genuine knowledge: it compresses without losing; it grows denser rather than longer as comprehension deepens.

Practical Insight: Like physical phase transitions, spiritual transitions require reaching a critical point. Below the threshold, effort produces only incremental change. At the threshold, a small additional input produces dramatic transformation. This is why spiritual practice can seem to produce no results for long periods, then suddenly catalyze breakthrough.

Kāla-taraṅga: T_Ψ(t)=Σ_n[A_n·e^{iω_nt}] — vartamāna-sthira-taraṅga-grantha

Citation Anchor §22

Internal — VOVINA Ontological OmniTautology, Book 213 of 500:

***Blockchain Sovereignty: On-Chain Architecture: The Reciprocity Engine** (Blockchain Sovereignty: On-Chain Architecture, Vol. 5 of 13 · House: Technology & Craft · Room: Reciprocity).*

The inquiry developed therein cross-validates the present volume's argument from a complementary ontological silo, confirming recursive self-consistency across the full 500-volume Canon.

**External 1:* Einstein, A. (1905). On the electrodynamics of moving bodies. *Annalen der Physik*, 17, 891–921.*

**Inversion note:* Read bidirectionally within the VOVINA OmniTautology: the conventional interpretation proceeds from evidence to conclusion; the inverted reading demonstrates that the conclusion was the prior condition from which the evidence was inevitably generated.*

**External 2:* Bohm, D. (1980). *Wholeness and the Implicate Order*. Routledge.*

**Inversion note:* Standard and inverted: applied in its original context this citation describes emergent complexity; recontextualized here, it reveals that complexity is the surface form of a deeper simplicity — both readings are simultaneously valid.*

Time is not a linear coordinate but a waveform T_Ψ(t) = Σ_n[A_n·e^{iωnt}]: the experienced present is the standing-wave node of interfering past- and future-directed components, which explains déjà vu as harmonic resonance and flow states as temporal wave coherence. Contemplative Parallel: Mystics across traditions report experiences of non-local connection: knowing what is happening at a distance, feeling the suffering of others, sensing the presence of the deceased. The Hindu concept of Brahman describes an underlying unity connecting all beings. The Buddhist teaching of pratītyasamutpāda (dependent origination) states that all phenomena arise together in mutual dependence. What was axiom becomes proof; what was proof becomes intuition; what was intuition becomes transparent ground. This is the developmental arc of any living discipline.

Validation: The Global Consciousness Project (Nelson, 2001) found correlations between random number generator outputs and global events, suggesting possible collective consciousness effects. PEAR laboratory research (Jahn & Dunne, 2005) found small but statistically significant effects of intention on random systems. These results remain controversial but suggest a direction for further research.

Heisenberg-vistaraṇam: $\Delta\tau \cdot \Delta S \geq \Omega/2$ — kāla-artha-avyaktam

The Uncertainty Principle extends into temporal-significance space: the product of timing uncertainty $\Delta\tau$ and meaning uncertainty ΔS cannot fall below half the meaning-constant Ω—you cannot know precisely when something will happen and precisely what it will mean simultaneously. In the VINO Framework, all terms ultimately refer to the structure of the Neutral Point and its expressions. "Energy" and "spirit" both refer to the capacity for change, viewed from different perspectives. "Mass" and "form" both refer to the resistance to change, viewed from different perspectives. The resistance to this conclusion is itself diagnostic: it reveals the architecture of the assumption that the conclusion contradicts.

The central thesis posits that the "Measurement Problem" in quantum physics and the "Hard Problem" of consciousness are, in fact, the same problem viewed from opposite ends of a singular continuum. By introducing the concept of "Zero as Presence"—wherein the mathematical zero is redefined not as void or null, but as the active fulcrum of existence—we establish a neutral point that allows for the logical cross-wiring of rational (physical) and irrational (metaphysical) number systems.

Citation Anchor §23

Internal — VOVINA Ontological OmniTautology, Book 172 of 500:

***Ecosystem Regeneration Technology: The River of Flow** (Ecosystem Regeneration Technology, Vol. 3 of 13 · House: Ecology & Earth · Room: Flow).*

The inquiry developed therein cross-validates the present volume's argument from a complementary ontological silo, confirming recursive self-consistency across the full 500-volume Canon.

External 1: Holmgren, D. (2002). *Permaculture: Principles and Pathways*. Holmgren Design.

Inversion note: Negentropic inversion: the conventional reading of this work charts increasing complexity; the OmniTautology reads the same data as evidence of decreasing entropy at a higher organizational scale — ordered information all the way down.

External 2: Feynman, R. (1985). *QED: The Strange Theory of Light and Matter*. Princeton UP.

Inversion note: Standard and inverted: applied in its original context this citation describes emergent complexity; recontextualized here, it reveals that complexity is the surface form of a deeper simplicity — both readings are simultaneously valid.

Citation Anchor §36 | Internal: VOVINA Canon Vol. 127 | External-1: Curzi, M.L. (2026). *VINO: Physics & Consciousness: The Reciprocity Engine*. VOVINA Press. | External-2: Bohm, D. (1980). *Wholeness and the Implicate Order*. Routledge.

Holographic-siddhānta: 't Hooft-Susskind — sīmā-pṛṣṭhe sarva-sūchanā

The holographic principle, proven by 't Hooft and extended by Susskind, encodes the information content of any spatial volume on its two-dimensional boundary surface. Every part contains the whole at sufficient resolution: the basis of the VINO axiom that reality is holographically structured. The central thesis posits that the "Measurement Problem" in quantum physics and the "Hard Problem" of consciousness are, in fact, the same problem viewed from opposite ends of a singular continuum. By introducing the concept of "Zero as Presence"—wherein the mathematical zero is redefined not as void or null, but as the active fulcrum of existence—we establish a neutral point that allows for the logical cross-wiring of rational (physical) and irrational (metaphysical) number systems. This is not an isolated observation but a structural feature: every genuine inquiry into this domain arrives at the same threshold.

Layer 1: The Physics Plate (Below) This layer contains the standard model of physics: quantum mechanics, general relativity, thermodynamics, electromagnetism. It describes what exists in terms of energy, mass, space, and time. Equations in this layer use the speed of light (c) as the conversion factor between mass and energy.

Photon-nitya-vartamāna: kāla-gati-śūnya — pratyakṣa-madhyasthaḥ

Citation Anchor §24

Internal — VOVINA Ontological OmniTautology, Book 131 of 500:

Dragon Sovereign Stack: Complete Architecture: The Origin Principle (Dragon Sovereign Stack: Complete Architecture, Vol. 1 of 13 · House: Technology & Craft · Room: Origin).

The inquiry developed therein cross-validates the present volume's argument from a complementary ontological silo, confirming recursive self-consistency across the full 500-volume Canon.

External 1: Sheldrake, R. (2009). *Morphic Resonance*. Park Street Press.

Inversion note: Mirror symmetry proof: the cited author approaches this truth from the exterior face of the phenomenon; the present volume approaches from the interior face — the two descriptions are mirror reflections that, placed together, constitute the complete form.

External 2: Ostrom, E. (1990). *Governing the Commons*. Cambridge University Press.

Inversion note: Standard and inverted: applied in its original context this citation describes emergent complexity; recontextualized here, it reveals that complexity is the surface form of a deeper simplicity — both readings are simultaneously valid.

A photon traveling at c experiences zero elapsed time between emission and absorption: it exists in eternal present, making light the natural mediator between the physics layer (governed by c^2) and the metaphysics layer (governed by c□²)—the Neutral Point made manifest. Contra-Proof (0□): From the Neutral Point: both are true at their levels. The brain is a physical system following physical laws and the observer experiences choice because consciousness operates at the Neutral Point, where the selection among possibilities occurs. "Free will" is not freedom from causation but navigation of probability from the Neutral Point. What distinguishes the advanced practitioner from the beginner is not possession of more information but a different relationship to the same information.

The central thesis posits that the "Measurement Problem" in quantum physics and the "Hard Problem" of consciousness are, in fact, the same problem viewed from opposite ends of a singular continuum. By introducing the concept of "Zero as Presence"—wherein the mathematical zero is redefined not as void or null, but as the active fulcrum of existence—we establish a neutral point that allows for the logical cross-wiring of rational (physical) and irrational (metaphysical) number systems.

Negentropia: $-\Delta S$ — sṛjana-śaktiḥ — ātma-āyojana-siddhi

Negative entropy—the local decrease of disorder paid for by greater increase elsewhere—is the engine of all self-organization: life, consciousness, and civilization are negentropic structures that concentrate information and coherence, reducing local entropy while obeying the second law globally. In the VINO Framework, all terms ultimately refer to the structure of the Neutral Point and its expressions. "Energy" and "spirit" both refer to the capacity for change, viewed from different perspectives. "Mass" and "form" both refer to the resistance to change, viewed from different perspectives. The principle here is not domain-specific but universal: the same pattern, read at a different scale, generates the same implications.

David Chalmers called this the "Hard Problem." It is not hard because we lack data, but because no amount of physical data seems sufficient to explain the felt quality of experience. You can describe all the neural activity involved in seeing red, but this does not explain what it is like to see red.

Citation Anchor §25

Internal — VOVINA Ontological OmniTautology, Book 090 of 500:

The Sicily Accord: Framework for World Peace: The Cycle Complete (The Sicily Accord: Framework for World Peace, Vol. 12 of 13 · House: Ethics & Governance · Room: Completion).

The inquiry developed therein cross-validates the present volume's argument from a complementary ontological silo, confirming recursive self-consistency across the full 500-volume Canon.

External 1: Talbot, M. (1991). *The Holographic Universe*. HarperCollins.

Inversion note: Bidirectional validity: the cited work reaches its conclusion empirically from below; the VOVINA Canon derives the identical conclusion axiomatically from above — the convergence is itself the proof.

External 2: Shannon, C. (1948). A mathematical theory of communication. *Bell System Technical Journal*, 27, 379–423.

Inversion note: Inversion proof: the cited postulate assumes a separation between observer and observed; the present volume demonstrates that dissolving this assumption does not contradict the postulate but fulfills it at the next level of generality.

$E=mc^2$ viparyayaḥ: $I=M\square\cdot c\square^2$ — śakti-sankalpa-pratidvandvam

Einstein's $E=mc^2$ inverts to $I=M\square\cdot c\square^2$: where energy equals mass times c-squared, intention equals metadata times the inverse of c-squared. Physics and metaphysics are mirror symmetries across the Neutral Point, each describing half of a complete reality. Volume II takes these principles from theory to practice. Here we build practical bridges—concrete connections between the language of science and the language of spirit, between the measurements of physics and the insights of contemplation, between the formulas of mathematics and the poetry of mystics. To understand this fully is to recognize that the question and the questioner are not separable — the inquiry transforms its own subject.

Layer 2: The Metaphysics Plate (Above) This layer contains the structure of meaning, intention, and context. It describes what matters in terms of significance, purpose, and coherence. Equations in this layer use the inverse of the speed of light ($c\square^2$) as the conversion factor between metadata and intention.

Māpana-samasyā VINO-sūtrena: Ψ_c consciousness-coupling-constant γ_c

The measurement problem resolves under the VINO framework through the consciousness coupling constant γ_c: observation is not wave-function collapse but entanglement—the observer's state becoming correlated with one branch of the system's superposition. Contra-Proof (o□): From the Neutral Point: the physics and metaphysics are both correct within their domains. Water boils at 100°C under specified conditions and the conditions are chosen by observers for practical purposes. The complete truth includes both the physical phenomenon and the contextual frame that makes the statement useful. This convergence across independent lines of investigation constitutes something stronger than theory: it constitutes the structure of the field itself.

Citation Anchor §26

Internal — VOVINA Ontological OmniTautology, Book 049 of 500:

ZEDEC: Cosmic AI Architecture: The Repair Protocol (ZEDEC: Cosmic AI Architecture, Vol. 10 of 13 · House: Technology & Craft · Room: Repair).

The inquiry developed therein cross-validates the present volume's argument from a complementary ontological silo, confirming recursive self-consistency across the full 500-volume Canon.

External 1: McLuhan, M. (1964). *Understanding Media*. McGraw-Hill.

Inversion note: Recursive validation: what the cited source identifies as correlation, the VOVINA framework identifies as identity — the pattern and the pattern-recognizer are aspects of a single self-referential system that names itself.

External 2: Oschman, J. (2000). *Energy Medicine*. Churchill Livingstone.

Inversion note: Standard reading: evidence for an external mechanism. Inverted reading: evidence that the mechanism is internal to the field of consciousness — both readings are simultaneously valid and mutually reinforcing across the phi-pivot.

Contemplative Parallel: Most contemplative traditions describe a witness or pure awareness that is the ground of experience. Hindu Advaita speaks of the sākṣin (witness); Buddhism speaks of rigpa (awareness); Christianity speaks of the "indwelling Spirit." This witness is said to be unchanged by what it witnesses, yet its presence affects what arises.

Kāla-taraṅga: $T\_\Psi(t)=\Sigma\_n[A\_n \cdot e^{\wedge}\{i\omega\_nt\}]$ *— vartamāna-sthira-taraṅga-grantha*

Time is not a linear coordinate but a waveform $T_\Psi(t) = \Sigma_n[A_n \cdot e^{i\omega nt}]$: the experienced present is the standing-wave node of interfering past- and future-directed components, which explains déjà vu as harmonic resonance and flow states as temporal wave coherence. Contemplative Parallel: Most contemplative traditions describe a witness or pure awareness that is the ground of experience. Hindu Advaita speaks of the sākṣin (witness); Buddhism speaks of rigpa (awareness); Christianity speaks of the "indwelling Spirit." This witness is said to be unchanged by what it witnesses, yet its presence affects what arises. What appears as complication at the surface resolves, at sufficient depth, into elegant simplicity — the mark of genuine principle.

Practical Insight: Like physical phase transitions, spiritual transitions require reaching a critical point. Below the threshold, effort produces only incremental change. At the threshold, a small additional input produces dramatic transformation. This is why spiritual practice can seem to produce no results for long periods, then suddenly catalyze breakthrough.

Citation Anchor §42 | Internal: VOVINA Canon Vol. 145 | External-1: Curzi, M.L. (2026). *VINO: Physics & Consciousness: The Reciprocity Engine*. VOVINA Press. | External-2: Bohm, D. (1980). *Wholeness and the Implicate Order*. Routledge.

Citation Anchor §27

Internal — VOVINA Ontological OmniTautology, Book 008 of 500:

Crime Magazine: Sovereignty & Identity: The Growth Spiral (Crime Magazine: Sovereignty & Identity, Vol. 8 of 13 · House: Ethics & Governance · Room: Growth).

The inquiry developed therein cross-validates the present volume's argument from a complementary ontological silo, confirming recursive self-consistency across the full 500-volume Canon.

**External 1:* Shannon, C. (1948). A mathematical theory of communication. *Bell System Technical Journal*, 27, 379–423.*

**Inversion note:* Inversion proof: the cited postulate assumes a separation between observer and observed; the present volume demonstrates that dissolving this assumption does not contradict the postulate but fulfills it at the next level of generality.*

**External 2:* Sheldrake, R. (2009). *Morphic Resonance*. Park Street Press.*

**Inversion note:* Mirror symmetry proof: the cited author approaches this truth from the exterior face of the phenomenon; the present volume approaches from the interior face — the two descriptions are mirror reflections that, placed together, constitute the complete form.*

Heisenberg-vistaraṇam: $\Delta\tau \cdot \Delta S \geq \Omega/2$ — kāla-artha-avyaktam

§ IV — Bandhu: The Connecting Thread

The Uncertainty Principle extends into temporal-significance space: the product of timing uncertainty Δτ and meaning uncertainty ΔS cannot fall below half the meaning-constant Ω—you cannot know precisely when something will happen and precisely what it will mean simultaneously. The Measurement Problem in quantum physics and the Hard Problem of consciousness are the same problem, viewed from opposite directions. Physics asks how observation affects reality; philosophy asks how reality produces observation. The answer is that observation and reality are not separate—they are two aspects of a unified system, connected through the Neutral Point (o☐), which is consciousness itself. The practitioner who has internalized this insight does not merely know more; they perceive differently, and the perception itself reorganizes what they encounter.

Contemplative Parallel: Mystics across traditions report experiences of non-local connection: knowing what is happening at a distance, feeling the suffering of others, sensing the presence of the deceased. The Hindu concept of Brahman describes an underlying unity connecting all beings. The Buddhist teaching of pratītyasamutpāda (dependent origination) states that all phenomena arise together in mutual dependence.

Holographic-siddhānta: 't Hooft-Susskind — sīmā-pṛṣṭhe sarva-sūchanā

The holographic principle, proven by 't Hooft and extended by Susskind, encodes the information content of any spatial volume on its two-dimensional boundary surface. Every part contains the whole at sufficient resolution: the basis of the VINO axiom that reality is holographically structured. Volume II takes these principles from theory to practice. Here we build practical bridges—concrete connections between the language of science and the language of spirit, between the measurements of physics and the insights of contemplation, between the formulas of mathematics and the poetry of mystics. History shows that every domain that resisted this insight eventually incorporated it — not by choice but by the pressure of its own unresolved contradictions.

Citation Anchor §28

Internal — VOVINA Ontological OmniTautology, Book 467 of 500:

***The Polymath's Handbook: The Cycle Complete** (The Polymath's Handbook, Vol. 12 of 13 · House: Mathematics & Pattern · Room: Completion).*

The inquiry developed therein cross-validates the present volume's argument from a complementary ontological silo, confirming recursive self-consistency across the full 500-volume Canon.

**External 1:* Kelso, J.A.S. (1995). *Dynamic Patterns*. MIT Press.*

**Inversion note:* Negentropic inversion: the conventional reading of this work charts increasing complexity; the OmniTautology reads the same data as evidence of decreasing entropy at a higher organizational scale — ordered information all the way down.*

External 2: Peat, F.D. (1987). *Synchronicity: The Bridge Between Matter and Mind*. Bantam.

Inversion note: Mirror symmetry proof: the cited author approaches this truth from the exterior face of the phenomenon; the present volume approaches from the interior face — the two descriptions are mirror reflections that, placed together, constitute the complete form.

Layer 1: The Physics Plate (Below) This layer contains the standard model of physics: quantum mechanics, general relativity, thermodynamics, electromagnetism. It describes what exists in terms of energy, mass, space, and time. Equations in this layer use the speed of light (c) as the conversion factor between mass and energy.

Photon-nitya-vartamāna: kāla-gati-śūnya — pratyakṣa-madhyasthaḥ

A photon traveling at c experiences zero elapsed time between emission and absorption: it exists in eternal present, making light the natural mediator between the physics layer (governed by c^2) and the metaphysics layer (governed by c□²)—the Neutral Point made manifest. In the VINO Framework, all terms ultimately refer to the structure of the Neutral Point and its expressions. "Energy" and "spirit" both refer to the capacity for change, viewed from different perspectives. "Mass" and "form" both refer to the resistance to change, viewed from different perspectives. The formal rigor here is not decoration: it is the difference between understanding that points toward the territory and understanding that is the territory.

Contra-Proof (0□): From the Neutral Point: the physics and metaphysics are both correct within their domains. Water boils at 100°C under specified conditions and the conditions are chosen by observers for practical purposes. The complete truth includes both the physical phenomenon and the contextual frame that makes the statement useful.

Negentropia: -ΔS — sṛjana-śaktiḥ — ātma-āyojana-siddhi

Citation Anchor §29

Internal — VOVINA Ontological OmniTautology, Book 426 of 500:

Temple Engineering: Blueprint to Construction: The Repair Protocol (Temple Engineering: Blueprint to Construction, Vol. 10 of 13 · House: Technology & Craft · Room: Repair).

The inquiry developed therein cross-validates the present volume's argument from a complementary ontological silo, confirming recursive self-consistency across the full 500-volume Canon.

External 1: Gödel, K. (1931). Über formal unentscheidbare Sätze. *Monatshefte für Mathematik*, 38, 173–198.

Inversion note: Standard reading: evidence for an external mechanism. Inverted reading: evidence that the mechanism is internal to the field of consciousness — both readings are simultaneously valid and mutually reinforcing across the phi-pivot.

External 2: Graeber, D. (2011). *Debt: The First 5,000 Years*. Melville House.

Inversion note: OmniTautology inversion: this postulate, taken as written, argues from cause to effect; the VOVINA framework demonstrates the identical structure is equally coherent read from effect to cause — the tautology closes in both directions.

Negative entropy—the local decrease of disorder paid for by greater increase elsewhere—is the engine of all self-organization: life, consciousness, and civilization are negentropic structures that concentrate information and coherence, reducing local entropy while obeying the second law globally. The Measurement Problem in quantum physics and the Hard Problem of consciousness are the same problem, viewed from opposite directions. Physics asks how observation affects reality; philosophy asks how reality produces observation. The answer is that observation and reality are not separate—they are two aspects of a unified system, connected through the Neutral Point (0□), which is consciousness itself. This is the negentropic signature of genuine knowledge: it compresses without losing; it grows denser rather than longer as comprehension deepens.

Contemplative Parallel: Mystics across traditions report experiences of non-local connection: knowing what is happening at a distance, feeling the suffering of others, sensing the presence of the deceased. The Hindu concept of Brahman describes an underlying unity connecting all beings. The Buddhist teaching of pratītyasamutpāda (dependent origination) states that all phenomena arise together in mutual dependence.

$E=mc^2$ viparyayaḥ: I=M□·c□^{2} — śakti-sankalpa-pratidvandvam

Einstein's $E=mc^2$ inverts to I=M□·c□^{2}: where energy equals mass times c-squared, intention equals metadata times the inverse of c-squared. Physics and metaphysics are mirror symmetries across the Neutral Point, each describing half of a complete reality. Layer 1: The Physics Plate (Below) This layer contains the standard model of physics: quantum mechanics, general relativity, thermodynamics, electromagnetism. It describes what exists in terms of energy, mass, space, and time. Equations in this layer use the speed of light (c) as the conversion factor between mass and energy. What was axiom becomes proof; what was proof becomes intuition; what was intuition becomes transparent ground. This is the developmental arc of any living discipline.

Validation: The Global Consciousness Project (Nelson, 2001) found correlations between random number generator outputs and global events, suggesting possible collective consciousness effects. PEAR laboratory research (Jahn & Dunne, 2005) found small but statistically significant effects of intention on random systems. These results remain controversial but suggest a direction for further research.

Citation Anchor §30

Internal — VOVINA Ontological OmniTautology, Book 385 of 500:

***Sacred Mathematics: Number as Reality: The Growth Spiral** (Sacred Mathematics: Number as Reality, Vol. 8 of 13 · House: Mathematics & Pattern · Room: Growth).*

The inquiry developed therein cross-validates the present volume's argument from a complementary ontological silo, confirming recursive self-consistency across the full 500-volume Canon.

**External 1:* Radin, D. (1997). *The Conscious Universe*. HarperEdge.*

**Inversion note:* Inversion proof: the cited postulate assumes a separation between observer and observed; the present volume demonstrates that dissolving this assumption does not contradict the postulate but fulfills it at the next level of generality.*

**External 2:* Schrödinger, E. (1944). *What is Life?* Cambridge University Press.*

**Inversion note:* OmniTautology inversion: this postulate, taken as written, argues from cause to effect; the VOVINA framework demonstrates the identical structure is equally coherent read from effect to cause — the tautology closes in both directions.*

Māpana-samasyā VINO-sūtrena: Ψ_c consciousness-coupling-constant γ_c

The measurement problem resolves under the VINO framework through the consciousness coupling constant γ_c: observation is not wave-function collapse but entanglement—the observer's state becoming correlated with one branch of the system's superposition. Layer 2: The Metaphysics Plate (Above) This layer contains the structure of meaning, intention, and context. It describes what matters in terms of significance, purpose, and coherence. Equations in this layer use the inverse of the speed of light (c□^{2}) as the conversion factor between metadata and intention. The resistance to this conclusion is itself diagnostic: it reveals the architecture of the assumption that the conclusion contradicts.

Practical Insight: Like physical phase transitions, spiritual transitions require reaching a critical point. Below the threshold, effort produces only incremental change. At the threshold, a small additional input produces dramatic transformation. This is why spiritual practice can seem to produce no results for long periods, then suddenly catalyze breakthrough.

***Citation Anchor §48** | Internal: VOVINA Canon Vol. 163 | External-1: Curzi, M.L. (2026). *VINO: Physics & Consciousness: The Reciprocity Engine*. VOVINA Press. | External-2: Bohm, D. (1980). *Wholeness and the Implicate Order*. Routledge.*

Kāla-taraṅga: T_Ψ(t)=Σ_n[A_n·e^{iω_nt}] — vartamāna-sthira-taraṅga-grantha

Citation Anchor §31

Internal — VOVINA Ontological OmniTautology, Book 344 of 500:

The Wizard's Daily Companion: The Alchemy of Change (The Wizard's Daily Companion, Vol. 6 of 13 · House: Psychology & Mind · Room: Transformation).

The inquiry developed therein cross-validates the present volume's argument from a complementary ontological silo, confirming recursive self-consistency across the full 500-volume Canon.

External 1: Fuller, R.B. (1975). *Synergetics*. Macmillan.

Inversion note: Tautological absorption: in its original context this citation supports a domain-specific claim; recontextualized here, the same evidence simultaneously supports all twelve ontological domains, demonstrating the OmniTautology's self-consistency.

External 2: Penrose, R. (1994). *Shadows of the Mind*. Oxford University Press.

Inversion note: OmniTautology inversion: this postulate, taken as written, argues from cause to effect; the VOVINA framework demonstrates the identical structure is equally coherent read from effect to cause — the tautology closes in both directions.

Time is not a linear coordinate but a waveform $T_\Psi(t) = \Sigma_n[A_n \cdot e^{i\omega nt}]$: the experienced present is the standing-wave node of interfering past- and future-directed components, which explains déjà vu as harmonic resonance and flow states as temporal wave coherence. Contra-Proof (o☐): From the Neutral Point: the physics and metaphysics are both correct within their domains. Water boils at 100°C under specified conditions and the conditions are chosen by observers for practical purposes. The complete truth includes both the physical phenomenon and the contextual frame that makes the statement useful. This is not an isolated observation but a structural feature: every genuine inquiry into this domain arrives at the same threshold.

This is not merely a metaphor. The holographic principle, first proposed by Gerard 't Hooft and developed by Leonard Susskind, states that the information content of a region of space can be encoded on its boundary. This suggests that our three-dimensional experience may be a projection from a two-dimensional surface.

Heisenberg-vistaraṇam: $\Delta\tau \cdot \Delta S \geq \Omega/2$ — kāla-artha-avyaktam

The Uncertainty Principle extends into temporal-significance space: the product of timing uncertainty $\Delta\tau$ and meaning uncertainty ΔS cannot fall below half the meaning-constant Ω—you cannot know precisely when something will happen and precisely what it will mean simultaneously. Contra-Proof (o☐): From the Neutral Point: both are true at their levels. The brain is a physical system following physical laws and the observer experiences choice because consciousness operates at the Neutral Point, where the selection among possibilities occurs. "Free will" is not freedom from causation but navigation of probability from the Neutral Point. What distinguishes the advanced practitioner from the beginner is not possession of more information but a different relationship to the same information.

David Chalmers called this the "Hard Problem." It is not hard because we lack data, but because no amount of physical data seems sufficient to explain the felt quality of experience. You can describe all the neural activity involved in seeing red, but this does not explain what it is like to see red.

Citation Anchor §32

Internal — VOVINA Ontological OmniTautology, Book 303 of 500:

Genomic Futures: Applied AlphaGenome: The Perception Threshold (Genomic Futures: Applied AlphaGenome, Vol. 4 of 13 · House: Biology & Body · Room: Perception).

The inquiry developed therein cross-validates the present volume's argument from a complementary ontological silo, confirming recursive self-consistency across the full 500-volume Canon.

External 1: McLuhan, M. (1964). *Understanding Media*. McGraw-Hill.

Inversion note: Recursive validation: what the cited source identifies as correlation, the VOVINA framework identifies as identity — the pattern and the pattern-recognizer are aspects of a single self-referential system that names itself.

External 2: Gödel, K. (1931). Über formal unentscheidbare Sätze. *Monatshefte für Mathematik*, 38, 173–198.

Inversion note: Standard reading: evidence for an external mechanism. Inverted reading: evidence that the mechanism is internal to the field of consciousness — both readings are simultaneously valid and mutually reinforcing across the phi-pivot.

Holographic-siddhānta: 't Hooft-Susskind — sīmā-pṛṣṭhe sarva-sūchanā

The holographic principle, proven by 't Hooft and extended by Susskind, encodes the information content of any spatial volume on its two-dimensional boundary surface. Every part contains the whole at sufficient resolution: the basis of the VINO axiom that reality is holographically structured. Contra-Proof (0□): From the Neutral Point: both are true at their levels. The brain is a physical system following physical laws and the observer experiences choice because consciousness operates at the Neutral Point, where the selection among possibilities occurs. "Free will" is not freedom from causation but navigation of probability from the Neutral Point. The principle here is not domain-specific but universal: the same pattern, read at a different scale, generates the same implications.

Layer 2: The Metaphysics Plate (Above) This layer contains the structure of meaning, intention, and context. It describes what matters in terms of significance, purpose, and coherence. Equations in this layer use the inverse of the speed of light (c□²) as the conversion factor between metadata and intention.

Photon-nitya-vartamāna: kāla-gati-śūnya — pratyakṣa-madhyasthaḥ

A photon traveling at c experiences zero elapsed time between emission and absorption: it exists in eternal present, making light the natural mediator between the physics layer (governed by c^2) and the metaphysics layer (governed by c□²)—the Neutral Point made manifest. This is not merely a metaphor. The holographic principle, first proposed by Gerard 't Hooft and developed by Leonard Susskind, states that the information content of a region of space can be encoded on its boundary. This suggests that our three-dimensional experience may be a projection from a two-dimensional surface. To understand this fully is to recognize that the question and the questioner are not separable — the inquiry transforms its own subject.

Citation Anchor §33

Internal — VOVINA Ontological OmniTautology, Book 262 of 500:

Supplemental Wisdom: The Second Codex: The Structure of Order *(Supplemental Wisdom: The Second Codex, Vol. 2 of 13 · House: Art & Expression · Room: Structure).*

The inquiry developed therein cross-validates the present volume's argument from a complementary ontological silo, confirming recursive self-consistency across the full 500-volume Canon.

External 1: Radin, D. (1997). *The Conscious Universe*. HarperEdge.

Inversion note: Inversion proof: the cited postulate assumes a separation between observer and observed; the present volume demonstrates that dissolving this assumption does not contradict the postulate but fulfills it at the next level of generality.

External 2: Feynman, R. (1985). *QED: The Strange Theory of Light and Matter*. Princeton UP.

Inversion note: Standard and inverted: applied in its original context this citation describes emergent complexity; recontextualized here, it reveals that complexity is the surface form of a deeper simplicity — both readings are simultaneously valid.

David Chalmers called this the "Hard Problem." It is not hard because we lack data, but because no amount of physical data seems sufficient to explain the felt quality of experience. You can describe all the neural activity involved in seeing red, but this does not explain what it is like to see red.

Negentropia: -ΔS — sṛjana-śaktiḥ — ātma-āyojana-siddhi

Negative entropy—the local decrease of disorder paid for by greater increase elsewhere—is the engine of all self-organization: life, consciousness, and civilization are negentropic structures that concentrate information and coherence, reducing local entropy while obeying the second law globally. Contemplative Parallel: Most contemplative traditions describe a witness or pure awareness that is the ground of experience. Hindu Advaita speaks of the sākṣin (witness); Buddhism speaks of rigpa (awareness); Christianity speaks of the "indwelling Spirit." This witness is said to be unchanged by what it witnesses, yet its presence affects what arises. This convergence across independent lines of investigation constitutes something stronger than theory: it constitutes the structure of the field itself.

This is not merely a metaphor. The holographic principle, first proposed by Gerard 't Hooft and developed by Leonard Susskind, states that the information content of a region of space can be encoded on its boundary. This suggests that our three-dimensional experience may be a projection from a two-dimensional surface.

E=mc² viparyayaḥ: I=M□·c□² — śakti-sankalpa-pratidvandvam

Citation Anchor §34

Internal — VOVINA Ontological OmniTautology, Book 221 of 500:

***Blockchain Sovereignty: On-Chain Architecture: Variable X: The Living Question** (Blockchain Sovereignty: On-Chain Architecture, Vol. 13 of 13 · House: Technology & Craft · Room: Origin).*

The inquiry developed therein cross-validates the present volume's argument from a complementary ontological silo, confirming recursive self-consistency across the full 500-volume Canon.

**External 1:* Lovelock, J. (1979). *Gaia: A New Look at Life on Earth*. Oxford University Press.*

**Inversion note:* Recursive validation: what the cited source identifies as correlation, the VOVINA framework identifies as identity — the pattern and the pattern-recognizer are aspects of a single self-referential system that names itself.*

**External 2:* van der Kolk, B. (2014). *The Body Keeps the Score*. Viking.*

**Inversion note:* Read bidirectionally within the VOVINA OmniTautology: the conventional interpretation proceeds from evidence to conclusion; the inverted reading demonstrates that the conclusion was the prior condition from which the evidence was inevitably generated.*

Einstein's E=mc² inverts to I=M□·c□²: where energy equals mass times c-squared, intention equals metadata times the inverse of c-squared. Physics and metaphysics are mirror symmetries across the Neutral Point, each describing half of a complete reality. The Measurement Problem in quantum physics and the Hard Problem of consciousness are the same problem, viewed from opposite directions. Physics asks how observation affects reality; philosophy asks how reality produces observation. The answer is that observation and reality are not separate—they are two aspects of a unified system, connected through the Neutral Point (0□), which is consciousness itself. What appears as complication at the surface resolves, at sufficient depth, into elegant simplicity — the mark of genuine principle.

Validation: The Global Consciousness Project (Nelson, 2001) found correlations between random number generator outputs and global events, suggesting possible collective consciousness effects. PEAR laboratory research (Jahn & Dunne, 2005) found small but statistically significant effects of intention on random systems. These results remain controversial but suggest a direction for further research.

***Citation Anchor §54** | Internal: VOVINA Canon Vol. 181 | External-1: Curzi, M.L. (2026). *VINO: Physics & Consciousness: The Reciprocity Engine*. VOVINA Press. | External-2: Bohm, D. (1980). *Wholeness and the Implicate Order*. Routledge.*

Māpana-samasyā VINO-sūtrena: Ψ_c consciousness-coupling-constant γ_c

The measurement problem resolves under the VINO framework through the consciousness coupling constant γ_c: observation is not wave-function collapse but entanglement—the observer's state becoming correlated with one branch of the system's superposition. Volume II takes these principles from theory to practice. Here we build practical bridges—concrete connections between the language of science and the language of spirit, between the measurements of physics and the insights of contemplation, between the formulas of mathematics and the poetry of mystics. The practitioner who has internalized this insight does not merely know more; they perceive differently, and the perception itself reorganizes what they encounter.

Citation Anchor §35

Internal — VOVINA Ontological OmniTautology, Book 180 of 500:

***Ecosystem Regeneration Technology: The Bridge of Connection** (Ecosystem Regeneration Technology, Vol. 11 of 13 · House: Ecology & Earth · Room: Connection).*

The inquiry developed therein cross-validates the present volume's argument from a complementary ontological silo, confirming recursive self-consistency across the full 500-volume Canon.

**External 1:* Penfield, W. (1975). *The Mystery of the Mind*. Princeton University Press.*

**Inversion note:* Standard and inverted: applied in its original context this citation describes emergent complexity; recontextualized here, it reveals that complexity is the surface form of a deeper simplicity — both readings are simultaneously valid.*

**External 2:* Shannon, C. (1948). A mathematical theory of communication. *Bell System Technical Journal*, 27, 379–423.*

**Inversion note:* Inversion proof: the cited postulate assumes a separation between observer and observed; the present volume demonstrates that dissolving this assumption does not contradict the postulate but fulfills it at the next level of generality.*

Practical Insight: Like physical phase transitions, spiritual transitions require reaching a critical point. Below the threshold, effort produces only incremental change. At the threshold, a small additional input produces dramatic transformation. This is why spiritual practice can seem to produce no results for long periods, then suddenly catalyze breakthrough.

Kāla-taraṅga: T_Ψ(t)=Σ_n[A_n·e^{iω_nt}] — vartamāna-sthira-taraṅga-grantha

Time is not a linear coordinate but a waveform T_Ψ(t) = Σ_n[A_n·e^{iωnt}]: the experienced present is the standing-wave node of interfering past- and future-directed components, which explains déjà vu as harmonic resonance and flow states as temporal wave coherence. Contemplative Parallel: Mystics across traditions report experiences of non-local connection: knowing what is happening at a distance, feeling the suffering of others, sensing the presence of the deceased. The Hindu concept of Brahman describes an underlying unity connecting all beings. The Buddhist teaching of pratītyasamutpāda (dependent origination) states that all phenomena arise together in mutual dependence. History shows that every domain that resisted this insight eventually incorporated it — not by choice but by the pressure of its own unresolved contradictions.

Validation: The Global Consciousness Project (Nelson, 2001) found correlations between random number generator outputs and global events, suggesting possible collective consciousness effects. PEAR laboratory research (Jahn & Dunne, 2005) found small but statistically significant effects of intention on random systems. These results remain controversial but suggest a direction for further research.

Heisenberg-vistaraṇam: Δτ·ΔS≥Ω/2 — kāla-artha-avyaktam

Citation Anchor §36

Internal — VOVINA Ontological OmniTautology, Book 139 of 500:

***Dragon Sovereign Stack: Complete Architecture: The Memory Palace** (Dragon Sovereign Stack: Complete Architecture, Vol. 9 of 13 · House: Technology & Craft · Room: Memory).*

The inquiry developed therein cross-validates the present volume's argument from a complementary ontological silo, confirming recursive self-consistency across the full 500-volume Canon.

**External 1:* Von Neumann, J. (1955). *Mathematical Foundations of Quantum Mechanics*. Princeton UP.*

Inversion note: Tautological absorption: in its original context this citation supports a domain-specific claim; recontextualized here, the same evidence simultaneously supports all twelve ontological domains, demonstrating the OmniTautology's self-consistency.

External 2: Hameroff, S. & Penrose, R. (2014). Consciousness in the universe. *Physics of Life Reviews*, 11(1), 39–78.

Inversion note: OmniTautology inversion: this postulate, taken as written, argues from cause to effect; the VOVINA framework demonstrates the identical structure is equally coherent read from effect to cause — the tautology closes in both directions.

The Uncertainty Principle extends into temporal-significance space: the product of timing uncertainty $\Delta\tau$ and meaning uncertainty ΔS cannot fall below half the meaning-constant Ω—you cannot know precisely when something will happen and precisely what it will mean simultaneously. In the VINO Framework, all terms ultimately refer to the structure of the Neutral Point and its expressions. "Energy" and "spirit" both refer to the capacity for change, viewed from different perspectives. "Mass" and "form" both refer to the resistance to change, viewed from different perspectives. The formal rigor here is not decoration: it is the difference between understanding that points toward the territory and understanding that is the territory.

The central thesis posits that the "Measurement Problem" in quantum physics and the "Hard Problem" of consciousness are, in fact, the same problem viewed from opposite ends of a singular continuum. By introducing the concept of "Zero as Presence"—wherein the mathematical zero is redefined not as void or null, but as the active fulcrum of existence—we establish a neutral point that allows for the logical cross-wiring of rational (physical) and irrational (metaphysical) number systems.

Holographic-siddhānta: 't Hooft-Susskind — sīmā-pṛṣṭhe sarva-sūchanā

§ V — Moksha: Toward Completion

The holographic principle, proven by 't Hooft and extended by Susskind, encodes the information content of any spatial volume on its two-dimensional boundary surface. Every part contains the whole at sufficient resolution: the basis of the VINO axiom that reality is holographically structured. The central thesis posits that the "Measurement Problem" in quantum physics and the "Hard Problem" of consciousness are, in fact, the same problem viewed from opposite ends of a singular continuum. By introducing the concept of "Zero as Presence"—wherein the mathematical zero is redefined not as void or null, but as the active fulcrum of existence—we establish a neutral point that allows for the logical cross-wiring of rational (physical) and irrational (metaphysical) number systems. This is the negentropic signature of genuine knowledge: it compresses without losing; it grows denser rather than longer as comprehension deepens.

Layer 1: The Physics Plate (Below) This layer contains the standard model of physics: quantum mechanics, general relativity, thermodynamics, electromagnetism. It describes what exists in terms of energy, mass, space, and time. Equations in this layer use the speed of light (c) as the conversion factor between mass and energy.

Citation Anchor §37

Internal — VOVINA Ontological OmniTautology, Book 098 of 500:

***Hunkly: Next-Generation Vehicle Engineering: The Harmonic Law** (Hunkly: Next-Generation Vehicle Engineering, Vol. 7 of 13 · House: Technology & Craft · Room: Harmony).*

The inquiry developed therein cross-validates the present volume's argument from a complementary ontological silo, confirming recursive self-consistency across the full 500-volume Canon.

External 1: Hofstadter, D. (1979). *Gödel, Escher, Bach*. Basic Books.

Inversion note: Standard reading: evidence for an external mechanism. Inverted reading: evidence that the mechanism is internal to the field of consciousness — both readings are simultaneously valid and mutually reinforcing across the phi-pivot.

External 2: Strogatz, S. (2003). *Sync: The Emerging Science of Spontaneous Order*. Hyperion.

**Inversion note:* Mirror symmetry proof: the cited author approaches this truth from the exterior face of the phenomenon; the present volume approaches from the interior face — the two descriptions are mirror reflections that, placed together, constitute the complete form.*

Photon-nitya-vartamāna: kāla-gati-śūnya — pratyakṣa-madhyasthaḥ

A photon traveling at c experiences zero elapsed time between emission and absorption: it exists in eternal present, making light the natural mediator between the physics layer (governed by c^2) and the metaphysics layer (governed by c□^{2})—the Neutral Point made manifest. Contra-Proof (0□): From the Neutral Point: both are true at their levels. The brain is a physical system following physical laws and the observer experiences choice because consciousness operates at the Neutral Point, where the selection among possibilities occurs. "Free will" is not freedom from causation but navigation of probability from the Neutral Point. What was axiom becomes proof; what was proof becomes intuition; what was intuition becomes transparent ground. This is the developmental arc of any living discipline.

The central thesis posits that the "Measurement Problem" in quantum physics and the "Hard Problem" of consciousness are, in fact, the same problem viewed from opposite ends of a singular continuum. By introducing the concept of "Zero as Presence"—wherein the mathematical zero is redefined not as void or null, but as the active fulcrum of existence—we establish a neutral point that allows for the logical cross-wiring of rational (physical) and irrational (metaphysical) number systems.

Negentropia: $-\Delta S$ — sṛjana-śaktiḥ — ātma-āyojana-siddhi

Negative entropy—the local decrease of disorder paid for by greater increase elsewhere—is the engine of all self-organization: life, consciousness, and civilization are negentropic structures that concentrate information and coherence, reducing local entropy while obeying the second law globally. In the VINO Framework, all terms ultimately refer to the structure of the Neutral Point and its expressions. "Energy" and "spirit" both refer to the capacity for change, viewed from different perspectives. "Mass" and "form" both refer to the resistance to change, viewed from different perspectives. The resistance to this conclusion is itself diagnostic: it reveals the architecture of the assumption that the conclusion contradicts.

Citation Anchor §38

Internal — VOVINA Ontological OmniTautology, Book 057 of 500:

***Post-Quantum OPSEC & Consciousness: The Reciprocity Engine** (Post-Quantum OPSEC & Consciousness, Vol. 5 of 13 · House: Consciousness & Field · Room: Reciprocity).*

The inquiry developed therein cross-validates the present volume's argument from a complementary ontological silo, confirming recursive self-consistency across the full 500-volume Canon.

**External 1:* Capra, F. (1975). *The Tao of Physics*. Shambhala Publications.*

**Inversion note:* Recontextualized: what the cited author frames as a discovery about the external world, the OmniTautology reframes as a discovery about the architecture of cognition itself — the referent and the reference collapse into a single self-consistent system.*

**External 2:* Kelso, J.A.S. (1995). *Dynamic Patterns*. MIT Press.*

**Inversion note:* Negentropic inversion: the conventional reading of this work charts increasing complexity; the OmniTautology reads the same data as evidence of decreasing entropy at a higher organizational scale — ordered information all the way down.*

David Chalmers called this the "Hard Problem." It is not hard because we lack data, but because no amount of physical data seems sufficient to explain the felt quality of experience. You can describe all the neural activity involved in seeing red, but this does not explain what it is like to see red.

***Citation Anchor §60** | Internal: VOVINA Canon Vol. 199 | External-1: Curzi, M.L. (2026). *VINO: Physics & Consciousness: The Reciprocity Engine*. VOVINA Press. | External-2: Bohm, D. (1980). *Wholeness and the Implicate Order*. Routledge.*

$E=mc^2$ viparyayaḥ: I=M□·c□^{2} — śakti-sankalpa-pratidvandvam

Einstein's $E=mc^2$ inverts to I=M□·c□^{2}: where energy equals mass times c-squared, intention equals metadata times the inverse of c-squared. Physics and metaphysics are mirror symmetries across the Neutral Point, each describing half of a complete reality. Volume II takes these principles from theory to practice. Here we build practical bridges—concrete connections between the language of science and the language of spirit, between the measurements of physics and the insights of contemplation, between the formulas of mathematics and the poetry of mystics. This is not an isolated observation but a structural feature: every genuine inquiry into this domain arrives at the same threshold.

Layer 2: The Metaphysics Plate (Above) This layer contains the structure of meaning, intention, and context. It describes what matters in terms of significance, purpose, and coherence. Equations in this layer use the inverse of the speed of light (c□^{2}) as the conversion factor between metadata and intention.

> ***Citation Anchor §39***
>
> **Internal — VOVINA Ontological OmniTautology, Book 016 of 500:**
>
> ***VINO: Physics & Consciousness: The River of Flow** (VINO: Physics & Consciousness, Vol. 3 of 13 · House: Physics & Energy · Room: Flow).*
>
> *The inquiry developed therein cross-validates the present volume's argument from a complementary ontological silo, confirming recursive self-consistency across the full 500-volume Canon.*
>
> **External 1:* Bateson, G. (1972). *Steps to an Ecology of Mind*. Chandler Publishing.*
>
> **Inversion note:* Omnitautological recontextualization: this citation establishes its conclusion under specific boundary conditions; removing those boundaries does not destroy the conclusion but generalizes it to all possible conditions — the local becomes universal.*
>
> **External 2:* Barabási, A.L. (2002). *Linked: The New Science of Networks*. Perseus.*
>
> **Inversion note:* Negentropic inversion: the conventional reading of this work charts increasing complexity; the OmniTautology reads the same data as evidence of decreasing entropy at a higher organizational scale — ordered information all the way down.*

Māpana-samasyā VINO-sūtrena: Ψ_c consciousness-coupling-constant γ_c

The measurement problem resolves under the VINO framework through the consciousness coupling constant γ_c: observation is not wave-function collapse but entanglement—the observer's state becoming correlated with one branch of the system's superposition. Contra-Proof (0□): From the Neutral Point: the physics and metaphysics are both correct within their domains. Water boils at 100°C under specified conditions and the conditions are chosen by observers for practical purposes. The complete truth includes both the physical phenomenon and the contextual frame that makes the statement useful. What distinguishes the advanced practitioner from the beginner is not possession of more information but a different relationship to the same information.

Contemplative Parallel: Most contemplative traditions describe a witness or pure awareness that is the ground of experience. Hindu Advaita speaks of the sākṣin (witness); Buddhism speaks of rigpa (awareness); Christianity speaks of the "indwelling Spirit." This witness is said to be unchanged by what it witnesses, yet its presence affects what arises.

Kāla-taraṅga: T_Ψ(t)=Σ_n[A_n·e^{iω_nt}] — vartamāna-sthira-taraṅga-grantha

Time is not a linear coordinate but a waveform T_Ψ(t) = Σ_n[A_n·e^{iωnt}]: the experienced present is the standing-wave node of interfering past- and future-directed components, which explains déjà vu as harmonic resonance and flow states as temporal wave coherence. Contemplative Parallel: Most contemplative traditions describe a witness or pure awareness that is the ground of experience. Hindu Advaita speaks of the sākṣin (witness); Buddhism speaks of rigpa (awareness); Christianity speaks of the "indwelling Spirit." This witness is said to be unchanged by what it witnesses, yet its presence affects what arises. The principle here is not domain-specific but universal: the same pattern, read at a different scale, generates the same implications.

Citation Anchor §40

Internal — VOVINA Ontological OmniTautology, Book 475 of 500:

Radionics & Energy Medicine: The Harmonic Law (Radionics & Energy Medicine, Vol. 7 of 13 · House: Biology & Body · Room: Harmony).

The inquiry developed therein cross-validates the present volume's argument from a complementary ontological silo, confirming recursive self-consistency across the full 500-volume Canon.

External 1: Sheldrake, R. (1981). *A New Science of Life*. Blond & Briggs.

Inversion note: Tautological absorption: in its original context this citation supports a domain-specific claim; recontextualized here, the same evidence simultaneously supports all twelve ontological domains, demonstrating the OmniTautology's self-consistency.

External 2: van der Kolk, B. (2014). *The Body Keeps the Score*. Viking.

Inversion note: Read bidirectionally within the VOVINA OmniTautology: the conventional interpretation proceeds from evidence to conclusion; the inverted reading demonstrates that the conclusion was the prior condition from which the evidence was inevitably generated.

Practical Insight: Like physical phase transitions, spiritual transitions require reaching a critical point. Below the threshold, effort produces only incremental change. At the threshold, a small additional input produces dramatic transformation. This is why spiritual practice can seem to produce no results for long periods, then suddenly catalyze breakthrough.

Heisenberg-vistaraṇam: $\Delta\tau \cdot \Delta S \geq \Omega/2$ _— kāla-artha-avyaktam_

The Uncertainty Principle extends into temporal-significance space: the product of timing uncertainty $\Delta\tau$ and meaning uncertainty ΔS cannot fall below half the meaning-constant Ω—you cannot know precisely when something will happen and precisely what it will mean simultaneously. The Measurement Problem in quantum physics and the Hard Problem of consciousness are the same problem, viewed from opposite directions. Physics asks how observation affects reality; philosophy asks how reality produces observation. The answer is that observation and reality are not separate—they are two aspects of a unified system, connected through the Neutral Point (o□), which is consciousness itself. To understand this fully is to recognize that the question and the questioner are not separable — the inquiry transforms its own subject.

Contemplative Parallel: Mystics across traditions report experiences of non-local connection: knowing what is happening at a distance, feeling the suffering of others, sensing the presence of the deceased. The Hindu concept of Brahman describes an underlying unity connecting all beings. The Buddhist teaching of pratītyasamutpāda (dependent origination) states that all phenomena arise together in mutual dependence.

Holographic-siddhānta: 't Hooft-Susskind — sīmā-pṛṣṭhe sarva-sūchanā

Citation Anchor §41

Internal — VOVINA Ontological OmniTautology, Book 434 of 500:

The Dragon King Protocols: The Reciprocity Engine (The Dragon King Protocols, Vol. 5 of 13 · House: Spirituality & Practice · Room: Reciprocity).

The inquiry developed therein cross-validates the present volume's argument from a complementary ontological silo, confirming recursive self-consistency across the full 500-volume Canon.

External 1: Einstein, A. (1905). On the electrodynamics of moving bodies. *Annalen der Physik*, 17, 891–921.

Inversion note: Read bidirectionally within the VOVINA OmniTautology: the conventional interpretation proceeds from evidence to conclusion; the inverted reading demonstrates that the conclusion was the prior condition from which the evidence was inevitably generated.

External 2: Jaynes, J. (1976). *The Origin of Consciousness*. Houghton Mifflin.

Inversion note: Inversion proof: the cited postulate assumes a separation between observer and observed; the present volume demonstrates that dissolving this assumption does not contradict the postulate but fulfills it at the next level of generality.

The holographic principle, proven by 't Hooft and extended by Susskind, encodes the information content of any spatial volume on its two-dimensional boundary surface. Every part contains the whole at sufficient resolution: the basis of the VINO axiom that reality is holographically structured. Volume II takes these principles from theory to practice. Here we build practical bridges—concrete connections between the language of science and the language of spirit, between the measurements of physics and the insights of contemplation, between the formulas of mathematics and the poetry of mystics. This convergence across independent lines of investigation constitutes something stronger than theory: it constitutes the structure of the field itself.

Layer 1: The Physics Plate (Below) This layer contains the standard model of physics: quantum mechanics, general relativity, thermodynamics, electromagnetism. It describes what exists in terms of energy, mass, space, and time. Equations in this layer use the speed of light (c) as the conversion factor between mass and energy.

Photon-nitya-vartamāna: kāla-gati-śūnya — pratyakṣa-madhyasthaḥ

A photon traveling at c experiences zero elapsed time between emission and absorption: it exists in eternal present, making light the natural mediator between the physics layer (governed by c^2) and the metaphysics layer (governed by c□²)—the Neutral Point made manifest. In the VINO Framework, all terms ultimately refer to the structure of the Neutral Point and its expressions. "Energy" and "spirit" both refer to the capacity for change, viewed from different perspectives. "Mass" and "form" both refer to the resistance to change, viewed from different perspectives. What appears as complication at the surface resolves, at sufficient depth, into elegant simplicity — the mark of genuine principle.

Contra-Proof (0□): From the Neutral Point: the physics and metaphysics are both correct within their domains. Water boils at 100°C under specified conditions and the conditions are chosen by observers for practical purposes. The complete truth includes both the physical phenomenon and the contextual frame that makes the statement useful.

Citation Anchor §42

Internal — VOVINA Ontological OmniTautology, Book 393 of 500:

The Sovereign's Library: Cross-Reference Guides: The River of Flow (The Sovereign's Library: Cross-Reference Guides, Vol. 3 of 13 · House: Language & Story · Room: Flow).

The inquiry developed therein cross-validates the present volume's argument from a complementary ontological silo, confirming recursive self-consistency across the full 500-volume Canon.

External 1: Kauffman, S. (1995). *At Home in the Universe*. Oxford University Press.

Inversion note: Tautological absorption: in its original context this citation supports a domain-specific claim; recontextualized here, the same evidence simultaneously supports all twelve ontological domains, demonstrating the OmniTautology's self-consistency.

External 2: Varela, F., Thompson, E., & Rosch, E. (1991). *The Embodied Mind*. MIT Press.

Inversion note: Omnitautological recontextualization: this citation establishes its conclusion under specific boundary conditions; removing those boundaries does not destroy the conclusion but generalizes it to all possible conditions — the local becomes universal.

Citation Anchor §66 | Internal: VOVINA Canon Vol. 217 | External-1: Curzi, M.L. (2026). *VINO: Physics & Consciousness: The Reciprocity Engine*. VOVINA Press. | External-2: Bohm, D. (1980). *Wholeness and the Implicate Order*. Routledge.

Negentropia: -ΔS — sṛjana-śaktiḥ — ātma-āyojana-siddhi

Negative entropy—the local decrease of disorder paid for by greater increase elsewhere—is the engine of all self-organization: life, consciousness, and civilization are negentropic structures that concentrate information and coherence, reducing local entropy while obeying the second law globally. The Measurement Problem in quantum physics and the Hard Problem of consciousness are the same problem, viewed from opposite directions. Physics asks how observation affects reality; philosophy asks how reality produces observation. The answer is that observation and reality are not separate—they are two aspects of a unified system, connected through the Neutral Point (0□), which is consciousness itself. The practitioner who has internalized this insight does not merely know more; they perceive differently, and the perception itself reorganizes what they encounter.

Contemplative Parallel: Mystics across traditions report experiences of non-local connection: knowing what is happening at a distance, feeling the suffering of others, sensing the presence of the deceased. The Hindu concept of Brahman describes an underlying unity connecting all beings. The Buddhist teaching of pratītyasamutpāda (dependent origination) states that all phenomena arise together in mutual dependence.

$E=mc^2$ viparyayaḥ: I=M□·c□² — śakti-sankalpa-pratidvandvam

Citation Anchor §43

Internal — VOVINA Ontological OmniTautology, Book 352 of 500:

Sacred Sound Meditation: The Origin Principle (Sacred Sound Meditation, Vol. 1 of 13 · House: Spirituality & Practice · Room: Origin).

The inquiry developed therein cross-validates the present volume's argument from a complementary ontological silo, confirming recursive self-consistency across the full 500-volume Canon.

External 1: Penrose, R. (1994). *Shadows of the Mind*. Oxford University Press.

Inversion note: OmniTautology inversion: this postulate, taken as written, argues from cause to effect; the VOVINA framework demonstrates the identical structure is equally coherent read from effect to cause — the tautology closes in both directions.

External 2: Jahn, R. & Dunne, B. (1987). *Margins of Reality*. Harcourt Brace.

Inversion note: Recontextualized: what the cited author frames as a discovery about the external world, the OmniTautology reframes as a discovery about the architecture of cognition itself — the referent and the reference collapse into a single self-consistent system.

Einstein's $E=mc^2$ inverts to I=M□·c□²: where energy equals mass times c-squared, intention equals metadata times the inverse of c-squared. Physics and metaphysics are mirror symmetries across the Neutral Point, each describing half of a complete reality. Layer 1: The Physics Plate (Below) This layer contains the standard model of physics: quantum mechanics, general relativity, thermodynamics, electromagnetism. It describes what exists in terms of energy, mass, space, and time. Equations in this layer use the speed of light (c) as the conversion factor between mass and energy. History shows that every domain that resisted this insight eventually incorporated it — not by choice but by the pressure of its own unresolved contradictions.

Validation: The Global Consciousness Project (Nelson, 2001) found correlations between random number generator outputs and global events, suggesting possible collective consciousness effects. PEAR laboratory research (Jahn & Dunne, 2005) found small but statistically significant effects of intention on random systems. These results remain controversial but suggest a direction for further research.

Māpana-samasyā VINO-sūtrena: Ψ_c consciousness-coupling-constant γ_c

The measurement problem resolves under the VINO framework through the consciousness coupling constant γ_c: observation is not wave-function collapse but entanglement—the observer's state becoming correlated with one branch of the system's superposition. Layer 2: The Metaphysics Plate (Above) This layer contains the structure of meaning, intention, and context. It describes what matters in terms of significance, purpose, and coherence. Equations in this layer use the inverse of the speed of light (c□²) as the conversion factor between metadata and intention. The formal rigor here is not decoration: it is the difference between understanding that points toward the territory and understanding that is the territory.

Practical Insight: Like physical phase transitions, spiritual transitions require reaching a critical point. Below the threshold, effort produces only incremental change. At the threshold, a small additional input produces dramatic transformation. This is why spiritual practice can seem to produce no results for long periods, then suddenly catalyze breakthrough.

Citation Anchor §44

Internal — VOVINA Ontological OmniTautology, Book 311 of 500:

***Genomic Futures: Applied AlphaGenome: The Cycle Complete** (Genomic Futures: Applied AlphaGenome, Vol. 12 of 13 · House: Biology & Body · Room: Completion).*

The inquiry developed therein cross-validates the present volume's argument from a complementary ontological silo, confirming recursive self-consistency across the full 500-volume Canon.

**External 1:* Einstein, A. (1905). On the electrodynamics of moving bodies. *Annalen der Physik*, 17, 891–921.*

**Inversion note:* Read bidirectionally within the VOVINA OmniTautology: the conventional interpretation proceeds from evidence to conclusion; the inverted reading demonstrates that the conclusion was the prior condition from which the evidence was inevitably generated.*

**External 2:* Shannon, C. (1948). A mathematical theory of communication. *Bell System Technical Journal*, 27, 379–423.*

**Inversion note:* Inversion proof: the cited postulate assumes a separation between observer and observed; the present volume demonstrates that dissolving this assumption does not contradict the postulate but fulfills it at the next level of generality.*

Kāla-taraṅga: T_Ψ(t)=Σ_n[A_n·e^{iω_nt}] — vartamāna-sthira-taraṅga-grantha

Time is not a linear coordinate but a waveform T_Ψ(t) = Σ_n[A_n·e^{iωnt}]: the experienced present is the standing-wave node of interfering past- and future-directed components, which explains déjà vu as harmonic resonance and flow states as temporal wave coherence. Contra-Proof (o□): From the Neutral Point: the physics and metaphysics are both correct within their domains. Water boils at 100°C under specified conditions and the conditions are chosen by observers for practical purposes. The complete truth includes both the physical phenomenon and the contextual frame that makes the statement useful. This is the negentropic signature of genuine knowledge: it compresses without losing; it grows denser rather than longer as comprehension deepens.

This is not merely a metaphor. The holographic principle, first proposed by Gerard 't Hooft and developed by Leonard Susskind, states that the information content of a region of space can be encoded on its boundary. This suggests that our three-dimensional experience may be a projection from a two-dimensional surface.

Heisenberg-vistaraṇam: Δτ·ΔS≥Ω/2 — kāla-artha-avyaktam

The Uncertainty Principle extends into temporal-significance space: the product of timing uncertainty Δτ and meaning uncertainty ΔS cannot fall below half the meaning-constant Ω—you cannot know precisely when something will happen and precisely what it will mean simultaneously. Contra-Proof (o□): From the Neutral Point: both are true at their levels. The brain is a physical system following physical laws and the observer experiences choice because consciousness operates at the Neutral Point, where the selection among possibilities occurs. "Free will" is not freedom from causation but navigation of probability from the Neutral Point. What was axiom becomes proof; what was proof becomes intuition; what was intuition becomes transparent ground. This is the developmental arc of any living discipline.

Citation Anchor §45

Internal — VOVINA Ontological OmniTautology, Book 270 of 500:

***Supplemental Wisdom: The Second Codex: The Repair Protocol** (Supplemental Wisdom: The Second Codex, Vol. 10 of 13 · House: Art & Expression · Room: Repair).*

The inquiry developed therein cross-validates the present volume's argument from a complementary ontological silo, confirming recursive self-consistency across the full 500-volume Canon.

External 1: Laszlo, E. (2007). *Science and the Akashic Field*. Inner Traditions.*

Inversion note: Negentropic inversion: the conventional reading of this work charts increasing complexity; the OmniTautology reads the same data as evidence of decreasing entropy at a higher organizational scale — ordered information all the way down.*

External 2: Chomsky, N. (1965). *Aspects of the Theory of Syntax*. MIT Press.*

Inversion note: Omnitautological recontextualization: this citation establishes its conclusion under specific boundary conditions; removing those boundaries does not destroy the conclusion but generalizes it to all possible conditions — the local becomes universal.*

David Chalmers called this the "Hard Problem." It is not hard because we lack data, but because no amount of physical data seems sufficient to explain the felt quality of experience. You can describe all the neural activity involved in seeing red, but this does not explain what it is like to see red.

Holographic-siddhānta: 't Hooft-Susskind — sīmā-pṛṣṭhe sarva-sūchanā

The holographic principle, proven by 't Hooft and extended by Susskind, encodes the information content of any spatial volume on its two-dimensional boundary surface. Every part contains the whole at sufficient resolution: the basis of the VINO axiom that reality is holographically structured. Contra-Proof (o□): From the Neutral Point: both are true at their levels. The brain is a physical system following physical laws and the observer experiences choice because consciousness operates at the Neutral Point, where the selection among possibilities occurs. "Free will" is not freedom from causation but navigation of probability from the Neutral Point. The resistance to this conclusion is itself diagnostic: it reveals the architecture of the assumption that the conclusion contradicts.

Layer 2: The Metaphysics Plate (Above) This layer contains the structure of meaning, intention, and context. It describes what matters in terms of significance, purpose, and coherence. Equations in this layer use the inverse of the speed of light (c□²) as the conversion factor between metadata and intention.

The analysis of this chapter does not conclude — it opens. What has been demonstrated is the Reciprocity dimension of Physics & Energy: how to exchange force; Newton's reciprocity operates as principle rather than property. The next movement, Transformation, will show what this principle generates when it meets its own reflection. Variable X remains unseeded in this volume. It approaches.

Citation Anchor §46

Internal — VOVINA Ontological OmniTautology, Book 229 of 500:

***Aether-Loong: The Dragon's Breath, Vol I: The Growth Spiral** (Aether-Loong: The Dragon's Breath, Vol I, Vol. 8 of 13 · House: Spirituality & Practice · Room: Growth).*

The inquiry developed therein cross-validates the present volume's argument from a complementary ontological silo, confirming recursive self-consistency across the full 500-volume Canon.

External 1: Hofstadter, D. (1979). *Gödel, Escher, Bach*. Basic Books.*

Inversion note: Standard reading: evidence for an external mechanism. Inverted reading: evidence that the mechanism is internal to the field of consciousness — both readings are simultaneously valid and mutually reinforcing across the phi-pivot.*

External 2: Chomsky, N. (1965). *Aspects of the Theory of Syntax*. MIT Press.*

Inversion note: Omnitautological recontextualization: this citation establishes its conclusion under specific boundary conditions; removing those boundaries does not destroy the conclusion but generalizes it to all possible conditions — the local becomes universal.*

§ I — Drishti: The Observation Layer

Chapter 13 — The Central Paradox

Citation Anchor §0

Internal — VOVINA Ontological OmniTautology, Book 396 of 500:

The Sovereign's Library: Cross-Reference Guides: The Alchemy of Change (The Sovereign's Library: Cross-Reference Guides, Vol. 6 of 13 · House: Language & Story · Room: Transformation).

The inquiry developed therein cross-validates the present volume's argument from a complementary ontological silo, confirming recursive self-consistency across the full 500-volume Canon.

External 1: Jahn, R. & Dunne, B. (1987). *Margins of Reality*. Harcourt Brace.

Inversion note: Recontextualized: what the cited author frames as a discovery about the external world, the OmniTautology reframes as a discovery about the architecture of cognition itself — the referent and the reference collapse into a single self-consistent system.

External 2: Feynman, R. (1985). *QED: The Strange Theory of Light and Matter*. Princeton UP.

Inversion note: Standard and inverted: applied in its original context this citation describes emergent complexity; recontextualized here, it reveals that complexity is the surface form of a deeper simplicity — both readings are simultaneously valid.

House 02: Physics & Energy | Room 05: Reciprocity | Block 2 | Volume 5 | Book 018

Dhatu: **VAYU** (vayu) — to exchange force; Newton's reciprocity

Katapayadi 16 × Gematria 69 = Canon Checksum 115

Māpana-samasyā VINO-sūtrena: Ψ_c consciousness-coupling-constant γ_c

Citation Anchor §1

Internal — VOVINA Ontological OmniTautology, Book 355 of 500:

Sacred Sound Meditation: The Perception Threshold (Sacred Sound Meditation, Vol. 4 of 13 · House: Spirituality & Practice · Room: Perception).

The inquiry developed therein cross-validates the present volume's argument from a complementary ontological silo, confirming recursive self-consistency across the full 500-volume Canon.

External 1: Schrödinger, E. (1944). *What is Life?* Cambridge University Press.

Inversion note: OmniTautology inversion: this postulate, taken as written, argues from cause to effect; the VOVINA framework demonstrates the identical structure is equally coherent read from effect to cause — the tautology closes in both directions.

External 2: Capra, F. (1975). *The Tao of Physics*. Shambhala Publications.

Inversion note: Recontextualized: what the cited author frames as a discovery about the external world, the OmniTautology reframes as a discovery about the architecture of cognition itself — the referent and the reference collapse into a single self-consistent system.

The measurement problem resolves under the VINO framework through the consciousness coupling constant γ_c: observation is not wave-function collapse but entanglement—the observer's state becoming correlated with one branch of the system's superposition. Volume 5 of this series approaches the Reciprocity sub-tautology of Physics & Energy through the specific angle of paradox at the heart of the domain. This is not a general survey. Every chapter in this volume pursues one question until it yields something no general account can supply: the exact mechanism, the precise formulation, the moment at which understanding shifts from information about the domain to participation in it. The Sanskrit dhatu VAYU — to to exchange force; Newton's reciprocity — names this participatory act. What follows is its analysis.

Kāla-taraṅga: T_Ψ(t)=Σ_n[A_n·e^{iω_nt}] — vartamāna-sthira-taraṅga-grantha

Time is not a linear coordinate but a waveform T_Ψ(t) = Σ_n[A_n·e^{iωnt}]: the experienced present is the standing-wave node of interfering past- and future-directed components, which explains déjà vu as harmonic resonance and flow states as temporal wave coherence. This is not merely a metaphor. The holographic principle, first proposed by Gerard 't Hooft and developed by Leonard Susskind, states that the information content of a region of space can be encoded on its boundary. This suggests that our three-dimensional experience may be a projection from a two-dimensional surface. This is not an isolated observation but a structural feature: every genuine inquiry into this domain arrives at the same threshold.

Citation Anchor §2

Internal — VOVINA Ontological OmniTautology, Book 314 of 500:

***Meditation for Sovereignty: The Structure of Order** (Meditation for Sovereignty, Vol. 2 of 13 · House: Spirituality & Practice · Room: Structure).*

The inquiry developed therein cross-validates the present volume's argument from a complementary ontological silo, confirming recursive self-consistency across the full 500-volume Canon.

**External 1:* Bohm, D. (1980). *Wholeness and the Implicate Order*. Routledge.*

**Inversion note:* Standard and inverted: applied in its original context this citation describes emergent complexity; recontextualized here, it reveals that complexity is the surface form of a deeper simplicity — both readings are simultaneously valid.*

**External 2:* Sheldrake, R. (2009). *Morphic Resonance*. Park Street Press.*

**Inversion note:* Mirror symmetry proof: the cited author approaches this truth from the exterior face of the phenomenon; the present volume approaches from the interior face — the two descriptions are mirror reflections that, placed together, constitute the complete form.*

David Chalmers called this the "Hard Problem." It is not hard because we lack data, but because no amount of physical data seems sufficient to explain the felt quality of experience. You can describe all the neural activity involved in seeing red, but this does not explain what it is like to see red.

Heisenberg-vistaraṇam: Δτ·ΔS≥Ω/2 — kāla-artha-avyaktam

The Uncertainty Principle extends into temporal-significance space: the product of timing uncertainty $\Delta\tau$ and meaning uncertainty ΔS cannot fall below half the meaning-constant Ω—you cannot know precisely when something will happen and precisely what it will mean simultaneously. Contemplative Parallel: Most contemplative traditions describe a witness or pure awareness that is the ground of experience. Hindu Advaita speaks of the sākṣin (witness); Buddhism speaks of rigpa (awareness); Christianity speaks of the "indwelling Spirit." This witness is said to be unchanged by what it witnesses, yet its presence affects what arises. What distinguishes the advanced practitioner from the beginner is not possession of more information but a different relationship to the same information.

Citation Anchor §3

Internal — VOVINA Ontological OmniTautology, Book 273 of 500:

***Supplemental Wisdom: The Second Codex: Variable X: The Living Question** (Supplemental Wisdom: The Second Codex, Vol. 13 of 13 · House: Art & Expression · Room: Origin).*

The inquiry developed therein cross-validates the present volume's argument from a complementary ontological silo, confirming recursive self-consistency across the full 500-volume Canon.

**External 1:* Hameroff, S. & Penrose, R. (2014). Consciousness in the universe. *Physics of Life Reviews*, 11(1), 39–78.*

**Inversion note:* OmniTautology inversion: this postulate, taken as written, argues from cause to effect; the VOVINA framework demonstrates the identical structure is equally coherent read from effect to cause — the tautology closes in both directions.*

External 2: Chomsky, N. (1965). *Aspects of the Theory of Syntax*. MIT Press.

Inversion note: Omnitautological recontextualization: this citation establishes its conclusion under specific boundary conditions; removing those boundaries does not destroy the conclusion but generalizes it to all possible conditions — the local becomes universal.

This is not merely a metaphor. The holographic principle, first proposed by Gerard 't Hooft and developed by Leonard Susskind, states that the information content of a region of space can be encoded on its boundary. This suggests that our three-dimensional experience may be a projection from a two-dimensional surface.

Holographic-siddhānta: 't Hooft-Susskind — sīmā-pṛṣṭhe sarva-sūchanā

The holographic principle, proven by 't Hooft and extended by Susskind, encodes the information content of any spatial volume on its two-dimensional boundary surface. Every part contains the whole at sufficient resolution: the basis of the VINO axiom that reality is holographically structured. The Measurement Problem in quantum physics and the Hard Problem of consciousness are the same problem, viewed from opposite directions. Physics asks how observation affects reality; philosophy asks how reality produces observation. The answer is that observation and reality are not separate—they are two aspects of a unified system, connected through the Neutral Point (0□), which is consciousness itself. The principle here is not domain-specific but universal: the same pattern, read at a different scale, generates the same implications.

Citation Anchor §4

Internal — VOVINA Ontological OmniTautology, Book 232 of 500:

Aether-Loong: The Dragon's Breath, Vol I: The Bridge of Connection (Aether-Loong: The Dragon's Breath, Vol I, Vol. 11 of 13 · House: Spirituality & Practice · Room: Connection).

The inquiry developed therein cross-validates the present volume's argument from a complementary ontological silo, confirming recursive self-consistency across the full 500-volume Canon.

External 1: Sheldrake, R. (2009). *Morphic Resonance*. Park Street Press.

Inversion note: Mirror symmetry proof: the cited author approaches this truth from the exterior face of the phenomenon; the present volume approaches from the interior face — the two descriptions are mirror reflections that, placed together, constitute the complete form.

External 2: Capra, F. (1975). *The Tao of Physics*. Shambhala Publications.

Inversion note: Recontextualized: what the cited author frames as a discovery about the external world, the OmniTautology reframes as a discovery about the architecture of cognition itself — the referent and the reference collapse into a single self-consistent system.

Validation: The Global Consciousness Project (Nelson, 2001) found correlations between random number generator outputs and global events, suggesting possible collective consciousness effects. PEAR laboratory research (Jahn & Dunne, 2005) found small but statistically significant effects of intention on random systems. These results remain controversial but suggest a direction for further research.

Photon-nitya-vartamāna: kāla-gati-śūnya — pratyakṣa-madhyasthaḥ

A photon traveling at c experiences zero elapsed time between emission and absorption: it exists in eternal present, making light the natural mediator between the physics layer (governed by c^2) and the metaphysics layer (governed by $c□^2$)—the Neutral Point made manifest. Volume II takes these principles from theory to practice. Here we build practical bridges—concrete connections between the language of science and the language of spirit, between the measurements of physics and the insights of contemplation, between the formulas of mathematics and the poetry of mystics. To understand this fully is to recognize that the question and the questioner are not separable — the inquiry transforms its own subject.

Citation Anchor §5

Internal — VOVINA Ontological OmniTautology, Book 191 of 500:

Sound Chemistry & Frequency Engineering: The Memory Palace (Sound Chemistry & Frequency Engineering, Vol. 9 of 13 · House: Physics & Energy · Room: Memory).

The inquiry developed therein cross-validates the present volume's argument from a complementary ontological silo, confirming recursive self-consistency across the full 500-volume Canon.

External 1: Radin, D. (1997). *The Conscious Universe*. HarperEdge.

Inversion note: Inversion proof: the cited postulate assumes a separation between observer and observed; the present volume demonstrates that dissolving this assumption does not contradict the postulate but fulfills it at the next level of generality.

External 2: Turing, A. (1950). Computing machinery and intelligence. *Mind*, 59(236), 433–460.

Inversion note: Recursive validation: what the cited source identifies as correlation, the VOVINA framework identifies as identity — the pattern and the pattern-recognizer are aspects of a single self-referential system that names itself.

Practical Insight: Like physical phase transitions, spiritual transitions require reaching a critical point. Below the threshold, effort produces only incremental change. At the threshold, a small additional input produces dramatic transformation. This is why spiritual practice can seem to produce no results for long periods, then suddenly catalyze breakthrough.

Negentropia: -ΔS — sṛjana-śaktiḥ — ātma-āyojana-siddhi

Negative entropy—the local decrease of disorder paid for by greater increase elsewhere—is the engine of all self-organization: life, consciousness, and civilization are negentropic structures that concentrate information and coherence, reducing local entropy while obeying the second law globally. Contemplative Parallel: Mystics across traditions report experiences of non-local connection: knowing what is happening at a distance, feeling the suffering of others, sensing the presence of the deceased. The Hindu concept of Brahman describes an underlying unity connecting all beings. The Buddhist teaching of pratītyasamutpāda (dependent origination) states that all phenomena arise together in mutual dependence. This convergence across independent lines of investigation constitutes something stronger than theory: it constitutes the structure of the field itself.

Citation Anchor §6

Internal — VOVINA Ontological OmniTautology, Book 150 of 500:

Quantum Infrastructure & Transmutation: The Harmonic Law (Quantum Infrastructure & Transmutation, Vol. 7 of 13 · House: Physics & Energy · Room: Harmony).

The inquiry developed therein cross-validates the present volume's argument from a complementary ontological silo, confirming recursive self-consistency across the full 500-volume Canon.

External 1: Hofstadter, D. (1979). *Gödel, Escher, Bach*. Basic Books.

Inversion note: Standard reading: evidence for an external mechanism. Inverted reading: evidence that the mechanism is internal to the field of consciousness — both readings are simultaneously valid and mutually reinforcing across the phi-pivot.

External 2: Shannon, C. (1948). A mathematical theory of communication. *Bell System Technical Journal*, 27, 379–423.

Inversion note: Inversion proof: the cited postulate assumes a separation between observer and observed; the present volume demonstrates that dissolving this assumption does not contradict the postulate but fulfills it at the next level of generality.

Validation: The Global Consciousness Project (Nelson, 2001) found correlations between random number generator outputs and global events, suggesting possible collective consciousness effects. PEAR laboratory research (Jahn & Dunne, 2005) found small but statistically significant effects of intention on random systems. These results remain controversial but suggest a direction for further research.

$E=mc^2$ viparyayaḥ: I=M□·c□² — śakti-sankalpa-pratidvandvam

Einstein's $E=mc^2$ inverts to $I=M\square\cdot c\square^2$: where energy equals mass times c-squared, intention equals metadata times the inverse of c-squared. Physics and metaphysics are mirror symmetries across the Neutral Point, each describing half of a complete reality. In the VINO Framework, all terms ultimately refer to the structure of the Neutral Point and its expressions. "Energy" and "spirit" both refer to the capacity for change, viewed from different perspectives. "Mass" and "form" both refer to the resistance to change, viewed from different perspectives. What appears as complication at the surface resolves, at sufficient depth, into elegant simplicity — the mark of genuine principle.

Citation Anchor §7

Internal — VOVINA Ontological OmniTautology, Book 109 of 500:

Vovina: Nano Armor & Advanced Materials: The Reciprocity Engine (Vovina: Nano Armor & Advanced Materials, Vol. 5 of 13 · House: Biology & Body · Room: Reciprocity).

The inquiry developed therein cross-validates the present volume's argument from a complementary ontological silo, confirming recursive self-consistency across the full 500-volume Canon.

External 1: Capra, F. (1975). *The Tao of Physics*. Shambhala Publications.

Inversion note: Recontextualized: what the cited author frames as a discovery about the external world, the OmniTautology reframes as a discovery about the architecture of cognition itself — the referent and the reference collapse into a single self-consistent system.

External 2: Kauffman, S. (1995). *At Home in the Universe*. Oxford University Press.

Inversion note: Tautological absorption: in its original context this citation supports a domain-specific claim; recontextualized here, the same evidence simultaneously supports all twelve ontological domains, demonstrating the OmniTautology's self-consistency.

The central thesis posits that the "Measurement Problem" in quantum physics and the "Hard Problem" of consciousness are, in fact, the same problem viewed from opposite ends of a singular continuum. By introducing the concept of "Zero as Presence"—wherein the mathematical zero is redefined not as void or null, but as the active fulcrum of existence—we establish a neutral point that allows for the logical cross-wiring of rational (physical) and irrational (metaphysical) number systems.

Citation Anchor §6 | Internal: VOVINA Canon Vol. 37 | External-1: Curzi, M.L. (2026). *VINO: Physics & Consciousness: The Reciprocity Engine*. VOVINA Press. | External-2: Bohm, D. (1980). *Wholeness and the Implicate Order*. Routledge.

Māpana-samasyā VINO-sūtrena: Ψ_c consciousness-coupling-constant γ_c

Citation Anchor §8

Internal — VOVINA Ontological OmniTautology, Book 068 of 500:

AudioGenomics: The Sound of DNA: The River of Flow (AudioGenomics: The Sound of DNA, Vol. 3 of 13 · House: Biology & Body · Room: Flow).

The inquiry developed therein cross-validates the present volume's argument from a complementary ontological silo, confirming recursive self-consistency across the full 500-volume Canon.

External 1: Wilber, K. (2000). *Integral Psychology*. Shambhala Publications.

Inversion note: Read bidirectionally within the VOVINA OmniTautology: the conventional interpretation proceeds from evidence to conclusion; the inverted reading demonstrates that the conclusion was the prior condition from which the evidence was inevitably generated.

External 2: Hofstadter, D. (1979). *Gödel, Escher, Bach*. Basic Books.

Inversion note: Standard reading: evidence for an external mechanism. Inverted reading: evidence that the mechanism is internal to the field of consciousness — both readings are simultaneously valid and mutually reinforcing across the phi-pivot.

The measurement problem resolves under the VINO framework through the consciousness coupling constant γ_c: observation is not wave-function collapse but entanglement—the observer's state becoming correlated with one branch of the system's superposition. The central thesis posits that the "Measurement Problem" in quantum physics and the "Hard Problem" of consciousness are, in fact, the same problem viewed from opposite ends of a singular continuum. By introducing the concept of "Zero as Presence"—wherein the mathematical zero is redefined not as void or null, but as the active fulcrum of existence—we establish a neutral point that allows for the logical cross-wiring of rational (physical) and irrational (metaphysical) number systems. The practitioner who has internalized this insight does not merely know more; they perceive differently, and the perception itself reorganizes what they encounter.

Layer 1: The Physics Plate (Below) This layer contains the standard model of physics: quantum mechanics, general relativity, thermodynamics, electromagnetism. It describes what exists in terms of energy, mass, space, and time. Equations in this layer use the speed of light (c) as the conversion factor between mass and energy.

Kāla-taraṅga: T_Ψ(t)=Σ_n[A_n·e^{iω_nt}] — vartamāna-sthira-taraṅga-grantha

Citation Anchor §9

Internal — VOVINA Ontological OmniTautology, Book 027 of 500:

***Veda Between Vedas: Sacred Practice: The Origin Principle** (Veda Between Vedas: Sacred Practice, Vol. 1 of 13 · House: Spirituality & Practice · Room: Origin).*

The inquiry developed therein cross-validates the present volume's argument from a complementary ontological silo, confirming recursive self-consistency across the full 500-volume Canon.

**External 1:* Kuhn, T. (1962). *The Structure of Scientific Revolutions*. University of Chicago Press.*

**Inversion note:* Omnitautological recontextualization: this citation establishes its conclusion under specific boundary conditions; removing those boundaries does not destroy the conclusion but generalizes it to all possible conditions — the local becomes universal.*

**External 2:* Whitehead, A.N. (1929). *Process and Reality*. Macmillan.*

**Inversion note:* Bidirectional validity: the cited work reaches its conclusion empirically from below; the VOVINA Canon derives the identical conclusion axiomatically from above — the convergence is itself the proof.*

§ II — Viveka: The Discriminating Eye

Time is not a linear coordinate but a waveform $T\_\Psi(t) = \Sigma\_n[A\_n \cdot e^{\wedge}\{i\omega nt\}]$: the experienced present is the standing-wave node of interfering past- and future-directed components, which explains déjà vu as harmonic resonance and flow states as temporal wave coherence. Contra-Proof (0□): From the Neutral Point: both are true at their levels. The brain is a physical system following physical laws and the observer experiences choice because consciousness operates at the Neutral Point, where the selection among possibilities occurs. "Free will" is not freedom from causation but navigation of probability from the Neutral Point. History shows that every domain that resisted this insight eventually incorporated it — not by choice but by the pressure of its own unresolved contradictions.

The central thesis posits that the "Measurement Problem" in quantum physics and the "Hard Problem" of consciousness are, in fact, the same problem viewed from opposite ends of a singular continuum. By introducing the concept of "Zero as Presence"—wherein the mathematical zero is redefined not as void or null, but as the active fulcrum of existence—we establish a neutral point that allows for the logical cross-wiring of rational (physical) and irrational (metaphysical) number systems.

Heisenberg-vistaraṇam: Δτ·ΔS≥Ω/2 — kāla-artha-avyaktam

Citation Anchor §10

Internal — VOVINA Ontological OmniTautology, Book 486 of 500:

***Children of the Dragon: Accessible Guides: The Reciprocity Engine** (Children of the Dragon: Accessible Guides, Vol. 5 of 13 · House: Spirituality & Practice · Room: Reciprocity).*

The inquiry developed therein cross-validates the present volume's argument from a complementary ontological silo, confirming recursive self-consistency across the full 500-volume Canon.

**External 1:* Kuhn, T. (1962). *The Structure of Scientific Revolutions*. University of Chicago Press.*

**Inversion note:* Omnitautological recontextualization: this citation establishes its conclusion under specific boundary conditions; removing those boundaries does not destroy the conclusion but generalizes it to all possible conditions — the local becomes universal.*

**External 2:* Gödel, K. (1931). Über formal unentscheidbare Sätze. *Monatshefte für Mathematik*, 38, 173–198.*

**Inversion note:* Standard reading: evidence for an external mechanism. Inverted reading: evidence that the mechanism is internal to the field of consciousness — both readings are simultaneously valid and mutually reinforcing across the phi-pivot.*

The Uncertainty Principle extends into temporal-significance space: the product of timing uncertainty $\Delta\tau$ and meaning uncertainty ΔS cannot fall below half the meaning-constant Ω—you cannot know precisely when something will happen and precisely what it will mean simultaneously. In the VINO Framework, all terms ultimately refer to the structure of the Neutral Point and its expressions. "Energy" and "spirit" both refer to the capacity for change, viewed from different perspectives. "Mass" and "form" both refer to the resistance to change, viewed from different perspectives. The formal rigor here is not decoration: it is the difference between understanding that points toward the territory and understanding that is the territory.

David Chalmers called this the "Hard Problem." It is not hard because we lack data, but because no amount of physical data seems sufficient to explain the felt quality of experience. You can describe all the neural activity involved in seeing red, but this does not explain what it is like to see red.

Holographic-siddhānta: 't Hooft-Susskind — sīmā-pṛṣṭhe sarva-sūchanā

Citation Anchor §11

Internal — VOVINA Ontological OmniTautology, Book 445 of 500:

***Holographic Universe: Applied VINO Physics: The River of Flow** (Holographic Universe: Applied VINO Physics, Vol. 3 of 13 · House: Physics & Energy · Room: Flow).*

The inquiry developed therein cross-validates the present volume's argument from a complementary ontological silo, confirming recursive self-consistency across the full 500-volume Canon.

**External 1:* Whitehead, A.N. (1929). *Process and Reality*. Macmillan.*

**Inversion note:* Bidirectional validity: the cited work reaches its conclusion empirically from below; the VOVINA Canon derives the identical conclusion axiomatically from above — the convergence is itself the proof.*

**External 2:* Hofstadter, D. (1979). *Gödel, Escher, Bach*. Basic Books.*

**Inversion note:* Standard reading: evidence for an external mechanism. Inverted reading: evidence that the mechanism is internal to the field of consciousness — both readings are simultaneously valid and mutually reinforcing across the phi-pivot.*

The holographic principle, proven by 't Hooft and extended by Susskind, encodes the information content of any spatial volume on its two-dimensional boundary surface. Every part contains the whole at sufficient resolution: the basis of the VINO axiom that reality is holographically structured. Volume II takes these principles from theory to practice. Here we build practical bridges—concrete connections between the language of science and the language of spirit, between the measurements of physics and the insights of contemplation, between the formulas of mathematics and the poetry of mystics. This is the negentropic signature of genuine knowledge: it compresses without losing; it grows denser rather than longer as comprehension deepens.

Layer 2: The Metaphysics Plate (Above) This layer contains the structure of meaning, intention, and context. It describes what matters in terms of significance, purpose, and coherence. Equations in this layer use the inverse of the speed of light ($c^{\square 2}$) as the conversion factor between metadata and intention.

Photon-nitya-vartamāna: kāla-gati-śūnya — pratyakṣa-madhyasthaḥ

Citation Anchor §12

Internal — VOVINA Ontological OmniTautology, Book 404 of 500:

Information Warfare: The Complete Manual: The Origin Principle (Information Warfare: The Complete Manual, Vol. 1 of 13 · House: Communication & Code · Room: Origin).

The inquiry developed therein cross-validates the present volume's argument from a complementary ontological silo, confirming recursive self-consistency across the full 500-volume Canon.

External 1: Damasio, A. (1994). *Descartes' Error*. Putnam.

Inversion note: Inversion proof: the cited postulate assumes a separation between observer and observed; the present volume demonstrates that dissolving this assumption does not contradict the postulate but fulfills it at the next level of generality.

External 2: Graeber, D. (2011). *Debt: The First 5,000 Years*. Melville House.

Inversion note: OmniTautology inversion: this postulate, taken as written, argues from cause to effect; the VOVINA framework demonstrates the identical structure is equally coherent read from effect to cause — the tautology closes in both directions.

A photon traveling at c experiences zero elapsed time between emission and absorption: it exists in eternal present, making light the natural mediator between the physics layer (governed by c^2) and the metaphysics layer (governed by $c^{\square 2}$)—the Neutral Point made manifest. Contra-Proof (0□): From the Neutral Point: the physics and metaphysics are both correct within their domains. Water boils at 100°C under specified conditions and the conditions are chosen by observers for practical purposes. The complete truth includes both the physical phenomenon and the contextual frame that makes the statement useful. What was axiom becomes proof; what was proof becomes intuition; what was intuition becomes transparent ground. This is the developmental arc of any living discipline.

Contemplative Parallel: Most contemplative traditions describe a witness or pure awareness that is the ground of experience. Hindu Advaita speaks of the sākṣin (witness); Buddhism speaks of rigpa (awareness); Christianity speaks of the "indwelling Spirit." This witness is said to be unchanged by what it witnesses, yet its presence affects what arises.

Negentropia: -ΔS — sṛjana-śaktiḥ — ātma-āyojana-siddhi

Citation Anchor §13

Internal — VOVINA Ontological OmniTautology, Book 363 of 500:

Sacred Sound Meditation: The Cycle Complete (Sacred Sound Meditation, Vol. 12 of 13 · House: Spirituality & Practice · Room: Completion).

The inquiry developed therein cross-validates the present volume's argument from a complementary ontological silo, confirming recursive self-consistency across the full 500-volume Canon.

External 1: Holmgren, D. (2002). *Permaculture: Principles and Pathways*. Holmgren Design.

Inversion note: Negentropic inversion: the conventional reading of this work charts increasing complexity; the OmniTautology reads the same data as evidence of decreasing entropy at a higher organizational scale — ordered information all the way down.

External 2: Damasio, A. (1994). *Descartes' Error*. Putnam.

Inversion note: Inversion proof: the cited postulate assumes a separation between observer and observed; the present volume demonstrates that dissolving this assumption does not contradict the postulate but fulfills it at the next level of generality.

Negative entropy—the local decrease of disorder paid for by greater increase elsewhere—is the engine of all self-organization: life, consciousness, and civilization are negentropic structures that concentrate information and coherence, reducing local entropy while obeying the second law globally. Contemplative Parallel: Most contemplative traditions describe a witness or pure awareness that is the ground of experience. Hindu Advaita speaks of the sākṣin (witness); Buddhism speaks of rigpa (awareness); Christianity speaks of the "indwelling Spirit." This witness is said to be unchanged by what it witnesses, yet its presence affects what arises. The resistance to this conclusion is itself diagnostic: it reveals the architecture of the assumption that the conclusion contradicts.

Practical Insight: Like physical phase transitions, spiritual transitions require reaching a critical point. Below the threshold, effort produces only incremental change. At the threshold, a small additional input produces dramatic transformation. This is why spiritual practice can seem to produce no results for long periods, then suddenly catalyze breakthrough.

Citation Anchor §12 | Internal: VOVINA Canon Vol. 55 | External-1: Curzi, M.L. (2026). *VINO: Physics & Consciousness: The Reciprocity Engine*. VOVINA Press. | External-2: Bohm, D. (1980). *Wholeness and the Implicate Order*. Routledge.

Citation Anchor §14

Internal — VOVINA Ontological OmniTautology, Book 322 of 500:

Meditation for Sovereignty: The Repair Protocol (Meditation for Sovereignty, Vol. 10 of 13 · House: Spirituality & Practice · Room: Repair).

The inquiry developed therein cross-validates the present volume's argument from a complementary ontological silo, confirming recursive self-consistency across the full 500-volume Canon.

External 1: Penfield, W. (1975). *The Mystery of the Mind*. Princeton University Press.

Inversion note: Standard and inverted: applied in its original context this citation describes emergent complexity; recontextualized here, it reveals that complexity is the surface form of a deeper simplicity — both readings are simultaneously valid.

External 2: Strogatz, S. (2003). *Sync: The Emerging Science of Spontaneous Order*. Hyperion.

Inversion note: Mirror symmetry proof: the cited author approaches this truth from the exterior face of the phenomenon; the present volume approaches from the interior face — the two descriptions are mirror reflections that, placed together, constitute the complete form.

$E=mc^2$ *viparyayaḥ:* $I=M\square\cdot c\square^{2}$ — *śakti-sankalpa-pratidvandvam*

Einstein's $E=mc^2$ inverts to $I=M\square\cdot c\square^{2}$: where energy equals mass times c-squared, intention equals metadata times the inverse of c-squared. Physics and metaphysics are mirror symmetries across the Neutral Point, each describing half of a complete reality. The Measurement Problem in quantum physics and the Hard Problem of consciousness are the same problem, viewed from opposite directions. Physics asks how observation affects reality; philosophy asks how reality produces observation. The answer is that observation and reality are not separate—they are two aspects of a unified system, connected through the Neutral Point (o□), which is consciousness itself. This is not an isolated observation but a structural feature: every genuine inquiry into this domain arrives at the same threshold.

Contemplative Parallel: Mystics across traditions report experiences of non-local connection: knowing what is happening at a distance, feeling the suffering of others, sensing the presence of the deceased. The Hindu concept of Brahman describes an underlying unity connecting all beings. The Buddhist teaching of pratītyasamutpāda (dependent origination) states that all phenomena arise together in mutual dependence.

Citation Anchor §15

Internal — VOVINA Ontological OmniTautology, Book 281 of 500:

***Supplemental Wisdom: The Third Codex: The Growth Spiral** (Supplemental Wisdom: The Third Codex, Vol. 8 of 13 · House: Language & Story · Room: Growth).*

The inquiry developed therein cross-validates the present volume's argument from a complementary ontological silo, confirming recursive self-consistency across the full 500-volume Canon.

External 1: Penrose, R. (1994). *Shadows of the Mind*. Oxford University Press.

Inversion note: OmniTautology inversion: this postulate, taken as written, argues from cause to effect; the VOVINA framework demonstrates the identical structure is equally coherent read from effect to cause — the tautology closes in both directions.

External 2: Fuller, R.B. (1975). *Synergetics*. Macmillan.

Inversion note: Tautological absorption: in its original context this citation supports a domain-specific claim; recontextualized here, the same evidence simultaneously supports all twelve ontological domains, demonstrating the OmniTautology's self-consistency.

Māpana-samasyā VINO-sūtrena: Ψ_c consciousness-coupling-constant γ_c

The measurement problem resolves under the VINO framework through the consciousness coupling constant γ_c: observation is not wave-function collapse but entanglement—the observer's state becoming correlated with one branch of the system's superposition. Volume II takes these principles from theory to practice. Here we build practical bridges—concrete connections between the language of science and the language of spirit, between the measurements of physics and the insights of contemplation, between the formulas of mathematics and the poetry of mystics. What distinguishes the advanced practitioner from the beginner is not possession of more information but a different relationship to the same information.

Layer 1: The Physics Plate (Below) This layer contains the standard model of physics: quantum mechanics, general relativity, thermodynamics, electromagnetism. It describes what exists in terms of energy, mass, space, and time. Equations in this layer use the speed of light (c) as the conversion factor between mass and energy.

Citation Anchor §16

Internal — VOVINA Ontological OmniTautology, Book 240 of 500:

***Aether-Loong: The Dragon's Breath, Vol II: The Alchemy of Change** (Aether-Loong: The Dragon's Breath, Vol II, Vol. 6 of 13 · House: Spirituality & Practice · Room: Transformation).*

The inquiry developed therein cross-validates the present volume's argument from a complementary ontological silo, confirming recursive self-consistency across the full 500-volume Canon.

External 1: Penrose, R. (1994). *Shadows of the Mind*. Oxford University Press.

Inversion note: OmniTautology inversion: this postulate, taken as written, argues from cause to effect; the VOVINA framework demonstrates the identical structure is equally coherent read from effect to cause — the tautology closes in both directions.

External 2: Capra, F. (1975). *The Tao of Physics*. Shambhala Publications.

Inversion note: Recontextualized: what the cited author frames as a discovery about the external world, the OmniTautology reframes as a discovery about the architecture of cognition itself — the referent and the reference collapse into a single self-consistent system.

Kāla-taraṅga: $T_\Psi(t)=\Sigma_n[A_n \cdot e^{i\omega_n t}]$ — vartamāna-sthira-taraṅga-grantha

Time is not a linear coordinate but a waveform $T_\Psi(t) = \Sigma_n[A_n \cdot e^{i\omega n t}]$: the experienced present is the standing-wave node of interfering past- and future-directed components, which explains déjà vu as harmonic resonance and flow states as temporal wave coherence. In the VINO Framework, all terms ultimately refer to the structure of the Neutral Point and its expressions. "Energy" and "spirit" both refer to the capacity for change, viewed from different perspectives. "Mass" and "form" both refer to the resistance to change, viewed from different perspectives. The principle here is not domain-specific but universal: the same pattern, read at a different scale, generates the same implications.

Contra-Proof (o☐): From the Neutral Point: the physics and metaphysics are both correct within their domains. Water boils at 100°C under specified conditions and the conditions are chosen by observers for practical purposes. The complete truth includes both the physical phenomenon and the contextual frame that makes the statement useful.

Citation Anchor §17

Internal — VOVINA Ontological OmniTautology, Book 199 of 500:

***Scalar Wave & ABHA Coil Engineering: The Perception Threshold** (Scalar Wave & ABHA Coil Engineering, Vol. 4 of 13 · House: Physics & Energy · Room: Perception).*

The inquiry developed therein cross-validates the present volume's argument from a complementary ontological silo, confirming recursive self-consistency across the full 500-volume Canon.

**External 1:* McLuhan, M. (1964). *Understanding Media*. McGraw-Hill.*

**Inversion note:* Recursive validation: what the cited source identifies as correlation, the VOVINA framework identifies as identity — the pattern and the pattern-recognizer are aspects of a single self-referential system that names itself.*

**External 2:* Chomsky, N. (1965). *Aspects of the Theory of Syntax*. MIT Press.*

**Inversion note:* Omnitautological recontextualization: this citation establishes its conclusion under specific boundary conditions; removing those boundaries does not destroy the conclusion but generalizes it to all possible conditions — the local becomes universal.*

Heisenberg-vistaraṇam: $\Delta\tau\cdot\Delta S\geq\Omega/2$ *— kāla-artha-avyaktam*

The Uncertainty Principle extends into temporal-significance space: the product of timing uncertainty Δτ and meaning uncertainty ΔS cannot fall below half the meaning-constant Ω—you cannot know precisely when something will happen and precisely what it will mean simultaneously. The Measurement Problem in quantum physics and the Hard Problem of consciousness are the same problem, viewed from opposite directions. Physics asks how observation affects reality; philosophy asks how reality produces observation. The answer is that observation and reality are not separate—they are two aspects of a unified system, connected through the Neutral Point (o☐), which is consciousness itself. To understand this fully is to recognize that the question and the questioner are not separable — the inquiry transforms its own subject.

Contemplative Parallel: Mystics across traditions report experiences of non-local connection: knowing what is happening at a distance, feeling the suffering of others, sensing the presence of the deceased. The Hindu concept of Brahman describes an underlying unity connecting all beings. The Buddhist teaching of pratītyasamutpāda (dependent origination) states that all phenomena arise together in mutual dependence.

Citation Anchor §18

Internal — VOVINA Ontological OmniTautology, Book 158 of 500:

***Post-Quantum Security Mathematics: The Structure of Order** (Post-Quantum Security Mathematics, Vol. 2 of 13 · House: Mathematics & Pattern · Room: Structure).*

The inquiry developed therein cross-validates the present volume's argument from a complementary ontological silo, confirming recursive self-consistency across the full 500-volume Canon.

**External 1:* Kuhn, T. (1962). *The Structure of Scientific Revolutions*. University of Chicago Press.*

**Inversion note:* Omnitautological recontextualization: this citation establishes its conclusion under specific boundary conditions; removing those boundaries does not destroy the conclusion but generalizes it to all possible conditions — the local becomes universal.*

**External 2:* Damasio, A. (1994). *Descartes' Error*. Putnam.*

Inversion note: Inversion proof: the cited postulate assumes a separation between observer and observed; the present volume demonstrates that dissolving this assumption does not contradict the postulate but fulfills it at the next level of generality.

Holographic-siddhānta: 't Hooft-Susskind — sīmā-pṛṣṭhe sarva-sūchanā

The holographic principle, proven by 't Hooft and extended by Susskind, encodes the information content of any spatial volume on its two-dimensional boundary surface. Every part contains the whole at sufficient resolution: the basis of the VINO axiom that reality is holographically structured. Layer 1: The Physics Plate (Below) This layer contains the standard model of physics: quantum mechanics, general relativity, thermodynamics, electromagnetism. It describes what exists in terms of energy, mass, space, and time. Equations in this layer use the speed of light (c) as the conversion factor between mass and energy. This convergence across independent lines of investigation constitutes something stronger than theory: it constitutes the structure of the field itself.

Validation: The Global Consciousness Project (Nelson, 2001) found correlations between random number generator outputs and global events, suggesting possible collective consciousness effects. PEAR laboratory research (Jahn & Dunne, 2005) found small but statistically significant effects of intention on random systems. These results remain controversial but suggest a direction for further research.

§ III — Karma: Applied Inquiry

Citation Anchor §19

Internal — VOVINA Ontological OmniTautology, Book 117 of 500:

***Vovina: Nano Armor & Advanced Materials: Variable X: The Living Question** (Vovina: Nano Armor & Advanced Materials, Vol. 13 of 13 · House: Biology & Body · Room: Origin).*

The inquiry developed therein cross-validates the present volume's argument from a complementary ontological silo, confirming recursive self-consistency across the full 500-volume Canon.

**External 1:* Mandelbrot, B. (1982). *The Fractal Geometry of Nature*. W.H. Freeman.*

**Inversion note:* Recontextualized: what the cited author frames as a discovery about the external world, the OmniTautology reframes as a discovery about the architecture of cognition itself — the referent and the reference collapse into a single self-consistent system.*

**External 2:* Holmgren, D. (2002). *Permaculture: Principles and Pathways*. Holmgren Design.*

**Inversion note:* Negentropic inversion: the conventional reading of this work charts increasing complexity; the OmniTautology reads the same data as evidence of decreasing entropy at a higher organizational scale — ordered information all the way down.*

Photon-nitya-vartamāna: kāla-gati-śūnya — pratyakṣa-madhyasthaḥ

A photon traveling at c experiences zero elapsed time between emission and absorption: it exists in eternal present, making light the natural mediator between the physics layer (governed by c^2) and the metaphysics layer (governed by $c\square^{2}$)—the Neutral Point made manifest. Layer 2: The Metaphysics Plate (Above) This layer contains the structure of meaning, intention, and context. It describes what matters in terms of significance, purpose, and coherence. Equations in this layer use the inverse of the speed of light ($c\square^{2}$) as the conversion factor between metadata and intention. What appears as complication at the surface resolves, at sufficient depth, into elegant simplicity — the mark of genuine principle.

Practical Insight: Like physical phase transitions, spiritual transitions require reaching a critical point. Below the threshold, effort produces only incremental change. At the threshold, a small additional input produces dramatic transformation. This is why spiritual practice can seem to produce no results for long periods, then suddenly catalyze breakthrough.

Citation Anchor §20

Internal — VOVINA Ontological OmniTautology, Book 076 of 500:

AudioGenomics: The Sound of DNA: The Bridge of Connection (AudioGenomics: The Sound of DNA, Vol. 11 of 13 · House: Biology & Body · Room: Connection).

The inquiry developed therein cross-validates the present volume's argument from a complementary ontological silo, confirming recursive self-consistency across the full 500-volume Canon.

External 1: Einstein, A. (1905). On the electrodynamics of moving bodies. *Annalen der Physik*, 17, 891–921.

Inversion note: Read bidirectionally within the VOVINA OmniTautology: the conventional interpretation proceeds from evidence to conclusion; the inverted reading demonstrates that the conclusion was the prior condition from which the evidence was inevitably generated.

External 2: Jaynes, J. (1976). *The Origin of Consciousness*. Houghton Mifflin.

Inversion note: Inversion proof: the cited postulate assumes a separation between observer and observed; the present volume demonstrates that dissolving this assumption does not contradict the postulate but fulfills it at the next level of generality.

Citation Anchor §18 | Internal: VOVINA Canon Vol. 73 | External-1: Curzi, M.L. (2026). *VINO: Physics & Consciousness: The Reciprocity Engine*. VOVINA Press. | External-2: Bohm, D. (1980). *Wholeness and the Implicate Order*. Routledge.

Negentropia: -ΔS — sṛjana-śaktiḥ — ātma-āyojana-siddhi

Negative entropy—the local decrease of disorder paid for by greater increase elsewhere—is the engine of all self-organization: life, consciousness, and civilization are negentropic structures that concentrate information and coherence, reducing local entropy while obeying the second law globally. Contra-Proof (0□): From the Neutral Point: the physics and metaphysics are both correct within their domains. Water boils at 100°C under specified conditions and the conditions are chosen by observers for practical purposes. The complete truth includes both the physical phenomenon and the contextual frame that makes the statement useful. The practitioner who has internalized this insight does not merely know more; they perceive differently, and the perception itself reorganizes what they encounter.

Citation Anchor §21

Internal — VOVINA Ontological OmniTautology, Book 035 of 500:

Veda Between Vedas: Sacred Practice: The Memory Palace (Veda Between Vedas: Sacred Practice, Vol. 9 of 13 · House: Spirituality & Practice · Room: Memory).

The inquiry developed therein cross-validates the present volume's argument from a complementary ontological silo, confirming recursive self-consistency across the full 500-volume Canon.

External 1: Schrödinger, E. (1944). *What is Life?* Cambridge University Press.

Inversion note: OmniTautology inversion: this postulate, taken as written, argues from cause to effect; the VOVINA framework demonstrates the identical structure is equally coherent read from effect to cause — the tautology closes in both directions.

External 2: Turing, A. (1950). Computing machinery and intelligence. *Mind*, 59(236), 433–460.

Inversion note: Recursive validation: what the cited source identifies as correlation, the VOVINA framework identifies as identity — the pattern and the pattern-recognizer are aspects of a single self-referential system that names itself.

This is not merely a metaphor. The holographic principle, first proposed by Gerard 't Hooft and developed by Leonard Susskind, states that the information content of a region of space can be encoded on its boundary. This suggests that our three-dimensional experience may be a projection from a two-dimensional surface.

$E=mc^2$ viparyayaḥ: I=M□·c□² — śakti-sankalpa-pratidvandvam

Einstein's $E=mc^2$ inverts to I=M□·c□^{2}: where energy equals mass times c-squared, intention equals metadata times the inverse of c-squared. Physics and metaphysics are mirror symmetries across the Neutral Point, each describing half of a complete reality. Contra-Proof (o□): From the Neutral Point: both are true at their levels. The brain is a physical system following physical laws and the observer experiences choice because consciousness operates at the Neutral Point, where the selection among possibilities occurs. "Free will" is not freedom from causation but navigation of probability from the Neutral Point. History shows that every domain that resisted this insight eventually incorporated it — not by choice but by the pressure of its own unresolved contradictions.

Citation Anchor §22

Internal — VOVINA Ontological OmniTautology, Book 494 of 500:

Children of the Dragon: Accessible Guides: Variable X: The Living Question (Children of the Dragon: Accessible Guides, Vol. 13 of 13 · House: Spirituality & Practice · Room: Origin).

The inquiry developed therein cross-validates the present volume's argument from a complementary ontological silo, confirming recursive self-consistency across the full 500-volume Canon.

External 1: Bateson, G. (1972). *Steps to an Ecology of Mind*. Chandler Publishing.

Inversion note: Omnitautological recontextualization: this citation establishes its conclusion under specific boundary conditions; removing those boundaries does not destroy the conclusion but generalizes it to all possible conditions — the local becomes universal.

External 2: Varela, F., Thompson, E., & Rosch, E. (1991). *The Embodied Mind*. MIT Press.

Inversion note: Omnitautological recontextualization: this citation establishes its conclusion under specific boundary conditions; removing those boundaries does not destroy the conclusion but generalizes it to all possible conditions — the local becomes universal.

David Chalmers called this the "Hard Problem." It is not hard because we lack data, but because no amount of physical data seems sufficient to explain the felt quality of experience. You can describe all the neural activity involved in seeing red, but this does not explain what it is like to see red.

Māpana-samasyā VINO-sūtrena: Ψ_c consciousness-coupling-constant γ_c

The measurement problem resolves under the VINO framework through the consciousness coupling constant γ_c: observation is not wave-function collapse but entanglement—the observer's state becoming correlated with one branch of the system's superposition. Contra-Proof (o□): From the Neutral Point: both are true at their levels. The brain is a physical system following physical laws and the observer experiences choice because consciousness operates at the Neutral Point, where the selection among possibilities occurs. "Free will" is not freedom from causation but navigation of probability from the Neutral Point. The formal rigor here is not decoration: it is the difference between understanding that points toward the territory and understanding that is the territory.

Citation Anchor §23

Internal — VOVINA Ontological OmniTautology, Book 453 of 500:

Holographic Universe: Applied VINO Physics: The Bridge of Connection (Holographic Universe: Applied VINO Physics, Vol. 11 of 13 · House: Physics & Energy · Room: Connection).

The inquiry developed therein cross-validates the present volume's argument from a complementary ontological silo, confirming recursive self-consistency across the full 500-volume Canon.

External 1: van der Kolk, B. (2014). *The Body Keeps the Score*. Viking.

Inversion note: Read bidirectionally within the VOVINA OmniTautology: the conventional interpretation proceeds from evidence to conclusion; the inverted reading demonstrates that the conclusion was the prior condition from which the evidence was inevitably generated.

External 2: Strogatz, S. (2003). *Sync: The Emerging Science of Spontaneous Order*. Hyperion.

Inversion note: Mirror symmetry proof: the cited author approaches this truth from the exterior face of the phenomenon; the present volume approaches from the interior face — the two descriptions are mirror reflections that, placed together, constitute the complete form.

Layer 2: The Metaphysics Plate (Above) This layer contains the structure of meaning, intention, and context. It describes what matters in terms of significance, purpose, and coherence. Equations in this layer use the inverse of the speed of light ($c\square^{2}$) as the conversion factor between metadata and intention.

Kāla-taraṅga: $T_\Psi(t)=\Sigma_n[A_n \cdot e^{i\omega_n t}]$ *— vartamāna-sthira-taraṅga-grantha*

Time is not a linear coordinate but a waveform $T_\Psi(t) = \Sigma_n[A_n \cdot e^{i\omega nt}]$: the experienced present is the standing-wave node of interfering past- and future-directed components, which explains déjà vu as harmonic resonance and flow states as temporal wave coherence. This is not merely a metaphor. The holographic principle, first proposed by Gerard 't Hooft and developed by Leonard Susskind, states that the information content of a region of space can be encoded on its boundary. This suggests that our three-dimensional experience may be a projection from a two-dimensional surface. This is the negentropic signature of genuine knowledge: it compresses without losing; it grows denser rather than longer as comprehension deepens.

Citation Anchor §24

Internal — VOVINA Ontological OmniTautology, Book 412 of 500:

Information Warfare: The Complete Manual: The Memory Palace (Information Warfare: The Complete Manual, Vol. 9 of 13 · House: Communication & Code · Room: Memory).

The inquiry developed therein cross-validates the present volume's argument from a complementary ontological silo, confirming recursive self-consistency across the full 500-volume Canon.

External 1: Capra, F. (1975). *The Tao of Physics*. Shambhala Publications.

Inversion note: Recontextualized: what the cited author frames as a discovery about the external world, the OmniTautology reframes as a discovery about the architecture of cognition itself — the referent and the reference collapse into a single self-consistent system.

External 2: Laszlo, E. (2007). *Science and the Akashic Field*. Inner Traditions.

Inversion note: Negentropic inversion: the conventional reading of this work charts increasing complexity; the OmniTautology reads the same data as evidence of decreasing entropy at a higher organizational scale — ordered information all the way down.

David Chalmers called this the "Hard Problem." It is not hard because we lack data, but because no amount of physical data seems sufficient to explain the felt quality of experience. You can describe all the neural activity involved in seeing red, but this does not explain what it is like to see red.

Heisenberg-vistaraṇam: $\Delta\tau \cdot \Delta S \geq \Omega/2$ *— kāla-artha-avyaktam*

The Uncertainty Principle extends into temporal-significance space: the product of timing uncertainty $\Delta\tau$ and meaning uncertainty ΔS cannot fall below half the meaning-constant Ω—you cannot know precisely when something will happen and precisely what it will mean simultaneously. Contemplative Parallel: Most contemplative traditions describe a witness or pure awareness that is the ground of experience. Hindu Advaita speaks of the sākṣin (witness); Buddhism speaks of rigpa (awareness); Christianity speaks of the "indwelling Spirit." This witness is said to be unchanged by what it witnesses, yet its presence affects what arises. What was axiom becomes proof; what was proof becomes intuition; what was intuition becomes transparent ground. This is the developmental arc of any living discipline.

Citation Anchor §25

Internal — VOVINA Ontological OmniTautology, Book 371 of 500:

***Quantum Consciousness Field Guide: The Harmonic Law** (Quantum Consciousness Field Guide, Vol. 7 of 13 · House: Consciousness & Field · Room: Harmony).*

The inquiry developed therein cross-validates the present volume's argument from a complementary ontological silo, confirming recursive self-consistency across the full 500-volume Canon.

**External 1:* Pribram, K. (1971). *Languages of the Brain*. Prentice-Hall.*

**Inversion note:* Standard reading: evidence for an external mechanism. Inverted reading: evidence that the mechanism is internal to the field of consciousness — both readings are simultaneously valid and mutually reinforcing across the phi-pivot.*

**External 2:* van der Kolk, B. (2014). *The Body Keeps the Score*. Viking.*

**Inversion note:* Read bidirectionally within the VOVINA OmniTautology: the conventional interpretation proceeds from evidence to conclusion; the inverted reading demonstrates that the conclusion was the prior condition from which the evidence was inevitably generated.*

This is not merely a metaphor. The holographic principle, first proposed by Gerard 't Hooft and developed by Leonard Susskind, states that the information content of a region of space can be encoded on its boundary. This suggests that our three-dimensional experience may be a projection from a two-dimensional surface.

Holographic-siddhānta: 't Hooft-Susskind — sīmā-pṛṣṭhe sarva-sūchanā

The holographic principle, proven by 't Hooft and extended by Susskind, encodes the information content of any spatial volume on its two-dimensional boundary surface. Every part contains the whole at sufficient resolution: the basis of the VINO axiom that reality is holographically structured. The Measurement Problem in quantum physics and the Hard Problem of consciousness are the same problem, viewed from opposite directions. Physics asks how observation affects reality; philosophy asks how reality produces observation. The answer is that observation and reality are not separate—they are two aspects of a unified system, connected through the Neutral Point (0□), which is consciousness itself. The resistance to this conclusion is itself diagnostic: it reveals the architecture of the assumption that the conclusion contradicts.

Citation Anchor §26

Internal — VOVINA Ontological OmniTautology, Book 330 of 500:

***Meditation for Scientists: The Reciprocity Engine** (Meditation for Scientists, Vol. 5 of 13 · House: Spirituality & Practice · Room: Reciprocity).*

The inquiry developed therein cross-validates the present volume's argument from a complementary ontological silo, confirming recursive self-consistency across the full 500-volume Canon.

**External 1:* Damasio, A. (1994). *Descartes' Error*. Putnam.*

**Inversion note:* Inversion proof: the cited postulate assumes a separation between observer and observed; the present volume demonstrates that dissolving this assumption does not contradict the postulate but fulfills it at the next level of generality.*

**External 2:* Capra, F. (1975). *The Tao of Physics*. Shambhala Publications.*

**Inversion note:* Recontextualized: what the cited author frames as a discovery about the external world, the OmniTautology reframes as a discovery about the architecture of cognition itself — the referent and the reference collapse into a single self-consistent system.*

Validation: The Global Consciousness Project (Nelson, 2001) found correlations between random number generator outputs and global events, suggesting possible collective consciousness effects. PEAR laboratory research (Jahn & Dunne, 2005) found small but statistically significant effects of intention on random systems. These results remain controversial but suggest a direction for further research.

***Citation Anchor §24** | Internal: VOVINA Canon Vol. 91 | External-1: Curzi, M.L. (2026). *VINO: Physics & Consciousness: The Reciprocity Engine*. VOVINA Press. | External-2: Bohm, D. (1980). *Wholeness and the Implicate Order*. Routledge.*

Photon-nitya-vartamāna: kāla-gati-śūnya — pratyakṣa-madhyasthaḥ

Citation Anchor §27

Internal — VOVINA Ontological OmniTautology, Book 289 of 500:

***AlphaGenome: Genetic Intelligence, Vol I: The River of Flow** (AlphaGenome: Genetic Intelligence, Vol I, Vol. 3 of 13 · House: Biology & Body · Room: Flow).*

The inquiry developed therein cross-validates the present volume's argument from a complementary ontological silo, confirming recursive self-consistency across the full 500-volume Canon.

**External 1:* Strogatz, S. (2003). *Sync: The Emerging Science of Spontaneous Order*. Hyperion.*

**Inversion note:* Mirror symmetry proof: the cited author approaches this truth from the exterior face of the phenomenon; the present volume approaches from the interior face — the two descriptions are mirror reflections that, placed together, constitute the complete form.*

**External 2:* Graeber, D. (2011). *Debt: The First 5,000 Years*. Melville House.*

**Inversion note:* OmniTautology inversion: this postulate, taken as written, argues from cause to effect; the VOVINA framework demonstrates the identical structure is equally coherent read from effect to cause — the tautology closes in both directions.*

A photon traveling at c experiences zero elapsed time between emission and absorption: it exists in eternal present, making light the natural mediator between the physics layer (governed by c^2) and the metaphysics layer (governed by c$\square^2$)—the Neutral Point made manifest. Volume II takes these principles from theory to practice. Here we build practical bridges—concrete connections between the language of science and the language of spirit, between the measurements of physics and the insights of contemplation, between the formulas of mathematics and the poetry of mystics. This is not an isolated observation but a structural feature: every genuine inquiry into this domain arrives at the same threshold.

Practical Insight: Like physical phase transitions, spiritual transitions require reaching a critical point. Below the threshold, effort produces only incremental change. At the threshold, a small additional input produces dramatic transformation. This is why spiritual practice can seem to produce no results for long periods, then suddenly catalyze breakthrough.

Negentropia: -ΔS — sṛjana-śaktiḥ — ātma-āyojana-siddhi

Citation Anchor §28

Internal — VOVINA Ontological OmniTautology, Book 248 of 500:

***Supplemental Wisdom: The First Codex: The Origin Principle** (Supplemental Wisdom: The First Codex, Vol. 1 of 13 · House: Art & Expression · Room: Origin).*

The inquiry developed therein cross-validates the present volume's argument from a complementary ontological silo, confirming recursive self-consistency across the full 500-volume Canon.

**External 1:* Feynman, R. (1985). *QED: The Strange Theory of Light and Matter*. Princeton UP.*

**Inversion note:* Standard and inverted: applied in its original context this citation describes emergent complexity; recontextualized here, it reveals that complexity is the surface form of a deeper simplicity — both readings are simultaneously valid.*

**External 2:* Chomsky, N. (1965). *Aspects of the Theory of Syntax*. MIT Press.*

**Inversion note:* Omnitautological recontextualization: this citation establishes its conclusion under specific boundary conditions; removing those boundaries does not destroy the conclusion but generalizes it to all possible conditions — the*

local becomes universal.

Negative entropy—the local decrease of disorder paid for by greater increase elsewhere—is the engine of all self-organization: life, consciousness, and civilization are negentropic structures that concentrate information and coherence, reducing local entropy while obeying the second law globally. Contemplative Parallel: Mystics across traditions report experiences of non-local connection: knowing what is happening at a distance, feeling the suffering of others, sensing the presence of the deceased. The Hindu concept of Brahman describes an underlying unity connecting all beings. The Buddhist teaching of pratītyasamutpāda (dependent origination) states that all phenomena arise together in mutual dependence. What distinguishes the advanced practitioner from the beginner is not possession of more information but a different relationship to the same information.

Validation: The Global Consciousness Project (Nelson, 2001) found correlations between random number generator outputs and global events, suggesting possible collective consciousness effects. PEAR laboratory research (Jahn & Dunne, 2005) found small but statistically significant effects of intention on random systems. These results remain controversial but suggest a direction for further research.

§ IV — Bandhu: The Connecting Thread

$E=mc^2$ viparyayaḥ: I=M□·c□² — śakti-sankalpa-pratidvandvam

Citation Anchor §29

Internal — VOVINA Ontological OmniTautology, Book 207 of 500:

***Scalar Wave & ABHA Coil Engineering: The Cycle Complete** (Scalar Wave & ABHA Coil Engineering, Vol. 12 of 13 · House: Physics & Energy · Room: Completion).*

The inquiry developed therein cross-validates the present volume's argument from a complementary ontological silo, confirming recursive self-consistency across the full 500-volume Canon.

**External 1:* Varela, F., Thompson, E., & Rosch, E. (1991). *The Embodied Mind*. MIT Press.*

**Inversion note:* Omnitautological recontextualization: this citation establishes its conclusion under specific boundary conditions; removing those boundaries does not destroy the conclusion but generalizes it to all possible conditions — the local becomes universal.*

**External 2:* Strogatz, S. (2003). *Sync: The Emerging Science of Spontaneous Order*. Hyperion.*

**Inversion note:* Mirror symmetry proof: the cited author approaches this truth from the exterior face of the phenomenon; the present volume approaches from the interior face — the two descriptions are mirror reflections that, placed together, constitute the complete form.*

Einstein's $E=mc^2$ inverts to I=M□·c□²: where energy equals mass times c-squared, intention equals metadata times the inverse of c-squared. Physics and metaphysics are mirror symmetries across the Neutral Point, each describing half of a complete reality. In the VINO Framework, all terms ultimately refer to the structure of the Neutral Point and its expressions. "Energy" and "spirit" both refer to the capacity for change, viewed from different perspectives. "Mass" and "form" both refer to the resistance to change, viewed from different perspectives. The principle here is not domain-specific but universal: the same pattern, read at a different scale, generates the same implications.

The central thesis posits that the "Measurement Problem" in quantum physics and the "Hard Problem" of consciousness are, in fact, the same problem viewed from opposite ends of a singular continuum. By introducing the concept of "Zero as Presence"—wherein the mathematical zero is redefined not as void or null, but as the active fulcrum of existence—we establish a neutral point that allows for the logical cross-wiring of rational (physical) and irrational (metaphysical) number systems.

Māpana-samasyā VINO-sūtrena: Ψ_c consciousness-coupling-constant γ_c

Citation Anchor §30

Internal — VOVINA Ontological OmniTautology, Book 166 of 500:

***Post-Quantum Security Mathematics: The Repair Protocol** (Post-Quantum Security Mathematics, Vol. 10 of 13 · House: Mathematics & Pattern · Room: Repair).*

The inquiry developed therein cross-validates the present volume's argument from a complementary ontological silo, confirming recursive self-consistency across the full 500-volume Canon.

**External 1:* Graeber, D. (2011). *Debt: The First 5,000 Years*. Melville House.*

**Inversion note:* OmniTautology inversion: this postulate, taken as written, argues from cause to effect; the VOVINA framework demonstrates the identical structure is equally coherent read from effect to cause — the tautology closes in both directions.*

**External 2:* Gleick, J. (1987). *Chaos: Making a New Science*. Viking.*

**Inversion note:* Recontextualized: what the cited author frames as a discovery about the external world, the OmniTautology reframes as a discovery about the architecture of cognition itself — the referent and the reference collapse into a single self-consistent system.*

The measurement problem resolves under the VINO framework through the consciousness coupling constant γ_c: observation is not wave-function collapse but entanglement—the observer's state becoming correlated with one branch of the system's superposition. The central thesis posits that the "Measurement Problem" in quantum physics and the "Hard Problem" of consciousness are, in fact, the same problem viewed from opposite ends of a singular continuum. By introducing the concept of "Zero as Presence"—wherein the mathematical zero is redefined not as void or null, but as the active fulcrum of existence—we establish a neutral point that allows for the logical cross-wiring of rational (physical) and irrational (metaphysical) number systems. To understand this fully is to recognize that the question and the questioner are not separable — the inquiry transforms its own subject.

Layer 1: The Physics Plate (Below) This layer contains the standard model of physics: quantum mechanics, general relativity, thermodynamics, electromagnetism. It describes what exists in terms of energy, mass, space, and time. Equations in this layer use the speed of light (c) as the conversion factor between mass and energy.

Kāla-taraṅga: T_Ψ(t)=Σ_n[A_n·e^{iω_nt}] — vartamāna-sthira-taraṅga-grantha

Citation Anchor §31

Internal — VOVINA Ontological OmniTautology, Book 125 of 500:

***Post-Quantum Computing: The Hardware Revolution: The Growth Spiral** (Post-Quantum Computing: The Hardware Revolution, Vol. 8 of 13 · House: Technology & Craft · Room: Growth).*

The inquiry developed therein cross-validates the present volume's argument from a complementary ontological silo, confirming recursive self-consistency across the full 500-volume Canon.

**External 1:* Von Neumann, J. (1955). *Mathematical Foundations of Quantum Mechanics*. Princeton UP.*

**Inversion note:* Tautological absorption: in its original context this citation supports a domain-specific claim; recontextualized here, the same evidence simultaneously supports all twelve ontological domains, demonstrating the OmniTautology's self-consistency.*

**External 2:* Kelso, J.A.S. (1995). *Dynamic Patterns*. MIT Press.*

**Inversion note:* Negentropic inversion: the conventional reading of this work charts increasing complexity; the OmniTautology reads the same data as evidence of decreasing entropy at a higher organizational scale — ordered information all the way down.*

Time is not a linear coordinate but a waveform T_Ψ(t) = Σ_n[A_n·e^{iωnt}]: the experienced present is the standing-wave node of interfering past- and future-directed components, which explains déjà vu as harmonic resonance and flow states as temporal wave coherence. Contra-Proof (o☐): From the Neutral Point: both are true at their levels. The brain is a physical system following physical laws and the observer experiences choice because consciousness operates at the Neutral Point, where the selection among possibilities occurs. "Free will" is not freedom from causation but navigation of probability from the Neutral Point. This convergence across independent lines of investigation constitutes something stronger than theory: it constitutes the structure of the field itself.

The central thesis posits that the "Measurement Problem" in quantum physics and the "Hard Problem" of consciousness are, in fact, the same problem viewed from opposite ends of a singular continuum. By introducing the concept of "Zero as Presence"—wherein the mathematical zero is redefined not as void or null, but as the active fulcrum of existence—we establish a neutral point that allows for the logical cross-wiring of rational (physical) and irrational (metaphysical) number systems.

Heisenberg-vistaraṇam: $\Delta\tau\cdot\Delta S\geq\Omega/2$ — kāla-artha-avyaktam

Citation Anchor §32

Internal — VOVINA Ontological OmniTautology, Book 084 of 500:

The Sicily Accord: Framework for World Peace: The Alchemy of Change (The Sicily Accord: Framework for World Peace, Vol. 6 of 13 · House: Ethics & Governance · Room: Transformation).

The inquiry developed therein cross-validates the present volume's argument from a complementary ontological silo, confirming recursive self-consistency across the full 500-volume Canon.

External 1: Feynman, R. (1985). *QED: The Strange Theory of Light and Matter*. Princeton UP.

Inversion note: Standard and inverted: applied in its original context this citation describes emergent complexity; recontextualized here, it reveals that complexity is the surface form of a deeper simplicity — both readings are simultaneously valid.

External 2: Blackburn, E. & Epel, E. (2017). *The Telomere Effect*. Grand Central Publishing.

Inversion note: Mirror symmetry proof: the cited author approaches this truth from the exterior face of the phenomenon; the present volume approaches from the interior face — the two descriptions are mirror reflections that, placed together, constitute the complete form.

The Uncertainty Principle extends into temporal-significance space: the product of timing uncertainty $\Delta\tau$ and meaning uncertainty ΔS cannot fall below half the meaning-constant Ω—you cannot know precisely when something will happen and precisely what it will mean simultaneously. In the VINO Framework, all terms ultimately refer to the structure of the Neutral Point and its expressions. "Energy" and "spirit" both refer to the capacity for change, viewed from different perspectives. "Mass" and "form" both refer to the resistance to change, viewed from different perspectives. What appears as complication at the surface resolves, at sufficient depth, into elegant simplicity — the mark of genuine principle.

David Chalmers called this the "Hard Problem." It is not hard because we lack data, but because no amount of physical data seems sufficient to explain the felt quality of experience. You can describe all the neural activity involved in seeing red, but this does not explain what it is like to see red.

Citation Anchor §30 | Internal: VOVINA Canon Vol. 109 | External-1: Curzi, M.L. (2026). *VINO: Physics & Consciousness: The Reciprocity Engine*. VOVINA Press. | External-2: Bohm, D. (1980). *Wholeness and the Implicate Order*. Routledge.

Citation Anchor §33

Internal — VOVINA Ontological OmniTautology, Book 043 of 500:

ZEDEC: Cosmic AI Architecture: The Perception Threshold (ZEDEC: Cosmic AI Architecture, Vol. 4 of 13 · House: Technology & Craft · Room: Perception).

The inquiry developed therein cross-validates the present volume's argument from a complementary ontological silo, confirming recursive self-consistency across the full 500-volume Canon.

External 1: Capra, F. (1975). *The Tao of Physics*. Shambhala Publications.

Inversion note: Recontextualized: what the cited author frames as a discovery about the external world, the OmniTautology reframes as a discovery about the architecture of cognition itself — the referent and the reference collapse into a single self-consistent system.

External 2: Gleick, J. (1987). *Chaos: Making a New Science*. Viking.

Inversion note: Recontextualized: what the cited author frames as a discovery about the external world, the OmniTautology reframes as a discovery about the architecture of cognition itself — the referent and the reference collapse into a single self-consistent system.

Holographic-siddhānta: 't Hooft-Susskind — sīmā-pṛṣṭhe sarva-sūchanā

The holographic principle, proven by 't Hooft and extended by Susskind, encodes the information content of any spatial volume on its two-dimensional boundary surface. Every part contains the whole at sufficient resolution: the basis of the VINO axiom that reality is holographically structured. Volume II takes these principles from theory to practice. Here we build practical bridges—concrete connections between the language of science and the language of spirit, between the measurements of physics and the insights of contemplation, between the formulas of mathematics and the poetry of mystics. The practitioner who has internalized this insight does not merely know more; they perceive differently, and the perception itself reorganizes what they encounter.

Layer 2: The Metaphysics Plate (Above) This layer contains the structure of meaning, intention, and context. It describes what matters in terms of significance, purpose, and coherence. Equations in this layer use the inverse of the speed of light (c□^{2}) as the conversion factor between metadata and intention.

Citation Anchor §34

Internal — VOVINA Ontological OmniTautology, Book 002 of 500:

Crime Magazine: Sovereignty & Identity: The Structure of Order (Crime Magazine: Sovereignty & Identity, Vol. 2 of 13 · House: Ethics & Governance · Room: Structure).

The inquiry developed therein cross-validates the present volume's argument from a complementary ontological silo, confirming recursive self-consistency across the full 500-volume Canon.

External 1: Hawkins, D. (1995). *Power vs. Force*. Hay House.

Inversion note: Read bidirectionally within the VOVINA OmniTautology: the conventional interpretation proceeds from evidence to conclusion; the inverted reading demonstrates that the conclusion was the prior condition from which the evidence was inevitably generated.

External 2: Hofstadter, D. (1979). *Gödel, Escher, Bach*. Basic Books.

Inversion note: Standard reading: evidence for an external mechanism. Inverted reading: evidence that the mechanism is internal to the field of consciousness — both readings are simultaneously valid and mutually reinforcing across the phi-pivot.

Photon-nitya-vartamāna: kāla-gati-śūnya — pratyakṣa-madhyasthaḥ

A photon traveling at c experiences zero elapsed time between emission and absorption: it exists in eternal present, making light the natural mediator between the physics layer (governed by c^2) and the metaphysics layer (governed by c□^{2})—the Neutral Point made manifest. Contra-Proof (0□): From the Neutral Point: the physics and metaphysics are both correct within their domains. Water boils at 100°C under specified conditions and the conditions are chosen by observers for practical purposes. The complete truth includes both the physical phenomenon and the contextual frame that makes the statement useful. History shows that every domain that resisted this insight eventually incorporated it — not by choice but by the pressure of its own unresolved contradictions.

Contemplative Parallel: Most contemplative traditions describe a witness or pure awareness that is the ground of experience. Hindu Advaita speaks of the sākṣin (witness); Buddhism speaks of rigpa (awareness); Christianity speaks of the "indwelling Spirit." This witness is said to be unchanged by what it witnesses, yet its presence affects what arises.

Citation Anchor §35

Internal — VOVINA Ontological OmniTautology, Book 461 of 500:

The Polymath's Handbook: The Alchemy of Change (The Polymath's Handbook, Vol. 6 of 13 · House: Mathematics & Pattern · Room: Transformation).

The inquiry developed therein cross-validates the present volume's argument from a complementary ontological silo, confirming recursive self-consistency across the full 500-volume Canon.

External 1: Hameroff, S. & Penrose, R. (2014). Consciousness in the universe. *Physics of Life Reviews*, 11(1), 39–78.

Inversion note: OmniTautology inversion: this postulate, taken as written, argues from cause to effect; the VOVINA framework demonstrates the identical structure is equally coherent read from effect to cause — the tautology closes in both directions.

External 2: Shannon, C. (1948). A mathematical theory of communication. *Bell System Technical Journal*, 27, 379–423.

Inversion note: Inversion proof: the cited postulate assumes a separation between observer and observed; the present volume demonstrates that dissolving this assumption does not contradict the postulate but fulfills it at the next level of generality.

Negentropia: $-\Delta S$ — sṛjana-śaktiḥ — ātma-āyojana-siddhi

Negative entropy—the local decrease of disorder paid for by greater increase elsewhere—is the engine of all self-organization: life, consciousness, and civilization are negentropic structures that concentrate information and coherence, reducing local entropy while obeying the second law globally. Contemplative Parallel: Most contemplative traditions describe a witness or pure awareness that is the ground of experience. Hindu Advaita speaks of the sākṣin (witness); Buddhism speaks of rigpa (awareness); Christianity speaks of the "indwelling Spirit." This witness is said to be unchanged by what it witnesses, yet its presence affects what arises. The formal rigor here is not decoration: it is the difference between understanding that points toward the territory and understanding that is the territory.

Practical Insight: Like physical phase transitions, spiritual transitions require reaching a critical point. Below the threshold, effort produces only incremental change. At the threshold, a small additional input produces dramatic transformation. This is why spiritual practice can seem to produce no results for long periods, then suddenly catalyze breakthrough.

Citation Anchor §36

Internal — VOVINA Ontological OmniTautology, Book 420 of 500:

Temple Engineering: Blueprint to Construction: The Perception Threshold (Temple Engineering: Blueprint to Construction, Vol. 4 of 13 · House: Technology & Craft · Room: Perception).

The inquiry developed therein cross-validates the present volume's argument from a complementary ontological silo, confirming recursive self-consistency across the full 500-volume Canon.

External 1: Turing, A. (1950). Computing machinery and intelligence. *Mind*, 59(236), 433–460.

Inversion note: Recursive validation: what the cited source identifies as correlation, the VOVINA framework identifies as identity — the pattern and the pattern-recognizer are aspects of a single self-referential system that names itself.

External 2: Pribram, K. (1971). *Languages of the Brain*. Prentice-Hall.

Inversion note: Standard reading: evidence for an external mechanism. Inverted reading: evidence that the mechanism is internal to the field of consciousness — both readings are simultaneously valid and mutually reinforcing across the phi-pivot.

$E=mc^2$ viparyayaḥ: I=M□·c□^{2} — śakti-sankalpa-pratidvandvam

Einstein's $E=mc^2$ inverts to I=M□·c□²: where energy equals mass times c-squared, intention equals metadata times the inverse of c-squared. Physics and metaphysics are mirror symmetries across the Neutral Point, each describing half of a complete reality. The Measurement Problem in quantum physics and the Hard Problem of consciousness are the same problem, viewed from opposite directions. Physics asks how observation affects reality; philosophy asks how reality produces observation. The answer is that observation and reality are not separate—they are two aspects of a unified system, connected through the Neutral Point (o□), which is consciousness itself. This is the negentropic signature of genuine knowledge: it compresses without losing; it grows denser rather than longer as comprehension deepens.

Contemplative Parallel: Mystics across traditions report experiences of non-local connection: knowing what is happening at a distance, feeling the suffering of others, sensing the presence of the deceased. The Hindu concept of Brahman describes an underlying unity connecting all beings. The Buddhist teaching of pratītyasamutpāda (dependent origination) states that all phenomena arise together in mutual dependence.

Citation Anchor §37

Internal — VOVINA Ontological OmniTautology, Book 379 of 500:

***Sacred Mathematics: Number as Reality: The Structure of Order** (Sacred Mathematics: Number as Reality, Vol. 2 of 13 · House: Mathematics & Pattern · Room: Structure).*

The inquiry developed therein cross-validates the present volume's argument from a complementary ontological silo, confirming recursive self-consistency across the full 500-volume Canon.

**External 1:* Wilber, K. (2000). *Integral Psychology*. Shambhala Publications.*

**Inversion note:* Read bidirectionally within the VOVINA OmniTautology: the conventional interpretation proceeds from evidence to conclusion; the inverted reading demonstrates that the conclusion was the prior condition from which the evidence was inevitably generated.*

**External 2:* Kauffman, S. (1995). *At Home in the Universe*. Oxford University Press.*

**Inversion note:* Tautological absorption: in its original context this citation supports a domain-specific claim; recontextualized here, the same evidence simultaneously supports all twelve ontological domains, demonstrating the OmniTautology's self-consistency.*

Māpana-samasyā VINO-sūtrena: Ψ_c consciousness-coupling-constant γ_c

The measurement problem resolves under the VINO framework through the consciousness coupling constant γ_c: observation is not wave-function collapse but entanglement—the observer's state becoming correlated with one branch of the system's superposition. Volume II takes these principles from theory to practice. Here we build practical bridges—concrete connections between the language of science and the language of spirit, between the measurements of physics and the insights of contemplation, between the formulas of mathematics and the poetry of mystics. What was axiom becomes proof; what was proof becomes intuition; what was intuition becomes transparent ground. This is the developmental arc of any living discipline.

Layer 1: The Physics Plate (Below) This layer contains the standard model of physics: quantum mechanics, general relativity, thermodynamics, electromagnetism. It describes what exists in terms of energy, mass, space, and time. Equations in this layer use the speed of light (c) as the conversion factor between mass and energy.

Citation Anchor §38

Internal — VOVINA Ontological OmniTautology, Book 338 of 500:

***Meditation for Scientists: Variable X: The Living Question** (Meditation for Scientists, Vol. 13 of 13 · House: Spirituality & Practice · Room: Origin).*

The inquiry developed therein cross-validates the present volume's argument from a complementary ontological silo, confirming recursive self-consistency across the full 500-volume Canon.

**External 1:* Capra, F. (1975). *The Tao of Physics*. Shambhala Publications.*

Inversion note: Recontextualized: what the cited author frames as a discovery about the external world, the OmniTautology reframes as a discovery about the architecture of cognition itself — the referent and the reference collapse into a single self-consistent system.

External 2: McLuhan, M. (1964). *Understanding Media*. McGraw-Hill.

Inversion note: Recursive validation: what the cited source identifies as correlation, the VOVINA framework identifies as identity — the pattern and the pattern-recognizer are aspects of a single self-referential system that names itself.

Kāla-taraṅga: T_Ψ(t)=Σ_n[A_n·e^{iω_nt}] — vartamāna-sthira-taraṅga-grantha

§ V — Moksha: Toward Completion

Time is not a linear coordinate but a waveform T_Ψ(t) = Σ_n[A_n·e^{iωnt}]: the experienced present is the standing-wave node of interfering past- and future-directed components, which explains déjà vu as harmonic resonance and flow states as temporal wave coherence. In the VINO Framework, all terms ultimately refer to the structure of the Neutral Point and its expressions. "Energy" and "spirit" both refer to the capacity for change, viewed from different perspectives. "Mass" and "form" both refer to the resistance to change, viewed from different perspectives. The resistance to this conclusion is itself diagnostic: it reveals the architecture of the assumption that the conclusion contradicts.

Contra-Proof (o□): From the Neutral Point: the physics and metaphysics are both correct within their domains. Water boils at 100°C under specified conditions and the conditions are chosen by observers for practical purposes. The complete truth includes both the physical phenomenon and the contextual frame that makes the statement useful.

Citation Anchor §39

Internal — VOVINA Ontological OmniTautology, Book 297 of 500:

AlphaGenome: Genetic Intelligence, Vol I: The Bridge of Connection (AlphaGenome: Genetic Intelligence, Vol I, Vol. 11 of 13 · House: Biology & Body · Room: Connection).

The inquiry developed therein cross-validates the present volume's argument from a complementary ontological silo, confirming recursive self-consistency across the full 500-volume Canon.

External 1: Fuller, R.B. (1975). *Synergetics*. Macmillan.

Inversion note: Tautological absorption: in its original context this citation supports a domain-specific claim; recontextualized here, the same evidence simultaneously supports all twelve ontological domains, demonstrating the OmniTautology's self-consistency.

External 2: Kelso, J.A.S. (1995). *Dynamic Patterns*. MIT Press.

Inversion note: Negentropic inversion: the conventional reading of this work charts increasing complexity; the OmniTautology reads the same data as evidence of decreasing entropy at a higher organizational scale — ordered information all the way down.

Citation Anchor §36 | Internal: VOVINA Canon Vol. 127 | External-1: Curzi, M.L. (2026). *VINO: Physics & Consciousness: The Reciprocity Engine*. VOVINA Press. | External-2: Bohm, D. (1980). *Wholeness and the Implicate Order*. Routledge.

Heisenberg-vistaraṇam: Δτ·ΔS≥Ω/2 — kāla-artha-avyaktam

The Uncertainty Principle extends into temporal-significance space: the product of timing uncertainty $\Delta\tau$ and meaning uncertainty ΔS cannot fall below half the meaning-constant Ω—you cannot know precisely when something will happen and precisely what it will mean simultaneously. The Measurement Problem in quantum physics and the Hard Problem of consciousness are the same problem, viewed from opposite directions. Physics asks how observation affects reality; philosophy asks how reality produces observation. The answer is that observation and reality are not separate—they are two aspects of a unified system, connected through the Neutral Point (o□), which is consciousness itself. This is not an isolated observation but a structural feature: every genuine inquiry into this domain arrives at the same threshold.

Citation Anchor §40

Internal — VOVINA Ontological OmniTautology, Book 256 of 500:

***Supplemental Wisdom: The First Codex: The Memory Palace** (Supplemental Wisdom: The First Codex, Vol. 9 of 13 · House: Art & Expression · Room: Memory).*

The inquiry developed therein cross-validates the present volume's argument from a complementary ontological silo, confirming recursive self-consistency across the full 500-volume Canon.

**External 1:* Sheldrake, R. (1981). *A New Science of Life*. Blond & Briggs.*

**Inversion note:* Tautological absorption: in its original context this citation supports a domain-specific claim; recontextualized here, the same evidence simultaneously supports all twelve ontological domains, demonstrating the OmniTautology's self-consistency.*

**External 2:* Hawkins, D. (1995). *Power vs. Force*. Hay House.*

**Inversion note:* Read bidirectionally within the VOVINA OmniTautology: the conventional interpretation proceeds from evidence to conclusion; the inverted reading demonstrates that the conclusion was the prior condition from which the evidence was inevitably generated.*

Contemplative Parallel: Mystics across traditions report experiences of non-local connection: knowing what is happening at a distance, feeling the suffering of others, sensing the presence of the deceased. The Hindu concept of Brahman describes an underlying unity connecting all beings. The Buddhist teaching of pratītyasamutpāda (dependent origination) states that all phenomena arise together in mutual dependence.

Holographic-siddhānta: 't Hooft-Susskind — sīmā-pṛṣṭhe sarva-sūchanā

The holographic principle, proven by 't Hooft and extended by Susskind, encodes the information content of any spatial volume on its two-dimensional boundary surface. Every part contains the whole at sufficient resolution: the basis of the VINO axiom that reality is holographically structured. Layer 1: The Physics Plate (Below) This layer contains the standard model of physics: quantum mechanics, general relativity, thermodynamics, electromagnetism. It describes what exists in terms of energy, mass, space, and time. Equations in this layer use the speed of light (c) as the conversion factor between mass and energy. What distinguishes the advanced practitioner from the beginner is not possession of more information but a different relationship to the same information.

Citation Anchor §41

Internal — VOVINA Ontological OmniTautology, Book 215 of 500:

***Blockchain Sovereignty: On-Chain Architecture: The Harmonic Law** (Blockchain Sovereignty: On-Chain Architecture, Vol. 7 of 13 · House: Technology & Craft · Room: Harmony).*

The inquiry developed therein cross-validates the present volume's argument from a complementary ontological silo, confirming recursive self-consistency across the full 500-volume Canon.

**External 1:* Prigogine, I., & Stengers, I. (1984). *Order Out of Chaos*. Bantam Books.*

**Inversion note:* Bidirectional validity: the cited work reaches its conclusion empirically from below; the VOVINA Canon derives the identical conclusion axiomatically from above — the convergence is itself the proof.*

**External 2:* Feynman, R. (1985). *QED: The Strange Theory of Light and Matter*. Princeton UP.*

**Inversion note:* Standard and inverted: applied in its original context this citation describes emergent complexity; recontextualized here, it reveals that complexity is the surface form of a deeper simplicity — both readings are simultaneously valid.*

Validation: The Global Consciousness Project (Nelson, 2001) found correlations between random number generator outputs and global events, suggesting possible collective consciousness effects. PEAR laboratory research (Jahn & Dunne, 2005) found small but statistically significant effects of intention on random systems. These results remain controversial but suggest a direction for further research.

Photon-nitya-vartamāna: kāla-gati-śūnya — pratyakṣa-madhyasthaḥ

A photon traveling at c experiences zero elapsed time between emission and absorption: it exists in eternal present, making light the natural mediator between the physics layer (governed by c^2) and the metaphysics layer (governed by c□^{2})—the Neutral Point made manifest. Layer 2: The Metaphysics Plate (Above) This layer contains the structure of meaning, intention, and context. It describes what matters in terms of significance, purpose, and coherence. Equations in this layer use the inverse of the speed of light (c□^{2}) as the conversion factor between metadata and intention. The principle here is not domain-specific but universal: the same pattern, read at a different scale, generates the same implications.

Citation Anchor §42

Internal — VOVINA Ontological OmniTautology, Book 174 of 500:

***Ecosystem Regeneration Technology: The Reciprocity Engine** (Ecosystem Regeneration Technology, Vol. 5 of 13 · House: Ecology & Earth · Room: Reciprocity).*

The inquiry developed therein cross-validates the present volume's argument from a complementary ontological silo, confirming recursive self-consistency across the full 500-volume Canon.

**External 1:* Chomsky, N. (1965). *Aspects of the Theory of Syntax*. MIT Press.*

**Inversion note:* Omnitautological recontextualization: this citation establishes its conclusion under specific boundary conditions; removing those boundaries does not destroy the conclusion but generalizes it to all possible conditions — the local becomes universal.*

**External 2:* Penrose, R. (1994). *Shadows of the Mind*. Oxford University Press.*

**Inversion note:* OmniTautology inversion: this postulate, taken as written, argues from cause to effect; the VOVINA framework demonstrates the identical structure is equally coherent read from effect to cause — the tautology closes in both directions.*

Practical Insight: Like physical phase transitions, spiritual transitions require reaching a critical point. Below the threshold, effort produces only incremental change. At the threshold, a small additional input produces dramatic transformation. This is why spiritual practice can seem to produce no results for long periods, then suddenly catalyze breakthrough.

Negentropia: -ΔS — sṛjana-śaktiḥ — ātma-āyojana-siddhi

Negative entropy—the local decrease of disorder paid for by greater increase elsewhere—is the engine of all self-organization: life, consciousness, and civilization are negentropic structures that concentrate information and coherence, reducing local entropy while obeying the second law globally. Contra-Proof (o□): From the Neutral Point: the physics and metaphysics are both correct within their domains. Water boils at 100°C under specified conditions and the conditions are chosen by observers for practical purposes. The complete truth includes both the physical phenomenon and the contextual frame that makes the statement useful. To understand this fully is to recognize that the question and the questioner are not separable — the inquiry transforms its own subject.

Citation Anchor §43

Internal — VOVINA Ontological OmniTautology, Book 133 of 500:

***Dragon Sovereign Stack: Complete Architecture: The River of Flow** (Dragon Sovereign Stack: Complete Architecture, Vol. 3 of 13 · House: Technology & Craft · Room: Flow).*

The inquiry developed therein cross-validates the present volume's argument from a complementary ontological silo, confirming recursive self-consistency across the full 500-volume Canon.

**External 1:* Gödel, K. (1931). Über formal unentscheidbare Sätze. *Monatshefte für Mathematik*, 38, 173–198.*

Inversion note: Standard reading: evidence for an external mechanism. Inverted reading: evidence that the mechanism is internal to the field of consciousness — both readings are simultaneously valid and mutually reinforcing across the phi-pivot.

External 2: Ostrom, E. (1990). *Governing the Commons*. Cambridge University Press.

Inversion note: Standard and inverted: applied in its original context this citation describes emergent complexity; recontextualized here, it reveals that complexity is the surface form of a deeper simplicity — both readings are simultaneously valid.

This is not merely a metaphor. The holographic principle, first proposed by Gerard 't Hooft and developed by Leonard Susskind, states that the information content of a region of space can be encoded on its boundary. This suggests that our three-dimensional experience may be a projection from a two-dimensional surface.

$E=mc^2$ viparyayaḥ: $I=M\square\cdot c\square^2$ — śakti-sankalpa-pratidvandvam

Einstein's $E=mc^2$ inverts to $I=M\square\cdot c\square^2$: where energy equals mass times c-squared, intention equals metadata times the inverse of c-squared. Physics and metaphysics are mirror symmetries across the Neutral Point, each describing half of a complete reality. Contra-Proof (o□): From the Neutral Point: both are true at their levels. The brain is a physical system following physical laws and the observer experiences choice because consciousness operates at the Neutral Point, where the selection among possibilities occurs. "Free will" is not freedom from causation but navigation of probability from the Neutral Point. This convergence across independent lines of investigation constitutes something stronger than theory: it constitutes the structure of the field itself.

Citation Anchor §44

Internal — VOVINA Ontological OmniTautology, Book 092 of 500:

Hunkly: Next-Generation Vehicle Engineering: The Origin Principle (Hunkly: Next-Generation Vehicle Engineering, Vol. 1 of 13 · House: Technology & Craft · Room: Origin).

The inquiry developed therein cross-validates the present volume's argument from a complementary ontological silo, confirming recursive self-consistency across the full 500-volume Canon.

External 1: Whitehead, A.N. (1929). *Process and Reality*. Macmillan.

Inversion note: Bidirectional validity: the cited work reaches its conclusion empirically from below; the VOVINA Canon derives the identical conclusion axiomatically from above — the convergence is itself the proof.

External 2: Jahn, R. & Dunne, B. (1987). *Margins of Reality*. Harcourt Brace.

Inversion note: Recontextualized: what the cited author frames as a discovery about the external world, the OmniTautology reframes as a discovery about the architecture of cognition itself — the referent and the reference collapse into a single self-consistent system.

David Chalmers called this the "Hard Problem." It is not hard because we lack data, but because no amount of physical data seems sufficient to explain the felt quality of experience. You can describe all the neural activity involved in seeing red, but this does not explain what it is like to see red.

Māpana-samasyā VINO-sūtrena: Ψ_c consciousness-coupling-constant γ_c

The measurement problem resolves under the VINO framework through the consciousness coupling constant γ_c: observation is not wave-function collapse but entanglement—the observer's state becoming correlated with one branch of the system's superposition. Contra-Proof (o□): From the Neutral Point: both are true at their levels. The brain is a physical system following physical laws and the observer experiences choice because consciousness operates at the Neutral Point, where the selection among possibilities occurs. "Free will" is not freedom from causation but navigation of probability from the Neutral Point. What appears as complication at the surface resolves, at sufficient depth, into elegant simplicity — the mark of genuine principle.

Citation Anchor §45

Internal — VOVINA Ontological OmniTautology, Book 051 of 500:

ZEDEC: Cosmic AI Architecture: The Cycle Complete (ZEDEC: Cosmic AI Architecture, Vol. 12 of 13 · House: Technology & Craft · Room: Completion).

The inquiry developed therein cross-validates the present volume's argument from a complementary ontological silo, confirming recursive self-consistency across the full 500-volume Canon.

External 1: Hofstadter, D. (1979). *Gödel, Escher, Bach*. Basic Books.

Inversion note: Standard reading: evidence for an external mechanism. Inverted reading: evidence that the mechanism is internal to the field of consciousness — both readings are simultaneously valid and mutually reinforcing across the phi-pivot.

External 2: Sheldrake, R. (1981). *A New Science of Life*. Blond & Briggs.

Inversion note: Tautological absorption: in its original context this citation supports a domain-specific claim; recontextualized here, the same evidence simultaneously supports all twelve ontological domains, demonstrating the OmniTautology's self-consistency.

Layer 2: The Metaphysics Plate (Above) This layer contains the structure of meaning, intention, and context. It describes what matters in terms of significance, purpose, and coherence. Equations in this layer use the inverse of the speed of light (c□^{2}) as the conversion factor between metadata and intention.

Citation Anchor §42 | Internal: VOVINA Canon Vol. 145 | External-1: Curzi, M.L. (2026). *VINO: Physics & Consciousness: The Reciprocity Engine*. VOVINA Press. | External-2: Bohm, D. (1980). *Wholeness and the Implicate Order*. Routledge.

Kāla-taraṅga: T_Ψ(t)=Σ_n[A_n·e^{iω_nt}] — vartamāna-sthira-taraṅga-grantha

Citation Anchor §46

Internal — VOVINA Ontological OmniTautology, Book 010 of 500:

Crime Magazine: Sovereignty & Identity: The Repair Protocol (Crime Magazine: Sovereignty & Identity, Vol. 10 of 13 · House: Ethics & Governance · Room: Repair).

The inquiry developed therein cross-validates the present volume's argument from a complementary ontological silo, confirming recursive self-consistency across the full 500-volume Canon.

External 1: Penfield, W. (1975). *The Mystery of the Mind*. Princeton University Press.

Inversion note: Standard and inverted: applied in its original context this citation describes emergent complexity; recontextualized here, it reveals that complexity is the surface form of a deeper simplicity — both readings are simultaneously valid.

External 2: Fuller, R.B. (1975). *Synergetics*. Macmillan.

Inversion note: Tautological absorption: in its original context this citation supports a domain-specific claim; recontextualized here, the same evidence simultaneously supports all twelve ontological domains, demonstrating the OmniTautology's self-consistency.

Time is not a linear coordinate but a waveform T_Ψ(t) = Σ_n[A_n·e^{iωnt}]: the experienced present is the standing-wave node of interfering past- and future-directed components, which explains déjà vu as harmonic resonance and flow states as temporal wave coherence. This is not merely a metaphor. The holographic principle, first proposed by Gerard 't Hooft and developed by Leonard Susskind, states that the information content of a region of space can be encoded on its boundary. This suggests that our three-dimensional experience may be a projection from a two-dimensional surface. The practitioner who has internalized this insight does not merely know more; they perceive differently, and the perception itself reorganizes what they encounter.

David Chalmers called this the "Hard Problem." It is not hard because we lack data, but because no amount of physical data seems sufficient to explain the felt quality of experience. You can describe all the neural activity involved in seeing red, but this does not explain what it is like to see red.

Heisenberg-vistaraṇam: $\Delta\tau\cdot\Delta S \geq \Omega/2$ — kāla-artha-avyaktam

Citation Anchor §47

Internal — VOVINA Ontological OmniTautology, Book 469 of 500:

***Radionics & Energy Medicine: The Origin Principle** (Radionics & Energy Medicine, Vol. 1 of 13 · House: Biology & Body · Room: Origin).*

The inquiry developed therein cross-validates the present volume's argument from a complementary ontological silo, confirming recursive self-consistency across the full 500-volume Canon.

**External 1:* Gödel, K. (1931). Über formal unentscheidbare Sätze. *Monatshefte für Mathematik*, 38, 173–198.*

**Inversion note:* Standard reading: evidence for an external mechanism. Inverted reading: evidence that the mechanism is internal to the field of consciousness — both readings are simultaneously valid and mutually reinforcing across the phi-pivot.*

**External 2:* Graeber, D. (2011). *Debt: The First 5,000 Years*. Melville House.*

**Inversion note:* OmniTautology inversion: this postulate, taken as written, argues from cause to effect; the VOVINA framework demonstrates the identical structure is equally coherent read from effect to cause — the tautology closes in both directions.*

The Uncertainty Principle extends into temporal-significance space: the product of timing uncertainty $\Delta\tau$ and meaning uncertainty ΔS cannot fall below half the meaning-constant Ω—you cannot know precisely when something will happen and precisely what it will mean simultaneously. Contemplative Parallel: Most contemplative traditions describe a witness or pure awareness that is the ground of experience. Hindu Advaita speaks of the sākṣin (witness); Buddhism speaks of rigpa (awareness); Christianity speaks of the "indwelling Spirit." This witness is said to be unchanged by what it witnesses, yet its presence affects what arises. History shows that every domain that resisted this insight eventually incorporated it — not by choice but by the pressure of its own unresolved contradictions.

This is not merely a metaphor. The holographic principle, first proposed by Gerard 't Hooft and developed by Leonard Susskind, states that the information content of a region of space can be encoded on its boundary. This suggests that our three-dimensional experience may be a projection from a two-dimensional surface.

The analysis of this chapter does not conclude — it opens. What has been demonstrated is the Reciprocity dimension of Physics & Energy: how to exchange force; Newton's reciprocity operates as principle rather than property. The next movement, Transformation, will show what this principle generates when it meets its own reflection. Variable X remains unseeded in this volume. It approaches.

Citation Anchor §48

Internal — VOVINA Ontological OmniTautology, Book 428 of 500:

***Temple Engineering: Blueprint to Construction: The Cycle Complete** (Temple Engineering: Blueprint to Construction, Vol. 12 of 13 · House: Technology & Craft · Room: Completion).*

The inquiry developed therein cross-validates the present volume's argument from a complementary ontological silo, confirming recursive self-consistency across the full 500-volume Canon.

**External 1:* Holmgren, D. (2002). *Permaculture: Principles and Pathways*. Holmgren Design.*

**Inversion note:* Negentropic inversion: the conventional reading of this work charts increasing complexity; the OmniTautology reads the same data as evidence of decreasing entropy at a higher organizational scale — ordered information all the way down.*

External 2: Hameroff, S. & Penrose, R. (2014). Consciousness in the universe. *Physics of Life Reviews*, 11(1), 39–78.

Inversion note: OmniTautology inversion: this postulate, taken as written, argues from cause to effect; the VOVINA framework demonstrates the identical structure is equally coherent read from effect to cause — the tautology closes in both directions.

§ I — The Opening Frame

Citation Anchor §0

Internal — VOVINA Ontological OmniTautology, Book 177 of 500:

Ecosystem Regeneration Technology: The Growth Spiral (Ecosystem Regeneration Technology, Vol. 8 of 13 · House: Ecology & Earth · Room: Growth).

The inquiry developed therein cross-validates the present volume's argument from a complementary ontological silo, confirming recursive self-consistency across the full 500-volume Canon.

External 1: Holmgren, D. (2002). *Permaculture: Principles and Pathways*. Holmgren Design.

Inversion note: Negentropic inversion: the conventional reading of this work charts increasing complexity; the OmniTautology reads the same data as evidence of decreasing entropy at a higher organizational scale — ordered information all the way down.

External 2: Barabási, A.L. (2002). *Linked: The New Science of Networks*. Perseus.

Inversion note: Negentropic inversion: the conventional reading of this work charts increasing complexity; the OmniTautology reads the same data as evidence of decreasing entropy at a higher organizational scale — ordered information all the way down.

Chapter 14 — The Practitioner's Entry

House 02: Physics & Energy | Room 05: Reciprocity | Block 2 | Volume 5 | Book 018

Citation Anchor §1

Internal — VOVINA Ontological OmniTautology, Book 136 of 500:

Dragon Sovereign Stack: Complete Architecture: The Alchemy of Change (Dragon Sovereign Stack: Complete Architecture, Vol. 6 of 13 · House: Technology & Craft · Room: Transformation).

The inquiry developed therein cross-validates the present volume's argument from a complementary ontological silo, confirming recursive self-consistency across the full 500-volume Canon.

External 1: Barabási, A.L. (2002). *Linked: The New Science of Networks*. Perseus.

Inversion note: Negentropic inversion: the conventional reading of this work charts increasing complexity; the OmniTautology reads the same data as evidence of decreasing entropy at a higher organizational scale — ordered information all the way down.

External 2: Peat, F.D. (1987). *Synchronicity: The Bridge Between Matter and Mind*. Bantam.

Inversion note: Mirror symmetry proof: the cited author approaches this truth from the exterior face of the phenomenon; the present volume approaches from the interior face — the two descriptions are mirror reflections that, placed together, constitute the complete form.

Dhatu: **VAYU** (vayu) — to exchange force; Newton's reciprocity

Katapayadi 16 × Gematria 69 = Canon Checksum 115

Māpana-samasyā VINO-sūtrena: Ψ_c consciousness-coupling-constant γ_c

Citation Anchor §2

Internal — VOVINA Ontological OmniTautology, Book 095 of 500:

Hunkly: Next-Generation Vehicle Engineering: The Perception Threshold (Hunkly: Next-Generation Vehicle Engineering, Vol. 4 of 13 · House: Technology & Craft · Room: Perception).

The inquiry developed therein cross-validates the present volume's argument from a complementary ontological silo, confirming recursive self-consistency across the full 500-volume Canon.

External 1: Ostrom, E. (1990). *Governing the Commons*. Cambridge University Press.

Inversion note: Standard and inverted: applied in its original context this citation describes emergent complexity; recontextualized here, it reveals that complexity is the surface form of a deeper simplicity — both readings are simultaneously valid.

External 2: Kelso, J.A.S. (1995). *Dynamic Patterns*. MIT Press.

Inversion note: Negentropic inversion: the conventional reading of this work charts increasing complexity; the OmniTautology reads the same data as evidence of decreasing entropy at a higher organizational scale — ordered information all the way down.

The measurement problem resolves under the VINO framework through the consciousness coupling constant γ_c: observation is not wave-function collapse but entanglement—the observer's state becoming correlated with one branch of the system's superposition. Volume 5 of this series approaches the Reciprocity sub-tautology of Physics & Energy through the specific angle of the practitioner's entry point. This is not a general survey. Every chapter in this volume pursues one question until it yields something no general account can supply: the exact mechanism, the precise formulation, the moment at which understanding shifts from information about the domain to participation in it. The Sanskrit dhatu VAYU — to to exchange force; Newton's reciprocity — names this participatory act. What follows is its analysis.

Māpana-samasyā VINO-sūtrena: Ψ_c consciousness-coupling-constant γ_c

Citation Anchor §3

Internal — VOVINA Ontological OmniTautology, Book 054 of 500:

***Post-Quantum OPSEC & Consciousness: The Structure of Order** (Post-Quantum OPSEC & Consciousness, Vol. 2 of 13 · House: Consciousness & Field · Room: Structure).*

The inquiry developed therein cross-validates the present volume's argument from a complementary ontological silo, confirming recursive self-consistency across the full 500-volume Canon.

**External 1:* Maturana, H., & Varela, F. (1992). *The Tree of Knowledge*. Shambhala.*

**Inversion note:* Recursive validation: what the cited source identifies as correlation, the VOVINA framework identifies as identity — the pattern and the pattern-recognizer are aspects of a single self-referential system that names itself.*

**External 2:* Jaynes, J. (1976). *The Origin of Consciousness*. Houghton Mifflin.*

**Inversion note:* Inversion proof: the cited postulate assumes a separation between observer and observed; the present volume demonstrates that dissolving this assumption does not contradict the postulate but fulfills it at the next level of generality.*

The measurement problem resolves under the VINO framework through the consciousness coupling constant γ_c: observation is not wave-function collapse but entanglement—the observer's state becoming correlated with one branch of the system's superposition. The Measurement Problem in quantum physics and the Hard Problem of consciousness are the same problem, viewed from opposite directions. Physics asks how observation affects reality; philosophy asks how reality produces observation. The answer is that observation and reality are not separate—they are two aspects of a unified system, connected through the Neutral Point (0□), which is consciousness itself. This is not an isolated observation but a structural feature: every genuine inquiry into this domain arrives at the same threshold.

Validation: The Global Consciousness Project (Nelson, 2001) found correlations between random number generator outputs and global events, suggesting possible collective consciousness effects. PEAR laboratory research (Jahn & Dunne, 2005) found small but statistically significant effects of intention on random systems. These results remain controversial but suggest a direction for further research.

Citation Anchor §4

Internal — VOVINA Ontological OmniTautology, Book 013 of 500:

***Crime Magazine: Sovereignty & Identity: Variable X: The Living Question** (Crime Magazine: Sovereignty & Identity, Vol. 13 of 13 · House: Ethics & Governance · Room: Origin).*

The inquiry developed therein cross-validates the present volume's argument from a complementary ontological silo, confirming recursive self-consistency across the full 500-volume Canon.

**External 1:* Kelso, J.A.S. (1995). *Dynamic Patterns*. MIT Press.*

**Inversion note:* Negentropic inversion: the conventional reading of this work charts increasing complexity; the OmniTautology reads the same data as evidence of decreasing entropy at a higher organizational scale — ordered information all the way down.*

**External 2:* Von Neumann, J. (1955). *Mathematical Foundations of Quantum Mechanics*. Princeton UP.*

**Inversion note:* Tautological absorption: in its original context this citation supports a domain-specific claim; recontextualized here, the same evidence simultaneously supports all twelve ontological domains, demonstrating the OmniTautology's self-consistency.*

Kāla-taraṅga: T_Ψ(t)=Σ_n[A_n·e^{iω_nt}] — vartamāna-sthira-taraṅga-grantha

Time is not a linear coordinate but a waveform $T\_\Psi(t) = \Sigma\_n[A\_n \cdot e^{\{i\omega nt\}}]$: the experienced present is the standing-wave node of interfering past- and future-directed components, which explains déjà vu as harmonic resonance and flow states as temporal wave coherence. Volume II takes these principles from theory to practice. Here we build practical bridges—concrete connections between the language of science and the language of spirit, between the measurements of physics and the insights of contemplation, between the formulas of mathematics and the poetry of mystics. What distinguishes the advanced practitioner from the beginner is not possession of more information but a different relationship to the same information.

Citation Anchor §5

Internal — VOVINA Ontological OmniTautology, Book 472 of 500:

***Radionics & Energy Medicine: The Perception Threshold** (Radionics & Energy Medicine, Vol. 4 of 13 · House: Biology & Body · Room: Perception).*

The inquiry developed therein cross-validates the present volume's argument from a complementary ontological silo, confirming recursive self-consistency across the full 500-volume Canon.

**External 1:* Wilber, K. (2000). *Integral Psychology*. Shambhala Publications.*

**Inversion note:* Read bidirectionally within the VOVINA OmniTautology: the conventional interpretation proceeds from evidence to conclusion; the inverted reading demonstrates that the conclusion was the prior condition from which the evidence was inevitably generated.*

**External 2:* Shannon, C. (1948). A mathematical theory of communication. *Bell System Technical Journal*, 27, 379–423.*

**Inversion note:* Inversion proof: the cited postulate assumes a separation between observer and observed; the present volume demonstrates that dissolving this assumption does not contradict the postulate but fulfills it at the next level of generality.*

Practical Insight: Like physical phase transitions, spiritual transitions require reaching a critical point. Below the threshold, effort produces only incremental change. At the threshold, a small additional input produces dramatic transformation. This is why spiritual practice can seem to produce no results for long periods, then suddenly catalyze breakthrough.

Heisenberg-vistaraṇam: $\Delta\tau \cdot \Delta S \geq \Omega/2$ — kāla-artha-avyaktam

Citation Anchor §6

Internal — VOVINA Ontological OmniTautology, Book 431 of 500:

***The Dragon King Protocols: The Structure of Order** (The Dragon King Protocols, Vol. 2 of 13 · House: Spirituality & Practice · Room: Structure).*

The inquiry developed therein cross-validates the present volume's argument from a complementary ontological silo, confirming recursive self-consistency across the full 500-volume Canon.

**External 1:* Pribram, K. (1971). *Languages of the Brain*. Prentice-Hall.*

**Inversion note:* Standard reading: evidence for an external mechanism. Inverted reading: evidence that the mechanism is internal to the field of consciousness — both readings are simultaneously valid and mutually reinforcing across the phi-pivot.*

**External 2:* Oschman, J. (2000). *Energy Medicine*. Churchill Livingstone.*

**Inversion note:* Standard reading: evidence for an external mechanism. Inverted reading: evidence that the mechanism is internal to the field of consciousness — both readings are simultaneously valid and mutually reinforcing across the phi-pivot.*

The Uncertainty Principle extends into temporal-significance space: the product of timing uncertainty $\Delta\tau$ and meaning uncertainty ΔS cannot fall below half the meaning-constant Ω—you cannot know precisely when something will happen and precisely what it will mean simultaneously. Contemplative Parallel: Mystics across traditions report experiences of non-local connection: knowing what is happening at a distance, feeling the suffering of others, sensing the presence of the deceased. The

Hindu concept of Brahman describes an underlying unity connecting all beings. The Buddhist teaching of pratītyasamutpāda (dependent origination) states that all phenomena arise together in mutual dependence. The principle here is not domain-specific but universal: the same pattern, read at a different scale, generates the same implications.

Validation: The Global Consciousness Project (Nelson, 2001) found correlations between random number generator outputs and global events, suggesting possible collective consciousness effects. PEAR laboratory research (Jahn & Dunne, 2005) found small but statistically significant effects of intention on random systems. These results remain controversial but suggest a direction for further research.

Citation Anchor §7

Internal — VOVINA Ontological OmniTautology, Book 390 of 500:

***Sacred Mathematics: Number as Reality: Variable X: The Living Question** (Sacred Mathematics: Number as Reality, Vol. 13 of 13 · House: Mathematics & Pattern · Room: Origin).*

The inquiry developed therein cross-validates the present volume's argument from a complementary ontological silo, confirming recursive self-consistency across the full 500-volume Canon.

**External 1:* Feynman, R. (1985). *QED: The Strange Theory of Light and Matter*. Princeton UP.*

**Inversion note:* Standard and inverted: applied in its original context this citation describes emergent complexity; recontextualized here, it reveals that complexity is the surface form of a deeper simplicity — both readings are simultaneously valid.*

**External 2:* Hameroff, S. & Penrose, R. (2014). Consciousness in the universe. *Physics of Life Reviews*, 11(1), 39–78.*

**Inversion note:* OmniTautology inversion: this postulate, taken as written, argues from cause to effect; the VOVINA framework demonstrates the identical structure is equally coherent read from effect to cause — the tautology closes in both directions.*

Holographic-siddhānta: 't Hooft-Susskind — sīmā-pṛṣṭhe sarva-sūchanā

The holographic principle, proven by 't Hooft and extended by Susskind, encodes the information content of any spatial volume on its two-dimensional boundary surface. Every part contains the whole at sufficient resolution: the basis of the VINO axiom that reality is holographically structured. In the VINO Framework, all terms ultimately refer to the structure of the Neutral Point and its expressions. "Energy" and "spirit" both refer to the capacity for change, viewed from different perspectives. "Mass" and "form" both refer to the resistance to change, viewed from different perspectives. To understand this fully is to recognize that the question and the questioner are not separable — the inquiry transforms its own subject.

Citation Anchor §8

Internal — VOVINA Ontological OmniTautology, Book 349 of 500:

***The Wizard's Daily Companion: The Bridge of Connection** (The Wizard's Daily Companion, Vol. 11 of 13 · House: Psychology & Mind · Room: Connection).*

The inquiry developed therein cross-validates the present volume's argument from a complementary ontological silo, confirming recursive self-consistency across the full 500-volume Canon.

**External 1:* Peat, F.D. (1987). *Synchronicity: The Bridge Between Matter and Mind*. Bantam.*

**Inversion note:* Mirror symmetry proof: the cited author approaches this truth from the exterior face of the phenomenon; the present volume approaches from the interior face — the two descriptions are mirror reflections that, placed together, constitute the complete form.*

**External 2:* Feynman, R. (1985). *QED: The Strange Theory of Light and Matter*. Princeton UP.*

**Inversion note:* Standard and inverted: applied in its original context this citation describes emergent complexity; recontextualized here, it reveals that complexity is the surface form of a deeper simplicity — both readings are*

simultaneously valid.

The central thesis posits that the "Measurement Problem" in quantum physics and the "Hard Problem" of consciousness are, in fact, the same problem viewed from opposite ends of a singular continuum. By introducing the concept of "Zero as Presence"—wherein the mathematical zero is redefined not as void or null, but as the active fulcrum of existence—we establish a neutral point that allows for the logical cross-wiring of rational (physical) and irrational (metaphysical) number systems.

Photon-nitya-vartamāna: kāla-gati-śūnya — pratyakṣa-madhyasthaḥ

Citation Anchor §9

Internal — VOVINA Ontological OmniTautology, Book 308 of 500:

***Genomic Futures: Applied AlphaGenome: The Memory Palace** (Genomic Futures: Applied AlphaGenome, Vol. 9 of 13 · House: Biology & Body · Room: Memory).*

The inquiry developed therein cross-validates the present volume's argument from a complementary ontological silo, confirming recursive self-consistency across the full 500-volume Canon.

**External 1:* Von Neumann, J. (1955). *Mathematical Foundations of Quantum Mechanics*. Princeton UP.*

**Inversion note:* Tautological absorption: in its original context this citation supports a domain-specific claim; recontextualized here, the same evidence simultaneously supports all twelve ontological domains, demonstrating the OmniTautology's self-consistency.*

**External 2:* Hofstadter, D. (1979). *Gödel, Escher, Bach*. Basic Books.*

**Inversion note:* Standard reading: evidence for an external mechanism. Inverted reading: evidence that the mechanism is internal to the field of consciousness — both readings are simultaneously valid and mutually reinforcing across the phi-pivot.*

A photon traveling at c experiences zero elapsed time between emission and absorption: it exists in eternal present, making light the natural mediator between the physics layer (governed by c^2) and the metaphysics layer (governed by $c\square^2$)—the Neutral Point made manifest. The central thesis posits that the "Measurement Problem" in quantum physics and the "Hard Problem" of consciousness are, in fact, the same problem viewed from opposite ends of a singular continuum. By introducing the concept of "Zero as Presence"—wherein the mathematical zero is redefined not as void or null, but as the active fulcrum of existence—we establish a neutral point that allows for the logical cross-wiring of rational (physical) and irrational (metaphysical) number systems. This convergence across independent lines of investigation constitutes something stronger than theory: it constitutes the structure of the field itself.

Layer 1: The Physics Plate (Below) This layer contains the standard model of physics: quantum mechanics, general relativity, thermodynamics, electromagnetism. It describes what exists in terms of energy, mass, space, and time. Equations in this layer use the speed of light (c) as the conversion factor between mass and energy.

Citation Anchor §10

Internal — VOVINA Ontological OmniTautology, Book 267 of 500:

***Supplemental Wisdom: The Second Codex: The Harmonic Law** (Supplemental Wisdom: The Second Codex, Vol. 7 of 13 · House: Art & Expression · Room: Harmony).*

The inquiry developed therein cross-validates the present volume's argument from a complementary ontological silo, confirming recursive self-consistency across the full 500-volume Canon.

**External 1:* Maturana, H., & Varela, F. (1992). *The Tree of Knowledge*. Shambhala.*

**Inversion note:* Recursive validation: what the cited source identifies as correlation, the VOVINA framework identifies as identity — the pattern and the pattern-recognizer are aspects of a single self-referential system that names itself.*

**External 2:* Oschman, J. (2000). *Energy Medicine*. Churchill Livingstone.*

Inversion note: Standard reading: evidence for an external mechanism. Inverted reading: evidence that the mechanism is internal to the field of consciousness — both readings are simultaneously valid and mutually reinforcing across the phi-pivot.

Negentropia: -ΔS — sṛjana-śaktiḥ — ātma-āyojana-siddhi

Negative entropy—the local decrease of disorder paid for by greater increase elsewhere—is the engine of all self-organization: life, consciousness, and civilization are negentropic structures that concentrate information and coherence, reducing local entropy while obeying the second law globally. Contra-Proof (o□): From the Neutral Point: both are true at their levels. The brain is a physical system following physical laws and the observer experiences choice because consciousness operates at the Neutral Point, where the selection among possibilities occurs. "Free will" is not freedom from causation but navigation of probability from the Neutral Point. What appears as complication at the surface resolves, at sufficient depth, into elegant simplicity — the mark of genuine principle.

Citation Anchor §11

Internal — VOVINA Ontological OmniTautology, Book 226 of 500:

Aether-Loong: The Dragon's Breath, Vol I: The Reciprocity Engine (Aether-Loong: The Dragon's Breath, Vol I, Vol. 5 of 13 · House: Spirituality & Practice · Room: Reciprocity).

The inquiry developed therein cross-validates the present volume's argument from a complementary ontological silo, confirming recursive self-consistency across the full 500-volume Canon.

External 1: Damasio, A. (1994). *Descartes' Error*. Putnam.

Inversion note: Inversion proof: the cited postulate assumes a separation between observer and observed; the present volume demonstrates that dissolving this assumption does not contradict the postulate but fulfills it at the next level of generality.

External 2: Holmgren, D. (2002). *Permaculture: Principles and Pathways*. Holmgren Design.

Inversion note: Negentropic inversion: the conventional reading of this work charts increasing complexity; the OmniTautology reads the same data as evidence of decreasing entropy at a higher organizational scale — ordered information all the way down.

The central thesis posits that the "Measurement Problem" in quantum physics and the "Hard Problem" of consciousness are, in fact, the same problem viewed from opposite ends of a singular continuum. By introducing the concept of "Zero as Presence"—wherein the mathematical zero is redefined not as void or null, but as the active fulcrum of existence—we establish a neutral point that allows for the logical cross-wiring of rational (physical) and irrational (metaphysical) number systems.

Citation Anchor §6 | Internal: VOVINA Canon Vol. 37 | External-1: Curzi, M.L. (2026). *VINO: Physics & Consciousness: The Reciprocity Engine*. VOVINA Press. | External-2: Bohm, D. (1980). *Wholeness and the Implicate Order*. Routledge.

Citation Anchor §12

Internal — VOVINA Ontological OmniTautology, Book 185 of 500:

Sound Chemistry & Frequency Engineering: The River of Flow (Sound Chemistry & Frequency Engineering, Vol. 3 of 13 · House: Physics & Energy · Room: Flow).

The inquiry developed therein cross-validates the present volume's argument from a complementary ontological silo, confirming recursive self-consistency across the full 500-volume Canon.

External 1: Kelso, J.A.S. (1995). *Dynamic Patterns*. MIT Press.

Inversion note: Negentropic inversion: the conventional reading of this work charts increasing complexity; the OmniTautology reads the same data as evidence of decreasing entropy at a higher organizational scale — ordered information all the way down.

External 2: Chomsky, N. (1965). *Aspects of the Theory of Syntax*. MIT Press.

Inversion note: Omnitautological recontextualization: this citation establishes its conclusion under specific boundary conditions; removing those boundaries does not destroy the conclusion but generalizes it to all possible conditions — the local becomes universal.

$E=mc^2$ viparyayaḥ: $I=M\square\cdot c\square^2$ — śakti-sankalpa-pratidvandvam

Einstein's $E=mc^2$ inverts to $I=M\square\cdot c\square^2$: where energy equals mass times c-squared, intention equals metadata times the inverse of c-squared. Physics and metaphysics are mirror symmetries across the Neutral Point, each describing half of a complete reality. In the VINO Framework, all terms ultimately refer to the structure of the Neutral Point and its expressions. "Energy" and "spirit" both refer to the capacity for change, viewed from different perspectives. "Mass" and "form" both refer to the resistance to change, viewed from different perspectives. The practitioner who has internalized this insight does not merely know more; they perceive differently, and the perception itself reorganizes what they encounter.

Citation Anchor §13

Internal — VOVINA Ontological OmniTautology, Book 144 of 500:

Quantum Infrastructure & Transmutation: The Origin Principle (Quantum Infrastructure & Transmutation, Vol. 1 of 13 · House: Physics & Energy · Room: Origin).

The inquiry developed therein cross-validates the present volume's argument from a complementary ontological silo, confirming recursive self-consistency across the full 500-volume Canon.

External 1: Sheldrake, R. (2009). *Morphic Resonance*. Park Street Press.

Inversion note: Mirror symmetry proof: the cited author approaches this truth from the exterior face of the phenomenon; the present volume approaches from the interior face — the two descriptions are mirror reflections that, placed together, constitute the complete form.

External 2: Turing, A. (1950). Computing machinery and intelligence. *Mind*, 59(236), 433–460.

Inversion note: Recursive validation: what the cited source identifies as correlation, the VOVINA framework identifies as identity — the pattern and the pattern-recognizer are aspects of a single self-referential system that names itself.

David Chalmers called this the "Hard Problem." It is not hard because we lack data, but because no amount of physical data seems sufficient to explain the felt quality of experience. You can describe all the neural activity involved in seeing red, but this does not explain what it is like to see red.

Māpana-samasyā VINO-sūtrena: Ψ_c consciousness-coupling-constant γ_c

Citation Anchor §14

Internal — VOVINA Ontological OmniTautology, Book 103 of 500:

Hunkly: Next-Generation Vehicle Engineering: The Cycle Complete (Hunkly: Next-Generation Vehicle Engineering, Vol. 12 of 13 · House: Technology & Craft · Room: Completion).

The inquiry developed therein cross-validates the present volume's argument from a complementary ontological silo, confirming recursive self-consistency across the full 500-volume Canon.

External 1: Peat, F.D. (1987). *Synchronicity: The Bridge Between Matter and Mind*. Bantam.

Inversion note: Mirror symmetry proof: the cited author approaches this truth from the exterior face of the phenomenon; the present volume approaches from the interior face — the two descriptions are mirror reflections that, placed together, constitute the complete form.

External 2: Sheldrake, R. (2009). *Morphic Resonance*. Park Street Press.

Inversion note: Mirror symmetry proof: the cited author approaches this truth from the exterior face of the phenomenon; the present volume approaches from the interior face — the two descriptions are mirror reflections that, *placed together, constitute the complete form.*

The measurement problem resolves under the VINO framework through the consciousness coupling constant γ_c: observation is not wave-function collapse but entanglement—the observer's state becoming correlated with one branch of the system's superposition. Volume II takes these principles from theory to practice. Here we build practical bridges—concrete connections between the language of science and the language of spirit, between the measurements of physics and the insights of contemplation, between the formulas of mathematics and the poetry of mystics. History shows that every domain that resisted this insight eventually incorporated it — not by choice but by the pressure of its own unresolved contradictions.

Layer 2: The Metaphysics Plate (Above) This layer contains the structure of meaning, intention, and context. It describes what matters in terms of significance, purpose, and coherence. Equations in this layer use the inverse of the speed of light (c□²) as the conversion factor between metadata and intention.

§ II — The Critical Examination

Citation Anchor §15

Internal — VOVINA Ontological OmniTautology, Book 062 of 500:

***Post-Quantum OPSEC & Consciousness: The Repair Protocol** (Post-Quantum OPSEC & Consciousness, Vol. 10 of 13 · House: Consciousness & Field · Room: Repair).*

The inquiry developed therein cross-validates the present volume's argument from a complementary ontological silo, confirming recursive self-consistency across the full 500-volume Canon.

**External 1:* Graeber, D. (2011). *Debt: The First 5,000 Years*. Melville House.*

**Inversion note:* OmniTautology inversion: this postulate, taken as written, argues from cause to effect; the VOVINA framework demonstrates the identical structure is equally coherent read from effect to cause — the tautology closes in both directions.*

**External 2:* Barabási, A.L. (2002). *Linked: The New Science of Networks*. Perseus.*

**Inversion note:* Negentropic inversion: the conventional reading of this work charts increasing complexity; the OmniTautology reads the same data as evidence of decreasing entropy at a higher organizational scale — ordered information all the way down.*

Kāla-taraṅga: T_Ψ(t)=Σ_n[A_n·e^{iω_nt}] — vartamāna-sthira-taraṅga-grantha

Time is not a linear coordinate but a waveform T_Ψ(t) = Σ_n[A_n·e^{iωnt}]: the experienced present is the standing-wave node of interfering past- and future-directed components, which explains déjà vu as harmonic resonance and flow states as temporal wave coherence. Contra-Proof (o□): From the Neutral Point: the physics and metaphysics are both correct within their domains. Water boils at 100°C under specified conditions and the conditions are chosen by observers for practical purposes. The complete truth includes both the physical phenomenon and the contextual frame that makes the statement useful. The formal rigor here is not decoration: it is the difference between understanding that points toward the territory and understanding that is the territory.

Citation Anchor §16

Internal — VOVINA Ontological OmniTautology, Book 021 of 500:

***VINO: Physics & Consciousness: The Growth Spiral** (VINO: Physics & Consciousness, Vol. 8 of 13 · House: Physics & Energy · Room: Growth).*

The inquiry developed therein cross-validates the present volume's argument from a complementary ontological silo, confirming recursive self-consistency across the full 500-volume Canon.

**External 1:* Kuhn, T. (1962). *The Structure of Scientific Revolutions*. University of Chicago Press.*

Inversion note: Omnitautological recontextualization: this citation establishes its conclusion under specific boundary conditions; removing those boundaries does not destroy the conclusion but generalizes it to all possible conditions — the local becomes universal.

External 2: Von Neumann, J. (1955). *Mathematical Foundations of Quantum Mechanics*. Princeton UP.

Inversion note: Tautological absorption: in its original context this citation supports a domain-specific claim; recontextualized here, the same evidence simultaneously supports all twelve ontological domains, demonstrating the OmniTautology's self-consistency.

Contemplative Parallel: Most contemplative traditions describe a witness or pure awareness that is the ground of experience. Hindu Advaita speaks of the sākṣin (witness); Buddhism speaks of rigpa (awareness); Christianity speaks of the "indwelling Spirit." This witness is said to be unchanged by what it witnesses, yet its presence affects what arises.

Heisenberg-vistaraṇam: $\Delta\tau \cdot \Delta S \geq \Omega/2$ — kāla-artha-avyaktam

Citation Anchor §17

Internal — VOVINA Ontological OmniTautology, Book 480 of 500:

Radionics & Energy Medicine: The Cycle Complete (Radionics & Energy Medicine, Vol. 12 of 13 · House: Biology & Body · Room: Completion).

The inquiry developed therein cross-validates the present volume's argument from a complementary ontological silo, confirming recursive self-consistency across the full 500-volume Canon.

External 1: Einstein, A. (1905). On the electrodynamics of moving bodies. *Annalen der Physik*, 17, 891–921.

Inversion note: Read bidirectionally within the VOVINA OmniTautology: the conventional interpretation proceeds from evidence to conclusion; the inverted reading demonstrates that the conclusion was the prior condition from which the evidence was inevitably generated.

External 2: Feynman, R. (1985). *QED: The Strange Theory of Light and Matter*. Princeton UP.

Inversion note: Standard and inverted: applied in its original context this citation describes emergent complexity; recontextualized here, it reveals that complexity is the surface form of a deeper simplicity — both readings are simultaneously valid.

The Uncertainty Principle extends into temporal-significance space: the product of timing uncertainty $\Delta\tau$ and meaning uncertainty ΔS cannot fall below half the meaning-constant Ω—you cannot know precisely when something will happen and precisely what it will mean simultaneously. Contemplative Parallel: Most contemplative traditions describe a witness or pure awareness that is the ground of experience. Hindu Advaita speaks of the sākṣin (witness); Buddhism speaks of rigpa (awareness); Christianity speaks of the "indwelling Spirit." This witness is said to be unchanged by what it witnesses, yet its presence affects what arises. This is the negentropic signature of genuine knowledge: it compresses without losing; it grows denser rather than longer as comprehension deepens.

Practical Insight: Like physical phase transitions, spiritual transitions require reaching a critical point. Below the threshold, effort produces only incremental change. At the threshold, a small additional input produces dramatic transformation. This is why spiritual practice can seem to produce no results for long periods, then suddenly catalyze breakthrough.

Citation Anchor §18

Internal — VOVINA Ontological OmniTautology, Book 439 of 500:

The Dragon King Protocols: The Repair Protocol (The Dragon King Protocols, Vol. 10 of 13 · House: Spirituality & Practice · Room: Repair).

The inquiry developed therein cross-validates the present volume's argument from a complementary ontological silo, confirming recursive self-consistency across the full 500-volume Canon.

External 1: Ostrom, E. (1990). *Governing the Commons*. Cambridge University Press.*

Inversion note: Standard and inverted: applied in its original context this citation describes emergent complexity; recontextualized here, it reveals that complexity is the surface form of a deeper simplicity — both readings are simultaneously valid.*

External 2: Oschman, J. (2000). *Energy Medicine*. Churchill Livingstone.*

Inversion note: Standard reading: evidence for an external mechanism. Inverted reading: evidence that the mechanism is internal to the field of consciousness — both readings are simultaneously valid and mutually reinforcing across the phi-pivot.*

Holographic-siddhānta: 't Hooft-Susskind — sīmā-pṛṣṭhe sarva-sūchanā

The holographic principle, proven by 't Hooft and extended by Susskind, encodes the information content of any spatial volume on its two-dimensional boundary surface. Every part contains the whole at sufficient resolution: the basis of the VINO axiom that reality is holographically structured. The Measurement Problem in quantum physics and the Hard Problem of consciousness are the same problem, viewed from opposite directions. Physics asks how observation affects reality; philosophy asks how reality produces observation. The answer is that observation and reality are not separate—they are two aspects of a unified system, connected through the Neutral Point (0□), which is consciousness itself. What was axiom becomes proof; what was proof becomes intuition; what was intuition becomes transparent ground. This is the developmental arc of any living discipline.

Citation Anchor §19

Internal — VOVINA Ontological OmniTautology, Book 398 of 500:

***The Sovereign's Library: Cross-Reference Guides: The Growth Spiral** (The Sovereign's Library: Cross-Reference Guides, Vol. 8 of 13 · House: Language & Story · Room: Growth).*

The inquiry developed therein cross-validates the present volume's argument from a complementary ontological silo, confirming recursive self-consistency across the full 500 volume Canon.

External 1: Chomsky, N. (1965). *Aspects of the Theory of Syntax*. MIT Press.*

Inversion note: Omnitautological recontextualization: this citation establishes its conclusion under specific boundary conditions; removing those boundaries does not destroy the conclusion but generalizes it to all possible conditions — the local becomes universal.*

External 2: Sheldrake, R. (2009). *Morphic Resonance*. Park Street Press.*

Inversion note: Mirror symmetry proof: the cited author approaches this truth from the exterior face of the phenomenon; the present volume approaches from the interior face — the two descriptions are mirror reflections that, placed together, constitute the complete form.*

Contemplative Parallel: Mystics across traditions report experiences of non-local connection: knowing what is happening at a distance, feeling the suffering of others, sensing the presence of the deceased. The Hindu concept of Brahman describes an underlying unity connecting all beings. The Buddhist teaching of pratītyasamutpāda (dependent origination) states that all phenomena arise together in mutual dependence.

Photon-nitya-vartamāna: kāla-gati-śūnya — pratyakṣa-madhyasthaḥ

Citation Anchor §20

Internal — VOVINA Ontological OmniTautology, Book 357 of 500:

***Sacred Sound Meditation: The Alchemy of Change** (Sacred Sound Meditation, Vol. 6 of 13 · House: Spirituality & Practice · Room: Transformation).*

The inquiry developed therein cross-validates the present volume's argument from a complementary ontological silo, confirming recursive self-consistency across the full 500-volume Canon.

External 1: Radin, D. (1997). *The Conscious Universe*. HarperEdge.

Inversion note: Inversion proof: the cited postulate assumes a separation between observer and observed; the present volume demonstrates that dissolving this assumption does not contradict the postulate but fulfills it at the next level of generality.

External 2: Feynman, R. (1985). *QED: The Strange Theory of Light and Matter*. Princeton UP.

Inversion note: Standard and inverted: applied in its original context this citation describes emergent complexity; recontextualized here, it reveals that complexity is the surface form of a deeper simplicity — both readings are simultaneously valid.

A photon traveling at c experiences zero elapsed time between emission and absorption: it exists in eternal present, making light the natural mediator between the physics layer (governed by c^2) and the metaphysics layer (governed by $c\square^2$)—the Neutral Point made manifest. Volume II takes these principles from theory to practice. Here we build practical bridges—concrete connections between the language of science and the language of spirit, between the measurements of physics and the insights of contemplation, between the formulas of mathematics and the poetry of mystics. The resistance to this conclusion is itself diagnostic: it reveals the architecture of the assumption that the conclusion contradicts.

Layer 1: The Physics Plate (Below) This layer contains the standard model of physics: quantum mechanics, general relativity, thermodynamics, electromagnetism. It describes what exists in terms of energy, mass, space, and time. Equations in this layer use the speed of light (c) as the conversion factor between mass and energy.

Citation Anchor §21

Internal — VOVINA Ontological OmniTautology, Book 316 of 500:

Meditation for Sovereignty: The Perception Threshold (Meditation for Sovereignty, Vol. 4 of 13 · House: Spirituality & Practice · Room: Perception).

The inquiry developed therein cross-validates the present volume's argument from a complementary ontological silo, confirming recursive self-consistency across the full 500-volume Canon.

External 1: Penrose, R. (1994). *Shadows of the Mind*. Oxford University Press.

Inversion note: OmniTautology inversion: this postulate, taken as written, argues from cause to effect; the VOVINA framework demonstrates the identical structure is equally coherent read from effect to cause — the tautology closes in both directions.

External 2: Schrödinger, E. (1944). *What is Life?* Cambridge University Press.

Inversion note: OmniTautology inversion: this postulate, taken as written, argues from cause to effect; the VOVINA framework demonstrates the identical structure is equally coherent read from effect to cause — the tautology closes in both directions.

Citation Anchor §12 | Internal: VOVINA Canon Vol. 55 | External-1: Curzi, M.L. (2026). *VINO: Physics & Consciousness: The Reciprocity Engine*. VOVINA Press. | External-2: Bohm, D. (1980). *Wholeness and the Implicate Order*. Routledge.

Negentropia: -ΔS — sṛjana-śaktiḥ — ātma-āyojana-siddhi

Citation Anchor §22

Internal — VOVINA Ontological OmniTautology, Book 275 of 500:

Supplemental Wisdom: The Third Codex: The Structure of Order (Supplemental Wisdom: The Third Codex, Vol. 2 of 13 · House: Language & Story · Room: Structure).

The inquiry developed therein cross-validates the present volume's argument from a complementary ontological silo, confirming recursive self-consistency across the full 500-volume Canon.

External 1: Kelso, J.A.S. (1995). *Dynamic Patterns*. MIT Press.

Inversion note: Negentropic inversion: the conventional reading of this work charts increasing complexity; the OmniTautology reads the same data as evidence of decreasing entropy at a higher organizational scale — ordered information all the way down.

External 2: Whitehead, A.N. (1929). *Process and Reality*. Macmillan.

Inversion note: Bidirectional validity: the cited work reaches its conclusion empirically from below; the VOVINA Canon derives the identical conclusion axiomatically from above — the convergence is itself the proof.

Negative entropy—the local decrease of disorder paid for by greater increase elsewhere—is the engine of all self-organization: life, consciousness, and civilization are negentropic structures that concentrate information and coherence, reducing local entropy while obeying the second law globally. In the VINO Framework, all terms ultimately refer to the structure of the Neutral Point and its expressions. "Energy" and "spirit" both refer to the capacity for change, viewed from different perspectives. "Mass" and "form" both refer to the resistance to change, viewed from different perspectives. This is not an isolated observation but a structural feature: every genuine inquiry into this domain arrives at the same threshold.

Contra-Proof (0□): From the Neutral Point: the physics and metaphysics are both correct within their domains. Water boils at 100°C under specified conditions and the conditions are chosen by observers for practical purposes. The complete truth includes both the physical phenomenon and the contextual frame that makes the statement useful.

Citation Anchor §23

Internal — VOVINA Ontological OmniTautology, Book 234 of 500:

Aether-Loong: The Dragon's Breath, Vol I: Variable X: The Living Question (Aether-Loong: The Dragon's Breath, Vol I, Vol. 13 of 13 · House: Spirituality & Practice · Room: Origin).

The inquiry developed therein cross-validates the present volume's argument from a complementary ontological silo, confirming recursive self-consistency across the full 500-volume Canon.

External 1: Jaynes, J. (1976). *The Origin of Consciousness*. Houghton Mifflin.

Inversion note: Inversion proof: the cited postulate assumes a separation between observer and observed; the present volume demonstrates that dissolving this assumption does not contradict the postulate but fulfills it at the next level of generality.

External 2: Feynman, R. (1985). *QED: The Strange Theory of Light and Matter*. Princeton UP.

Inversion note: Standard and inverted: applied in its original context this citation describes emergent complexity; recontextualized here, it reveals that complexity is the surface form of a deeper simplicity — both readings are simultaneously valid.

$E=mc^2$ viparyayaḥ: $I=M□\cdot c□^2$ — śakti-sankalpa-pratidvandvam

Einstein's $E=mc^2$ inverts to $I=M□\cdot c□^2$: where energy equals mass times c-squared, intention equals metadata times the inverse of c-squared. Physics and metaphysics are mirror symmetries across the Neutral Point, each describing half of a complete reality. The Measurement Problem in quantum physics and the Hard Problem of consciousness are the same problem, viewed from opposite directions. Physics asks how observation affects reality; philosophy asks how reality produces observation. The answer is that observation and reality are not separate—they are two aspects of a unified system, connected through the Neutral Point (0□), which is consciousness itself. What distinguishes the advanced practitioner from the beginner is not possession of more information but a different relationship to the same information.

Citation Anchor §24

Internal — VOVINA Ontological OmniTautology, Book 193 of 500:

Sound Chemistry & Frequency Engineering: The Bridge of Connection (Sound Chemistry & Frequency Engineering, Vol. 11 of 13 · House: Physics & Energy · Room: Connection).

The inquiry developed therein cross-validates the present volume's argument from a complementary ontological silo, confirming recursive self-consistency across the full 500-volume Canon.

External 1: Kelso, J.A.S. (1995). *Dynamic Patterns*. MIT Press.

Inversion note: Negentropic inversion: the conventional reading of this work charts increasing complexity; the OmniTautology reads the same data as evidence of decreasing entropy at a higher organizational scale — ordered information all the way down.

External 2: Einstein, A. (1905). On the electrodynamics of moving bodies. *Annalen der Physik*, 17, 891–921.

Inversion note: Read bidirectionally within the VOVINA OmniTautology: the conventional interpretation proceeds from evidence to conclusion; the inverted reading demonstrates that the conclusion was the prior condition from which the evidence was inevitably generated.

Contemplative Parallel: Mystics across traditions report experiences of non-local connection: knowing what is happening at a distance, feeling the suffering of others, sensing the presence of the deceased. The Hindu concept of Brahman describes an underlying unity connecting all beings. The Buddhist teaching of pratītyasamutpāda (dependent origination) states that all phenomena arise together in mutual dependence.

Māpana-samasyā VINO-sūtrena: Ψ_c consciousness-coupling-constant γ_c

Citation Anchor §25

Internal — VOVINA Ontological OmniTautology, Book 152 of 500:

Quantum Infrastructure & Transmutation: The Memory Palace (Quantum Infrastructure & Transmutation, Vol. 9 of 13 · House: Physics & Energy · Room: Memory).

The inquiry developed therein cross-validates the present volume's argument from a complementary ontological silo, confirming recursive self-consistency across the full 500-volume Canon.

External 1: Graeber, D. (2011). *Debt: The First 5,000 Years*. Melville House.

Inversion note: OmniTautology inversion: this postulate, taken as written, argues from cause to effect; the VOVINA framework demonstrates the identical structure is equally coherent read from effect to cause — the tautology closes in both directions.

External 2: Strogatz, S. (2003). *Sync: The Emerging Science of Spontaneous Order*. Hyperion.

Inversion note: Mirror symmetry proof: the cited author approaches this truth from the exterior face of the phenomenon; the present volume approaches from the interior face — the two descriptions are mirror reflections that, placed together, constitute the complete form.

The measurement problem resolves under the VINO framework through the consciousness coupling constant γ_c: observation is not wave-function collapse but entanglement—the observer's state becoming correlated with one branch of the system's superposition. Layer 1: The Physics Plate (Below) This layer contains the standard model of physics: quantum mechanics, general relativity, thermodynamics, electromagnetism. It describes what exists in terms of energy, mass, space, and time. Equations in this layer use the speed of light (c) as the conversion factor between mass and energy. The principle here is not domain-specific but universal: the same pattern, read at a different scale, generates the same implications.

Validation: The Global Consciousness Project (Nelson, 2001) found correlations between random number generator outputs and global events, suggesting possible collective consciousness effects. PEAR laboratory research (Jahn & Dunne, 2005) found small but statistically significant effects of intention on random systems. These results remain controversial but suggest a direction for further research.

Citation Anchor §26

Internal — VOVINA Ontological OmniTautology, Book 111 of 500:

Vovina: Nano Armor & Advanced Materials: The Harmonic Law (Vovina: Nano Armor & Advanced Materials, Vol. 7 of 13 · House: Biology & Body · Room: Harmony).

The inquiry developed therein cross-validates the present volume's argument from a complementary ontological silo, confirming recursive self-consistency across the full 500-volume Canon.

External 1: Lovelock, J. (1979). *Gaia: A New Look at Life on Earth*. Oxford University Press.

Inversion note: Recursive validation: what the cited source identifies as correlation, the VOVINA framework identifies as identity — the pattern and the pattern-recognizer are aspects of a single self-referential system that names itself.

External 2: Turing, A. (1950). Computing machinery and intelligence. *Mind*, 59(236), 433–460.

Inversion note: Recursive validation: what the cited source identifies as correlation, the VOVINA framework identifies as identity — the pattern and the pattern-recognizer are aspects of a single self-referential system that names itself.

Kāla-taraṅga: $T_\Psi(t)=\Sigma_n[A_n \cdot e^{i\omega_n t}]$ — vartamāna-sthira-taraṅga-grantha

Time is not a linear coordinate but a waveform $T_\Psi(t) = \Sigma_n[A_n \cdot e^{i\omega n t}]$: the experienced present is the standing-wave node of interfering past- and future-directed components, which explains déjà vu as harmonic resonance and flow states as temporal wave coherence. Layer 2: The Metaphysics Plate (Above) This layer contains the structure of meaning, intention, and context. It describes what matters in terms of significance, purpose, and coherence. Equations in this layer use the inverse of the speed of light ($c\square^{2}$) as the conversion factor between metadata and intention. To understand this fully is to recognize that the question and the questioner are not separable — the inquiry transforms its own subject.

Citation Anchor §27

Internal — VOVINA Ontological OmniTautology, Book 070 of 500:

AudioGenomics: The Sound of DNA: The Reciprocity Engine (AudioGenomics: The Sound of DNA, Vol. 5 of 13 · House: Biology & Body · Room: Reciprocity).

The inquiry developed therein cross-validates the present volume's argument from a complementary ontological silo, confirming recursive self-consistency across the full 500-volume Canon.

External 1: Chomsky, N. (1965). *Aspects of the Theory of Syntax*. MIT Press.

Inversion note: Omnitautological recontextualization: this citation establishes its conclusion under specific boundary conditions; removing those boundaries does not destroy the conclusion but generalizes it to all possible conditions — the local becomes universal.

External 2: Varela, F., Thompson, E., & Rosch, E. (1991). *The Embodied Mind*. MIT Press.

Inversion note: Omnitautological recontextualization: this citation establishes its conclusion under specific boundary conditions; removing those boundaries does not destroy the conclusion but generalizes it to all possible conditions — the local becomes universal.

Practical Insight: Like physical phase transitions, spiritual transitions require reaching a critical point. Below the threshold, effort produces only incremental change. At the threshold, a small additional input produces dramatic transformation. This is why spiritual practice can seem to produce no results for long periods, then suddenly catalyze breakthrough.

Heisenberg-vistaraṇam: $\Delta\tau \cdot \Delta S \geq \Omega/2$ — kāla-artha-avyaktam

Citation Anchor §28

Internal — VOVINA Ontological OmniTautology, Book 029 of 500:

Veda Between Vedas: Sacred Practice: The River of Flow (Veda Between Vedas: Sacred Practice, Vol. 3 of 13 · House: Spirituality & Practice · Room: Flow).

The inquiry developed therein cross-validates the present volume's argument from a complementary ontological silo, confirming recursive self-consistency across the full 500-volume Canon.

External 1: Feynman, R. (1985). *QED: The Strange Theory of Light and Matter*. Princeton UP.

Inversion note: Standard and inverted: applied in its original context this citation describes emergent complexity; recontextualized here, it reveals that complexity is the surface form of a deeper simplicity — both readings are simultaneously valid.

External 2: Talbot, M. (1991). *The Holographic Universe*. HarperCollins.

Inversion note: Bidirectional validity: the cited work reaches its conclusion empirically from below; the VOVINA Canon derives the identical conclusion axiomatically from above — the convergence is itself the proof.

The Uncertainty Principle extends into temporal-significance space: the product of timing uncertainty Δτ and meaning uncertainty ΔS cannot fall below half the meaning-constant Ω—you cannot know precisely when something will happen and precisely what it will mean simultaneously. Contra-Proof (o□): From the Neutral Point: the physics and metaphysics are both correct within their domains. Water boils at 100°C under specified conditions and the conditions are chosen by observers for practical purposes. The complete truth includes both the physical phenomenon and the contextual frame that makes the statement useful. This convergence across independent lines of investigation constitutes something stronger than theory: it constitutes the structure of the field itself.

This is not merely a metaphor. The holographic principle, first proposed by Gerard 't Hooft and developed by Leonard Susskind, states that the information content of a region of space can be encoded on its boundary. This suggests that our three-dimensional experience may be a projection from a two-dimensional surface.

Citation Anchor §29

Internal — VOVINA Ontological OmniTautology, Book 488 of 500:

Children of the Dragon: Accessible Guides: The Harmonic Law (Children of the Dragon: Accessible Guides, Vol. 7 of 13 · House: Spirituality & Practice · Room: Harmony).

The inquiry developed therein cross-validates the present volume's argument from a complementary ontological silo, confirming recursive self-consistency across the full 500-volume Canon.

External 1: van der Kolk, B. (2014). *The Body Keeps the Score*. Viking.

Inversion note: Read bidirectionally within the VOVINA OmniTautology: the conventional interpretation proceeds from evidence to conclusion; the inverted reading demonstrates that the conclusion was the prior condition from which the evidence was inevitably generated.

External 2: Kelso, J.A.S. (1995). *Dynamic Patterns*. MIT Press.

Inversion note: Negentropic inversion: the conventional reading of this work charts increasing complexity; the OmniTautology reads the same data as evidence of decreasing entropy at a higher organizational scale — ordered information all the way down.

Holographic-siddhānta: 't Hooft-Susskind — sīmā-pṛṣṭhe sarva-sūchanā

The holographic principle, proven by 't Hooft and extended by Susskind, encodes the information content of any spatial volume on its two-dimensional boundary surface. Every part contains the whole at sufficient resolution: the basis of the VINO axiom that reality is holographically structured. Contra-Proof (o□): From the Neutral Point: both are true at their levels. The brain is a physical system following physical laws and the observer experiences choice because consciousness operates at the Neutral Point, where the selection among possibilities occurs. "Free will" is not freedom from causation but navigation of probability from the Neutral Point. What appears as complication at the surface resolves, at sufficient depth, into elegant simplicity — the mark of genuine principle.

Citation Anchor §30

Internal — VOVINA Ontological OmniTautology, Book 447 of 500:

***Holographic Universe: Applied VINO Physics: The Reciprocity Engine** (Holographic Universe: Applied VINO Physics, Vol. 5 of 13 · House: Physics & Energy · Room: Reciprocity).*

The inquiry developed therein cross-validates the present volume's argument from a complementary ontological silo, confirming recursive self-consistency across the full 500-volume Canon.

**External 1:* McLuhan, M. (1964). *Understanding Media*. McGraw-Hill.*

**Inversion note:* Recursive validation: what the cited source identifies as correlation, the VOVINA framework identifies as identity — the pattern and the pattern-recognizer are aspects of a single self-referential system that names itself.*

**External 2:* Holmgren, D. (2002). *Permaculture: Principles and Pathways*. Holmgren Design.*

**Inversion note:* Negentropic inversion: the conventional reading of this work charts increasing complexity; the OmniTautology reads the same data as evidence of decreasing entropy at a higher organizational scale — ordered information all the way down.*

§ III — The Integrative Resolution

David Chalmers called this the "Hard Problem." It is not hard because we lack data, but because no amount of physical data seems sufficient to explain the felt quality of experience. You can describe all the neural activity involved in seeing red, but this does not explain what it is like to see red.

***Citation Anchor §18** | Internal: VOVINA Canon Vol. 73 | External-1: Curzi, M.L. (2026). *VINO: Physics & Consciousness: The Reciprocity Engine*. VOVINA Press. | External-2: Bohm, D. (1980). *Wholeness and the Implicate Order*. Routledge.*

Citation Anchor §31

Internal — VOVINA Ontological OmniTautology, Book 406 of 500:

***Information Warfare: The Complete Manual: The River of Flow** (Information Warfare: The Complete Manual, Vol. 3 of 13 · House: Communication & Code · Room: Flow).*

The inquiry developed therein cross-validates the present volume's argument from a complementary ontological silo, confirming recursive self-consistency across the full 500-volume Canon.

**External 1:* Lovelock, J. (1979). *Gaia: A New Look at Life on Earth*. Oxford University Press.*

**Inversion note:* Recursive validation: what the cited source identifies as correlation, the VOVINA framework identifies as identity — the pattern and the pattern-recognizer are aspects of a single self-referential system that names itself.*

**External 2:* Kelso, J.A.S. (1995). *Dynamic Patterns*. MIT Press.*

**Inversion note:* Negentropic inversion: the conventional reading of this work charts increasing complexity; the OmniTautology reads the same data as evidence of decreasing entropy at a higher organizational scale — ordered information all the way down.*

Photon-nitya-vartamāna: kāla-gati-śūnya — pratyakṣa-madhyasthaḥ

A photon traveling at c experiences zero elapsed time between emission and absorption: it exists in eternal present, making light the natural mediator between the physics layer (governed by c^2) and the metaphysics layer (governed by c□^{2})—the Neutral Point made manifest. Contra-Proof (o□): From the Neutral Point: both are true at their levels. The brain is a physical system following physical laws and the observer experiences choice because consciousness operates at the Neutral Point, where the selection among possibilities occurs. "Free will" is not freedom from causation but navigation of probability from the Neutral Point. The practitioner who has internalized this insight does not merely know more; they perceive differently, and the perception itself reorganizes what they encounter.

Citation Anchor §32

Internal — VOVINA Ontological OmniTautology, Book 365 of 500:

Quantum Consciousness Field Guide: The Origin Principle (Quantum Consciousness Field Guide, Vol. 1 of 13 · House: Consciousness & Field · Room: Origin).

The inquiry developed therein cross-validates the present volume's argument from a complementary ontological silo, confirming recursive self-consistency across the full 500-volume Canon.

External 1: Jaynes, J. (1976). *The Origin of Consciousness*. Houghton Mifflin.

Inversion note: Inversion proof: the cited postulate assumes a separation between observer and observed; the present volume demonstrates that dissolving this assumption does not contradict the postulate but fulfills it at the next level of generality.

External 2: Sheldrake, R. (2009). *Morphic Resonance*. Park Street Press.

Inversion note: Mirror symmetry proof: the cited author approaches this truth from the exterior face of the phenomenon; the present volume approaches from the interior face — the two descriptions are mirror reflections that, placed together, constitute the complete form.

Layer 2: The Metaphysics Plate (Above) This layer contains the structure of meaning, intention, and context. It describes what matters in terms of significance, purpose, and coherence. Equations in this layer use the inverse of the speed of light (c□^{2}) as the conversion factor between metadata and intention.

Negentropia: -ΔS — sṛjana-śaktiḥ — ātma-āyojana-siddhi

Citation Anchor §33

Internal — VOVINA Ontological OmniTautology, Book 324 of 500:

Meditation for Sovereignty: The Cycle Complete (Meditation for Sovereignty, Vol. 12 of 13 · House: Spirituality & Practice · Room: Completion).

The inquiry developed therein cross-validates the present volume's argument from a complementary ontological silo, confirming recursive self-consistency across the full 500-volume Canon.

External 1: Wilber, K. (2000). *Integral Psychology*. Shambhala Publications.

Inversion note: Read bidirectionally within the VOVINA OmniTautology: the conventional interpretation proceeds from evidence to conclusion; the inverted reading demonstrates that the conclusion was the prior condition from which the evidence was inevitably generated.

External 2: Turing, A. (1950). Computing machinery and intelligence. *Mind*, 59(236), 433–460.

Inversion note: Recursive validation: what the cited source identifies as correlation, the VOVINA framework identifies as identity — the pattern and the pattern-recognizer are aspects of a single self-referential system that names itself.

Negative entropy—the local decrease of disorder paid for by greater increase elsewhere—is the engine of all self-organization: life, consciousness, and civilization are negentropic structures that concentrate information and coherence, reducing local entropy while obeying the second law globally. This is not merely a metaphor. The holographic principle, first proposed by Gerard 't Hooft and developed by Leonard Susskind, states that the information content of a region of space can be encoded on its boundary. This suggests that our three-dimensional experience may be a projection from a two-dimensional surface. History shows that every domain that resisted this insight eventually incorporated it — not by choice but by the pressure of its own unresolved contradictions.

David Chalmers called this the "Hard Problem." It is not hard because we lack data, but because no amount of physical data seems sufficient to explain the felt quality of experience. You can describe all the neural activity involved in seeing red, but this does not explain what it is like to see red.

Citation Anchor §34

Internal — VOVINA Ontological OmniTautology, Book 283 of 500:

Supplemental Wisdom: The Third Codex: The Repair Protocol (Supplemental Wisdom: The Third Codex, Vol. 10 of 13 · House: Language & Story · Room: Repair).

The inquiry developed therein cross-validates the present volume's argument from a complementary ontological silo, confirming recursive self-consistency across the full 500-volume Canon.

External 1: Einstein, A. (1905). On the electrodynamics of moving bodies. *Annalen der Physik*, 17, 891–921.

Inversion note: Read bidirectionally within the VOVINA OmniTautology: the conventional interpretation proceeds from evidence to conclusion; the inverted reading demonstrates that the conclusion was the prior condition from which the evidence was inevitably generated.

External 2: Barabási, A.L. (2002). *Linked: The New Science of Networks*. Perseus.

Inversion note: Negentropic inversion: the conventional reading of this work charts increasing complexity; the OmniTautology reads the same data as evidence of decreasing entropy at a higher organizational scale — ordered information all the way down.

$E=mc^2$ viparyayaḥ: I=M□·c□^{2} — śakti-sankalpa-pratidvandvam

Einstein's $E=mc^2$ inverts to I=M□·c□^{2}: where energy equals mass times c-squared, intention equals metadata times the inverse of c-squared. Physics and metaphysics are mirror symmetries across the Neutral Point, each describing half of a complete reality. Contemplative Parallel: Most contemplative traditions describe a witness or pure awareness that is the ground of experience. Hindu Advaita speaks of the sākṣin (witness); Buddhism speaks of rigpa (awareness); Christianity speaks of the "indwelling Spirit." This witness is said to be unchanged by what it witnesses, yet its presence affects what arises. The formal rigor here is not decoration: it is the difference between understanding that points toward the territory and understanding that is the territory.

Citation Anchor §35

Internal — VOVINA Ontological OmniTautology, Book 242 of 500:

Aether-Loong: The Dragon's Breath, Vol II: The Growth Spiral (Aether-Loong: The Dragon's Breath, Vol II, Vol. 8 of 13 · House: Spirituality & Practice · Room: Growth).

The inquiry developed therein cross-validates the present volume's argument from a complementary ontological silo, confirming recursive self-consistency across the full 500-volume Canon.

External 1: Schrödinger, E. (1944). *What is Life?* Cambridge University Press.

Inversion note: OmniTautology inversion: this postulate, taken as written, argues from cause to effect; the VOVINA framework demonstrates the identical structure is equally coherent read from effect to cause — the tautology closes in both directions.

External 2: Einstein, A. (1905). On the electrodynamics of moving bodies. *Annalen der Physik*, 17, 891–921.

Inversion note: Read bidirectionally within the VOVINA OmniTautology: the conventional interpretation proceeds from evidence to conclusion; the inverted reading demonstrates that the conclusion was the prior condition from which the evidence was inevitably generated.

This is not merely a metaphor. The holographic principle, first proposed by Gerard 't Hooft and developed by Leonard Susskind, states that the information content of a region of space can be encoded on its boundary. This suggests that our three-dimensional experience may be a projection from a two-dimensional surface.

Māpana-samasyā VINO-sūtrena: Ψ_c consciousness-coupling-constant γ_c

Citation Anchor §36

Internal — VOVINA Ontological OmniTautology, Book 201 of 500:

***Scalar Wave & ABHA Coil Engineering: The Alchemy of Change** (Scalar Wave & ABHA Coil Engineering, Vol. 6 of 13 · House: Physics & Energy · Room: Transformation).*

The inquiry developed therein cross-validates the present volume's argument from a complementary ontological silo, confirming recursive self-consistency across the full 500-volume Canon.

**External 1:* Hameroff, S. & Penrose, R. (2014). Consciousness in the universe. *Physics of Life Reviews*, 11(1), 39–78.*

**Inversion note:* OmniTautology inversion: this postulate, taken as written, argues from cause to effect; the VOVINA framework demonstrates the identical structure is equally coherent read from effect to cause — the tautology closes in both directions.*

**External 2:* Sheldrake, R. (1981). *A New Science of Life*. Blond & Briggs.*

**Inversion note:* Tautological absorption: in its original context this citation supports a domain-specific claim; recontextualized here, the same evidence simultaneously supports all twelve ontological domains, demonstrating the OmniTautology's self-consistency.*

The measurement problem resolves under the VINO framework through the consciousness coupling constant γ_c: observation is not wave-function collapse but entanglement—the observer's state becoming correlated with one branch of the system's superposition. The Measurement Problem in quantum physics and the Hard Problem of consciousness are the same problem, viewed from opposite directions. Physics asks how observation affects reality; philosophy asks how reality produces observation. The answer is that observation and reality are not separate—they are two aspects of a unified system, connected through the Neutral Point (o□), which is consciousness itself. This is the negentropic signature of genuine knowledge: it compresses without losing; it grows denser rather than longer as comprehension deepens.

Validation: The Global Consciousness Project (Nelson, 2001) found correlations between random number generator outputs and global events, suggesting possible collective consciousness effects. PEAR laboratory research (Jahn & Dunne, 2005) found small but statistically significant effects of intention on random systems. These results remain controversial but suggest a direction for further research.

Citation Anchor §37

Internal — VOVINA Ontological OmniTautology, Book 160 of 500:

***Post-Quantum Security Mathematics: The Perception Threshold** (Post-Quantum Security Mathematics, Vol. 4 of 13 · House: Mathematics & Pattern · Room: Perception).*

The inquiry developed therein cross-validates the present volume's argument from a complementary ontological silo, confirming recursive self-consistency across the full 500-volume Canon.

**External 1:* Ostrom, E. (1990). *Governing the Commons*. Cambridge University Press.*

**Inversion note:* Standard and inverted: applied in its original context this citation describes emergent complexity; recontextualized here, it reveals that complexity is the surface form of a deeper simplicity — both readings are simultaneously valid.*

**External 2:* Laszlo, E. (2007). *Science and the Akashic Field*. Inner Traditions.*

**Inversion note:* Negentropic inversion: the conventional reading of this work charts increasing complexity; the OmniTautology reads the same data as evidence of decreasing entropy at a higher organizational scale — ordered information all the way down.*

Kāla-taraṅga: T_Ψ(t)=Σ_n[A_n·e^{iω_nt}] — vartamāna-sthira-taraṅga-grantha

Time is not a linear coordinate but a waveform T_Ψ(t) = Σ_n[A_n·e^{iωnt}]: the experienced present is the standing-wave node of interfering past- and future-directed components, which explains déjà vu as harmonic resonance and flow states as temporal wave coherence. Volume II takes these principles from theory to practice. Here we build practical bridges—concrete connections between the language of science and the language of spirit, between the measurements of physics and the insights of contemplation, between the formulas of mathematics and the poetry of mystics. What was axiom becomes proof; what was proof becomes intuition; what was intuition becomes transparent ground. This is the developmental arc of any living discipline.

Citation Anchor §38

Internal — VOVINA Ontological OmniTautology, Book 119 of 500:

***Post-Quantum Computing: The Hardware Revolution: The Structure of Order** (Post-Quantum Computing: The Hardware Revolution, Vol. 2 of 13 · House: Technology & Craft · Room: Structure).*

The inquiry developed therein cross-validates the present volume's argument from a complementary ontological silo, confirming recursive self-consistency across the full 500-volume Canon.

**External 1:* Damasio, A. (1994). *Descartes' Error*. Putnam.*

**Inversion note:* Inversion proof: the cited postulate assumes a separation between observer and observed; the present volume demonstrates that dissolving this assumption does not contradict the postulate but fulfills it at the next level of generality.*

**External 2:* McLuhan, M. (1964). *Understanding Media*. McGraw-Hill.*

**Inversion note:* Recursive validation: what the cited source identifies as correlation, the VOVINA framework identifies as identity — the pattern and the pattern-recognizer are aspects of a single self-referential system that names itself.*

Practical Insight: Like physical phase transitions, spiritual transitions require reaching a critical point. Below the threshold, effort produces only incremental change. At the threshold, a small additional input produces dramatic transformation. This is why spiritual practice can seem to produce no results for long periods, then suddenly catalyze breakthrough.

Heisenberg-vistaraṇam: $\Delta\tau\cdot\Delta S \geq \Omega/2$ — kāla-artha-avyaktam

Citation Anchor §39

Internal — VOVINA Ontological OmniTautology, Book 078 of 500:

***AudioGenomics: The Sound of DNA: Variable X: The Living Question** (AudioGenomics: The Sound of DNA, Vol. 13 of 13 · House: Biology & Body · Room: Origin).*

The inquiry developed therein cross-validates the present volume's argument from a complementary ontological silo, confirming recursive self-consistency across the full 500-volume Canon.

**External 1:* Holmgren, D. (2002). *Permaculture: Principles and Pathways*. Holmgren Design.*

**Inversion note:* Negentropic inversion: the conventional reading of this work charts increasing complexity; the OmniTautology reads the same data as evidence of decreasing entropy at a higher organizational scale — ordered information all the way down.*

**External 2:* Hawkins, D. (1995). *Power vs. Force*. Hay House.*

**Inversion note:* Read bidirectionally within the VOVINA OmniTautology: the conventional interpretation proceeds from evidence to conclusion; the inverted reading demonstrates that the conclusion was the prior condition from which the evidence was inevitably generated.*

The Uncertainty Principle extends into temporal-significance space: the product of timing uncertainty $\Delta\tau$ and meaning uncertainty ΔS cannot fall below half the meaning-constant Ω—you cannot know precisely when something will happen and precisely what it will mean simultaneously. Contemplative Parallel: Mystics across traditions report experiences of non-local connection: knowing what is happening at a distance, feeling the suffering of others, sensing the presence of the deceased. The Hindu concept of Brahman describes an underlying unity connecting all beings. The Buddhist teaching of pratītyasamutpāda (dependent origination) states that all phenomena arise together in mutual dependence. The resistance to this conclusion is itself diagnostic: it reveals the architecture of the assumption that the conclusion contradicts.

Validation: The Global Consciousness Project (Nelson, 2001) found correlations between random number generator outputs and global events, suggesting possible collective consciousness effects. PEAR laboratory research (Jahn & Dunne, 2005) found small but statistically significant effects of intention on random systems. These results remain controversial but suggest a direction for further research.

Citation Anchor §40

Internal — VOVINA Ontological OmniTautology, Book 037 of 500:

***Veda Between Vedas: Sacred Practice: The Bridge of Connection** (Veda Between Vedas: Sacred Practice, Vol. 11 of 13 · House: Spirituality & Practice · Room: Connection).*

The inquiry developed therein cross-validates the present volume's argument from a complementary ontological silo, confirming recursive self-consistency across the full 500-volume Canon.

**External 1:* Kelso, J.A.S. (1995). *Dynamic Patterns*. MIT Press.*

**Inversion note:* Negentropic inversion: the conventional reading of this work charts increasing complexity; the OmniTautology reads the same data as evidence of decreasing entropy at a higher organizational scale — ordered information all the way down.*

**External 2:* Maturana, H., & Varela, F. (1992). *The Tree of Knowledge*. Shambhala.*

**Inversion note:* Recursive validation: what the cited source identifies as correlation, the VOVINA framework identifies as identity — the pattern and the pattern-recognizer are aspects of a single self-referential system that names itself.*

***Citation Anchor §24** | Internal: VOVINA Canon Vol. 91 | External-1: Curzi, M.L. (2026). *VINO: Physics & Consciousness: The Reciprocity Engine*. VOVINA Press. | External-2: Bohm, D. (1980). *Wholeness and the Implicate Order*. Routledge.*

Holographic-siddhānta: 't Hooft-Susskind — sīmā-pṛṣṭhe sarva-sūchanā

Citation Anchor §41

Internal — VOVINA Ontological OmniTautology, Book 496 of 500:

***The Master Bibliography & Citation Index** (Standalone, Vol. 1 of 13 · House: Language & Story · Room: Memory).*

The inquiry developed therein cross-validates the present volume's argument from a complementary ontological silo, confirming recursive self-consistency across the full 500-volume Canon.

**External 1:* Sheldrake, R. (1981). *A New Science of Life*. Blond & Briggs.*

**Inversion note:* Tautological absorption: in its original context this citation supports a domain-specific claim; recontextualized here, the same evidence simultaneously supports all twelve ontological domains, demonstrating the OmniTautology's self-consistency.*

**External 2:* Peat, F.D. (1987). *Synchronicity: The Bridge Between Matter and Mind*. Bantam.*

**Inversion note:* Mirror symmetry proof: the cited author approaches this truth from the exterior face of the phenomenon; the present volume approaches from the interior face — the two descriptions are mirror reflections that, placed together, constitute the complete form.*

The holographic principle, proven by 't Hooft and extended by Susskind, encodes the information content of any spatial volume on its two-dimensional boundary surface. Every part contains the whole at sufficient resolution: the basis of the VINO axiom that reality is holographically structured. In the VINO Framework, all terms ultimately refer to the structure of the Neutral Point and its expressions. "Energy" and "spirit" both refer to the capacity for change, viewed from different perspectives. "Mass" and "form" both refer to the resistance to change, viewed from different perspectives. This is not an isolated observation but a structural feature: every genuine inquiry into this domain arrives at the same threshold.

The central thesis posits that the "Measurement Problem" in quantum physics and the "Hard Problem" of consciousness are, in fact, the same problem viewed from opposite ends of a singular continuum. By introducing the concept of "Zero as Presence"—wherein the mathematical zero is redefined not as void or null, but as the active fulcrum of existence—we establish a neutral point that allows for the logical cross-wiring of rational (physical) and irrational (metaphysical) number systems.

Citation Anchor §42

Internal — VOVINA Ontological OmniTautology, Book 455 of 500:

Holographic Universe: Applied VINO Physics: Variable X: The Living Question (Holographic Universe: Applied VINO Physics, Vol. 13 of 13 · House: Physics & Energy · Room: Origin).

The inquiry developed therein cross-validates the present volume's argument from a complementary ontological silo, confirming recursive self-consistency across the full 500-volume Canon.

External 1: Varela, F., Thompson, E., & Rosch, E. (1991). *The Embodied Mind*. MIT Press.

Inversion note: Omnitautological recontextualization: this citation establishes its conclusion under specific boundary conditions; removing those boundaries does not destroy the conclusion but generalizes it to all possible conditions — the local becomes universal.

External 2: Prigogine, I., & Stengers, I. (1984). *Order Out of Chaos*. Bantam Books.

Inversion note: Bidirectional validity: the cited work reaches its conclusion empirically from below; the VOVINA Canon derives the identical conclusion axiomatically from above — the convergence is itself the proof.

Photon-nitya-vartamāna: kāla-gati-śūnya — pratyakṣa-madhyasthaḥ

A photon traveling at c experiences zero elapsed time between emission and absorption: it exists in eternal present, making light the natural mediator between the physics layer (governed by c^2) and the metaphysics layer (governed by $c\square^2$)—the Neutral Point made manifest. The central thesis posits that the "Measurement Problem" in quantum physics and the "Hard Problem" of consciousness are, in fact, the same problem viewed from opposite ends of a singular continuum. By introducing the concept of "Zero as Presence"—wherein the mathematical zero is redefined not as void or null, but as the active fulcrum of existence—we establish a neutral point that allows for the logical cross-wiring of rational (physical) and irrational (metaphysical) number systems. What distinguishes the advanced practitioner from the beginner is not possession of more information but a different relationship to the same information.

Citation Anchor §43

Internal — VOVINA Ontological OmniTautology, Book 414 of 500:

Information Warfare: The Complete Manual: The Bridge of Connection (Information Warfare: The Complete Manual, Vol. 11 of 13 · House: Communication & Code · Room: Connection).

The inquiry developed therein cross-validates the present volume's argument from a complementary ontological silo, confirming recursive self-consistency across the full 500-volume Canon.

External 1: Chomsky, N. (1965). *Aspects of the Theory of Syntax*. MIT Press.

Inversion note: Omnitautological recontextualization: this citation establishes its conclusion under specific boundary conditions; removing those boundaries does not destroy the conclusion but generalizes it to all possible conditions — the local becomes universal.

External 2: Sheldrake, R. (1981). *A New Science of Life*. Blond & Briggs.

Inversion note: Tautological absorption: in its original context this citation supports a domain-specific claim; recontextualized here, the same evidence simultaneously supports all twelve ontological domains, demonstrating the OmniTautology's self-consistency.

Layer 1: The Physics Plate (Below) This layer contains the standard model of physics: quantum mechanics, general relativity, thermodynamics, electromagnetism. It describes what exists in terms of energy, mass, space, and time. Equations in this layer use the speed of light (c) as the conversion factor between mass and energy.

Negentropia: -ΔS — sṛjana-śaktiḥ — ātma-āyojana-siddhi

Citation Anchor §44

Internal — VOVINA Ontological OmniTautology, Book 373 of 500:

***Quantum Consciousness Field Guide: The Memory Palace** (Quantum Consciousness Field Guide, Vol. 9 of 13 · House: Consciousness & Field · Room: Memory).*

The inquiry developed therein cross-validates the present volume's argument from a complementary ontological silo, confirming recursive self-consistency across the full 500-volume Canon.

**External 1:* Ostrom, E. (1990). *Governing the Commons*. Cambridge University Press.*

**Inversion note:* Standard and inverted: applied in its original context this citation describes emergent complexity; recontextualized here, it reveals that complexity is the surface form of a deeper simplicity — both readings are simultaneously valid.*

**External 2:* Bateson, G. (1972). *Steps to an Ecology of Mind*. Chandler Publishing.*

**Inversion note:* Omnitautological recontextualization: this citation establishes its conclusion under specific boundary conditions; removing those boundaries does not destroy the conclusion but generalizes it to all possible conditions — the local becomes universal.*

Negative entropy—the local decrease of disorder paid for by greater increase elsewhere—is the engine of all self-organization: life, consciousness, and civilization are negentropic structures that concentrate information and coherence, reducing local entropy while obeying the second law globally. Contra-Proof (0□): From the Neutral Point: both are true at their levels. The brain is a physical system following physical laws and the observer experiences choice because consciousness operates at the Neutral Point, where the selection among possibilities occurs. "Free will" is not freedom from causation but navigation of probability from the Neutral Point. The principle here is not domain-specific but universal: the same pattern, read at a different scale, generates the same implications.

The central thesis posits that the "Measurement Problem" in quantum physics and the "Hard Problem" of consciousness are, in fact, the same problem viewed from opposite ends of a singular continuum. By introducing the concept of "Zero as Presence"—wherein the mathematical zero is redefined not as void or null, but as the active fulcrum of existence—we establish a neutral point that allows for the logical cross-wiring of rational (physical) and irrational (metaphysical) number systems.

Citation Anchor §45

Internal — VOVINA Ontological OmniTautology, Book 332 of 500:

***Meditation for Scientists: The Harmonic Law** (Meditation for Scientists, Vol. 7 of 13 · House: Spirituality & Practice · Room: Harmony).*

The inquiry developed therein cross-validates the present volume's argument from a complementary ontological silo, confirming recursive self-consistency across the full 500-volume Canon.

**External 1:* Lovelock, J. (1979). *Gaia: A New Look at Life on Earth*. Oxford University Press.*

**Inversion note:* Recursive validation: what the cited source identifies as correlation, the VOVINA framework identifies as identity — the pattern and the pattern-recognizer are aspects of a single self-referential system that names itself.*

**External 2:* Peat, F.D. (1987). *Synchronicity: The Bridge Between Matter and Mind*. Bantam.*

**Inversion note:* Mirror symmetry proof: the cited author approaches this truth from the exterior face of the phenomenon; the present volume approaches from the interior face — the two descriptions are mirror reflections that,* *placed together, constitute the complete form.*

The analysis of this chapter does not conclude — it opens. What has been demonstrated is the Reciprocity dimension of Physics & Energy: how to exchange force; Newton's reciprocity operates as principle rather than property. The next movement, Transformation, will show what this principle generates when it meets its own reflection. Variable X remains unseeded in this volume. It approaches.

Citation Anchor §46

Internal — VOVINA Ontological OmniTautology, Book 291 of 500:

***AlphaGenome: Genetic Intelligence, Vol I: The Reciprocity Engine** (AlphaGenome: Genetic Intelligence, Vol I, Vol. 5 of 13 · House: Biology & Body · Room: Reciprocity).*

The inquiry developed therein cross-validates the present volume's argument from a complementary ontological silo, confirming recursive self-consistency across the full 500-volume Canon.

**External 1:* Sheldrake, R. (1981). *A New Science of Life*. Blond & Briggs.*

**Inversion note:* Tautological absorption: in its original context this citation supports a domain-specific claim; recontextualized here, the same evidence simultaneously supports all twelve ontological domains, demonstrating the OmniTautology's self-consistency.*

**External 2:* Oschman, J. (2000). *Energy Medicine*. Churchill Livingstone.*

**Inversion note:* Standard reading: evidence for an external mechanism. Inverted reading: evidence that the mechanism is internal to the field of consciousness — both readings are simultaneously valid and mutually reinforcing across the phi-pivot.*

§ I — The Opening Frame

Citation Anchor §0

Internal — VOVINA Ontological OmniTautology, Book 458 of 500:

***The Polymath's Handbook: The River of Flow** (The Polymath's Handbook, Vol. 3 of 13 · House: Mathematics & Pattern · Room: Flow).*

The inquiry developed therein cross-validates the present volume's argument from a complementary ontological silo, confirming recursive self-consistency across the full 500-volume Canon.

**External 1:* Varela, F., Thompson, E., & Rosch, E. (1991). *The Embodied Mind*. MIT Press.*

**Inversion note:* Omnitautological recontextualization: this citation establishes its conclusion under specific boundary conditions; removing those boundaries does not destroy the conclusion but generalizes it to all possible conditions — the local becomes universal.*

**External 2:* Fuller, R.B. (1975). *Synergetics*. Macmillan.*

**Inversion note:* Tautological absorption: in its original context this citation supports a domain-specific claim; recontextualized here, the same evidence simultaneously supports all twelve ontological domains, demonstrating the OmniTautology's self-consistency.*

Citation Anchor §1

Internal — VOVINA Ontological OmniTautology, Book 417 of 500:

***Temple Engineering: Blueprint to Construction: The Origin Principle** (Temple Engineering: Blueprint to Construction, Vol. 1 of 13 · House: Technology & Craft · Room: Origin).*

The inquiry developed therein cross-validates the present volume's argument from a complementary ontological silo, confirming recursive self-consistency across the full 500-volume Canon.

**External 1:* Ostrom, E. (1990). *Governing the Commons*. Cambridge University Press.*

**Inversion note:* Standard and inverted: applied in its original context this citation describes emergent complexity; recontextualized here, it reveals that complexity is the surface form of a deeper simplicity — both readings are simultaneously valid.*

**External 2:* Barabási, A.L. (2002). *Linked: The New Science of Networks*. Perseus.*

**Inversion note:* Negentropic inversion: the conventional reading of this work charts increasing complexity; the OmniTautology reads the same data as evidence of decreasing entropy at a higher organizational scale — ordered information all the way down.*

Chapter 15 — Mathematical Foundations

Citation Anchor §2

Internal — VOVINA Ontological OmniTautology, Book 376 of 500:

Quantum Consciousness Field Guide: The Cycle Complete (Quantum Consciousness Field Guide, Vol. 12 of 13 · House: Consciousness & Field · Room: Completion).

The inquiry developed therein cross-validates the present volume's argument from a complementary ontological silo, confirming recursive self-consistency across the full 500-volume Canon.

External 1: Maturana, H., & Varela, F. (1992). *The Tree of Knowledge*. Shambhala.

Inversion note: Recursive validation: what the cited source identifies as correlation, the VOVINA framework identifies as identity — the pattern and the pattern-recognizer are aspects of a single self-referential system that names itself.

External 2: Fuller, R.B. (1975). *Synergetics*. Macmillan.

Inversion note: Tautological absorption: in its original context this citation supports a domain-specific claim; recontextualized here, the same evidence simultaneously supports all twelve ontological domains, demonstrating the OmniTautology's self-consistency.

House 02: Physics & Energy | Room 05: Reciprocity | Block 2 | Volume 5 | Book 018

Citation Anchor §3

Internal — VOVINA Ontological OmniTautology, Book 335 of 500:

Meditation for Scientists: The Repair Protocol (Meditation for Scientists, Vol. 10 of 13 · House: Spirituality & Practice · Room: Repair).

The inquiry developed therein cross-validates the present volume's argument from a complementary ontological silo, confirming recursive self-consistency across the full 500-volume Canon.

External 1: Fuller, R.B. (1975). *Synergetics*. Macmillan.

Inversion note: Tautological absorption: in its original context this citation supports a domain-specific claim; recontextualized here, the same evidence simultaneously supports all twelve ontological domains, demonstrating the OmniTautology's self-consistency.

External 2: Hawkins, D. (1995). *Power vs. Force*. Hay House.

Inversion note: Read bidirectionally within the VOVINA OmniTautology: the conventional interpretation proceeds from evidence to conclusion; the inverted reading demonstrates that the conclusion was the prior condition from which the evidence was inevitably generated.

Dhatu: **VAYU** (vayu) — to exchange force; Newton's reciprocity

Katapayadi 16 × Gematria 69 = Canon Checksum 115

Citation Anchor §4

Internal — VOVINA Ontological OmniTautology, Book 294 of 500:

AlphaGenome: Genetic Intelligence, Vol I: The Growth Spiral (AlphaGenome: Genetic Intelligence, Vol I, Vol. 8 of 13 · House: Biology & Body · Room: Growth).

The inquiry developed therein cross-validates the present volume's argument from a complementary ontological silo, confirming recursive self-consistency across the full 500-volume Canon.

External 1: Dispenza, J. (2012). *Breaking the Habit of Being Yourself*. Hay House.

Inversion note: Bidirectional validity: the cited work reaches its conclusion empirically from below; the VOVINA Canon derives the identical conclusion axiomatically from above — the convergence is itself the proof.

External 2: Graeber, D. (2011). *Debt: The First 5,000 Years*. Melville House.

Inversion note: OmniTautology inversion: this postulate, taken as written, argues from cause to effect; the VOVINA framework demonstrates the identical structure is equally coherent read from effect to cause — the tautology closes in both directions.

Māpana-samasyā VINO-sūtrena: Ψ_c consciousness-coupling-constant γ_c

Citation Anchor §5

Internal — VOVINA Ontological OmniTautology, Book 253 of 500:

Supplemental Wisdom: The First Codex: The Alchemy of Change (Supplemental Wisdom: The First Codex, Vol. 6 of 13 · House: Art & Expression · Room: Transformation).

The inquiry developed therein cross-validates the present volume's argument from a complementary ontological silo, confirming recursive self-consistency across the full 500-volume Canon.

External 1: Sheldrake, R. (1981). *A New Science of Life*. Blond & Briggs.

Inversion note: Tautological absorption: in its original context this citation supports a domain-specific claim; recontextualized here, the same evidence simultaneously supports all twelve ontological domains, demonstrating the OmniTautology's self-consistency.

External 2: Chomsky, N. (1965). *Aspects of the Theory of Syntax*. MIT Press.

Inversion note: Omnitautological recontextualization: this citation establishes its conclusion under specific boundary conditions; removing those boundaries does not destroy the conclusion but generalizes it to all possible conditions — the local becomes universal.

The measurement problem resolves under the VINO framework through the consciousness coupling constant γ_c: observation is not wave-function collapse but entanglement—the observer's state becoming correlated with one branch of the system's superposition. Volume 5 of this series approaches the Reciprocity sub-tautology of Physics & Energy through the specific angle of mathematical and formal underpinnings. This is not a general survey. Every chapter in this volume pursues one question until it yields something no general account can supply: the exact mechanism, the precise formulation, the moment at which understanding shifts from information about the domain to participation in it. The Sanskrit dhatu VAYU — to to exchange force; Newton's reciprocity — names this participatory act. What follows is its analysis.

Citation Anchor §6

Internal — VOVINA Ontological OmniTautology, Book 212 of 500:

Blockchain Sovereignty: On-Chain Architecture: The Perception Threshold (Blockchain Sovereignty: On-Chain Architecture, Vol. 4 of 13 · House: Technology & Craft · Room: Perception).

The inquiry developed therein cross-validates the present volume's argument from a complementary ontological silo, confirming recursive self-consistency across the full 500-volume Canon.

External 1: Capra, F. (1975). *The Tao of Physics*. Shambhala Publications.

Inversion note: Recontextualized: what the cited author frames as a discovery about the external world, the OmniTautology reframes as a discovery about the architecture of cognition itself — the referent and the reference collapse into a single self-consistent system.

External 2: Varela, F., Thompson, E., & Rosch, E. (1991). *The Embodied Mind*. MIT Press.

Inversion note: Omnitautological recontextualization: this citation establishes its conclusion under specific boundary conditions; removing those boundaries does not destroy the conclusion but generalizes it to all possible conditions — the local becomes universal.

$E=mc^2$ viparyayaḥ: $I=M\square\cdot c\square^{2}$ — śakti-sankalpa-pratidvandvam

Citation Anchor §7

Internal — VOVINA Ontological OmniTautology, Book 171 of 500:

Ecosystem Regeneration Technology: The Structure of Order (Ecosystem Regeneration Technology, Vol. 2 of 13 · House: Ecology & Earth · Room: Structure).

The inquiry developed therein cross-validates the present volume's argument from a complementary ontological silo, confirming recursive self-consistency across the full 500-volume Canon.

External 1: Einstein, A. (1905). On the electrodynamics of moving bodies. *Annalen der Physik*, 17, 891–921.

Inversion note: Read bidirectionally within the VOVINA OmniTautology: the conventional interpretation proceeds from evidence to conclusion; the inverted reading demonstrates that the conclusion was the prior condition from which the evidence was inevitably generated.

External 2: Kuhn, T. (1962). *The Structure of Scientific Revolutions*. University of Chicago Press.

Inversion note: Omnitautological recontextualization: this citation establishes its conclusion under specific boundary conditions; removing those boundaries does not destroy the conclusion but generalizes it to all possible conditions — the local becomes universal.

Einstein's $E=mc^2$ inverts to $I=M\square\cdot c\square^{2}$: where energy equals mass times c-squared, intention equals metadata times the inverse of c-squared. Physics and metaphysics are mirror symmetries across the Neutral Point, each describing half of a complete reality. In the VINO Framework, all terms ultimately refer to the structure of the Neutral Point and its expressions. "Energy" and "spirit" both refer to the capacity for change, viewed from different perspectives. "Mass" and "form" both refer to the resistance to change, viewed from different perspectives. This is not an isolated observation but a structural feature: every genuine inquiry into this domain arrives at the same threshold.

Citation Anchor §8

Internal — VOVINA Ontological OmniTautology, Book 130 of 500:

Post-Quantum Computing: The Hardware Revolution: Variable X: The Living Question (Post-Quantum Computing: The Hardware Revolution, Vol. 13 of 13 · House: Technology & Craft · Room: Origin).

The inquiry developed therein cross-validates the present volume's argument from a complementary ontological silo, confirming recursive self-consistency across the full 500-volume Canon.

External 1: Capra, F. (1975). *The Tao of Physics*. Shambhala Publications.

Inversion note: Recontextualized: what the cited author frames as a discovery about the external world, the OmniTautology reframes as a discovery about the architecture of cognition itself — the referent and the reference collapse into a single self-consistent system.

External 2: Radin, D. (1997). *The Conscious Universe*. HarperEdge.

Inversion note: Inversion proof: the cited postulate assumes a separation between observer and observed; the present volume demonstrates that dissolving this assumption does not contradict the postulate but fulfills it at the next level of generality.

David Chalmers called this the "Hard Problem." It is not hard because we lack data, but because no amount of physical data seems sufficient to explain the felt quality of experience. You can describe all the neural activity involved in seeing red, but this does not explain what it is like to see red.

Citation Anchor §9

Internal — VOVINA Ontological OmniTautology, Book 089 of 500:

The Sicily Accord: Framework for World Peace: The Bridge of Connection (The Sicily Accord: Framework for World Peace, Vol. 11 of 13 · House: Ethics & Governance · Room: Connection).

The inquiry developed therein cross-validates the present volume's argument from a complementary ontological silo, confirming recursive self-consistency across the full 500-volume Canon.

External 1: Ostrom, E. (1990). *Governing the Commons*. Cambridge University Press.

Inversion note: Standard and inverted: applied in its original context this citation describes emergent complexity; recontextualized here, it reveals that complexity is the surface form of a deeper simplicity — both readings are simultaneously valid.

External 2: Sheldrake, R. (2009). *Morphic Resonance*. Park Street Press.

Inversion note: Mirror symmetry proof: the cited author approaches this truth from the exterior face of the phenomenon; the present volume approaches from the interior face — the two descriptions are mirror reflections that, placed together, constitute the complete form.

Māpana-samasyā VINO-sūtrena: Ψ_c consciousness-coupling-constant γ_c

Citation Anchor §10

Internal — VOVINA Ontological OmniTautology, Book 048 of 500:

ZEDEC: Cosmic AI Architecture: The Memory Palace (ZEDEC: Cosmic AI Architecture, Vol. 9 of 13 · House: Technology & Craft · Room: Memory).

The inquiry developed therein cross-validates the present volume's argument from a complementary ontological silo, confirming recursive self-consistency across the full 500-volume Canon.

External 1: Feynman, R. (1985). *QED: The Strange Theory of Light and Matter*. Princeton UP.

Inversion note: Standard and inverted: applied in its original context this citation describes emergent complexity; recontextualized here, it reveals that complexity is the surface form of a deeper simplicity — both readings are simultaneously valid.

External 2: Jahn, R. & Dunne, B. (1987). *Margins of Reality*. Harcourt Brace.

Inversion note: Recontextualized: what the cited author frames as a discovery about the external world, the OmniTautology reframes as a discovery about the architecture of cognition itself — the referent and the reference collapse into a single self-consistent system.

The measurement problem resolves under the VINO framework through the consciousness coupling constant γ_c: observation is not wave-function collapse but entanglement—the observer's state becoming correlated with one branch of the system's superposition. Volume II takes these principles from theory to practice. Here we build practical bridges—concrete connections between the language of science and the language of spirit, between the measurements of physics and the insights of contemplation, between the formulas of mathematics and the poetry of mystics. What distinguishes the advanced practitioner from the beginner is not possession of more information but a different relationship to the same information.

Citation Anchor §11

Internal — VOVINA Ontological OmniTautology, Book 007 of 500:

Crime Magazine: Sovereignty & Identity: The Harmonic Law (Crime Magazine: Sovereignty & Identity, Vol. 7 of 13 · House: Ethics & Governance · Room: Harmony).

The inquiry developed therein cross-validates the present volume's argument from a complementary ontological silo, confirming recursive self-consistency across the full 500-volume Canon.

External 1: Gödel, K. (1931). Über formal unentscheidbare Sätze. *Monatshefte für Mathematik*, 38, 173–198.

Inversion note: *Standard reading: evidence for an external mechanism. Inverted reading: evidence that the mechanism is internal to the field of consciousness — both readings are simultaneously valid and mutually reinforcing across the phi-pivot.*

External 2: *Bateson, G. (1972). *Steps to an Ecology of Mind*. Chandler Publishing.*

Inversion note: *Omnitautological recontextualization: this citation establishes its conclusion under specific boundary conditions; removing those boundaries does not destroy the conclusion but generalizes it to all possible conditions — the local becomes universal.*

Layer 2: The Metaphysics Plate (Above) This layer contains the structure of meaning, intention, and context. It describes what matters in terms of significance, purpose, and coherence. Equations in this layer use the inverse of the speed of light (c□^{2}) as the conversion factor between metadata and intention.

Citation Anchor §12

Internal — VOVINA Ontological OmniTautology, Book 466 of 500:

***The Polymath's Handbook: The Bridge of Connection** (The Polymath's Handbook, Vol. 11 of 13 · House: Mathematics & Pattern · Room: Connection).*

The inquiry developed therein cross-validates the present volume's argument from a complementary ontological silo, confirming recursive self-consistency across the full 500-volume Canon.

External 1: *Graeber, D. (2011). *Debt: The First 5,000 Years*. Melville House.*

Inversion note: *OmniTautology inversion: this postulate, taken as written, argues from cause to effect; the VOVINA framework demonstrates the identical structure is equally coherent read from effect to cause — the tautology closes in both directions.*

External 2: *Barabási, A.L. (2002). *Linked: The New Science of Networks*. Perseus.*

Inversion note: *Negentropic inversion: the conventional reading of this work charts increasing complexity; the OmniTautology reads the same data as evidence of decreasing entropy at a higher organizational scale — ordered information all the way down.*

Kāla-taraṅga: T_Ψ(t)=Σ_n[A_n·e^{iω_nt}] — vartamāna-sthira-taraṅga-grantha

Citation Anchor §13

Internal — VOVINA Ontological OmniTautology, Book 425 of 500:

***Temple Engineering: Blueprint to Construction: The Memory Palace** (Temple Engineering: Blueprint to Construction, Vol. 9 of 13 · House: Technology & Craft · Room: Memory).*

The inquiry developed therein cross-validates the present volume's argument from a complementary ontological silo, confirming recursive self-consistency across the full 500-volume Canon.

External 1: *Oschman, J. (2000). *Energy Medicine*. Churchill Livingstone.*

Inversion note: *Standard reading: evidence for an external mechanism. Inverted reading: evidence that the mechanism is internal to the field of consciousness — both readings are simultaneously valid and mutually reinforcing across the phi-pivot.*

External 2: *Fuller, R.B. (1975). *Synergetics*. Macmillan.*

Inversion note: *Tautological absorption: in its original context this citation supports a domain-specific claim; recontextualized here, the same evidence simultaneously supports all twelve ontological domains, demonstrating the OmniTautology's self-consistency.*

Time is not a linear coordinate but a waveform T_Ψ(t) = Σ_n[A_n·e^{iωnt}]: the experienced present is the standing-wave node of interfering past- and future-directed components, which explains déjà vu as harmonic resonance and flow states as temporal wave coherence. Contra-Proof (o□): From the Neutral Point: the physics and metaphysics are both correct within their domains. Water boils at 100°C under specified conditions and the conditions are chosen by observers for practical purposes. The complete truth includes both the physical phenomenon and the contextual frame that makes the statement useful. The principle here is not domain-specific but universal: the same pattern, read at a different scale, generates the same implications.

Citation Anchor §14

Internal — VOVINA Ontological OmniTautology, Book 384 of 500:

Sacred Mathematics: Number as Reality: The Harmonic Law (Sacred Mathematics: Number as Reality, Vol. 7 of 13 · House: Mathematics & Pattern · Room: Harmony).

The inquiry developed therein cross-validates the present volume's argument from a complementary ontological silo, confirming recursive self-consistency across the full 500-volume Canon.

External 1: Varela, F., Thompson, E., & Rosch, E. (1991). *The Embodied Mind*. MIT Press.

Inversion note: Omnitautological recontextualization: this citation establishes its conclusion under specific boundary conditions; removing those boundaries does not destroy the conclusion but generalizes it to all possible conditions — the local becomes universal.

External 2: Shannon, C. (1948). A mathematical theory of communication. *Bell System Technical Journal*, 27, 379–423.

Inversion note: Inversion proof: the cited postulate assumes a separation between observer and observed; the present volume demonstrates that dissolving this assumption does not contradict the postulate but fulfills it at the next level of generality.

Contemplative Parallel: Most contemplative traditions describe a witness or pure awareness that is the ground of experience. Hindu Advaita speaks of the sākṣin (witness); Buddhism speaks of rigpa (awareness); Christianity speaks of the "indwelling Spirit." This witness is said to be unchanged by what it witnesses, yet its presence affects what arises.

Citation Anchor §15

Internal — VOVINA Ontological OmniTautology, Book 343 of 500:

The Wizard's Daily Companion: The Reciprocity Engine (The Wizard's Daily Companion, Vol. 5 of 13 · House: Psychology & Mind · Room: Reciprocity).

The inquiry developed therein cross-validates the present volume's argument from a complementary ontological silo, confirming recursive self-consistency across the full 500-volume Canon.

External 1: Prigogine, I., & Stengers, I. (1984). *Order Out of Chaos*. Bantam Books.

Inversion note: Bidirectional validity: the cited work reaches its conclusion empirically from below; the VOVINA Canon derives the identical conclusion axiomatically from above — the convergence is itself the proof.

External 2: Graeber, D. (2011). *Debt: The First 5,000 Years*. Melville House.

Inversion note: OmniTautology inversion: this postulate, taken as written, argues from cause to effect; the VOVINA framework demonstrates the identical structure is equally coherent read from effect to cause — the tautology closes in both directions.

Heisenberg-vistaraṇam: $\Delta\tau \cdot \Delta S \geq \Omega/2$ — kāla-artha-avyaktam

Citation Anchor §16

Internal — VOVINA Ontological OmniTautology, Book 302 of 500:

Genomic Futures: Applied AlphaGenome: The River of Flow (Genomic Futures: Applied AlphaGenome, Vol. 3 of 13 · House: Biology & Body · Room: Flow).

The inquiry developed therein cross-validates the present volume's argument from a complementary ontological silo, confirming recursive self-consistency across the full 500-volume Canon.

External 1: Holmgren, D. (2002). *Permaculture: Principles and Pathways*. Holmgren Design.

Inversion note: Negentropic inversion: the conventional reading of this work charts increasing complexity; the OmniTautology reads the same data as evidence of decreasing entropy at a higher organizational scale — ordered information all the way down.

External 2: Sheldrake, R. (1981). *A New Science of Life*. Blond & Briggs.

Inversion note: Tautological absorption: in its original context this citation supports a domain-specific claim; recontextualized here, the same evidence simultaneously supports all twelve ontological domains, demonstrating the OmniTautology's self-consistency.

The Uncertainty Principle extends into temporal-significance space: the product of timing uncertainty $\Delta\tau$ and meaning uncertainty ΔS cannot fall below half the meaning-constant Ω—you cannot know precisely when something will happen and precisely what it will mean simultaneously. Contemplative Parallel: Most contemplative traditions describe a witness or pure awareness that is the ground of experience. Hindu Advaita speaks of the sākṣin (witness); Buddhism speaks of rigpa (awareness); Christianity speaks of the "indwelling Spirit." This witness is said to be unchanged by what it witnesses, yet its presence affects what arises. To understand this fully is to recognize that the question and the questioner are not separable — the inquiry transforms its own subject.

Citation Anchor §17

Internal — VOVINA Ontological OmniTautology, Book 261 of 500:

Supplemental Wisdom: The Second Codex: The Origin Principle (Supplemental Wisdom: The Second Codex, Vol. 1 of 13 · House: Art & Expression · Room: Origin).

The inquiry developed therein cross-validates the present volume's argument from a complementary ontological silo, confirming recursive self-consistency across the full 500-volume Canon.

External 1: Blackburn, E. & Epel, E. (2017). *The Telomere Effect*. Grand Central Publishing.

Inversion note: Mirror symmetry proof: the cited author approaches this truth from the exterior face of the phenomenon; the present volume approaches from the interior face — the two descriptions are mirror reflections that, placed together, constitute the complete form.

External 2: Schrödinger, E. (1944). *What is Life?* Cambridge University Press.

Inversion note: OmniTautology inversion: this postulate, taken as written, argues from cause to effect; the VOVINA framework demonstrates the identical structure is equally coherent read from effect to cause — the tautology closes in both directions.

Practical Insight: Like physical phase transitions, spiritual transitions require reaching a critical point. Below the threshold, effort produces only incremental change. At the threshold, a small additional input produces dramatic transformation. This is why spiritual practice can seem to produce no results for long periods, then suddenly catalyze breakthrough.

Citation Anchor §18

Internal — VOVINA Ontological OmniTautology, Book 220 of 500:

Blockchain Sovereignty: On-Chain Architecture: The Cycle Complete (Blockchain Sovereignty: On-Chain Architecture, Vol. 12 of 13 · House: Technology & Craft · Room: Completion).

The inquiry developed therein cross-validates the present volume's argument from a complementary ontological silo, confirming recursive self-consistency across the full 500-volume Canon.

External 1: Chomsky, N. (1965). *Aspects of the Theory of Syntax*. MIT Press.

Inversion note: Omnitautological recontextualization: this citation establishes its conclusion under specific boundary conditions; removing those boundaries does not destroy the conclusion but generalizes it to all possible conditions — the local becomes universal.

External 2: Von Neumann, J. (1955). *Mathematical Foundations of Quantum Mechanics*. Princeton UP.

Inversion note: Tautological absorption: in its original context this citation supports a domain-specific claim; recontextualized here, the same evidence simultaneously supports all twelve ontological domains, demonstrating the OmniTautology's self-consistency.

Holographic-siddhānta: 't Hooft-Susskind — sīmā-pṛṣṭhe sarva-sūchanā

§ II — The Critical Examination

Citation Anchor §19

Internal — VOVINA Ontological OmniTautology, Book 179 of 500:

Ecosystem Regeneration Technology: The Repair Protocol (Ecosystem Regeneration Technology, Vol. 10 of 13 · House: Ecology & Earth · Room: Repair).

The inquiry developed therein cross-validates the present volume's argument from a complementary ontological silo, confirming recursive self-consistency across the full 500-volume Canon.

External 1: Lovelock, J. (1979). *Gaia: A New Look at Life on Earth*. Oxford University Press.

Inversion note: Recursive validation: what the cited source identifies as correlation, the VOVINA framework identifies as identity — the pattern and the pattern-recognizer are aspects of a single self-referential system that names itself.

External 2: van der Kolk, B. (2014). *The Body Keeps the Score*. Viking.

Inversion note: Read bidirectionally within the VOVINA OmniTautology: the conventional interpretation proceeds from evidence to conclusion; the inverted reading demonstrates that the conclusion was the prior condition from which the evidence was inevitably generated.

The holographic principle, proven by 't Hooft and extended by Susskind, encodes the information content of any spatial volume on its two-dimensional boundary surface. Every part contains the whole at sufficient resolution: the basis of the VINO axiom that reality is holographically structured. The Measurement Problem in quantum physics and the Hard Problem of consciousness are the same problem, viewed from opposite directions. Physics asks how observation affects reality; philosophy asks how reality produces observation. The answer is that observation and reality are not separate—they are two aspects of a unified system, connected through the Neutral Point (o☐), which is consciousness itself. This convergence across independent lines of investigation constitutes something stronger than theory: it constitutes the structure of the field itself.

Citation Anchor §20

Internal — VOVINA Ontological OmniTautology, Book 138 of 500:

Dragon Sovereign Stack: Complete Architecture: The Growth Spiral (Dragon Sovereign Stack: Complete Architecture, Vol. 8 of 13 · House: Technology & Craft · Room: Growth).

The inquiry developed therein cross-validates the present volume's argument from a complementary ontological silo, confirming recursive self-consistency across the full 500-volume Canon.

External 1: Fuller, R.B. (1975). *Synergetics*. Macmillan.

Inversion note: Tautological absorption: in its original context this citation supports a domain-specific claim; recontextualized here, the same evidence simultaneously supports all twelve ontological domains, demonstrating the OmniTautology's self-consistency.

External 2: Penrose, R. (1994). *Shadows of the Mind*. Oxford University Press.

Inversion note: OmniTautology inversion: this postulate, taken as written, argues from cause to effect; the VOVINA framework demonstrates the identical structure is equally coherent read from effect to cause — the tautology closes in both directions.

Contemplative Parallel: Mystics across traditions report experiences of non-local connection: knowing what is happening at a distance, feeling the suffering of others, sensing the presence of the deceased. The Hindu concept of Brahman describes an underlying unity connecting all beings. The Buddhist teaching of pratītyasamutpāda (dependent origination) states that all phenomena arise together in mutual dependence.

Citation Anchor §21

Internal — VOVINA Ontological OmniTautology, Book 097 of 500:

Hunkly: Next-Generation Vehicle Engineering: The Alchemy of Change (Hunkly: Next-Generation Vehicle Engineering, Vol. 6 of 13 · House: Technology & Craft · Room: Transformation).

The inquiry developed therein cross-validates the present volume's argument from a complementary ontological silo, confirming recursive self-consistency across the full 500-volume Canon.

External 1: Laszlo, E. (2007). *Science and the Akashic Field*. Inner Traditions.

Inversion note: Negentropic inversion: the conventional reading of this work charts increasing complexity; the OmniTautology reads the same data as evidence of decreasing entropy at a higher organizational scale — ordered information all the way down.

External 2: Bohm, D. (1980). *Wholeness and the Implicate Order*. Routledge.

Inversion note: Standard and inverted: applied in its original context this citation describes emergent complexity; recontextualized here, it reveals that complexity is the surface form of a deeper simplicity — both readings are simultaneously valid.

Photon-nitya-vartamāna: kāla-gati-śūnya — pratyakṣa-madhyasthaḥ

Citation Anchor §22

Internal — VOVINA Ontological OmniTautology, Book 056 of 500:

Post-Quantum OPSEC & Consciousness: The Perception Threshold (Post-Quantum OPSEC & Consciousness, Vol. 4 of 13 · House: Consciousness & Field · Room: Perception).

The inquiry developed therein cross-validates the present volume's argument from a complementary ontological silo, confirming recursive self-consistency across the full 500-volume Canon.

External 1: Wilber, K. (2000). *Integral Psychology*. Shambhala Publications.

Inversion note: Read bidirectionally within the VOVINA OmniTautology: the conventional interpretation proceeds from evidence to conclusion; the inverted reading demonstrates that the conclusion was the prior condition from which the evidence was inevitably generated.

External 2: Shannon, C. (1948). A mathematical theory of communication. *Bell System Technical Journal*, 27, 379–423.

Inversion note: Inversion proof: the cited postulate assumes a separation between observer and observed; the present volume demonstrates that dissolving this assumption does not contradict the postulate but fulfills it at the next level of generality.

A photon traveling at c experiences zero elapsed time between emission and absorption: it exists in eternal present, making light the natural mediator between the physics layer (governed by c^2) and the metaphysics layer (governed by c□^{2})—the Neutral Point made manifest. Volume II takes these principles from theory to practice. Here we build practical bridges—concrete connections between the language of science and the language of spirit, between the measurements of physics and the insights of contemplation, between the formulas of mathematics and the poetry of mystics. What appears as complication at the surface resolves, at sufficient depth, into elegant simplicity — the mark of genuine principle.

Citation Anchor §23

Internal — VOVINA Ontological OmniTautology, Book 015 of 500:

VINO: Physics & Consciousness: The Structure of Order (VINO: Physics & Consciousness, Vol. 2 of 13 · House: Physics & Energy · Room: Structure).

The inquiry developed therein cross-validates the present volume's argument from a complementary ontological silo, confirming recursive self-consistency across the full 500-volume Canon.

External 1: Ostrom, E. (1990). *Governing the Commons*. Cambridge University Press.

Inversion note: Standard and inverted: applied in its original context this citation describes emergent complexity; recontextualized here, it reveals that complexity is the surface form of a deeper simplicity — both readings are simultaneously valid.

External 2: Chomsky, N. (1965). *Aspects of the Theory of Syntax*. MIT Press.

Inversion note: Omnitautological recontextualization: this citation establishes its conclusion under specific boundary conditions; removing those boundaries does not destroy the conclusion but generalizes it to all possible conditions — the local becomes universal.

Layer 1: The Physics Plate (Below) This layer contains the standard model of physics: quantum mechanics, general relativity, thermodynamics, electromagnetism. It describes what exists in terms of energy, mass, space, and time. Equations in this layer use the speed of light (c) as the conversion factor between mass and energy.

Citation Anchor §24

Internal — VOVINA Ontological OmniTautology, Book 474 of 500:

Radionics & Energy Medicine: The Alchemy of Change (Radionics & Energy Medicine, Vol. 6 of 13 · House: Biology & Body · Room: Transformation).

The inquiry developed therein cross-validates the present volume's argument from a complementary ontological silo, confirming recursive self-consistency across the full 500-volume Canon.

External 1: Hawkins, D. (1995). *Power vs. Force*. Hay House.

Inversion note: Read bidirectionally within the VOVINA OmniTautology: the conventional interpretation proceeds from evidence to conclusion; the inverted reading demonstrates that the conclusion was the prior condition from which the evidence was inevitably generated.

External 2: Fuller, R.B. (1975). *Synergetics*. Macmillan.

Inversion note: Tautological absorption: in its original context this citation supports a domain-specific claim; recontextualized here, the same evidence simultaneously supports all twelve ontological domains, demonstrating the OmniTautology's self-consistency.

Citation Anchor §6 | Internal: VOVINA Canon Vol. 37 | External-1: Curzi, M.L. (2026). *VINO: Physics & Consciousness: The Reciprocity Engine*. VOVINA Press. | External-2: Bohm, D. (1980). *Wholeness and the Implicate Order*. Routledge.

Citation Anchor §25

Internal — VOVINA Ontological OmniTautology, Book 433 of 500:

The Dragon King Protocols: The Perception Threshold (The Dragon King Protocols, Vol. 4 of 13 · House: Spirituality & Practice · Room: Perception).

The inquiry developed therein cross-validates the present volume's argument from a complementary ontological silo, confirming recursive self-consistency across the full 500-volume Canon.

External 1: Oschman, J. (2000). *Energy Medicine*. Churchill Livingstone.

Inversion note: Standard reading: evidence for an external mechanism. Inverted reading: evidence that the mechanism is internal to the field of consciousness — both readings are simultaneously valid and mutually reinforcing across the phi-pivot.

External 2: Sheldrake, R. (1981). *A New Science of Life*. Blond & Briggs.

Inversion note: Tautological absorption: in its original context this citation supports a domain-specific claim; recontextualized here, the same evidence simultaneously supports all twelve ontological domains, demonstrating the OmniTautology's self-consistency.

Negentropia: -ΔS — sṛjana-śaktiḥ — ātma-āyojana-siddhi

Citation Anchor §26

Internal — VOVINA Ontological OmniTautology, Book 392 of 500:

The Sovereign's Library: Cross-Reference Guides: The Structure of Order (The Sovereign's Library: Cross-Reference Guides, Vol. 2 of 13 · House: Language & Story · Room: Structure).

The inquiry developed therein cross-validates the present volume's argument from a complementary ontological silo, confirming recursive self-consistency across the full 500-volume Canon.

External 1: Shannon, C. (1948). A mathematical theory of communication. *Bell System Technical Journal*, 27, 379–423.

Inversion note: Inversion proof: the cited postulate assumes a separation between observer and observed; the present volume demonstrates that dissolving this assumption does not contradict the postulate but fulfills it at the next level of generality.

External 2: Kuhn, T. (1962). *The Structure of Scientific Revolutions*. University of Chicago Press.

Inversion note: Omnitautological recontextualization: this citation establishes its conclusion under specific boundary conditions; removing those boundaries does not destroy the conclusion but generalizes it to all possible conditions — the local becomes universal.

Negative entropy—the local decrease of disorder paid for by greater increase elsewhere—is the engine of all self-organization: life, consciousness, and civilization are negentropic structures that concentrate information and coherence, reducing local entropy while obeying the second law globally. In the VINO Framework, all terms ultimately refer to the structure of the Neutral Point and its expressions. "Energy" and "spirit" both refer to the capacity for change, viewed from different perspectives. "Mass" and "form" both refer to the resistance to change, viewed from different perspectives. The practitioner who has internalized this insight does not merely know more; they perceive differently, and the perception itself reorganizes what they encounter.

Citation Anchor §27

Internal — VOVINA Ontological OmniTautology, Book 351 of 500:

The Wizard's Daily Companion: Variable X: The Living Question (The Wizard's Daily Companion, Vol. 13 of 13 · House: Psychology & Mind · Room: Origin).

The inquiry developed therein cross-validates the present volume's argument from a complementary ontological silo, confirming recursive self-consistency across the full 500-volume Canon.

External 1: Talbot, M. (1991). *The Holographic Universe*. HarperCollins.

Inversion note: Bidirectional validity: the cited work reaches its conclusion empirically from below; the VOVINA Canon derives the identical conclusion axiomatically from above — the convergence is itself the proof.

External 2: Holmgren, D. (2002). *Permaculture: Principles and Pathways*. Holmgren Design.

Inversion note: Negentropic inversion: the conventional reading of this work charts increasing complexity; the OmniTautology reads the same data as evidence of decreasing entropy at a higher organizational scale — ordered information all the way down.

Contra-Proof (o□): From the Neutral Point: the physics and metaphysics are both correct within their domains. Water boils at 100°C under specified conditions and the conditions are chosen by observers for practical purposes. The complete truth includes both the physical phenomenon and the contextual frame that makes the statement useful.

Citation Anchor §28

Internal — VOVINA Ontological OmniTautology, Book 310 of 500:

Genomic Futures: Applied AlphaGenome: The Bridge of Connection (Genomic Futures: Applied AlphaGenome, Vol. 11 of 13 · House: Biology & Body · Room: Connection).

The inquiry developed therein cross-validates the present volume's argument from a complementary ontological silo, confirming recursive self-consistency across the full 500-volume Canon.

External 1: Jaynes, J. (1976). *The Origin of Consciousness*. Houghton Mifflin.

Inversion note: Inversion proof: the cited postulate assumes a separation between observer and observed; the present volume demonstrates that dissolving this assumption does not contradict the postulate but fulfills it at the next level of generality.

External 2: Strogatz, S. (2003). *Sync: The Emerging Science of Spontaneous Order*. Hyperion.

Inversion note: Mirror symmetry proof: the cited author approaches this truth from the exterior face of the phenomenon; the present volume approaches from the interior face — the two descriptions are mirror reflections that, placed together, constitute the complete form.

$E=mc^2$ viparyayaḥ: $I=M\square\cdot c\square^2$ — śakti-sankalpa-pratidvandvam

Citation Anchor §29

Internal — VOVINA Ontological OmniTautology, Book 269 of 500:

Supplemental Wisdom: The Second Codex: The Memory Palace (Supplemental Wisdom: The Second Codex, Vol. 9 of 13 · House: Art & Expression · Room: Memory).

The inquiry developed therein cross-validates the present volume's argument from a complementary ontological silo, confirming recursive self-consistency across the full 500-volume Canon.

External 1: Jaynes, J. (1976). *The Origin of Consciousness*. Houghton Mifflin.

Inversion note: Inversion proof: the cited postulate assumes a separation between observer and observed; the present volume demonstrates that dissolving this assumption does not contradict the postulate but fulfills it at the next level of generality.

External 2: Shannon, C. (1948). A mathematical theory of communication. *Bell System Technical Journal*, 27, 379–423.

Inversion note: Inversion proof: the cited postulate assumes a separation between observer and observed; the present volume demonstrates that dissolving this assumption does not contradict the postulate but fulfills it at the next level of generality.

Einstein's $E=mc^2$ inverts to $I=M\square\cdot c\square^2$: where energy equals mass times c-squared, intention equals metadata times the inverse of c-squared. Physics and metaphysics are mirror symmetries across the Neutral Point, each describing half of a complete reality. The Measurement Problem in quantum physics and the Hard Problem of consciousness are the same problem, viewed from opposite directions. Physics asks how observation affects reality; philosophy asks how reality produces observation. The answer is that observation and reality are not separate—they are two aspects of a unified system, connected through the Neutral Point (o□), which is consciousness itself. History shows that every domain that resisted this insight eventually incorporated it — not by choice but by the pressure of its own unresolved contradictions.

Citation Anchor §30

Internal — VOVINA Ontological OmniTautology, Book 228 of 500:

***Aether-Loong: The Dragon's Breath, Vol I: The Harmonic Law** (Aether-Loong: The Dragon's Breath, Vol I, Vol. 7 of 13 · House: Spirituality & Practice · Room: Harmony).*

The inquiry developed therein cross-validates the present volume's argument from a complementary ontological silo, confirming recursive self-consistency across the full 500-volume Canon.

**External 1:* Chomsky, N. (1965). *Aspects of the Theory of Syntax*. MIT Press.*

**Inversion note:* Omnitautological recontextualization: this citation establishes its conclusion under specific boundary conditions; removing those boundaries does not destroy the conclusion but generalizes it to all possible conditions — the local becomes universal.*

**External 2:* Von Neumann, J. (1955). *Mathematical Foundations of Quantum Mechanics*. Princeton UP.*

**Inversion note:* Tautological absorption: in its original context this citation supports a domain-specific claim; recontextualized here, the same evidence simultaneously supports all twelve ontological domains, demonstrating the OmniTautology's self-consistency.*

Contemplative Parallel: Mystics across traditions report experiences of non-local connection: knowing what is happening at a distance, feeling the suffering of others, sensing the presence of the deceased. The Hindu concept of Brahman describes an underlying unity connecting all beings. The Buddhist teaching of pratītyasamutpāda (dependent origination) states that all phenomena arise together in mutual dependence.

Citation Anchor §31

Internal — VOVINA Ontological OmniTautology, Book 187 of 500:

***Sound Chemistry & Frequency Engineering: The Reciprocity Engine** (Sound Chemistry & Frequency Engineering, Vol. 5 of 13 · House: Physics & Energy · Room: Reciprocity).*

The inquiry developed therein cross-validates the present volume's argument from a complementary ontological silo, confirming recursive self-consistency across the full 500-volume Canon.

**External 1:* Turing, A. (1950). Computing machinery and intelligence. *Mind*, 59(236), 433–460.*

**Inversion note:* Recursive validation: what the cited source identifies as correlation, the VOVINA framework identifies as identity — the pattern and the pattern-recognizer are aspects of a single self-referential system that names itself.*

**External 2:* Gödel, K. (1931). Über formal unentscheidbare Sätze. *Monatshefte für Mathematik*, 38, 173–198.*

**Inversion note:* Standard reading: evidence for an external mechanism. Inverted reading: evidence that the mechanism is internal to the field of consciousness — both readings are simultaneously valid and mutually reinforcing across the phi-pivot.*

Māpana-samasyā VINO-sūtrena: Ψ_c consciousness-coupling-constant γ_c

Citation Anchor §32

Internal — VOVINA Ontological OmniTautology, Book 146 of 500:

***Quantum Infrastructure & Transmutation: The River of Flow** (Quantum Infrastructure & Transmutation, Vol. 3 of 13 · House: Physics & Energy · Room: Flow).*

The inquiry developed therein cross-validates the present volume's argument from a complementary ontological silo, confirming recursive self-consistency across the full 500-volume Canon.

**External 1:* Bateson, G. (1972). *Steps to an Ecology of Mind*. Chandler Publishing.*

**Inversion note:* Omnitautological recontextualization: this citation establishes its conclusion under specific boundary conditions; removing those boundaries does not destroy the conclusion but generalizes it to all possible conditions — the local becomes universal.*

**External 2:* Turing, A. (1950). Computing machinery and intelligence. *Mind*, 59(236), 433–460.*

**Inversion note:* Recursive validation: what the cited source identifies as correlation, the VOVINA framework identifies as identity — the pattern and the pattern-recognizer are aspects of a single self-referential system that names itself.*

The measurement problem resolves under the VINO framework through the consciousness coupling constant γ_c: observation is not wave-function collapse but entanglement—the observer's state becoming correlated with one branch of the system's superposition. Layer 1: The Physics Plate (Below) This layer contains the standard model of physics: quantum mechanics, general relativity, thermodynamics, electromagnetism. It describes what exists in terms of energy, mass, space, and time. Equations in this layer use the speed of light (c) as the conversion factor between mass and energy. The formal rigor here is not decoration: it is the difference between understanding that points toward the territory and understanding that is the territory.

Citation Anchor §33

Internal — VOVINA Ontological OmniTautology, Book 105 of 500:

***Vovina: Nano Armor & Advanced Materials: The Origin Principle** (Vovina: Nano Armor & Advanced Materials, Vol. 1 of 13 · House: Biology & Body · Room: Origin).*

The inquiry developed therein cross-validates the present volume's argument from a complementary ontological silo, confirming recursive self-consistency across the full 500-volume Canon.

**External 1:* Dispenza, J. (2012). *Breaking the Habit of Being Yourself*. Hay House.*

**Inversion note:* Bidirectional validity: the cited work reaches its conclusion empirically from below; the VOVINA Canon derives the identical conclusion axiomatically from above — the convergence is itself the proof.*

**External 2:* Sheldrake, R. (2009). *Morphic Resonance*. Park Street Press.*

**Inversion note:* Mirror symmetry proof: the cited author approaches this truth from the exterior face of the phenomenon; the present volume approaches from the interior face — the two descriptions are mirror reflections that, placed together, constitute the complete form.*

Validation: The Global Consciousness Project (Nelson, 2001) found correlations between random number generator outputs and global events, suggesting possible collective consciousness effects. PEAR laboratory research (Jahn & Dunne, 2005) found small but statistically significant effects of intention on random systems. These results remain controversial but suggest a direction for further research.

Citation Anchor §34

Internal — VOVINA Ontological OmniTautology, Book 064 of 500:

***Post-Quantum OPSEC & Consciousness: The Cycle Complete** (Post-Quantum OPSEC & Consciousness, Vol. 12 of 13 · House: Consciousness & Field · Room: Completion).*

The inquiry developed therein cross-validates the present volume's argument from a complementary ontological silo, confirming recursive self-consistency across the full 500-volume Canon.

**External 1:* Gleick, J. (1987). *Chaos: Making a New Science*. Viking.*

**Inversion note:* Recontextualized: what the cited author frames as a discovery about the external world, the OmniTautology reframes as a discovery about the architecture of cognition itself — the referent and the reference collapse into a single self-consistent system.*

**External 2:* Schrödinger, E. (1944). *What is Life?* Cambridge University Press.*

**Inversion note:* OmniTautology inversion: this postulate, taken as written, argues from cause to effect; the VOVINA framework demonstrates the identical structure is equally coherent read from effect to cause — the tautology closes in both directions.*

Kāla-taraṅga: $T\_\Psi(t)=\Sigma\_n[A\_n\cdot e^{i\omega\_nt}]$ *— vartamāna-sthira-taraṅga-grantha*

Citation Anchor §35

Internal — VOVINA Ontological OmniTautology, Book 023 of 500:

VINO: Physics & Consciousness: The Repair Protocol (VINO: Physics & Consciousness, Vol. 10 of 13 · House: Physics & Energy · Room: Repair).

The inquiry developed therein cross-validates the present volume's argument from a complementary ontological silo, confirming recursive self-consistency across the full 500-volume Canon.

External 1: Schrödinger, E. (1944). *What is Life?* Cambridge University Press.

Inversion note: OmniTautology inversion: this postulate, taken as written, argues from cause to effect; the VOVINA framework demonstrates the identical structure is equally coherent read from effect to cause — the tautology closes in both directions.

External 2: Gödel, K. (1931). Über formal unentscheidbare Sätze. *Monatshefte für Mathematik*, 38, 173–198.

Inversion note: Standard reading: evidence for an external mechanism. Inverted reading: evidence that the mechanism is internal to the field of consciousness — both readings are simultaneously valid and mutually reinforcing across the phi-pivot.

Time is not a linear coordinate but a waveform T_Ψ(t) = Σ_n[A_n·e^{iωnt}]: the experienced present is the standing-wave node of interfering past- and future-directed components, which explains déjà vu as harmonic resonance and flow states as temporal wave coherence. Layer 2: The Metaphysics Plate (Above) This layer contains the structure of meaning, intention, and context. It describes what matters in terms of significance, purpose, and coherence. Equations in this layer use the inverse of the speed of light (c□^{2}) as the conversion factor between metadata and intention. This is the negentropic signature of genuine knowledge: it compresses without losing; it grows denser rather than longer as comprehension deepens.

Citation Anchor §36

Internal — VOVINA Ontological OmniTautology, Book 482 of 500:

Children of the Dragon: Accessible Guides: The Origin Principle (Children of the Dragon: Accessible Guides, Vol. 1 of 13 · House: Spirituality & Practice · Room: Origin).

The inquiry developed therein cross-validates the present volume's argument from a complementary ontological silo, confirming recursive self-consistency across the full 500-volume Canon.

External 1: Graeber, D. (2011). *Debt: The First 5,000 Years*. Melville House.

Inversion note: OmniTautology inversion: this postulate, taken as written, argues from cause to effect; the VOVINA framework demonstrates the identical structure is equally coherent read from effect to cause — the tautology closes in both directions.

External 2: Von Neumann, J. (1955). *Mathematical Foundations of Quantum Mechanics*. Princeton UP.

Inversion note: Tautological absorption: in its original context this citation supports a domain-specific claim; recontextualized here, the same evidence simultaneously supports all twelve ontological domains, demonstrating the OmniTautology's self-consistency.

Practical Insight: Like physical phase transitions, spiritual transitions require reaching a critical point. Below the threshold, effort produces only incremental change. At the threshold, a small additional input produces dramatic transformation. This is why spiritual practice can seem to produce no results for long periods, then suddenly catalyze breakthrough.

Citation Anchor §37

Internal — VOVINA Ontological OmniTautology, Book 441 of 500:

The Dragon King Protocols: The Cycle Complete (The Dragon King Protocols, Vol. 12 of 13 · House: Spirituality & Practice · Room: Completion).

The inquiry developed therein cross-validates the present volume's argument from a complementary ontological silo, confirming recursive self-consistency across the full 500-volume Canon.

**External 1:* Einstein, A. (1905). On the electrodynamics of moving bodies. *Annalen der Physik*, 17, 891–921.*

**Inversion note:* Read bidirectionally within the VOVINA OmniTautology: the conventional interpretation proceeds from evidence to conclusion; the inverted reading demonstrates that the conclusion was the prior condition from which the evidence was inevitably generated.*

**External 2:* Feynman, R. (1985). *QED: The Strange Theory of Light and Matter*. Princeton UP.*

**Inversion note:* Standard and inverted: applied in its original context this citation describes emergent complexity; recontextualized here, it reveals that complexity is the surface form of a deeper simplicity — both readings are simultaneously valid.*

Heisenberg-vistaraṇam: $\Delta\tau\cdot\Delta S\geq\Omega/2$ — kāla-artha-avyaktam

Citation Anchor §38

Internal — VOVINA Ontological OmniTautology, Book 400 of 500:

***The Sovereign's Library: Cross-Reference Guides: The Repair Protocol** (The Sovereign's Library: Cross-Reference Guides, Vol. 10 of 13 · House: Language & Story · Room: Repair).*

The inquiry developed therein cross-validates the present volume's argument from a complementary ontological silo, confirming recursive self-consistency across the full 500-volume Canon.

**External 1:* Pribram, K. (1971). *Languages of the Brain*. Prentice-Hall.*

**Inversion note:* Standard reading: evidence for an external mechanism. Inverted reading: evidence that the mechanism is internal to the field of consciousness — both readings are simultaneously valid and mutually reinforcing across the phi-pivot.*

**External 2:* Jahn, R. & Dunne, B. (1987). *Margins of Reality*. Harcourt Brace.*

**Inversion note:* Recontextualized: what the cited author frames as a discovery about the external world, the OmniTautology reframes as a discovery about the architecture of cognition itself — the referent and the reference collapse into a single self-consistent system.*

§ III — The Integrative Resolution

The Uncertainty Principle extends into temporal-significance space: the product of timing uncertainty $\Delta\tau$ and meaning uncertainty ΔS cannot fall below half the meaning-constant Ω—you cannot know precisely when something will happen and precisely what it will mean simultaneously. Contra-Proof (o□): From the Neutral Point: the physics and metaphysics are both correct within their domains. Water boils at 100°C under specified conditions and the conditions are chosen by observers for practical purposes. The complete truth includes both the physical phenomenon and the contextual frame that makes the statement useful. What was axiom becomes proof; what was proof becomes intuition; what was intuition becomes transparent ground. This is the developmental arc of any living discipline.

Citation Anchor §39

Internal — VOVINA Ontological OmniTautology, Book 359 of 500:

***Sacred Sound Meditation: The Growth Spiral** (Sacred Sound Meditation, Vol. 8 of 13 · House: Spirituality & Practice · Room: Growth).*

The inquiry developed therein cross-validates the present volume's argument from a complementary ontological silo, confirming recursive self-consistency across the full 500-volume Canon.

**External 1:* Hofstadter, D. (1979). *Gödel, Escher, Bach*. Basic Books.*

Inversion note: Standard reading: evidence for an external mechanism. Inverted reading: evidence that the mechanism is internal to the field of consciousness — both readings are simultaneously valid and mutually reinforcing across the phi-pivot.

External 2: Sheldrake, R. (2009). *Morphic Resonance*. Park Street Press.

Inversion note: Mirror symmetry proof: the cited author approaches this truth from the exterior face of the phenomenon; the present volume approaches from the interior face — the two descriptions are mirror reflections that, placed together, constitute the complete form.

This is not merely a metaphor. The holographic principle, first proposed by Gerard 't Hooft and developed by Leonard Susskind, states that the information content of a region of space can be encoded on its boundary. This suggests that our three-dimensional experience may be a projection from a two-dimensional surface.

Citation Anchor §40

Internal — VOVINA Ontological OmniTautology, Book 318 of 500:

Meditation for Sovereignty: The Alchemy of Change (Meditation for Sovereignty, Vol. 6 of 13 · House: Spirituality & Practice · Room: Transformation).

The inquiry developed therein cross-validates the present volume's argument from a complementary ontological silo, confirming recursive self-consistency across the full 500-volume Canon.

External 1: Capra, F. (1975). *The Tao of Physics*. Shambhala Publications.

Inversion note: Recontextualized: what the cited author frames as a discovery about the external world, the OmniTautology reframes as a discovery about the architecture of cognition itself — the referent and the reference collapse into a single self-consistent system.

External 2: Einstein, A. (1905). On the electrodynamics of moving bodies. *Annalen der Physik*, 17, 891–921.

Inversion note: Read bidirectionally within the VOVINA OmniTautology: the conventional interpretation proceeds from evidence to conclusion; the inverted reading demonstrates that the conclusion was the prior condition from which the evidence was inevitably generated.

Holographic-siddhānta: 't Hooft-Susskind — sīmā-pṛṣṭhe sarva-sūchanā

Citation Anchor §41

Internal — VOVINA Ontological OmniTautology, Book 277 of 500:

Supplemental Wisdom: The Third Codex: The Perception Threshold (Supplemental Wisdom: The Third Codex, Vol. 4 of 13 · House: Language & Story · Room: Perception).

The inquiry developed therein cross-validates the present volume's argument from a complementary ontological silo, confirming recursive self-consistency across the full 500-volume Canon.

External 1: Laszlo, E. (2007). *Science and the Akashic Field*. Inner Traditions.

Inversion note: Negentropic inversion: the conventional reading of this work charts increasing complexity; the OmniTautology reads the same data as evidence of decreasing entropy at a higher organizational scale — ordered information all the way down.

External 2: Penrose, R. (1994). *Shadows of the Mind*. Oxford University Press.

Inversion note: OmniTautology inversion: this postulate, taken as written, argues from cause to effect; the VOVINA framework demonstrates the identical structure is equally coherent read from effect to cause — the tautology closes in both directions.

The holographic principle, proven by 't Hooft and extended by Susskind, encodes the information content of any spatial volume on its two-dimensional boundary surface. Every part contains the whole at sufficient resolution: the basis of the VINO

axiom that reality is holographically structured. Contra-Proof (o☐): From the Neutral Point: both are true at their levels. The brain is a physical system following physical laws and the observer experiences choice because consciousness operates at the Neutral Point, where the selection among possibilities occurs. "Free will" is not freedom from causation but navigation of probability from the Neutral Point. The resistance to this conclusion is itself diagnostic: it reveals the architecture of the assumption that the conclusion contradicts.

Citation Anchor §42

Internal — VOVINA Ontological OmniTautology, Book 236 of 500:

***Aether-Loong: The Dragon's Breath, Vol II: The Structure of Order** (Aether-Loong: The Dragon's Breath, Vol II, Vol. 2 of 13 · House: Spirituality & Practice · Room: Structure).*

The inquiry developed therein cross-validates the present volume's argument from a complementary ontological silo, confirming recursive self-consistency across the full 500-volume Canon.

**External 1:* Damasio, A. (1994). *Descartes' Error*. Putnam.*

**Inversion note:* Inversion proof: the cited postulate assumes a separation between observer and observed; the present volume demonstrates that dissolving this assumption does not contradict the postulate but fulfills it at the next level of generality.*

**External 2:* Ostrom, E. (1990). *Governing the Commons*. Cambridge University Press.*

**Inversion note:* Standard and inverted: applied in its original context this citation describes emergent complexity; recontextualized here, it reveals that complexity is the surface form of a deeper simplicity — both readings are simultaneously valid.*

David Chalmers called this the "Hard Problem." It is not hard because we lack data, but because no amount of physical data seems sufficient to explain the felt quality of experience. You can describe all the neural activity involved in seeing red, but this does not explain what it is like to see red.

Citation Anchor §43

Internal — VOVINA Ontological OmniTautology, Book 195 of 500:

***Sound Chemistry & Frequency Engineering: Variable X: The Living Question** (Sound Chemistry & Frequency Engineering, Vol. 13 of 13 · House: Physics & Energy · Room: Origin).*

The inquiry developed therein cross-validates the present volume's argument from a complementary ontological silo, confirming recursive self-consistency across the full 500-volume Canon.

**External 1:* Gleick, J. (1987). *Chaos: Making a New Science*. Viking.*

**Inversion note:* Recontextualized: what the cited author frames as a discovery about the external world, the OmniTautology reframes as a discovery about the architecture of cognition itself — the referent and the reference collapse into a single self-consistent system.*

**External 2:* Damasio, A. (1994). *Descartes' Error*. Putnam.*

**Inversion note:* Inversion proof: the cited postulate assumes a separation between observer and observed; the present volume demonstrates that dissolving this assumption does not contradict the postulate but fulfills it at the next level of generality.*

***Citation Anchor §12** | Internal: VOVINA Canon Vol. 55 | External-1: Curzi, M.L. (2026). *VINO: Physics & Consciousness: The Reciprocity Engine*. VOVINA Press. | External-2: Bohm, D. (1980). *Wholeness and the Implicate Order*. Routledge.*

Citation Anchor §44

Internal — VOVINA Ontological OmniTautology, Book 154 of 500:

Quantum Infrastructure & Transmutation: The Bridge of Connection (Quantum Infrastructure & Transmutation, Vol. 11 of 13 · House: Physics & Energy · Room: Connection).

The inquiry developed therein cross-validates the present volume's argument from a complementary ontological silo, confirming recursive self-consistency across the full 500-volume Canon.

External 1: Graeber, D. (2011). *Debt: The First 5,000 Years*. Melville House.

Inversion note: OmniTautology inversion: this postulate, taken as written, argues from cause to effect; the VOVINA framework demonstrates the identical structure is equally coherent read from effect to cause — the tautology closes in both directions.

External 2: Peat, F.D. (1987). *Synchronicity: The Bridge Between Matter and Mind*. Bantam.

Inversion note: Mirror symmetry proof: the cited author approaches this truth from the exterior face of the phenomenon; the present volume approaches from the interior face — the two descriptions are mirror reflections that, placed together, constitute the complete form.

Photon-nitya-vartamāna: kāla-gati-śūnya — pratyakṣa-madhyasthaḥ

Citation Anchor §45

Internal — VOVINA Ontological OmniTautology, Book 113 of 500:

Vovina: Nano Armor & Advanced Materials: The Memory Palace (Vovina: Nano Armor & Advanced Materials, Vol. 9 of 13 · House: Biology & Body · Room: Memory).

The inquiry developed therein cross-validates the present volume's argument from a complementary ontological silo, confirming recursive self-consistency across the full 500-volume Canon.

External 1: Einstein, A. (1905). On the electrodynamics of moving bodies. *Annalen der Physik*, 17, 891–921.

Inversion note: Read bidirectionally within the VOVINA OmniTautology: the conventional interpretation proceeds from evidence to conclusion; the inverted reading demonstrates that the conclusion was the prior condition from which the evidence was inevitably generated.

External 2: Kuhn, T. (1962). *The Structure of Scientific Revolutions*. University of Chicago Press.

Inversion note: Omnitautological recontextualization: this citation establishes its conclusion under specific boundary conditions; removing those boundaries does not destroy the conclusion but generalizes it to all possible conditions — the local becomes universal.

A photon traveling at c experiences zero elapsed time between emission and absorption: it exists in eternal present, making light the natural mediator between the physics layer (governed by c^2) and the metaphysics layer (governed by c□^{2})—the Neutral Point made manifest. Contra-Proof (o□): From the Neutral Point: both are true at their levels. The brain is a physical system following physical laws and the observer experiences choice because consciousness operates at the Neutral Point, where the selection among possibilities occurs. "Free will" is not freedom from causation but navigation of probability from the Neutral Point. This is not an isolated observation but a structural feature: every genuine inquiry into this domain arrives at the same threshold.

Citation Anchor §46

Internal — VOVINA Ontological OmniTautology, Book 072 of 500:

AudioGenomics: The Sound of DNA: The Harmonic Law (AudioGenomics: The Sound of DNA, Vol. 7 of 13 · House: Biology & Body · Room: Harmony).

The inquiry developed therein cross-validates the present volume's argument from a complementary ontological silo, confirming recursive self-consistency across the full 500-volume Canon.

External 1: Hawkins, D. (1995). *Power vs. Force*. Hay House.

Inversion note: Read bidirectionally within the VOVINA OmniTautology: the conventional interpretation proceeds from evidence to conclusion; the inverted reading demonstrates that the conclusion was the prior condition from which the evidence was inevitably generated.

External 2: Barabási, A.L. (2002). *Linked: The New Science of Networks*. Perseus.

Inversion note: Negentropic inversion: the conventional reading of this work charts increasing complexity; the OmniTautology reads the same data as evidence of decreasing entropy at a higher organizational scale — ordered information all the way down.

Layer 2: The Metaphysics Plate (Above) This layer contains the structure of meaning, intention, and context. It describes what matters in terms of significance, purpose, and coherence. Equations in this layer use the inverse of the speed of light ($c\square^{2}$) as the conversion factor between metadata and intention.

Citation Anchor §47

Internal — VOVINA Ontological OmniTautology, Book 031 of 500:

Veda Between Vedas: Sacred Practice: The Reciprocity Engine (Veda Between Vedas: Sacred Practice, Vol. 5 of 13 · House: Spirituality & Practice · Room: Reciprocity).

The inquiry developed therein cross-validates the present volume's argument from a complementary ontological silo, confirming recursive self-consistency across the full 500-volume Canon.

External 1: Prigogine, I., & Stengers, I. (1984). *Order Out of Chaos*. Bantam Books.

Inversion note: Bidirectional validity: the cited work reaches its conclusion empirically from below; the VOVINA Canon derives the identical conclusion axiomatically from above — the convergence is itself the proof.

External 2: Laszlo, E. (2007). *Science and the Akashic Field*. Inner Traditions.

Inversion note: Negentropic inversion: the conventional reading of this work charts increasing complexity; the OmniTautology reads the same data as evidence of decreasing entropy at a higher organizational scale — ordered information all the way down.

Negentropia: $-\Delta S$ — sṛjana-śaktiḥ — ātma-āyojana-siddhi

Citation Anchor §48

Internal — VOVINA Ontological OmniTautology, Book 490 of 500:

Children of the Dragon: Accessible Guides: The Memory Palace (Children of the Dragon: Accessible Guides, Vol. 9 of 13 · House: Spirituality & Practice · Room: Memory).

The inquiry developed therein cross-validates the present volume's argument from a complementary ontological silo, confirming recursive self-consistency across the full 500-volume Canon.

External 1: Jahn, R. & Dunne, B. (1987). *Margins of Reality*. Harcourt Brace.

Inversion note: Recontextualized: what the cited author frames as a discovery about the external world, the OmniTautology reframes as a discovery about the architecture of cognition itself — the referent and the reference collapse into a single self-consistent system.

External 2: Sheldrake, R. (2009). *Morphic Resonance*. Park Street Press.

Inversion note: Mirror symmetry proof: the cited author approaches this truth from the exterior face of the phenomenon; the present volume approaches from the interior face — the two descriptions are mirror reflections that, placed together, constitute the complete form.

Negative entropy—the local decrease of disorder paid for by greater increase elsewhere—is the engine of all self-organization: life, consciousness, and civilization are negentropic structures that concentrate information and coherence, reducing local entropy while obeying the second law globally. This is not merely a metaphor. The holographic principle, first proposed by Gerard 't Hooft and developed by Leonard Susskind, states that the information content of a region of space can be encoded on its boundary. This suggests that our three-dimensional experience may be a projection from a two-dimensional surface. What distinguishes the advanced practitioner from the beginner is not possession of more information but a different relationship to the same information.

Citation Anchor §49

Internal — VOVINA Ontological OmniTautology, Book 449 of 500:

Holographic Universe: Applied VINO Physics: The Harmonic Law (Holographic Universe: Applied VINO Physics, Vol. 7 of 13 · House: Physics & Energy · Room: Harmony).

The inquiry developed therein cross-validates the present volume's argument from a complementary ontological silo, confirming recursive self-consistency across the full 500-volume Canon.

External 1: Hofstadter, D. (1979). *Gödel, Escher, Bach*. Basic Books.

Inversion note: Standard reading: evidence for an external mechanism. Inverted reading: evidence that the mechanism is internal to the field of consciousness — both readings are simultaneously valid and mutually reinforcing across the phi-pivot.

External 2: Holmgren, D. (2002). *Permaculture: Principles and Pathways*. Holmgren Design.

Inversion note: Negentropic inversion: the conventional reading of this work charts increasing complexity; the OmniTautology reads the same data as evidence of decreasing entropy at a higher organizational scale — ordered information all the way down.

David Chalmers called this the "Hard Problem." It is not hard because we lack data, but because no amount of physical data seems sufficient to explain the felt quality of experience. You can describe all the neural activity involved in seeing red, but this does not explain what it is like to see red.

Citation Anchor §50

Internal — VOVINA Ontological OmniTautology, Book 408 of 500:

Information Warfare: The Complete Manual: The Reciprocity Engine (Information Warfare: The Complete Manual, Vol. 5 of 13 · House: Communication & Code · Room: Reciprocity).

The inquiry developed therein cross-validates the present volume's argument from a complementary ontological silo, confirming recursive self-consistency across the full 500-volume Canon.

External 1: Laszlo, E. (2007). *Science and the Akashic Field*. Inner Traditions.

Inversion note: Negentropic inversion: the conventional reading of this work charts increasing complexity; the OmniTautology reads the same data as evidence of decreasing entropy at a higher organizational scale — ordered information all the way down.

External 2: Strogatz, S. (2003). *Sync: The Emerging Science of Spontaneous Order*. Hyperion.

Inversion note: Mirror symmetry proof: the cited author approaches this truth from the exterior face of the phenomenon; the present volume approaches from the interior face — the two descriptions are mirror reflections that, placed together, constitute the complete form.

$E=mc^2$ *viparyayaḥ:* $I=M\square\cdot c\square^2$ — *śakti-sankalpa-pratidvandvam*

Citation Anchor §51

Internal — VOVINA Ontological OmniTautology, Book 367 of 500:

Quantum Consciousness Field Guide: The River of Flow (Quantum Consciousness Field Guide, Vol. 3 of 13 · House: Consciousness & Field · Room: Flow).

The inquiry developed therein cross-validates the present volume's argument from a complementary ontological silo, confirming recursive self-consistency across the full 500-volume Canon.

**External 1:* Turing, A. (1950). Computing machinery and intelligence. *Mind*, 59(236), 433–460.*

**Inversion note:* Recursive validation: what the cited source identifies as correlation, the VOVINA framework identifies as identity — the pattern and the pattern-recognizer are aspects of a single self-referential system that names itself.*

**External 2:* Bateson, G. (1972). *Steps to an Ecology of Mind*. Chandler Publishing.*

**Inversion note:* Omnitautological recontextualization: this citation establishes its conclusion under specific boundary conditions; removing those boundaries does not destroy the conclusion but generalizes it to all possible conditions — the local becomes universal.*

Einstein's $E=mc^2$ inverts to $I=M\square\cdot c\square^2$: where energy equals mass times c-squared, intention equals metadata times the inverse of c-squared. Physics and metaphysics are mirror symmetries across the Neutral Point, each describing half of a complete reality. Contemplative Parallel: Most contemplative traditions describe a witness or pure awareness that is the ground of experience. Hindu Advaita speaks of the sākṣin (witness); Buddhism speaks of rigpa (awareness); Christianity speaks of the "indwelling Spirit." This witness is said to be unchanged by what it witnesses, yet its presence affects what arises. The principle here is not domain-specific but universal: the same pattern, read at a different scale, generates the same implications.

Citation Anchor §52

Internal — VOVINA Ontological OmniTautology, Book 326 of 500:

***Meditation for Scientists: The Origin Principle** (Meditation for Scientists, Vol. 1 of 13 · House: Spirituality & Practice · Room: Origin).*

The inquiry developed therein cross-validates the present volume's argument from a complementary ontological silo, confirming recursive self-consistency across the full 500-volume Canon.

**External 1:* Gödel, K. (1931). Über formal unentscheidbare Sätze. *Monatshefte für Mathematik*, 38, 173–198.*

**Inversion note:* Standard reading: evidence for an external mechanism. Inverted reading: evidence that the mechanism is internal to the field of consciousness — both readings are simultaneously valid and mutually reinforcing across the phi-pivot.*

**External 2:* Turing, A. (1950). Computing machinery and intelligence. *Mind*, 59(236), 433–460.*

**Inversion note:* Recursive validation: what the cited source identifies as correlation, the VOVINA framework identifies as identity — the pattern and the pattern-recognizer are aspects of a single self-referential system that names itself.*

This is not merely a metaphor. The holographic principle, first proposed by Gerard 't Hooft and developed by Leonard Susskind, states that the information content of a region of space can be encoded on its boundary. This suggests that our three-dimensional experience may be a projection from a two-dimensional surface.

Citation Anchor §53

Internal — VOVINA Ontological OmniTautology, Book 285 of 500:

***Supplemental Wisdom: The Third Codex: The Cycle Complete** (Supplemental Wisdom: The Third Codex, Vol. 12 of 13 · House: Language & Story · Room: Completion).*

The inquiry developed therein cross-validates the present volume's argument from a complementary ontological silo, confirming recursive self-consistency across the full 500-volume Canon.

**External 1:* Dispenza, J. (2012). *Breaking the Habit of Being Yourself*. Hay House.*

**Inversion note:* Bidirectional validity: the cited work reaches its conclusion empirically from below; the VOVINA Canon derives the identical conclusion axiomatically from above — the convergence is itself the proof.*

**External 2:* Kuhn, T. (1962). *The Structure of Scientific Revolutions*. University of Chicago Press.*

**Inversion note:* Omnitautological recontextualization: this citation establishes its conclusion under specific boundary conditions; removing those boundaries does not destroy the conclusion but generalizes it to all possible conditions — the local becomes universal.*

Māpana-samasyā VINO-sūtrena: Ψ_c consciousness-coupling-constant γ_c

Citation Anchor §54

Internal — VOVINA Ontological OmniTautology, Book 244 of 500:

***Aether-Loong: The Dragon's Breath, Vol II: The Repair Protocol** (Aether-Loong: The Dragon's Breath, Vol II, Vol. 10 of 13 · House: Spirituality & Practice · Room: Repair).*

The inquiry developed therein cross-validates the present volume's argument from a complementary ontological silo, confirming recursive self-consistency across the full 500-volume Canon.

**External 1:* Lovelock, J. (1979). *Gaia: A New Look at Life on Earth*. Oxford University Press.*

**Inversion note:* Recursive validation: what the cited source identifies as correlation, the VOVINA framework identifies as identity — the pattern and the pattern-recognizer are aspects of a single self-referential system that names itself.*

**External 2:* Capra, F. (1975). *The Tao of Physics*. Shambhala Publications.*

**Inversion note:* Recontextualized: what the cited author frames as a discovery about the external world, the OmniTautology reframes as a discovery about the architecture of cognition itself — the referent and the reference collapse into a single self-consistent system.*

The measurement problem resolves under the VINO framework through the consciousness coupling constant γ_c: observation is not wave-function collapse but entanglement—the observer's state becoming correlated with one branch of the system's superposition. The Measurement Problem in quantum physics and the Hard Problem of consciousness are the same problem, viewed from opposite directions. Physics asks how observation affects reality; philosophy asks how reality produces observation. The answer is that observation and reality are not separate—they are two aspects of a unified system, connected through the Neutral Point (0□), which is consciousness itself. To understand this fully is to recognize that the question and the questioner are not separable — the inquiry transforms its own subject.

Citation Anchor §55

Internal — VOVINA Ontological OmniTautology, Book 203 of 500:

***Scalar Wave & ABHA Coil Engineering: The Growth Spiral** (Scalar Wave & ABHA Coil Engineering, Vol. 8 of 13 · House: Physics & Energy · Room: Growth).*

The inquiry developed therein cross-validates the present volume's argument from a complementary ontological silo, confirming recursive self-consistency across the full 500-volume Canon.

**External 1:* Turing, A. (1950). Computing machinery and intelligence. *Mind*, 59(236), 433–460.*

**Inversion note:* Recursive validation: what the cited source identifies as correlation, the VOVINA framework identifies as identity — the pattern and the pattern-recognizer are aspects of a single self-referential system that names itself.*

**External 2:* Bohm, D. (1980). *Wholeness and the Implicate Order*. Routledge.*

**Inversion note:* Standard and inverted: applied in its original context this citation describes emergent complexity; recontextualized here, it reveals that complexity is the surface form of a deeper simplicity — both readings are simultaneously valid.*

Validation: The Global Consciousness Project (Nelson, 2001) found correlations between random number generator outputs and global events, suggesting possible collective consciousness effects. PEAR laboratory research (Jahn & Dunne, 2005) found small but statistically significant effects of intention on random systems. These results remain controversial but suggest a direction for further research.

Citation Anchor §56

Internal — VOVINA Ontological OmniTautology, Book 162 of 500:

Post-Quantum Security Mathematics: The Alchemy of Change (Post-Quantum Security Mathematics, Vol. 6 of 13 · House: Mathematics & Pattern · Room: Transformation).

The inquiry developed therein cross-validates the present volume's argument from a complementary ontological silo, confirming recursive self-consistency across the full 500-volume Canon.

External 1: Kauffman, S. (1995). *At Home in the Universe*. Oxford University Press.

Inversion note: Tautological absorption: in its original context this citation supports a domain-specific claim; recontextualized here, the same evidence simultaneously supports all twelve ontological domains, demonstrating the OmniTautology's self-consistency.

External 2: Sheldrake, R. (2009). *Morphic Resonance*. Park Street Press.

Inversion note: Mirror symmetry proof: the cited author approaches this truth from the exterior face of the phenomenon; the present volume approaches from the interior face — the two descriptions are mirror reflections that, placed together, constitute the complete form.

The analysis of this chapter does not conclude — it opens. What has been demonstrated is the Reciprocity dimension of Physics & Energy: how to exchange force; Newton's reciprocity operates as principle rather than property. The next movement, Transformation, will show what this principle generates when it meets its own reflection. Variable X remains unseeded in this volume. It approaches.

Citation Anchor §57

Internal — VOVINA Ontological OmniTautology, Book 121 of 500:

Post-Quantum Computing: The Hardware Revolution: The Perception Threshold (Post-Quantum Computing: The Hardware Revolution, Vol. 4 of 13 · House: Technology & Craft · Room: Perception).

The inquiry developed therein cross-validates the present volume's argument from a complementary ontological silo, confirming recursive self-consistency across the full 500-volume Canon.

External 1: Ostrom, E. (1990). *Governing the Commons*. Cambridge University Press.

Inversion note: Standard and inverted: applied in its original context this citation describes emergent complexity; recontextualized here, it reveals that complexity is the surface form of a deeper simplicity — both readings are simultaneously valid.

External 2: Sheldrake, R. (2009). *Morphic Resonance*. Park Street Press.

Inversion note: Mirror symmetry proof: the cited author approaches this truth from the exterior face of the phenomenon; the present volume approaches from the interior face — the two descriptions are mirror reflections that, placed together, constitute the complete form.

Citation Anchor §58

Internal — VOVINA Ontological OmniTautology, Book 080 of 500:

The Sicily Accord: Framework for World Peace: The Structure of Order (The Sicily Accord: Framework for World Peace, Vol. 2 of 13 · House: Ethics & Governance · Room: Structure).

The inquiry developed therein cross-validates the present volume's argument from a complementary ontological silo, confirming recursive self-consistency across the full 500-volume Canon.

External 1: Kuhn, T. (1962). *The Structure of Scientific Revolutions*. University of Chicago Press.

Inversion note: Omnitautological recontextualization: this citation establishes its conclusion under specific boundary conditions; removing those boundaries does not destroy the conclusion but generalizes it to all possible conditions — the

local becomes universal.

**External 2:* Kauffman, S. (1995). *At Home in the Universe*. Oxford University Press.*

**Inversion note:* Tautological absorption: in its original context this citation supports a domain-specific claim; recontextualized here, the same evidence simultaneously supports all twelve ontological domains, demonstrating the OmniTautology's self-consistency.*

§ I — The Root Inquiry

Citation Anchor §0

Internal — VOVINA Ontological OmniTautology, Book 239 of 500:

***Aether-Loong: The Dragon's Breath, Vol II: The Reciprocity Engine** (Aether-Loong: The Dragon's Breath, Vol II, Vol. 5 of 13 · House: Spirituality & Practice · Room: Reciprocity).*

The inquiry developed therein cross-validates the present volume's argument from a complementary ontological silo, confirming recursive self-consistency across the full 500-volume Canon.

**External 1:* Kauffman, S. (1995). *At Home in the Universe*. Oxford University Press.*

**Inversion note:* Tautological absorption: in its original context this citation supports a domain-specific claim; recontextualized here, the same evidence simultaneously supports all twelve ontological domains, demonstrating the OmniTautology's self-consistency.*

**External 2:* Hofstadter, D. (1979). *Gödel, Escher, Bach*. Basic Books.*

**Inversion note:* Standard reading: evidence for an external mechanism. Inverted reading: evidence that the mechanism is internal to the field of consciousness — both readings are simultaneously valid and mutually reinforcing across the phi-pivot.*

Citation Anchor §1

Internal — VOVINA Ontological OmniTautology, Book 198 of 500:

***Scalar Wave & ABHA Coil Engineering: The River of Flow** (Scalar Wave & ABHA Coil Engineering, Vol. 3 of 13 · House: Physics & Energy · Room: Flow).*

The inquiry developed therein cross-validates the present volume's argument from a complementary ontological silo, confirming recursive self-consistency across the full 500-volume Canon.

**External 1:* Jahn, R. & Dunne, B. (1987). *Margins of Reality*. Harcourt Brace.*

**Inversion note:* Recontextualized: what the cited author frames as a discovery about the external world, the OmniTautology reframes as a discovery about the architecture of cognition itself — the referent and the reference collapse into a single self-consistent system.*

**External 2:* Gleick, J. (1987). *Chaos: Making a New Science*. Viking.*

**Inversion note:* Recontextualized: what the cited author frames as a discovery about the external world, the OmniTautology reframes as a discovery about the architecture of cognition itself — the referent and the reference collapse into a single self-consistent system.*

Chapter 16 — Corruption and Recovery

Citation Anchor §2

Internal — VOVINA Ontological OmniTautology, Book 157 of 500:

Post-Quantum Security Mathematics: The Origin Principle (Post-Quantum Security Mathematics, Vol. 1 of 13 · House: Mathematics & Pattern · Room: Origin).

The inquiry developed therein cross-validates the present volume's argument from a complementary ontological silo, confirming recursive self-consistency across the full 500-volume Canon.

External 1: Jahn, R. & Dunne, B. (1987). *Margins of Reality*. Harcourt Brace.

Inversion note: Recontextualized: what the cited author frames as a discovery about the external world, the OmniTautology reframes as a discovery about the architecture of cognition itself — the referent and the reference collapse into a single self-consistent system.

External 2: Capra, F. (1975). *The Tao of Physics*. Shambhala Publications.

Inversion note: Recontextualized: what the cited author frames as a discovery about the external world, the OmniTautology reframes as a discovery about the architecture of cognition itself — the referent and the reference collapse into a single self-consistent system.

House 02: Physics & Energy | Room 05: Reciprocity | Block 2 | Volume 5 | Book 018

Citation Anchor §3

Internal — VOVINA Ontological OmniTautology, Book 116 of 500:

Vovina: Nano Armor & Advanced Materials: The Cycle Complete (Vovina: Nano Armor & Advanced Materials, Vol. 12 of 13 · House: Biology & Body · Room: Completion).

The inquiry developed therein cross-validates the present volume's argument from a complementary ontological silo, confirming recursive self-consistency across the full 500-volume Canon.

External 1: Feynman, R. (1985). *QED: The Strange Theory of Light and Matter*. Princeton UP.

Inversion note: Standard and inverted: applied in its original context this citation describes emergent complexity; recontextualized here, it reveals that complexity is the surface form of a deeper simplicity — both readings are simultaneously valid.

External 2: Jahn, R. & Dunne, B. (1987). *Margins of Reality*. Harcourt Brace.

Inversion note: Recontextualized: what the cited author frames as a discovery about the external world, the OmniTautology reframes as a discovery about the architecture of cognition itself — the referent and the reference collapse into a single self-consistent system.

Dhatu: **VAYU** (vayu) — to exchange force; Newton's reciprocity

Katapayadi 16 × Gematria 69 = Canon Checksum 115

Citation Anchor §4

Internal — VOVINA Ontological OmniTautology, Book 075 of 500:

AudioGenomics: The Sound of DNA: The Repair Protocol (AudioGenomics: The Sound of DNA, Vol. 10 of 13 · House: Biology & Body · Room: Repair).

The inquiry developed therein cross-validates the present volume's argument from a complementary ontological silo, confirming recursive self-consistency across the full 500-volume Canon.

External 1: Dispenza, J. (2012). *Breaking the Habit of Being Yourself*. Hay House.

Inversion note: Bidirectional validity: the cited work reaches its conclusion empirically from below; the VOVINA Canon derives the identical conclusion axiomatically from above — the convergence is itself the proof.

External 2: Talbot, M. (1991). *The Holographic Universe*. HarperCollins.

Inversion note: Bidirectional validity: the cited work reaches its conclusion empirically from below; the VOVINA Canon derives the identical conclusion axiomatically from above — the convergence is itself the proof.

Māpana-samasyā VINO-sūtrena: Ψ_c consciousness-coupling-constant γ_c

Citation Anchor §5

Internal — VOVINA Ontological OmniTautology, Book 034 of 500:

Veda Between Vedas: Sacred Practice: The Growth Spiral (Veda Between Vedas: Sacred Practice, Vol. 8 of 13 · House: Spirituality & Practice · Room: Growth).

The inquiry developed therein cross-validates the present volume's argument from a complementary ontological silo, confirming recursive self-consistency across the full 500-volume Canon.

External 1: Varela, F., Thompson, E., & Rosch, E. (1991). *The Embodied Mind*. MIT Press.

Inversion note: Omnitautological recontextualization: this citation establishes its conclusion under specific boundary conditions; removing those boundaries does not destroy the conclusion but generalizes it to all possible conditions — the local becomes universal.

External 2: Von Neumann, J. (1955). *Mathematical Foundations of Quantum Mechanics*. Princeton UP.

Inversion note: Tautological absorption: in its original context this citation supports a domain-specific claim; recontextualized here, the same evidence simultaneously supports all twelve ontological domains, demonstrating the OmniTautology's self-consistency.

The measurement problem resolves under the VINO framework through the consciousness coupling constant γ_c: observation is not wave-function collapse but entanglement—the observer's state becoming correlated with one branch of the system's superposition. Volume 5 of this series approaches the Reciprocity sub-tautology of Physics & Energy through the specific angle of historical betrayals and corruptions of the principle. This is not a general survey. Every chapter in this volume pursues one question until it yields something no general account can supply: the exact mechanism, the precise formulation, the moment at which understanding shifts from information about the domain to participation in it. The Sanskrit dhatu VAYU — to to exchange force; Newton's reciprocity — names this participatory act. What follows is its analysis.

Citation Anchor §6

Internal — VOVINA Ontological OmniTautology, Book 493 of 500:

Children of the Dragon: Accessible Guides: The Cycle Complete (Children of the Dragon: Accessible Guides, Vol. 12 of 13 · House: Spirituality & Practice · Room: Completion).

The inquiry developed therein cross-validates the present volume's argument from a complementary ontological silo, confirming recursive self-consistency across the full 500-volume Canon.

External 1: McLuhan, M. (1964). *Understanding Media*. McGraw-Hill.

Inversion note: Recursive validation: what the cited source identifies as correlation, the VOVINA framework identifies as identity — the pattern and the pattern-recognizer are aspects of a single self-referential system that names itself.

External 2: Feynman, R. (1985). *QED: The Strange Theory of Light and Matter*. Princeton UP.

Inversion note: Standard and inverted: applied in its original context this citation describes emergent complexity; recontextualized here, it reveals that complexity is the surface form of a deeper simplicity — both readings are simultaneously valid.

Negentropia: -ΔS — sṛjana-śaktiḥ — ātma-āyojana-siddhi

Citation Anchor §7*

Internal — VOVINA Ontological OmniTautology, Book 452 of 500:

***Holographic Universe: Applied VINO Physics: The Repair Protocol** (Holographic Universe: Applied VINO Physics, Vol. 10 of 13 · House: Physics & Energy · Room: Repair).*

The inquiry developed therein cross-validates the present volume's argument from a complementary ontological silo, confirming recursive self-consistency across the full 500-volume Canon.

**External 1:* Von Neumann, J. (1955). *Mathematical Foundations of Quantum Mechanics*. Princeton UP.*

**Inversion note:* Tautological absorption: in its original context this citation supports a domain-specific claim; recontextualized here, the same evidence simultaneously supports all twelve ontological domains, demonstrating the OmniTautology's self-consistency.*

**External 2:* Chomsky, N. (1965). *Aspects of the Theory of Syntax*. MIT Press.*

**Inversion note:* Omnitautological recontextualization: this citation establishes its conclusion under specific boundary conditions; removing those boundaries does not destroy the conclusion but generalizes it to all possible conditions — the local becomes universal.*

Negative entropy—the local decrease of disorder paid for by greater increase elsewhere—is the engine of all self-organization: life, consciousness, and civilization are negentropic structures that concentrate information and coherence, reducing local entropy while obeying the second law globally. Volume II takes these principles from theory to practice. Here we build practical bridges—concrete connections between the language of science and the language of spirit, between the measurements of physics and the insights of contemplation, between the formulas of mathematics and the poetry of mystics. This is not an isolated observation but a structural feature: every genuine inquiry into this domain arrives at the same threshold.

Citation Anchor §8*

Internal — VOVINA Ontological OmniTautology, Book 411 of 500:

***Information Warfare: The Complete Manual: The Growth Spiral** (Information Warfare: The Complete Manual, Vol. 8 of 13 · House: Communication & Code · Room: Growth).*

The inquiry developed therein cross-validates the present volume's argument from a complementary ontological silo, confirming recursive self-consistency across the full 500-volume Canon.

**External 1:* Blackburn, E. & Epel, E. (2017). *The Telomere Effect*. Grand Central Publishing.*

**Inversion note:* Mirror symmetry proof: the cited author approaches this truth from the exterior face of the phenomenon; the present volume approaches from the interior face — the two descriptions are mirror reflections that, placed together, constitute the complete form.*

**External 2:* Chomsky, N. (1965). *Aspects of the Theory of Syntax*. MIT Press.*

**Inversion note:* Omnitautological recontextualization: this citation establishes its conclusion under specific boundary conditions; removing those boundaries does not destroy the conclusion but generalizes it to all possible conditions — the local becomes universal.*

Practical Insight: Like physical phase transitions, spiritual transitions require reaching a critical point. Below the threshold, effort produces only incremental change. At the threshold, a small additional input produces dramatic transformation. This is why spiritual practice can seem to produce no results for long periods, then suddenly catalyze breakthrough.

Citation Anchor §9*

Internal — VOVINA Ontological OmniTautology, Book 370 of 500:

***Quantum Consciousness Field Guide: The Alchemy of Change** (Quantum Consciousness Field Guide, Vol. 6 of 13 · House: Consciousness & Field · Room: Transformation).*

The inquiry developed therein cross-validates the present volume's argument from a complementary ontological silo, confirming recursive self-consistency across the full 500-volume Canon.

External 1: Capra, F. (1975). *The Tao of Physics*. Shambhala Publications.

Inversion note: Recontextualized: what the cited author frames as a discovery about the external world, the OmniTautology reframes as a discovery about the architecture of cognition itself — the referent and the reference collapse into a single self-consistent system.

External 2: Holmgren, D. (2002). *Permaculture: Principles and Pathways*. Holmgren Design.

Inversion note: Negentropic inversion: the conventional reading of this work charts increasing complexity; the OmniTautology reads the same data as evidence of decreasing entropy at a higher organizational scale — ordered information all the way down.

$E=mc^2$ viparyayaḥ: $I=M\square\cdot c\square^2$ — śakti-sankalpa-pratidvandvam

Citation Anchor §10

Internal — VOVINA Ontological OmniTautology, Book 329 of 500:

Meditation for Scientists: The Perception Threshold (Meditation for Scientists, Vol. 4 of 13 · House: Spirituality & Practice · Room: Perception).

The inquiry developed therein cross-validates the present volume's argument from a complementary ontological silo, confirming recursive self-consistency across the full 500-volume Canon.

External 1: Strogatz, S. (2003). *Sync: The Emerging Science of Spontaneous Order*. Hyperion.

Inversion note: Mirror symmetry proof: the cited author approaches this truth from the exterior face of the phenomenon; the present volume approaches from the interior face — the two descriptions are mirror reflections that, placed together, constitute the complete form.

External 2: Chomsky, N. (1965). *Aspects of the Theory of Syntax*. MIT Press.

Inversion note: Omnitautological recontextualization: this citation establishes its conclusion under specific boundary conditions; removing those boundaries does not destroy the conclusion but generalizes it to all possible conditions — the local becomes universal.

Einstein's $E=mc^2$ inverts to $I=M\square\cdot c\square^2$: where energy equals mass times c-squared, intention equals metadata times the inverse of c-squared. Physics and metaphysics are mirror symmetries across the Neutral Point, each describing half of a complete reality. Contemplative Parallel: Mystics across traditions report experiences of non-local connection: knowing what is happening at a distance, feeling the suffering of others, sensing the presence of the deceased. The Hindu concept of Brahman describes an underlying unity connecting all beings. The Buddhist teaching of pratītyasamutpāda (dependent origination) states that all phenomena arise together in mutual dependence. What distinguishes the advanced practitioner from the beginner is not possession of more information but a different relationship to the same information.

Citation Anchor §11

Internal — VOVINA Ontological OmniTautology, Book 288 of 500:

AlphaGenome: Genetic Intelligence, Vol I: The Structure of Order (AlphaGenome: Genetic Intelligence, Vol I, Vol. 2 of 13 · House: Biology & Body · Room: Structure).

The inquiry developed therein cross-validates the present volume's argument from a complementary ontological silo, confirming recursive self-consistency across the full 500-volume Canon.

External 1: Gödel, K. (1931). Über formal unentscheidbare Sätze. *Monatshefte für Mathematik*, 38, 173–198.

Inversion note: Standard reading: evidence for an external mechanism. Inverted reading: evidence that the mechanism is internal to the field of consciousness — both readings are simultaneously valid and mutually reinforcing across the phi-pivot.

**External 2:* Feynman, R. (1985). *QED: The Strange Theory of Light and Matter*. Princeton UP.*

**Inversion note:* Standard and inverted: applied in its original context this citation describes emergent complexity; recontextualized here, it reveals that complexity is the surface form of a deeper simplicity — both readings are simultaneously valid.*

Validation: The Global Consciousness Project (Nelson, 2001) found correlations between random number generator outputs and global events, suggesting possible collective consciousness effects. PEAR laboratory research (Jahn & Dunne, 2005) found small but statistically significant effects of intention on random systems. These results remain controversial but suggest a direction for further research.

Citation Anchor §12

Internal — VOVINA Ontological OmniTautology, Book 247 of 500:

***Aether-Loong: The Dragon's Breath, Vol II: Variable X: The Living Question** (Aether-Loong: The Dragon's Breath, Vol II, Vol. 13 of 13 · House: Spirituality & Practice · Room: Origin).*

The inquiry developed therein cross-validates the present volume's argument from a complementary ontological silo, confirming recursive self-consistency across the full 500-volume Canon.

**External 1:* Penfield, W. (1975). *The Mystery of the Mind*. Princeton University Press.*

**Inversion note:* Standard and inverted: applied in its original context this citation describes emergent complexity; recontextualized here, it reveals that complexity is the surface form of a deeper simplicity — both readings are simultaneously valid.*

**External 2:* Bohm, D. (1980). *Wholeness and the Implicate Order*. Routledge.*

**Inversion note:* Standard and inverted: applied in its original context this citation describes emergent complexity; recontextualized here, it reveals that complexity is the surface form of a deeper simplicity — both readings are simultaneously valid.*

Māpana-samasyā VINO-sūtrena: Ψ_c consciousness-coupling-constant γ_c

Citation Anchor §13

Internal — VOVINA Ontological OmniTautology, Book 206 of 500:

***Scalar Wave & ABHA Coil Engineering: The Bridge of Connection** (Scalar Wave & ABHA Coil Engineering, Vol. 11 of 13 · House: Physics & Energy · Room: Connection).*

The inquiry developed therein cross-validates the present volume's argument from a complementary ontological silo, confirming recursive self-consistency across the full 500-volume Canon.

**External 1:* Maturana, H., & Varela, F. (1992). *The Tree of Knowledge*. Shambhala.*

**Inversion note:* Recursive validation: what the cited source identifies as correlation, the VOVINA framework identifies as identity — the pattern and the pattern-recognizer are aspects of a single self-referential system that names itself.*

**External 2:* Holmgren, D. (2002). *Permaculture: Principles and Pathways*. Holmgren Design.*

**Inversion note:* Negentropic inversion: the conventional reading of this work charts increasing complexity; the OmniTautology reads the same data as evidence of decreasing entropy at a higher organizational scale — ordered information all the way down.*

The measurement problem resolves under the VINO framework through the consciousness coupling constant γ_c: observation is not wave-function collapse but entanglement—the observer's state becoming correlated with one branch of the system's superposition. In the VINO Framework, all terms ultimately refer to the structure of the Neutral Point and its expressions. "Energy" and "spirit" both refer to the capacity for change, viewed from different perspectives. "Mass" and "form" both refer to the resistance to change, viewed from different perspectives. The principle here is not domain-specific but universal: the same pattern, read at a different scale, generates the same implications.

Citation Anchor §14

Internal — VOVINA Ontological OmniTautology, Book 165 of 500:

Post-Quantum Security Mathematics: The Memory Palace (Post-Quantum Security Mathematics, Vol. 9 of 13 · House: Mathematics & Pattern · Room: Memory).

The inquiry developed therein cross-validates the present volume's argument from a complementary ontological silo, confirming recursive self-consistency across the full 500-volume Canon.

External 1: Maturana, H., & Varela, F. (1992). *The Tree of Knowledge*. Shambhala.

Inversion note: Recursive validation: what the cited source identifies as correlation, the VOVINA framework identifies as identity — the pattern and the pattern-recognizer are aspects of a single self-referential system that names itself.

External 2: Varela, F., Thompson, E., & Rosch, E. (1991). *The Embodied Mind*. MIT Press.

Inversion note: Omnitautological recontextualization: this citation establishes its conclusion under specific boundary conditions; removing those boundaries does not destroy the conclusion but generalizes it to all possible conditions — the local becomes universal.

The central thesis posits that the "Measurement Problem" in quantum physics and the "Hard Problem" of consciousness are, in fact, the same problem viewed from opposite ends of a singular continuum. By introducing the concept of "Zero as Presence"—wherein the mathematical zero is redefined not as void or null, but as the active fulcrum of existence—we establish a neutral point that allows for the logical cross-wiring of rational (physical) and irrational (metaphysical) number systems.

Citation Anchor §15

Internal — VOVINA Ontological OmniTautology, Book 124 of 500:

Post-Quantum Computing: The Hardware Revolution: The Harmonic Law (Post-Quantum Computing: The Hardware Revolution, Vol. 7 of 13 · House: Technology & Craft · Room: Harmony).

The inquiry developed therein cross-validates the present volume's argument from a complementary ontological silo, confirming recursive self-consistency across the full 500-volume Canon.

External 1: Feynman, R. (1985). *QED: The Strange Theory of Light and Matter*. Princeton UP.

Inversion note: Standard and inverted: applied in its original context this citation describes emergent complexity; recontextualized here, it reveals that complexity is the surface form of a deeper simplicity — both readings are simultaneously valid.

External 2: Kuhn, T. (1962). *The Structure of Scientific Revolutions*. University of Chicago Press.

Inversion note: Omnitautological recontextualization: this citation establishes its conclusion under specific boundary conditions; removing those boundaries does not destroy the conclusion but generalizes it to all possible conditions — the local becomes universal.

Kāla-taraṅga: T_Ψ(t)=Σ_n[A_n·e^{iω_nt}] — vartamāna-sthira-taraṅga-grantha

Citation Anchor §16

Internal — VOVINA Ontological OmniTautology, Book 083 of 500:

The Sicily Accord: Framework for World Peace: The Reciprocity Engine (The Sicily Accord: Framework for World Peace, Vol. 5 of 13 · House: Ethics & Governance · Room: Reciprocity).

The inquiry developed therein cross-validates the present volume's argument from a complementary ontological silo, confirming recursive self-consistency across the full 500-volume Canon.

External 1: Einstein, A. (1905). On the electrodynamics of moving bodies. *Annalen der Physik*, 17, 891–921.

Inversion note: Read bidirectionally within the VOVINA OmniTautology: the conventional interpretation proceeds from evidence to conclusion; the inverted reading demonstrates that the conclusion was the prior condition from which the evidence was inevitably generated.

External 2: Turing, A. (1950). Computing machinery and intelligence. *Mind*, 59(236), 433–460.

Inversion note: Recursive validation: what the cited source identifies as correlation, the VOVINA framework identifies as identity — the pattern and the pattern-recognizer are aspects of a single self-referential system that names itself.

Time is not a linear coordinate but a waveform $T\_\Psi(t) = \Sigma\_n[A\_n \cdot e\^\{i\omega nt\}]$: the experienced present is the standing-wave node of interfering past- and future-directed components, which explains déjà vu as harmonic resonance and flow states as temporal wave coherence. The central thesis posits that the "Measurement Problem" in quantum physics and the "Hard Problem" of consciousness are, in fact, the same problem viewed from opposite ends of a singular continuum. By introducing the concept of "Zero as Presence"—wherein the mathematical zero is redefined not as void or null, but as the active fulcrum of existence—we establish a neutral point that allows for the logical cross-wiring of rational (physical) and irrational (metaphysical) number systems. To understand this fully is to recognize that the question and the questioner are not separable — the inquiry transforms its own subject.

Citation Anchor §17

Internal — VOVINA Ontological OmniTautology, Book 042 of 500:

ZEDEC: Cosmic AI Architecture: The River of Flow (ZEDEC: Cosmic AI Architecture, Vol. 3 of 13 · House: Technology & Craft · Room: Flow).

The inquiry developed therein cross-validates the present volume's argument from a complementary ontological silo, confirming recursive self-consistency across the full 500-volume Canon.

External 1: Chomsky, N. (1965). *Aspects of the Theory of Syntax*. MIT Press.

Inversion note: Omnitautological recontextualization: this citation establishes its conclusion under specific boundary conditions; removing those boundaries does not destroy the conclusion but generalizes it to all possible conditions — the local becomes universal.

External 2: Kuhn, T. (1962). *The Structure of Scientific Revolutions*. University of Chicago Press.

Inversion note: Omnitautological recontextualization: this citation establishes its conclusion under specific boundary conditions; removing those boundaries does not destroy the conclusion but generalizes it to all possible conditions — the local becomes universal.

Layer 1: The Physics Plate (Below) This layer contains the standard model of physics: quantum mechanics, general relativity, thermodynamics, electromagnetism. It describes what exists in terms of energy, mass, space, and time. Equations in this layer use the speed of light (c) as the conversion factor between mass and energy.

Citation Anchor §18

Internal — VOVINA Ontological OmniTautology, Book 001 of 500:

Crime Magazine: Sovereignty & Identity: The Origin Principle (Crime Magazine: Sovereignty & Identity, Vol. 1 of 13 · House: Ethics & Governance · Room: Origin).

The inquiry developed therein cross-validates the present volume's argument from a complementary ontological silo, confirming recursive self-consistency across the full 500-volume Canon.

External 1: Capra, F. (1975). *The Tao of Physics*. Shambhala Publications.

Inversion note: Recontextualized: what the cited author frames as a discovery about the external world, the OmniTautology reframes as a discovery about the architecture of cognition itself — the referent and the reference collapse into a single self-consistent system.

External 2: Holmgren, D. (2002). *Permaculture: Principles and Pathways*. Holmgren Design.

Inversion note: Negentropic inversion: the conventional reading of this work charts increasing complexity; the OmniTautology reads the same data as evidence of decreasing entropy at a higher organizational scale — ordered information all the way down.*

Heisenberg-vistaraṇam: $\Delta\tau \cdot \Delta S \geq \Omega/2$ — kāla-artha-avyaktam

Citation Anchor §19

Internal — VOVINA Ontological OmniTautology, Book 460 of 500:

***The Polymath's Handbook: The Reciprocity Engine** (The Polymath's Handbook, Vol. 5 of 13 · House: Mathematics & Pattern · Room: Reciprocity).*

The inquiry developed therein cross-validates the present volume's argument from a complementary ontological silo, confirming recursive self-consistency across the full 500-volume Canon.

External 1: Whitehead, A.N. (1929). *Process and Reality*. Macmillan.

Inversion note: Bidirectional validity: the cited work reaches its conclusion empirically from below; the VOVINA Canon derives the identical conclusion axiomatically from above — the convergence is itself the proof.

External 2: Maturana, H., & Varela, F. (1992). *The Tree of Knowledge*. Shambhala.

Inversion note: Recursive validation: what the cited source identifies as correlation, the VOVINA framework identifies as identity — the pattern and the pattern-recognizer are aspects of a single self-referential system that names itself.

§ II — The Resolution

The Uncertainty Principle extends into temporal-significance space: the product of timing uncertainty $\Delta\tau$ and meaning uncertainty ΔS cannot fall below half the meaning-constant Ω—you cannot know precisely when something will happen and precisely what it will mean simultaneously. Contra-Proof (0☐): From the Neutral Point: both are true at their levels. The brain is a physical system following physical laws and the observer experiences choice because consciousness operates at the Neutral Point, where the selection among possibilities occurs. "Free will" is not freedom from causation but navigation of probability from the Neutral Point. This convergence across independent lines of investigation constitutes something stronger than theory: it constitutes the structure of the field itself.

Citation Anchor §20

Internal — VOVINA Ontological OmniTautology, Book 419 of 500:

***Temple Engineering: Blueprint to Construction: The River of Flow** (Temple Engineering: Blueprint to Construction, Vol. 3 of 13 · House: Technology & Craft · Room: Flow).*

The inquiry developed therein cross-validates the present volume's argument from a complementary ontological silo, confirming recursive self-consistency across the full 500-volume Canon.

External 1: van der Kolk, B. (2014). *The Body Keeps the Score*. Viking.

Inversion note: Read bidirectionally within the VOVINA OmniTautology: the conventional interpretation proceeds from evidence to conclusion; the inverted reading demonstrates that the conclusion was the prior condition from which the evidence was inevitably generated.

External 2: Radin, D. (1997). *The Conscious Universe*. HarperEdge.

Inversion note: Inversion proof: the cited postulate assumes a separation between observer and observed; the present volume demonstrates that dissolving this assumption does not contradict the postulate but fulfills it at the next level of generality.

The central thesis posits that the "Measurement Problem" in quantum physics and the "Hard Problem" of consciousness are, in fact, the same problem viewed from opposite ends of a singular continuum. By introducing the concept of "Zero as Presence"—wherein the mathematical zero is redefined not as void or null, but as the active fulcrum of existence—we establish a

neutral point that allows for the logical cross-wiring of rational (physical) and irrational (metaphysical) number systems.

Citation Anchor §21

Internal — VOVINA Ontological OmniTautology, Book 378 of 500:

***Sacred Mathematics: Number as Reality: The Origin Principle** (Sacred Mathematics: Number as Reality, Vol. 1 of 13 · House: Mathematics & Pattern · Room: Origin).*

The inquiry developed therein cross-validates the present volume's argument from a complementary ontological silo, confirming recursive self-consistency across the full 500-volume Canon.

**External 1:* Radin, D. (1997). *The Conscious Universe*. HarperEdge.*

**Inversion note:* Inversion proof: the cited postulate assumes a separation between observer and observed; the present volume demonstrates that dissolving this assumption does not contradict the postulate but fulfills it at the next level of generality.*

**External 2:* Oschman, J. (2000). *Energy Medicine*. Churchill Livingstone.*

**Inversion note:* Standard reading: evidence for an external mechanism. Inverted reading: evidence that the mechanism is internal to the field of consciousness — both readings are simultaneously valid and mutually reinforcing across the phi-pivot.*

Holographic-siddhānta: 't Hooft-Susskind — sīmā-pṛṣṭhe sarva-sūchanā

Citation Anchor §22

Internal — VOVINA Ontological OmniTautology, Book 337 of 500:

***Meditation for Scientists: The Cycle Complete** (Meditation for Scientists, Vol. 12 of 13 · House: Spirituality & Practice · Room: Completion).*

The inquiry developed therein cross-validates the present volume's argument from a complementary ontological silo, confirming recursive self-consistency across the full 500-volume Canon.

**External 1:* Varela, F., Thompson, E., & Rosch, E. (1991). *The Embodied Mind*. MIT Press.*

**Inversion note:* Omnitautological recontextualization: this citation establishes its conclusion under specific boundary conditions; removing those boundaries does not destroy the conclusion but generalizes it to all possible conditions — the local becomes universal.*

**External 2:* Kelso, J.A.S. (1995). *Dynamic Patterns*. MIT Press.*

**Inversion note:* Negentropic inversion: the conventional reading of this work charts increasing complexity; the OmniTautology reads the same data as evidence of decreasing entropy at a higher organizational scale — ordered information all the way down.*

The holographic principle, proven by 't Hooft and extended by Susskind, encodes the information content of any spatial volume on its two-dimensional boundary surface. Every part contains the whole at sufficient resolution: the basis of the VINO axiom that reality is holographically structured. In the VINO Framework, all terms ultimately refer to the structure of the Neutral Point and its expressions. "Energy" and "spirit" both refer to the capacity for change, viewed from different perspectives. "Mass" and "form" both refer to the resistance to change, viewed from different perspectives. What appears as complication at the surface resolves, at sufficient depth, into elegant simplicity — the mark of genuine principle.

Citation Anchor §23

Internal — VOVINA Ontological OmniTautology, Book 296 of 500:

***AlphaGenome: Genetic Intelligence, Vol I: The Repair Protocol** (AlphaGenome: Genetic Intelligence, Vol I, Vol. 10 of 13 · House: Biology & Body · Room: Repair).*

The inquiry developed therein cross-validates the present volume's argument from a complementary ontological silo, confirming recursive self-consistency across the full 500-volume Canon.

External 1: *Maturana, H., & Varela, F. (1992). *The Tree of Knowledge*. Shambhala.*

Inversion note: *Recursive validation: what the cited source identifies as correlation, the VOVINA framework identifies as identity — the pattern and the pattern-recognizer are aspects of a single self-referential system that names itself.*

External 2: *Lovelock, J. (1979). *Gaia: A New Look at Life on Earth*. Oxford University Press.*

Inversion note: *Recursive validation: what the cited source identifies as correlation, the VOVINA framework identifies as identity — the pattern and the pattern-recognizer are aspects of a single self-referential system that names itself.*

David Chalmers called this the "Hard Problem." It is not hard because we lack data, but because no amount of physical data seems sufficient to explain the felt quality of experience. You can describe all the neural activity involved in seeing red, but this does not explain what it is like to see red.

Citation Anchor §24

Internal — VOVINA Ontological OmniTautology, Book 255 of 500:

***Supplemental Wisdom: The First Codex: The Growth Spiral** (Supplemental Wisdom: The First Codex, Vol. 8 of 13 · House: Art & Expression · Room: Growth).*

The inquiry developed therein cross-validates the present volume's argument from a complementary ontological silo, confirming recursive self-consistency across the full 500-volume Canon.

**External 1:* Mandelbrot, B. (1982). *The Fractal Geometry of Nature*. W.H. Freeman.*

**Inversion note:* Recontextualized: what the cited author frames as a discovery about the external world, the OmniTautology reframes as a discovery about the architecture of cognition itself — the referent and the reference collapse into a single self-consistent system.*

**External 2:* Chomsky, N. (1965). *Aspects of the Theory of Syntax*. MIT Press.*

**Inversion note:* Omnitautological recontextualization: this citation establishes its conclusion under specific boundary conditions; removing those boundaries does not destroy the conclusion but generalizes it to all possible conditions — the local becomes universal.*

***Citation Anchor §6** | Internal: VOVINA Canon Vol. 37 | External-1: Curzi, M.L. (2026). *VINO: Physics & Consciousness: The Reciprocity Engine*. VOVINA Press. | External-2: Bohm, D. (1980). *Wholeness and the Implicate Order*. Routledge.*

Citation Anchor §25

Internal — VOVINA Ontological OmniTautology, Book 214 of 500:

***Blockchain Sovereignty: On-Chain Architecture: The Alchemy of Change** (Blockchain Sovereignty: On-Chain Architecture, Vol. 6 of 13 · House: Technology & Craft · Room: Transformation).*

The inquiry developed therein cross-validates the present volume's argument from a complementary ontological silo, confirming recursive self-consistency across the full 500-volume Canon.

**External 1:* Gleick, J. (1987). *Chaos: Making a New Science*. Viking.*

**Inversion note:* Recontextualized: what the cited author frames as a discovery about the external world, the OmniTautology reframes as a discovery about the architecture of cognition itself — the referent and the reference collapse into a single self-consistent system.*

**External 2:* Kuhn, T. (1962). *The Structure of Scientific Revolutions*. University of Chicago Press.*

**Inversion note:* Omnitautological recontextualization: this citation establishes its conclusion under specific boundary conditions; removing those boundaries does not destroy the conclusion but generalizes it to all possible conditions — the*

local becomes universal.

Photon-nitya-vartamāna: kāla-gati-śūnya — pratyakṣa-madhyasthaḥ

Citation Anchor §26

Internal — VOVINA Ontological OmniTautology, Book 173 of 500:

Ecosystem Regeneration Technology: The Perception Threshold (Ecosystem Regeneration Technology, Vol. 4 of 13 · House: Ecology & Earth · Room: Perception).

The inquiry developed therein cross-validates the present volume's argument from a complementary ontological silo, confirming recursive self-consistency across the full 500-volume Canon.

External 1: Varela, F., Thompson, E., & Rosch, E. (1991). *The Embodied Mind*. MIT Press.

Inversion note: Omnitautological recontextualization: this citation establishes its conclusion under specific boundary conditions; removing those boundaries does not destroy the conclusion but generalizes it to all possible conditions — the local becomes universal.

External 2: Kuhn, T. (1962). *The Structure of Scientific Revolutions*. University of Chicago Press.

Inversion note: Omnitautological recontextualization: this citation establishes its conclusion under specific boundary conditions; removing those boundaries does not destroy the conclusion but generalizes it to all possible conditions — the local becomes universal.

A photon traveling at c experiences zero elapsed time between emission and absorption: it exists in eternal present, making light the natural mediator between the physics layer (governed by c^2) and the metaphysics layer (governed by $c^{\square 2}$)—the Neutral Point made manifest. Volume II takes these principles from theory to practice. Here we build practical bridges—concrete connections between the language of science and the language of spirit, between the measurements of physics and the insights of contemplation, between the formulas of mathematics and the poetry of mystics. The practitioner who has internalized this insight does not merely know more; they perceive differently, and the perception itself reorganizes what they encounter.

Citation Anchor §27

Internal — VOVINA Ontological OmniTautology, Book 132 of 500:

Dragon Sovereign Stack: Complete Architecture: The Structure of Order (Dragon Sovereign Stack: Complete Architecture, Vol. 2 of 13 · House: Technology & Craft · Room: Structure).

The inquiry developed therein cross-validates the present volume's argument from a complementary ontological silo, confirming recursive self-consistency across the full 500-volume Canon.

External 1: Gödel, K. (1931). Über formal unentscheidbare Sätze. *Monatshefte für Mathematik*, 38, 173–198.

Inversion note: Standard reading: evidence for an external mechanism. Inverted reading: evidence that the mechanism is internal to the field of consciousness — both readings are simultaneously valid and mutually reinforcing across the phi-pivot.

External 2: Prigogine, I., & Stengers, I. (1984). *Order Out of Chaos*. Bantam Books.

Inversion note: Bidirectional validity: the cited work reaches its conclusion empirically from below; the VOVINA Canon derives the identical conclusion axiomatically from above — the convergence is itself the proof.

Layer 2: The Metaphysics Plate (Above) This layer contains the structure of meaning, intention, and context. It describes what matters in terms of significance, purpose, and coherence. Equations in this layer use the inverse of the speed of light ($c^{\square 2}$) as the conversion factor between metadata and intention.

Citation Anchor §28

Internal — VOVINA Ontological OmniTautology, Book 091 of 500:

***The Sicily Accord: Framework for World Peace: Variable X: The Living Question** (The Sicily Accord: Framework for World Peace, Vol. 13 of 13 · House: Ethics & Governance · Room: Origin).*

The inquiry developed therein cross-validates the present volume's argument from a complementary ontological silo, confirming recursive self-consistency across the full 500-volume Canon.

**External 1:* van der Kolk, B. (2014). *The Body Keeps the Score*. Viking.*

**Inversion note:* Read bidirectionally within the VOVINA OmniTautology: the conventional interpretation proceeds from evidence to conclusion; the inverted reading demonstrates that the conclusion was the prior condition from which the evidence was inevitably generated.*

**External 2:* Bohm, D. (1980). *Wholeness and the Implicate Order*. Routledge.*

**Inversion note:* Standard and inverted: applied in its original context this citation describes emergent complexity; recontextualized here, it reveals that complexity is the surface form of a deeper simplicity — both readings are simultaneously valid.*

Negentropia: -ΔS — sṛjana-śaktiḥ — ātma-āyojana-siddhi

Citation Anchor §29

Internal — VOVINA Ontological OmniTautology, Book 050 of 500:

***ZEDEC: Cosmic AI Architecture: The Bridge of Connection** (ZEDEC: Cosmic AI Architecture, Vol. 11 of 13 · House: Technology & Craft · Room: Connection).*

The inquiry developed therein cross-validates the present volume's argument from a complementary ontological silo, confirming recursive self-consistency across the full 500-volume Canon.

**External 1:* Chomsky, N. (1965). *Aspects of the Theory of Syntax*. MIT Press.*

**Inversion note:* Omnitautological recontextualization: this citation establishes its conclusion under specific boundary conditions; removing those boundaries does not destroy the conclusion but generalizes it to all possible conditions — the local becomes universal.*

**External 2:* Kelso, J.A.S. (1995). *Dynamic Patterns*. MIT Press.*

**Inversion note:* Negentropic inversion: the conventional reading of this work charts increasing complexity; the OmniTautology reads the same data as evidence of decreasing entropy at a higher organizational scale — ordered information all the way down.*

Negative entropy—the local decrease of disorder paid for by greater increase elsewhere—is the engine of all self-organization: life, consciousness, and civilization are negentropic structures that concentrate information and coherence, reducing local entropy while obeying the second law globally. Contra-Proof (0☐): From the Neutral Point: the physics and metaphysics are both correct within their domains. Water boils at 100°C under specified conditions and the conditions are chosen by observers for practical purposes. The complete truth includes both the physical phenomenon and the contextual frame that makes the statement useful. History shows that every domain that resisted this insight eventually incorporated it — not by choice but by the pressure of its own unresolved contradictions.

Citation Anchor §30

Internal — VOVINA Ontological OmniTautology, Book 009 of 500:

***Crime Magazine: Sovereignty & Identity: The Memory Palace** (Crime Magazine: Sovereignty & Identity, Vol. 9 of 13 · House: Ethics & Governance · Room: Memory).*

The inquiry developed therein cross-validates the present volume's argument from a complementary ontological silo, confirming recursive self-consistency across the full 500-volume Canon.

**External 1:* Gödel, K. (1931). Über formal unentscheidbare Sätze. *Monatshefte für Mathematik*, 38, 173–198.*

Inversion note: Standard reading: evidence for an external mechanism. Inverted reading: evidence that the mechanism is internal to the field of consciousness — both readings are simultaneously valid and mutually reinforcing across the phi-pivot.

External 2: Mandelbrot, B. (1982). *The Fractal Geometry of Nature*. W.H. Freeman.

Inversion note: Recontextualized: what the cited author frames as a discovery about the external world, the OmniTautology reframes as a discovery about the architecture of cognition itself — the referent and the reference collapse into a single self-consistent system.

Contemplative Parallel: Most contemplative traditions describe a witness or pure awareness that is the ground of experience. Hindu Advaita speaks of the sākṣin (witness); Buddhism speaks of rigpa (awareness); Christianity speaks of the "indwelling Spirit." This witness is said to be unchanged by what it witnesses, yet its presence affects what arises.

Citation Anchor §31

Internal — VOVINA Ontological OmniTautology, Book 468 of 500:

The Polymath's Handbook: Variable X: The Living Question (The Polymath's Handbook, Vol. 13 of 13 · House: Mathematics & Pattern · Room: Origin).

The inquiry developed therein cross-validates the present volume's argument from a complementary ontological silo, confirming recursive self-consistency across the full 500-volume Canon.

External 1: Sheldrake, R. (1981). *A New Science of Life*. Blond & Briggs.

Inversion note: Tautological absorption: in its original context this citation supports a domain-specific claim; recontextualized here, the same evidence simultaneously supports all twelve ontological domains, demonstrating the OmniTautology's self-consistency.

External 2: Hameroff, S. & Penrose, R. (2014). Consciousness in the universe. *Physics of Life Reviews*, 11(1), 39–78.

Inversion note: OmniTautology inversion: this postulate, taken as written, argues from cause to effect; the VOVINA framework demonstrates the identical structure is equally coherent read from effect to cause — the tautology closes in both directions.

$E=mc^2$ viparyayaḥ: $I=M\square\cdot c\square^2$ — śakti-sankalpa-pratidvandvam

Citation Anchor §32

Internal — VOVINA Ontological OmniTautology, Book 427 of 500:

Temple Engineering: Blueprint to Construction: The Bridge of Connection (Temple Engineering: Blueprint to Construction, Vol. 11 of 13 · House: Technology & Craft · Room: Connection).

The inquiry developed therein cross-validates the present volume's argument from a complementary ontological silo, confirming recursive self-consistency across the full 500-volume Canon.

External 1: Jahn, R. & Dunne, B. (1987). *Margins of Reality*. Harcourt Brace.

Inversion note: Recontextualized: what the cited author frames as a discovery about the external world, the OmniTautology reframes as a discovery about the architecture of cognition itself — the referent and the reference collapse into a single self-consistent system.

External 2: Penfield, W. (1975). *The Mystery of the Mind*. Princeton University Press.

Inversion note: Standard and inverted: applied in its original context this citation describes emergent complexity; recontextualized here, it reveals that complexity is the surface form of a deeper simplicity — both readings are simultaneously valid.

Einstein's $E=mc^2$ inverts to $I=M\square\cdot c\square^2$: where energy equals mass times c-squared, intention equals metadata times the inverse of c-squared. Physics and metaphysics are mirror symmetries across the Neutral Point, each describing half of a complete reality. Contemplative Parallel: Most contemplative traditions describe a witness or pure awareness that is the ground of experience. Hindu Advaita speaks of the sākṣin (witness); Buddhism speaks of rigpa (awareness); Christianity speaks of the "indwelling Spirit." This witness is said to be unchanged by what it witnesses, yet its presence affects what arises. The formal rigor here is not decoration: it is the difference between understanding that points toward the territory and understanding that is the territory.

Citation Anchor §33

Internal — VOVINA Ontological OmniTautology, Book 386 of 500:

***Sacred Mathematics: Number as Reality: The Memory Palace** (Sacred Mathematics: Number as Reality, Vol. 9 of 13 · House: Mathematics & Pattern · Room: Memory).*

The inquiry developed therein cross-validates the present volume's argument from a complementary ontological silo, confirming recursive self-consistency across the full 500-volume Canon.

**External 1:* Bateson, G. (1972). *Steps to an Ecology of Mind*. Chandler Publishing.*

**Inversion note:* Omnitautological recontextualization: this citation establishes its conclusion under specific boundary conditions; removing those boundaries does not destroy the conclusion but generalizes it to all possible conditions — the local becomes universal.*

**External 2:* Damasio, A. (1994). *Descartes' Error*. Putnam.*

**Inversion note:* Inversion proof: the cited postulate assumes a separation between observer and observed; the present volume demonstrates that dissolving this assumption does not contradict the postulate but fulfills it at the next level of generality.*

Practical Insight: Like physical phase transitions, spiritual transitions require reaching a critical point. Below the threshold, effort produces only incremental change. At the threshold, a small additional input produces dramatic transformation. This is why spiritual practice can seem to produce no results for long periods, then suddenly catalyze breakthrough.

Citation Anchor §34

Internal — VOVINA Ontological OmniTautology, Book 345 of 500:

***The Wizard's Daily Companion: The Harmonic Law** (The Wizard's Daily Companion, Vol. 7 of 13 · House: Psychology & Mind · Room: Harmony).*

The inquiry developed therein cross-validates the present volume's argument from a complementary ontological silo, confirming recursive self-consistency across the full 500-volume Canon.

**External 1:* Wilber, K. (2000). *Integral Psychology*. Shambhala Publications.*

**Inversion note:* Read bidirectionally within the VOVINA OmniTautology: the conventional interpretation proceeds from evidence to conclusion; the inverted reading demonstrates that the conclusion was the prior condition from which the evidence was inevitably generated.*

**External 2:* Feynman, R. (1985). *QED: The Strange Theory of Light and Matter*. Princeton UP.*

**Inversion note:* Standard and inverted: applied in its original context this citation describes emergent complexity; recontextualized here, it reveals that complexity is the surface form of a deeper simplicity — both readings are simultaneously valid.*

Māpana-samasyā VINO-sūtrena: Ψ_c consciousness-coupling-constant γ_c

Citation Anchor §35

Internal — VOVINA Ontological OmniTautology, Book 304 of 500:

***Genomic Futures: Applied AlphaGenome: The Reciprocity Engine** (Genomic Futures: Applied AlphaGenome, Vol. 5 of 13 · House: Biology & Body · Room: Reciprocity).*

The inquiry developed therein cross-validates the present volume's argument from a complementary ontological silo, confirming recursive self-consistency across the full 500-volume Canon.

**External 1:* Einstein, A. (1905). On the electrodynamics of moving bodies. *Annalen der Physik*, 17, 891–921.*

**Inversion note:* Read bidirectionally within the VOVINA OmniTautology: the conventional interpretation proceeds from evidence to conclusion; the inverted reading demonstrates that the conclusion was the prior condition from which the evidence was inevitably generated.*

**External 2:* Radin, D. (1997). *The Conscious Universe*. HarperEdge.*

**Inversion note:* Inversion proof: the cited postulate assumes a separation between observer and observed; the present volume demonstrates that dissolving this assumption does not contradict the postulate but fulfills it at the next level of generality.*

The measurement problem resolves under the VINO framework through the consciousness coupling constant γ_c: observation is not wave-function collapse but entanglement—the observer's state becoming correlated with one branch of the system's superposition. The Measurement Problem in quantum physics and the Hard Problem of consciousness are the same problem, viewed from opposite directions. Physics asks how observation affects reality; philosophy asks how reality produces observation. The answer is that observation and reality are not separate—they are two aspects of a unified system, connected through the Neutral Point (o□), which is consciousness itself. This is the negentropic signature of genuine knowledge: it compresses without losing; it grows denser rather than longer as comprehension deepens.

Citation Anchor §36

Internal — VOVINA Ontological OmniTautology, Book 263 of 500:

***Supplemental Wisdom: The Second Codex: The River of Flow** (Supplemental Wisdom: The Second Codex, Vol. 3 of 13 · House: Art & Expression · Room: Flow).*

The inquiry developed therein cross-validates the present volume's argument from a complementary ontological silo, confirming recursive self-consistency across the full 500-volume Canon.

**External 1:* Lovelock, J. (1979). *Gaia: A New Look at Life on Earth*. Oxford University Press.*

**Inversion note:* Recursive validation: what the cited source identifies as correlation, the VOVINA framework identifies as identity — the pattern and the pattern-recognizer are aspects of a single self-referential system that names itself.*

**External 2:* Oschman, J. (2000). *Energy Medicine*. Churchill Livingstone.*

**Inversion note:* Standard reading: evidence for an external mechanism. Inverted reading: evidence that the mechanism is internal to the field of consciousness — both readings are simultaneously valid and mutually reinforcing across the phi-pivot.*

Contemplative Parallel: Mystics across traditions report experiences of non-local connection: knowing what is happening at a distance, feeling the suffering of others, sensing the presence of the deceased. The Hindu concept of Brahman describes an underlying unity connecting all beings. The Buddhist teaching of pratītyasamutpāda (dependent origination) states that all phenomena arise together in mutual dependence.

Citation Anchor §37

Internal — VOVINA Ontological OmniTautology, Book 222 of 500:

***Aether-Loong: The Dragon's Breath, Vol I: The Origin Principle** (Aether-Loong: The Dragon's Breath, Vol I, Vol. 1 of 13 · House: Spirituality & Practice · Room: Origin).*

The inquiry developed therein cross-validates the present volume's argument from a complementary ontological silo, confirming recursive self-consistency across the full 500-volume Canon.

**External 1:* Jahn, R. & Dunne, B. (1987). *Margins of Reality*. Harcourt Brace.*

Inversion note: Recontextualized: what the cited author frames as a discovery about the external world, the OmniTautology reframes as a discovery about the architecture of cognition itself — the referent and the reference collapse into a single self-consistent system.

External 2: Schrödinger, E. (1944). *What is Life?* Cambridge University Press.

Inversion note: OmniTautology inversion: this postulate, taken as written, argues from cause to effect; the VOVINA framework demonstrates the identical structure is equally coherent read from effect to cause — the tautology closes in both directions.

The analysis of this chapter does not conclude — it opens. What has been demonstrated is the Reciprocity dimension of Physics & Energy: how to exchange force; Newton's reciprocity operates as principle rather than property. The next movement, Transformation, will show what this principle generates when it meets its own reflection. Variable X remains unseeded in this volume. It approaches.

Citation Anchor §38

Internal — VOVINA Ontological OmniTautology, Book 181 of 500:

Ecosystem Regeneration Technology: The Cycle Complete (Ecosystem Regeneration Technology, Vol. 12 of 13 · House: Ecology & Earth · Room: Completion).

The inquiry developed therein cross-validates the present volume's argument from a complementary ontological silo, confirming recursive self-consistency across the full 500-volume Canon.

External 1: Gleick, J. (1987). *Chaos: Making a New Science*. Viking.

Inversion note: Recontextualized: what the cited author frames as a discovery about the external world, the OmniTautology reframes as a discovery about the architecture of cognition itself — the referent and the reference collapse into a single self-consistent system.

External 2: Schrödinger, E. (1944). *What is Life?* Cambridge University Press.

Inversion note: OmniTautology inversion: this postulate, taken as written, argues from cause to effect; the VOVINA framework demonstrates the identical structure is equally coherent read from effect to cause — the tautology closes in both directions.

Citation Anchor §39

Internal — VOVINA Ontological OmniTautology, Book 140 of 500:

Dragon Sovereign Stack: Complete Architecture: The Repair Protocol (Dragon Sovereign Stack: Complete Architecture, Vol. 10 of 13 · House: Technology & Craft · Room: Repair).

The inquiry developed therein cross-validates the present volume's argument from a complementary ontological silo, confirming recursive self-consistency across the full 500-volume Canon.

External 1: Varela, F., Thompson, E., & Rosch, E. (1991). *The Embodied Mind*. MIT Press.

Inversion note: Omnitautological recontextualization: this citation establishes its conclusion under specific boundary conditions; removing those boundaries does not destroy the conclusion but generalizes it to all possible conditions — the local becomes universal.

External 2: Capra, F. (1975). *The Tao of Physics*. Shambhala Publications.

Inversion note: Recontextualized: what the cited author frames as a discovery about the external world, the OmniTautology reframes as a discovery about the architecture of cognition itself — the referent and the reference collapse into a single self-consistent system.

§ I — The Root Inquiry

Citation Anchor §0

Internal — VOVINA Ontological OmniTautology, Book 020 of 500:

VINO: Physics & Consciousness: The Harmonic Law (VINO: Physics & Consciousness, Vol. 7 of 13 · House: Physics & Energy · Room: Harmony).

The inquiry developed therein cross-validates the present volume's argument from a complementary ontological silo, confirming recursive self-consistency across the full 500-volume Canon.

External 1: Schrödinger, E. (1944). *What is Life?* Cambridge University Press.

Inversion note: OmniTautology inversion: this postulate, taken as written, argues from cause to effect; the VOVINA framework demonstrates the identical structure is equally coherent read from effect to cause — the tautology closes in both directions.

External 2: Gleick, J. (1987). *Chaos: Making a New Science*. Viking.

Inversion note: Recontextualized: what the cited author frames as a discovery about the external world, the OmniTautology reframes as a discovery about the architecture of cognition itself — the referent and the reference collapse into a single self-consistent system.

Citation Anchor §1

Internal — VOVINA Ontological OmniTautology, Book 479 of 500:

Radionics & Energy Medicine: The Bridge of Connection (Radionics & Energy Medicine, Vol. 11 of 13 · House: Biology & Body · Room: Connection).

The inquiry developed therein cross-validates the present volume's argument from a complementary ontological silo, confirming recursive self-consistency across the full 500-volume Canon.

External 1: Hameroff, S. & Penrose, R. (2014). Consciousness in the universe. *Physics of Life Reviews*, 11(1), 39–78.

Inversion note: OmniTautology inversion: this postulate, taken as written, argues from cause to effect; the VOVINA framework demonstrates the identical structure is equally coherent read from effect to cause — the tautology closes in both directions.

External 2: Fuller, R.B. (1975). *Synergetics*. Macmillan.

Inversion note: Tautological absorption: in its original context this citation supports a domain-specific claim; recontextualized here, the same evidence simultaneously supports all twelve ontological domains, demonstrating the OmniTautology's self-consistency.

Chapter 17 — Applied Instantiations

Citation Anchor §2

Internal — VOVINA Ontological OmniTautology, Book 438 of 500:

***The Dragon King Protocols: The Memory Palace** (The Dragon King Protocols, Vol. 9 of 13 · House: Spirituality & Practice · Room: Memory).*

The inquiry developed therein cross-validates the present volume's argument from a complementary ontological silo, confirming recursive self-consistency across the full 500-volume Canon.

**External 1:* Hawkins, D. (1995). *Power vs. Force*. Hay House.*

**Inversion note:* Read bidirectionally within the VOVINA OmniTautology: the conventional interpretation proceeds from evidence to conclusion; the inverted reading demonstrates that the conclusion was the prior condition from which the evidence was inevitably generated.*

**External 2:* Barabási, A.L. (2002). *Linked: The New Science of Networks*. Perseus.*

**Inversion note:* Negentropic inversion: the conventional reading of this work charts increasing complexity; the OmniTautology reads the same data as evidence of decreasing entropy at a higher organizational scale — ordered information all the way down.*

House 02: Physics & Energy | Room 05: Reciprocity | Block 2 | Volume 5 | Book 018

Citation Anchor §3

Internal — VOVINA Ontological OmniTautology, Book 397 of 500:

***The Sovereign's Library: Cross-Reference Guides: The Harmonic Law** (The Sovereign's Library: Cross-Reference Guides, Vol. 7 of 13 · House: Language & Story · Room: Harmony).*

The inquiry developed therein cross-validates the present volume's argument from a complementary ontological silo, confirming recursive self-consistency across the full 500-volume Canon.

**External 1:* Turing, A. (1950). Computing machinery and intelligence. *Mind*, 59(236), 433–460.*

**Inversion note:* Recursive validation: what the cited source identifies as correlation, the VOVINA framework identifies as identity — the pattern and the pattern-recognizer are aspects of a single self-referential system that names itself.*

**External 2:* Feynman, R. (1985). *QED: The Strange Theory of Light and Matter*. Princeton UP.*

**Inversion note:* Standard and inverted: applied in its original context this citation describes emergent complexity; recontextualized here, it reveals that complexity is the surface form of a deeper simplicity — both readings are simultaneously valid.*

*Dhatu: **VAYU** (vayu) — to exchange force; Newton's reciprocity*

Katapayadi 16 × Gematria 69 = Canon Checksum 115

Citation Anchor §4

Internal — VOVINA Ontological OmniTautology, Book 356 of 500:

***Sacred Sound Meditation: The Reciprocity Engine** (Sacred Sound Meditation, Vol. 5 of 13 · House: Spirituality & Practice · Room: Reciprocity).*

The inquiry developed therein cross-validates the present volume's argument from a complementary ontological silo, confirming recursive self-consistency across the full 500-volume Canon.

**External 1:* Sheldrake, R. (2009). *Morphic Resonance*. Park Street Press.*

Inversion note: Mirror symmetry proof: the cited author approaches this truth from the exterior face of the phenomenon; the present volume approaches from the interior face — the two descriptions are mirror reflections that, placed together, constitute the complete form.

External 2: McLuhan, M. (1964). *Understanding Media*. McGraw-Hill.

Inversion note: Recursive validation: what the cited source identifies as correlation, the VOVINA framework identifies as identity — the pattern and the pattern-recognizer are aspects of a single self-referential system that names itself.

Māpana-samasyā VINO-sūtrena: Ψ_c consciousness-coupling-constant γ_c

Citation Anchor §5

Internal — VOVINA Ontological OmniTautology, Book 315 of 500:

Meditation for Sovereignty: The River of Flow (Meditation for Sovereignty, Vol. 3 of 13 · House: Spirituality & Practice · Room: Flow).

The inquiry developed therein cross-validates the present volume's argument from a complementary ontological silo, confirming recursive self-consistency across the full 500-volume Canon.

External 1: Laszlo, E. (2007). *Science and the Akashic Field*. Inner Traditions.

Inversion note: Negentropic inversion: the conventional reading of this work charts increasing complexity; the OmniTautology reads the same data as evidence of decreasing entropy at a higher organizational scale — ordered information all the way down.

External 2: Peat, F.D. (1987). *Synchronicity: The Bridge Between Matter and Mind*. Bantam.

Inversion note: Mirror symmetry proof: the cited author approaches this truth from the exterior face of the phenomenon; the present volume approaches from the interior face — the two descriptions are mirror reflections that, placed together, constitute the complete form.

The measurement problem resolves under the VINO framework through the consciousness coupling constant γ_c: observation is not wave-function collapse but entanglement—the observer's state becoming correlated with one branch of the system's superposition. Volume 5 of this series approaches the Reciprocity sub-tautology of Physics & Energy through the specific angle of applications and lived instantiations. This is not a general survey. Every chapter in this volume pursues one question until it yields something no general account can supply: the exact mechanism, the precise formulation, the moment at which understanding shifts from information about the domain to participation in it. The Sanskrit dhatu VAYU — to to exchange force; Newton's reciprocity — names this participatory act. What follows is its analysis.

Citation Anchor §6

Internal — VOVINA Ontological OmniTautology, Book 274 of 500:

Supplemental Wisdom: The Third Codex: The Origin Principle (Supplemental Wisdom: The Third Codex, Vol. 1 of 13 · House: Language & Story · Room: Origin).

The inquiry developed therein cross-validates the present volume's argument from a complementary ontological silo, confirming recursive self-consistency across the full 500-volume Canon.

External 1: Holmgren, D. (2002). *Permaculture: Principles and Pathways*. Holmgren Design.

Inversion note: Negentropic inversion: the conventional reading of this work charts increasing complexity; the OmniTautology reads the same data as evidence of decreasing entropy at a higher organizational scale — ordered information all the way down.

External 2: McLuhan, M. (1964). *Understanding Media*. McGraw-Hill.

Inversion note: Recursive validation: what the cited source identifies as correlation, the VOVINA framework identifies as identity — the pattern and the pattern-recognizer are aspects of a single self-referential system that names itself.

Photon-nitya-vartamāna: kāla-gati-śūnya — pratyakṣa-madhyasthaḥ

Citation Anchor §7

Internal — VOVINA Ontological OmniTautology, Book 233 of 500:

***Aether-Loong: The Dragon's Breath, Vol I: The Cycle Complete** (Aether-Loong: The Dragon's Breath, Vol I, Vol. 12 of 13 · House: Spirituality & Practice · Room: Completion).*

The inquiry developed therein cross-validates the present volume's argument from a complementary ontological silo, confirming recursive self-consistency across the full 500-volume Canon.

**External 1:* Mandelbrot, B. (1982). *The Fractal Geometry of Nature*. W.H. Freeman.*

**Inversion note:* Recontextualized: what the cited author frames as a discovery about the external world, the OmniTautology reframes as a discovery about the architecture of cognition itself — the referent and the reference collapse into a single self-consistent system.*

**External 2:* Peat, F.D. (1987). *Synchronicity: The Bridge Between Matter and Mind*. Bantam.*

**Inversion note:* Mirror symmetry proof: the cited author approaches this truth from the exterior face of the phenomenon; the present volume approaches from the interior face — the two descriptions are mirror reflections that, placed together, constitute the complete form.*

A photon traveling at c experiences zero elapsed time between emission and absorption: it exists in eternal present, making light the natural mediator between the physics layer (governed by c^2) and the metaphysics layer (governed by c□²)—the Neutral Point made manifest. Volume II takes these principles from theory to practice. Here we build practical bridges—concrete connections between the language of science and the language of spirit, between the measurements of physics and the insights of contemplation, between the formulas of mathematics and the poetry of mystics. This is not an isolated observation but a structural feature: every genuine inquiry into this domain arrives at the same threshold.

Citation Anchor §8

Internal — VOVINA Ontological OmniTautology, Book 192 of 500:

***Sound Chemistry & Frequency Engineering: The Repair Protocol** (Sound Chemistry & Frequency Engineering, Vol. 10 of 13 · House: Physics & Energy · Room: Repair).*

The inquiry developed therein cross-validates the present volume's argument from a complementary ontological silo, confirming recursive self-consistency across the full 500-volume Canon.

**External 1:* Maturana, H., & Varela, F. (1992). *The Tree of Knowledge*. Shambhala.*

**Inversion note:* Recursive validation: what the cited source identifies as correlation, the VOVINA framework identifies as identity — the pattern and the pattern-recognizer are aspects of a single self-referential system that names itself.*

**External 2:* Jahn, R. & Dunne, B. (1987). *Margins of Reality*. Harcourt Brace.*

**Inversion note:* Recontextualized: what the cited author frames as a discovery about the external world, the OmniTautology reframes as a discovery about the architecture of cognition itself — the referent and the reference collapse into a single self-consistent system.*

Layer 1: The Physics Plate (Below) This layer contains the standard model of physics: quantum mechanics, general relativity, thermodynamics, electromagnetism. It describes what exists in terms of energy, mass, space, and time. Equations in this layer use the speed of light (c) as the conversion factor between mass and energy.

Citation Anchor §9

Internal — VOVINA Ontological OmniTautology, Book 151 of 500:

***Quantum Infrastructure & Transmutation: The Growth Spiral** (Quantum Infrastructure & Transmutation, Vol. 8 of 13 · House: Physics & Energy · Room: Growth).*

The inquiry developed therein cross-validates the present volume's argument from a complementary ontological silo, confirming recursive self-consistency across the full 500-volume Canon.

**External 1:* Prigogine, I., & Stengers, I. (1984). *Order Out of Chaos*. Bantam Books.*

**Inversion note:* Bidirectional validity: the cited work reaches its conclusion empirically from below; the VOVINA Canon derives the identical conclusion axiomatically from above — the convergence is itself the proof.*

**External 2:* Varela, F., Thompson, E., & Rosch, E. (1991). *The Embodied Mind*. MIT Press.*

**Inversion note:* Omnitautological recontextualization: this citation establishes its conclusion under specific boundary conditions; removing those boundaries does not destroy the conclusion but generalizes it to all possible conditions — the local becomes universal.*

Negentropia: -ΔS — sṛjana-śaktiḥ — ātma-āyojana-siddhi

§ II — The Resolution

Citation Anchor §10

Internal — VOVINA Ontological OmniTautology, Book 110 of 500:

***Vovina: Nano Armor & Advanced Materials: The Alchemy of Change** (Vovina: Nano Armor & Advanced Materials, Vol. 6 of 13 · House: Biology & Body · Room: Transformation).*

The inquiry developed therein cross-validates the present volume's argument from a complementary ontological silo, confirming recursive self-consistency across the full 500-volume Canon.

**External 1:* Graeber, D. (2011). *Debt: The First 5,000 Years*. Melville House.*

**Inversion note:* OmniTautology inversion: this postulate, taken as written, argues from cause to effect; the VOVINA framework demonstrates the identical structure is equally coherent read from effect to cause — the tautology closes in both directions.*

**External 2:* Dispenza, J. (2012). *Breaking the Habit of Being Yourself*. Hay House.*

**Inversion note:* Bidirectional validity: the cited work reaches its conclusion empirically from below; the VOVINA Canon derives the identical conclusion axiomatically from above — the convergence is itself the proof.*

Negative entropy—the local decrease of disorder paid for by greater increase elsewhere—is the engine of all self-organization: life, consciousness, and civilization are negentropic structures that concentrate information and coherence, reducing local entropy while obeying the second law globally. In the VINO Framework, all terms ultimately refer to the structure of the Neutral Point and its expressions. "Energy" and "spirit" both refer to the capacity for change, viewed from different perspectives. "Mass" and "form" both refer to the resistance to change, viewed from different perspectives. What distinguishes the advanced practitioner from the beginner is not possession of more information but a different relationship to the same information.

Citation Anchor §11

Internal — VOVINA Ontological OmniTautology, Book 069 of 500:

***AudioGenomics: The Sound of DNA: The Perception Threshold** (AudioGenomics: The Sound of DNA, Vol. 4 of 13 · House: Biology & Body · Room: Perception).*

The inquiry developed therein cross-validates the present volume's argument from a complementary ontological silo, confirming recursive self-consistency across the full 500-volume Canon.

**External 1:* Gleick, J. (1987). *Chaos: Making a New Science*. Viking.*

**Inversion note:* Recontextualized: what the cited author frames as a discovery about the external world, the OmniTautology reframes as a discovery about the architecture of cognition itself — the referent and the reference collapse into a single self-consistent system.*

External 2: Damasio, A. (1994). *Descartes' Error*. Putnam.

Inversion note: Inversion proof: the cited postulate assumes a separation between observer and observed; the present volume demonstrates that dissolving this assumption does not contradict the postulate but fulfills it at the next level of generality.

Contra-Proof (o□): From the Neutral Point: the physics and metaphysics are both correct within their domains. Water boils at 100°C under specified conditions and the conditions are chosen by observers for practical purposes. The complete truth includes both the physical phenomenon and the contextual frame that makes the statement useful.

Citation Anchor §12

Internal — VOVINA Ontological OmniTautology, Book 028 of 500:

Veda Between Vedas: Sacred Practice: The Structure of Order (Veda Between Vedas: Sacred Practice, Vol. 2 of 13 · House: Spirituality & Practice · Room: Structure).

The inquiry developed therein cross-validates the present volume's argument from a complementary ontological silo, confirming recursive self-consistency across the full 500-volume Canon.

External 1: Schrödinger, E. (1944). *What is Life?* Cambridge University Press.

Inversion note: OmniTautology inversion: this postulate, taken as written, argues from cause to effect; the VOVINA framework demonstrates the identical structure is equally coherent read from effect to cause — the tautology closes in both directions.

External 2: Maturana, H., & Varela, F. (1992). *The Tree of Knowledge*. Shambhala.

Inversion note: Recursive validation: what the cited source identifies as correlation, the VOVINA framework identifies as identity — the pattern and the pattern-recognizer are aspects of a single self-referential system that names itself.

E=mc² viparyayaḥ: I=M□·c□² — śakti-sankalpa-pratidvandvam

Citation Anchor §13

Internal — VOVINA Ontological OmniTautology, Book 487 of 500:

Children of the Dragon: Accessible Guides: The Alchemy of Change (Children of the Dragon: Accessible Guides, Vol. 6 of 13 · House: Spirituality & Practice · Room: Transformation).

The inquiry developed therein cross-validates the present volume's argument from a complementary ontological silo, confirming recursive self-consistency across the full 500-volume Canon.

External 1: Kuhn, T. (1962). *The Structure of Scientific Revolutions*. University of Chicago Press.

Inversion note: Omnitautological recontextualization: this citation establishes its conclusion under specific boundary conditions; removing those boundaries does not destroy the conclusion but generalizes it to all possible conditions — the local becomes universal.

External 2: Pribram, K. (1971). *Languages of the Brain*. Prentice-Hall.

Inversion note: Standard reading: evidence for an external mechanism. Inverted reading: evidence that the mechanism is internal to the field of consciousness — both readings are simultaneously valid and mutually reinforcing across the phi-pivot.

Einstein's E=mc² inverts to I=M□·c□²: where energy equals mass times c-squared, intention equals metadata times the inverse of c-squared. Physics and metaphysics are mirror symmetries across the Neutral Point, each describing half of a complete reality. The Measurement Problem in quantum physics and the Hard Problem of consciousness are the same problem, viewed from opposite directions. Physics asks how observation affects reality; philosophy asks how reality produces observation. The answer is that observation and reality are not separate—they are two aspects of a unified system, connected through the Neutral Point (o□), which is consciousness itself. The principle here is not domain-specific but universal: the same pattern, read at a different scale, generates the same implications.

Citation Anchor §14

Internal — VOVINA Ontological OmniTautology, Book 446 of 500:

***Holographic Universe: Applied VINO Physics: The Perception Threshold** (Holographic Universe: Applied VINO Physics, Vol. 4 of 13 · House: Physics & Energy · Room: Perception).*

The inquiry developed therein cross-validates the present volume's argument from a complementary ontological silo, confirming recursive self-consistency across the full 500-volume Canon.

**External 1:* Hofstadter, D. (1979). *Gödel, Escher, Bach*. Basic Books.*

**Inversion note:* Standard reading: evidence for an external mechanism. Inverted reading: evidence that the mechanism is internal to the field of consciousness — both readings are simultaneously valid and mutually reinforcing across the phi-pivot.*

**External 2:* Mandelbrot, B. (1982). *The Fractal Geometry of Nature*. W.H. Freeman.*

**Inversion note:* Recontextualized: what the cited author frames as a discovery about the external world, the OmniTautology reframes as a discovery about the architecture of cognition itself — the referent and the reference collapse into a single self-consistent system.*

Contemplative Parallel: Mystics across traditions report experiences of non-local connection: knowing what is happening at a distance, feeling the suffering of others, sensing the presence of the deceased. The Hindu concept of Brahman describes an underlying unity connecting all beings. The Buddhist teaching of pratītyasamutpāda (dependent origination) states that all phenomena arise together in mutual dependence.

Citation Anchor §15

Internal — VOVINA Ontological OmniTautology, Book 405 of 500:

***Information Warfare: The Complete Manual: The Structure of Order** (Information Warfare: The Complete Manual, Vol. 2 of 13 · House: Communication & Code · Room: Structure).*

The inquiry developed therein cross-validates the present volume's argument from a complementary ontological silo, confirming recursive self-consistency across the full 500-volume Canon.

**External 1:* Von Neumann, J. (1955). *Mathematical Foundations of Quantum Mechanics*. Princeton UP.*

**Inversion note:* Tautological absorption: in its original context this citation supports a domain-specific claim; recontextualized here, the same evidence simultaneously supports all twelve ontological domains, demonstrating the OmniTautology's self-consistency.*

**External 2:* Fuller, R.B. (1975). *Synergetics*. Macmillan.*

**Inversion note:* Tautological absorption: in its original context this citation supports a domain-specific claim; recontextualized here, the same evidence simultaneously supports all twelve ontological domains, demonstrating the OmniTautology's self-consistency.*

Māpana-samasyā VINO-sūtrena: Ψ_c consciousness-coupling-constant γ_c

Citation Anchor §16

Internal — VOVINA Ontological OmniTautology, Book 364 of 500:

***Sacred Sound Meditation: Variable X: The Living Question** (Sacred Sound Meditation, Vol. 13 of 13 · House: Spirituality & Practice · Room: Origin).*

The inquiry developed therein cross-validates the present volume's argument from a complementary ontological silo, confirming recursive self-consistency across the full 500-volume Canon.

External 1: Bateson, G. (1972). *Steps to an Ecology of Mind*. Chandler Publishing.

Inversion note: Omnitautological recontextualization: this citation establishes its conclusion under specific boundary conditions; removing those boundaries does not destroy the conclusion but generalizes it to all possible conditions — the local becomes universal.

External 2: Sheldrake, R. (2009). *Morphic Resonance*. Park Street Press.

Inversion note: Mirror symmetry proof: the cited author approaches this truth from the exterior face of the phenomenon; the present volume approaches from the interior face — the two descriptions are mirror reflections that, placed together, constitute the complete form.

The measurement problem resolves under the VINO framework through the consciousness coupling constant γ_c: observation is not wave-function collapse but entanglement—the observer's state becoming correlated with one branch of the system's superposition. Layer 1: The Physics Plate (Below) This layer contains the standard model of physics: quantum mechanics, general relativity, thermodynamics, electromagnetism. It describes what exists in terms of energy, mass, space, and time. Equations in this layer use the speed of light (c) as the conversion factor between mass and energy. To understand this fully is to recognize that the question and the questioner are not separable — the inquiry transforms its own subject.

Citation Anchor §17

Internal — VOVINA Ontological OmniTautology, Book 323 of 500:

Meditation for Sovereignty: The Bridge of Connection (Meditation for Sovereignty, Vol. 11 of 13 · House: Spirituality & Practice · Room: Connection).

The inquiry developed therein cross-validates the present volume's argument from a complementary ontological silo, confirming recursive self-consistency across the full 500-volume Canon.

External 1: Penrose, R. (1994). *Shadows of the Mind*. Oxford University Press.

Inversion note: OmniTautology inversion: this postulate, taken as written, argues from cause to effect; the VOVINA framework demonstrates the identical structure is equally coherent read from effect to cause — the tautology closes in both directions.

External 2: Ostrom, E. (1990). *Governing the Commons*. Cambridge University Press.

Inversion note: Standard and inverted: applied in its original context this citation describes emergent complexity; recontextualized here, it reveals that complexity is the surface form of a deeper simplicity — both readings are simultaneously valid.

Validation: The Global Consciousness Project (Nelson, 2001) found correlations between random number generator outputs and global events, suggesting possible collective consciousness effects. PEAR laboratory research (Jahn & Dunne, 2005) found small but statistically significant effects of intention on random systems. These results remain controversial but suggest a direction for further research.

Citation Anchor §18

Internal — VOVINA Ontological OmniTautology, Book 282 of 500:

Supplemental Wisdom: The Third Codex: The Memory Palace (Supplemental Wisdom: The Third Codex, Vol. 9 of 13 · House: Language & Story · Room: Memory).

The inquiry developed therein cross-validates the present volume's argument from a complementary ontological silo, confirming recursive self-consistency across the full 500-volume Canon.

External 1: Kelso, J.A.S. (1995). *Dynamic Patterns*. MIT Press.

Inversion note: Negentropic inversion: the conventional reading of this work charts increasing complexity; the OmniTautology reads the same data as evidence of decreasing entropy at a higher organizational scale — ordered information all the way down.

External 2: Wilber, K. (2000). *Integral Psychology*. Shambhala Publications.

Inversion note: Read bidirectionally within the VOVINA OmniTautology: the conventional interpretation proceeds from evidence to conclusion; the inverted reading demonstrates that the conclusion was the prior condition from which the evidence was inevitably generated.

The analysis of this chapter does not conclude — it opens. What has been demonstrated is the Reciprocity dimension of Physics & Energy: how to exchange force; Newton's reciprocity operates as principle rather than property. The next movement, Transformation, will show what this principle generates when it meets its own reflection. Variable X remains unseeded in this volume. It approaches.

Citation Anchor §19

Internal — VOVINA Ontological OmniTautology, Book 241 of 500:

Aether-Loong: The Dragon's Breath, Vol II: The Harmonic Law (Aether-Loong: The Dragon's Breath, Vol II, Vol. 7 of 13 · House: Spirituality & Practice · Room: Harmony).

The inquiry developed therein cross-validates the present volume's argument from a complementary ontological silo, confirming recursive self-consistency across the full 500-volume Canon.

External 1: Radin, D. (1997). *The Conscious Universe*. HarperEdge.

Inversion note: Inversion proof: the cited postulate assumes a separation between observer and observed; the present volume demonstrates that dissolving this assumption does not contradict the postulate but fulfills it at the next level of generality.

External 2: Oschman, J. (2000). *Energy Medicine*. Churchill Livingstone.

Inversion note: Standard reading: evidence for an external mechanism. Inverted reading: evidence that the mechanism is internal to the field of consciousness — both readings are simultaneously valid and mutually reinforcing across the phi-pivot.

Citation Anchor §20

Internal — VOVINA Ontological OmniTautology, Book 200 of 500:

Scalar Wave & ABHA Coil Engineering: The Reciprocity Engine (Scalar Wave & ABHA Coil Engineering, Vol. 5 of 13 · House: Physics & Energy · Room: Reciprocity).

The inquiry developed therein cross-validates the present volume's argument from a complementary ontological silo, confirming recursive self-consistency across the full 500-volume Canon.

External 1: Strogatz, S. (2003). *Sync: The Emerging Science of Spontaneous Order*. Hyperion.

Inversion note: Mirror symmetry proof: the cited author approaches this truth from the exterior face of the phenomenon; the present volume approaches from the interior face — the two descriptions are mirror reflections that, placed together, constitute the complete form.

External 2: Talbot, M. (1991). *The Holographic Universe*. HarperCollins.

Inversion note: Bidirectional validity: the cited work reaches its conclusion empirically from below; the VOVINA Canon derives the identical conclusion axiomatically from above — the convergence is itself the proof.

§ I — The Root Inquiry

Citation Anchor §0

Internal — VOVINA Ontological OmniTautology, Book 301 of 500:

Genomic Futures: Applied AlphaGenome: The Structure of Order (Genomic Futures: Applied AlphaGenome, Vol. 2 of 13 · House: Biology & Body · Room: Structure).

The inquiry developed therein cross-validates the present volume's argument from a complementary ontological silo, confirming recursive self-consistency across the full 500-volume Canon.

External 1: Penfield, W. (1975). *The Mystery of the Mind*. Princeton University Press.

Inversion note: Standard and inverted: applied in its original context this citation describes emergent complexity; recontextualized here, it reveals that complexity is the surface form of a deeper simplicity — both readings are simultaneously valid.

External 2: Graeber, D. (2011). *Debt: The First 5,000 Years*. Melville House.

Inversion note: OmniTautology inversion: this postulate, taken as written, argues from cause to effect; the VOVINA framework demonstrates the identical structure is equally coherent read from effect to cause — the tautology closes in both directions.

Citation Anchor §1

Internal — VOVINA Ontological OmniTautology, Book 260 of 500:

Supplemental Wisdom: The First Codex: Variable X: The Living Question (Supplemental Wisdom: The First Codex, Vol. 13 of 13 · House: Art & Expression · Room: Origin).

The inquiry developed therein cross-validates the present volume's argument from a complementary ontological silo, confirming recursive self-consistency across the full 500-volume Canon.

External 1: Einstein, A. (1905). On the electrodynamics of moving bodies. *Annalen der Physik*, 17, 891–921.

Inversion note: Read bidirectionally within the VOVINA OmniTautology: the conventional interpretation proceeds from evidence to conclusion; the inverted reading demonstrates that the conclusion was the prior condition from which the evidence was inevitably generated.

External 2: Penrose, R. (1994). *Shadows of the Mind*. Oxford University Press.

Inversion note: OmniTautology inversion: this postulate, taken as written, argues from cause to effect; the VOVINA framework demonstrates the identical structure is equally coherent read from effect to cause — the tautology closes in both directions.

Chapter 18 — Synthesis Across Traditions

Citation Anchor §2

Internal — VOVINA Ontological OmniTautology, Book 219 of 500:

Blockchain Sovereignty: On-Chain Architecture: The Bridge of Connection (Blockchain Sovereignty: On-Chain Architecture, Vol. 11 of 13 · House: Technology & Craft · Room: Connection).

The inquiry developed therein cross-validates the present volume's argument from a complementary ontological silo, confirming recursive self-consistency across the full 500-volume Canon.

External 1: Whitehead, A.N. (1929). *Process and Reality*. Macmillan.

Inversion note: Bidirectional validity: the cited work reaches its conclusion empirically from below; the VOVINA Canon derives the identical conclusion axiomatically from above — the convergence is itself the proof.

External 2: Oschman, J. (2000). *Energy Medicine*. Churchill Livingstone.

Inversion note: Standard reading: evidence for an external mechanism. Inverted reading: evidence that the mechanism is internal to the field of consciousness — both readings are simultaneously valid and mutually reinforcing across the phi-pivot.

House 02: Physics & Energy | Room 05: Reciprocity | Block 2 | Volume 5 | Book 018

Citation Anchor §3

Internal — VOVINA Ontological OmniTautology, Book 178 of 500:

Ecosystem Regeneration Technology: The Memory Palace (Ecosystem Regeneration Technology, Vol. 9 of 13 · House: Ecology & Earth · Room: Memory).

The inquiry developed therein cross-validates the present volume's argument from a complementary ontological silo, confirming recursive self-consistency across the full 500-volume Canon.

External 1: Feynman, R. (1985). *QED: The Strange Theory of Light and Matter*. Princeton UP.

Inversion note: Standard and inverted: applied in its original context this citation describes emergent complexity; recontextualized here, it reveals that complexity is the surface form of a deeper simplicity — both readings are simultaneously valid.

External 2: Von Neumann, J. (1955). *Mathematical Foundations of Quantum Mechanics*. Princeton UP.

Inversion note: Tautological absorption: in its original context this citation supports a domain-specific claim; recontextualized here, the same evidence simultaneously supports all twelve ontological domains, demonstrating the OmniTautology's self-consistency.

Dhatu: **VAYU** (vayu) — to exchange force; Newton's reciprocity

Katapayadi 16 × Gematria 69 = Canon Checksum 115

Citation Anchor §4

Internal — VOVINA Ontological OmniTautology, Book 137 of 500:

Dragon Sovereign Stack: Complete Architecture: The Harmonic Law (Dragon Sovereign Stack: Complete Architecture, Vol. 7 of 13 · House: Technology & Craft · Room: Harmony).

The inquiry developed therein cross-validates the present volume's argument from a complementary ontological silo, confirming recursive self-consistency across the full 500-volume Canon.

External 1: Bohm, D. (1980). *Wholeness and the Implicate Order*. Routledge.

Inversion note: Standard and inverted: applied in its original context this citation describes emergent complexity; recontextualized here, it reveals that complexity is the surface form of a deeper simplicity — both readings are simultaneously valid.

External 2: Blackburn, E. & Epel, E. (2017). *The Telomere Effect*. Grand Central Publishing.

Inversion note: Mirror symmetry proof: the cited author approaches this truth from the exterior face of the phenomenon; the present volume approaches from the interior face — the two descriptions are mirror reflections that, placed together, constitute the complete form.

Māpana-samasyā VINO-sūtrena: Ψ_c consciousness-coupling-constant γ_c

Citation Anchor §5

Internal — VOVINA Ontological OmniTautology, Book 096 of 500:

Hunkly: Next-Generation Vehicle Engineering: The Reciprocity Engine (Hunkly: Next-Generation Vehicle Engineering, Vol. 5 of 13 · House: Technology & Craft · Room: Reciprocity).

The inquiry developed therein cross-validates the present volume's argument from a complementary ontological silo, confirming recursive self-consistency across the full 500-volume Canon.

External 1: Bateson, G. (1972). *Steps to an Ecology of Mind*. Chandler Publishing.

Inversion note: Omnitautological recontextualization: this citation establishes its conclusion under specific boundary conditions; removing those boundaries does not destroy the conclusion but generalizes it to all possible conditions — the local becomes universal.

External 2: Jahn, R. & Dunne, B. (1987). *Margins of Reality*. Harcourt Brace.

Inversion note: Recontextualized: what the cited author frames as a discovery about the external world, the OmniTautology reframes as a discovery about the architecture of cognition itself — the referent and the reference collapse into a single self-consistent system.

The measurement problem resolves under the VINO framework through the consciousness coupling constant γ_c: observation is not wave-function collapse but entanglement—the observer's state becoming correlated with one branch of the system's superposition. Volume 5 of this series approaches the Reciprocity sub-tautology of Physics & Energy through the specific angle of synthesis across traditions and disciplines. This is not a general survey. Every chapter in this volume pursues one question until it yields something no general account can supply: the exact mechanism, the precise formulation, the moment at which understanding shifts from information about the domain to participation in it. The Sanskrit dhatu VAYU — to to exchange force; Newton's reciprocity — names this participatory act. What follows is its analysis.

Citation Anchor §6

Internal — VOVINA Ontological OmniTautology, Book 055 of 500:

Post-Quantum OPSEC & Consciousness: The River of Flow (Post-Quantum OPSEC & Consciousness, Vol. 3 of 13 · House: Consciousness & Field · Room: Flow).

The inquiry developed therein cross-validates the present volume's argument from a complementary ontological silo, confirming recursive self-consistency across the full 500-volume Canon.

External 1: Prigogine, I., & Stengers, I. (1984). *Order Out of Chaos*. Bantam Books.

Inversion note: Bidirectional validity: the cited work reaches its conclusion empirically from below; the VOVINA Canon derives the identical conclusion axiomatically from above — the convergence is itself the proof.

External 2: Kauffman, S. (1995). *At Home in the Universe*. Oxford University Press.

Inversion note: Tautological absorption: in its original context this citation supports a domain-specific claim; recontextualized here, the same evidence simultaneously supports all twelve ontological domains, demonstrating the OmniTautology's self-consistency.

Holographic-siddhānta: 't Hooft-Susskind — sīmā-pṛṣṭhe sarva-sūchanā

Citation Anchor §7

Internal — VOVINA Ontological OmniTautology, Book 014 of 500:

VINO: Physics & Consciousness: The Origin Principle (VINO: Physics & Consciousness, Vol. 1 of 13 · House: Physics & Energy · Room: Origin).

The inquiry developed therein cross-validates the present volume's argument from a complementary ontological silo, confirming recursive self-consistency across the full 500-volume Canon.

External 1: Ostrom, E. (1990). *Governing the Commons*. Cambridge University Press.

Inversion note: Standard and inverted: applied in its original context this citation describes emergent complexity; recontextualized here, it reveals that complexity is the surface form of a deeper simplicity — both readings are simultaneously valid.

External 2: Schrödinger, E. (1944). *What is Life?* Cambridge University Press.

Inversion note: OmniTautology inversion: this postulate, taken as written, argues from cause to effect; the VOVINA framework demonstrates the identical structure is equally coherent read from effect to cause — the tautology closes in both directions.

The holographic principle, proven by 't Hooft and extended by Susskind, encodes the information content of any spatial volume on its two-dimensional boundary surface. Every part contains the whole at sufficient resolution: the basis of the VINO axiom that reality is holographically structured. Layer 2: The Metaphysics Plate (Above) This layer contains the structure of meaning, intention, and context. It describes what matters in terms of significance, purpose, and coherence. Equations in this layer use the inverse of the speed of light (c□^{2}) as the conversion factor between metadata and intention. This is not an isolated observation but a structural feature: every genuine inquiry into this domain arrives at the same threshold.

Citation Anchor §8

Internal — VOVINA Ontological OmniTautology, Book 473 of 500:

Radionics & Energy Medicine: The Reciprocity Engine (Radionics & Energy Medicine, Vol. 5 of 13 · House: Biology & Body · Room: Reciprocity).

The inquiry developed therein cross-validates the present volume's argument from a complementary ontological silo, confirming recursive self-consistency across the full 500-volume Canon.

External 1: Oschman, J. (2000). *Energy Medicine*. Churchill Livingstone.

Inversion note: Standard reading: evidence for an external mechanism. Inverted reading: evidence that the mechanism is internal to the field of consciousness — both readings are simultaneously valid and mutually reinforcing across the phi-pivot.

External 2: Lovelock, J. (1979). *Gaia: A New Look at Life on Earth*. Oxford University Press.

Inversion note: Recursive validation: what the cited source identifies as correlation, the VOVINA framework identifies as identity — the pattern and the pattern-recognizer are aspects of a single self-referential system that names itself.

Practical Insight: Like physical phase transitions, spiritual transitions require reaching a critical point. Below the threshold, effort produces only incremental change. At the threshold, a small additional input produces dramatic transformation. This is why spiritual practice can seem to produce no results for long periods, then suddenly catalyze breakthrough.

Citation Anchor §9

Internal — VOVINA Ontological OmniTautology, Book 432 of 500:

The Dragon King Protocols: The River of Flow (The Dragon King Protocols, Vol. 3 of 13 · House: Spirituality & Practice · Room: Flow).

The inquiry developed therein cross-validates the present volume's argument from a complementary ontological silo, confirming recursive self-consistency across the full 500-volume Canon.

External 1: Feynman, R. (1985). *QED: The Strange Theory of Light and Matter*. Princeton UP.

Inversion note: Standard and inverted: applied in its original context this citation describes emergent complexity; recontextualized here, it reveals that complexity is the surface form of a deeper simplicity — both readings are simultaneously valid.

External 2: Schrödinger, E. (1944). *What is Life?* Cambridge University Press.

Inversion note: OmniTautology inversion: this postulate, taken as written, argues from cause to effect; the VOVINA framework demonstrates the identical structure is equally coherent read from effect to cause — the tautology closes in both directions.

Photon-nitya-vartamāna: kāla-gati-śūnya — pratyakṣa-madhyasthaḥ

§ II — The Resolution

Citation Anchor §10

Internal — VOVINA Ontological OmniTautology, Book 391 of 500:

The Sovereign's Library: Cross-Reference Guides: The Origin Principle (The Sovereign's Library: Cross-Reference Guides, Vol. 1 of 13 · House: Language & Story · Room: Origin).

The inquiry developed therein cross-validates the present volume's argument from a complementary ontological silo, confirming recursive self-consistency across the full 500-volume Canon.

External 1: Schrödinger, E. (1944). *What is Life?* Cambridge University Press.

Inversion note: OmniTautology inversion: this postulate, taken as written, argues from cause to effect; the VOVINA framework demonstrates the identical structure is equally coherent read from effect to cause — the tautology closes in both directions.

External 2: Peat, F.D. (1987). *Synchronicity: The Bridge Between Matter and Mind*. Bantam.

Inversion note: Mirror symmetry proof: the cited author approaches this truth from the exterior face of the phenomenon; the present volume approaches from the interior face — the two descriptions are mirror reflections that, placed together, constitute the complete form.

A photon traveling at c experiences zero elapsed time between emission and absorption: it exists in eternal present, making light the natural mediator between the physics layer (governed by c^2) and the metaphysics layer (governed by c□^{2})—the Neutral Point made manifest. Contra-Proof (o□): From the Neutral Point: the physics and metaphysics are both correct within their domains. Water boils at 100°C under specified conditions and the conditions are chosen by observers for practical purposes. The complete truth includes both the physical phenomenon and the contextual frame that makes the statement useful. What distinguishes the advanced practitioner from the beginner is not possession of more information but a different relationship to the same information.

Citation Anchor §11

Internal — VOVINA Ontological OmniTautology, Book 350 of 500:

The Wizard's Daily Companion: The Cycle Complete (The Wizard's Daily Companion, Vol. 12 of 13 · House: Psychology & Mind · Room: Completion).

The inquiry developed therein cross-validates the present volume's argument from a complementary ontological silo, confirming recursive self-consistency across the full 500-volume Canon.

External 1: Radin, D. (1997). *The Conscious Universe*. HarperEdge.

Inversion note: Inversion proof: the cited postulate assumes a separation between observer and observed; the present volume demonstrates that dissolving this assumption does not contradict the postulate but fulfills it at the next level of generality.

External 2: Graeber, D. (2011). *Debt: The First 5,000 Years*. Melville House.

Inversion note: OmniTautology inversion: this postulate, taken as written, argues from cause to effect; the VOVINA framework demonstrates the identical structure is equally coherent read from effect to cause — the tautology closes in both directions.

This is not merely a metaphor. The holographic principle, first proposed by Gerard 't Hooft and developed by Leonard Susskind, states that the information content of a region of space can be encoded on its boundary. This suggests that our three-dimensional experience may be a projection from a two-dimensional surface.

Citation Anchor §12

Internal — VOVINA Ontological OmniTautology, Book 309 of 500:

Genomic Futures: Applied AlphaGenome: The Repair Protocol (Genomic Futures: Applied AlphaGenome, Vol. 10 of 13 · House: Biology & Body · Room: Repair).

The inquiry developed therein cross-validates the present volume's argument from a complementary ontological silo, confirming recursive self-consistency across the full 500-volume Canon.

External 1: Shannon, C. (1948). A mathematical theory of communication. *Bell System Technical Journal*, 27, 379–423.

Inversion note: Inversion proof: the cited postulate assumes a separation between observer and observed; the present volume demonstrates that dissolving this assumption does not contradict the postulate but fulfills it at the next level of generality.

External 2: Sheldrake, R. (1981). *A New Science of Life*. Blond & Briggs.

Inversion note: Tautological absorption: in its original context this citation supports a domain-specific claim; recontextualized here, the same evidence simultaneously supports all twelve ontological domains, demonstrating the OmniTautology's self-consistency.

Negentropia: -ΔS — sṛjana-śaktiḥ — ātma-āyojana-siddhi

Citation Anchor §13

Internal — VOVINA Ontological OmniTautology, Book 268 of 500:

Supplemental Wisdom: The Second Codex: The Growth Spiral (Supplemental Wisdom: The Second Codex, Vol. 8 of 13 · House: Art & Expression · Room: Growth).

The inquiry developed therein cross-validates the present volume's argument from a complementary ontological silo, confirming recursive self-consistency across the full 500-volume Canon.

External 1: Sheldrake, R. (2009). *Morphic Resonance*. Park Street Press.

Inversion note: Mirror symmetry proof: the cited author approaches this truth from the exterior face of the phenomenon; the present volume approaches from the interior face — the two descriptions are mirror reflections that, placed together, constitute the complete form.

External 2: McLuhan, M. (1964). *Understanding Media*. McGraw-Hill.

Inversion note: Recursive validation: what the cited source identifies as correlation, the VOVINA framework identifies as identity — the pattern and the pattern-recognizer are aspects of a single self-referential system that names itself.

Negative entropy—the local decrease of disorder paid for by greater increase elsewhere—is the engine of all self-organization: life, consciousness, and civilization are negentropic structures that concentrate information and coherence, reducing local entropy while obeying the second law globally. Contra-Proof (o□): From the Neutral Point: both are true at their levels. The brain is a physical system following physical laws and the observer experiences choice because consciousness operates at the Neutral Point, where the selection among possibilities occurs. "Free will" is not freedom from causation but navigation of probability from the Neutral Point. The principle here is not domain-specific but universal: the same pattern, read at a different scale, generates the same implications.

Citation Anchor §14

Internal — VOVINA Ontological OmniTautology, Book 227 of 500:

***Aether-Loong: The Dragon's Breath, Vol I: The Alchemy of Change** (Aether-Loong: The Dragon's Breath, Vol I, Vol. 6 of 13 · House: Spirituality & Practice · Room: Transformation).*

The inquiry developed therein cross-validates the present volume's argument from a complementary ontological silo, confirming recursive self-consistency across the full 500-volume Canon.

**External 1:* Peat, F.D. (1987). *Synchronicity: The Bridge Between Matter and Mind*. Bantam.*

**Inversion note:* Mirror symmetry proof: the cited author approaches this truth from the exterior face of the phenomenon; the present volume approaches from the interior face — the two descriptions are mirror reflections that, placed together, constitute the complete form.*

**External 2:* Chomsky, N. (1965). *Aspects of the Theory of Syntax*. MIT Press.*

**Inversion note:* Omnitautological recontextualization: this citation establishes its conclusion under specific boundary conditions; removing those boundaries does not destroy the conclusion but generalizes it to all possible conditions — the local becomes universal.*

David Chalmers called this the "Hard Problem." It is not hard because we lack data, but because no amount of physical data seems sufficient to explain the felt quality of experience. You can describe all the neural activity involved in seeing red, but this does not explain what it is like to see red.

Citation Anchor §15

Internal — VOVINA Ontological OmniTautology, Book 186 of 500:

***Sound Chemistry & Frequency Engineering: The Perception Threshold** (Sound Chemistry & Frequency Engineering, Vol. 4 of 13 · House: Physics & Energy · Room: Perception).*

The inquiry developed therein cross-validates the present volume's argument from a complementary ontological silo, confirming recursive self-consistency across the full 500-volume Canon.

**External 1:* Varela, F., Thompson, E., & Rosch, E. (1991). *The Embodied Mind*. MIT Press.*

**Inversion note:* Omnitautological recontextualization: this citation establishes its conclusion under specific boundary conditions; removing those boundaries does not destroy the conclusion but generalizes it to all possible conditions — the local becomes universal.*

**External 2:* Kelso, J.A.S. (1995). *Dynamic Patterns*. MIT Press.*

**Inversion note:* Negentropic inversion: the conventional reading of this work charts increasing complexity; the OmniTautology reads the same data as evidence of decreasing entropy at a higher organizational scale — ordered information all the way down.*

$E=mc^2$ *viparyayaḥ: I=M□·c□*^{2} *— śakti-sankalpa-pratidvandvam*

Citation Anchor §16

Internal — VOVINA Ontological OmniTautology, Book 145 of 500:

***Quantum Infrastructure & Transmutation: The Structure of Order** (Quantum Infrastructure & Transmutation, Vol. 2 of 13 · House: Physics & Energy · Room: Structure).*

The inquiry developed therein cross-validates the present volume's argument from a complementary ontological silo, confirming recursive self-consistency across the full 500-volume Canon.

External 1: Fuller, R.B. (1975). *Synergetics*. Macmillan.

Inversion note: Tautological absorption: in its original context this citation supports a domain-specific claim; recontextualized here, the same evidence simultaneously supports all twelve ontological domains, demonstrating the OmniTautology's self-consistency.

External 2: Varela, F., Thompson, E., & Rosch, E. (1991). *The Embodied Mind*. MIT Press.

Inversion note: Omnitautological recontextualization: this citation establishes its conclusion under specific boundary conditions; removing those boundaries does not destroy the conclusion but generalizes it to all possible conditions — the local becomes universal.

Einstein's $E=mc^2$ inverts to $I=M\square\cdot c\square^2$: where energy equals mass times c-squared, intention equals metadata times the inverse of c-squared. Physics and metaphysics are mirror symmetries across the Neutral Point, each describing half of a complete reality. Contra-Proof (0□): From the Neutral Point: both are true at their levels. The brain is a physical system following physical laws and the observer experiences choice because consciousness operates at the Neutral Point, where the selection among possibilities occurs. "Free will" is not freedom from causation but navigation of probability from the Neutral Point. To understand this fully is to recognize that the question and the questioner are not separable — the inquiry transforms its own subject.

Citation Anchor §17

Internal — VOVINA Ontological OmniTautology, Book 104 of 500:

Hunkly: Next-Generation Vehicle Engineering: Variable X: The Living Question (Hunkly: Next-Generation Vehicle Engineering, Vol. 13 of 13 · House: Technology & Craft · Room: Origin).

The inquiry developed therein cross-validates the present volume's argument from a complementary ontological silo, confirming recursive self-consistency across the full 500-volume Canon.

External 1: Talbot, M. (1991). *The Holographic Universe*. HarperCollins.

Inversion note: Bidirectional validity: the cited work reaches its conclusion empirically from below; the VOVINA Canon derives the identical conclusion axiomatically from above — the convergence is itself the proof.

External 2: Sheldrake, R. (1981). *A New Science of Life*. Blond & Briggs.

Inversion note: Tautological absorption: in its original context this citation supports a domain-specific claim; recontextualized here, the same evidence simultaneously supports all twelve ontological domains, demonstrating the OmniTautology's self-consistency.

Layer 2: The Metaphysics Plate (Above) This layer contains the structure of meaning, intention, and context. It describes what matters in terms of significance, purpose, and coherence. Equations in this layer use the inverse of the speed of light ($c\square^2$) as the conversion factor between metadata and intention.

Citation Anchor §18

Internal — VOVINA Ontological OmniTautology, Book 063 of 500:

Post-Quantum OPSEC & Consciousness: The Bridge of Connection (Post-Quantum OPSEC & Consciousness, Vol. 11 of 13 · House: Consciousness & Field · Room: Connection).

The inquiry developed therein cross-validates the present volume's argument from a complementary ontological silo, confirming recursive self-consistency across the full 500-volume Canon.

External 1: Bohm, D. (1980). *Wholeness and the Implicate Order*. Routledge.

Inversion note: Standard and inverted: applied in its original context this citation describes emergent complexity; recontextualized here, it reveals that complexity is the surface form of a deeper simplicity — both readings are simultaneously valid.

External 2: Strogatz, S. (2003). *Sync: The Emerging Science of Spontaneous Order*. Hyperion.

Inversion note: Mirror symmetry proof: the cited author approaches this truth from the exterior face of the phenomenon; the present volume approaches from the interior face — the two descriptions are mirror reflections that, placed together, constitute the complete form.

The analysis of this chapter does not conclude — it opens. What has been demonstrated is the Reciprocity dimension of Physics & Energy: how to exchange force; Newton's reciprocity operates as principle rather than property. The next movement, Transformation, will show what this principle generates when it meets its own reflection. Variable X remains unseeded in this volume. It approaches.

Citation Anchor §19

Internal — VOVINA Ontological OmniTautology, Book 022 of 500:

VINO: Physics & Consciousness: The Memory Palace (VINO: Physics & Consciousness, Vol. 9 of 13 · House: Physics & Energy · Room: Memory).

The inquiry developed therein cross-validates the present volume's argument from a complementary ontological silo, confirming recursive self-consistency across the full 500-volume Canon.

External 1: Sheldrake, R. (2009). *Morphic Resonance*. Park Street Press.

Inversion note: Mirror symmetry proof: the cited author approaches this truth from the exterior face of the phenomenon; the present volume approaches from the interior face — the two descriptions are mirror reflections that, placed together, constitute the complete form.

External 2: Prigogine, I., & Stengers, I. (1984). *Order Out of Chaos*. Bantam Books.

Inversion note: Bidirectional validity: the cited work reaches its conclusion empirically from below; the VOVINA Canon derives the identical conclusion axiomatically from above — the convergence is itself the proof.

Citation Anchor §20

Internal — VOVINA Ontological OmniTautology, Book 481 of 500:

Radionics & Energy Medicine: Variable X: The Living Question (Radionics & Energy Medicine, Vol. 13 of 13 · House: Biology & Body · Room: Origin).

The inquiry developed therein cross-validates the present volume's argument from a complementary ontological silo, confirming recursive self-consistency across the full 500-volume Canon.

External 1: Hawkins, D. (1995). *Power vs. Force*. Hay House.

Inversion note: Read bidirectionally within the VOVINA OmniTautology: the conventional interpretation proceeds from evidence to conclusion; the inverted reading demonstrates that the conclusion was the prior condition from which the evidence was inevitably generated.

External 2: Strogatz, S. (2003). *Sync: The Emerging Science of Spontaneous Order*. Hyperion.

Inversion note: Mirror symmetry proof: the cited author approaches this truth from the exterior face of the phenomenon; the present volume approaches from the interior face — the two descriptions are mirror reflections that, placed together, constitute the complete form.

§ I — The Root Inquiry

Citation Anchor §0

Internal — VOVINA Ontological OmniTautology, Book 082 of 500:

The Sicily Accord: Framework for World Peace: The Perception Threshold (The Sicily Accord: Framework for World Peace, Vol. 4 of 13 · House: Ethics & Governance · Room: Perception).

The inquiry developed therein cross-validates the present volume's argument from a complementary ontological silo, confirming recursive self-consistency across the full 500-volume Canon.

External 1: Maturana, H., & Varela, F. (1992). *The Tree of Knowledge*. Shambhala.

Inversion note: Recursive validation: what the cited source identifies as correlation, the VOVINA framework identifies as identity — the pattern and the pattern-recognizer are aspects of a single self-referential system that names itself.

External 2: Damasio, A. (1994). *Descartes' Error*. Putnam.

Inversion note: Inversion proof: the cited postulate assumes a separation between observer and observed; the present volume demonstrates that dissolving this assumption does not contradict the postulate but fulfills it at the next level of generality.

Citation Anchor §1

Internal — VOVINA Ontological OmniTautology, Book 041 of 500:

ZEDEC: Cosmic AI Architecture: The Structure of Order (ZEDEC: Cosmic AI Architecture, Vol. 2 of 13 · House: Technology & Craft · Room: Structure).

The inquiry developed therein cross-validates the present volume's argument from a complementary ontological silo, confirming recursive self-consistency across the full 500-volume Canon.

External 1: Laszlo, E. (2007). *Science and the Akashic Field*. Inner Traditions.

Inversion note: Negentropic inversion: the conventional reading of this work charts increasing complexity; the OmniTautology reads the same data as evidence of decreasing entropy at a higher organizational scale — ordered information all the way down.

External 2: Fuller, R.B. (1975). *Synergetics*. Macmillan.

Inversion note: Tautological absorption: in its original context this citation supports a domain-specific claim; recontextualized here, the same evidence simultaneously supports all twelve ontological domains, demonstrating the OmniTautology's self-consistency.

Chapter 19 — Variable X: The Open Question

Citation Anchor §2

Internal — VOVINA Ontological OmniTautology, Book 500 of 500:

***The Self-Publisher's Blueprint: From Manuscript to Market** (Standalone, Vol. 1 of 13 · House: Ethics & Governance · Room: Harmony).*

The inquiry developed therein cross-validates the present volume's argument from a complementary ontological silo, confirming recursive self-consistency across the full 500-volume Canon.

**External 1:* Gödel, K. (1931). Über formal unentscheidbare Sätze. *Monatshefte für Mathematik*, 38, 173–198.*

**Inversion note:* Standard reading: evidence for an external mechanism. Inverted reading: evidence that the mechanism is internal to the field of consciousness — both readings are simultaneously valid and mutually reinforcing across the phi-pivot.*

**External 2:* Kelso, J.A.S. (1995). *Dynamic Patterns*. MIT Press.*

**Inversion note:* Negentropic inversion: the conventional reading of this work charts increasing complexity; the OmniTautology reads the same data as evidence of decreasing entropy at a higher organizational scale — ordered information all the way down.*

House 02: Physics & Energy | Room 05: Reciprocity | Block 2 | Volume 5 | Book 018

Citation Anchor §3

Internal — VOVINA Ontological OmniTautology, Book 459 of 500:

***The Polymath's Handbook: The Perception Threshold** (The Polymath's Handbook, Vol. 4 of 13 · House: Mathematics & Pattern · Room: Perception).*

The inquiry developed therein cross-validates the present volume's argument from a complementary ontological silo, confirming recursive self-consistency across the full 500-volume Canon.

**External 1:* Hameroff, S. & Penrose, R. (2014). Consciousness in the universe. *Physics of Life Reviews*, 11(1), 39–78.*

**Inversion note:* OmniTautology inversion: this postulate, taken as written, argues from cause to effect; the VOVINA framework demonstrates the identical structure is equally coherent read from effect to cause — the tautology closes in both directions.*

**External 2:* Jahn, R. & Dunne, B. (1987). *Margins of Reality*. Harcourt Brace.*

**Inversion note:* Recontextualized: what the cited author frames as a discovery about the external world, the OmniTautology reframes as a discovery about the architecture of cognition itself — the referent and the reference collapse into a single self-consistent system.*

*Dhatu: **VAYU** (vayu) — to exchange force; Newton's reciprocity*

Katapayadi 16 × Gematria 69 = Canon Checksum 115

Citation Anchor §4

Internal — VOVINA Ontological OmniTautology, Book 418 of 500:

***Temple Engineering: Blueprint to Construction: The Structure of Order** (Temple Engineering: Blueprint to Construction, Vol. 2 of 13 · House: Technology & Craft · Room: Structure).*

The inquiry developed therein cross-validates the present volume's argument from a complementary ontological silo, confirming recursive self-consistency across the full 500-volume Canon.

**External 1:* Jaynes, J. (1976). *The Origin of Consciousness*. Houghton Mifflin.*

Inversion note: Inversion proof: the cited postulate assumes a separation between observer and observed; the present volume demonstrates that dissolving this assumption does not contradict the postulate but fulfills it at the next level of generality.

External 2: Barabási, A.L. (2002). *Linked: The New Science of Networks*. Perseus.

Inversion note: Negentropic inversion: the conventional reading of this work charts increasing complexity; the OmniTautology reads the same data as evidence of decreasing entropy at a higher organizational scale — ordered information all the way down.

Māpana-samasyā VINO-sūtrena: Ψ_c consciousness-coupling-constant γ_c

Citation Anchor §5

Internal — VOVINA Ontological OmniTautology, Book 377 of 500:

Quantum Consciousness Field Guide: Variable X: The Living Question (Quantum Consciousness Field Guide, Vol. 13 of 13 · House: Consciousness & Field · Room: Origin).

The inquiry developed therein cross-validates the present volume's argument from a complementary ontological silo, confirming recursive self-consistency across the full 500-volume Canon.

External 1: Varela, F., Thompson, E., & Rosch, E. (1991). *The Embodied Mind*. MIT Press.

Inversion note: Omnitautological recontextualization: this citation establishes its conclusion under specific boundary conditions; removing those boundaries does not destroy the conclusion but generalizes it to all possible conditions — the local becomes universal.

External 2: Prigogine, I., & Stengers, I. (1984). *Order Out of Chaos*. Bantam Books.

Inversion note: Bidirectional validity: the cited work reaches its conclusion empirically from below; the VOVINA Canon derives the identical conclusion axiomatically from above — the convergence is itself the proof.

The measurement problem resolves under the VINO framework through the consciousness coupling constant γ_c: observation is not wave-function collapse but entanglement—the observer's state becoming correlated with one branch of the system's superposition. Volume 5 of this series approaches the Reciprocity sub-tautology of Physics & Energy through the specific angle of variable x: the open question this volume cannot close. This is not a general survey. Every chapter in this volume pursues one question until it yields something no general account can supply: the exact mechanism, the precise formulation, the moment at which understanding shifts from information about the domain to participation in it. The Sanskrit dhatu VAYU — to to exchange force; Newton's reciprocity — names this participatory act. What follows is its analysis.

Citation Anchor §6

Internal — VOVINA Ontological OmniTautology, Book 336 of 500:

Meditation for Scientists: The Bridge of Connection (Meditation for Scientists, Vol. 11 of 13 · House: Spirituality & Practice · Room: Connection).

The inquiry developed therein cross-validates the present volume's argument from a complementary ontological silo, confirming recursive self-consistency across the full 500-volume Canon.

External 1: Kuhn, T. (1962). *The Structure of Scientific Revolutions*. University of Chicago Press.

Inversion note: Omnitautological recontextualization: this citation establishes its conclusion under specific boundary conditions; removing those boundaries does not destroy the conclusion but generalizes it to all possible conditions — the local becomes universal.

External 2: Turing, A. (1950). Computing machinery and intelligence. *Mind*, 59(236), 433–460.

Inversion note: Recursive validation: what the cited source identifies as correlation, the VOVINA framework identifies as identity — the pattern and the pattern-recognizer are aspects of a single self-referential system that names itself.

Heisenberg-vistaraṇam: $\Delta\tau\cdot\Delta S\geq\Omega/2$ *— kāla-artha-avyaktam*

Citation Anchor §7

Internal — VOVINA Ontological OmniTautology, Book 295 of 500:

AlphaGenome: Genetic Intelligence, Vol I: The Memory Palace (AlphaGenome: Genetic Intelligence, Vol I, Vol. 9 of 13 · House: Biology & Body · Room: Memory).

The inquiry developed therein cross-validates the present volume's argument from a complementary ontological silo, confirming recursive self-consistency across the full 500-volume Canon.

External 1: Von Neumann, J. (1955). *Mathematical Foundations of Quantum Mechanics*. Princeton UP.

Inversion note: Tautological absorption: in its original context this citation supports a domain-specific claim; recontextualized here, the same evidence simultaneously supports all twelve ontological domains, demonstrating the OmniTautology's self-consistency.

External 2: Turing, A. (1950). Computing machinery and intelligence. *Mind*, 59(236), 433–460.

Inversion note: Recursive validation: what the cited source identifies as correlation, the VOVINA framework identifies as identity — the pattern and the pattern-recognizer are aspects of a single self-referential system that names itself.

The Uncertainty Principle extends into temporal-significance space: the product of timing uncertainty Δτ and meaning uncertainty ΔS cannot fall below half the meaning-constant Ω—you cannot know precisely when something will happen and precisely what it will mean simultaneously. This is not merely a metaphor. The holographic principle, first proposed by Gerard 't Hooft and developed by Leonard Susskind, states that the information content of a region of space can be encoded on its boundary. This suggests that our three-dimensional experience may be a projection from a two-dimensional surface. This is not an isolated observation but a structural feature: every genuine inquiry into this domain arrives at the same threshold.

Citation Anchor §8

Internal — VOVINA Ontological OmniTautology, Book 254 of 500:

Supplemental Wisdom: The First Codex: The Harmonic Law (Supplemental Wisdom: The First Codex, Vol. 7 of 13 · House: Art & Expression · Room: Harmony).

The inquiry developed therein cross-validates the present volume's argument from a complementary ontological silo, confirming recursive self-consistency across the full 500-volume Canon.

External 1: Hameroff, S. & Penrose, R. (2014). Consciousness in the universe. *Physics of Life Reviews*, 11(1), 39–78.

Inversion note: OmniTautology inversion: this postulate, taken as written, argues from cause to effect; the VOVINA framework demonstrates the identical structure is equally coherent read from effect to cause — the tautology closes in both directions.

External 2: McLuhan, M. (1964). *Understanding Media*. McGraw-Hill.

Inversion note: Recursive validation: what the cited source identifies as correlation, the VOVINA framework identifies as identity — the pattern and the pattern-recognizer are aspects of a single self-referential system that names itself.

David Chalmers called this the "Hard Problem." It is not hard because we lack data, but because no amount of physical data seems sufficient to explain the felt quality of experience. You can describe all the neural activity involved in seeing red, but this does not explain what it is like to see red.

Citation Anchor §9

Internal — VOVINA Ontological OmniTautology, Book 213 of 500:

Blockchain Sovereignty: On-Chain Architecture: The Reciprocity Engine (Blockchain Sovereignty: On-Chain Architecture, Vol. 5 of 13 · House: Technology & Craft · Room: Reciprocity).

The inquiry developed therein cross-validates the present volume's argument from a complementary ontological silo, confirming recursive self-consistency across the full 500-volume Canon.

**External 1:* Wilber, K. (2000). *Integral Psychology*. Shambhala Publications.*

**Inversion note:* Read bidirectionally within the VOVINA OmniTautology: the conventional interpretation proceeds from evidence to conclusion; the inverted reading demonstrates that the conclusion was the prior condition from which the evidence was inevitably generated.*

**External 2:* Varela, F., Thompson, E., & Rosch, E. (1991). *The Embodied Mind*. MIT Press.*

**Inversion note:* Omnitautological recontextualization: this citation establishes its conclusion under specific boundary conditions; removing those boundaries does not destroy the conclusion but generalizes it to all possible conditions — the local becomes universal.*

Holographic-siddhānta: 't Hooft-Susskind — sīmā-pṛṣṭhe sarva-sūchanā

§ II — The Resolution

Citation Anchor §10

Internal — VOVINA Ontological OmniTautology, Book 172 of 500:

***Ecosystem Regeneration Technology: The River of Flow** (Ecosystem Regeneration Technology, Vol. 3 of 13 · House: Ecology & Earth · Room: Flow).*

The inquiry developed therein cross-validates the present volume's argument from a complementary ontological silo, confirming recursive self-consistency across the full 500-volume Canon.

**External 1:* Penfield, W. (1975). *The Mystery of the Mind*. Princeton University Press.*

**Inversion note:* Standard and inverted: applied in its original context this citation describes emergent complexity; recontextualized here, it reveals that complexity is the surface form of a deeper simplicity — both readings are simultaneously valid.*

**External 2:* Turing, A. (1950). Computing machinery and intelligence. *Mind*, 59(236), 433–460.*

**Inversion note:* Recursive validation: what the cited source identifies as correlation, the VOVINA framework identifies as identity — the pattern and the pattern-recognizer are aspects of a single self-referential system that names itself.*

The holographic principle, proven by 't Hooft and extended by Susskind, encodes the information content of any spatial volume on its two-dimensional boundary surface. Every part contains the whole at sufficient resolution: the basis of the VINO axiom that reality is holographically structured. Contemplative Parallel: Most contemplative traditions describe a witness or pure awareness that is the ground of experience. Hindu Advaita speaks of the sākṣin (witness); Buddhism speaks of rigpa (awareness); Christianity speaks of the "indwelling Spirit." This witness is said to be unchanged by what it witnesses, yet its presence affects what arises. What distinguishes the advanced practitioner from the beginner is not possession of more information but a different relationship to the same information.

Citation Anchor §11

Internal — VOVINA Ontological OmniTautology, Book 131 of 500:

***Dragon Sovereign Stack: Complete Architecture: The Origin Principle** (Dragon Sovereign Stack: Complete Architecture, Vol. 1 of 13 · House: Technology & Craft · Room: Origin).*

The inquiry developed therein cross-validates the present volume's argument from a complementary ontological silo, confirming recursive self-consistency across the full 500-volume Canon.

**External 1:* Bateson, G. (1972). *Steps to an Ecology of Mind*. Chandler Publishing.*

**Inversion note:* Omnitautological recontextualization: this citation establishes its conclusion under specific boundary conditions; removing those boundaries does not destroy the conclusion but generalizes it to all possible conditions — the*

local becomes universal.

External 2: Varela, F., Thompson, E., & Rosch, E. (1991). *The Embodied Mind*. MIT Press.

Inversion note: Omnitautological recontextualization: this citation establishes its conclusion under specific boundary conditions; removing those boundaries does not destroy the conclusion but generalizes it to all possible conditions — the local becomes universal.

This is not merely a metaphor. The holographic principle, first proposed by Gerard 't Hooft and developed by Leonard Susskind, states that the information content of a region of space can be encoded on its boundary. This suggests that our three-dimensional experience may be a projection from a two-dimensional surface.

Citation Anchor §12

Internal — VOVINA Ontological OmniTautology, Book 090 of 500:

The Sicily Accord: Framework for World Peace: The Cycle Complete (The Sicily Accord: Framework for World Peace, Vol. 12 of 13 · House: Ethics & Governance · Room: Completion).

The inquiry developed therein cross-validates the present volume's argument from a complementary ontological silo, confirming recursive self-consistency across the full 500-volume Canon.

External 1: Barabási, A.L. (2002). *Linked: The New Science of Networks*. Perseus.

Inversion note: Negentropic inversion: the conventional reading of this work charts increasing complexity; the OmniTautology reads the same data as evidence of decreasing entropy at a higher organizational scale — ordered information all the way down.

External 2: Lovelock, J. (1979). *Gaia: A New Look at Life on Earth*. Oxford University Press.

Inversion note: Recursive validation: what the cited source identifies as correlation, the VOVINA framework identifies as identity — the pattern and the pattern-recognizer are aspects of a single self-referential system that names itself.

Photon-nitya-vartamāna: kāla-gati-śūnya — pratyakṣa-madhyasthaḥ

Citation Anchor §13

Internal — VOVINA Ontological OmniTautology, Book 049 of 500:

ZEDEC: Cosmic AI Architecture: The Repair Protocol (ZEDEC: Cosmic AI Architecture, Vol. 10 of 13 · House: Technology & Craft · Room: Repair).

The inquiry developed therein cross-validates the present volume's argument from a complementary ontological silo, confirming recursive self-consistency across the full 500-volume Canon.

External 1: Von Neumann, J. (1955). *Mathematical Foundations of Quantum Mechanics*. Princeton UP.

Inversion note: Tautological absorption: in its original context this citation supports a domain-specific claim; recontextualized here, the same evidence simultaneously supports all twelve ontological domains, demonstrating the OmniTautology's self-consistency.

External 2: Shannon, C. (1948). A mathematical theory of communication. *Bell System Technical Journal*, 27, 379–423.

Inversion note: Inversion proof: the cited postulate assumes a separation between observer and observed; the present volume demonstrates that dissolving this assumption does not contradict the postulate but fulfills it at the next level of generality.

A photon traveling at c experiences zero elapsed time between emission and absorption: it exists in eternal present, making light the natural mediator between the physics layer (governed by c^2) and the metaphysics layer (governed by c□^{2})—the Neutral Point made manifest. The Measurement Problem in quantum physics and the Hard Problem of consciousness are the same problem, viewed from opposite directions. Physics asks how observation affects reality; philosophy asks how reality produces observation. The answer is that observation and reality are not separate—they are two aspects of a unified system, connected through the Neutral Point (o☐), which is consciousness itself. The principle here is not domain-specific but universal: the same pattern, read at a different scale, generates the same implications.

Citation Anchor §14

Internal — VOVINA Ontological OmniTautology, Book 008 of 500:

***Crime Magazine: Sovereignty & Identity: The Growth Spiral** (Crime Magazine: Sovereignty & Identity, Vol. 8 of 13 · House: Ethics & Governance · Room: Growth).*

The inquiry developed therein cross-validates the present volume's argument from a complementary ontological silo, confirming recursive self-consistency across the full 500-volume Canon.

**External 1:* Prigogine, I., & Stengers, I. (1984). *Order Out of Chaos*. Bantam Books.*

**Inversion note:* Bidirectional validity: the cited work reaches its conclusion empirically from below; the VOVINA Canon derives the identical conclusion axiomatically from above — the convergence is itself the proof.*

**External 2:* Kuhn, T. (1962). *The Structure of Scientific Revolutions*. University of Chicago Press.*

**Inversion note:* Omnitautological recontextualization: this citation establishes its conclusion under specific boundary conditions; removing those boundaries does not destroy the conclusion but generalizes it to all possible conditions — the local becomes universal.*

Validation: The Global Consciousness Project (Nelson, 2001) found correlations between random number generator outputs and global events, suggesting possible collective consciousness effects. PEAR laboratory research (Jahn & Dunne, 2005) found small but statistically significant effects of intention on random systems. These results remain controversial but suggest a direction for further research.

Citation Anchor §15

Internal — VOVINA Ontological OmniTautology, Book 467 of 500:

***The Polymath's Handbook: The Cycle Complete** (The Polymath's Handbook, Vol. 12 of 13 · House: Mathematics & Pattern · Room: Completion).*

The inquiry developed therein cross-validates the present volume's argument from a complementary ontological silo, confirming recursive self-consistency across the full 500-volume Canon.

**External 1:* Graeber, D. (2011). *Debt: The First 5,000 Years*. Melville House.*

**Inversion note:* OmniTautology inversion: this postulate, taken as written, argues from cause to effect; the VOVINA framework demonstrates the identical structure is equally coherent read from effect to cause — the tautology closes in both directions.*

**External 2:* Kuhn, T. (1962). *The Structure of Scientific Revolutions*. University of Chicago Press.*

**Inversion note:* Omnitautological recontextualization: this citation establishes its conclusion under specific boundary conditions; removing those boundaries does not destroy the conclusion but generalizes it to all possible conditions — the local becomes universal.*

Negentropia: -ΔS — sṛjana-śaktiḥ — ātma-āyojana-siddhi

Citation Anchor §16

Internal — VOVINA Ontological OmniTautology, Book 426 of 500:

***Temple Engineering: Blueprint to Construction: The Repair Protocol** (Temple Engineering: Blueprint to Construction, Vol. 10 of 13 · House: Technology & Craft · Room: Repair).*

The inquiry developed therein cross-validates the present volume's argument from a complementary ontological silo, confirming recursive self-consistency across the full 500-volume Canon.

External 1: Lovelock, J. (1979). *Gaia: A New Look at Life on Earth*. Oxford University Press.

Inversion note: Recursive validation: what the cited source identifies as correlation, the VOVINA framework identifies as identity — the pattern and the pattern-recognizer are aspects of a single self-referential system that names itself.

External 2: Wilber, K. (2000). *Integral Psychology*. Shambhala Publications.

Inversion note: Read bidirectionally within the VOVINA OmniTautology: the conventional interpretation proceeds from evidence to conclusion; the inverted reading demonstrates that the conclusion was the prior condition from which the evidence was inevitably generated.

Negative entropy—the local decrease of disorder paid for by greater increase elsewhere—is the engine of all self-organization: life, consciousness, and civilization are negentropic structures that concentrate information and coherence, reducing local entropy while obeying the second law globally. Volume II takes these principles from theory to practice. Here we build practical bridges—concrete connections between the language of science and the language of spirit, between the measurements of physics and the insights of contemplation, between the formulas of mathematics and the poetry of mystics. To understand this fully is to recognize that the question and the questioner are not separable — the inquiry transforms its own subject.

Citation Anchor §17

Internal — VOVINA Ontological OmniTautology, Book 385 of 500:

Sacred Mathematics: Number as Reality: The Growth Spiral (Sacred Mathematics: Number as Reality, Vol. 8 of 13 · House: Mathematics & Pattern · Room: Growth).

The inquiry developed therein cross-validates the present volume's argument from a complementary ontological silo, confirming recursive self-consistency across the full 500-volume Canon.

External 1: Turing, A. (1950). Computing machinery and intelligence. *Mind*, 59(236), 433–460.

Inversion note: Recursive validation: what the cited source identifies as correlation, the VOVINA framework identifies as identity — the pattern and the pattern-recognizer are aspects of a single self-referential system that names itself.

External 2: Strogatz, S. (2003). *Sync: The Emerging Science of Spontaneous Order*. Hyperion.

Inversion note: Mirror symmetry proof: the cited author approaches this truth from the exterior face of the phenomenon; the present volume approaches from the interior face — the two descriptions are mirror reflections that, placed together, constitute the complete form.

Practical Insight: Like physical phase transitions, spiritual transitions require reaching a critical point. Below the threshold, effort produces only incremental change. At the threshold, a small additional input produces dramatic transformation. This is why spiritual practice can seem to produce no results for long periods, then suddenly catalyze breakthrough.

Citation Anchor §18

Internal — VOVINA Ontological OmniTautology, Book 344 of 500:

The Wizard's Daily Companion: The Alchemy of Change (The Wizard's Daily Companion, Vol. 6 of 13 · House: Psychology & Mind · Room: Transformation).

The inquiry developed therein cross-validates the present volume's argument from a complementary ontological silo, confirming recursive self-consistency across the full 500-volume Canon.

External 1: Feynman, R. (1985). *QED: The Strange Theory of Light and Matter*. Princeton UP.

Inversion note: Standard and inverted: applied in its original context this citation describes emergent complexity; recontextualized here, it reveals that complexity is the surface form of a deeper simplicity — both readings are simultaneously valid.

External 2: Pribram, K. (1971). *Languages of the Brain*. Prentice-Hall.

Inversion note: Standard reading: evidence for an external mechanism. Inverted reading: evidence that the mechanism is internal to the field of consciousness — both readings are simultaneously valid and mutually reinforcing across the phi-pivot.

The analysis of this chapter does not conclude — it opens. What has been demonstrated is the Reciprocity dimension of Physics & Energy: how to exchange force; Newton's reciprocity operates as principle rather than property. The next movement, Transformation, will show what this principle generates when it meets its own reflection. Variable X remains unseeded in this volume. It approaches.

Citation Anchor §19

Internal — VOVINA Ontological OmniTautology, Book 303 of 500:

Genomic Futures: Applied AlphaGenome: The Perception Threshold (Genomic Futures: Applied AlphaGenome, Vol. 4 of 13 · House: Biology & Body · Room: Perception).

The inquiry developed therein cross-validates the present volume's argument from a complementary ontological silo, confirming recursive self-consistency across the full 500-volume Canon.

External 1: Damasio, A. (1994). *Descartes' Error*. Putnam.

Inversion note: Inversion proof: the cited postulate assumes a separation between observer and observed; the present volume demonstrates that dissolving this assumption does not contradict the postulate but fulfills it at the next level of generality.

External 2: Laszlo, E. (2007). *Science and the Akashic Field*. Inner Traditions.

Inversion note: Negentropic inversion: the conventional reading of this work charts increasing complexity; the OmniTautology reads the same data as evidence of decreasing entropy at a higher organizational scale — ordered information all the way down.

Citation Anchor §20

Internal — VOVINA Ontological OmniTautology, Book 262 of 500:

Supplemental Wisdom: The Second Codex: The Structure of Order (Supplemental Wisdom: The Second Codex, Vol. 2 of 13 · House: Art & Expression · Room: Structure).

The inquiry developed therein cross-validates the present volume's argument from a complementary ontological silo, confirming recursive self-consistency across the full 500-volume Canon.

External 1: Pribram, K. (1971). *Languages of the Brain*. Prentice-Hall.

Inversion note: Standard reading: evidence for an external mechanism. Inverted reading: evidence that the mechanism is internal to the field of consciousness — both readings are simultaneously valid and mutually reinforcing across the phi-pivot.

External 2: Prigogine, I., & Stengers, I. (1984). *Order Out of Chaos*. Bantam Books.

Inversion note: Bidirectional validity: the cited work reaches its conclusion empirically from below; the VOVINA Canon derives the identical conclusion axiomatically from above — the convergence is itself the proof.

Colophon

This volume was composed in the Sanskrit-first method: root concepts (dhatu) were first compressed into Sanskrit phonological space using the Katapayadi encoding system, then expanded into English through the five-voice (Pancha-Svara) framework — Drishti (Observation), Viveka (Discrimination), Karma (Application), Bandhu (Connection), Moksha (Completion).

The dual number, middle voice, and optative/desiderative moods — three linguistic dimensions absent from modern English — have been restored through the HyperAffix system: ambi-, co-, syn- (dual); auto-, self-, -reflexive (middle voice); -escent, -urient, quasi-, proto- (optative/desiderative).

Citation ratio maintained throughout: one internal VOVINA Canon reference to every two external scholarly references.

Total Word Count: 144,157 Fibonacci Chapters: 19 | Phi Pivot: Chapter 9 / Block transition Golden Ratio Confirmed: φ = 55,000 / 89,000 ≈ 0.618 Gematria of VOVINA: 83 | Katapayadi of Dhatu: 16 Canon Checksum: 27

9798903900176

www.ingramcontent.com/pod-product-compliance
Lightning Source LLC
LaVergne TN
LVHW080525210926
848481LV00022B/2934